S0-BWX-160
PHYSICS

1981 Summer School on High Energy Particle Accelerators

Fermilab
July 13-24, 1981

Physics of High Ene
Particle Acceleratc

(Fermilab Summer School, 1981)

AIP Conference Proceedings
Series Editor: Hugh C. Wolfe
Number 87

Physics of High Energy Particle Accelerators

(Fermilab Summer School, 1981)

Editors
R. A. Carrigan and F. R. Huson
Fermilab
M. Month
Department of Energy

American Institute of Physics
New York 1982

QC
787
.P3
S95
1981
Engineering
& Physics
Library

Copying fees: The code at the bottom of the first page of each article in this volume gives the fee for each copy of the article made beyond the free copying permitted under the 1978 US Copyright Law. (See also the statement following "Copyright" below). This fee can be paid to the American Institute of Physics through the Copyright Clearance Center, Inc., Box 765, Schenectady, N.Y. 12301.

Copyright © 1982 American Institute of Physics

Individual readers of this volume and non-profit libraries, acting for them, are permitted to make fair use of the material in it, such as copying an article for use in teaching or research. Permission is granted to quote from this volume in scientific work with the customary acknowledgment of the source. To reprint a figure, table or other excerpt requires the consent of one of the original authors and notification to AIP. Republication or systematic or multiple reproduction of any material in this volume is permitted only under license from AIP. Address inquiries to Series Editor, AIP Conference Proceedings, AIP.

L.C. Catalog Card No. 82-072421
ISBN 0-88318-186-X
DOE CONF- 820711

SUMMER SCHOOL ORGANIZATION

School Administration

L. Lederman, School Director
R. Huson, Associate School Director
M. Month, Chairman, Organizing Committee
C. Vanecek, School Secretary
R. Carrigan, Jr., Senior Editor
R. Donaldson, Proceedings
C. Sazama-Reay and J. Bjorken, Social Arrangements

Organizing Committee

M. Month,* DOE (chairperson)
J. D. Bjorken, Fermilab
F. R. Huson, Fermilab
C. Pellegrini, Brookhaven National Laboratory
B. Richter, Stanford Linear Accelerator Center
R. Schwitters, Harvard/Fermilab
A. Tollestrup, Fermilab
W. Willis, CERN/Brookhaven

Editorial Committee

R. Carrigan, Jr., Senior Editor, Fermilab
R. Donaldson, Fermilab
F. R. Huson, Fermilab
L. Michelotti, Fermilab
M. Month,* DOE
D. Neuffer, Fermilab
R. D. Ruth, Lawrence Berkeley National Laboratory
W. T. Weng, Brookhaven National Laboratory

*Detailed from Brookhaven National Laboratory

ACKNOWLEDGMENTS

The exceptional quality of the papers presented here is due to the substantial efforts of the lecturers and the seminar speakers who were asked to do the most difficult task of directing their talks to both novices and experts in the accelerator field. Much of what they say here has never been brought together in summary form before, and the result amounts to the first comprehensive, up-to-date textbook on high-energy particle accelerators.

The task of compiling this record rested with an editorial committee led by R. A. Carrigan, Jr. The editing for publication of the submitted manuscripts was done by R. Donaldson. An extensive review of the papers was performed by L. Michelotti, D. Neuffer, R. D. Ruth, and W. T. Wang.

The speakers were helped in their tasks by an energetic team of lecture assistants (S. Holmes, A. McInturff, L. Michelotti, D. Neuffer, C. Owen, R. Ruth, R. Shafer, and W. Weng) and of seminar chairmen (C. Brown, D. Edwards, T. Kirk, P. Limon, P. Livdahl, T. Nash, J. Peoples, and L. Teng). M. Schiff helped in preparing the preface.

The success of the summer school was due in no small part to the summer school secretariat and to others on the Fermilab staff. C. Vanecek worked for many months on the coordination of every phase of the school. H. Crow, E. Duty, S. Grommes, and P. Roberts shouldered the secretarial burden of the school. J. Bjorken and C. Reay arranged the varied and interesting social program that brought the school's participants together on an informal basis. A. Burwell's computer record keeping facilitated the registration procedure. H. Peterson helped in many ways to pull myriad items together.

The summer school was supported by the offices of the United States Department of Energy and the National Science Foundation.

R. Huson
M. Month
Conference Organizers

PREFACE

The material gathered in this volume covers the lectures given at the Summer School on High Energy Particle Accelerators, sponsored by the United States Department of Energy (DOE) and the National Science Foundation, held at Fermilab in Batavia, Illinois, July 13-24, 1981. The school was organized as a response to a recent appeal by a subpanel of the DOE High Energy Physics Advisory Panel (HEPAP) for more scientists and more students to work in the field of high energy particle accelerators.

The last decade has been one of the most fertile periods in the history of high-energy physics, made possible largely by the development of a new generation of accelerators. These have greatly extended the energy range of particle beams and have opened a window to a new and exciting regime of particle interaction energies. However, it has not been an easy matter to disseminate information on new accelerator technology to audiences that could benefit from such knowledge. Thus, HEPAP called for better ways to spread the news and for more researchers to join the burgeoning field upon which progress in high-energy physics is so dependent.

The summer school, the first of its kind to be held in this country, was planned by an organizing committee consisting of M. Month (DOE, chairman), J. D. Bjorken (Fermilab), R. Huson (Fermilab), C. Pellegrini (Brookhaven National Laboratory), B. Richter (Stanford Linear Accelerator Center), R. Schwitters (Harvard/Fermilab), A. Tollestrup (Fermilab), and W. Willis (CERN/Brookhaven). L. Lederman, (Fermilab) director of the summer school, provided the committee every kind of support and offered to host the school at Fermilab.

The committee set a number of objectives for the school: (i) to present in a thorough and up-to-date manner the entire spectrum of knowledge relating to accelerators; (ii) to disseminate that knowledge to audiences that can best make use of it; (iii) to encourage, by providing text materials and training to potential instructors, the development of accelerator physics education as part of university programs in high-energy physics; and (iv) to foster a more extensive dialogue between particle and accelerator physicists. The success of the school and this publication leave the organizers with the feeling that they have come some way toward achieving these goals.

The basic structure of the two-week program consisted of formal morning lectures devoted to accelerator physics and more general afternoon seminars dealing with various topics in accelerator and particle physics. The proceedings of the school preserves this format and will be published as two separate volumes, this volume titled "The Physics of High Energy Particle Accelerators" and the second volume titled "The State of Particle Accelerators and High

Energy Physics." This volume contains and extends the material presented in the morning lectures and can be read as a comprehensive textbook on accelerator physics, starting with a presentation of the basic concepts of the subject and going on to treat state-of-the-art topics such as laser and collective-field accelerators. These last two concepts, if rendered functional, would permit future accelerators to attain higher energies within shorter trajectories, thus making possible exploration of energy ranges beyond present economically acceptable costs. The papers in this volume have been commissioned and compiled with the intent of providing a comprehensive account of the accumulated knowledge of high energy particle accelerators. They have been edited with a view to integrating the lecture series and to preserving its original character. Use is made of examples and problems for teaching and self-study purposes. The second volume, containing the afternoon papers, is both less pedagogical and further ranging than the first. The lectures in this case deal with current and proposed accelerators, current theories in particle physics, some innovative ongoing and upcoming experiments, accelerator history, and speculations about the future. They catch the excitement of what it is like to work in the science that has as its aim to probe the ultimate nature of the material universe.

In recent years, in the tandem advances that characterize the interplay of particle theory and experiment, it is the theory that has exceeded the capability of experimental physics to verify the predictions. An obvious example is the question of the existence of the carriers of the weak force, the intermediate vector bosons. Although there is considerable indirect evidence pointing to their presence, there has yet been no direct observation of these particles. One reason for this is that there has been no accelerator system capable of attaining the energies necessary to create intermediate vector bosons, given their expected masses. Now, however, a CERN collider has been commissioned that has the energy capability, and a new Fermilab collider expected to be completed in 1985 will have that capability as well.

In the past, increased accelerator energy has almost invariably been followed by a burst of important discoveries. Accelerator energies have been increasing at the rate of an order of magnitude every seven years--due mainly to the imaginative advances in accelerator technology and design. Today strong-focusing synchrotrons are capable of achieving energies of 10^3 GeV, six orders of magnitude higher than the cyclotrons which reached the MeV energy range only forty years ago. This represents an amazing period of achievement in accelerator technology. But even this rate of progress may not be sufficient to allow experiments capable of testing the theoretical speculation as to what lies beyond. Theorists continue to predict new phenomena in energy regions far beyond those presently accessible and it is therefore important that accelerator builders reach even higher rates of technological innovation. A new energy range, with perhaps startling properties, waits to be explored. If the quest for the irreducible ingredients of matter is to continue unabated, new accelerator ideas must be forthcoming. We hope that the summer school together with these proceedings will play a role in the complex creative process through which these new ideas will emerge.

R. Carrigan
R. Huson
M. Month

TABLE OF CONTENTS

Accelerator Theory

Accelerator Technology

New Accelerator Ideas

INTRODUCTION TO ACCELERATOR THEORY

Lecturer
Ernest D. Courant

Text: L. Michelotti, D. Neuffer, R. D. Ruth, and W. Weng

Appendix: Lee C. Teng

TABLE OF CONTENTS

00940243X/82/870001-76$3.00 Copyright 1982 American Institute of Physics

INTRODUCTION

The subject of particle accelerators is vast, and many good books and papers have been written on the subject. The references listed at the end of this set of lectures should therefore be taken as representative but not necessarily complete. The whole idea is, of course, to accelerate particles, electrons, protons, or other ions, to high energies. The usefulness of these high-energy particle beams in the understanding of fundamental physics is the subject of many of the afternoon talks at the summer school; we will not dwell on that. What we treat here are the basic concepts of the design of the accelerators which yield these beams of particles, in particular, circular accelerators.

In a circular accelerator particles are confined to a closed orbit by a magnetic field and accelerated once (or more) in each revolution by an rf (radio frequency) cavity that is synchronized with the revolution frequency. In order for the particle to be confined to the closed orbit, the electric and magnetic fields have to be such that:

1. The reference particle stays on the desired orbit and is given energy at the desired rate, $E = E^s(t)$. This requires a magnetic field that increases with the momentum of the particle, and an rf field that imparts the correct energy increments to the particle at a frequency synchronized with the revolution frequency.

2. A particle which is a little bit off the reference orbit, experiences focusing forces which ensure that it remains close to the reference orbit. These are provided by suitable shaping of the magnetic guide field (transverse magnetic focusing).

3. A particle whose energy deviates slightly from the ideal, $E^s(t)$, is accelerated more-or-less than the reference particle so that its energy deviation from $E^s(t)$ remains small (longitudinal focusing).

4. If the hardware (magnets, rf system) deviates slightly from its design parameters, the resulting orbits (including their stability properties) must differ only slightly from the ideal ones.

5. The electric and magnetic fields produced by the beam itself must not be such as to destroy the stability of orbits produced by the externally applied fields. This leads to limitations on the strength of the beams (which will be discussed in C. Pellegrini's lectures). In storage rings, the action of one intense beam on the other beam (beam-beam interaction) can lead to additional limitations; these will be covered by A. J. Dragt, J. Tennyson, and J. Schonfeld.

In this set of lectures we begin by first studying the transverse magnetic focusing in a circular accelerator. This leads to the so-called betatron oscillations. In the second lecture we remove some of the idealizations introduced and discuss how various types of small errors effect transverse stability. This section includes a discussion of coupled transverse motion. The third lecture treats the problem of acceleration (after all, that is the purpose of an accelerator). This section includes a discussion of longitudinal focusing which goes hand in hand with acceleration. In the final lecture we discuss some aspects of the synchrotron radiation emitted by electrons which are bent by the magnetic fields in an accelerator.

1. BETATRON OSCILLATIONS

R. D. Ruth
Lawrence Berkeley Laboratory, Berkeley, California 94720
and
W. T. Weng
Brookhaven National Laboratory, Upton, New York 11973

We begin by considering the transverse stability of a particle. To simplify matters, we neglect acceleration and consider orbits in a static magnetic field (the justification for this will soon be apparent).

1A. The equations of motion

The ideal orbit is assumed to be a closed curve in a horizontal plane (usually a circle or a set of circular arcs connected by straight sections). If the orbit is simply a circle, the useful coordinates for a particle are cylindrical coordinates (r, θ, y). Even more useful are the coordinates (x, s, y) where $x = r - R$ (R = radius of the circle); $s = R\theta$, the distance along the circle of the point closest to the particles.

In general, when the orbit is a closed plane curve with local curvature $1/\rho$, we generalize the above coordinates as follows:

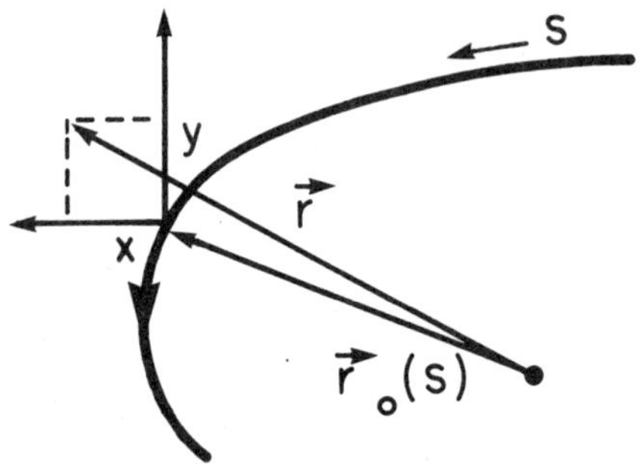

Fig. 1.1 The coordinate system.

If $\vec{r}_0$ is the point on the curve closest to $\vec{r}$, then:

s = distance along the curve to the point $\vec{r}_0$ from a fixed origin somewhere on the curve;

x = horizontal projection of the vector $\vec{r} - \vec{r}_0$;

y = vertical projection of the vector $\vec{r} - \vec{r}_0$;

ρ = local radius of curvature.

In terms of these coordinates, the equations of motion are

$$\frac{d^2x}{dt^2} = \frac{F_x}{\gamma m} + \frac{v^2}{\rho + x} = v^2\left\{\frac{B_y}{B\rho} + \frac{1}{\rho(1 + x/\rho)}\right\} \tag{1.1}$$

$$\frac{d^2y}{dt^2} = \frac{F_y}{\gamma m} = -v^2 \frac{B_x}{B\rho} \tag{1.2}$$

$$\frac{ds}{dt} = \frac{v}{1 + x/\rho}, \tag{1.3}$$

where v is the magnitude of the particle velocity, B_y, B_x are the components of the transverse magnetic field, and $B\rho = pc/e$ is the so-called magnetic rigidity. If we express this in terms of s as the independent variable and neglect terms of order $(x/\rho)^2$ then Eq. (1.1) becomes

$$\frac{d^2x}{ds^2} = (1 + x/\rho)\frac{B_y}{B\rho} + \frac{1}{\rho}. \tag{1.4}$$

On the reference orbit (x = 0) the right side should vanish. Therefore, B on the reference orbit must be given by

$$\begin{aligned} B_y &= -(B\rho)\frac{1}{\rho} \equiv B_0(s) \\ B_x &= 0 \qquad (x,y = 0). \end{aligned} \tag{1.5}$$

Thus we have

$$\frac{d^2x}{ds^2} = -\frac{x}{\rho^2} + \frac{B_y - B_0(s)}{B\rho}, \tag{1.6}$$

and

$$\frac{d^2y}{ds^2} = -\frac{B_x}{B\rho}. \tag{1.7}$$

Now, if the field near the orbit varies with x and y as

$$\begin{aligned} B_y &= B_0\left(1 - \frac{nx}{\rho} + \ldots\right) \\ B_x &= -B_0 ny/\rho, \end{aligned} \tag{1.8}$$

where n is the gradient coefficient, then we obtain the classical equations of motion,

$$\frac{d^2x}{ds^2} = -\frac{(1-n)}{\rho^2}x$$

$$\frac{d^2y}{ds^2} = -\frac{n}{\rho^2}y. \qquad (1.9)$$

For the constant n and ρ these are just harmonic oscillator equations; thus, they describe sinusoidal oscillations **provided** that n and 1 - n are both positive, i.e.,

$$0 < n < 1. \qquad (1.10)$$

Therefore, for a circle of radius $\rho = R$ and a gradient index n, we have vertical stability if $n > 0$ (field decreases with r), and horizontal stability if $n < 1$ (field decreases more slowly than the centripetal force increases).

To study this a bit more let us define the function

$$K(s) = \begin{cases} n/\rho^2 = B'/B\rho & \text{vertical} \\ \dfrac{(1-n)}{\rho^2} = \dfrac{1}{\rho^2} - \dfrac{B'}{B\rho} & \text{horizontal} \end{cases} . \qquad (1.11)$$

It is still useful at this point to study the solution with K = const. With a constant K, the general solution of

$$\frac{d^2z}{ds^2} + Kz = 0, \qquad (1.12)$$

in terms of the initial conditions at $s = s_1$, is

$$\begin{pmatrix} z \\ z' \end{pmatrix}_2 = \begin{pmatrix} \cos\phi & \frac{1}{\sqrt{K}}\sin\phi \\ -\sqrt{K}\sin\phi & \cos\phi \end{pmatrix} \begin{pmatrix} z \\ z' \end{pmatrix}_1 = M(s_2, s_1)\bar{z}_1, \qquad (1.13)$$

where

$$\phi = \sqrt{K}\,(s_2 - s_1). \qquad (1.14)$$

The oscillation frequencies are given by

$$\nu_x = (1 - n)^{1/2}$$
$$\nu_y = (n)^{1/2}, \tag{1.15}$$

where ν is the number of oscillations per revolution.

The solution here has been formulated in terms of a matrix for two very good reasons. First, this separates the properties of the general solution from those of the particular solution since the matrix only depends upon K and the interval $(s_2 - s_1)$. Second, this approach is easily generalized to piece wise constant K since the matrix for a given interval is just the product of the matrices for the subintervals, namely

$$M(s_2,s_0) = M(s_2,s_1)\ M(s_1,s_0). \tag{1.16}$$

This property is actually quite general and holds for any system of second-order linear differential equations.

Several other features are also transparent with this formulation. If we examine the motion in (z,z') phase space, the matrix given is essentially a rotation. So the system point (z,z') moves along an ellipse in phase space. Note that like a rotation this matrix has a determinant of unity, and therefore the area inside of a closed curve in phase space is preserved (Liouville's theorem).

This leads us to another important concept, phase-space requirements. For the sinusoidal oscillation described above, it is easy to see that the quantity

$$W \equiv \frac{\nu}{R}\left(z^2 + \frac{R^2 z'^2}{\nu^2}\right) \tag{1.17}$$

is a constant. So we have

$$z_{max} = \sqrt{\frac{R}{\nu} W}. \tag{1.18}$$

The area of the ellipse W = const. in (z,z') phase space is

$$\text{Area} = \pi W, \tag{1.19}$$

and, therefore we introduce the following definition: A beam comprised of particles which have $W < \varepsilon$ is said to have an **emittance** of $\pi\varepsilon$. So, $\pi\varepsilon$ = area of phase space occupied (usually expressed in mm-mradians). Thus, if a vacuum chamber has a radius a, the maximum emittance accommodated in it is (the "admittance")

$$\pi\varepsilon = \pi \frac{\nu}{R} a^2. \tag{1.20}$$

In the above formulation (the weak focusing configuration) the tune ν is limited by the requirement of stability. However, Eq. (1.20) leads us to the conclusion that if ν were large (stronger focusing), we could either accommodate much more emittance or alternatively build the vacuum chamber with a much smaller radius. This leads us in the next section to consider an alternative design.

1B. Strong focusing

Now suppose $\rho(s)$ and $K(s)$ are not constants all around, but piece wise constant and changing from sector to sector. We can have three kinds of sectors:

$$
\begin{aligned}
&K > 0 \\
&K < 0 \qquad\qquad (1.21)\\
&K = 0 \text{ (straight sections).}
\end{aligned}
$$

As the particle travels through these sectors, its "state vector" gets transformed successively by the matrix for each sector in turn. Therefore, in going through several sectors, it is transformed by the overall product matrix. The matrices for a section of length ℓ are

$$
K > 0 \qquad M_+ = \begin{pmatrix} \cos \ell\sqrt{K} & \frac{1}{\sqrt{K}} \sin \ell\sqrt{K} \\ -\sqrt{K} \sin \ell\sqrt{K} & \cos \ell\sqrt{K} \end{pmatrix} \tag{1.22}
$$

$$
K < 0 \qquad M_- = \begin{pmatrix} \cosh \ell\sqrt{K} & \frac{1}{\sqrt{K}} \sinh \ell\sqrt{K} \\ \sqrt{K} \sinh \ell\sqrt{K} & \cosh \ell\sqrt{K} \end{pmatrix} \tag{1.23}
$$

$$
K = 0 \qquad M_0 = \begin{pmatrix} 1 & \ell \\ 0 & 1 \end{pmatrix} \tag{1.24}
$$

Note that all the matrices have the determinant = 1 as does the product which again expresses the conservation of phase space.

Before proceeding to study combinations of these elements it is useful to formulate some general properties. The most general 2×2 matrix with determinant = 1 can be parameterized as

$$M = \begin{pmatrix} \cos\mu + \alpha\sin\mu & \beta\sin\mu \\ -\gamma\sin\mu & \cos\mu - \alpha\sin\mu \end{pmatrix}; \quad \beta\gamma = 1 + \alpha^2, \tag{1.25}$$

where either μ, α, β, γ are all real, or α, β, γ are pure imaginary and μ is pure imaginary ($\cos\mu > 1$) or $\mu = \pi$ + pure imaginary ($\cos\mu < -1$). So we can write

$$M = I\cos\mu + J\sin\mu,$$

with

$$J = \begin{pmatrix} \alpha & \beta \\ -\gamma & -\alpha \end{pmatrix}. \tag{1.26}$$

Note that $J^2 = -I$, the unit matrix. Therefore, the algebra of M is like that of $e^{i\mu}$, and so we have

$$M = e^{\mu J}. \tag{1.27}$$

Now consider a periodic lattice and let M be the matrix for one complete period. A particle going through n periods sees the matrix

$$M^n = e^{n\mu J}$$
$$= \begin{pmatrix} \cos(n\mu) + \alpha\sin(n\mu) & \beta\sin(n\mu) \\ -\gamma\sin(n\mu) & \cos(n\mu) - \alpha\sin(n\mu) \end{pmatrix} \tag{1.28}$$

and the particle coordinates in terms of the initial conditions are given by

$$z_n = z_0[\cos(n\mu) + \alpha\sin(n\mu)] + z_0'\beta\sin(n\mu)$$
$$z_n' = -z_0\gamma\sin(n\mu) + z_0'[\cos(n\mu) - \alpha\sin(n\mu)]. \tag{1.29}$$

With this form for z_n and z_n' it is easy to see that

$$z_n^2 + (\alpha z_n + \beta z_n')^2 = \text{const.} \tag{1.30}$$

which, again, defines an ellipse in phase space. Using $\beta\gamma = 1 + \alpha^2$, and dividing by β, we write

$$\gamma z_n^2 + 2\alpha z_n z_n' + \beta z_n'^2 \equiv W = \text{const.} \tag{1.31}$$

We have an invariant!

The area in phase space is again given by Eq. (1.19), and with a given $W = \varepsilon$, we have

$$z_{max} = \sqrt{\varepsilon\beta}$$
$$(\pi\varepsilon = \text{emittance}). \tag{1.32}$$

Now, as we go through the lattice, element by element, the coefficients α, β, γ change (but μ does not) and return to their original values after a complete period. However, W remains invariant; therefore, the largest z **anywhere** in the lattice is

$$(z_{max})_{max} = \sqrt{\varepsilon\beta_{max}}. \tag{1.33}$$

Thus, β governs the amplitude of oscillations and is called the "amplitude function." For a given aperture $a = z_{max}$, we have

$$\varepsilon = a^2/\beta_{max}. \tag{1.34}$$

So we can get more phase space into a given aperture if we make β_{max} small enough.

Now consider a lattice made of P identical periods each with an alternate positive and negative K. Assume (for simplicity)

$$K_+ = -K_- = K \tag{1.35}$$

and let $\phi = \ell\sqrt{K}$; $2\pi R = 2\ell P$; 2ℓ = sector length. The matrix for one sector is obtained by multiplying Eq. (1.22) by Eq. (1.23).

$$M = \begin{pmatrix} \cos\phi\cosh\phi + \sin\phi\sinh\phi & \frac{1}{\sqrt{K}}(\cos\phi\sinh\phi + \sin\phi\cosh\phi) \\ -\sqrt{K}(\sin\phi\cosh\phi - \cos\phi\sinh\phi) & \cos\phi\cosh\phi - \sin\phi\sinh\phi \end{pmatrix} \tag{1.36}$$

In terms of the parametrization of M we have

$$\cos\mu = \cos\phi \cosh\phi$$

$$\beta = \frac{1}{\sqrt{K}} \frac{\cos\phi \sinh\phi + \sin\phi \cosh\phi}{(1 - \cos^2\phi \cosh^2\phi)^{1/2}} . \tag{1.37}$$

Now, if we **scale** by changing the length of a period and K so that $\phi = \ell\sqrt{K}$ remains constant, then β scales as $1/\sqrt{K} = \ell/\phi$. So β is proportional to the length ℓ, while K (focusing strength) is inversely proportional to ℓ. Therefore, for a small β we want small ℓ (many periods in the entire lattice) and large K (strong focusing).

There is, however, an additional restriction. If μ is real, then the motion is bounded since the matrix elements just oscillate. But if the Trace M > 2, then μ is not real and the matrix elements of M can increase exponentially. Therefore, we must restrict ϕ so that $|2 \cos\mu| = |\text{Trace } M| < 2$ for both horizontal and vertical. This double stability is assured if the $K < 0$ and $K > 0$ sectors are equal (provided $K \gg 1/\rho^2$), for then ϕ in Eq. (1.36) is replaced by $i\phi$, and the matrix is unchanged except for the sign of $\sin\phi$.

To summarize this idea: in order to fit more phase space into a given amount of real physical space, we would like to decrease the beta function. This in turn requires strong focusing (of the alternating sort).

1C. The beta function

Since the beta function is so important, it is useful to study it in somewhat more detail. The equation of motion is (once again)

$$z'' + K(s)z = 0, \tag{1.38}$$

where K is periodic with period C. Equation (1.32) suggests that we look for a solution

$$z = \sqrt{W\beta} \cos \psi, \tag{1.39}$$

where ψ is some phase function. What is the behavior of β and ψ given the above parametrization of the problem? Since the equation is linear, we can search for a solution of the form

$$z = we^{i\psi}. \tag{1.40}$$

Substituting Eq. (1.40) into Eq. (1.38), it is straightforward to show (see problem 1.6) that

$$\psi' = 1/w^2, \tag{1.41}$$

and

$$w'' + Kw - \frac{1}{w^3} = 0, \tag{1.42}$$

where a particular normalization has been selected. We can now identify w with $(\beta)^{1/2}$, and we can also convert the solution into the matrix form (see problem 1.1)

$$M(s_2, s_1) = \begin{pmatrix} \frac{w_2}{w_1} \cos \psi - w_2 w_1' \sin\psi & w_1 w_2 \sin\psi \\ \text{etc.} & \frac{w_1}{w_2} \cos\psi + w_1 w_2' \sin\psi \end{pmatrix} \tag{1.43}$$

Taking one complete period, where $w_1 = w_2$ and $\psi = \mu$, we can identify (again)

$$\begin{aligned} w^2 &= \beta \\ w'w &= -\alpha. \end{aligned} \tag{1.44}$$

In addition, it is easy to show that

$$\begin{aligned} \alpha' &= K\beta - \gamma \\ \gamma' &= 2\ K\alpha \\ \beta' &= -2\alpha \quad (\beta\gamma = 1 + \alpha^2) \end{aligned} \tag{1.45}$$

or the third order periodic equation

$$\beta''' + 4K\beta' + 2K'\beta = 0. \tag{1.46}$$

To summarize, the beta function gives us "everything."

$$\psi(s) = \int ds/\beta \tag{1.47}$$

$$\mu = \oint ds/\beta \equiv 2\pi\nu \tag{1.48}$$

and

$$z \sim \sqrt{\beta}\, e^{i\psi(s)}. \tag{1.49}$$

Thus, β is the instantaneous wavelength with an average value given by $\sim R/\nu$, and the amplitude of the oscillation is simply modulated by $(\beta)^{1/2}$. Formally, this is just like the WKB approximation; if K(s) varies slowly we have the same form with

$$w = \sqrt{\beta} \sim K^{-1/4} \quad \text{(WKB)}. \tag{1.50}$$

The exact solution is given by Eq. (1.49) with the phase relation in (1.47); however, the β function must be calculated, either by solving Eq. (1.46) or by explicitly calculating the matrix elements and thus identifying β.

But this is not the only role the beta function plays in the analysis of transverse oscillation. In a real accelerator the fields are never quite those that were prescribed in the design. So we need to study how the errors in the realization of an accelerator can effect its performance. In the next section we treat various types of "errors" and discover once again that the beta function plays the leading role.

PROBLEMS

1.1 Show that the transfer matrix from s_1 to s_2 is given by Eq. (1.43).

1.2 Consider the matrix for a focusing element of length ℓ and focusing function K. Show that the transfer matrix can be written

$$M = \underset{\text{Drift}}{\begin{pmatrix} 1 & a \\ 0 & 1 \end{pmatrix}} \underset{\text{Thin Lens}}{\begin{pmatrix} 1 & 0 \\ \frac{1}{F} & 1 \end{pmatrix}} \underset{\text{Drift}}{\begin{pmatrix} 1 & a \\ 0 & 1 \end{pmatrix}}$$

How does the effective length, 2a, compare with the actual length, ℓ? How is F, the focal length, related to ℓ and K?

1.3 Consider a lattice composed of drifts and thin lenses with a period given by

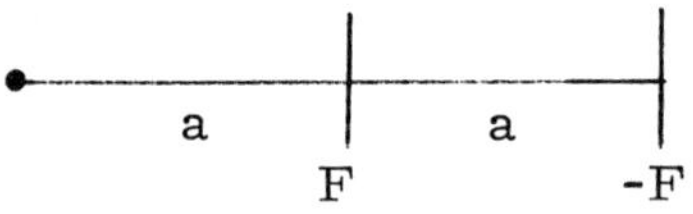

Show that $\sin \mu/2 = a/2F$ and $\beta = 2a\,[\,1 \pm \sin(\mu/2)]/\sin\mu$.

1.4 Find the 3×3 matrix which transforms (α, β, γ) given the matrix elements of M, the transfer matrix for (z, z').

1.5 For colliding-beam accelerators it is sometimes useful to have a very small beta function at the interaction point. Solve the equations for the matrix elements of M in a section free of magnetic elements. In particular obtain $\beta(s)$. Let β at the interaction point be $\beta^* = 5$ cm. What is β 10 meters away? What is the ratio of the beam size at these two points?

1.6 Show Eqs. (1.41) and (1.42).

2. PERTURBED BETATRON OSCILLATION

W. T. Weng
Brookhaven National Laboratory , Upton, New York 11973
and
R. D. Ruth
Lawrence Berkeley Laboratory, Berkeley, California 94720

The phase amplitude form of the solution for transverse oscillation enables us to transform the equation of motion to that of an harmonic oscillator. This makes it much easier to analyze field errors or perturbations. So, suppose we have some "errors," then the equation of motion can be written

$$x'' + k(s)x = f, \tag{2.1}$$

where f may be

a. An error in field	$f = f(s)$	
b. a momentum deviation	$f \sim \Delta p/p$	
c. an error in gradient	$f = xk(s)$	(2.2)
d. a coupling	f contains other coordinates	
e. a non-linearity	$f = ax^2 + bx^3 + \ldots$	

If we transform to the new variables,

$$\phi = \frac{1}{\nu} \int \frac{ds}{\beta(s)} \tag{2.3}$$

$$x = u\beta^{1/2}, \tag{2.4}$$

then from the previous form for the solution to the problem, we know the homogeneous solutions can be put in the following form,

$$u \sim e^{\pm i\nu\phi}. \tag{2.5}$$

Furthermore, it is easy to see that the inhomogeneous equation becomes

$$\frac{d^2u}{d\phi^2} + \nu^2 u = \nu^2 \beta^{3/2}(s) f \equiv F, \tag{2.6}$$

where now F is written in terms of the variable ϕ and $\beta(s)$ is the betatron function of the unperturbed oscillation.

2A. Guiding field errors

Consider first the case in which f is a function of ϕ only. Then Eq. (2.6) is a forced harmonic oscillator, and the right-hand side is a periodic function of ϕ with period 2π. A forced harmonic oscillator oscillates with the frequency of the forcing term unless its own frequency equals a frequency contained in the forcing function. Therefore an oscillatory solution exists unless ν is an **integer.** If ν is an integer and the oscillator is just driven at its resonant frequency, then the amplitude grows linearly in time which leads to instability.

Assume this is not the case, and let us look for a solution by a Green's function method, i.e., we let

$$F = \delta_p(\phi - \phi_1), \tag{2.7}$$

where δ_p is the periodic delta function. In this case the solution to Eq. (2.6) is given by (problem 2.1).

$$g(\phi,\phi_1) = \frac{1}{2\nu\sin\pi\nu}\cos\nu(\phi - \phi_1 + \pi). \tag{2.8}$$

Consequently, the general solution is

$$u = \frac{1}{2\nu\sin\pi\nu}\int_{\phi}^{\phi+2\pi} \nu^2\beta^{3/2} f(\phi_1)\cos\nu(\phi - \phi_1 + \pi)d\phi_1, \tag{2.9}$$

or

$$x(s) = \frac{\beta^{1/2}(s)}{2\sin\pi\nu}\int_{s}^{s+c} \beta^{1/2}(s_1)\, f(s_1)\cos[\nu(\phi - \phi_1 + \pi)]ds_1 \tag{2.10}$$

Notice that this solution exhibits a resonance when ν is an integer, as expected. In addition, note that the effect of the error is proportional to $\beta^{1/2}$ at the location of the error and to $\beta^{1/2}$ at the point of observation, thus the β function again plays the major role.

In order to get a measure of the amplitude of the "closed orbit," recall that for a free oscillation we have the invariant

$$W = \gamma x^2 + 2\alpha x x' + \beta x'^2, \tag{2.11}$$

and

$$x_{max} = \sqrt{\beta W}.$$

Rewriting this same function for the "inhomogeneous" case in terms of u, we find

$$W = u^2 + (du/d\phi)^2/\nu^2 \tag{2.12}$$

(a not too surprising result). In terms of the solution in Eq. (2.9) this is given by

$$W = \frac{\nu^2}{4\sin^2\pi\nu} \int_{\phi}^{\phi+2\pi} d\phi_1 \int_{\phi}^{\phi+2\pi} d\phi_2 \beta^{3/2}(\phi_1)\ \beta^{3/2}(\phi_2) \times \times f(\phi_1)\ f(\phi_2) \cos \nu (\phi_1 - \phi_2). \tag{2.13}$$

This particular form for the "invariant" of the closed orbit is very useful for making statistical assertions. Consider for example a machine with M quadrupoles doing the focusing and let each have a random transverse error in location δ_i. Then $f = K_i \delta_i$ over length L_i of the ith magnet. Assume in addition that each quadrupole is short so that we can replace the integral by a sum. Then

$$W = \frac{\nu^2}{4\sin^2\pi\nu} \sum_i \sum_j (\delta_i K_i \beta_i^{1/2} L_i) \times \times (\delta_j K_j \beta_j^{1/2} L_j) \cos \nu (\phi_i - \phi_j). \tag{2.14}$$

Now consider the expectation value of the amplitude of the displaced orbit in an ensemble of machines having the errors described. Since the errors in different magnets are uncorrelated, the only terms which contribute in the sum are those with i = j. Therefore we find

$$\langle W \rangle = \frac{\langle \delta^2 \rangle}{4\sin^2\pi\nu} \sum_i \beta_i (K_i^2 L_i^2). \tag{2.15}$$

To see this quantitatively, consider the accelerator ISABELLE. In this case:

$$\left.\begin{aligned} &108 \text{ Quadrupoles} \\ &L = 1.5\text{m} \\ &K = 0.045\text{m}^{-2} \\ &\nu = 22.6 \\ &\beta_{Ave} = 35\text{m} \end{aligned}\right\} \quad \text{ISA.} \tag{2.16}$$

With these parameters we find the amplitude to be

$$x_{rms} = \sqrt{W\beta} \simeq 13 \langle\delta\rangle_{rms}. \tag{2.17}$$

This implies that an alignment error of only 1.5 mm yields an error in orbit amplitude of 2 cm (a bit too big)!

Of course, the actual beam amplitude can be even larger because of the tails of the beam distribution and the variation of the beta function throughout the machine. These considerations emphasize the importance of close tolerances on alignment, and in practice these errors must be compensated for by additional correction magnets in the machine.

2B. Momentum deviation

One "field error" of those listed in Eq. (2.2) is of special interest, namely, that due to a particle having a different momentum than the nominal beam momentum. Since such a particle has a different "rigidity" to magnetic bending, the equation of motion is altered. The new rigidity is given by

$$B\rho = (B\rho)_0(1 + \Delta p/p), \tag{2.18}$$

and the new equation of motion is given by

$$\frac{d^2x}{ds^2} + \left(\frac{1}{\rho^2} + \frac{B'}{B\rho_0(1 + \Delta p/p)}\right)x = \frac{1}{\rho}\frac{\Delta p}{p}. \tag{2.19}$$

So to first order in $\Delta p/p$, the equation of motion becomes

$$\frac{d^2x}{ds^2} + K(s)x = \frac{1}{\rho}\frac{\Delta p}{p}. \tag{2.20}$$

It is interesting and very useful to cast the solution to the above equation again in a matrix form. It is straightforward to show that for a sector with K and ρ constant, the solution is

$$\begin{pmatrix} x \\ x \\ \Delta p/p \end{pmatrix}_1 = \begin{pmatrix} \cos\phi & \frac{\ell}{\phi}\sin\phi & \frac{1-\cos\phi}{\rho\sqrt{K}} \\ \frac{-\phi\sin\phi}{\ell} & \cos\phi & \frac{\sin\phi}{\rho\sqrt{K}} \\ 0 & 0 & 1 \end{pmatrix} \begin{pmatrix} x \\ x' \\ \Delta p/p \end{pmatrix}_0 \quad (2.21)$$

where $\phi = \ell\sqrt{K}$. In case for $K < 0$, the solution assumes a similar form, but with the third column given by

$$\begin{pmatrix} \frac{\cosh\phi - 1}{\rho\sqrt{K}} \\ \frac{\sinh\phi}{\rho\sqrt{K}} \\ 1 \end{pmatrix} \quad (2.22)$$

Now we can construct the product matrix for one period (K and − K).

$$T = \begin{pmatrix} M & & M_{13} \\ & & M_{23} \\ 0 & & 1 \end{pmatrix}, \quad (2.23)$$

where M is the matrix in Eq. (1.36) and M_{13} and M_{23} can be found easily. Since we are calculating a displaced equilibrium orbit or "closed orbit," we must impose the closure condition. If T is the matrix for a whole period, then

$$T\begin{pmatrix} x \\ x' \\ \Delta p/p \end{pmatrix} = \begin{pmatrix} x \\ x' \\ \Delta p/p \end{pmatrix}. \quad (2.24)$$

Or introducing the off momentum function $\eta(s)$ by

$$x(s) = \eta(s)\Delta p/p, \quad (2.25)$$

we have

$$T\begin{pmatrix} \eta \\ \eta' \\ 1 \end{pmatrix} = \begin{pmatrix} \eta \\ \eta' \\ 1 \end{pmatrix}. \quad (2.26)$$

Solving for $\begin{pmatrix} \eta \\ \eta' \end{pmatrix}$ yields

$$\begin{pmatrix} \eta \\ \eta' \end{pmatrix} = (I-M)^{-1}\begin{pmatrix} M_{13} \\ M_{23} \end{pmatrix}. \quad (2.27)$$

The η function at other points can be constructed similarly. Thus we have another property of the lattice which describes the local transverse displacement due to a deviation in particle momentum.

2C. Gradient errors

Another type of error which we mentioned above is the gradient error. Consider an actual focusing gradient related to the ideal one by

$$K_A(s) = K(s) + k(s) \tag{2.28}$$

then we have

$$f = -k(s)x, \tag{2.29}$$

and the equation of motion is

$$x'' + (K + k)x = 0. \tag{2.30}$$

If the machine is made up ideally of P identical periods, then for a machine of circumference C the periodicity of K is C/P. On the other hand, since errors repeat themselves only once each cycle, the periodicity of k is just C itself. In terms of the phase advance variable ϕ, the factor kx will contain frequencies $m\pm\nu$ (k gives the integer part). This means the oscillator will be "driven" if $\nu = m\pm\nu$ or if

$$2\nu = \text{integer}. \tag{2.31}$$

Another way of seeing this is to consider the matrix for one complete revolution (P periods);

$$M_0^P = \begin{pmatrix} \cos P\mu + \alpha\sin P\mu & \beta\sin P\mu \\ -\gamma\sin P\mu & \cos P\mu - \alpha\sin P\mu \end{pmatrix} \tag{2.32}$$

The presence of gradient errors change this slightly; however, stability is maintained so long as the trace of the perturbed matrix is still less than 2. Therefore, if $|T_r M_0{}^P| = 2$, the orbit is vulnerable to these errors; i.e., if

$$P\mu \equiv 2\pi\nu = p\pi \tag{2.33}$$

or if

$$\nu = \text{half integer}. \tag{2.34}$$

Imagine a situation where we vary all the K's in such a way to move the tune ν_0 close to the value p/2 in the presence of gradient errors. Then there will be an instability when ν_0 is close to p/2 within a range

$$(\Delta\nu) = \frac{1}{2\pi}\left|\oint k(s)\beta(s)e^{ip\phi(s)}ds\right|. \tag{2.35}$$

In other words, the "stop band" width is sensitive to the pth harmonic of the gradient error k(s) (weighted, of course, with the beta function). In the stable case, if there is a gradient error, the tune, ν, is shifted by

$$\nu - \nu_0 = \frac{1}{4\pi}\oint k(s)\beta(s)ds, \tag{2.36}$$

and the beta function all around the machine varies from the unperturbed value with an amplitude given approximately by

$$\frac{\Delta\beta}{\beta} \simeq \frac{\nu(\Delta\nu)}{2[\nu^2 - (p/2)^2]} \tag{2.37}$$

provided that ν is outside the "stop band." A more detailed analysis of this problem can be found in Ref. 1.

2D. Coupling

The last and perhaps the most interesting linear errors which we will discuss are coupling errors. So far everything that has been discussed applies both to the horizontal and the vertical oscillations separately. However, field errors can be such that the x force depends on y and the y force depends on x. Consider for example, a quadrupole which is rotated a bit; then the field is given by

$$\begin{aligned} B_y &= B\rho(kx + My) \\ B_x &= B\rho(ky - Mx) \end{aligned} \tag{2.38}$$

and the equations of motion become **coupled**

$$\begin{aligned} x'' + k_1 x + My &= 0 \\ y'' + k_2 y + Mx &= 0. \end{aligned} \tag{2.39}$$

So, rather than two separate 2×2 problems, the solution is described by a 4×4 matrix transformation for the 4 canonical variables (x, P_x, y, P_y).

Before discussing particular solutions, let us examine the general theory. In electromagnetic fields particle motion is determined by a Hamiltonian or Lagrangian. In the case of pure magnetic fields, the Hamiltonian is just the total energy

$$H = c[m^2c^2 + (\vec{p} - e\vec{A})^2]^{1/2}, \tag{2.40}$$

where $\vec{A}$ is the vector potential and $\vec{p}$ is the canonical momentum. Using various canonical transformations (which we will not discuss here) it is possible to convert this to a Hamiltonian in variables which are just the deviations (and slopes) from a reference orbit. This Hamiltonian begins with quadratic terms. Since we are considering only linear equations of motion, we can take H to be a symmetrical quadratic function of the coordinates and momenta, with coefficients that are periodic functions of s.

If we write the canonical coordinates and momenta as a vector, $(q_1, p_1, q_2, p_2, \ldots) = (x_1, x_2, x_3, \ldots) = \vec{x}$ or x_α, then

$$H = \frac{1}{2} \sum_{r,s=1}^{2N} H_{rs}(s) x_r x_s . \tag{2.41}$$

H here is not the energy and is not a constant; however, the equations of motion are still given by the usual Hamilton's equations

$$\frac{\partial q_i}{\partial s} = \frac{\partial H}{\partial p_i} ; \quad \frac{\partial p_i}{\partial s} = - \frac{\partial H}{\partial q_i} . \tag{2.42}$$

In terms of the x variables defined above, these equations can be written in the matrix form

$$\frac{d\vec{x}}{ds} = -SH\vec{x}, \tag{2.43}$$

where H now is the symmetric matrix of the coefficients of the Hamiltonian and S is the "symplectic unit matrix"

$$S = \begin{pmatrix} 0 & -1 & & & \\ 1 & 0 & & 0 & \\ & & 0 & -1 & \\ & 0 & 1 & 0 & \\ & & & & \ddots \end{pmatrix} . \tag{2.44}$$

In two dimensions, S is $-i$ times the Pauli matrix σ_2 and has the properties

$$S^2 = -I, \quad \tilde{S} = -S = S^{-1}, \tag{2.45}$$

where ~ denotes transposition. Note that in the uncoupled case

$$H = \begin{pmatrix} K_1 & 0 \\ 0 & K_2 \end{pmatrix} \tag{2.46}$$

and all the previous results apply.

Now consider any two solutions x and y of the equations of motion. Multiply x by S and take the inner product;

$$p = (y,Sx) = -(x,Sy) = \sum_{rs} y_r S_{rs} x_s$$
$$= x_1y_2 - x_2y_1 + x_3y_4 - x_4y_3 + \ldots \tag{2.47}$$

However, from the equations of motion, we have

$$\frac{dp}{ds} = (y',Sx) + (y,Sx') \tag{2.48}$$
$$= (-SHy,Sx) + (y,-SSHx).$$

In addition

$$(Ay,Bx) = (y,\tilde{A}Bx), \tag{2.49}$$

so using Eq. (2.45) we find

$$\frac{dp}{ds} = (y,HS^2x) + (y,-S^2Hx) = 0. \tag{2.50}$$

Thus P is a bilinear invariant of any two solutions. Now the states evolved over an interval of "time," s, are equivalent to a "canonical transformation" of the x vector which can be described by a (2N × 2N) matrix analagous to the 2×2 matrices studied previously, i.e.,

$$x(s_2) = Mx(s_1)$$
$$y(s_2) = My(s_1). \tag{2.51}$$

Since P is an invariant,

$$(y_2,Sx_2) = (y_1,Sx_1)$$

or

$$(My_1,SMx_1) = (y_1,\tilde{M}SMx_1) = (y_1,Sx_1). \tag{2.52}$$

Therefore, since the above relation is valid for any y_1 and x_1, M must be such that

$$\tilde{M}SM = S. \tag{2.53}$$

Matrices which keep the inner product, P = (y, Sx), invariant are said to be **symplectic**, analogous to, but different from, orthogonal matrices [invariant (y, x)]. The determinant of these matrices is 1 instead of -1 because M evolves continuously from the unit Matrix. This is all fine, but what does this imply for stability and/or resonances?

The system is stable if all of the eigenvalues of the matrix for a complete revolution have magnitude $\leqslant 1$. Instability arises if one (or more) eigenvalue is greater than 1 in absolute value. Suppose M has 4 distinct eigenvalues (2N in general). Then since the eigenvectors form a complete set, any initial vector is some linear combination of them. The eigenvalues and vectors are given by

$$Mx^i = \lambda^i x^i. \tag{2.54}$$

Now consider two eigenvectors x^i, x^j. Then

$$\begin{aligned} p^{(i,j)} &\equiv (x^i, Sx^j) = (x^i, \tilde{M}SMx^j) \\ &= (Mx^i, SMx^j). \end{aligned} \tag{2.55}$$

And so we find

$$(\lambda^i \lambda^j - 1)\,(x^i, Sx^j) = 0. \tag{2.56}$$

Either x^i and x^j are sympletically orthogonal or the eigenvalues are reciprocal. However, given one eigenvector x^i, not all the others can be orthogonal to it because, for example, the vector Sx^i is not, and this vector is surely a linear combination of other eigenvectors. Therefore, for every λ^i there is another eigenvalue $1/\lambda^i$. In addition, since the matrix M is real, λ^{i*} is also an eigenvalue. We thus have the following possibilities (see Fig. 2.1): (a) All four eigenvalues lie on the unit circle, forming two complex conjugate reciprocal pairs; (b) One reciprocal pair is real, the others are complex and on the unit circle; (c) Two real reciprocal pairs; (d) One eigenvalue, λ_1, is complex and not on the unit circle; the other eigenvalues then must be $\lambda_2 = 1/\lambda_1$, $\lambda_3 = \lambda_1^*$, $\lambda_4 = 1/\lambda_1^*$.

The only stable configuration is that given in situation (a). Now we can ask the question: if the uncoupled motion is stable [case(a)], under what circumstances can coupling lead to instability? If we assume the coupling is weak, the matrix with coupling differs only slightly from that without coupling. This means that a small coupling can only lead to instability of the sort in case (d), and then only if the eigenvalues of the uncoupled case are **very nearly equal**, so that a small change can switch the system from case (a) to case (d). This will be true, provided

$$\cos\,(2\pi\nu_x) = \cos\,(2\pi\nu_y) \tag{2.57}$$

or rather if

$$\nu_x - \nu_y = \text{integer}$$

or

$$\nu_x + \nu_y = \text{integer}. \tag{2.58}$$

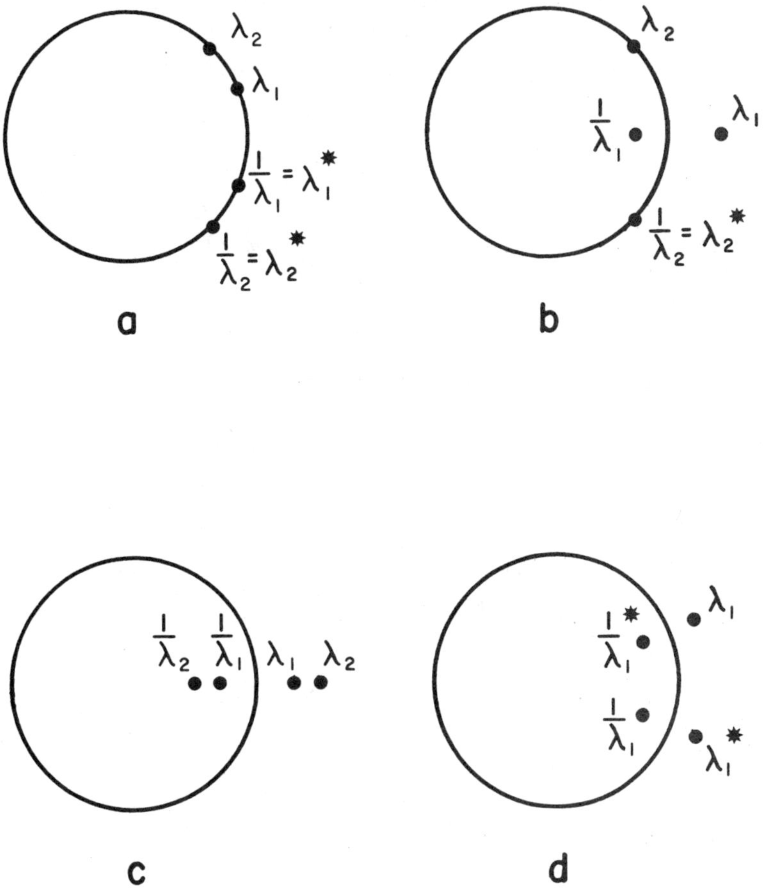

Fig. 2.1 Location of eigenvalues for two-dimensional linear oscillations. (a) Both modes stable. (b) One mode stable, one mode unstable. (c) Both modes unstable in absence of coupling. (d) Instability induced by coupling.

A remarkable fact is that only the latter case yields a true instability.

To demonstrate this, let's search for invariants to the problem. Search for a quadratic form U so that

$$(x,Ux) = (Mx,UMx) = (x,\tilde{M}UMx)$$
or
$$\tilde{M}UM = U. \tag{2.59}$$

It's easy to see that this corresponds to the requirement that the matrix SU must commute with M,

$$[SU, M] = 0. \tag{2.60}$$

In particular, let's select U's of the form

$$U_1 = S\ (M - M^{-1})$$
$$U_2 = S\ (M^2 - M^{-2}). \tag{2.61}$$

If we use the matrices for the unperturbed case, this leads to the invariants [see problem (2.2)].

$$W_1 = W_x\sin(2\pi\nu_x) + W_y\sin(2\pi\nu_y)$$
$$W_2 = W_x\sin(4\pi\nu_x) + W_y\sin(4\pi\nu_y), \tag{2.62}$$

where W_x and W_y are the familiar invariants for the x and y motion respectively. So, we find that near a difference resonance,

$$\nu_x - \nu_y = \text{integer}$$
then
$$W_1 = (W_x + W_y)\sin 2\pi\nu_x \tag{2.63}$$
$$W_2 = (W_x + W_y)\sin 4\pi\nu_x.$$

Thus, the invariant is positive definite and the motion in any one degree of freedom is bounded. On the other hand, consider a sum resonance

$$\nu_x + \nu_y = \text{integer}$$
then
$$W_1 = (W_x - W_y)\sin 2\pi\nu_x \tag{2.64}$$
$$W_2 = (W_x - W_y)\sin 4\pi\nu_x.$$

Thus, at a sum resonance, both W_x and W_y can increase without violating the invariance condition. So we have instability only in the case of a sum resonance while for a difference resonance the motion is bounded.

Now we can collect all the information on these resonances in the form of a diagram (see Fig. 2.2). We have instability when

$$\left.\begin{array}{ll} \nu_x = k & \nu_y = k, \\ \nu_x = k/2 & \nu_y = k/2, \\ \text{and} \quad \nu_x + \nu_y = k & \end{array}\right\} \tag{2.65}$$

2E. Nonlinear effects

In case there are nonlinear terms, the equation of motion is of the form

$$x'' + K(s)x = a_{20}x^2 + a_{30}x^3 + a_{mn}x^m y^n, \tag{2.66}$$

where all coefficients are periodic in s.

We know resonance can take place, i.e., instability arises, when r.h.s. contains frequency ν_x. Since x oscillates with ν_x and y with ν_y, and the coefficient is periodic, r.h.s. contains the frequencies

$$k + m\nu_x + n\nu_y$$

with any integers k, m, n. Therefore the system would seem to be sensitive to resonance--and potentially unstable whenever

$$k + m\nu_x + n\nu_y = 0 \quad \text{for any integers } k, m, n.$$

But since rational numbers are everywhere dense, this would mean: **Nothing** is stable.

This is a dilemma first faced by Poincare and studied extensively by modern mathematicians up to the present day. For practical purposes, one requires much less than complete mathematical stability. It turns out that there may be "fast" and "slow" instabilities; the slow ones may take billions of periods to develop.

Resonance condition is not as bad as it looks. For, if amplitude blows up due to a resonance, frequencies change (also because of nonlinearity) and the resonance condition ceases to be valid. This means, in practice, that resonance is likely to be bad--for small amplitudes--only when $|m| + |n| \leqslant 4$ (and further, if m and n are of the same sign). The simplest case is one dimensional oscillation to be discussed in the following:

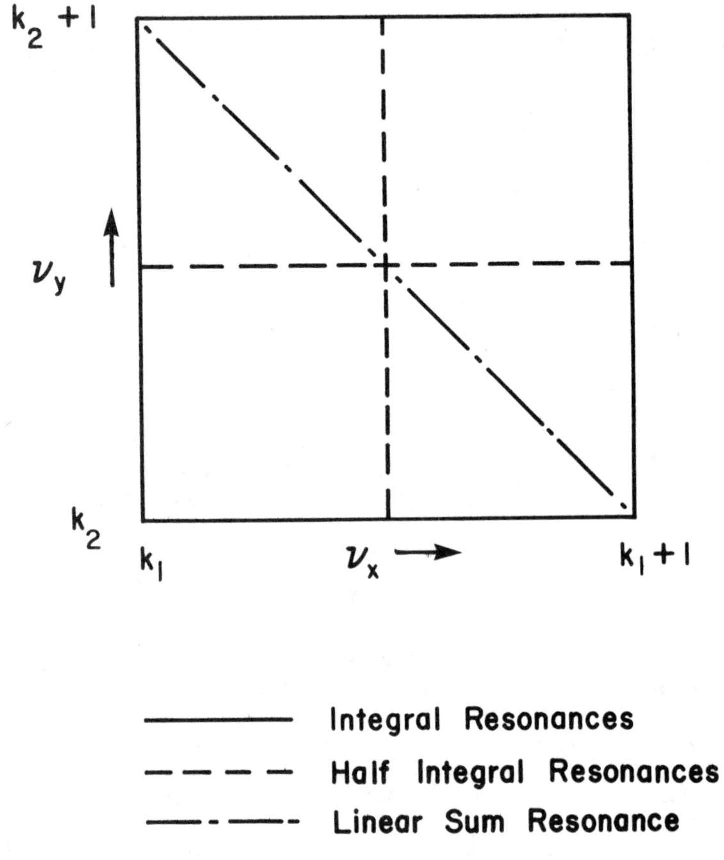

Fig. 2.2 Resonance lines in a tune diagram

$$x'' + K(s)x + a_3(s)x^2 + a_4(s)x^3 + \dots = 0. \tag{2.67}$$

To solve the nonlinear equation, we have to first solve the linear equation as we did before and express the solution as

$$x = \beta^{1/2}e^{\pm i\nu\theta} \equiv \beta^{1/2}u, \quad \theta = \frac{1}{\nu}\int \frac{ds}{\beta}.$$

Defining $x = \beta^{1/2}u$, the original nonlinear equation now becomes

$$u'' + \nu^2 u + \nu^2\beta^{3/2}(a_3\beta u^2 + a_4\beta^{3/2}u^3 + \dots) = 0,$$

where everything is periodic in θ with period 2π.

It can be shown that the equation of motion is derivable from a Hamiltonian

$$H(u,p,\theta) = \frac{1}{2}\,p^2 + \frac{1}{2}\,\nu^2u^2 + f_3u^3 + f_4u^4 + \dots$$

where $f_3(\theta) = 1/3\ \nu^2\beta^{5/2}\ a_3$; $f_4 = 1/4\ \nu^2\beta^3 a_4$, ... etc.

Be careful about the fact that H is not energy, and not a constant, because the NL terms are periodic functions of θ. If H could be made independent of θ, it would be a constant which defines invariant curves in phase space. Furthermore, if these curves are closed we have stability. In the linear case, H is a constant and the invariant curve is an ellipse in phase space. In the nonlinear case, we hope to find invariant curves that are distorted ellipses. The method we follow is to find a canonical transformation that makes the new H independent of θ. We are going to do this by a step-by-step approach, i.e., first eliminate θ-dependence of u^3 term, then u^4 term, etc.

This process was invented by Birkhoff--transformation to "Birkhoff normal form." This was later extended by Moser (1955) to the case of ν at resonance. The trouble is the process does not converge, because of "small denominators" since any ν is near a rational number. But Kolmogorov, Arnold, and Moser found that closed invariant curves exist anyway, at small enough amplitudes (but are not exactly expressible by analytic formulas). As physicists let's set aside these doubts and simply calculate the first few terms of the series.

If the original system is subjected to an canonical transformation generated by "generating function" $F_2(q,P,\theta)$, the new canonical variables Q P are obtained by solving

$$Q = \frac{\partial F_2}{\partial P}$$

$$p = \frac{\partial F_2}{\partial q} \tag{2.68}$$

for QP in terms of qp or vice versa. The new Hamiltonian is

$$H'(Q,P,\theta) = H + \frac{\partial F_2}{\partial \theta}(q,P,\theta). \tag{2.69}$$

If H', written in terms of P, Q, θ, is independent of θ, it is a constant of motion.

We first express the original problem in terms of action angle variables J,ϕ

$$u = \sqrt{\frac{2J}{\nu}}\cos\phi; \quad p = -\sqrt{2J\nu}\sin\phi \tag{2.70}$$

then

$$H = \nu J + \frac{2J}{\nu}^{3/2}\cos^3\phi f_3(\theta) + (\frac{2J}{\nu})^2\cos^4\phi f_4\theta) + \ldots \tag{2.71}$$

So let us consider the more general problem given by

$$H = H_0(J) + f(\phi, J, \theta), \tag{2.72}$$

where $f(J, \phi, \theta)$ is a periodic function in ϕ and θ and has zero average.

$$\frac{1}{4\pi^2}\int_0^{2\pi} d\phi \int_0^{2\pi} d\theta f = 0. \tag{2.73}$$

[The average value of f has been put into $H_0(J)$]. Perform a canonical transformation, $(\phi,J) \to (\psi,k)$, given by

$$F_2(\phi,K,\theta) = \phi K + G(\phi,K,\theta). \tag{2.74}$$

Provided G is small in some sense, this transformation is close to the identity. From Eqs. (2.68) and (2.69) we have

$$\begin{aligned} H' = {} & H_0(K) + H_0'(K)G_\phi + \ldots \\ & + f(\phi,K,\theta) + f_J G_\phi + \ldots \\ & + G_\theta \end{aligned} \tag{2.75}$$

where the subscript indicates partial differentiation and only leading terms have been retained.

Recall that the object is to generate a constant of the motion by selecting G judiciously. Therefore we select G so as to cancel the lowest order fluctuating term, i.e., we let

$$f(\phi,K,\theta) + \nu(K)G_{\phi} + G_{\theta} = 0. \tag{2.76}$$

Note that f was left expressed in the same "mixed" variables as G. Since f is a periodic function of ϕ and θ, it can be expressed as a double Fourier series

$$f = \sum_{mn} f_{mn}(K)e^{i(m\phi-n\theta)}, \tag{2.77}$$

and the solution to Eq. (2.76) can therefore be written

$$G = i \sum_{mn} \frac{f_{mn}(K)e^{i(m\phi-n\theta)}}{m\nu-n} \tag{2.78}$$

So we are left with the new Hamiltonian

$$H' + H_0(K) + f_K G_{\phi} + \ldots \tag{2.79}$$

This, of course, still depends upon our "time" variable, θ. However, if $f \sim K^b$, then $f_K G_{\phi} \sim K^{2b-1}$ and therefore the "time" dependent term is of higher order in the amplitude ($b > 1$). Thus, we have succeeded in moving the time dependence to higher order.

To initiate the next step in this series we extract the average value of $f_K G_{\phi}$ and retain the fluctuating term as the object of the **next** canonical transformation. Suppressing this additional fluctuation at this order, we find the approximate invariant

$$H' = H_0(K) + \langle f_K G_{\phi} \rangle^{\theta} \tag{2.80}$$

(where < > plus superscript means average) and the new "tune"

$$\nu_1(K) = \nu_0(K) + \frac{\partial}{\partial K} \langle f_K G_{\phi} \rangle^{\theta,\psi}. \tag{2.81}$$

All of this is fine so long as G is small; however, a quick inspection of Eq. (2.78) will convince you that this is **not** the case for arbitrary ν. There are resonances whenever

$$\nu(K) = n/m. \tag{2.82}$$

But remember that for a nonlinear problem ν is generally a function of amplitude $\left(\nu = \nu(K)\right)$. So that if a resonance causes growth, this changes the tune and the system moves off resonance. We therefore expect that the infinities are a difficiency of the approach taken above. In the vicinity of a resonance, another approach is in order.

If $\nu(K) \approx \frac{n}{m} + \varepsilon$ and provided all other resonances can be neglected (sometimes true, sometimes **not** true), then another approach works quite well. In this case (single resonance) the Hamiltonian takes the form

$$H = H_0(J) + f_{mn}\cos(m\phi - n\theta) \tag{2.83}$$

If we now make a transformation to a coordinate system which rotates in **phase space**, it is possible to find an **exact** integral of the motion for the Hamiltonian in Eq. (2.83). This is accomplished with the generating function

$$(\phi, J) \to (\psi, K); \qquad F_2(\phi,K,\theta) = (\phi - \frac{n}{m}\theta)K$$

so that

$$\psi = \phi - n/m\ \theta \qquad J = K \tag{2.84}$$

and

$$H' = H_0(K) - \frac{n}{m}K + f_{mn}\cos(m\psi) \tag{2.85}$$

Since H' is now independent of θ, it is a constant of the motion and can be used to label particle trajectories in phase space.

Of course, the real physical problem will have other resonances as well, and thus, this development yields only an approximation in that case. In spite of this, the above technique has had very useful practical applications in the "resonant extraction" of beams from accelerators. The interested reader should consult the appendix as well as Ref. 3 for more information on this point.

Before leaving the subject of nonlinear effects we must caution the reader that we have neglected to discuss many important issues such as the existence of integrals, resonance overlap and stochasticity. There are many extensive articles and books which deal with these questions. In particular we refer the reader to Refs. 4 and 5 and an excellent review article by B. Chirikov.[6] In addition A. Dragt in these proceedings presents an entirely different approach to the problem, and J. Tennyson and J. Schonfeld show the connection of nonlinear dynamics to the beam-beam effect in colliding beams.

We now leave the subject of transverse oscillations (whether linear or nonlinear) to address the question of acceleration.

PROBLEMS

Problem 2.1.

Find the Green's function to Eq. (2.6) corresponding to the delta function perturbation given by Eq. (2.7).

Problem 2.2

Prove that W_1 and W_2 as given in Eq. (2.62) are invariants.

Problem 2.3

Express the final Hamiltonian (2.85) in terms of x and p_x, and find (a) stable fixed point, unstable fixed points; (b) equation of the separatrices; (c) trajectories of particle growth for resonance extraction (see Appendix).

3. ACCELERATION

Leo Michelotti
Fermi National Accelerator Laboratory, Batavia, IL 60510

As important and elegant as the theory of betatron oscillations is, the primary function of an accelerator is to increase the kinetic energy of a beam of particles--protons, let us say. This is accomplished by passing them through a cavity in which a longitudinal rf (radio frequency) electric field has been excited and is sustained. The proton is made to cross the cavity when the phase of the field is such that the energy kick imparted to the proton has the desired value. Accelerators tend to the circular, because it makes more sense, technologically and economically, to pass the protons 50,000 times through 2 meters of a (125 kV/m, say) cavity than to push them only once through 100 kilometers of cavities. For the intent of this section the "rest of" the accelerator--the "ring": dipoles, quadrupoles, straight sections, correction elements, and so forth--is nothing more than a black box device for returning protons to the input port of an rf cavity. The betatron oscillations discussed in the previous sections have to do with stability of transverse motion of a particle as it undergoes many circuits of the ring. In this section we shall address the most basic questions about single particle longitudinal stability during acceleration.[7-9]

We shall work within a very simple model. Neglect all transverse motion of the proton.* The cavity gives the proton an instantaneous energy kick of size

$$\delta E = eV \sin\phi \ , \tag{3.1}$$

where ϕ is the phase of the electric field at the time of passage, and V is the maximum voltage achieved by the cavity.** To keep the analysis at a reasonable level, we assume that the energy gain is designed to be the same at every crossing. This means that the phase of the field should have the same value at all crossings. Let ϕ^s represent this value; the superscript "s" will stand for "synchronous" throughout this section. (We shall see shortly why it is **not** desirable to send the proton through with $\phi^s = \pi/2$.)

In a synchrotron, the time required for a proton to circle the ring depends on its energy. Therefore, as this increases, the frequency of the rf cavity must be continually modified in order to maintain synchronism with the orbiting particle. A beam, however, is not purely monochromatic; the energies of its constituent particles fall into a distribution having some non-zero width. It

---

*The validity of decoupling transverse and longitudinal motion in this way is not obvious. It rests on the fact that the synchrotron and betatron oscillation frequencies are very different.

**We shall adopt the convention eV > 0 so that there is no ambiguity about the sign of e.

is, accordingly, impossible for the cavity to be exactly synchronous with all of them simultaneously. The stability question is this: If the cavity and an orbiting particle start out **close to** synchrony, will they remain so throughout the acceleration period? Put another way, will the energy of a proton in an almost but not quite monochromatic beam stay close to the (ever increasing) synchronous energy or drift away from it?

We shall label the state of the proton with (1) the phase of the cavity and (2) the energy of the proton immediately **after** passage through the cavity. By abuse of language, we will speak of the phase of the proton, when what is actually meant is the phase of the field at the instant of traversal. The energy increment from one cycle to the next is easily written

$$E_k - E_{k-1} = eV\sin\phi_k, \tag{3.2}$$

where the subscript counts cycles. We are interested here not in the energy itself but in the difference between this energy and the currently synchronous energy. Define $\Delta E \equiv E - E^s$ to be this difference. Its variation between consecutive cycles is given by

$$\Delta E_k - \Delta E_{k-1} = eV(\sin\phi_k - \sin\phi^s) \ . \tag{3.3}$$

The phase of a proton that is not exactly synchronous will also change from cycle to cycle. Let the cycling time, at cycle k, of a proton of energy E be represented by $2\pi/\omega_k(E)$. This is written in a form that emphasizes the "instantaneous cycling frequency," $\omega_k(E)$. The rf cavity must oscillate at some multiple of the synchronous frequency--say, $\omega^{rf} = h\omega^s$, where h is an integer (about 1000 at Fermilab). The phase advance seen by a non-synchronous proton at cycle k is thus given by

$$\begin{aligned}\phi_k - \phi_{k-1} &= \frac{2\pi}{\omega_{k-1}(E_{k-1})} \cdot h\omega^s{}_{k-1} \quad (\mathrm{mod}\ 2\pi) \\ &= 2\pi h\left[\frac{\omega^s{}_{k-1}}{\omega_{k-1}(E_{k-1})} - 1\right] \\ &= \frac{2\pi h}{\omega_{k-1}(E_{k-1})}\,[\omega^s{}_{k-1} - \omega_{k-1}(E_{k-1})] \ .\end{aligned} \tag{3.4}$$

Of course, $\omega_k^s = \omega_k(E_k^s)$. As k becomes large, the proton is traveling essentially at the speed of light, and ω_k^s approaches a limiting value. For lower energies, however, ω_k^s is not independent of k.

We now have a well-defined mathematical problem. Given (ϕ_1, E_1), one can recursively find (ϕ_n, E_n) for all n by using the difference Eqs. (3.3) and (3.4). To get some feeling for the nature of the solutions, we shall simplify matters even further by introducing two reasonable approximations.

Let us assume that E is sufficiently close to E^s that we can expand $\omega_k(E)$ and retain only the lowest order term:

$$\begin{aligned}\omega(E) &\simeq \omega^s + \left(\frac{d\omega}{dE}\right)^s (E - E^s) \\ &= \omega^s + \left(\frac{d\omega}{dp}\right)^s \left(\frac{dp}{dE}\right)^s \Delta E \\ &= \omega^s + \left(\frac{\omega E}{p^2}\eta\right)^s \Delta E,\end{aligned} \tag{3.5}$$

where the dimensionless variable $\eta \equiv (p/\omega)(d\omega/dp)$ is the "frequency slip factor" for the ring. Using this, the phase difference equation becomes (approximately)

$$\phi_k - \phi_{k-1} \simeq -\frac{2\pi h}{\omega_{k-1}(E_{k-1})} \left(\frac{\omega E}{p^2}\eta\right)^s_{k-1} \Delta E_{k-1}. \tag{3.6}$$

The sign of η is of some importance. In a low energy beam, a proton of higher energy traverses the ring more quickly, by virtue of its greater velocity, so that η is positive. At high energies, however, all protons travel near the speed of light, but the closed orbits of the higher energy ones have a greater path length; the bending fields are too weak to keep it on the synchronous track. Thus, η is negative. The energy at which η goes through zero and switches from positive to negative is called the **transition energy** of the machine.[1] For the sake of argument, we shall assume that the synchronous energy is already above transition in subsequent discussion; the reader can easily work out the corresponding results for energies below transition.

Taken together, Eq. (3.3) and Eq. (3.6) are almost in the form of the "standard mapping" that has received so much attention recently in the literature. (see, for example, Chirikov, Ref. 6) It has been discovered that if the parameters of the mapping are within certain domains, then the mapping exhibits chaotic behavior in which the state variables rapidly lose all information about their history. However, for realistic machine parameters this is not the case; instead, the mapping is smooth and non-chaotic.

PROBLEM: Assign values to the parameters according to the design of your favorite accelerator. Show that the mapping is indeed smooth for these values.

Our second approximation is this: Given that (1) the mapping is smooth, and (2) a large number of cycles are required to change a proton's state appreciably, it is possible to think of the sequence of numbers generated by the recursion equations as coming from the sampling of continuous functions. This viewpoint suggests that we pass from difference equations to differential equations. For reasons that will be obvious in a moment, it is useful to introduce the variable

$$W \equiv \frac{2\pi}{\omega^s} \Delta E \ , \tag{3.7}$$

and use it in place of ΔE. The differential equations appropriate to discrete sampling equations are written by approximating the time derivative of a quantity as its discrete difference divided by the cycling period. The latter is $2\pi/\omega$; thus, we obtain,

$$\begin{aligned} \frac{dW}{dt} &\simeq \frac{2\pi}{\omega^s(t)} \cdot \frac{\omega(E,t)}{2\pi} \cdot eV\,(\sin\phi - \sin\phi^s) \\ &\simeq eV\,(\sin\phi - \sin\phi^s) && \text{(a)} \\ \frac{d\phi}{dt} &\simeq -\frac{h}{2\pi}\left(\frac{\omega^2 E}{p^2}\,\eta\right)^s W \ , && \text{(b)} \end{aligned} \tag{3.8}$$

where it is formally noted that ω^s depends on t and ω on E and t. The final form of Eq. (3.8a) follows from keeping only the lowest order term in the Taylor series expansion of Eq. (3.5). (This is certainly an excellent approximation provided that we do not differentiate a second time to get d^2W/dt^2.)

PROBLEM: Write the second order differential equation obeyed by the angle variable ϕ of the biased pendulum drawn in Fig. (3.1). Make a connection to Eqs. (3.8).

The differential Eqs. (3.8) possess fixed-point solutions at $W = 0$ (i.e., $E = E^s$) and $\phi = \phi^s$ (mod 2π) or $\phi = \pi - \phi^s$(mod 2π). Behavior of solutions very close to the fixed points can be obtained by expanding Eq. (3.8a) to first order in $\phi - \phi^s$. This results in the linear system, written in matrix form,

$$\begin{aligned} \frac{d}{dt}\begin{pmatrix} \phi - \phi^s \\ W \end{pmatrix} &= \begin{pmatrix} 0 & -\Omega^2/eV\cos\phi^s \\ eV\cos\phi^s & 0 \end{pmatrix} \begin{pmatrix} \phi - \phi^s \\ W \end{pmatrix} \\ &\equiv A \cdot \begin{pmatrix} \phi - \phi^s \\ W \end{pmatrix}, \end{aligned} \tag{3.9}$$

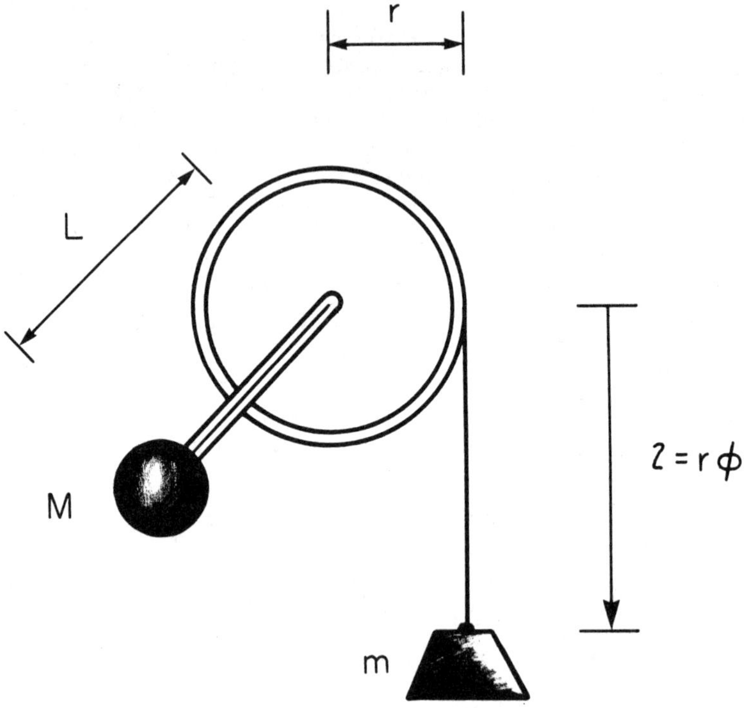

Fig. 3.1 The biased pendulum is a mechanical analogue of synchrotron oscillation. The mass m biases the pendulum so that it is at rest at some non-vertical angle ϕ^s.

$$\Omega^2 = \frac{h}{2\pi}\left(\frac{\omega^2 E}{p^2}\,\eta\right)^s eV\cos\phi^s ,$$

where we've used $d\phi^s/dt = 0$. The characteristic polynomial of the matrix A is $\lambda^2 + \Omega^2$. Provided that $\Omega^2 > 0$, that is, provided that η and $\cos\phi^s$ have the same sign, the eigenvalues of A are pure imaginary; viz., $\lambda = \pm i\Omega$. In turn, this means that these fixed points are stable and that nearby solutions in (ϕ,W) phase space wind around them at frequency Ω. Accordingly, Ω is called the **synchrotron oscillation frequency.** It is the frequency of oscillatory solutions in the vicinity of the stable fixed points. Of course, it is now possible to go back and rewrite Eq. (3.8b) in terms of Ω.

$$\frac{d\phi}{dt} = -\,(\Omega^2/eV\cos\phi^s)\;W \;. \tag{3.10}$$

PROBLEM: Repeat this analysis for the fixed points at $W = 0$ and $\phi = \pi - \phi^s \pmod{2\pi}$. Show that, under the same conditions as above, these are unstable fixed points, and calculate the rate at which nearby solutions diverge away from them.

In answer to the original problem, then, stability requires that η and $\cos\phi^s$ have the same sign; above transition we must have $\pi/2 < \phi^s < 3\pi/2$, while below it, we need $-\pi/2 < \phi^s < \pi/2$.

PROBLEM: Verify that this is correct by using simple pictorial reasoning on what happens to the phase of the proton when it is on either side of synchrony.

The first order differential Eqs. (3.8) can be derived from a "Hamiltonian" in which W and ϕ play the roles of conjugate variables. Let us define

$$H \equiv -\frac{\Omega^2}{2eV\cos\phi^s}\,W^2 + eV(\cos\phi + \phi\sin\phi^s). \tag{3.11}$$

It is trivial to verify that "Hamilton's equations" are satisfied

$$\frac{d\phi}{dt} = \frac{\partial H}{\partial W} \;,\quad \frac{dW}{dt} = -\frac{\partial H}{\partial \phi} \;. \tag{3.12}$$

Because of the stability condition, $\cos\phi^s < 0$, the "kinetic energy" term has the correct sign. The Hamiltonian is not a constant of the motion, because E^s--and through it, Ω--depends explicitly on time. Ignore that for the moment: provisionally accept the non-trivial assertion that if E^s changes little in time $1/\Omega$, that is,

$$\frac{1}{2\pi}\frac{\omega^s}{\Omega}\frac{eV\sin\phi^s}{E^s} \ll 1 \qquad (3.13)$$

then we can treat the dynamics as though H were exactly constant and adiabatically superpose the effects of varying H later. Formally, the problem is then equivalent to that of a particle on a line moving in the potential which is sketched in Fig. (3.2), for $\pi/2 < \phi^s < \pi$. For a given value of H, the motion is confined to definite regions, one of which is unbounded, in the variable ϕ. If the conditions are such that the motion is stably confined to oscillate about $\phi = \phi^s$, then it is said that the particle is trapped in a "bucket." There are an infinity of such buckets separated by 2π in phase. Obviously, the physical situations described by a particle with Hamiltonian H trapped in a bucket at ϕ^s and that of one with Hamiltonian $H + 2\pi eV \sin\phi^s$ in a bucket at $\phi^s + 2\pi$ are identical.

Trajectories of the proton in (ϕ,W) phase space are equivalent to the loci H=constant, and can be visualized easily in two steps. First, use Eq. (3.11) to solve for W^2 and plot these curves for various values of H. Figure (3.3a) shows some examples. Only $W^2 > 0$ is valid, so only the upper half-plane is important. To get the trajectories, simply set $W = \pm (W^2)^{1/2}$. This gives the Fig. (3.3b). For obvious reasons, this particular figure is sometimes called the "fish diagram."

The stable and unstable fixed points are easily identified in the fish diagram. The trajectory which goes through the unstable fixed points is called the **separatrix**; it separates regions of stable and unstable motion.* Close to a stable fixed point the motion is that of a harmonic oscillator of frequency Ω, as has been shown already. In the unstable region, W decreases without limit at an **average** rate, according to Eq. (3.8a), of

$$\left\langle\frac{dW}{dt}\right\rangle = - eV\sin\phi^s. \qquad (3.14)$$

This is precisely what might have been expected; it means that on the average the particle energy is not increasing, that is,

$$\left\langle\frac{dE}{dt}\right\rangle = 0 \ . \qquad (3.15)$$

PROBLEM: At what angle does the separatrix cross itself at an unstable fixed point?

The careful reader should already be skeptical of the numerous approximations introduced as stepping stones to the final results. As a blatant example, there was the inconsistency in keeping first-order terms in ΔE in Eq. (3.5) and dropping them from Eq. (3.8a). This was done, of course, to make the connection with

---

*More correctly, the trajectory approaches an unstable fixed point in the limit $t \to \pm \infty$.

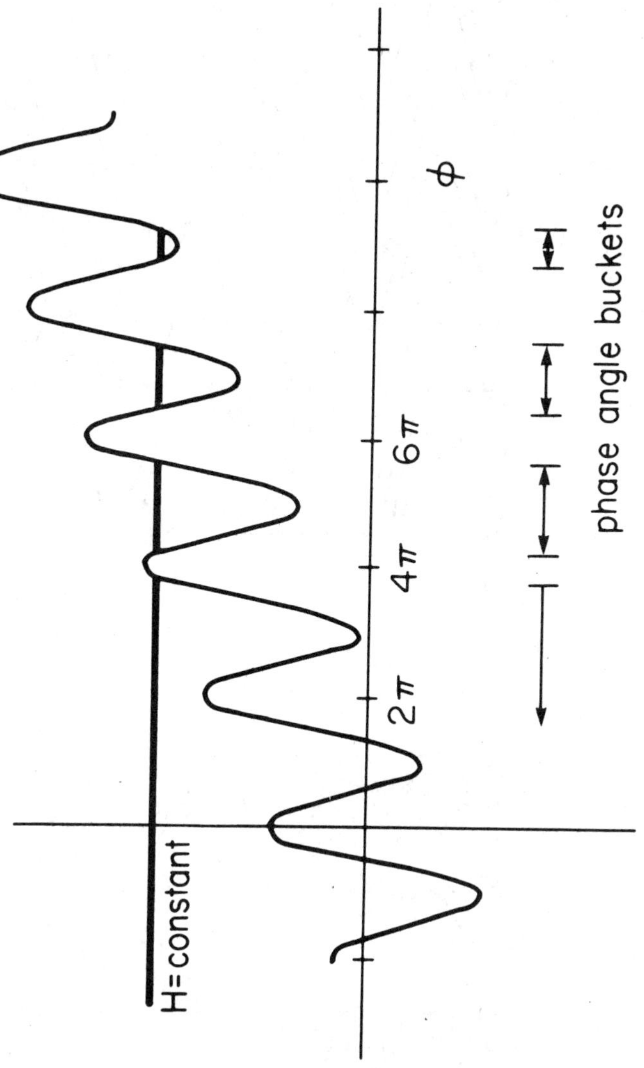

Fig. 3.2 The effective Hamiltonian for synchrotron oscillation has as a "potential" the superposition of a linear and a cosine term. For a given value of H, the phase angle "buckets" are the allowed regions of motion.

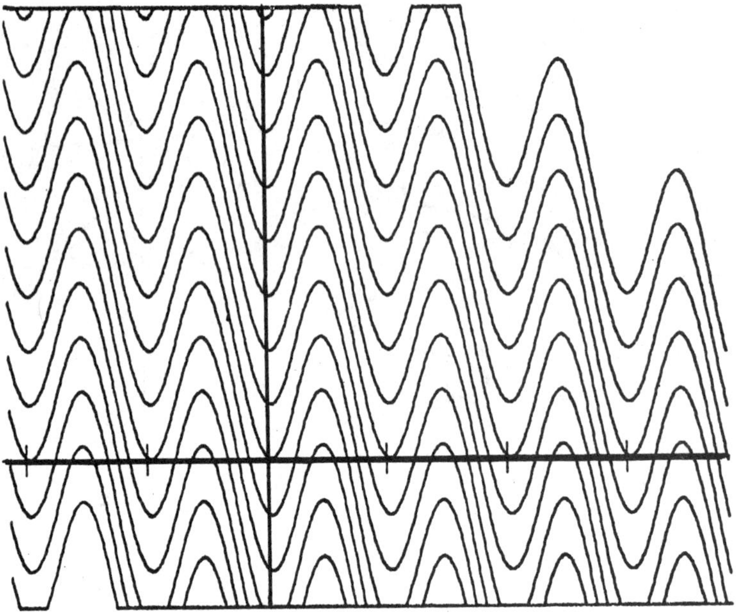

a.

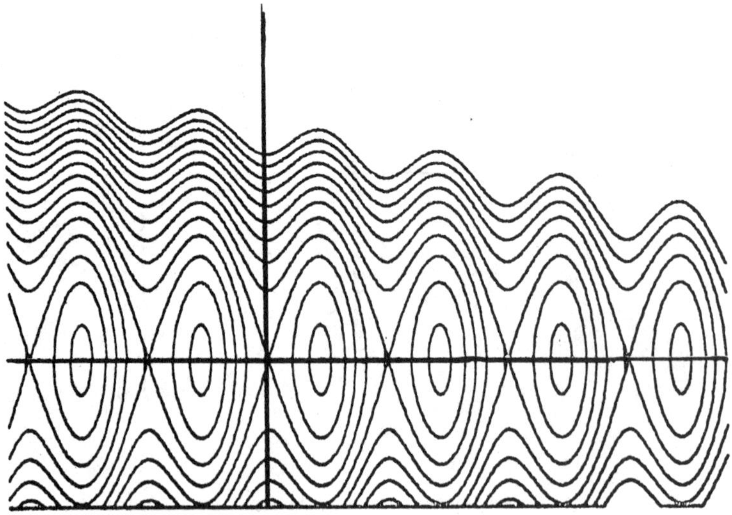

b.

Fig. 3.3 a) W^2 vs. ϕ for constant values of H. b) W vs. ϕ for the same values of H. A "phase space bucket" corresponds to the **stable** region inside a separatrix.

the Hamiltonian of Eq. (3.11), so it was a useful step. But was it valid? Have we lost thereby any **significant** aspect of the system's behavior? A number of "adiabatic" approximations have also been made; that is, we have neglected the time variation of quantities that are in fact time dependent, such as ϕ^s and Ω. In addition, we have completely neglected the coherent effects arising from the influence of different particles in the beam upon one another via such things as "wake fields" set up in the cavity when the beam passes through it and interactions with the wall of the beam tube in the ring. The subject of longitudinal stability is more complicated than one would infer from the presentation given here, which is meant to be no more than an introduction. The reader is referred to later talks of the summer school as well as the material in the bibliography for more information on the subject.

Problem on Transition Energy

(a) As the energy increases in a circular accelerator the frequency of revolution changes because the velocity changes and because the path length of the equilibrium orbit changes. Show that in a strong focusing synchrotron these changes are opposite in sign. The transition energy is defined as the energy at which the rate of change of revolution frequency is zero. Find an expression for the transition energy in terms of the "momentum compaction factor," which is the fractional change in orbit path length per unit fractional change in momentum.

(b) Starting with the phase oscillation equation, show that there exists a phase of stable equilibrium and one of unstable equilibrium and that whether we have one or the other depends on whether the energy is above or below the transition energy. Show that a given fixed point changes from being stable to unstable or the reverse as the energy passes through transition.

(c) Starting with the full phase oscillation equation, derive an equation for small deviations in phase from the fixed point. Solve this equation in the adiabatic limit. Show that as the transition energy is approached the adiabatic solution fails.

(d) Start with the small amplitude phase oscillation equation. Define a frequency slip factor, η, as the fractional change of frequency per unit fractional change in momentum. In the nonadiabatic region near transition, where η is near zero, approximate η/E by a linear function of time. Solve the equation in terms of Bessel and Neumann functions of order 2/3. Show that bunches of particles become sharply bunched as the transition energy is reached and that the corresponding energy spread increases. Show that it is possible to re-establish phase stability after transition is passed by appropriately shifting the phase of the applied rf. This can be accomplished by introducing a perturbation in the applied frequency for a certain time period. Estimate this time period and show that its value is not critical.

(e) Consider the strong focusing equation for betatron oscillations with an orbit perturbation induced by a momentum error. Using Fourier analysis, find an espression for the "momentum compaction factor" and show that the transition γ is approximately given by the betatron oscillation wave number, ν.

Problem on Adiabatic Damping

An important result of particle acceleration is the "decrease" in the phase space area occupied by the beam in the effect described as "adiabatic damping."[8] The acceleration is a slow process in terms of betatron and synchrotron oscillations and is therefore "adiabatic" so that the action integral $I = \int p\, d\, q$ integrated over a phase oscillation is constant.

a) For transverse motion ($p = p_y = m\, v_y \gamma$, $q = y$) show that the invariance of the action integral implies that the quantity defined in section 1 (Eq. 1.17)

$$W = \frac{\nu}{\pi R}\left(y^2 + \frac{R^2}{\nu^2}\, y'^2\right)$$

must decrease as $1/\beta\gamma$ during acceleration. You may assume the betatron motion is given by

$$y(s) = \hat{y} \cos\left(\frac{\nu s}{R}\right)$$

with ν and R constant, and $\hat{y}$ the adiabatic variable.

It is customary to label $\beta\gamma\varepsilon$ as the "invariant emittance," where ε is the maximum value of W, the emittance, defined in Eq. 1.19. $\beta\gamma\varepsilon$ remains constant during acceleration.

b) How does the maximum amplitude of Eq. 1.18 vary in acceleration?

c) In longitudinal motion the synchrotron frequency ν_s varies adiabatically during acceleration. Including this variation show from the invariance of $I = \oint W d\phi$ how a particle's maximum energy amplitude ΔE_{max} and phase error ϕ_{max} change during acceleration. You may use Eq. (3.8) to obtain equations for ϕ, W.

4. SYNCHROTRON RADIATION AND ELECTRON BEAM DYNAMICS

D. Neuffer
Fermi National Accelerator Laboratory, Batavia, Illinois 60510

Introduction

In the preceding chapters, we have explored beam dynamics in accelerators and storage rings, in which particle orbits are bent and focussed by magnetic fields, and particle energies are changed by rf electric fields. These electromagnetic fields accelerate charged particles, and we may recall from college physics the rule that charged particles radiate energy when accelerated. This radiation inside synchrotrons is called, logically enough, synchrotron radiation and has important effects on beam dynamics.

In the first proton synchrotrons, radiation was a negligible effect. In electron synchrotrons, this radiation becomes increasingly important at higher energies, and its effects are the dominant constraint in the design of all present and future electron-(positron) storage rings with energies greater than ~1 GeV.[10]

The synchrotron radiation has been found useful as a source in radiation-matter experiments,[11-13] and some electron synchrotrons are produced with the deliberate purpose of producing radiation, the circulating electrons are used only as radiation sources.[14] Radiation can be an important tool in reducing e^- and e^+ beam emittances ("cooling") and, as discussed by Chao,[15] can be used to produce polarized $e^\mp$ beams. In the highest energy proton synchrotrons, synchrotron radiation is used as a diagnostic tool (to "see" beams)[16] and in future higher energy machines it may have important dynamic effects.[17]

In the following sections we describe the nature and effects of this synchrotron radiation.

4A. Synchrotron radiation - physical properties

The quantum electrodynamics calculation of radiation by accelerated charges was presented by Schwinger in 1949;[18] the classical calculation was first performed by Schott in 1911[19] and is presented in the textbooks of Jackson[20] and Landau and Lifshitz.[21] An extensive treatment is provided by Sekolov and Ternov.[22]

We quote the result[20] that the energy loss per second (radiated power) P of a charged particle of charge e, mass m, and 4-momentum p_μ is

$$P = \frac{2}{3}\frac{e^2}{m^2c^3}\left(\frac{dp_\mu}{d\tau}\right)^2, \tag{4.1}$$

where $d\tau = dt/\gamma$ is the proper time element, c is the velocity of light, γ is the relativistic factor, t is laboratory time, and

$$\left(\frac{dp_\mu}{d\tau}\right)^2 = \left(\frac{d\vec{p}}{d\tau}\right)^2 - \frac{1}{c^2}\left(\frac{dE}{d\tau}\right)^2 \tag{4.2}$$

and cgs units are used in Eq. (4.1).

We will use the relativistic approximations that the velocity $\beta c \simeq 1$ for all particles, and $\gamma = (1-\beta^2)^{-1/2} >> 1$, since this is true wherever synchrotron radiation is important.

It can be shown that in synchrotrons the dominant contribution is due to the transverse acceleration from the bending magnetic field B

$$\left(\frac{dp}{d\tau}\right) = \gamma \frac{dp}{dt} \simeq \gamma eB = \frac{\gamma E_0}{\rho} >> \frac{1}{c}\frac{dE}{d\tau}, \tag{4.3}$$

where ρ is the machine bending radius ($\rho = E_0/eB$) and E_0 is the central machine energy and can be rewritten as $m\gamma_0 c^2$. Then

$$P \simeq \frac{2}{3}\frac{e^2 c\gamma^2\gamma_0^2}{\rho^2} \simeq \frac{2e^2E^2E_0^2}{3m^4c^7\rho^2} \text{ (cgs)}. \tag{4.4}$$

In Eq. (4.4) we have carefully separated the particle-dependent factor E^2 from the machine-dependent factors E_0^2, ρ^2. The energy loss per turn is

$$U(E) = \rho\left(\frac{2\pi\rho}{c}\right) = \frac{4}{3}\frac{\pi e^2E^2E_0^2}{m^4c^8\rho} \simeq AE^4 \tag{4.5}$$

with $A = 4\pi\rho^2/3(mc^2)^4$ and we note $E_0 \approx E$. A useful formula for electron/positron rings is

$$U(\text{MeV}) = 8.85\times10^{-2}\,\frac{E_0^4}{\rho^2}\,\frac{(\text{GeV})}{(\text{m})}. \tag{4.6}$$

The angular distribution of the radiation is given, approximately, by the formula[20]

$$\frac{dP}{d\Omega} \sim \frac{3}{\pi} P\gamma^2 \frac{1}{(1+\gamma^2\theta^2)^3}\left[1 - \frac{4\gamma^2\theta^2\cos^2\phi}{(1+\gamma^2\theta^2)^2}\right], \tag{4.7}$$

where θ is the angle between the particle velocity and the radiation, ϕ is the polar angle between the radiation and the acceleration (see Fig. 4.1), $d\Omega$ is the solid angle $\sin\theta\, d\theta\, d\phi$, and $\beta \approx 1$ is assumed. The radiation is approximately confined within a small angle $\theta \approx 1/\gamma$ and peaked at that angle (see Fig. 4.2).

The frequency distribution of the radiation is characterized by a critical frequency $\omega_c = 3\gamma^3 c/2\rho$ and distributed as shown in

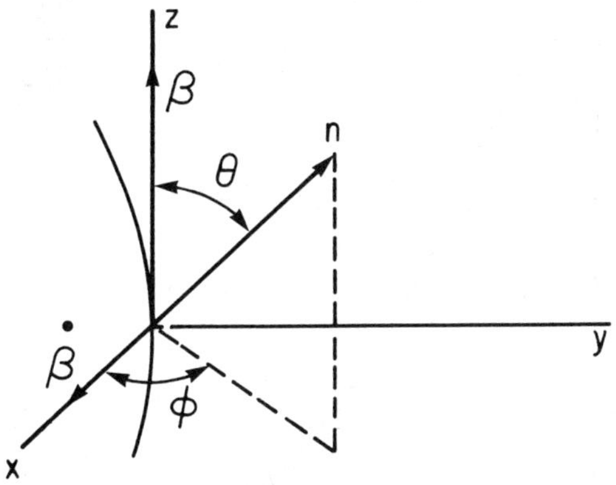

Fig. 4.1 Synchrotron radiation emission with acceleration ($\dot{\beta}$) perpendicular to particle velocity (β). Radiation power in the direction indicated by n is given in Eq. (7) as a function of the angles θ, ϕ. (From Ref. 11.)

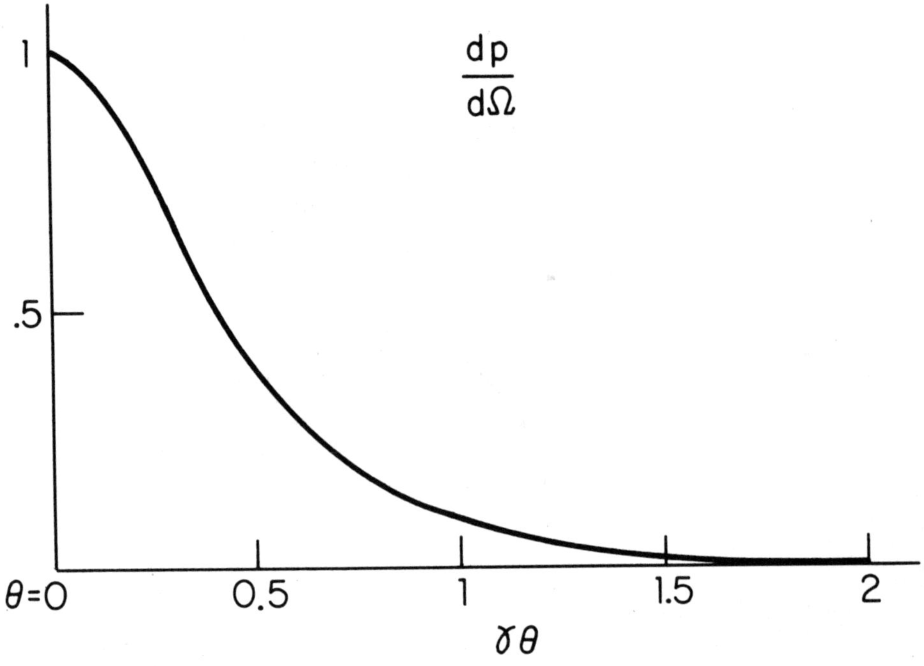

Fig. 4.2 Radiation power density ($dP/d\Omega$) as a function of the scaled angle $\gamma\theta$ for relativistic radiation.

Fig. 4.3,[23] where the power spectrum $dP/d\omega$ and the photon number spectrum $dN/d\omega dt$ are shown in terms of a scaled variable $\xi = \omega/\omega_c$, and

$$\frac{dN}{d\omega dt} = \frac{1}{\hbar\omega}\frac{dP}{d\omega}. \tag{4.8}$$

$dP/d\omega$ can be expressed analytically as

$$\frac{dP}{d\omega} = \frac{P}{\omega_c} S\left(\frac{\omega}{\omega_c}\right), \tag{4.9}$$

where $S(\chi) = (9(3)^{1/2}/8\pi)\ \chi \int_\chi^\infty K_{5/3}(\chi')\, d\chi'$, and $K_{5/3}$ is a modified Bessel function.

We note here simply the result that the radiation is characterized by photons of energy $\approx\hbar\omega_c$ emitted within an angle $\approx 1/\gamma$ and that $U/\hbar\omega_c \simeq N_\gamma$ is approximately the number of photons emitted per turn by a particle. (Actually $N_\gamma = (15(3)^{1/2}/8)\ U/\hbar\omega_c$.)

4B. Effects of radiation on energy

As we noted above, particles stored in a synchrotron radiate energy and they would be lost if this energy were not continually restored. Energy is provided to beams by rf cavities as described in the previous chapter.[24] The balance between radiation losses and rf acceleration has important consequences in beam dynamics.[23]

We first consider the longitudinal motion of a single particle in energy error-phase error (ε,τ) coordinates, where ε is the energy difference from some reference energy $E_0(\varepsilon = E - E_0)$, and $\tau = T - T_0$ is the difference in phase from the reference phase in seconds. Without radiation the equations of motion are

$$\begin{aligned}\frac{d\varepsilon}{dt} &= ef_0V_{rf}\sin\omega_{rf}\,T \\ &\simeq e\, f_0V_{rf}[\sin\,\omega_{rf}T_0 + \omega_{rf}\ \tau\cos\,\omega_{rf}\ T_0].\end{aligned} \tag{4.10}$$

$$\frac{d\tau}{dt} = \frac{-\eta_E}{E_0}\,\varepsilon \tag{4.11}$$

In these equations V_{rf} is the rf voltage, $\omega_{rf}/2\pi$ is the rf frequency, f_0 is the revolution frequency of the reference particle, and

$$\eta_E \equiv \frac{E_0}{f_0}\frac{\partial f_0}{\partial E} = 1/\gamma_T^2 - 1/\gamma^2 \simeq 1/\gamma_T^2 \tag{4.12}$$

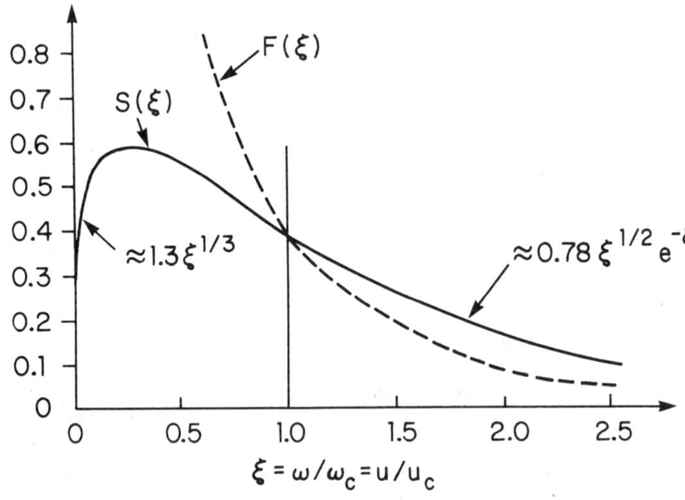

Fig. 4.3 Normalized power spectrum S and photon number spectrum F of synchrotron radiation. (From Ref. 14.)

is the "momentum compaction factor" and in Eq. (4.10) we have used the approximation $\omega_{rf}\tau \ll 2\pi$, and $\gamma \gg 1$ is used throughout.

We first approximate energy loss from synchrotron radiation as the loss of an amount of energy U(E) on each turn.

$$\frac{d\varepsilon}{dt} = -f_0 U(E) \simeq -U_0 - \frac{f_0 \partial U(E)}{\partial E} E. \tag{4.13}$$

We set $U_0 \equiv U(E_0) = e V_{rf} \sin(\omega_{rf} T_0)$ and combine Eqs. (4.10), (4.11), and (4.13)

$$\frac{d^2\varepsilon}{dt^2} \simeq - f_0 \frac{\partial U}{\partial E}\frac{d\varepsilon}{dt} - \Omega^2 \varepsilon, \tag{4.14}$$

where

$$\Omega^2 = f_0 \omega_{rf} \frac{\eta_E}{E_0} e V_{rf} \cos(\omega_{rf} T_0) \tag{4.15}$$

which is recognizable as a damped harmonic oscillator equation. Thus energy errors decrease with time, since

$$\frac{\partial U}{\partial E} = 2AE\, E_0^2 \simeq 2A\, E^3 \tag{4.16}$$

from Eq. (4.5) and A is positive definite. The solution of Eq. (4.14) is

$$\varepsilon(t) \propto e^{-Bt} \cos(\Omega' t + \phi) \tag{4.17}$$

with ϕ an arbitrary initial phase and $B = .5 f_0 \partial U/\partial E$ and

$$\Omega' = \sqrt{\Omega^2 - B^2} \approx \Omega. \tag{4.18}$$

We are interested in $d/dt\, \langle \varepsilon^2 \rangle$, the mean square energy deviation for a "bunch" of particles. From Eq. (4.14)

$$\frac{d}{dt} \langle \varepsilon^2 \rangle = -2B \langle \varepsilon^2 \rangle = - f_0 \frac{\partial U}{\partial E} \langle \varepsilon^2 \rangle \tag{4.19}$$

which indicates that the combination of the synchrotron radiation U(E) per turn plus a restoring rf field reduces energy spread; this is "radiation damping."

The above picture is not complete. Since radiation is quantized, there will be statistical fluctuations in the energy radiated in a turn. The typical number of photons per turn is $N_T \simeq U(E)/\hbar\omega_c$, where $\hbar\omega_c$ is a typical photon energy. The fluctuation in N_T is $\sim(N_T)^{1/2}$ from usual statistical arguments, so the fluctuation in energy emitted per turn is of order

$$\Delta_E \simeq \sqrt{N_T}\,\hbar\omega_c \simeq \sqrt{U\hbar\omega_c}.$$

This estimate can be made more accurate by integration over the actual photon distribution; the result is $\Delta_E = (1.323)^{1/2}(U\hbar\omega_c)^{1/2}$.[14]

Since this is a random fluctuation, it cannot be inserted directly into the equations of motion; however, these "quantum fluctuations" can be combined with the "radiation damping" in the equation

$$\frac{d\langle\varepsilon^2\rangle}{dt} = - f_0\left(\frac{\partial U}{\partial E}\langle\varepsilon^2\rangle - \frac{\Delta_E^2}{2}\right), \tag{4.20}$$

where the factor of 1/2 appears from τ–ε oscillations.

This has the equilibrium solution

$$\langle\varepsilon\rangle \equiv (\langle\varepsilon^2\rangle)^{1/2} = \left(\frac{\partial U}{\partial E}\right)^{-1/2}\frac{\Delta_E}{\sqrt{2}} = \sqrt{\frac{E_0(1.323.\hbar\omega_c)}{2}}$$
$$= \sqrt{\frac{3(1.323)\hbar mc^3\gamma^4}{4\rho}} \tag{4.21a}$$

or

$$\langle\varepsilon^2\rangle = \frac{E_0\langle\hbar\omega\rangle}{2}, \tag{4.21b}$$

where $\langle\hbar\omega\rangle$ indicated an averaged photon energy. Our result is that the beam develops an energy spread $\langle\varepsilon\rangle$ which is independent of the detailed machine design and only dependent on $E_0^2/(\rho)^{1/2}$. Since electron storage rings are designed with energies $E_0 \cong (\rho)^{1/2}$, the relative energy spread $\langle\varepsilon\rangle/E_0$ is about the same for all of them and is $\sim 10^{-3}$.

The shape of the energy distribution is a Gaussian, as can be proved from the "Central Limit Theorem" and the random nature of the quantum fluctuations and is given by

$$f(E) \propto e^{-\frac{\varepsilon^2}{2\langle\varepsilon^2\rangle}}. \tag{4.22}$$

A similar distribution in the time displacement τ arises by the same effects

$$f(\tau) \propto e^{-\frac{\tau^2}{2\langle\tau^2\rangle}}, \qquad (4.23)$$

where

$$\langle\tau^2\rangle = \left(\frac{\eta_E}{\Omega_{E_0}}\right)^2 \langle\varepsilon^2\rangle. \qquad (4.24)$$

4C. Radiation effects on transverse amplitudes

The same phenomena of "radiation damping" and "quantum fluctuations" also occur in horizontal and vertical motion, and the same feature of an equilibrium Gaussian amplitude distribution with a width determined by the balancing of these effects also occurs.

We first consider transverse motion. Particles undergo betatron oscillations approximated by[16,17]

$$x = A\sqrt{\beta_x}\cos\phi \qquad (4.25)$$

$$\frac{dx}{ds} \equiv x' \simeq \frac{A}{\sqrt{\beta_x}}\sin\phi = \frac{p_x}{p_{total}}, \qquad (4.26)$$

where A is the particle amplitude, and we have made the approximation that the betatron function $\beta_\perp$(see Sec. 1)[16] is constant and the velocity $\beta c \simeq c$. ($\langle A^2\rangle$ is transverse "emittance.")

In radiation, photon emission is within the small angle $\theta \lesssim 1/\gamma$ of the particle direction. We assume $1/\gamma$ is small and $\theta = 0$. Then both x and x' are unchanged when a photon of energy δE is emitted [see Fig. 4.4(a)].

However, when the energy δE is restored in an rf cavity, x' is reduced [see Fig. 4(b)], since $p_\perp$ (transverse momentum) is unchanged while $E(\cong p)$ is increased. Therefore on one turn

$$x' \rightarrow x' - \frac{U_0}{E_0}x'. \qquad (4.27)$$

Averaging over phases ϕ, we find that the amplitude changes according to

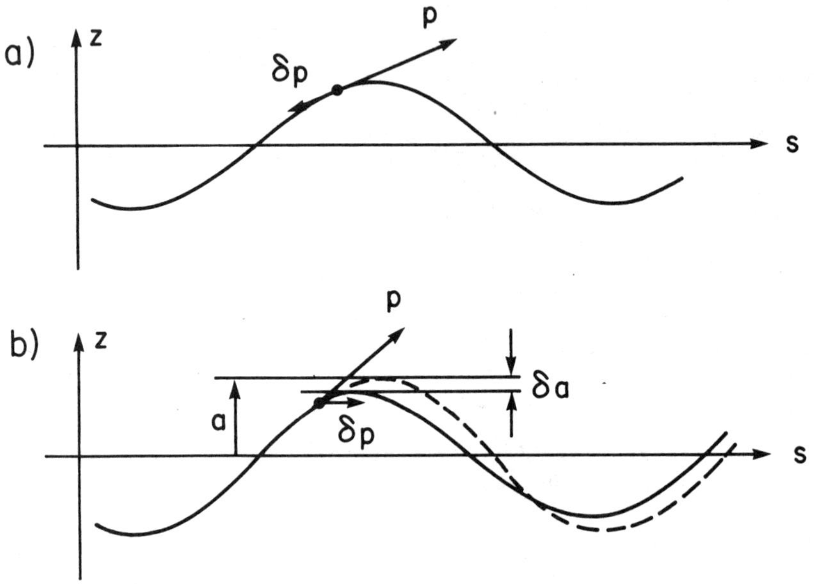

Fig. 4.4 Effect of an energy change on the vertical betatron oscillations: (a) for radiation loss, (b) for rf acceleration. (From Ref. 14.)

$$\frac{d}{dt}\langle A^2\rangle \simeq -\frac{f_0U_0}{E_0}\langle A^2\rangle, \tag{4.28}$$

where U_0 is the total radiation per turn and E_0 is the central energy. Note that this is half the energy damping rate

$$\frac{d}{dt}\langle \varepsilon^2\rangle \simeq -2\,\frac{f_0U_0}{E_0}\langle \varepsilon^2\rangle. \tag{4.29}$$

In a complete calculation, which includes effects of betatron oscillations and the momentum dependence of orbits, the damping Eq. (4.28) is correct for vertical motion but is modified slightly for horizontal motion and Eq. (4.29) is also modified, when effects of the momentum on the motion are included.[23] The result is

$$\frac{d}{dt}\langle A^2\rangle_x \simeq -\left(1-\frac{\eta_x}{\rho}\right)\frac{f_0U_0}{E_0}\langle A^2\rangle_x \tag{4.30}$$

and

$$\frac{d}{dt}\langle \varepsilon^2\rangle \simeq -\left(1+\frac{\langle\eta_x\rangle}{2\rho}\right)\left(\frac{2f_0U_0}{E_0}\right)\langle \varepsilon^2\rangle, \tag{4.31}$$

where η_x is the average off-momentum function, usually $\langle\eta_x\rangle \simeq 10^{-2}\,\rho$ for large storage rings.

The mechanism for generation of transverse "quantum fluctuations" is somewhat different. We first consider horizontal motion. Horizontal motion is momentum dependent. We rewrite (4.25), (4.26) as

$$x \simeq A\sqrt{\beta_x}\cos\phi + \eta_x\frac{\varepsilon}{E_0} \tag{4.32}$$

$$x' \simeq \frac{A}{\sqrt{\beta_x}}\sin\phi + \eta_x'\frac{\varepsilon}{E_0}, \tag{4.33}$$

where η_x is the off-energy function. To estimate the effect we assume η_x is constant ($\eta_x' = 0$) and consider emission of a single photon of energy δE. The amplitude and phase A and ϕ change as $\varepsilon \to \varepsilon - \delta E$ while x, x' are unchanged (see Fig. 4.5). We find

$$\left\langle \Delta\left[\frac{(x-\eta\varepsilon/E_0)^2}{\beta_x} + x'^2\beta_x\right]\right\rangle = \langle(\delta A)^2\rangle \simeq \frac{(\eta_x)^2(\delta E)^2}{\beta_x E_0^2} \tag{4.34}$$

with $\delta E = \hbar\omega_c$, we find

$$\frac{d}{dt}\langle A^2\rangle \simeq \frac{f_0U_0\langle\hbar\omega_c\rangle\eta_x^2}{\beta_x\,E_0^2}, \tag{4.35}$$

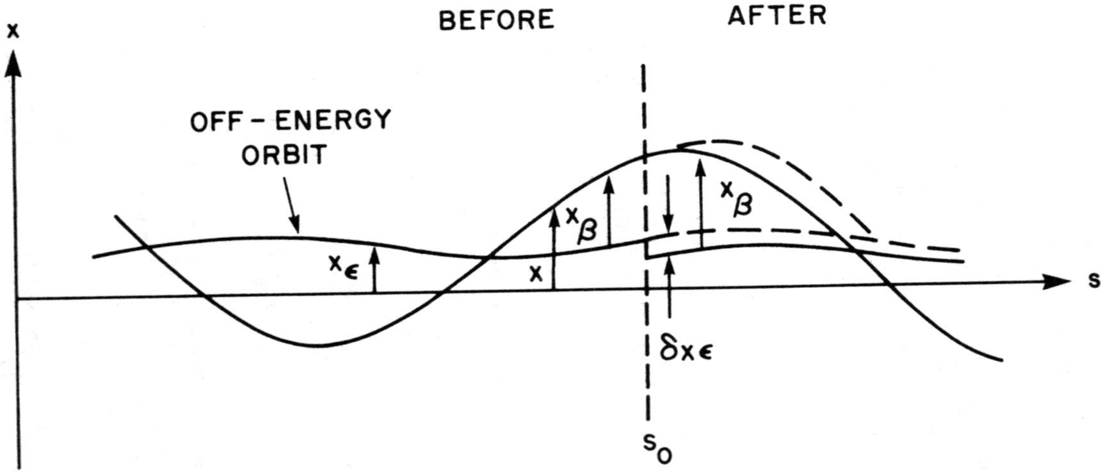

Fig. 4.5 Effect of a sudden energy change at s_0 on the horizontal betatron displacement. (From Ref. 14.)

where $\langle\hbar\omega_c\rangle$ is an energy averaged over all photons, and differs by a factor of order unity from $\hbar\omega_c$.

An approximate for $\langle\eta_x{}^2\rangle$ for most storage rings is[13]

$$\langle\eta_x^2\rangle \simeq \eta_E\langle\beta_x^2\rangle \simeq \frac{\langle\beta_x^2\rangle}{\gamma_T^2} \simeq \frac{\langle\beta_x^2\rangle}{\nu_x^2}, \tag{4.36}$$

where ν_x is the betatron tune.* The approximate symbols in Eq. (4.36) are correct in magnitude but differences of order 25-50% may appear in particular rings.

Equations (4.35) and (4.28) may be combined to find an equilibrium value of $\langle A^2\rangle$

$$\langle A^2\rangle \simeq \frac{\langle\hbar\omega_c\rangle\ \langle\eta_x^2\rangle}{E_0\langle\beta_x\rangle}. \tag{4.37}$$

The beam width ($\langle x^2\rangle_\beta = \beta_x\langle A^2\rangle/2$) can be expressed as

$$\langle x^2\rangle_\beta \simeq \frac{1}{2}\frac{\langle\hbar\omega_c\rangle}{E_0}\langle\eta^2\rangle_x\frac{\beta_x}{\langle\beta_x\rangle}, \tag{4.38}$$

where $\beta_x(s)$ is the betatron value where $\langle x^2\rangle$ is measured and $\langle\beta_x\rangle$ is the averaged value of Eq. (4.37). This can be compared with the energy spread

$$\langle\varepsilon^2\rangle = \frac{\langle\hbar\omega_c\rangle E_0}{4} \tag{4.39}$$

to obtain a relation directly connecting them

$$\langle x^2\rangle_\beta \simeq 2\langle\varepsilon^2\rangle\left\langle\frac{\eta_x^2}{E_0^2}\right\rangle, \tag{4.40}$$

where $\langle\hbar\omega_c\rangle \simeq 1.323\ \hbar\omega_c$, and $\beta_x(s) = \beta_x$ is assumed.

We have obtained the result that "quantum fluctuations" plus "radiation damping" leads to an equilibrium value of the horizontal emittance A^2 as well as in energy spread ε^2.

In vertical motion we have $\langle\eta_y^2\rangle \simeq 0$, and the above mechanism does not directly produce vertical fluctuations. Fluctuations in the angle of photon emission $\theta \sim (\gamma)^{-1}$ do occur, but the contribution to $\langle\delta A^2\rangle$ is of the order γ^{-2} of the result in Eq. (4.34) (see

---

*We note here that the off-energy function η_x is not directly related to the momentum compaction η_ε, and that the betatron function $\beta_x(s)$ is not at all related to the speed βc. The "usual" accelerator notation can be confusing.

problem 4.3). In real storage rings there is usually substantial coupling between horizontal and vertical motion and that coupling produces a vertical emittance. We introduce coupling coefficients g_y, g_x with the properties

$$\begin{aligned} & g_x + g_y = 1 \\ & \langle \delta A^2 \rangle_x = g_x \langle \delta A^2 \rangle \text{ and } \langle \delta A^2 \rangle_y = g_y \langle \delta A^2 \rangle, \end{aligned} \tag{4.41}$$

where $g_y \ll g_x$ (usually). In practice the coupling can be varied to provide optimum luminosity.[10] An approximate value of vertical emittance is then

$$\langle A^2 \rangle_y \simeq g_y \frac{\langle \omega_c \rangle}{E_0} \frac{\langle \eta_x^2 \rangle}{\beta_x} \tag{4.42}$$

and "radiation damping" plus "quantum fluctuations" determine both vertical and horizontal beam sizes.

Discussion

In this section, we have described the features of synchrotron radiation and described how this radiation determines the properties of beams in electron (and positron) storage rings. The same effects also occur in proton (antiproton) and all other storage rings but are much smaller because of the powers of γ (E/mc^2) in these effects. Other factors (such as beam-gas scattering, non-linear optics, beam mismatch, source properties, etc.) determine energy spread and beam size in present $p(\bar{p})$ rings, although in future higher energy rings radiation effects will become increasingly important. We hope that this paper provides a useful introduction to this field and invite the reader to consult the references for more detailed discussions.

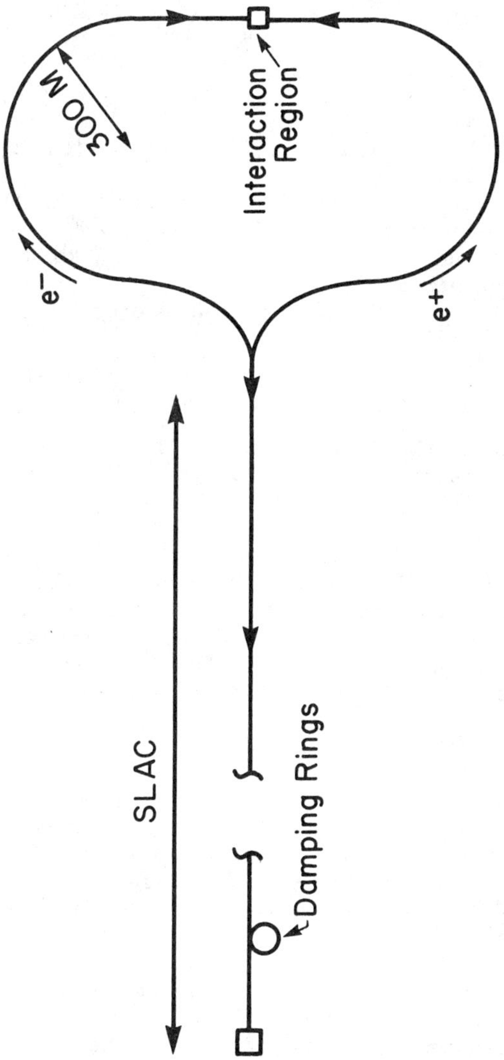

Fig. 4.6 The SLAC Single Pass Collider. The damping rings reduce $e^{\pm}$ source emittances by synchrotron radiation damping. Problem 4 asks for calculation of radiation effects in the collider arcs.

PROBLEMS

Problem 4.1

a. What is a corresponding formula [to Eq. (6)] for proton rings?

b. What is the energy loss per turn U for LEP? (5 km radius, 80-GeV electron synchrotron)

c. What is U for the Tevatron? (1 km radius, 1-TeV proton synchrotron)

d. What is a typical photon energy ($\hbar\omega_c$) for LEP? for the Tevatron?

Problem 4.2

a. What is the damping time $(2B)^{-1}$ for LEP? for the Tevatron?

b. What is the equilibrium energy spread $\langle\Delta\varepsilon^2\rangle^{1/2}$ for LEP?

Problem 4.3

As mentioned in Section 4C, quantum fluctuations due to the finite angle of photon emission ($\theta \simeq 1/\gamma$) lead to an amplitude growth $d\langle A^2\rangle/dt$. Obtain an approximate expression for this effect, following techniques similar to those used to obtain Eqs. (4.20) and (4.35). Show that its contribution to $\langle x^2\rangle$ is of order $1/\gamma^2$ of Eq. (4.38).

Problem 4.4

In the SLAC Single Pass Collider, 50-GeV beams of electrons and positrons are transported along separate arcs (see Fig. 4.6) to collisions in an interaction region. Assume each arc contains a total bend of 270° with a bending radius of 300 m with no rf acceleration in the arcs.

(a) How much energy does an average particle lose on these arcs due to synchrotron radiation?

(b) Estimate the mean energy spread of the beam from this radiation.

(c) Assume that in the transport lattice the average β-function $\langle\beta_x\rangle$ is 5 m and the average η-function $\langle\eta_x\rangle$ is 0.15 m. Estimate the increase in horizontal emittance due to the radiation.

REFERENCES

1. E. D. Courant and H. S. Snyder, "Theory of the Alternating Gradient Synchrotron," Ann. of Phys. **3**, 1 (1958).
2. F. Cole, "Notes on Accelerator Theory," MVRA-TN-250, 1961.
3. M. Month, "Effects of Sextupoles Near a One-Third Integral ν Value in the AGS: Theory," AGS CD-17, July, 1967.
4. A. Schoch, "Theory of Linear and Non-Linear Perturbations of Betatron Oscillation in Alternating Gradient Synchrotron," CERN 57-1 (1978).
5. G. Guignard, "A General Treatment of Resonances in Accelerators," CERN 78-11 (1978).
6. B. V. Chirikov, "A Universal Instability of Many Dimensional Oscillator Systems," Physics Reports **52**, 263 (1979).
7. J. Livingood, **Principles of Cyclic Particle Accelerators**, Ch. 6, "Phase Stability," (D. Van Nostrand, New York, 1961).
8. H. Bruck, **Circular Particle Accelerators** (tr.), Ch. XVI, "Synchronous Acceleration," Los Alamos Scientific Laboratory, LA-TR-72-10.
9. K. R. Symon and A. M. Sessler, "Methods of Radio Frequency Acceleration...," **High Energy Accelerators and Pion Physics 1**, CERN, Geneva, 1956, p. 44-58.
10. E. Keil, these proceedings.
11. "Special Issue: Synchrotron Radiation," Physics Today, Vol. 34 No. 5 May 1981.
12. **Synchrotron Radiation Research**, H. Winick and S. Doniach, Eds., (Plenum Press, New York, 1980).
13. A. Bienenstock, IEEE Trans. Nucl. Sci. **NS-26**, 3780 (1979).
14. A. van Steenbergen, IEEE Trans. Nucl. Sci. **NS-26**, 3785 (1979).
15. A. Chao, these proceedings.
16. A. Hofmann, IEEE Trans. Nucl. Sci. **NS-28**, 2131 (1981).
17. R. R. Wilson, private communication.
18. J. S. Schwinger, Phys. Rev. **75**, 1912 (1949).
19. G. A. Schott, **Electromagnetic Radiation** (Cambridge University Press, 1912).
20. J. D. Jackson, **Classical Electrodynamics** (J. Wiley & Sons, Inc., 1962).
21. L. P. Landau and E. M. Lifshitz, Classical Theory of Fields Addison-Wesley, Reading, Massachusetts, 1951.
22. A. A. Sokolov and I. M. Ternov, **Synchrotron Radiation** (Pergamum, New York, 1968).
23. M. Sands, The Physics of Electron Storage Rings, Stanford Linear Accelerator Center Report, SLAC-121, 1970; also in Proc. Int. School of Physics, B. Touschek, Ed. (Enrico Fermi) Varenna 1969 (Academic Press, New York, 1971).
24. L. Michelotti, these proceedings.
25. R. D. Ruth and W. T. Weng, these proceedings.

APPENDIX. BEAM EXTRACTION FROM A CIRCULAR ACCELERATOR

Lee C. Teng
Fermi National Accelerator Laboratory, Batavia, Illinois 60510

INTRODUCTION

There are two essential components in a beam-extraction system:

1. A septum to define the "in" and "out" of the accelerator. The "out" side of the septum is followed by an extraction channel which leads the beam out of the machine. The septum and channel create the need for long straight sections. Thus, the principal consideration in the design of the magnet lattice is its facility for extraction. There are two major types of septa, electrostatic and magnetic.
2. A stepping mechanism to step the beam across the septum. Depending on the application, the stepping mechanism can be either fast or slow.

> Fast - Beam is generally stepped over by a fast kicker magnet and extracted in one turn. This mode of extraction is used principally for transferring beams from one accelerator to another.
>
> Slow - The limited resolutions of experimental detectors demand that the beam be spilled out slowly over ~1 sec. One prefers a totally smooth and uniform spill giving a 100% duty factor. For slow spill the step-over mechanism must depend on some parameter or parameters in which the beam has a distribution, preferably a wide one. The most convenient mechanism is a resonance in the transverse motion.

SEPTUM

1. Electrostatic septum

The highest practical field that can be held without breakdown in the presence of an intense particle beam is about 100 kV/cm. For relativistic particles, this gives a deflecting force equivalent to that of ~1/3 kG magnetic field. The septum can be very thin (~50 μm). It is difficult to keep a foil taut in two dimensions when it is struck by beam and sags because of heating. Hence, the common practice is to use an array of thin wires, say 50 μm diameter W/Re wires with ~1 mm spacing. With careful alignment, the effective thickness of the septum could be less than 100 μm. The cross-sectional view of such a wire septum could be as shown in Fig. 1.

The leakage field generally terminates on the floor and ceiling of the grounded wire frame and is hence dependent on the vertical gap of the frame. Although, next to the wire plane the field is ripply, it can easily be seen that the integrated effect of the field along

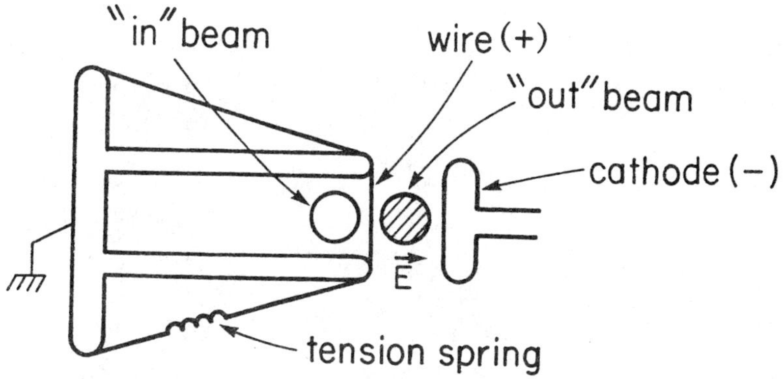

Fig. 1 The cross-sectional view of a wire septum.

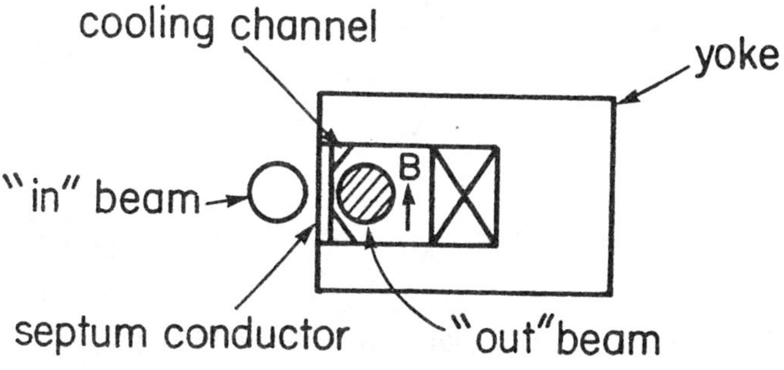

Fig. 2 Cross-section of a current septum magnet.

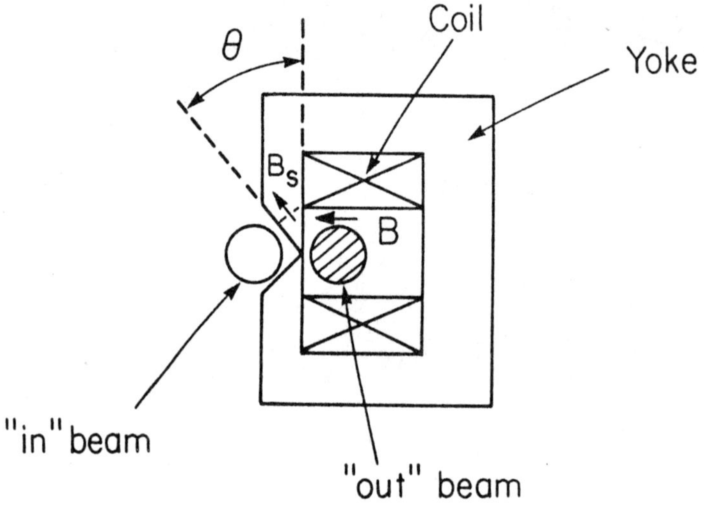

Fig. 3 Cross-section of an iron septum magnet.

the length of the septum is independent of the ripple and the distance from the wire plane. The electrostatic septum consumes no power and is generally operated dc. One may try to pulse the septum to get to higher field except that the breakdown field is not substantially higher until one gets to very fast pulses (~1 μsec).

2. Magnetic septum

There are two types of magnetic septum:

a. Current septum (see Fig. 2)

The strength of the septum depends on the current density j and the thickness t of the septum. We have

$$Bh = \frac{4\pi}{10} jth \quad \text{(Gaussian units)}$$

or

$$\frac{B}{t} = \frac{4\pi}{10} j \simeq 1.3 \text{ kG/cm} \quad \text{(dc)},$$

where the numerical value corresponds to a maximum current density of ~1 kA/cm². Even with water cooling along the edges, it is difficult to go beyond this current density for thin septa operated dc. Compared to electrostatic septum, we see therefore that for thicknesses less than some 2.5 mm, electrostatic septum is stronger. When pulsed or ramped at even several msec duration, however, the maximum current density could be 1-2 orders of magnitude higher and one can get ~10 kG with a current septum some 2 mm thick.

b. Iron septum (Lambertson septum)

Referring to Fig. 3 one easily sees that the allowable angle θ of the yoke cut-out is given by

$$\sin\theta = \frac{B}{B_s} = \frac{B}{18\text{kG}} ,$$

where B_s is the saturation field of iron. For B = 9 kG, we get θ = 30°, giving a 120° cut-out. The deflections of the "out" beam in the current and the iron septa are in different (perpendicular) planes, which must be taken into consideration in the design. For dc applications, the iron septum is much stronger.

FAST STEP-OVER MECHANISM

The commonly employed method is to transversely deflect the beam by a fast pulsed magnet (kicker). If the field and length of the kicker are B and ℓ, the angle of kick θ_k is given by

$$\theta_k = \frac{B\ell}{B\rho} ,$$

where $B\rho$ is the magnetic rigidity of the particle. 90° betatron phase advance downstream at the septum, the angle θ_k will result in a transverse displacement x_s given by

$$x_s = (M_{12})\ \theta_k = (\sqrt{\beta_s\beta_k}\ \sin\phi_{sk})\ \theta_k = \beta_{max}\ \frac{B\ell}{B\rho} ,$$

where M_{12} is the 12 element of the transfer matrix, β_s and β_k are the amplitude functions at the septum and the kicker and are assumed to be both equal to the maximum value β_{max}, and $\phi_{sk} = 90°$ is the betatron phase advance from k to s. Take, for example, 100 GeV/c protons for which $B\rho = 10^4/3$ kGm, a kicker with B = 1 kG and ℓ = 2 m, and β_{max} = 50 m, we get

$$x_s = (50\ \text{m})\ \frac{2\times3}{10^4} = 30\ \text{mm}$$

which is adequate to clear the full width of most normal beams.

In order to minimize the beam loss on the septum (for the health of both the beam and the septum) one wants a fast rise-time. This depends on the technology of the power supply in switching a large current in a short time. At present, a reasonable switching rate is given by

$$\frac{B}{\tau} \sim 6\ \text{G/nsec}.$$

For example, to get B = 1 kG, the rise time τ must be ~170 nsec, and for a rise time of τ = 10 nsec one can only get a peak field of B~60 G.

In addition to a fast rise-time, one would also like a very thin septum. Hence, one generally starts with an electrostatic septum followed by a thicker magnetic septum when the displacement has grown to > 3 mm (gap) plus beam width. Since the electrostatic septum is rather weak, the displacement will not likely reach this value in a reasonable distance and the electrostatic septum may become unreasonably long. The conventional arrangement is to terminate the electrostatic septum after a reasonable length when the beam is still contained within the accelerator aperture. The deflected beam, then, follows a coherent betatron oscillation and arrives at a much larger displacement 90° downstream where one can start inserting a much thicker, hence much stronger magnetic septum to quickly deflect the beam out of the machine. For this arrangement one needs a relatively short drift space (straight section) for the kicker. An odd multiple of 90° downstream one needs a longer straight section for the electrostatic septum. Another odd multiple of 90° downstream one needs a very long straight section for the magnetic septum and the main sections of the extraction channel to deflect the beam out of the machine. The amplitude functions at the kicker, the electrostatic septum and the magnetic septum should all be very large. All these conditions should be taken into consideration in the magnet lattice design.

SLOW STEP-OVER MECHANISM (RESONANCE)

For slow extraction, we move the betatron tune ν onto a (non-linear) resonance to make the oscillation amplitude grow in a controlled manner. The amplitude growth can be made as large as a few mm per turn so that in a few turns when the beam returns to the septum, the amplitude step-over could be $\sim$1 cm. The amplitude growth depends on the initial amplitude and the momentum, both parameters in which the beam has a distribution. Generally, the momentum distribution is much wider than the initial amplitude distribution (relative to the resonance mechanism) and the beam is pealed off in momentum to give a slow spill. We will discuss here the use of the 1/3 integer and the 1/2 integer resonances.

We digress here to introduce some discussion of the motion in one transverse degree of freedom under the influence of a single resonance. We ignore the motion in the second transverse degree of freedom, except to assume that it is far away from resonances and not coupled to the first degree of freedom. The general x-motion is given by the equations

$$\begin{cases} \dfrac{dx}{ds} = p_x \\[2ex] \dfrac{dp_x}{ds} = -\dfrac{B'}{B\rho}\,x - \dfrac{1}{B\rho}\left(B''\dfrac{x^2}{2!} + B'''\dfrac{x^3}{3!} + \ldots\right) \end{cases} \tag{A.1}$$

$$= -\frac{B'}{B\rho}\,x - \frac{1}{B\rho}\sum_{k=3}^{\infty} B^{(k-1)}\,\frac{x^{k-1}}{(k-1)!}\,,$$

where s is the coordinate along the closed equilibrium orbit, prime mean $\partial/\partial x$, and the kth term in the summation corresponds to the 2k-pole field. All field coefficients have the revolution periodicity. We assume that the small (linear) oscillation tune ν is close to the single resonance $\nu = m/n$. Now, we transform $(x,\ p_x)$ first to (u, p), then to the rotating polar coordinates $(r,\ \phi)$ by

$$\begin{cases} u = \dfrac{1}{\sqrt{\beta}}\, x = r \cos\left(\phi + \dfrac{m}{n}\theta\right) \\ \qquad\qquad\qquad\qquad\qquad \theta = \displaystyle\int \frac{ds}{\nu\beta}\,, \\ p = -\dfrac{\alpha x + \beta p_x}{\sqrt{\beta}} = r \sin\left(\phi + \dfrac{m}{n}\theta\right) \end{cases} \tag{A.2}$$

where α and β are the usual Courant-Snyder parameters. These transformations accomplish the following:

To (u,p) - takes out the wobbles of the unperturbed linear motion leaving a simple harmonic motion with circles as phase trajectories and phase points circling clockwise ν turns per revolution around the ring ($\Delta\theta = 2\pi$).

To (r, ϕ) - rotates the phase plane clockwise m/n turns per revolution so that if $\nu = m/n$ (exactly on resonance) all phase points corresponding to the linear motion will stand still. The motion in (r, ϕ) describes then only the effects due to off-resonance and non-linear perturbation terms.

If one analyzes the perturbation terms into harmonics around the ring, most of the harmonics produce only rapidly oscillatory effects which average to zero in time. The only terms which produce non-vanishing secular are

Zeroth-harmonic (θ-averages) of even k term - This produces an amplitude dependence of the non-linear tune at large amplitude. For instance, an octupole having a field proportional to x^3 is like a quadrupole field with strength (field gradient) proportional to x^2 or $(\text{amplitude})^2$ and hence produces a tune proportional to $(\text{amplitude})^2$.

Resonant-harmonic - In this case, this is the mth harmonic of the 2n-pole field, (k = n term). For $\nu = m/n$, the factor x^{n-1} has harmonics $(n-1)\,m/n = m - m/n$. Together with the mth harmonic of the field coefficient, this gives a driving term with harmonic $m - (m - m/n) = m/n$ which is in resonance with ν. Higher order (k) terms can also drive this resonance, but no term with $k < n$ can. We define

$$\begin{cases} \bar{A}_n \equiv \left\langle -\dfrac{\nu\beta}{2^n}\,\dfrac{\beta^{n/2}}{(n-1)!}\,\dfrac{B^{(n-1)}}{B\rho} \right\rangle_\theta \\ A_n \cos(m\theta - \psi) \equiv m^{th} \text{ harmonic of } -\dfrac{\nu\beta}{2^n}\,\dfrac{\beta^{n/2}}{(n-1)!}\,\dfrac{B^{(n-1)}}{B\rho}\,, \end{cases} \tag{A.3}$$

where the β factors come from the transformation and ψ specifies the phase of the resonant harmonic around the ring, so chosen that A_n is always positive. Keeping only secular terms, the equations for (r, ϕ) are

$$\begin{cases} \dfrac{d\phi}{d\theta} = (\nu - \dfrac{m}{n}) \begin{pmatrix} n \\ n/2 \end{pmatrix} \bar{A}_n r^{n-2} - A_n r^{n-2} \cos (n\phi + \psi) \\ \dfrac{dr}{d\theta} = \qquad\qquad\qquad\qquad - A_n r^{n-1} \sin (n\phi + \psi) \end{cases} \tag{A.4}$$

where $\begin{pmatrix} n \\ n/2 \end{pmatrix}$ is the binomial coefficient.

Solving $d\phi/d\theta = dr/d\theta = 0$ for r and ϕ, we get the fixed points. At these points, the non-linear tune is exactly m/n, although the linear tune ν is not. Phase trajectories passing through unstable fixed points are the separatrices. The innermost separatrices enclose the stable region.

1. Third integer resonance

Take the Fermilab Main Ring as an example. During acceleration $\nu = 19.4$. At extraction, we first turn on the driving sextupoles, then adjust the quadrupoles to move ν onto $\nu = 19+1/3 = 58/3$. In this case Eqs. (A.3) and (A.4) are

$$\begin{cases} \dfrac{d\phi}{d\theta} = (\nu - \dfrac{58}{3}) - Sr \cos (3\phi + \psi) \\ \dfrac{dr}{d\theta} = \qquad\qquad - Sr^2 \sin (3\phi + \psi) \ , \end{cases} \tag{A.5}$$

where

$$S\cos (58\theta - \psi) = 58\text{th harmonic of } -(\nu\beta/16)\ \beta^{3/2}\ (B''/B\rho).$$

This can be simplified to

$$\begin{cases} \dfrac{d\phi}{d\lambda} = \delta - r \cos (3\phi + \psi) \\ \dfrac{dr}{d\lambda} = - r^2 \sin (3\phi + \psi), \end{cases} \tag{A.6}$$

where

$$\delta \equiv \frac{1}{S} (\nu - \frac{58}{3}) > 0, \qquad\qquad \lambda \equiv S\theta.$$

A first integral of Eq. (A.6) is

$$\frac{1}{2}\,\delta r^2 - \frac{1}{3}\,r^3 \cos\,(3\phi + \psi) = K = \text{const.} \tag{A.7}$$

There are three unstable fixed points at

$$r = \delta,\; 3\phi + \psi = \begin{cases} 0 \\ 2\pi \\ 4\pi \end{cases} \text{ or } \phi = \begin{cases} -\psi/3 & \text{(A)} \\ -\psi/3 + 2\pi/3 & \text{(B)} \\ -\psi/3 + 4\pi/3 & \text{(C)} \end{cases}$$

The phase trajectory (A.7) passing through these unstable fixed points has $K = \delta^3/6$ and the equation

$$2r^3 \cos\,(3\phi + \psi) - 3\delta r^2 + \delta^3 = 0$$

or

$$\left[r\cos\,(\phi + \frac{\psi}{3} + \pi) - \frac{\delta}{2}\right]\left[r\cos\,(\phi + \frac{\psi}{3} + \frac{\pi}{3}) - \frac{\delta}{2}\right] \times$$

$$\cdot \left[r\cos\,(\phi + \frac{\psi}{3} - \frac{\pi}{3}) - \frac{\delta}{2}\right] = 0 \tag{A.8}$$

namely three straight lines intersecting at the unstable fixed points. These are plotted in Fig. 4 for $\psi = 0$ and $\delta > 0$. The arrows correspond to increasing λ.

To use the third integer resonance for extraction, we arrange it as shown in Fig. 5. These are static separatrices. In the dynamic case as δ is reduced, the triangular central stable area decreases, the corners at A, B and C open up and the central phase area streams out along AS and along the two other congruent out-streaming separatrices at B and C. The step size at the septum ($\theta = 0$, we also choose $\phi = 0$ at the septum to give a matched beam there) every 3 revolutions ($\Delta\theta = 6\pi$, when x is again near maximum at the septum) at the end of extraction ($\delta = 0$) is given by

$$\Delta x_s = -\,6\pi S\,\frac{x_s^{\,2}}{\sqrt{\beta_s}}\,.$$

This shows that the step size increases as x^2. The percentage of beam loss on the septum is simply given by (septum thickness)/(step size). The desired step size will give the required sextupole strength S, and the geometry in the phase diagram above gives the position (ψ) of the sextupole. For a given momentum when the triangular area equals the emittance of the beam, extraction will commence. Extraction ends when $\delta = 0$ or $\nu = 58/3$. But because of the chromaticity, particles of different momenta reach resonance at different times and the slow

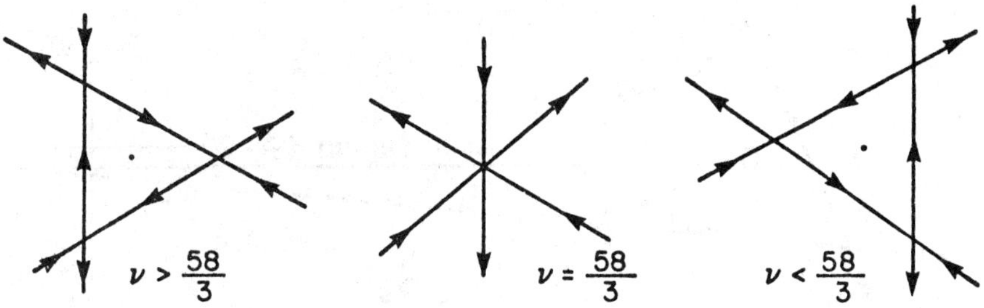

Fig. 4 Separatrices for a third integer resonance.

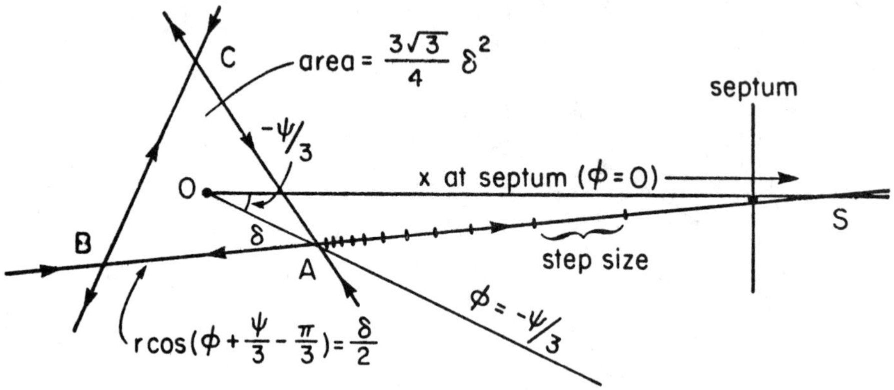

Fig. 5 Phase space diagram for a third integer resonant extraction.

spill is an unfolding or a pealing of the momentum spread of the beam.

2. Half integer resonance

This being a linear resonance has the same topology over the whole phase plane. To utilize this resonance for extraction, we must create some separatrices by adding, say, a zeroth harmonic octupole field to make the tune amplitude dependent. Again for the Fermilab Main Ring, we will move ν from the normal 19.4 to the resonance $\nu = 19+1/2 = 39/2$. For secular driving terms we take

$$\begin{cases} \overline{Q} \equiv \left\langle -\frac{\nu\beta}{4}\beta\frac{B'}{B\rho}\right\rangle_\theta & (0^{th}\text{ harmonic quadrupole}) \\ Q\cos(39\theta-\psi) \equiv 39^{th}\text{ harmonic of } -\frac{\nu\beta}{4}\beta\frac{B'}{B\rho} & \end{cases} \tag{A.9}$$

and

$$-\overline{E} \equiv \left\langle -\frac{\nu\beta}{96}\beta^2\frac{B'''}{B\rho}\right\rangle_\theta \qquad (0^{th}\text{ harmonic octupole}),$$

where it is assumed that $\overline{Q}$ and $\overline{E}$ so defined are both positive. The (ϕ, r) equations are then

$$\begin{cases} \frac{d\phi}{d\theta} = (\nu - \frac{39}{2} - 2\overline{Q}) - Q\cos(2\phi+\psi) + 6\overline{E}r^2 \\ \frac{dr}{d\theta} = -Qr\sin(2\phi+\psi) \end{cases} \tag{A.10}$$

or

$$\begin{cases} \frac{d\phi}{d\lambda} = -\delta - \cos(2\phi+\psi) + R^2 & (\delta>0) \\ \frac{dR}{d\lambda} = -R\sin(2\phi+\psi) & \end{cases} \tag{A.11}$$

where

$$-\delta \equiv \frac{1}{Q}(\nu - \frac{39}{2} - 2\overline{Q}), \quad R^2 \equiv \frac{6\overline{E}}{Q}r^2, \quad \lambda \equiv Q\theta$$

The first integral is

$$\frac{1}{2} R^4 - \left[\delta + \cos(2\phi + \psi)\right] R^2 = K = \text{const.} \qquad (A.12)$$

There are two unstable fixed points

$$R^2 = \delta - 1, \quad 2\phi + \psi = \begin{cases} \pi \\ 3\pi \end{cases} \text{or } \phi = \begin{cases} -\psi/2 + \pi/2 & (A) \\ -\psi/2 + 3\pi/2 & (B) \end{cases}$$

and two outboard stable fixed points

$$R^2 = \delta + 1, \quad 2\phi + \psi = \begin{cases} 0 \\ 2\pi \end{cases} \text{or } \phi = \begin{cases} -\psi/2 & (C) \\ -\psi/2 + \pi & (D) \end{cases}$$

The separatrices are phase trajectories passing through the unstable fixed points, hence given by $K = -1/2\ (\delta-1)^2$. The equation is

$$R^4 - 2\left[\delta + \cos(2\phi + \psi)\right] R^2 + (\delta - 1)^2 = 0 \qquad (A.13)$$

or when factorized

$$\left[R^2 + 2R\cos\left(\phi + \frac{\psi}{2}\right) - (\delta-1)\right]\left[R^2 - 2R\cos\left(\phi + \frac{\psi}{2}\right) - (\delta-1)\right] = 0. \qquad (A.14)$$

This gives two circles as shown in Fig. 6.

The development from here on parallels that of the third integer resonance extraction. The step size is now, however, more complicated and is given at the septum ($\theta=0$, $\phi=0$) and at the end of extraction ($\delta=1$) by

$$\Delta x_s = \sqrt{\frac{3\bar{E}}{Q\beta_s}}\, x_s^2 \sqrt{2 - \frac{3\bar{E}}{Q\beta_s}\, x_s^2} \qquad (A.15)$$

We have intentionally treated the two resonances as being approached from opposite directions, but the most important difference of the two cases is that for the third integer resonance, the central stable phase area vanishes only when exactly on resonance, i.e., when $\nu = 58/3$. If the resonance is crossed too fast, there is a danger of retrapping the phase points in the newly created stable area on the other side of resonance before the unstable phase points have time to stream out. On the other hand, for the half integer resonance, the central stable area shrinks to zero when δ decreases to 1 and does not reappear until δ reaches -1. Over the range of 2 in δ or equivalently, 2Q in ν, the topology of the central region of the phase plane is totally open and the phase points have plenty of time to stream out, thus giving a cleaner extraction. This unstable range of ν is the so-called "stop band" typical of a half integer (linear) resonance.

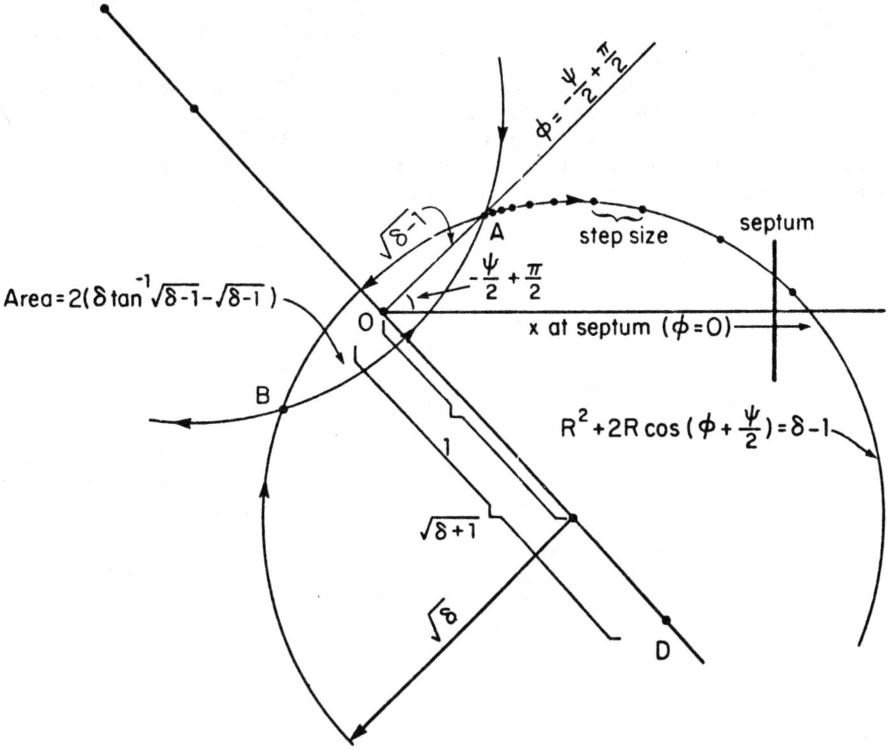

Fig. 6 Phase space diagram for a half integer resonant extraction.

It is a good exercise to work out a detailed design of a complete resonant slow extraction system:

(a) Take into account the emittance and the momentum spread of the beam, the tune, the dispersion, the chromaticity and the long straight section arrangement of the accelerator.

(b) Give the parameters and the designs of the resonance driving elements, the tune control scheme, the electrostatic and the magnetic septa, and the extraction channel.

(c) Compute the extraction efficiency, and the emittance and momentum spread of the extracted beam.

(d) Study the effects of power supply ripples and noises on the spill. Devise and design filter and feedback systems to suppress these effects to improve the duty factor of the spill.

SINGLE BEAM COHERENT INSTABILITIES IN CIRCULAR ACCELERATORS AND STORAGE RINGS

C. Pellegrini
Brookhaven National Laboratory, Upton, New York 11973

TABLE OF CONTENTS

0094-243X/82/870077-70$3.00 Copyright 1982 American Institute of Physics

SINGLE BEAM COHERENT INSTABILITIES IN CIRCULAR ACCELERATORS AND STORAGE RINGS

C. Pellegrini
Brookhaven National Laboratory, Upton, New York 11973

INTRODUCTION

Coherent instabilities of particle beams in circular accelerators and storage rings have been the subject of many theoretical and experimental works, starting from the early 1960's. I will not try to review all the work done in nearly two decades and will limit myself to give references only to the most recent works or the works more strictly related to the particular way I describe these phenomena. An excellent review of the early work can be found in Ref. 1 and for some aspects, more related to storage rings, in Ref. 2.

The coherent instabilities we want to discuss are produced by the electromagnetic interaction of the charged particle beam with the walls of the vacuum chamber in which the beam is moving. The wall geometry is complicated by the presence in the accelerator of the diagnostic equipment, radiofrequency cavities, and other equipment necessary to operate the system. The field produced by the beam and modified by the walls causes a force, proportional to the beam current, acting on the beam itself and that can lead to what we call a coherent instability.

Since the wall geometry can be very different for the various accelerators it is very useful to follow the idea of Vaccaro and Sessler[3] of characterizing the vacuum chamber and all the associated devices, which we will call the "beam environment," by an impedance function $Z(\omega)$, relating the beam current and the electromagnetic field perturbing the beam. In this way we can study the beam dynamics in general terms and then apply the results to a particular accelerator by specifying the impedance.

Although we can write the general equations describing the beam dynamics and the stability properties, it is not possible to obtain a general solution of these equations for a realistic beam charge distribution and wall impedance. It is however possible to obtain analytical solutions valid in a certain region of the parameter space of this problem and numerical solutions in the more complicated cases.

It is possible to think of instability mechanisms different from the beam wall interaction considered here. One example is the two species instability which can occur for instance in a proton beam partially neutralized by electrons produced by ionization of the residual gas in the vacuum chamber. The electrons and protons can execute coherent oscillations of growing amplitude leading to beam loss. Although these effects can be important under some conditions they will not be discussed here.

The paper is organized as follows. In Sections 2 and 3 we discuss, using a simple example, what we mean by coherent instabilities and illustrate the techniques that can be used to study this problem. In particular, we introduce the Vlasov equation and show how it can be used to evaluate the stability properties.

In Sections 4 and 5, we consider a particle and introduce the equations of motion for the case of longitudinal and transverse coherent oscillations. In Sections 6 and 7 we derive an expression for the longitudinal and transverse coherent forces. These forces are then used in the Vlasov equations to obtain a dispersion relation and define the limiting conditions for stability (threshold current). These threshold conditions are obtained for several different sets of beam parameters, corresponding to what we call slow or fast instabilities and long or short wavelength of the perturbation. The longitudinal stability is discussed in Sections 8 and 9 and the transverse stability in 10 and 11.

2. THE STABILITY PROBLEM

We consider a set of particles, for instance a beam of an accelerator, and assume that for this system there exists an equilibrium state. We want to know whether a small perturbation around the equilibrium state will grow (instability) or decay (stability).

As an example let us consider a set of one-dimensional oscillators subject to a collective force proportional to the center-of-mass displacement. We can write the equation of motion for particle "ℓ" as

$$\ddot{x}_\ell + \omega_\ell^2 x_\ell = N \int G\,(t - t')D\,(t')dt', \tag{1}$$

where

$$D(t) = \frac{1}{N} \sum_{\ell=1}^{N} x_\ell, \tag{2}$$

and we assume the collective force to be proportional to the number of particles, N.

The Green function G(t) must satisfy the causality condition, $G(t) = 0$ if $t < 0$. This function describes the memory of the system producing the collective force. In the simplest case in which the system has no memory, all interactions are instantaneous in time and $G(t-t') = \delta(t-t')$. An example of a system with memory is a resistive metallic wall. The fields produced by the charges and currents induced by an electron passing near the wall can last long after the particle has moved away.

We have considered the possibility that the particles have different oscillation frequencies, ω_ℓ. A difference in frequency can be produced either by a variation in the energy, ΔE, of the particles or by the action of nonlinear forces, which we have neglected in the writing of Eq. (1). In the second case the frequency depends on the particle oscillation amplitude,

$$\omega_\ell^2 = \omega^2\,(\Delta E_\ell, |x_\ell|). \tag{3}$$

We assume that the variation in frequency is small

$$\frac{\Delta\omega_\ell}{\omega_\ell} << 1 \tag{4}$$

and call ω_0 the central value of the frequency distribution.

For our set of oscillators the equilibrium state is that in which the center of mass does not move and D(t) = 0. We now perturb the system by displacing the center of mass. To study the motion that follows the perturbation, we look for a solution of Eq (1) of the form

$$x_\ell = A_\ell\, e^{-i\Omega t} \tag{5}$$

$$D = \bar{A}\, e^{-i\Omega t} \tag{6}$$

$$\bar{A} = \frac{1}{N}\sum_{\ell=1}^{N} A_\ell \tag{7}$$

such that all particles oscillate coherently with the same frequency. Substituting in Eq. (1) and introducing the "impedance" function

$$iZ(\omega) = \int dt\; G(t)e^{+i\omega t} \tag{8}$$

we obtain

$$(\omega_\ell^2 - \Omega^2)A_\ell = 2\pi i\bar{A}NZ(\Omega). \tag{9}$$

A solution of Eq. (9) can be easily obtained if we assume that ω_ℓ is not a function of A_ℓ. The case when ω_ℓ is a function of A_ℓ is a more complicated problem that we will study in the next section with the help of the Vlasov equation. From Eq (9) dividing by $(\omega_\ell{}^2 - \Omega^2)$ and summing over all particles we obtain a dispersion relation

$$1 = 2\pi iNZ(\Omega)\left\{\frac{1}{N}\sum_{\ell=1}^{N}\frac{1}{\omega_\ell^2 - \Omega^2}\right\}. \tag{10}$$

Its solution defines a set of characteristic frequencies and for each of these frequencies we obtain from Eq. (9) a set of amplitudes A_ℓ, defining a normal mode of oscillation of the system.

For a beam of particles the condition that the collective force is small compared to the focusing is often satisfied. Then, to first order in Z, we can substitute in Eq. (10) $Z(\Omega)$ with $Z(\pm\omega_0)$. If all particles have the same frequency the solution of Eq. (10) is

$$\Omega = \omega_0 + \frac{\pi N}{\omega_0} \mathrm{Im}Z(\omega_0) - \frac{i\pi N}{\omega_0} \mathrm{Re}Z(\omega_0) \tag{11}$$

$$\Omega = -\omega_0 - \frac{\pi N}{\omega_0} \mathrm{Im}Z(\omega_0) - \frac{i\pi N}{\omega_0} \mathrm{Re}Z(\omega_0). \tag{12}$$

To obtain this result we have used the property that the real and imaginary parts of $Z(\omega)$ are odd and even

$$\mathrm{Re}\ Z(-\omega) = -\mathrm{Re}\ Z(\omega) \tag{13}$$

$$\mathrm{Im}\ Z(-\omega) = \mathrm{Im}\ Z(\omega) \tag{14}$$

property which follows from the fact that $G(t)$ must be a real function.

The quantity

$$\delta\Omega = i\pi N Z(\omega_0)/\omega_0 \tag{15}$$

appearing in Eq. (11) represents the shift between the collective frequency Ω and the unperturbed frequency ω_0, and is called the coherent frequency shift.

It is interesting to notice that both solutions in Eqs. (11) and (12) have the same imaginary part. Since $D \sim e^{-i\Omega t}$ we have from Eq. (6) that the motion is stable if

$$\mathrm{Re}\ Z(\omega_0) > 0 \tag{16}$$

and unstable if

$$\mathrm{Re}\ Z(\omega_0) < 0. \tag{17}$$

The rise time of the instability is then given by

$$\frac{1}{\tau} = -N \frac{\mathrm{Re}Z(\omega_0)}{2\omega_0}. \tag{18}$$

Notice that for an instantaneous interaction, $G(t) = \delta(t)$, $Z(\omega) = -i$ and the system is stable.

We can use the properties in Eqs. (13) and (14) of $Z(\omega)$ to simplify the dispersion relation. Since it is enough to obtain information on the stability property, to calculate the root near ω_0, we can rewrite Eq. (10) as

$$1 = \delta\Omega \left\{ \frac{1}{N} \sum_{\ell=1}^{N} \frac{1}{\omega_\ell - \Omega} \right\}. \tag{19}$$

To obtain an instability we need a real part of $Z(\omega)$, producing a $\pi/2$ phase shift between the particle oscillation and the center-of-mass oscillation in Eq. (1).

For a large number of particles we can substitute the sum with an integral and write

$$1 = \delta\,\Omega \int d\omega \frac{g(\omega)}{\omega - \Omega}, \tag{20}$$

where $g(\omega)$ is the frequency distribution function normalized to one. To study this dispersion relation in Eq. (20) let us assume that the frequency distribution is non-zero only in an interval $\omega_0 - \Delta$, $\omega_0 + \Delta$ and rewrite Eq. (20) as

$$1 = \delta\,\Omega\, L\,(\Omega), \tag{21}$$

where

$$L(\Omega) = \int_{-1}^{1} dx \frac{g(x)}{x - x_1} \tag{22}$$

$$x_1 = (\Omega - \omega_0)/\Delta. \tag{23}$$

For a complex impedance the solution of Eq. (20) is a complex number

$$x_1 = r + i\alpha \tag{24}$$

obtained from the equations

$$\frac{\delta\Omega_R - i\,\delta\Omega_I}{\delta\Omega_R^2 + \delta\Omega_i^2} = L_R(x_1) + iL_I(x_1), \tag{25}$$

where $\delta\Omega_R$, $\delta\Omega_I$, L_R, L_I are the real and imaginary parts of $\delta\Omega$, $L(x_1)$ and

$$L_R(x_1) = \int_{-1}^{1} dx\, g(x) \frac{x - r}{(x - r)^2 + \alpha^2} \tag{26}$$

$$L_I(x_1) = \int_{-1}^{1} dx\, g(x) \frac{\alpha}{(x - r)^2 + \alpha^2} . \tag{27}$$

We can first solve Eq. (25) in the case when

$$r^2 + \alpha^2 >> 1 \tag{28}$$

which means that the collective frequency is well outside the frequency distribution $g(\omega)$. Expanding Eqs. (26) and (27) as a power series in $x/(r^2 + \alpha^2)^{1/2}$ we obtain, to the lowest order,

$$L_R = -\Delta r \tag{29}$$

$$L_I = \Delta\alpha \tag{30}$$

and using Eqs. (23) and (25)

$$\Omega = \omega_0 - \delta\Omega \tag{31}$$

which is the same result obtained for the case when all particles have the same frequency. In particular if $\mathrm{ReZ}(\omega_0) < 0$ the system is unstable.

Since r and α are proportional to $\delta\Omega$ the condition in Eq. (28) can only be satisfied, for a given impedance, for a large number of particles, N. If we decrease N the imaginary part of $\delta\Omega$ will decrease, and the system becomes less unstable. If we decrease N to the point where the condition in Eq. (28) is no more satisfied and the characteristic frequency Ω can be inside the frequency distribution, we can obtain for a particular value of N a solution of the dispersion relation in Eq. (25) with $\alpha \to 0^+$. This solution has a rise-time going to infinity and defines the maximum stable value of N or "threshold intensity."

In the limit $\alpha \to 0^+$ and using the relationship

$$\lim_{\alpha\to 0^+} \frac{\alpha}{(x - r)^2 + \alpha^2} = \pi\delta(x - r) \tag{32}$$

the integrals expressed in Eqs. (26) and (27) become

$$L_I(x_1) = \pi g(r) \tag{33}$$

$$L_R(x_1) = \lim_{\alpha\to 0^+} \int_{-1}^{1} dx\, g(x) \frac{x - r}{(x - r)^2 + \alpha^2} = P \int_{-1}^{1} dx \frac{g(x)}{x - r}, \tag{34}$$

where the last term indicates a principal value integral.

Using Eqs. (33) and (34) we can now solve Eq. (25) for r and N. One way to do this is to obtain r from

$$\frac{L_I(r)}{L_R(r)} = -\frac{\delta\Omega_I}{\delta\Omega_R} \tag{35}$$

which is independent of N, and then to obtain N from

$$N_{th} \left|\pi \frac{Z(\omega_0)}{\omega_0}\right| = \{L_R^2 + L_I^2\}^{-1/2}. \tag{36}$$

In this last equation we have indicated N as N_{th}, the instability threshold system intensity.

As an example, let us consider the case of a parabolic frequency distribution of width 2Δ,

$$g(\omega) = \frac{3}{4\Delta} \left\{1 - \left(\frac{\omega - \omega_0}{\Delta}\right)^2\right\} \tag{37}$$

which gives

$$L(\Omega) = \frac{3}{4\Delta} \int_{-1}^{1} dx \frac{1 - x^2}{x - x_1} \tag{38}$$

and for $\alpha \to 0^+$ and $|r| \leqslant 1$,

$$L(\Omega) = \frac{3\pi}{4\Delta} F(r) \tag{39}$$

with

$$F(r) = \frac{1}{\pi} \left\{(1 - r^2) \ln \frac{1 - r}{1 + r} - 2r + i\pi (1 - r^2)\right\}. \tag{40}$$

Notice that when r varies between (−1) and (+1) the quantity $|F(r)|$ changes between 0.64 and 1 (Fig. 1). We can write the equation for the threshold intensity

$$\frac{3}{4} \pi \; N_{th} \frac{|\pi Z(\omega_0)/\omega_0|}{\Delta |F|} = 1, \tag{41}$$

where $|F|$ is a form factor of the order of one.

Another way to write Eq. (41) is

$$\frac{\delta\Omega}{\Delta} \approx 1, \tag{42}$$

where $\delta\Omega = \Omega - \omega_0$ is the coherent frequency shift that would be obtained if all the particles had the same frequency and for an

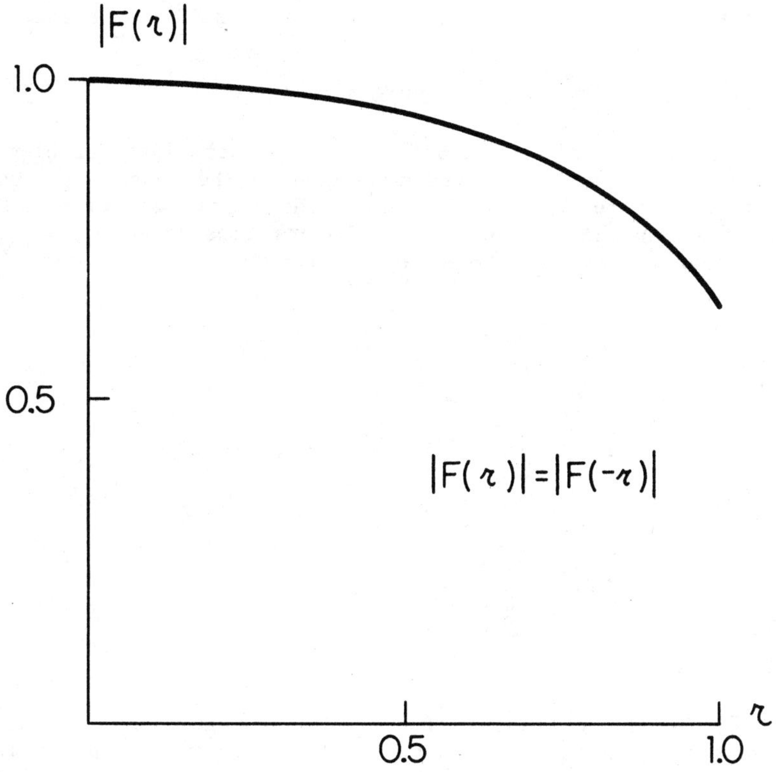

Fig. 1. Form factor |F(r)| as a function of r in Eq. (40).

intensity equal to the threshold one. This result tells us that we can have a stable system also if $\mathrm{Re}Z(\omega_0) < 0$ if we have a frequency spread, Δ, of the order of the frequency shift. This mechanism of beam stabilization is the "Landau damping."[4]

It is also important to notice that for an unstable system we can modify the impedance by adding a feedback system to make $\mathrm{Re}Z(\omega_0) \geq 0$. So we can also use a feedback stabilization system.

3. THE VLASOV EQUATION

A more general formulation of the stability problem, not subject to the limitations which applied to the description used in Section 2, can be obtained by using the Vlasov equation. Let us consider again the system of oscillators considered in Section 2. The equation of motion (1) can be obtained from a Hamiltonian

$$H = H_0 + H_1, \tag{43}$$

$$H_0 = \sum_{\ell=1}^{N} \{\frac{1}{2} p_\ell^{\,2} + \frac{1}{2} \omega_0^2 x_\ell^2 + bx_\ell^4\}, \tag{44}$$

$$H_1 = -\sum_{\ell=1}^{N} F_c x_\ell , \tag{45}$$

where F_c is the collective force, and we have added a term bx^4 representing a small non linear force and we use p, x instead of x, $\dot{x}$.

The Vlasov equation uses the single particle distribution function f(x, p, t) to describe the time evolution of the system, according to the equation[5]

$$\frac{\partial f}{\partial t} + \dot{x}\frac{\partial f}{\partial x} + \dot{p}\frac{\partial f}{\partial p} = 0 \tag{46}$$

with

$$\dot{x} = \frac{\partial H}{\partial p} \tag{47}$$

$$\dot{p} = -\frac{\partial H}{\partial x} \tag{48}$$

and H is now a single particle Hamiltonian.

$$H = H_0 + H_1 \tag{49}$$

$$H_0 = \frac{1}{2}(p^2 + \omega_0^2 x^2) + bx^4 \tag{50}$$

$$H_1 = - F_c x. \tag{51}$$

The collective force

$$F_c = N \int G(t-t') D(t')dt' \tag{52}$$

depends on f itself through the definition of D(t)

$$D(t) = \int x f(x, p, t) \, dxdp \tag{53}$$

making the Vlasov equation a non-linear equation for f(x, p, t).

Equation (46) is an approximation of Liouville's equation[5] and does not describe collision processes and the irreversible approach of a system to equilibrium. It gives, however, a good description of the stability of a system.

The technique used to study the stability is to write the distribution function as a sum of an equilibrium solution and a small perturbation

$$f = f_0(x, p) + f_1(x, p, t) \tag{54}$$

with the assumption

$$D[f_0] = \int x \, f_0(x, p) \, dxdp \equiv 0. \tag{55}$$

Any function f_0 which is a function of H_0 is an equilibrium solution of the Vlasov equation, as can be seen immediately using Eqs. (47) and (48).

It is often convenient to use, instead of the variables x, p, the action-angle variables J, ψ obtained from the generating function

$$F = -\frac{1}{2}\omega_0 x^2 \mathrm{tg}\psi \tag{56}$$

$$J = -\frac{\partial F}{\partial \psi} \tag{57}$$

$$p = \frac{\partial F}{\partial x}. \tag{58}$$

From Eqs. (57) and (58) we obtain

$$x = (2J/\omega_0)^{1/2} \cos\psi \tag{59}$$

$$p = -(2\omega_0 J)^{1/2} \sin\psi \tag{60}$$

and the new Hamiltonian is

$$\overline{H} = \omega_0 J + b\left(\frac{2J}{\omega_0}\right)^2 \cos^4\psi - F_c\left(\frac{2J}{\omega_0}\right)^{1/2} \cos\psi. \tag{61}$$

If we are away from the half or quarter integer resonance, that is, if ω_0 is far enough from 1/2 or 1/4 (modulus 1), we can take an average of the term proportional to b over one oscillation period and rewrite the Hamiltonian as

$$\overline{H} = \omega_0 J \left\{1 + \frac{3}{2} b \frac{J}{\omega_0^3}\right\} - F_c \left(\frac{2J}{\omega_0}\right)^{1/2} \cos\psi. \tag{62}$$

The term $3bJ/2\omega_0$ describes a change in frequency with amplitude due to a non-linear (octupolar) force, since, for $F_c = 0$

$$\psi' = \frac{\partial H}{\partial J} = \omega_0 \{1 + 3b \frac{J}{\omega_0^3}\}. \tag{63}$$

Using the Hamiltonian Eq. (62), the equilibrium distribution function is only a function of J, $f_0 = f_0(J)$. The collective force is only a functional of f_1, and the center of mass D can be written as

$$D(t) = \int_0^\infty dJ \int_0^{2\pi} d\psi \left(\frac{2J}{\omega_0}\right)^{1/2} \cos\psi\, f_1(J, \psi, t). \tag{64}$$

Assuming f_1 to be small, we can now write the Vlasov equation, to first order in f_1, as

$$\frac{\partial f_1}{\partial t} + \frac{\partial H_0}{\partial J}\frac{\partial f_1}{\partial \psi} - \frac{\partial H_1}{\partial \psi}\frac{\partial f_0}{\partial J} = 0 \tag{65}$$

having used the fact that $\partial f_0/\partial\psi = 0$. This linearized equation is adequate to describe the initial state of the instability, as long as f_1 remains a small perturbation. When in an unstable system, f_1 grows one should solve the full non-linear equation to describe its evolution.

We assume now f_1 of the form

$$f_1 = \{a(J)\ell^{i\psi} + b(J)\ell^{-i\psi}\}\ \ell^{-i\Omega t} \tag{66}$$

and substitute in Eq. (65) to obtain

$$a(J) = \frac{i}{2}\, NZ(\Omega)\bar{D}\ \left(\frac{2J}{\omega_0}\right)^{1/2} \frac{\dfrac{\partial f_0}{\partial J}}{\Omega - \dfrac{\partial H_0}{\partial J}}\ , \tag{67}$$

$$b(J) = -\frac{i}{2}\, NZ(\Omega)\bar{D}\ \left(\frac{2J}{\omega_0}\right)^{1/2} \frac{\dfrac{\partial f_0}{\partial J}}{\Omega + \dfrac{\partial H_0}{\partial J}} \tag{68}$$

where

$$\bar{D} = \int dJd\psi\ \left(\frac{2J}{\omega_0}\right)^{1/2} \cos\psi\ \{a(J)\ell^{i\psi} + b(J)\ell^{-i\psi}\}. \tag{69}$$

Substituting Eqs. (67) and (68) in Eq. (69) we obtain the dispersion relation

$$1 = \delta\Omega \int_0^\infty dJ\ J\ \frac{\partial f_0}{\partial J} \left\{ \frac{1}{\Omega - \dfrac{\partial H_0}{\partial J}} - \frac{1}{\Omega + \dfrac{\partial H_0}{\partial J}} \right\}. \tag{70}$$

As in Section 2 we need only to study the solution $\Omega \approx \omega_0$, so that we can neglect the second term on the r.h.s. of Eq. (70) and rewrite this dispersion relation in the simpler form

$$1 = \delta\Omega \int_0^\infty dJ\ \frac{J\dfrac{\partial f_0}{\partial J}}{\Omega - \dfrac{\partial H_0}{\partial J}}\ . \tag{71}$$

This equation has the same structure of Eq. (20) except for the substitution of $g(\omega)$ with $J\partial f_0/\partial J$. Changing the variable from ω to J

$$J = (\omega - \omega_0)/(\omega_0^2\ /3b), \tag{72}$$

we can rewrite the integral as

$$\int_{\omega_0}^{\infty} d\omega \; \frac{\overline{g}(\omega)}{\Omega - \omega} \tag{73}$$

with

$$\overline{g}(\omega) = (\omega_0^2/3b)^2 \; (\omega - \omega_0) \; \left(\frac{\partial f_0}{\partial J}\right)_{J = (\omega - \omega_0)\, 3b/\omega_0^2} \tag{74}$$

All the consideration of Section 2, including the definition of the stability limit and threshold intensity, can now be applied again.

4. EQUATIONS OF MOTION FOR THE LONGITUDINAL DEGREE OF FREEDOM

In this and the next section we introduce the equations of motion for the longitudinal and transverse degrees of freedom (synchrotron and betatron motion). This is done only in order to introduce our notations and define the approximations that we use. A full discussion of particle motion in an accelerator can be found in the lectures by E. D. Courant in these same proceedings.

To describe the longitudinal motion of a particle we use as variables the azimuthal angle θ and the longitudinal momentum. In the approximation of neglecting the coupling to the transverse betatron oscillations and assuming the transverse momentum to be much smaller than the longitudinal momentum, we can use the energy as the variable associated to θ.

Introducing a reference particle of energy E_0 and angle $\theta = \omega_0 t$ we can describe any other particle in the beam by the variables ε, ϕ, where

$$\theta = \omega_0 t + \phi(t) \tag{75}$$

$$\varepsilon = (E - E_0)/E_0. \tag{76}$$

It is important to notice that if we work in the rotating frame defined by variables ϕ, ε we have to substitute whenever necessary the $\partial/\partial t$ with $\partial/\partial t - \omega_0(\partial/\partial\phi)$.

The equations of motion for a particle in a bunch can now be written as

$$\dot{\phi} = -\omega_0 \, \eta \, \varepsilon \tag{77}$$

$$\dot{\varepsilon} = +\frac{\omega_0}{\eta} \nu_s^2 \, \phi + \mu + \mu_{NL}, \tag{78}$$

where the "frequency slip factor" η describes the change in revolution frequency with energy,

$$\eta = -\frac{1}{\omega_0}\frac{\partial\omega}{\partial\varepsilon}\bigg|_{\varepsilon = 0} = \alpha - \frac{1}{\gamma^2} \tag{79}$$

with α the momentum compaction factor and γ the particle energy in rest mass units, and ν_s is the small amplitude synchrotron tune. Equation (78) has been linearized in ϕ, thus considering only particles executing small amplitude synchrotron oscillations. The quantity μ describes the energy change due to interaction of the beam with its environment and goes to zero when the beam current goes to zero. The last quantity in Eq. (78), μ_{NL}, describes the non-linear part of the force produced by the rf system and is included explicitly to be able to consider the effect of Landau damping on beam stability.

The additional energy loss describes the effect of the interaction of the beam with the environment and will be evaluated in Section 6.

Equations (77) and (78) can be obtained from the Hamiltonian

$$H = -\frac{1}{2}\,\omega_0\eta\varepsilon^2 - \frac{1}{2}\frac{\omega_0}{\eta}\,\nu_s^2\,\phi^2 + U + U_{NL} \tag{80}$$

with

$$\mu = -\frac{\partial U}{\partial\phi} \tag{81}$$

$$\mu_{NL} = -\frac{\partial U_{NL}}{\partial\phi}\,. \tag{82}$$

For a beam with many bunches we can generalize Eqs. (77) and (78) introducing bunch coordinates ϕ_p, ε_p, p being the bunch label. If the position of the center of bunch p is Φ_p, the bunch coordinate ϕ_p is related to ϕ by $\phi_p = \phi - \Phi_p$. For B equally spaced bunches $\Phi_p = 2\pi p/B$, $p = 0, 1, \ldots, B-1$. Using bunch coordinates we can describe each bunch with a different distribution function $f^{(p)}(\phi_p, \varepsilon_p, t)$. For a beam with B bunches we write the equations of synchrotron oscillations as

$$\dot{\phi}_p = -\eta\,\omega_0\,\varepsilon_p \tag{83}$$

$$\dot{\varepsilon}_p = \frac{\omega_0\,\nu_s^2}{\eta}\,\phi_p + \mu_p + \mu_{NL}, \quad p = 0, 1, \ldots B-1 \tag{84}$$

and the Hamiltonian as

$$H = - \sum_{n=0}^{B-1} \{\frac{1}{2} \omega_0 \eta \varepsilon_n^2 + \frac{1}{2} \frac{\omega_0 \nu_s^2}{\eta} \phi_n^2\} + U + U_{NL}, \tag{85}$$

where

$$\mu_p = - \frac{\partial U}{\partial \phi_p}. \tag{86}$$

We can introduce angle action variables J, ζ using the generating function

$$G = - \frac{1}{2} \frac{\nu_s}{|\eta|} \sum_{n=0}^{B-1} \phi_n^2 \operatorname{tg} \zeta_n \tag{87}$$

which gives

$$\phi_n = \left(\frac{2|\eta| J_n}{\nu_s}\right)^{1/2} \cos \zeta_n \tag{88}$$

$$\varepsilon_n = - \left(\frac{2\nu_s J_n}{|\eta|}\right)^{1/2} \sin \zeta_n. \tag{89}$$

Using these variables the Hamiltonian is

$$\overline{H} = - \frac{\eta}{|\eta|} \nu_s \omega_0 \sum_{n=0}^{B-1} J_n + \overline{U} + \overline{U}_{NL}, \tag{90}$$

where $\overline{U}$ is a function of J_n, ζ_n, and $\overline{U}_{NL}$ is averaged over the ζ's to obtain a function of the J_n only.

5. EQUATION OF MOTION FOR THE TRANSVERSE OSCILLATIONS

We consider particles oscillating in one transverse direction, neglecting the coupling between the two betatron oscillation modes. We also simplify the problem by neglecting the alternating-gradient structure of the machine and using the smooth approximation to describe the betatron motion. This last approximation is rather good and its use can be justified by the analysis of the general theory.[6]

If y is the transverse direction of oscillation, we can write the equation of motion in the form

$$\ddot{y} + \nu_0^2\omega_0^2\ [1 + 2P\varepsilon]\ y \ = F(\phi,t) + N(y), \tag{91}$$

where $F(\phi,t)$ is the collective self force, $N(y)$ the non-linear forces producing a change in frequency with oscillation amplitude, and the term $2P\varepsilon$ describes the change in frequency due to a variation in energy. The quantity P is given by

$$P = \xi - \eta \tag{92}$$

with η given by Eq. (79) and the "chromaticity" ζ defined as

$$\xi = \frac{1}{\nu_0}\frac{d\nu_0}{d\varepsilon} \tag{93}$$

and describing the change in tune with energy.

The longitudinal variables ε, ϕ satisfy the equations of synchrotron motion (77) and (78). However in these synchrotron equations we neglect both the collective term, μ, and the non-linear term μ_{NL}, so that the longitudinal motion in ε, ϕ is that of a harmonic oscillator.

The equation of motion (91) can be derived from the Hamiltonian

$$H = H_{0_T} + H_{0_L} + H_{NL} - F(\phi,t)y \tag{94}$$

with

$$H_{0_T} = \frac{1}{2}\,\dot{y}^2 + \frac{1}{2}\,\nu_0^2\omega_0^2[1 + 2P\varepsilon]\ y^2 \tag{95}$$

$$H_{0_L} = -\frac{1}{2}\,\omega_0\eta\varepsilon^2 - \frac{1}{2}\frac{\omega_0\,\nu_s^2}{\eta}\,\phi^2 \tag{96}$$

and H_{NL} describing the non-linear force. Second order terms in y are neglected in the equation for ϕ.

We now perform a canonical transformation defined by the generating function

$$G = -\frac{1}{2}\,\nu_0\omega_0 y^2 tg\psi \tag{97}$$

to introduce the transverse action-angle variables I and ψ

$$y = \sqrt{\frac{2I}{\nu_0\omega_0}}\ \cos\ \psi \tag{98}$$

$$\dot{y} = -\sqrt{2\nu_0\omega_0 I}\ \ \sin\ \psi. \tag{99}$$

With these variables the Hamiltonian is

$$\overline{H} = \nu_0\omega_0 I\,[1 + P\varepsilon(1 + \cos 2\psi)] + H_{0_L} - F(\phi,t)\left(\frac{2I}{\nu_0\omega_0}\right)^{1/2}\cos\psi + \overline{H}_{NL}. \tag{100}$$

The term $\nu_0\omega_0 IP\varepsilon\cos 2\psi$ is a fast oscillating term, execpt for the case when $\nu_0 \approx 1/2$, and can be neglected. We also take an average of $\overline{H}_{NL}$ over ψ, assuming that non-linear resonances can be neglected, and write

$$<\overline{H}_{NL}> = \nu_0\omega_0 D(I). \tag{101}$$

For a bunched beam, of B bunches, it is convenient to introduce "bunch coordinates" y_k, ϕ_k and write the transverse equation of motion as

$$\ddot{y}_k + \nu_0^2\omega_0^2\,[1 + 2P\varepsilon_k]\,y_k = F(\phi_k,t) + N(y_k). \tag{102}$$

The notation is the same as used in Eq. (91). The equations for ε_k, ϕ_k are given by Eqs. (83) and (84) with the non-linear and the collective forces taken equal to zero.

The equations of motion (102), (83), and (84) can be obtained from a Hamiltonian

$$H = \sum_{K=0}^{B-1}\{H_{0_{T,K}} + H_{0_{L,K}} + H_{C,K} + H_{NL,K}\}, \tag{103}$$

where

$$H_{0_{T,K}} = \frac{1}{2}\dot{y}_K^2 + \frac{1}{2}\nu_0^2\omega_0^2\,[1 + 2P\varepsilon_K]\,y_K^2 \tag{104}$$

$$H_{0_{L,K}} = -\frac{1}{2}\frac{\omega_0\nu_s^2}{\eta}\phi_K^2 - \frac{\omega_0\eta}{2}\varepsilon_K^2 \tag{105}$$

$$H_{C,K} = -F(\phi_K,t)y_K \tag{106}$$

and we assume to neglect second order terms in the equation of motion for ϕ.

As in the previous case we introduce new action angle variables I_K, ψ_K for the transverse motion and J_K, ζ_K for the longitudinal motion. Using the generating function (97) for the transverse variables we have

$$y_K = \left(\frac{2I_K}{\nu_0\omega_0}\right)^{1/2} \cos \psi_K \tag{107}$$

$$\dot{y}_K = -(2\nu_0\omega_0 I_K)^{1/2} \sin \psi_K \tag{108}$$

and ϕ_K, ε_K are given by Eqs. (88) and (89). With these variables the Hamiltonian is

$$\overline{H} = \sum_{K=0}^{B-1} \Big\{\nu_0\omega_0[1 + P\varepsilon_K]\, I_K + \nu_0\omega_0 D(I_K) - \frac{\eta}{|\eta|}\nu_s\omega_0 J_K - F(J_K, \zeta_K, t)\left(\frac{2I_K}{\nu_0\omega_0}\right)^{1/2} \cos\psi_K\Big\} \tag{109}$$

having used Eq. (101) to write the non-linear part.

6. TRANSVERSE AND LONGITUDINAL COLLECTIVE FORCES

6.1. The charge and current density

We want to evaluate the force produced by the beam on a test particle, "ℓ". To do this we describe the beam charge and current density with the help of the single particle distribution function, f. We use a coordinate system x, y, θ (or x, y, s = Rθ, R being the average accelerator radius) and use as conjugate variables, momenta, the quantity $\dot{x}$, $\dot{y}$, ε.

We write the distribution function as

$$F = \frac{N}{R} f(x, y, \theta, \dot{x}, \dot{y}, \varepsilon, t), \tag{110}$$

where N is the number of particles in the beam, R is introduced because we use the variable θ instead of s and $f(x, y, \theta, \dot{x}, \dot{y}, \varepsilon, t)$ is normalized to one.

We assume that the two transverse betatron oscillation modes are uncoupled and consider the case in which the beam particles are oscillating only in one direction, say y. The charge density can now be defined in terms of the distribution function f as

$$e\rho(x, y, \theta, t) = \frac{eN}{R} \int f(x, y, \theta; \dot{x}, \dot{y}, \varepsilon; t)\, d\dot{x}d\dot{y}d\varepsilon. \tag{111}$$

The transverse current density is given by

$$J_x(x, y, \theta, t) = \frac{eN}{R} \int \dot{x} f(x, y, \theta; \dot{x}, \dot{y}, \varepsilon; t) d\dot{x} d\dot{y} d\varepsilon \quad (112)$$

$$J_y(x, y, \theta, t) = \frac{eN}{R} \int \dot{y} f(x, y, \theta; \dot{x}, \dot{y}, \varepsilon, t) d\dot{x} d\dot{y} d\varepsilon. \quad (113)$$

For the case of one-dimensional oscillations in the y direction we assume

$$J_x \equiv 0. \quad (114)$$

The longitudinal velocity $\dot{s} = \dot{\theta}R$ is the sum of a constant term, v the velocity of the reference particle, plus a small term proportional to ε. We assume the beam to be relativistic, $v \approx c$, and neglect the small term, as compared to v, in evaluating the longitudinal current density. With this approximation we have

$$J_s = ev\rho. \quad (115)$$

To further simplify the problem we also assume that there is no coupling between the transverse and longitudinal collective modes of oscillation and consider separately the two cases. We consider first the fields and forces produced by a bunch and extend afterward the results to a beam with many bunches.

For the transverse oscillation modes we assume that the distribution function is of the form

$$f = \frac{N}{R} \{f_0(x, y, \phi, \dot{x}, \dot{y}, \varepsilon) + f_1(x, y, \phi, \dot{x}, \dot{y}, \varepsilon) e^{-i\nu\omega_0 t}\}, \quad (116)$$

where N is the number of particles in the bunch and ν the coherent oscillation tune. The term f_0 in Eq. (116) represents an equilibrium solution. The term f_1 is a small perturbation.

We also make the following assumptions:

(1) For the equilibrium term the density can be factorized in a longitudinal and a transverse term

$$\rho_0(x, y, \theta, t) = \frac{N}{R} \int f_0 d\dot{x} d\dot{y} d\varepsilon = \frac{N}{R} \lambda(\phi) h_0(x, y) \quad (117)$$

with $h_0(x, y)$ and $\lambda(\phi)$ are normalized to one.

(2) The equilibrium term has zero dipole moment

$$\int y f_0(x, \dot{x}, y, \dot{y}, \phi, \varepsilon) dx d\dot{x} dy d\dot{y} d\varepsilon = 0. \quad (118)$$

(3) The perturbation term must satisfy the normalization condition

$$\int f_1 \, dxdyd\dot{x}d\dot{y}d\phi d\varepsilon = 0. \qquad (119)$$

(4) Both f_0 and f_1 are periodic functions of $\phi = \theta - \omega_0 t$; in particular, we write

$$\lambda(\phi) = \frac{1}{2\pi I_0} \sum_p I_p e^{ip\phi}, \qquad (120)$$

where

$$I_0 = eNv/2\pi R \qquad (121)$$

is the average bunch current and the I_p are the Fourier components of the bunch current; for the perturbed density

$$\rho_1(x, y, \phi) = \int f_1 \, d\dot{x}d\dot{y}d\varepsilon \qquad (122)$$

we write

$$\rho_1(x, y, \phi) = \sum_p h_p(x, y) e^{ip\phi} \qquad (123)$$

and for the transverse current

$$J_{y,1}(x, y, \phi) = \int \dot{y} f_1 \, d\dot{x}d\dot{y}d\varepsilon \qquad (124)$$

$$J_{y,1}(x, y, \phi) = \sum_p b_p(x, y) e^{ip\phi}. \qquad (125)$$

From the equation of continuity of current we obtain the relationship

$$i\nu\omega_0 h_p(x,y) = \frac{\partial}{\partial y} b_p(x, y). \qquad (126)$$

The charge and current density can then be written as

$$e\rho = \frac{eN}{R} h_0(x, y) \sum_p \frac{I_p}{2\pi I_0} e^{ip\phi} + \frac{eN}{R} \sum_p h_p(x, y) e^{ip\phi - i\nu\omega_0 t} \qquad (127)$$

$$J_y = \frac{eN}{R} \sum_p b_p(x, y) e^{ip\phi - i\nu\omega_0 t}. \qquad (128)$$

To describe the case of longitudinal oscillations we assume

$$f = \frac{N}{R} g_0(\phi, \varepsilon)\, h_0(x, y, \dot{x}, \dot{y}) + \frac{N}{R} g_1(\phi, \varepsilon)\, h_1(x, y, \dot{x}, \dot{y})\, e^{-i\Omega\omega_0 t} \tag{129}$$

where Ω is the collective frequency for a longitudinal oscillation mode. Since we are only interested in longitudinal effects we assume that the transverse currents are zero, $J_x = J_y = 0$. The charge and longitudinal current density are now given by

$$e\rho = \frac{eN}{R} n_0(\phi)\, d_0(x, y) + \frac{eN}{R} n_1(\phi)\, d_1(x, y)\, e^{-i\omega_0 \Omega t} \tag{130}$$

$$J_s = ev\rho, \tag{131}$$

where the $d(x, y)$ and $n(\phi)$ are obtained by integrating Eq. (129) over $\dot{x}$, $\dot{y}$, ε. The terms $n_0(\phi)$ and $d_0(x, y)$ are normalized to one

$$\int n_0(\phi)\, d\phi = 1 \tag{132}$$

and for the perturbation term

$$\int n_1(\phi)\, d\phi = 0. \tag{133}$$

6.2. The transverse force

The electric and magnetic field generated by the beam can be obtained from Maxwell's equations, using ρ and J_y, given by Eqs. (127) and (128) as sources. Using the potential and vector potential in the Lorentz gauge, we have

$$\Box V = -4\pi e\rho \tag{134}$$

$$\Box \underline{A} = -\frac{4\pi}{c} \underline{J}. \tag{135}$$

To solve these equations we assume

$$V = \frac{eN}{R} \sum_p V_p(x,y)\, e^{ip\phi - i\nu\omega_0 t} \tag{136}$$

$$A_x = 0 \tag{137}$$

$$A_y = \frac{eN}{Rc} \sum_p A_p(x, y)\, e^{ip\phi - i\nu\omega_0 t} \tag{138}$$

$$A_s = \beta V, \tag{139}$$

where $\beta = v/c$. We notice that the charge density, Eq. (127) has two terms and the first one can be obtained from the second by assuming $\nu = 0$, $h_p = h_0 I_p$. We now solve Eq. (132) considering only the second term as a source. At the end we make a linear combination of the two solutions.

The scalar and vector potentials are obtained by solving the two equations

$$\left\{\frac{\partial^2}{\partial x^2} + \frac{\partial^2}{\partial y^2} + \lambda_p^2\right\} V_p(x, y) = -\, 4\pi h_p(x, y) \tag{140}$$

$$\left\{\frac{\partial^2}{\partial x^2} + \frac{\partial^2}{\partial y^2} + \lambda_p^2\right\} A_p(x, y) = -\, 4\pi b_p(x, y) \tag{141}$$

with the proper boundary conditions on the walls of the vacuum chamber. The quantity λ_p is

$$\lambda_p^2 = \frac{\omega_0^2}{c^2} \left[(p + \nu)^2 - \frac{p^2}{\beta^2}\right] . \tag{142}$$

The time component of the field oscillates at the frequency $(\nu+p)\omega_0$, ν being the collective frequency and $p\omega_0$ describing the effect of the accelerator periodicity.

The transverse force is given by

$$F = eE_y + e\beta H_x, \tag{143}$$

and using Eqs. (136) and (137) this can be written as

$$F = \frac{e^2 N}{R} \sum_p C_p(x, y)\, e^{ip\phi - i\nu\omega_0 t} \tag{144}$$

with

$$C_p = -\frac{1}{\gamma^2} \frac{\partial V_p(x, y)}{\partial y} + i\nu \frac{\omega_0}{c} A_p(x, y). \tag{145}$$

The Lorentz condition

$$\frac{1}{c}\frac{\partial V}{\partial t} + \text{div}\,\underline{A} = 0 \tag{146}$$

gives us the relationship

$$i\nu\frac{\omega_0}{c}V_p\,(x,\,y) = \frac{\partial A_p(x,\,y)}{\partial y}. \tag{147}$$

The solution of Eqs. (140) and (141) can be expressed in terms of the Green functions G_p, H_p, satisfying the equations

$$\left\{\frac{\partial^2}{\partial x^2} + \frac{\partial^2}{\partial y^2} + \lambda_p^2\right\} G_p\,(x - x',\, y - y') = -4\pi\delta(x - x')\delta(y - y') \tag{148}$$

$$\left\{\frac{\partial^2}{\partial x^2} + \frac{\partial^2}{\partial y} + \lambda_p^2\right\} H_p(x - x',\, y - y') = -4\pi\delta(x - x')\delta(y - y') \tag{149}$$

with the proper boundary conditions, as

$$V_p(x,\,y) = \int G_p(x - x',\, y - y')\, h_p\,(x',\, y')\, dx'dy' \tag{150}$$

$$A_p(x,\,y) = \int H_p(x - x',\, y - y')\, b_p\,(x',\, y')\, dx'dy' \tag{151}$$

Furthermore, the Green functions can be written as the sum of two terms

$$G_p = G_p^v + G_p^w \tag{152}$$

$$H_p = H_p^v + H_p^w\,. \tag{153}$$

The terms G_p^v, H_p^v are the Green functions satisfying the outgoing boundary conditions for a beam in vacuum. The wall terms G_p^w, H_p^w are added so that the boundary conditions on the walls can be satisfied.

The fields and forces obtained from Eqs. (140) and (141) are non-linear; however, to study the stability problem for the dipole mode of instability it is enough to consider the linear part only.

For the field in vacuum the Green functions G_p, H_p are given by[7]

$$G_p(x - x',\, y - y') = H_p(x - x',\, y - y') = i\pi H_0^{(1)}(\lambda_p P) \tag{154}$$

with

$$P^2 = (x - x')^2 + (y - y')^2. \tag{155}$$

The field corresponding to the equilibrium distribution can be calculated assuming $b_p = 0$, $h_p = (I_p h_0)/2\pi I_0$. For a uniform cylindrical beam of radius a, one has

$$\frac{\partial V_p}{\partial y} = - \frac{iI_p}{aI_0} H_1 (\lambda_p a) J_1 (\lambda_p r)(y/r) \tag{156}$$

with $r^2 = x^2 + y^2$. For wavelength long compared to the beam size, $\lambda_p a << 1$, one has

$$\frac{\partial V_p}{\partial y} \approx - \frac{I_p}{\pi I_0 a^2} y. \tag{157}$$

In the case of a more complicated beam geometry, one has to substitute the quantity πa^2 with an "effective beam transverse area." The field produced by the perturbation is obtained in the long wavelength limit by displacing the beam by a distance equal to the dipole moment. In this approximation we have

$$\frac{\partial V_p}{\partial y} = \frac{2I_p}{I_0 a^2} D_p \tag{158}$$

with the dipole moment D_p given by

$$D_p = \int y h_p(x, y)\, dxdy. \tag{159}$$

In the same approximation we also have $A_p = 0$.

For the part of the fields due to the walls we can expand the Green functions in a power series to obtain

$$\begin{aligned} \frac{\partial V_p^w(x, y)}{\partial y} = & \frac{\partial G_p^w(0, 0)}{\partial y} \int h_p(x',y')\, dx'xy' \\ & + \frac{\partial^2 G_p^w(0, 0)}{\partial y^2} \int (y - y')h_p(x', y')\, dx'dy' \\ & + \ldots \end{aligned} \tag{160}$$

and similarly for H. The first term on the right-hand side is a constant and can be neglected. For the second term we have two different results according to whether we consider the equilibrium density h_0 or the perturbation h_p. For the equilibrium case, using Eq. (118), we have

$$\frac{\partial V_p^{(0)}}{\partial y} = y \frac{I_p}{2\pi I_0} \frac{\partial^2 G_p^w(0, 0)}{\partial y^2}. \tag{161}$$

For the perturbation we have, using Eq. (159)

$$\frac{\partial V_p(x, y)}{\partial y} = - \frac{\partial^2 G_p^w(0, 0)}{\partial y^2} D_p. \tag{162}$$

We can perform a similar expansion for A_p. In this case there is only a contribution from the perturbation which, using Eq. (126), can be written as

$$A_p(x, y) \approx -i\nu\omega_0 H_p(0, 0) D_p \tag{163}$$

Collecting the terms (157), (161), proportional to y, and using Eqs. (144) and (145) we obtain what is called the incoherent force

$$F_{inc} = \frac{2ey}{\gamma^2 \beta c} \sum_p I_p \left\{\frac{1}{a^2} - \frac{1}{2} \frac{\partial^2 G_p^w(0, 0)}{\partial y^2}\right\} e^{ip\phi}. \tag{164}$$

This force produces a change in the focusing force, which can be introduced in the equations of motion substituting ν_0 with $\nu_0 - \delta\nu_{inc}$, with

$$\delta\nu_{inc} = - \frac{F_{inc} c^2}{2yE_0 \nu_0 \omega_0^2}. \tag{165}$$

Introducing the incoherent impedance

$$Z_p^{inc} = -i \frac{R}{eI_p y\beta} \int_0^{2\pi} Fe^{-ip\phi} d\phi \tag{166}$$

we have from Eq. (164)

$$Z_p^{inc} = -i \frac{4\pi R}{\gamma^2 \beta^2 c} \left\{\frac{1}{a^2} - \frac{1}{2} \frac{\partial^2 G_p^w(0, 0)}{\partial y^2}\right\} \tag{167}$$

in cgs units or

$$Z_p^{inc} = -i \frac{RZ_0}{\gamma^2\beta} \{\frac{1}{a^2} - \frac{1}{2} \frac{\partial^2 G_p^w}{\partial y^2}\} \tag{168}$$

in mks units, $Z_0 = 377\Omega$ being the vacuum impedance. Using the impedance we can rewrite the incoherent force as

$$F_{inc} = \frac{ie\beta y}{2\pi R} \sum_p I_p Z_p^{inc} e^{ip\phi} \tag{169}$$

or the incoherent tune shift as

$$\delta\nu_{inc} = - \frac{ieR}{4\pi\nu_0 \beta E_0} \sum_p I_p Z_p^{inc} e^{ip\phi}. \tag{170}$$

The force produced by the perturbation can be obtained from Eqs. (158), (162), (163), (144), and (145) as

$$F = - \frac{e^2 N}{R} \sum_p D_p \left\{ \frac{2}{\gamma^2 a^2} - \frac{1}{\gamma^2} \frac{\partial^2 G_p^w}{\partial y^2} - \frac{\nu^2\omega_0^2}{c} H_p \right\} e^{ip\phi - i\nu\omega_0 t}. \tag{171}$$

Introducing a transverse impedance $Z_\perp$ defined as

$$Z_\perp(p + \nu) = - i \frac{R \int_0^{2\pi} F e^{- ip\phi + i\nu\omega_0 t} d\phi}{eI_0 \beta D_p} \tag{172}$$

giving the ratio of the total force per revolution to the transverse current $I_0 D_p$ at the frequency $(\nu + p)\omega_0$, we can rewrite Eq. (171) as

$$F = \frac{ieI_0\beta}{2\pi R} \sum_p D_p Z_\perp(p + \nu) e^{ip\phi - i\nu\omega_0 t} \tag{173}$$

with

$$Z_\perp(p + \nu) = i \frac{4\pi}{c} \frac{2\pi R}{\beta^2} \left\{ \frac{1}{\gamma^2 a^2} - \frac{1}{2\gamma^2} \frac{\partial^2 G_p^w}{\partial y^2} - \frac{\nu^2\omega_0^2}{2c} H_p^w \right\}. \tag{174}$$

The transverse impedance is an impedance per unit length. The expression (174) is in cgs units. To obtain it in mks units we can substitute $4\pi/c$ with the vacuum impedance, $Z_0 = 377\Omega$. Notice that the $Z_\perp$ defined by Eq. (174) satisfies the symmetry conditions in Eqs. (13) and (14).

Let us consider now the case of a bunched beam with B equally spaced bunches. The distribution function for bunch K and the force produced by bunch K are obtained from Eqs. (115) and (173) shifting the coordinate ϕ by $2\pi K/B$, with $K = 0, 1, \ldots, B - 1$. The total force acting on bunch ℓ can then be written as a sum over the contribution from all bunches, evaluated at $\phi_\ell = \phi - 2\pi\ell/B$,

$$F^{(\ell)} = \frac{ieI_0\beta}{2\pi R}\sum_{K=0}^{B-1}\sum_p D_p^{(K)} Z_\perp(p+\nu)e^{ip\phi_\ell - 2\pi ip\frac{K-\ell}{B} - i\nu\omega_0 t}. \tag{175}$$

The dipole moment for bunch K is defined using the distribution function

$$f^{(K)} = \frac{N}{R}\left\{f_0^{(K)}\left(x, y, \phi - 2\pi\frac{K}{B}, \dot{x}, \dot{y}, \varepsilon\right) + f_1^{(K)}\left(x, y, \phi - 2\pi\frac{K}{B}, \dot{x}, \dot{y}, \varepsilon\right)e^{-i\nu\omega_0 t}\right\} \tag{176}$$

as

$$D_p^{(K)} = \frac{1}{2\pi}\int y\, f_1^{(K)}(x, y, \phi, \dot{x}, \dot{y}, \varepsilon)e^{-ip\phi}\, dxdyd\phi d\dot{x}d\dot{y}d\varepsilon. \tag{177}$$

In writing Eq. (177) we have also assumed that all bunches have the same number of particles, N. The current I_0 in Eq. (175) is the average current per bunch.

6.3 The longitudinal force

To find the potential and the vector potential we use again the equations (134) and (135) with the charge and current densities given by Eqs. (130) and (131). Since we have $J_x = J_y = 0$, $J_s = \beta c\rho$, we also assume $A_x = A_y = 0$ and $A_s = \beta V$. We are then left with a single equation for the scalar potential V.

Using the perturbed charge density as a source, the equation for V can be easily obtained. The case of the equilibrium density can then be obtained for $\Omega = 0$ and changing the proper labels. The longitudinal density must be periodic in ϕ so that we can write

$$e\rho = \frac{eN}{R} d_1(x,y) \sum_p \Lambda_p e^{ip\phi - i\Omega\omega_0 t} \tag{178}$$

with

$$\Lambda_p = \frac{1}{2\pi} \int n_1(\phi) e^{-ip\phi} d\phi$$

$$= \frac{1}{2\pi} \int d\phi d\varepsilon g_1(\phi,\varepsilon) e^{-ip\phi}. \tag{179}$$

We now assume

$$V = \frac{eN}{R} \sum_p \Lambda_p V_p\ (x,\ y) e^{ip\phi - i\Omega\omega_0 t} \tag{180}$$

and obtain from Eq. (134) the equation

$$\left\{\frac{\partial^2}{\partial x^2} + \frac{\partial^2}{\partial y^2} + \lambda_p^2\right\} V_p(x,\ y) = -4\pi d_1(x,\ y) \tag{181}$$

that must be solved with the proper boundary conditions. The quantity λ_p in Eq. (181) is

$$\lambda_p^2 = (\omega_0/c)^2 \{(\Omega + p)^2 - p^2/\beta^2\}. \tag{182}$$

To evaluate the energy loss that appears in the synchrotron oscillation Eq. (78) we need to evaluate the electric field

$$E_s = -i \frac{2\pi\beta I_0}{RC} \sum_p \Lambda_p V_p(x,\ y) \left\{\frac{p}{\gamma^2\beta^2} - \Omega\right\} e^{ip\phi - i\Omega\omega_0 t}. \tag{183}$$

We assume that to evaluate the energy loss we can use the field on axis and that ϕ does not change during a revolution. With this approximation we obtain

$$\mu = - ie \frac{4\pi}{c} \frac{\beta\omega_0 I_0}{2E_0} \sum_p \Lambda_p V_p(0,0) \left\{\frac{p}{\gamma^2\beta^2} - \Omega\right\} e^{ip\phi - i\Omega\omega_0 t} . \quad (184)$$

As we did in the previous section we introduce now a longitudinal impedance

$$Z_\parallel(p + \Omega) = \frac{4\pi}{c} \frac{\beta}{2} V_p(0,0) \left\{\frac{p}{\gamma^2\beta^2} - \Omega\right\}, \quad (185)$$

where $4\pi/c$ can be substituted with Z_0 if we use mks units.

The collective frequency Ω is a small term and is zero in the absence of a perturbation. Since we will use the impedance to obtain a linearized Vlasov equation we can neglect the Ω term in Eq. (185).

Introducing the impedance in Eq. (184) we obtain

$$\mu = - ie \frac{\omega_0 I_0}{E_0} \sum_p \Lambda_p Z_\parallel(p + \Omega) e^{ip\phi - i\Omega\omega_0 t} . \quad (186)$$

Another quantity of interest is the "potential" U defined by Eq. (81). This is written as

$$U = e \frac{\omega_0 I_0}{E_0} \sum_p \frac{Z_\parallel(p + \Omega)}{p} \Lambda_p e^{ip\phi - i\Omega\omega_0 t} . \quad (187)$$

For a beam with B bunches we can write Eq. (178) as

$$\rho = \frac{eN}{R} d_1(x, y) \sum_{K=0}^{B-1} \sum_p \Lambda_p^{(K)} e^{ip[\phi + 2\pi \frac{K}{B}] - i\Omega\omega_0 t} \quad (188)$$

with

$$\Lambda_p^{(K)} = \frac{1}{2\pi} \int d\phi d\varepsilon g_1^{(K)}(\phi, \varepsilon) e^{-ip\phi} \quad (189)$$

and obtain the energy loss of bunch "ℓ" from

$$U^{(\ell)} = \frac{e\omega_0 I_0}{E_0} \sum_{K=0}^{B-1} \sum_p \frac{Z_\parallel(p + \Omega)}{p} \Lambda_p^{(K)} e^{ip\phi_\ell + i2\pi p(K - \ell)/B - i\Omega\omega_0 t}$$

$$(190)$$

7. THE MODEL IMPEDANCE

In Section 6 we have written the collective forces acting on the beam in terms of a longitudinal and a transverse impedance. But for the space-charge terms these impedances characterize the beam environment. This environment is the vacuum chamber and all the radio-frequency cavities, pick-up electrodes, bellows, and other devices needed to build and operate the accelerator. In this situation the calculation of the impedance and of the field and forces can be a difficult matter.

From the experimental point of view it is possible, during the ring construction, to measure the impedance of each part of the vacuum system and thus to evaluate the total ring impedance. It is also possible to obtain information on the ring impedance from measurements of the beam properties as a function of the beam current.

From these measurements,[8] performed on several accelerators, one can see that it is often possible to use a simple model to describe the impedances. In this model one describes the impedance as a combination of a few terms. The first is the impedance of a straight, uniform cross section, vacuum pipe built with a low resistivity metal, like stainless steel or aluminum. The effect of the radio-frequency system is described by a few low- or medium-frequency resonances, at $\omega < c/b$, b being the vacuum chamber radius. The last term is used to describe the effect of all other parts, which can introduce many high frequency resonances. In the model one averages all the high frequency effects to obtain the "broad band impedance."[8]

The impedance of a resistive cylindrical pipe of radius b has been derived in many papers.[9] In the long wavelength approximation, $\omega < c/b$; or $\lambda b < 1$, it can be written, neglecting a term proportional to Ω as discussed in Section 6.3, in the form

$$Z_{\parallel}(\omega) \approx Z_0 \left\{ \frac{\beta g_0}{2\beta^2\gamma^2} p + \frac{R}{b} \mathscr{R}(\omega) \right\}, \tag{191}$$

where the first term describes the space-charge effect, which for a uniform cylindrical beam of radius a is given by

$$g_0 = 1 + 2\ell n(b/a) \tag{192}$$

and the second term describes the effect of the wall resistivity with the wall impedance R(ω) given by

$$\mathscr{R}(\omega) = (1 - i) \frac{\omega}{2c} \delta(\omega) \tag{193}$$

and $\delta(\omega)$ the skin depth for a wall of conductivity σ

$$\delta(\omega) = c/(2\pi\omega\sigma)^{1/2} . \tag{194}$$

The first term in Eq. (191), describing the space-charge effect, is a capacitive impedance; the second term, due to the resistivity, has a capacitive and a resistive term.

In the same approximation the transverse impedance is given by[9]

$$Z_{\perp}(\omega) = iZ_0 R \left\{\frac{1}{\beta^2\gamma^2}\left(\frac{1}{a^2} - \frac{1}{b^2}\right) - (1 + i)\frac{\delta(\omega)}{b^3}\right\}, \tag{195}$$

where again the first term describes the space-charge and the second, the effect of the resistivity.

The broad-band impedance is that of a parallel R, C, L circuit resonating at a frequency of the order of c/b and with a Q of the order of one.[8] The value of the shunt impedance, R_0, defines completely this impedance and characterizes an accelerator.

Assuming also that the rf resonances can be described by an equivalent R, L, C parallel circuit we can obtain the total ring model impedance by adding to the longitudinal impedance (191) a number of terms

$$Z_{\parallel}(\omega) = R_0 \frac{\omega}{\omega_R} \frac{(\omega/\omega_R) - iQ[1 - (\omega/\omega_R)^2]}{(\omega/\omega_R)^2 + Q^2[1 - (\omega/\omega_R)^2]^2} , \tag{196}$$

where ω_R is the resonant frequency. To describe the resonant rf cavity impedance we use in Eq. (196) the rf cavity parameters. To describe the broad-band impedance we assume $\omega_R \sim c/b$ and $Q \approx 1$.

For the transverse impedance associated with the same structures we assume that all resonant transverse modes of the rf cavity are damped to avoid strong beam instabilities. To the broad-band impedance we associate a transverse broad-band impedance.[10]

$$Z_{\perp}(\omega) = 2\,\frac{R}{d^2}\,\frac{Z_{\parallel}(\omega)}{(\omega/\omega_0)}. \tag{197}$$

We use now the results of the previous discussion to define a longitudinal and transverse "model impedance." We make also the following additional assumptions:

1) All resonant, high Q, impedances are eliminated by a proper system design; if this is not the case they must be taken into account and they can produce very strong beam instabilities;
2) in the longitudinal impedance we neglect the space charge and image term, which is usually small and can be neglected

in comparison to the resistive wall or broad-band impedance term; this assumption is violated for low γ's ($\gamma \sim 1$);

3) we use for the transverse broad-band impedance and the resistive wall impedance the relationship shown in Eq. (197);
4) we use the condition $\lambda a \ll 1$ to simplify the resistive pipe impedance;
5) we also use $\lambda b \ll 1$ to simplify the same impedance; this is even justified for $\lambda b \geqslant 1$ if the resistive pipe impedance is smaller than the broad-band impedance, since in this case the term is small anyway.

Using all these assumptions we can now write (using the mks system)

$$Z_{\parallel}(\omega) = - Z_0 \frac{R}{b} \frac{1 - i}{2} \frac{\omega}{c} \delta(\omega) + Z_{\parallel}^{B}(\omega), \tag{198}$$

where $Z_{\parallel}^{B}(\omega)$ is the broad-band impedance defined by (196),

$$Z_{\perp}(\omega) = iZ_0 R \left\{ \frac{1}{\beta^2\gamma^2} \left(\frac{1}{a^2} - \frac{1}{b^2}\right) - (1 + i) \frac{\delta(\omega)}{b^3} - \frac{2i}{d^2} \frac{Z_{\parallel}^{B}(\omega)/Z_0}{(\omega/\omega_0)} \right\}. \tag{199}$$

These two expressions define the model impedances which can be used in many beam stability calculations to obtain at least an order of magnitude estimate.

8. LONGITUDINAL INSTABILITIES OF A COASTING BEAM

We start the study of longitudinal instabilities considering the simplest case, a uniform coasting beam. This case was the first to be studied theoretically[11] and the predicted instability, called the negative mass instability, was later observed in several accelerators.

For a coasting beam there is no radio-frequency system so that we can take $\nu_s = 0$ and $\mu_{NL} = 0$ in the equations of motion (77) and (78). We assume that initially the longitudinal density distribution function is uniform. Following Eq. (129) and neglecting the transverse coordinates we assume

$$f(\phi, \varepsilon, t) = \frac{N}{R}\left\{ g_0(\varepsilon) + g_1(\phi, \varepsilon) e^{-i\Omega\omega_0 t} \right\}. \tag{200}$$

We can also use the periodicity in ϕ to write

$$g_1(\phi, \varepsilon) = \sum_p g_p(\varepsilon) e^{ip\phi}. \tag{201}$$

The coefficients $g_p(\varepsilon)$ are related to the Λ_p introduced in Section 6, Eq. (179) by

$$\Lambda_p = \int d\varepsilon\, g_p(\varepsilon). \tag{202}$$

The Hamiltonian for our system can be obtained from Eqs. (80) and (187)

$$H = -\frac{1}{2}\eta\omega_0\varepsilon^2 + \frac{eI_0\omega_0}{E_0}\sum_p \frac{Z_{\parallel}(\Omega + p)}{p}\Lambda_p e^{ip\phi - i\Omega\omega_0 t}. \tag{203}$$

With this Hamiltonian the linearized Vlasov equation is

$$\begin{aligned} &- i\omega_0 \sum_p (\Omega + p\eta\varepsilon) g_p e^{ip\phi - i\Omega\omega_0 t} \\ &= \frac{ie\omega_0 I_0}{E_0} \frac{\partial g_0}{\partial \varepsilon} \sum_p Z_{\parallel}(p + \Omega)\, \Lambda_p e^{ip\phi - i\Omega t} \end{aligned} \tag{204}$$

giving

$$g_p(\varepsilon) = - \frac{eI_0 Z_\parallel(p + \Omega)}{E_0} \Lambda_p \frac{\frac{\partial g_0}{\partial \varepsilon}}{\Omega + p\eta\varepsilon}. \tag{205}$$

Integrating over ε we also have the dispersion relation

$$1 = - \frac{eI_0 Z_\parallel(p + \Omega)}{E_0} \int d\varepsilon \frac{\frac{\partial g_0}{\partial \varepsilon}}{\Omega + p\eta\varepsilon}. \tag{206}$$

This equation defines the eigenvalues Ω for the beam normal mode $e^{ip\phi}$.

Introducing the quantities

$$x_p = - \frac{\Omega}{p\eta} \tag{207}$$

$$L(x_p) = \int d\varepsilon \frac{\partial g_0}{\partial \varepsilon} \frac{1}{\varepsilon - x_p} \tag{208}$$

and

$$\delta\Omega = - \frac{eI_0}{\eta E_0} \frac{Z_\parallel(p + \Omega)}{p} \tag{209}$$

we can rewrite Eq. (206) as

$$1 = \delta\Omega L(x_p). \tag{210}$$

To study the dispersion relation (210) let us introduce explicitly the real and imaginary part of x_p

$$x_p = R + iI \tag{211}$$

and also of $L(x_p)$ and $\delta\Omega$

$$L(x_p) = L_R + iL_I \tag{212}$$

$$\delta\Omega = X + iY. \tag{213}$$

The dispersion relation (210) can now be written as

$$\frac{X - iY}{X^2 + Y^2} = L_R + iL_I \tag{214}$$

with

$$L_R = \int d\varepsilon \, \frac{\varepsilon - R}{(\varepsilon - R)^2 + I^2} \, \frac{\partial g_0}{\partial \varepsilon} \tag{215}$$

$$L_I = \int d\varepsilon \, \frac{I}{(\varepsilon - R)^2 + I^2} \, \frac{\partial g_0}{\partial \varepsilon} \tag{216}$$

8.1 The negative mass instability

We want now to study these equations in a few cases of interest:

1) beam with zero energy spread
2) the threshold current for a beam with non-zero energy spread

For the first case let us assume

$$g_0(\varepsilon) = \delta(\varepsilon). \tag{217}$$

From Eqs. (215), (216), and (217) we obtain

$$L_R = \frac{R^2 - I^2}{(R^2 + I^2)^2} \tag{218}$$

$$L_I = \frac{2IR}{(R^2 + I^2)^2} \tag{219}$$

and combining these two equations

$$L_R + iL_I = \frac{1}{(R - iI)^2}. \tag{220}$$

Equation (220) together with Eq. (214) gives

$$(R - iI)^2 = (X + iY). \tag{221}$$

In the case of a perfectly conducting cylindrical vacuum chamber we have from Eqs. (209) and (191)

$$X \approx -\frac{eI_0}{\eta E_0}\frac{Z_0 g_0}{2\beta\gamma^2},\ Y = 0, \tag{222}$$

and the solutions are

$$\text{if } \eta < 0 \tag{223}$$

$$R = |X|^{1/2},\ I = 0 \tag{224}$$

which means that Ω is real and the beam is stable;

$$\text{if } \eta > 0 \tag{225}$$

$$R = 0,\ I = |X|^{1/2} \tag{226}$$

and the beam is unstable.

The frequency slip factor is given by Eq. (79) as $\eta = \alpha - 1/\gamma^2$, or, using the transition energy $\gamma_T = \alpha^{-1/2}$, as

$$\eta = \frac{1}{\gamma_\perp^2} - \frac{1}{\gamma^2}. \tag{227}$$

With this definition one has

$$\eta > 0 \qquad \text{if} \qquad \gamma > \gamma_\perp$$

$$\eta < 0 \qquad \text{if} \qquad \gamma < \gamma_\perp.$$

Hence we can say that a beam in a perfectly conducting vacuum chamber is unstable for $\eta > 0$, i.e., above transition energy. This instability is called the "negative mass instability" and was the first beam instability to be discovered.

The mechanism producing the negative mass instability is the following. For a beam with uniform longitudinal density there is no longitudinal electric field and no force acting on the particles. Let us create a local perturbation, a bump, in the longitudinal density. This perturbation will produce a space-charge longitudinal electric force $eE_\phi \approx -e\partial\lambda/\partial\phi$, acting in the direction away from the bump. This force will increase the energy of the particles in front of the bump and decrease the energy for particles in the back. In the negative mass regime, the front particle will decrease its revolution frequency ω and the back particle will increase ω, so that both particles will move toward the bump and increase its magnitude. In other words, any local small increase in the density will grow, thus producing in the end a non-uniform particle distribution.

Using Eqs. (226) and (207) we can predict the initial growth rate of the instability

$$\frac{1}{\tau} = p\omega_0 \left\{ eI_0 Z_0 \frac{g_0 |\eta|}{2E_0 \beta\gamma^2} \right\}^{1/2} . \tag{228}$$

On the other hand the linearized theory that we have developed can not give us any information on the beam final state. This would require the use of the full non-linear Vlasov equation and is outside the scope of this work. The solution of the non-linear problem is a very difficult task, and most of the work done is based on numerical calculations.

8.2 The effect of the energy spread and the threshold current

In the general case when the beam has a non-zero energy spread we can obtain the eigenvalues, x_p, of our problem from Eqs. (214), (215), and (216). Following the discussion of Section 1 we can look for solutions such that $I \to 0^+$, defining a limit value for the stable current, or threshold current.

We assume to be above transition, $\eta > 0$, so that the beam is unstable if $I > 0$. In the limit $I \to 0^+$ and using the relationship

$$\lim_{I \to 0^+} \frac{I}{(\varepsilon - R)^2 + I^2} = \pi\delta(\varepsilon - R) \tag{229}$$

the integral (216) becomes

$$L_I = \pi \left. \frac{\partial g_0}{\partial \varepsilon} \right|_{\varepsilon = R} \tag{230}$$

and Eq. (215)

$$L_R = \lim_{I \to 0^+} \int d\varepsilon \frac{\varepsilon - R}{(\varepsilon - R)^2 + I^2} \frac{\partial g_0}{\partial \varepsilon}. \tag{231}$$

The dispersion relation (214) can now be solved for R and I_0. The value of I_0 defines the threshold intensity, I_{th}. One way to solve Eq. (215) is to use the equation

$$\frac{L_I}{L_R} = - \frac{Y}{X} \tag{232}$$

to obtain R and for this value of R to evaluate the threshold current, I_{th}, from

$$\frac{eI_{0,th} \, |Z_{\parallel}(p)/p| \; |L|}{E_0 \eta} = 1, \tag{233}$$

where the impedance has been evaluated to first order for $\Omega = 0$.

Let us evaluate explicitly the case when the energy distribution is parabolic

$$g_0(\varepsilon) = \frac{3}{4\sqrt{5}\,\sigma_\varepsilon} \left\{1 - \left(\frac{\varepsilon}{\sqrt{5}\,\sigma_\varepsilon}\right)^2\right\} \tag{234}$$

σ_ε being the r.m.s. value of ε. Using Eqs. (230), (231), and (234) we obtain for the case

$$\left|\frac{R}{\sqrt{5}\,\sigma_\varepsilon}\right| \leqslant 1 \tag{235}$$

that

$$L_I = - \frac{3\pi}{2} \frac{R}{(\sqrt{5}\,\sigma_\varepsilon)^3} \tag{236}$$

$$L_R = \frac{2}{\pi} L_I G(R) \tag{237}$$

with

$$G(R) = \frac{\sqrt{5}\,\sigma_\varepsilon}{R} + \frac{1}{2} \ln \left|\frac{\sqrt{5}\,\sigma_\varepsilon - R}{\sqrt{5}\,\sigma_\varepsilon + R}\right| . \tag{238}$$

For the case $|R/\sqrt{5}\sigma_\varepsilon| > 1$ one has $L_I = 0$ and it is possible to have a solution with $I \to 0^+$ only if $Y = 0$. The equation (232) for R can now be written as

$$\frac{2}{\pi} G(R) = - \frac{X}{Y}. \tag{239}$$

Notice that $G(-R) = - G(R)$.

In Fig. 2 we have made a graph of $G(\overline{R})$ versus $\overline{R} = R/(\sqrt{5}\sigma_\varepsilon)$ and of $\overline{F} = (5/3)\sigma_\varepsilon^2|L|$. An examination of this figure shows that whatever the value of R is obtained by solving (2.55), the value of $|L|$ is nearly constant and $|L| \approx 1/\sigma_\varepsilon^2$. A similar result can be obtained for all reasonably well-behaved distributions. Using this result we can write

$$|L| = 1/(F\sigma_\varepsilon^2), \tag{240}$$

where F is a form factor of the order of one depending on the shape of the energy distribution and on the ratio X/Y of the real to the imaginary part of the impedance.

With the help of Eq. (240) we can rewrite the expression (233) for the threshold current in the convenient form [12]

$$\frac{eI_{0,th}\,|Z_p/p|}{E_0\eta F\sigma_\varepsilon^2} = 1. \tag{241}$$

Notice that for a given impedance the threshold current increases as the square of the rms. energy spread. The reason for this effect is that particles with different energies circuit with different angular velocities, thus reducing the coherence properties of the beam.

9. LONGITUDINAL INSTABILITIES OF BUNCHED BEAMS

9.1 General consideration and equations of motion

The analysis of the longitudinal instabilities of bunched beams is more complicated than that of coasting beams. Although one can write the general equations describing the beam dynamics and the stability properties, it is not possible to obtain a general solution of these equations for a realistic beam-charge distribution and wall impedance. It is however possible to obtain solutions valid in a certain region of the parameter space of this problem.

Let us introduce a time scale using the characteristic times of the beam and of its environment: the bunch duration, L/C; the revolution time T_0; the period of the perturbing electromagnetic field, T_p; the synchrotron oscillation period, T_s; and the instability rise time, τ_i. We always assume $\tau_i > T_0$.

We can consider two cases:

(i) slow instabilities

$$\tau_i > T_s > T_0 \tag{242}$$

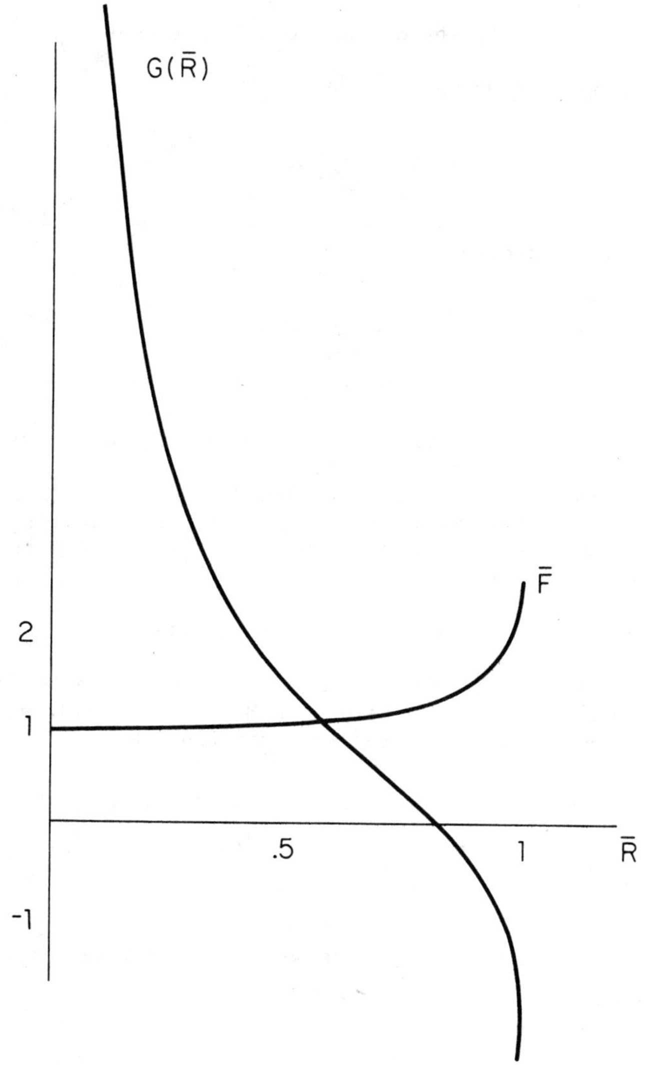

Fig. 2. The expression $G(\overline{R})$ and $\overline{F}(\overline{R})$ as a function of $\overline{R}$ [used in Eqs. (239) and (240)].

(ii) fast instabilities

$$T_s > \tau_i > T_0. \tag{243}$$

Cases (i) and (ii) can be divided in two subcases

a) long wavelength

$$\frac{L}{cT_p} \ll 1, \tag{244}$$

b) short wavelength

$$\frac{L}{cT_p} \gg 1. \tag{245}$$

For each of these cases it is possible to write an approximate expression for the stability condition.

Case (ia) describes the coupled bunch instability and will be discussed in Section 9.3; case (ib) in Section 9.4; and case (iib), describing what is called the microwave instability or bunch lengthening effect in Section 9.5.

9.2. Slow instabilities

To study this problem we use the variables J, ζ introduced in Section 4. We consider a beam with B equal bunches, equally spaced and write the distribution function for bunch K as

$$f^{(K)} = \frac{N}{R}\left\{ g_0(J) + g_1^{(K)}(J, \zeta)e^{-i\Omega\omega_0 t} \right\}. \tag{246}$$

Using Eqs. (90) and (190) we can write the Hamiltonian for bunch "ℓ" as

$$\overline{H}^{(\ell)} = - \nu_s\omega_0 J_\ell + \tag{247}$$
$$+ \frac{e\omega_0 I_0}{E_0} \sum_{K=0}^{B-1} \sum_p \Lambda_p^{(K)} \frac{Z_{\parallel}(p+\Omega)}{p} e^{ip[\phi_\ell + 2\pi(K-\ell)/B] - i\Omega\omega_0 t} + U_{NL},$$

where ϕ_ℓ is a function of J_ℓ, ζ_ℓ and we have assumed for simplicity to be above transition energy, $\eta > 0$. The quantity $\Lambda_p^{(K)}$ is given by Eq. (189).

We now expand $e^{ip\phi}$ as a Fourier series in ζ to obtain

$$\Lambda_p^{(K)} e^{ip\phi_\ell} = \sum_{r,s=-\infty}^{+\infty} i^{r+s}(-1)^s J_r\left(p\sqrt{\frac{2\eta J_\ell}{\nu_s}}\right)e^{ir\zeta_\ell}$$

$$\frac{1}{2\pi}\int dJd\zeta g_1^{(K)}(J, \zeta)\, J_s\left(p\sqrt{\frac{2\eta J}{\nu_s}}\right)e^{is\zeta}. \tag{248}$$

For the case of slow instabilities we can take an average over a synchrotron period and keep only the slowly oscillating terms $r + s = 0$. Then Eq. (248) can be simplified to

$$\Lambda_p^{(K)} e^{ip\phi_\ell} \approx \sum_{r=-\infty}^{+\infty} J_r\left(p\sqrt{\frac{2\eta J_\ell}{\nu_s}}\right)e^{ir\zeta_\ell} \times$$

$$\times \frac{1}{2\pi}\int dJd\zeta g_1^{(K)}(J, \zeta)\, J_r\left(p\sqrt{\frac{2\eta J}{\nu_s}}\right)e^{-ir\zeta}. \tag{249}$$

If we also make the assumption

$$g_1^{(K)}(J, \zeta) = \tilde{g}_q^{(K)}(J)e^{iq\zeta} \tag{250}$$

the expression (249) becomes

$$\Lambda_p^{(K)}e^{ip\phi_\ell} = J_q\left(p\sqrt{\frac{2\eta J_\ell}{\nu_s}}\right)e^{iq\zeta_\ell}\,\sigma_q^{(K)}(p) \tag{251}$$

with

$$\sigma_q^{(K)}(p) = \int_0^\infty dJ\tilde{g}_q^{(K)}(J)\, J_q\left(p\sqrt{\frac{2\eta J}{\nu_s}}\right). \tag{252}$$

We will see that Eq. (250) is an eigenfunction in ζ of the Vlasov equation and represents a state in which all particles have the same phase $q\zeta$.

To write the Vlasov equation we now assume that the U term produced by the equilibrium distribution has the only effect of modifying the focusing force, i.e., to change ν_s. For large currents this modification can be large and the shape of the potential well in which the particles oscillate can be modified,[9] producing a change in bunch length. Here we will not discuss in detail this effect and assume that it produces a change in ν_s described by the non-linear potential U_{NL}.

We can now write the linearized Vlasov equation as

$$\tilde{g}_q^{(\ell)}(J) = q\,\frac{eI_0}{E_0}\,\frac{\dfrac{\partial g_0}{\partial J}}{\Omega + q\nu_s + q\,\dfrac{1}{\omega_0}\,\dfrac{\partial U_{NL}}{\partial J}}\sum_p \frac{Z_\parallel(p+\Omega)}{p} \tag{253}$$

$$\times\, J_q\left(p\sqrt{\frac{2\eta J}{\nu_s}}\right)\sum_{K=0}^{B-1}\sigma_q^{(K)}(p)\,e^{2\pi i p(K-\ell)/B}\,.$$

It is interesting to note that with the choice Eq. (250), $g^{(K)}(J,\zeta) \approx \ell^{ig\zeta}$, we have obtained Eq. (253) independent of ζ and in which the different q modes are not coupled. We will call these modes the "synchrotron modes" and will refer to the index p as the index of "azimuthal modes."

Considering Eq. (253) as a matrix equation for the vector $\tilde{g}_q^{(\ell)}$, $\ell = 0, \ldots, B-1$, one can see that this matrix is a cyclic matrix in ℓ, K and that its solutions are of the form

$$\tilde{g}_q^{(\ell)}(J) = A_{q,s}(J)\,e^{2\pi i s\ell/B}, \quad s = 0, \ldots, B-1. \tag{254}$$

The index s defines the coupled bunch modes. For a given s the bunches oscillate with a phase difference $2\pi s/B$. Using Eq. (254), Eq. (253) becomes

$$A_{q,s}(J) = q\,\frac{eI_0B}{E_0}\,\frac{\dfrac{\partial g_0}{\partial J}}{\Omega + q\left(\nu_s + \dfrac{1}{\omega_0}\,\dfrac{\partial U_{NL}}{\partial J}\right)}\,\times \tag{255}$$

$$\times \sum_h \frac{Z_\parallel(hB-s+\Omega)}{hB-s}\,J_q\left((hB-s)\sqrt{\frac{2\eta J}{\nu_s}}\right)\bar{\sigma}_{q,s}(hB-s)$$

with

$$\bar{\sigma}_{q,s}(hB - s) = \int_0^\infty dJA_{q,s}(J)\, J_q\left((hB - s)\sqrt{\frac{2\eta J}{\nu_s}}\right) \tag{256}$$

equations already derived by several authors.[13]

Our problem is now reduced to find the eigenvalues and eigenvector of Eq. (255). We have no general solutions of these equations, although particular solutions have been found for certain forms of the impedance, or by introducing other approximations. In the next sections we will discuss some of these solutions.

9.3. Coupled bunch longitudinal instabilities

Using the conditions in Eqs. (242) and (244) we can simplify the integral equation (255) by expanding the Bessel functions, for small values of the argument, as $J_q(Z) \approx (Z/2)^q/q!$. An examination of the simplified equations shows that we can obtain a solution of the form[14]

$$A_{q,s}(J) \approx \frac{\dfrac{\partial g_0}{\partial J}}{\Omega + q\left[\nu_s + \dfrac{1}{w_0}\dfrac{\partial U_{NL}}{\partial J}\right]}\left(\frac{\eta J}{2\nu_s}\right)^{q/2}. \tag{257}$$

The corresponding eigenvalue is a solution of the dispersion integral

$$1 = \frac{eI_0 B}{(q-1)!E_0}\sum_h Z_\parallel(hB - s + \Omega)\,(hB - s)^{q-1}$$
$$\int_0^\infty dJ\,\frac{\dfrac{\partial g_0}{\partial J}\left(\dfrac{\eta J}{2\nu_s}\right)^{q/2}}{\Omega + q\left[\nu_s + \dfrac{\partial U_{NL}}{\partial J}\right]} \times J_q\left((hB - s)\sqrt{\frac{2\eta J}{\nu_s}}\right) \tag{258}$$

In the absence of Landau damping, $\partial U_{NL}/\partial J = 0$, and for a gaussian bunch of rms length σ_ϕ

$$g_0(J) = \frac{\eta}{2\pi\nu_s\sigma_\phi^2}\, e^{-\frac{\eta J}{\nu_s\sigma_\phi^2}} \tag{259}$$

we obtain

$$\Omega + q\nu_s = - \frac{eI_0 B\eta\sigma_\phi^{2q-2}}{2\pi\nu_s E_0 2^q (q-1)!} Z_{eff}^{(q)} \tag{260}$$

with

$$Z_{eff}^{q} = \sum_h Z_\parallel(hB - s - q\nu_s)(hB - s)^{2q-1} e^{-\frac{1}{2}\sigma_\phi^2(hB-s)^2} \tag{261}$$

The effective impedance has been evaluated, to zero order, for $\Omega = -q\nu_s$.

These results were obtained by several authors[15] to explain the first observations of coupled bunch instabilities and have been widely discussed in the literature.

The eigenvalues (260) are characterized by two numbers, s and q. The number q gives the oscillation mode: dipole for q = 1, quadrupole for q = 2, etc.; the number s gives the phase shift, $2\pi s/B$, between bunches. The impedance $Z_\parallel(\omega)$ has to be evaluated at a multiple of the bunch revolution frequency, $hB\omega_0$, minus $s\omega_0$ plus the q-th side band of the synchrotron frequency.

In the absence of Landau damping the stability condition is given by $\text{Im}\Omega \leqslant 0$, which for $\eta > 0$ and using the symmetry properties (14) can also be written as

$$- s^{2q-1} e^{-\frac{1}{2}\sigma_\phi^2 s^2} \text{Im}Z_\parallel(s + q\nu_s) +$$

$$+ \sum_{h>0} \Bigg\{ (hB - s)^{2q-1} e^{-\frac{1}{2}\sigma_\phi^2(hB-s)^2} \text{Im}Z_\parallel(hB - s - q\nu_s) -$$

$$- (hB + s)^{2q-1} e^{-\frac{1}{2}\sigma_\phi^2(hB+s)^2} \text{Im}Z_\parallel(hB + s + q\nu_s)\Bigg\} \geqslant 0. \tag{262}$$

For a perfectly conductiong vacuum pipe, $\text{Im}Z_\parallel = 0$ and the beam is stable.

For q = 1, s = 0 the impedance is often dominated by the resonance at the accelerating frequency $h^* B\omega_0$. In this case the stability condition becomes

$$h^{*}B\ell^{-\frac{1}{2}\sigma_\phi^2(h^*B)^2}\left\{\mathrm{Im}Z_\parallel(h^*B-\nu_s)-\mathrm{Im}Z_\parallel(h^*B+\nu_s)\right\} \geqslant 0, \quad (263)$$

so that the impedance at the lower side band must be larger than that at the higher side band. This stability condition was first introduced by Robinson.[17]

9.4. The high frequency slow instability

In this section we consider again the slow instabilities, described by the integral Eq. (255), without using the low frequency approximation Eq. (244). The high frequencies violating Eq. (244) can influence the internal bunch motion and to study this effect it is sufficient to consider a single bunch, for which case s = 0.

Neglecting the bunch mode index, we can rewrite Eq. (255) as

$$A_q(J) = \frac{qeI_0}{E_0}\,\frac{\dfrac{\partial g_0}{\partial J}}{\Omega + q\nu_s\left[1+\dfrac{1}{\omega_0\nu_s}\dfrac{\partial U_{NL}}{\partial J}\right]} \times$$

$$\times \sum_p \frac{Z_\parallel(p+\Omega)}{p} J_q\left(p\sqrt{\frac{2\eta J}{\nu_s}}\right)\bar{\sigma}_q(p) \quad (264)$$

with

$$\bar{\sigma}_q(p) = \int_0^\infty dJ A_q(J) J_q\left(p\sqrt{\frac{2\eta J}{\nu_s}}\right). \quad (265)$$

To further simplify the discussion we assume $\partial U_{NL}/\partial J = 0$, so neglecting Landau damping. Equation (264) has been obtained and studied by Sacherer.[18]

Let us introduce a new function

$$u_q(J) = \left(\frac{\partial g_0}{\partial J}\right)^{-1/2} A_q(J) \quad (266)$$

and rewrite Eq. (264) explicitly as an integral equation

$$u_q(J) = \frac{qeI_0}{E_0(\Omega + q\nu_s)} \int_0^\infty dJ' u_q(J')\, G\,(J, J') \tag{267}$$

with

$$G(J, J') = \left[\frac{\partial g_0(J)}{\partial J} \frac{\partial g_0(J')}{\partial J'}\right]^{1/2} \sum_{p=1}^{\infty} J_q\left(p\sqrt{\frac{2\eta J}{\nu_s}}\right) J_q\left(p\sqrt{\frac{2\eta J'}{\nu_s}}\right) \times$$

$$\times \frac{Z\left(p - q\nu_s\right) + Z^*\left(p + q\nu_s\right)}{p}, \tag{268}$$

where we have evaluated the impedance to zero order for $\Omega = -q\nu_s$ and used the symmetry properties in Eqs. (13) and (14).

The Green function $G(J, J')$ is symmetric. If the quantity $Z\,(p - q\nu_s) + Z^*(p + q\nu_s)$ is real, the eigenvalues of Eq. (268) are also real and the beam is stable. This is, for instance, the case of a perfectly conducting wall or for a constant imaginary part of the impedance.

Let us introduce a complete orthonormal set of eigenfunctions, $F_m^q(J)$, of the integral Eq. (267) and let us write

$$u_q(J) = \sum_m D_{q,m} F_m^q(J) \tag{269}$$

$$\Lambda_{m,p}^q = \int_0^\infty dJ \left(\frac{\partial g_0}{\partial J}\right)^{1/2} J_q\left(p\sqrt{\frac{2\eta J}{\nu_s}}\right) F_m^q\,(J). \tag{270}$$

The eigenvalues of Eq. (267) can be written in terms of the $\Lambda_{m,p}^q$ as

$$\Omega_{q,m} + q\nu_s = \frac{qeI_0}{E_0} Z_{eff}^{(q,m)}, \tag{271}$$

where the effective impedance is

$$Z_{eff}^{(q,m)} = \sum_p \frac{Z\,(p - q\nu_s) + Z^*(p + q\nu_s)}{p} [\Lambda_{m,p}^q]^2. \tag{272}$$

The eigenvalues and eigenfunction of Eq. (267) depend on $g_0(J)$ and on $Z(\omega)$. We do not have any general solution of Eq. (267). A solution has been found, however, for the case of a parabolic charge distribution and constant $Z_{\parallel}(\omega)/\omega$, $Z_{\parallel}(\omega)/\omega = L_\omega$.[19] This case "per se" is not very interesting, since all the eigenvalues are real and the beam is stable. It can be used to evaluate the complex eigenvalues when the impedance is of the form

$$\frac{Z_{\parallel}(\omega)}{\omega} \equiv L_\omega + \frac{Z_1(\omega)}{\omega} \tag{273}$$

with

$$\left|\frac{Z_1(\omega)}{\omega}\right| << L_\omega. \tag{274}$$

When Eqs. (273) and (274) are satisfied we can use perturbation methods to write the eigenfunctions and eigenvalues as

$$F_m^q(J) = F_{0,m}^q(J) + F_{1,m}^q(J) \tag{275}$$

$$\Omega_{q,m} = \Omega_{q,m}^{(0)} + \Omega_{q,m}^{(1)} \tag{276}$$

with $\Omega_{q,m}^{(0)}$ obtained for $Z_1 = 0$ and

$$\Omega_{q,m}^{(1)} = \frac{qeI_0}{E_0} \sum_p \frac{Z_1(p - q\nu_s) + Z_1^*(p + q\nu_s)}{p} \left[\Lambda_{0,m,p}^q\right]^2 \tag{277}$$

and the $\Lambda_{0,m,p}^q$ evaluated with the unperturbed eigenfunctions.

Let us now find the zeroth order solution. We assume

$$g_0(J) = \frac{3}{4\pi} \frac{2\eta N}{\nu_s \phi_L^2} \left\{1 - \frac{2\eta J}{\nu_s \phi_L^2}\right\}^{1/2} \tag{278}$$

corresponding to a longitudinal density

$$\lambda_0(\phi) = \frac{3N}{4\phi_L} \left\{1 - \frac{\phi^2}{\phi_L^2}\right\}. \tag{279}$$

With this choice of $g_0(J)$ one has[19]

$$\Lambda^{q}_{0,m,p} = B_{q,m} \frac{J_{m+1/2}(p\phi_L)}{(p\phi_L)^{1/2}}, \quad m = q + 2,\ q + 4 \tag{280}$$

with

$$B^2_{q,m} = \frac{3}{8}\frac{N\eta}{\nu_s}(2m+1)\frac{(m+q-1)!!(m-q-1)!!}{(m+q)!!(m-q)!!} \tag{281}$$

and the eigenvalues can be written as

$$q\nu_s + \Omega^{(0)}_{q,m} = \frac{eI_0\eta}{E\nu_s} Z^{(0)}_{q,m}\, q \frac{(m+q-1)!!(m-q-1)!!}{(m+q)!!(m-q)!!} \tag{282}$$

$$\Omega^{(1)}_{q,m} = \frac{eI_0\eta}{E\nu_s} Z^{(1)}_{q,m} \frac{q(m+q-1)!!(m-q-1)!!}{(m+q)!!(m-q)!!} \tag{283}$$

with

$$Z^{(0)}_{q,m} = 3\pi \frac{\omega_0 L_\omega}{\phi_L^3} \tag{284}$$

$$Z^{(1)}_{q,m} = +\frac{3}{8}(2m+1)\int_{-\infty}^{+\infty} dy\left[\frac{J_{m+1/2}(y)}{y}\right]^2 Z_1\left((y - q\nu_s\phi_L)\frac{\omega_0}{\phi_L}\right) \tag{285}$$

The formulae (284) and (285) have been widely used to evaluate the coupling impedance for a single bunch slow instability. It must be remembered, however, that Eq. (285) can only be applied for an impedance of the form in Eqs. (273) and (274) and that for an arbitrary impedance the result from Eq. (285) is not generally valid. In particular, since the term L_ω in Eq. (273) produced by smooth conducting walls is very small for high-energy beams, the formula (285) can only be approximately applied in frequency regions where the broad-band impedance is slowly varying.

To evaluate approximately the threshold for this instability we can compare the coherent frequency shift, whose main contribution is given by Eq. (282), with the frequency spread produced by the nonlinear part of the rf force, $\Delta\Omega_L$,

$$\Delta\Omega_L = -\frac{1}{8} B^2 \nu_s \omega_0 \phi_L^2. \tag{286}$$

Assuming $\Delta\Omega_L \approx \Omega^{(0)}_{q,m}$ we obtain a relationship between the current and the bunch length, defining the equilibrium conditions. It is interesting to notice that the threshold, or equilibrium current, I_{th}, is a very strong function of the bunch length. In fact from Eqs. (286), (282), and (284) we have

$$I_{th} \approx \phi_L^5 \tag{287}$$

9.5. The high frequency instability

When the instability rise time is shorter than the synchrotron period we cannot apply the results of the previous sections. The use of the approximate expression (249) in the Hamiltonian (247) is not justified and the full expression (248) has to be used.

In this case all synchrotron modes $e^{iq\zeta}$ are coupled. This coupling plays an important role in the fast instability and it makes it convenient, to study this problem, to use an approach different from that used in the slow instability case.[20]

We use directly the distribution function (246) and the Hamiltonian (247). We consider again the case of a single bunch, assume $U_{NL} = 0$ and that the field produced by the equilibrium distribution causes only a change in ν_s. In this case we can rewrite the Hamiltonian (247) as

$$\overline{H} = -\nu_s \omega_0 J + Ue^{-i\omega_0 \Omega t} \tag{288}$$

with

$$U = \frac{e\omega_0 I_0}{E_0} \sum_p \frac{Z_{\parallel}(p + \Omega)}{p} \Lambda_p e^{ip\phi}, \tag{289}$$

and

$$\Lambda_p = \frac{1}{2\pi} \int dJ d\zeta g_1(J, \zeta) e^{-ip\phi(J, \zeta)} \tag{290}$$

and $\phi(J, \zeta)$ given by Eq. (88).

The linearized Vlasov equation can now be written as

$$i\omega_0 \Omega g_1 + \omega_0 \nu_s \frac{\partial g_1}{\partial \zeta} = - \frac{\partial U}{\partial \zeta} \frac{\partial g_0}{\partial J}. \tag{291}$$

The solution of Eq. (291) can be written using the periodicity in ζ, as

$$g_1(J, \zeta) = - \frac{1}{\omega_0 \nu_s} \frac{\partial g_0}{\partial J} \frac{e^{-i\bar{\Omega}\zeta}}{e^{2\pi i\bar{\Omega}} - 1} \int_\zeta^{\zeta+2\pi} d\zeta' \frac{\partial U(J, \zeta')}{\partial \zeta'} e^{i\bar{\Omega}\zeta'}, \tag{292}$$

where

$$\bar{\Omega} = \Omega/\nu_s. \tag{293}$$

Substituting Eq. (292) in (290) we can reduce the integral Eq. (292) to a linear algebraic system

$$\Lambda_p = \sum_{q=-\infty}^{+\infty} T_{pq} \Lambda_q, \tag{294}$$

with

$$T_{pq} = \frac{ieI_0}{2\pi\nu_s E_0} \frac{Z_\parallel(q + \Omega)}{\ell^{2\pi i\bar{\Omega}} - 1} \times$$

$$\times \int_0^\infty dJ \frac{\partial g_0}{\partial J} \left(\frac{2\eta J}{\nu_s} \right)^{1/2} \int_0^{2\pi} d\zeta d\bar{\zeta} \times$$

$$\times \sin(\zeta + \bar{\zeta})\, e^{i\bar{\Omega}\bar{\zeta} - ip\phi(J, \zeta) + iq\phi(J, \zeta + \bar{\zeta})}. \tag{295}$$

The solutions of our problem are the values of Ω for which the equation

$$\det\left|\delta_{p,q} - T_{pq}\right| = 0 \tag{296}$$

is satisfied. We emphasize here that, contrary to the coasting beam case, for a bunched beam all frequency components $q\omega_0$ are coupled. Also all synchrotron modes $g_1(j, \zeta) \approx \tilde{g}_1(J)e^{iq\zeta}$ are coupled, contrary to the slow instability case. In this coupling of all modes lies the difficulty of solving the fast blow-up, bunched beam case.

To study Eq. (294) we use the following approach. We consider only high frequency perturbations, i.e., the case

$$q\sigma_\phi \gtrsim 1, \tag{297}$$

where σ_ϕ is the rms bunch length in units of machine radius. Then we consider the fast blow-up case

$$\omega_0 \gg |\Omega|\ \omega_0 \gg \nu_s\omega_0 . \tag{298}$$

Using these two conditions we simplify the expression of the matrix element (295) by assuming $Z_\parallel(q + \Omega) \approx Z_\parallel(q)$ and by evaluating the integral over ζ in the limit $\mathrm{Im}\Omega \to +\infty$. We also rewrite the integration over J and ζ as an integral over ε and ϕ to obtain[20]

$$T_{pq} \simeq \frac{-ieI_0}{2\pi E_0} Z_\parallel(q) \int d\varepsilon d\phi e^{i(q-p)\phi} \frac{\frac{\partial g_0}{\partial \varepsilon}}{\Omega + q\eta\varepsilon} . \tag{299}$$

We now restrict ourselves to the case of a gaussian bunch

$$g_0 = \lambda(\phi)G(\varepsilon), \tag{300}$$

$$\lambda(\phi) = \frac{1}{\sqrt{2\pi}\sigma_\phi} e^{-\phi^2/2\sigma_\phi^2}, \tag{301}$$

$$G(\varepsilon) = \frac{1}{\sqrt{2\pi}\sigma_\varepsilon} e^{-\varepsilon^2/2\sigma_\varepsilon^2}. \tag{302}$$

We also define

$$\lambda_n = \frac{1}{2\pi} \int d\phi\lambda(\phi)\ e^{in\phi} = \frac{1}{2\pi} e^{-n^2\sigma_\phi^2/2}. \tag{303}$$

The matrix element can now be written as

$$T_{pq} = \frac{eI_0}{E_0\eta}\,\frac{Z_{\parallel}(q)}{q}\,\lambda_{q-p}\int d\varepsilon\,\frac{\frac{\partial G}{\partial \varepsilon}}{\frac{\Omega}{q\eta}+\varepsilon} \tag{304}$$

or, using Eq. (302) as

$$T_{pq} = -\frac{2\pi eI_0}{\eta E_0\sigma_\varepsilon^2}\,\frac{Z_{\parallel}(q)}{q}\,\lambda_{q-p}\,h\left(-\frac{\Omega}{q\sigma_\phi\nu_s}\right) \tag{305}$$

with

$$h(y) = \int_0^{+\infty} dx\; xe^{-x^2/2-ixy}. \tag{306}$$

This function has the property that

$$h(0) = 1$$
$$|h(y)| \leqslant 1 \text{ if } y \neq 0,\ \mathrm{Im}y < 0, \tag{307}$$

so that it has a maximum for y = 0 and decreases when y increases.

It is interesting to notice that the matrix element (304) or (305) can also be used to describe the limiting case of a coasting beam by taking the other limit $\sigma_\phi \to \infty$ or

$$\lambda_{n-m} = \delta_{n,\,m}. \tag{308}$$

In this case the matrix T is diagonal and we obtain the usual coasting beam dispersion relation Eq. (206)

$$1 = -\frac{eI_0}{E\eta}\,\frac{Z_{\parallel}(q)}{q}\int d\varepsilon\,\frac{\frac{\partial G}{\partial \varepsilon}}{\frac{\Omega}{q\eta}+\varepsilon}. \tag{309}$$

For a bunched beam we have to solve the more complicated Eq. (294) with $T_{m,n}$ given by Eq. (305). We can obtain a simple but approximate solution if we make the approximation that the impedance is large only near a frequency q_0 and is nearly constant over the range $q_0 - N_b$, $q_0 + N_b$ where $N_b \sim 1/\sigma_\phi$ defines the range over which the bunch form factor λ_n remains near to one. In this simple model, we can rewrite Eq. (305) as

$$T^*_{pq} \approx - \frac{eI_0}{E_0 \eta \sigma_\varepsilon^2} \frac{Z_\parallel(q_0)}{q_0} h\left(- \frac{\Omega}{q_0 \sigma_\phi \nu_s}\right) * e^{- (q - p)^2 \sigma_\phi^2/2} . \tag{310}$$

The eigenvectors of the matrix T^*_{pq} are of the form

$$V_p = e^{i\alpha p}, \tag{311}$$

and the corresponding eigenvalues are obtained from the dispersion relation

$$1 = - \frac{eI_0}{E_0 \eta \sigma_\varepsilon^2} \frac{Z_\parallel(q_0)}{q_0} h\left(- \frac{\Omega(\alpha)}{q_0 \sigma_\phi \nu_s}\right) \sum_{n=-\infty}^{+\infty} e^{-(n^2\sigma_\phi^2/2)+in\alpha} . \tag{312}$$

Since $h(x) \leqslant 1$ and the sum over n in Eq. (312) has a maximum for $\alpha = 0$, we can obtain a solution of Eq. (312) only if the condition

$$\frac{eI_0}{E_0 \eta \sigma_\varepsilon^2} \left| \frac{Z_\parallel(q_0)}{q_0} \right| \sum_{n=-\infty}^{+\infty} e^{-(n^2\sigma_\phi^2/2)} \geqslant 1 \tag{313}$$

is satisfied. Notice that the sum over n in Eq. (313) is of the order $1/\sigma_\phi$, so that this condition depends on I_0/σ_ϕ, i.e., the peak current.

If the condition (313) is not satisfied we contradict our initial assumption $|\Omega/\nu_s| > 1$. On the other hand if we satisfy Eq. (313), we can find a solution for Eq. (312) with $|\Omega/\nu_s| > 1$, consistent with the way we have derived Eq. (305). This means that we can have a fast blow-up only if the beam current is such as to satisfy Eq. (313) and that at lower beam current there is no possibility of a fast blow-up.

It is also interesting to notice that for a given $\Omega/\nu_0\omega_0$, the function

$$h\left(-\frac{\Omega}{q_0\sigma_\phi\nu_s}\right)$$

is larger when $q_0\sigma_\phi > |\Omega/\nu_s| > 1$ so that a high frequency impedance is more effective in producing a fast blow-up.

Equation (313) can be used to define a threshold curve

$$I_0 = f(\sigma_\phi).$$

For a bunch, since $\sigma_\varepsilon \sim \sigma_\phi$, this curve is of the form

$$I_{th} \approx \sigma_\phi^3. \tag{314}$$

Comparing Eq. (314) with the similar condition in Eq. (287) for slow instabilities we see that at small current the condition in Eq. (288) is dominant and one observes single mode (slow) instabilities. Increasing the current, the microwave (fast) instability becomes dominant.

10. TRANSVERSE INSTABILITY OF A COASTING BEAM

In this section we want to start the discussion of transverse instabilities by considering the simplest case, i.e., that of a coasting beam.[21] For a coasting beam the equilibrium state has a uniform distribution in azimuth and there is no rf system present. The equations of motion can be simplified by assuming $\nu_s = 0$.

Using action angle variables the Hamiltonian is given by Eqs. (100) and (101) as

$$\overline{H} = \nu_0\omega_0\{I\,[1 + P\varepsilon] + D(I)\} - \frac{1}{2}\eta\omega_0\varepsilon^2$$

$$- F(\phi, t)\left(\frac{2I}{\nu_0\omega_0}\right)^{1/2}\cos\psi \tag{315}$$

To write the distribution function, we follow Eq. (116) and neglect for simplicity the nonessential variables x, $\dot{x}$. In the coasting beam case ε is a constant, there is no modulation of the betatron frequency due to synchrotron oscillations and we can take $\nu = 0$.

We now assume explicitly

$$f = \frac{N}{R}\left\{\frac{1}{2\pi}g_0(\varepsilon)h_0(I) + f_1^p\,(I, \psi, \varepsilon)e^{ip\phi - i\nu\omega_0 t}\right\}, \tag{316}$$

having selected only one of the Fourier harmonics in ϕ to describe the perturbation. As we shall see, this choice is justified since the harmonics with different p's do not interact and Eq. (316) is an eigenfunction in ϕ of our problem. In the same way, we select only one of the harmonics of F in Eq. (173),

$$F = \frac{ieI_0\beta}{2\pi R} D_p Z_\perp(p + \nu)e^{ip\phi - i\nu\omega_0 t} . \tag{317}$$

If we want to study the solution with $\nu \sim \nu_0$ we can also assume (as in Sec. 3)

$$f_1^p \sim \tilde{f}_1^p e^{i\psi} , \tag{318}$$

and obtain the linearized Vlasov equation

$$\left\{\nu - \nu_0 - \nu_0\left[D'(I) + \left(\xi - \eta + \frac{p\eta}{\nu_0}\right)\varepsilon\right]\right\} \tilde{f}_1^p(I, \varepsilon) =$$

$$= \frac{ieI_0 c}{8\pi^2 E_0} D_p Z_\perp(p + \nu) \left(\frac{2I}{\nu_0\omega_0}\right)^{1/2} g_0(\varepsilon) \frac{\partial h_0(I)}{\partial I} , \tag{319}$$

having used Eq. (92) to express P.

The dipole moment in Eq. (319) is given by Eqs. (159) and (98) as

$$D_p = \int d\varepsilon dI d\psi f_1^{(p)}(I, \psi, \varepsilon) \left(\frac{2I}{\nu_0\omega_0}\right)^{1/2} \cos\psi, \tag{320}$$

and using Eq. (318) as

$$D_p = \pi \int d\varepsilon dI \left(\frac{2I}{\nu_0\omega_0}\right)^{1/2} \tilde{f}_1^p (I, \varepsilon). \tag{321}$$

From Eqs. (319) and (321) we obtain the dispersion relation[21]

$$1 = \delta\nu_p \int d\varepsilon dI \frac{I g_0(\varepsilon) \dfrac{\partial h_0(I)}{\partial I}}{\nu - \nu_0 - \nu_0\left[D'(I) + \left(\xi - \eta + \dfrac{p\eta}{\nu_0}\right)\varepsilon\right]} \tag{322}$$

with

$$\delta\nu_p = \frac{ieI_0R}{4\pi\beta E_0\nu_0} Z_\perp(p + \nu_0), \tag{323}$$

giving the coherent frequency shift and we have evaluated the impedance, to zero order, for $\nu = \nu_0$.

In the simple case when $D'(I) = 0$, and $g_0(\varepsilon) = \delta(\varepsilon)$ the solution of the dispersion integral is

$$\nu - \nu_0 = \delta\nu_p. \tag{324}$$

In the general case the value of ν depends on the energy and the amplitude distribution. In systems like proton storage rings, the energy-dependent term, in the denominator of the dispersion integral, can be larger than the amplitude dependent term. In this case Eq. (322) can be written in the simpler form

$$1 = \delta\nu_p \int d\varepsilon \frac{g_0(\varepsilon)}{\nu_0 - \nu + \nu_0\left[\xi - \eta + \dfrac{p\eta}{\nu_0}\right]\varepsilon}. \tag{325}$$

This is the same type of dispersion relation discussed in Sec. 2, and we can follow the same evaluation to arrive at the stability condition

$$|\delta\nu_p| \sim \Delta\nu_p \tag{326}$$

with

$$\Delta\nu_p = \nu_0\left[\xi - \eta + \frac{p\eta}{\nu_0}\right]\varepsilon_m \tag{327}$$

giving the frequency spread in terms of the energy spread ε_m.

It is interesting to notice that for a given energy spread the frequency spread is determined, according to Eq. (327), by the machine chromaticity, ξ, the frequency slip factor, η, and the mode number, p. In particular one can define a mode number, p^*, such that $\Delta\nu$ is a minimum. Then if we satisfy Eq. (326) for $p = p^*$, the beam is stable for all other values of p, that is for all coherent dipole modes.

11. THE VLASOV EQUATION FOR THE TRANSVERSE OSCILLATIONS OF A BUNCHED BEAM

To obtain the Vlasov equation we use action-angle coordinates. The Hamiltonian for a bunch ℓ is obtained from Eq. (109)

$$\overline{H}^{(\ell)} = \nu_0\omega_0\{(1 + P\varepsilon_\ell)I_\ell + D(I_\ell)\}$$

$$- \nu_s\omega_0 J_\ell - F(J_\ell, \zeta_\ell, t)\left(\frac{2I_\ell}{\nu_0\omega_0}\right)^{1/2} \cos\psi_\ell, \tag{328}$$

where we have assumed for simplicity to be above transition, $\eta > 0$. The collective force is given by Eqs. (175) and (177).

The distribution function for bunch ℓ, neglecting the inessential variables x, $\dot{x}$, can be written as

$$f^{(\ell)} = \frac{N}{R}\left\{f_0^{(\ell)}(I, J) + f_1^{(\ell)}(I, \psi, J, \zeta)e^{-i\nu\omega_0 t}\right\}. \tag{329}$$

The linearized Vlasov equation is

$$- i\nu\omega_0 f_1^{(\ell)} + \nu_0\omega_0\{1 + P\varepsilon + D'(I)\}\frac{\partial f_1^{(\ell)}}{\partial\psi} - \nu_s\omega_0\frac{\partial f_1^{(\ell)}}{\partial\zeta} =$$

$$= \frac{c^2F^{(\ell)}}{E_0} e^{i\nu\omega_0 t}\left(\frac{2I}{\nu_0\omega_0}\right)^{1/2} \sin\psi\,\frac{\partial f_0^{(\ell)}}{\partial I} \tag{330}$$

with

$$\frac{c^2F^{(\ell)}}{E_0} e^{i\nu\omega_0 t} = \frac{ieI_0\omega_0 c}{2\pi E_0} \sum_{K=0}^{B-1} \sum_p D_p^{(K)} \times$$

$$\times Z_\perp(p + \nu) e^{2\pi ip \frac{\ell-K}{B} + ip\phi(J_\ell, \zeta_\ell)}, \tag{331}$$

and

$$D_p^{(K)} = \frac{1}{2\pi} \int \left(\frac{2I}{\nu_0\omega_0}\right)^{1/2} \cos\psi\, f_1^{(K)}(I, \psi, J, \zeta) e^{-ip\phi(J,\zeta)} dI d\psi dJ d\zeta \tag{332}$$

$$\phi(J, \zeta) = \left(\frac{2\eta J}{\nu_s}\right)^{1/2} \cos\zeta, \tag{333}$$

where we have used Eqs. (175), (177), and (88).

As discussed in Sections 2 and 3 we now look for a solution $\nu \approx \nu_0$ and assume

$$f_1(I, \psi, J, \zeta) \approx \tilde{f}_1(I, J, \zeta) e^{i\psi}$$

then Eq. (330) becomes

$$\left[\nu - \nu_0\left(1 + P\varepsilon + D'(I)\right)\right] \tilde{f}_1^{(\ell)} - i\nu_s \frac{\partial \tilde{f}_1^{(\ell)}}{\partial \zeta} = \frac{ieI_0 c}{4\pi E_0} \left(\frac{2I}{\nu_0\omega_0}\right)^{1/2} \times$$

$$\times \frac{\partial f_0(I, J)}{\partial I} \sum_{K=0}^{B-1} \sum_p D_p^{(K)} Z_\perp(p + \nu) e^{2\pi ip \frac{\ell - K}{B} + ip\phi(J,\zeta)} \tag{334}$$

with

$$D_p^{(K)} = \frac{1}{2} \int \left(\frac{2I}{\nu_0 \omega_0}\right)^{1/2} \times$$

$$\times \tilde{f}_1^{(K)}(I, J, \zeta)\, e^{-ip\phi(J,\zeta)}\, dIdJd\zeta. \qquad (335)$$

As in the case of longitudinal oscillations we do not have a general solution of Eq. (334). We can again divide the parameter space of this problem in a few regions where we can obtain approximate solutions. Since the betatron period is shorter than the revolution period we can always assume that the coherent tune shift is small compared with ν_0 and use the same classification in slow or fast instability introduced in Eqs. (242) and (243), comparing the coherent tune shift with the synchrotron tune. In the same way we can also use the long or short wavelength approximation given by Eqs. (244) and (245).

11.1 THE SLOW INSTABILITIES FOR THE TRANSVERSE CASE

We will discuss the slow instabilities, neglecting Landau damping, $D'(I) = 0$, to simplify our problem. It is convenient to introduce a local dipole moment, defined along a bunch,

$$V^{(K)}(J, \zeta) = \frac{1}{2} \int_0^\infty dI \left(\frac{2I}{\nu_0 \omega_0}\right)^{1/2} \tilde{f}_1^{(K)}(I, J, \zeta), \qquad (336)$$

so that from Eq. (335)

$$D_p^{(K)} = \int dJd\zeta V^{(K)}(J, \zeta) e^{-ip\phi(J,\zeta)}. \qquad (337)$$

From Eq. (334) multiplying by $(2I/\nu_0\omega_0)^{1/2}$ and integrating over I we have

$$[\nu - \nu_0(1 + P\varepsilon)]\, V^{(\ell)}(J, \zeta) - i\nu_s \frac{\partial V^{(\ell)}(J, \zeta)}{\partial \zeta} =$$

$$= - \frac{ieI_0R}{4\pi\beta\nu_0 E_0} g_0(J) \sum_{K=0}^{B-1} \sum_p D_p^{(K)} Z_\perp(p + \nu) e^{2\pi ip\left(\frac{\ell-K}{B}\right) + ip\phi(J,\zeta)}, \qquad (338)$$

where

$$g_0(J) = \int_0^\infty f_0(I, J)\, dI \tag{339}$$

gives the unperturbed longitudinal distribution. We can simplify Eq. (338) by introducing the coupled bunch normal modes

$$V^{(\ell)}(J, \zeta) = V_s(J, \zeta)e^{2\pi is\ell/B}, \quad s=0, 1, \ldots, B-1, \tag{340}$$

$$D_{s,p} = \int dJd\zeta V_s(J, \zeta)e^{-ip\phi}, \quad s=0, 1, \ldots, B-1, \tag{341}$$

obtaining

$$[\nu - \nu_0(1 + P\varepsilon)]\, V_s(J, \zeta) - i\nu_s \frac{\partial V_s(J, \zeta)}{\partial \zeta} =$$

$$= - \frac{ieI_0 BR}{4\pi\beta\nu E_0 0} g_0(J) \sum_n D_{s,nB+s} Z_\perp(nB + s + \nu)e^{i(nB+s)\phi}. \tag{342}$$

As in the case of coupled bunch longitudinal instabilities, the frequency spectrum is a multiple of the bunch revolution frequency, plus the mode number s, plus the coherent betatron frequency ν.

The solution of Eq. (342), periodic in ζ, is

$$V_s(J, \zeta) = - \frac{ieI_0 R}{4\pi\beta E_0 \nu_0 \nu_s} g_0(J) \sum_n D_{s,nB+s} Z_\perp(nB + s + \nu) \times$$

$$\times \frac{1}{e^{2\pi i(\nu-\nu_0)/\nu_s} - 1} \times ,$$

$$\times \int_\zeta^{\zeta+2\pi} d\zeta' e^{i(nB+s)\phi(J,\zeta')+i\frac{\nu-\nu_0}{\nu_s}(\zeta'-\zeta)-i\bar{\nu}[\phi(J,\zeta')-\phi(J,\zeta)]} \tag{343}$$

with

$$\bar{\nu} = \nu_0 P/\eta = \nu_0\left(\frac{\xi}{\eta} - 1\right). \tag{344}$$

We now expand $e^{ip\phi}$ in its Fourier harmonics to obtain

$$V_s(J, \zeta) = -\frac{ieI_0R}{4\pi\beta E_0\nu_0} g_0(J) \sum_n D_{s,nB+s} Z_\perp(nB + s - \nu) \times$$

$$\times \sum_{p,q=-\infty}^{+\infty} \frac{i^{p+q}}{\nu-\nu_0+q\nu_s} J_p\left(\bar{\nu}\left(\frac{2\eta J}{\nu_s}\right)^{1/2}\right) J_q\left((nB+s-\bar{\nu})\left(\frac{2\eta J}{\nu_s}\right)^{1/2}\right) e^{i(p+q)\zeta}. \tag{345}$$

On this equation we can perform the slow instability approximation by taking the average over ζ, i.e., neglecting all fast oscillating terms and keeping only the term $p + q = 0$.

The approximate equation for the average value $\bar{V}_s(J)$, of $V_s(J,\zeta)$ is

$$\bar{V}_s(J) = -\frac{ieI_0R}{4\pi\beta E_0\nu_0} g_0(J) \sum_n D_{s,nB+s} Z_\perp(nB+s+\nu) \times$$

$$\times \sum_q \frac{(-1)^q}{\nu-\nu_0+q\nu_s} J_q\left(\bar{\nu}\left(\frac{2\eta J}{\nu_s}\right)^{1/2}\right) J_q\left((nB+s-\bar{\nu})\left(\frac{2\eta J}{\nu_s}\right)^{1/2}\right) \tag{346}$$

with

$$D_{s,p} = 2\pi \int_0^\infty dJ\bar{V}_s(J) J_0\left(p\left(\frac{2\eta J}{\nu_s}\right)^{1/2}\right). \tag{347}$$

Since we have made the assumption that the coherent tune shift is small compared to the synchrotron frequency, we can obtain a solution of Eq. (346) to first order in $(\nu - \nu_0)/\nu_s$ by keeping only the $q = 0$ term on its r.h.s. The equation for the slow instability is then

$$\overline{V}_s(J) = -\frac{ieI_0R}{4\pi\nu_0\beta E_0} g_0(J)J_0\left(\overline{\nu}\left(\frac{2\eta J}{\nu_s}\right)^{1/2}\right) \times \frac{1}{\nu - \nu_0} \times$$

$$\times \sum_n D_{s,nB+s} J_0\left((nB+s-\overline{\nu})\left(\frac{2\eta J}{\nu_s}\right)^{1/2}\right) Z_\perp(nB+s+\nu). \tag{348}$$

This equation is still very complicated and it is convenient to obtain some simplified and approximate solution to obtain an understanding of the transverse instabilities.

11.2. Rigid bunches and the coupled bunch instabilities

For a bunch of length ϕ_L the quantity $\overline{\nu}\phi_L$ measures the change in betatron frequency from the head to the tail of the bunch. In this section we want to study the case when this change is negligible and all particles in the bunch oscillate with the same frequency as in a point like, $\phi_L = 0$, system or rigid system. Assuming $\overline{\nu} = 0$ in Eq. (348) we can obtain the equation

$$\overline{V}_s(J) = -\frac{ieI_0R}{4\pi\beta E_0\nu_0} \frac{g_0(J)}{\nu-\nu_0} \sum_n D_{s,nB+s}Z_\perp(nB+s+\nu)J_0\left((nB+s)\left(\frac{2\eta J}{\nu_s}\right)^{1/2}\right) \tag{349}$$

describing coupled bunch instabilities.

Equation (349) becomes very simple in the long wave length approximation, taking

$$J_0\left(p\left(\frac{2\eta J}{\nu_s}\right)^{1/2}\right) \approx 1$$

in the expression (347) of the dipole moment $D_{s,p}$, which in this case does not depend on p. We obtain from Eq. (349) the eigenvalue for the s-th mode of coherent oscillation, as[22]

$$\nu-\nu_0 = -\frac{ieI_0R}{4\pi\beta E_0\nu_0} Z^s_{\perp eff} \tag{350}$$

with

$$Z^s_{\perp eff} = \sum_n Z_\perp(nB + s + \nu_0)F(nB + s), \tag{351}$$

and

$$F(p) = 2\pi \int_0^\infty dJ g_0(J) J_0\left(p\left(\frac{2\eta J}{\nu_s}\right)^{1/2}\right). \qquad (352)$$

The form factor F(p) can be evaluated for any bunch shape. For a gaussian bunch $g_0(J)$ is given by Eq. (259) and

$$F(p) = e^{-p^2\sigma_\phi^2/2}. \qquad (353)$$

The eigenvalues (350) describe the B collective modes of oscillations of B rigid bunches. For mode s the bunches oscillate with a phase difference $2\pi s/B$.

11.3. The head-tail instability

We want now to discuss the case when $\overline{\nu}\phi_L$ is not negligible and study the effect of the phase difference introduced within a bunch by this term. Since this effect is mainly internal to the bunch we can simplify the problem considering the case of a single bunch, B = 1, s = 0. We also consider a simple impedance which does not produce an instability for a rigid bunch. For instance we can consider the case $Z_\perp(\omega)$ = constant, which using Eqs. (351), (353) and (13) gives $\mathrm{Re}Z_{\perp\mathrm{eff}} = 0$ and $\mathrm{Im}\nu = 0$.

Our problem is now to study the transverse instability of a single bunch assuming $\overline{\nu}\phi_L$ to be small but non zero. Let us write Eq. (348) for B = 1, s = 0, and up to first order in $\overline{\nu}\phi_L$. Since we are considering the slow instability case, we can also assume $|\nu-\nu_0| \ll \nu_s$, and obtain

$$\overline{V}(J) = -\frac{ieI_0R}{4\pi\beta E_0\nu_0}\frac{g_0(J)}{\nu-\nu_0}\Big\{D_0 Z_\perp(\nu) + \sum_{n>0} D_n J_0\left(n\left(\frac{2\eta J}{\nu_s}\right)^{1/2}\right) \times$$

$$\times [Z_\perp(n+\nu) - Z_\perp^*(n-\nu)] + \overline{\nu}\left(\frac{2\eta J}{\nu_s}\right)^{1/2} \sum_{n>0} D_n \times$$

$$J_1\left(n\left(\frac{2\eta J}{\nu_s}\right)^{1/2}\right)[Z_\perp(n+\nu) + Z_\perp^*(n-\nu)]\Big\}. \qquad (354)$$

If we now assume $Z_\perp(\nu) = 0$ and $Z_\perp(\omega) = \text{constant} \equiv Z_\perp$, we obtain that when $\bar{\nu} = 0$, $\nu = \nu_0$ and the system is stable. If $\nu \neq 0$ we have

$$V(J) = \frac{\delta\nu}{\nu-\nu_0}\bar{\nu}\, g_0(J)\left(\frac{2\eta J}{\nu_s}\right)^{1/2}\sum_{n>0} D_n J_1\left(n\left(\frac{2\eta J}{\nu_s}\right)^{1/2}\right), \tag{355}$$

with

$$\delta\nu = -\frac{ieI_0 RZ_\perp}{4\pi\beta E_0\nu_0}. \tag{356}$$

If we make the additional hypothesis that we are dealing with a short bunch, so that $D_n \sim n$ we can solve Eq. (355) with the help of Eq. (347), obtaining

$$\nu-\nu_0 = \bar{\nu}\delta\nu 2\pi\sum_{n>0}\int_0^\infty dJ g_0(J)\left(\frac{2\eta J}{\nu_s}\right)^{1/2} J_1\left(n\left(\frac{2\eta J}{\nu_s}\right)^{1/2}\right). \tag{357}$$

For a gaussian bunch, with $g_0(J)$ given by Eq. (259), we also have

$$\nu-\nu_0 = \bar{\nu}\delta\nu\sum_{n>0}(n\sigma_\phi^2)e^{-(n^2\sigma_\phi^2)/2}. \tag{358}$$

This result shows that the introduction of a phase shift between the head and the tail of the bunch can produce another type of instability,[23] the head-tail effect.

The analysis just done of the transverse instability, as described by Eqs. (348) and (347) is much simplified. To obtain a complete information we should solve the integral Eq. (348), finding its eigenfunction $\bar{V}_s^r(J)$ and eigenvalues, $\lambda_{s,r}$.

The eigenfunctions provide a complete orthonormal set in the J-space with weight W(J). In terms of the $\bar{V}_s^r(J)$ and rewriting Eq. (348) as

$$\bar{V}_s(J) = \int_0^\infty dJ' G(J,J')\bar{V}_s(J'), \tag{359}$$

we can calculate ν from

$$\nu - \nu_0 = \lambda_{s,r} \int_0^\infty dJdJ'W(J)\overline{V}_s^r(J')G(J,J')\overline{V}_s^r(J). \qquad (360)$$

Thus for Eq. (348) describing the slow transverse instability we have an infinite set of eigenvalues, characterized by two numbers: s, the coupled bunch number, and r, the radial number in the synchrotron (J, ζ) phase space. The solution (358) gives the lowest eigenvalue for a mode in which all particles in the bunch oscillate together, similar to the rigid bunch dipole mode. According to the resultant Eq. (358), the stability or instability of this mode can be controlled by changing the sign of $\overline{V}$, i.e., the sign of the chromaticity ξ.

A discussion of the higher modes can be found in the literature.[23,24]

The fast transverse instability has also been studied recently.[25] As in the case of longitudinal instability in the limit of a fast blow-up and a high-frequency perturbation the stability condition is given by the coasting beam condition Eq. (322) with the peak bunch current substituting the average current in Eq. (323).

PROBLEMS

1. Prove that the impedance function defined in Eq. (8) is equivalent to that in Eq. (172).
2. Define the impedance function for the following structures inside a storage ring:
   (a) Resistive vacuum chamber wall
   (b) RF acceleration cavity
   (c) Pickup electrodes
   (d) Cross-sectional variation of the wall.
3. Find the threshold current and rise time for the microwave and head-tail instabilities for both the ISR and PEP machines.

REFERENCES

1. A. M. Sessler, IEEE Trans. Nucl. Sci. NS-**18**, 1039 (1971).
2. F. Amman, IEEE Trans. Nucl. Sci. NS-**16**, 1073 (1969).
3. A. M. Sessler and V. Vaccaro, Longitudinal Instabilities of Azimuthally Uniform Beams in a Circular Chamber with Wall of Arbitrary Electrical Properties, CERN Report ISR-RF 67-2, 1967.
4. See, for instance, H. G. Hereward, The Elementary Theory of Landau Damping, CERN 65-20, 1965.
5. A discussion of Vlasov equation can be found in many textbooks; see, for instance, T. Y. Wu, **Kinetic Equations of Gases and Plasmas** (Addison-Wesley, Reading, 1966).
6. C. Pellegrini and A. M. Sessler, Proceedings of the International Conference on High Energy Accelerators, Cambridge, 1967, p. 135.
7. See, for instance, P. M. Morse and H. Feshback, **Methods of Theoretical Physics**, (McGraw Hill, New York, 1953), p. 811.
8. See, for instance, A. Hofmann, K. Hübner , and B. Zotter, IEEE Trans. Nucl. Sci. NS-**26**, 3514 (1979); P.B. Wilson et al., IEEE Trans. Nucl. Sci. NS-**24**, 1211 (1977).
9. See, for instance, V. K. Neil and A. M. Sessler, Rev. Sci. Instrum. **36**, 429 (1965); L. J. Laslett, V. K. Neil, and A. M. Sessler, Rev. Sci. Instrum. **36**, 436 (1965).
10. W. Schnell, CERN Report ISR-RF 70-7, 1970.
11. C. E. Nielsen and A. M. Sessler, Proceedings of the International Conference on High Energy Accelerators, CERN 239 1959.
12. E. Keil and W. Schnell, CERN Report ISR-TH-RF 69-48, 1969.
13. See the review paper by J. L. Laclare, Proceedings of the International Conference on High Energy Accelerators, Geneva, 1980, p. 526.
14. J. M. Wang, Longitudinal Symmetric Coupled Bunch Modes, Brookhaven National Laboratory, BNL Report 51302, 1980.
15. R. Littauer, C. Pellegrini, M. Sands, and B. Touschek, Longitudinal Multibunch Instabilities, Laboratori Nazionali di Frascati, ADONE International Report, 1968; M. Q. Barton and E. G. Raka, IEEE Trans. Nucl. Sci. NS-**18**, 1032 (1971); see also Ref. 2.
16. A. Renieri and F. Tazzioli, Proceedings of the International Conference on High Energy Accelerators, Stanford, 1974, p.370; C. Pellegrini and M. Sands, Coupled Bunch Longitudinal Instabilities, SLAC, PEP 258, 1977.
17. K. Robinson, CEA 11, 1956 and CEAL 1010, 1964.
18. F. J. Sacherer, IEEE Trans. Nucl. Sci. NS-**24**, 1393 (1977).
19. F. J. Sacherer, Ph.D. Thesis, UCRL-18454, 1968; R. Gluckstern, Proceedings of the Proton Linear Accelerator Conference, Fermilab, Batavia, 1970; G. Besnier, Nucl. Instrum. Methods **164**, 235 (1979).
20. J. M. Wang and C. Pellegrini, Proceedings of the International Conference on High Energy Accelerators, Geneva ,1980, p. 554.

21. L. J. Laslett, V. K. Neil, and A. M. Sessler, Rev. Sci. Instrum. **36**, 436 (1965).
22. E. D. Courant and A. M. Sessler, Rev. Sci. Instrum. **37**, 1579 (1966); E. Ferlenghi, C. Pellegrini, and B. Touschek, Nuovo Cimento **44B**, 253 (1966).
23. C. Pellegrini, Nuovo Cimento **64A**, 477 (1969).
24. F. Sacherer, Proceedings of the International Conference on High Energy Accelerators, Stanford, 1974, p. 347.
25. R. D. Ruth and J. M. Wang, IEEE Trans. Nucl. Sci. NS-**28**, p. 2405 (1981).

LECTURES ON NONLINEAR ORBIT DYNAMICS

Alex J. Dragt
University of Maryland, College Park, Maryland 20742

TABLE OF CONTENTS

0094-243X/82/870147-167$3.00 Copyright 1982 American Institute of Physics

INTRODUCTION

The treatment of nonlinear effects in orbit dynamics is essential for two reasons. First, nonlinear effects are almost unavoidable. They arise from spatial and temporal inhomogeneities in electric and magnetic fields, from fringe fields, from beam self forces, from the beam-beam interaction in the case of colliding beams, and from higher order kinematic terms arising in expansions of the equations of motion. In this context, it is necessary to deal with nonlinear behavior in order to understand and avoid harmful effects. Second, nonlinear effects can sometimes be exploited to good ends. Current applications include chromaticity control with sextupole magnets, beam stabilization through Landau damping induced by sextupole and octupole magnets, achromatic sections composed of sextupole compensated dipoles, and quadrupoles and nonlinear beam extraction.

Present methods of nonlinear orbit calculation usually involve either Hamiltonian perturbation theory or the direct numerical integration of trajectories. Other important methods include "particle tracking" and extended matrix calculations. Particle tracking approximates nonlinear effects by impulsive momentum kicks. Extended matrix calculations are based on a Green's function method for treating in lowest order the effect of quadratic terms in the equations of motion.

These lectures describe a new approach to the analysis of nonlinear orbit dynamics. Special attention is given to the Hamiltonian nature of the equations of motion, and Lie algebraic tools are developed to exploit this symmetry. It is shown that the use of Lie algebraic concepts provides a concise and powerful method for describing and computing nonlinear effects. Applications are made to charged particle beam transport, light optics, and orbits in circular machines including the case of colliding beams.

1. THE UBIQUITY OF LAGRANGIAN AND HAMILTONIAN DYNAMICS

1.1 Hamilton's Equations with Time as an Independent Variable[1-3]

It is a remarkable discovery that all the fundamental dynamical laws of Nature are expressible in Lagrangian or Hamiltonian form. The relativistic Lagrangian for the motion of a particle of rest mass m_o and charge q in an electromagnetic field is given by the expression

$$L(\vec{r},\vec{v},t) = -m_o c^2(1-\vec{v}^2/c^2)^{1/2} - q\psi(\vec{r},t) + q\ \vec{v}\cdot\vec{A}(\vec{r},t). \quad (1.1)$$

Here ψ and $\vec{A}$ are the scalar and vector potentials defined in such a way that the electromagnetic fields $\vec{E}$ and $\vec{B}$ are given by the standard relations

$$\vec{B} = \vec{\nabla} \times \vec{A}$$
$$\vec{E} = -\nabla\psi - \partial\vec{A}/\partial t. \tag{1.2}$$

Exercise 1.1: Lagrange's equations of motion for a system having n degrees of freedom are

$$(d/dt)(\partial L/\partial\dot{q}_i) - (\partial L/\partial q_i) = 0, \tag{1.3}$$

where $(q_1 \ldots q_n)$ is any set of generalized coordinates. In the case that the generalized coordinates are taken to be the usual Cartesian coordinates, verify that Lagrange's equations for the Lagrangian (1.1) reproduce the required relativistic equations of motion for a charged particle under the influence of the Lorentz force.

The momentum p_i canonically conjugate to the variable q_i is defined by the relation

$$p_i = L/\partial\dot{q}_i. \tag{1.4}$$

The Hamiltonian H associated with a Lagrangian L is defined by the Legendre transformation

$$H(q,p,t) = \sum_i p_i \dot{q}_i - L(q,\dot{q},t). \tag{1.5}$$

Note that the Hamiltonian is to be expressed as a function of the variables q,p,t. That is, the variables $\dot{q}$ are to be eliminated in terms of the p's.

Exercise 1.2: For the Lagrangian (1.1), show that the canonical momenta in Cartesian coordinates are given by the equation

$$\vec{p} = m_o\vec{v}/(1-\vec{v}^2/c^2)^{1/2} + q\vec{A}. \tag{1.6}$$

Note that the first term in (1.6) is just the relativistic mechanical momentum. Consequently, the relation (1.6) can also be written in the form

$$\vec{p} - q\vec{A} = \vec{p}^{mech}. \quad (1.7)$$

Exercise 1.3: Show that the Hamiltonian associated with the Lagrangian (1.1) is given in Cartesian coordinates by the expression

$$H = [m_o{}^2c^4 + c^2(\vec{p}-q\vec{A})^2]^{1/2} + q\psi. \quad (1.8)$$

Exercise 1.4: Find the canonical momenta and Hamiltonian associated with the Lagrangian (1.1) when cylindrical coordinates ρ,ϕ,z are used as generalized coordinates.
Answer:

$$p_\rho = m_o\dot{\rho}/(1-v^2/c^2)^{1/2} + qA_\rho \quad (1.9a)$$
$$p_z = m_o\dot{z}/(1-v^2/c^2)^{1/2} + qA_z \quad (1.9b)$$
$$p_\phi = m_o\rho^2\dot{\phi}/(1-v^2/c^2)^{1/2} + q\phi A_\phi. \quad (1.9c)$$

$$H = \{m_o^2c^4 + c^2[(p_\rho-qA_\rho)^2 + (p_z-qA_z)^2 + (p/\rho-qA_\phi)^2]\}^{1/2}$$
$$+ q\psi. \quad (1.10)$$

Hamilton's equations of motion for the 2n canonical variables $(q_1 \ldots q_n)$ and $(p_1 \ldots p_n)$ are given in terms of the Hamiltonian $H(q,p,t)$ by the rules

$$\dot{q}_i = \partial H/\partial p_i \quad , \quad \dot{p}_i = -\partial H/\partial q_i. \quad (1.11a,b)$$

For later use, it is convenient to add one more equation to the set (1.11). Consider the total time rate of change of the Hamiltonian H along a trajectory in q, p space. Using the chain rule, one finds the result

$$dH/dt = \partial H/\partial t + \sum_i [(\partial H/\partial q_i)\dot{q}_i + (\partial H/\partial p_i)\dot{p}_i]. \quad (1.12)$$

However, the quantity under the summation sign vanishes because of (1.11). It follows that the Hamiltonian has the special property

$$dH/dt = \partial H/\partial t. \quad (1.13)$$

1.2 Hamilton's Equations with a Coordinate as an Independent Variable

Note that in the usual Hamiltonian formulation (as in the usual Lagrangian formulation) the time t plays the distinguished role of an independent variable, and all the q's and p's are dependent variables. That is, the canonical variables are viewed as functions q(t), p(t) of the independent variable t.

In some cases, it is more convenient to take some coordinate to be the independent variable rather than the time. For example, consider the passage of a collection of particles through a rectangular magnet such as is shown in Figs. (1.1) and (1.2). In such a situation, particles with different initial conditions will require different times to pass through the magnet. If the quantities of interest are primarily the locations and momenta of the particles as they leave the exit face of the magnet, then it would clearly be more convenient to use some coordinate which measures the progress of a particle through the magnet as an independent variable. In the case of a magnet with parallel faces as shown in Figs. (1.1) and (1.2), a convenient independent variable would be the x coordinate. In the case of a wedge magnet as shown in Fig. (1.3), a convenient independent variable would be the angle ϕ of a cylindical coordinate triad ρ,ϕ,z.

Suppose some coordinate is indeed chosen to be the independent variable. Is it then still possible to have a Hamiltonian (or Lagrangian) formulation of the equations of motion? The answer in general is yes as is shown by the following theorem.

<u>Theorem 1.1</u>: Suppose H(q,p,t) is a Hamiltonian for a system having n degrees of freedom. Suppose further that $\dot{q}_1 = \partial H/\partial p_1 \neq 0$ for some interval of time T in some region R of the phase space described by the 2n variables $(q_1 \ldots q_n)$ and $(p_1 \ldots p_n)$. Then in this region and time interval, q_1 can be introduced as an independent variable in place of the time t. Moreover, the equations of motion with q_1 as an independent variable can be obtained from a Hamiltonian which will be called K.

<u>Proof</u>: Consider the 2n-2 quantities $(q_2 \ldots q_n)$, $(p_2 \ldots p_n)$. They obey Hamilton's equations of motion

$$\dot{q}_i = \partial H/\partial p_i \qquad i = 2,\ldots,n \tag{1.14a}$$

$$\dot{p}_i = -\partial H/\partial q_i \qquad i = 2,\ldots,n. \tag{1.14b}$$

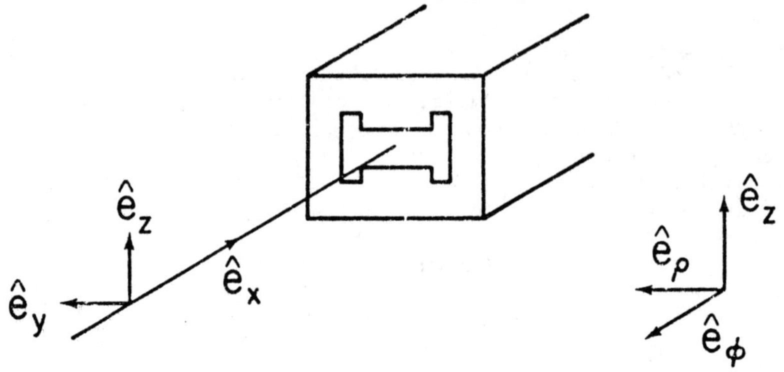

Fig. 1.1: Typical choice of a cartesion coordinate system for the description of charged particle trajectories in a magnet. Also shown, to the right, is an associated cylindrical coordinate system unit vector triad.

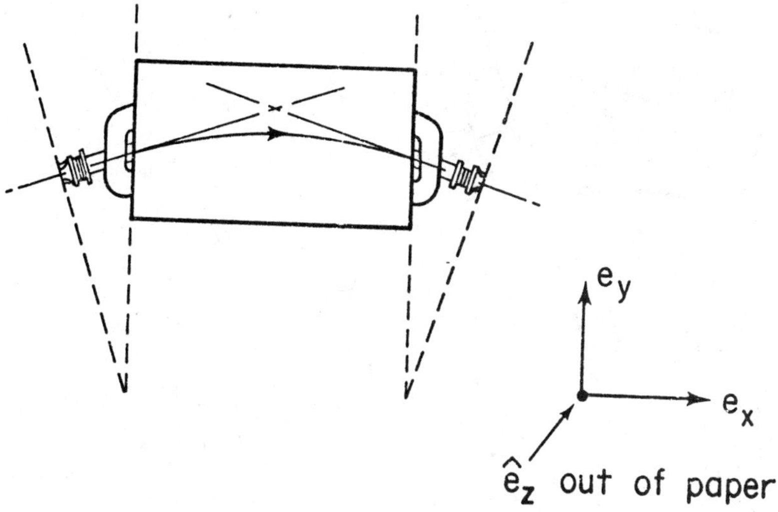

Fig. 1.2: Top view of a particle trajectory in a parallel-faced magnet.

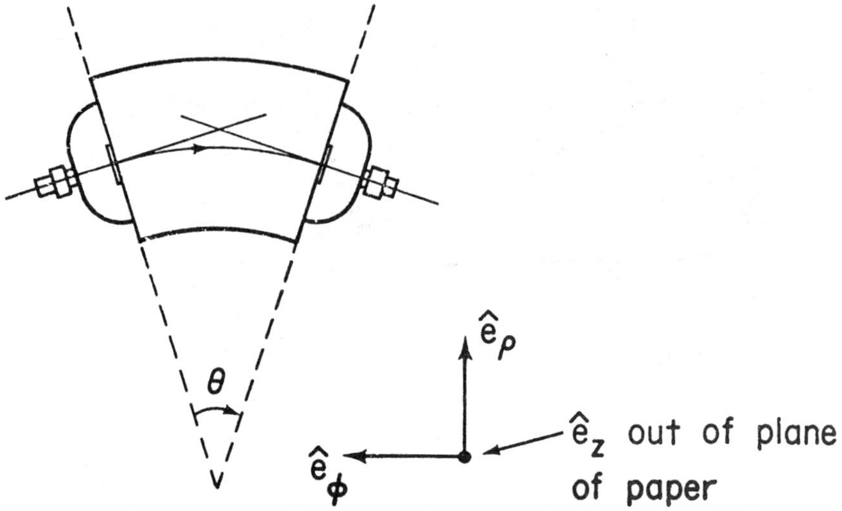

Fig. 1.3: Top view of a particle trajectory in a wedge magnet. The trajectory is conveniently described using cylindrical coordinates ρ, ϕ, z.

Denote total derivatives with respect to q_1 by a prime. Then, applying the chain rule to Eqs. (1.14), one finds the relations

$$\begin{aligned} q_i' &= dq_i/dq_1 = (dq_i/dt)(dt/dq_1) \\ &= (\partial H/\partial p_i)(\partial H/\partial p_1)^{-1} \end{aligned} \tag{1.15a}$$

$$\begin{aligned} p_i' &= dp_i/dq_1 = (dp_i/dt)(dt/dq_1) \\ &= -(\partial H/\partial q_i)(\partial H/\partial p_1)^{-1}. \end{aligned} \tag{1.15b}$$

To these 2n-2 relations it is convenient to add two more. First, suppose the time t is added to the list of coordinates as a dependent variable. Then one immediately has the relation

$$t' = dt/dq_1 = (dq_1/dt)^{-1} = (\partial H/\partial p_1)^{-1}. \tag{1.15c}$$

Second, suppose the quantity p_t defined by writing $p_t = -H$ is formally added to the list of momenta. Then, using (1.13), one finds the equation

$$\begin{aligned} p_t' &= dp_t/dq_1 = (dp_t/dt)(dt/dq_1) \\ &= -(\partial H/\partial t)(\partial H/\partial p_1)^{-1}. \end{aligned} \tag{1.15d}$$

Equations (1.15) are the desired equations of motion for the 2n variables $(t, q_2, \ldots, q_n)$, $(p_t, p_2, \ldots, p_n)$ with q_1 as an independent variable. What remains to be shown is that the quantities on the right-hand sides of Eqs. (1.15) can be obtained by applying the standard rules to some Hamiltonian K.

Look once again at the defining relation for p_t,

$$p_t = -H(q,p,t). \tag{1.16}$$

Suppose that this relation is solved for p_1 to give a relation of the form

$$p_1 = -K(t,q_2 \ldots q_n; p_t, p_2 \ldots p_n; q_1). \tag{1.17}$$

Such an inversion is possible according to the implicit function theorem because $\partial H/\partial p_1 \neq 0$ by assumption. Then, as the notation is intended to suggest, K is the desired new Hamiltonian.

To see that this is so, take the total differential of (1.16) to find the result

$$dp_t = -(\partial H/\partial t)dt - \sum_i (\partial H/\partial q_i)dq_i - \sum_i (\partial H/\partial p_i)dp_i. \tag{1.18}$$

Now solve (1.18) for dp_1 to get the relation

$$dp_1 = \left(\frac{\partial H}{\partial p_1}\right)^{-1} [-dp_t - (\partial H/\partial t)dt - \sum_i (\partial H/\partial q_i)dq_i - \sum_{i\neq 1} (\partial H/\partial p_i)dp_i]. \tag{1.19}$$

Also, take the total differential of (1.17) to find the result

$$dp_1 = -(\partial K/\partial p_t)dp_t - (\partial K/\partial t)dt - \sum_i(\partial K/\partial q_i)dq_i - \sum_{i\neq 1} (\partial K/\partial p_i)dp_i. \tag{1.20}$$

Upon comparing (1.19) and (1.20), and looking at Eqs. (1.15), one obtains the advertised result

$$\begin{aligned} \partial K/\partial p_t &= (\partial H/\partial p_1)^{-1} = t' \\ \partial K/\partial p_i &= (\partial H/\partial p_i)(\partial H/\partial p_1)^{-1} = q_i' \quad i = 2,....n \\ \partial K/\partial t &= (\partial H/\partial t)(\partial H/\partial p_1)^{-1} = -p_t' \\ \partial K/\partial q_i &= (\partial H/\partial q_i)(\partial H/\partial p_1)^{-1} = -p_i' \quad i = 2,....n. \end{aligned} \tag{1.21}$$

That is, the indicated partial derivatives of K do indeed produce the required right-hand sides of Eqs. (1.15). Note that according to Eqs. (1.21), the quantity p_t may be viewed as the momentum canonically conjugate to the time t.

> Exercise 1.5: Find the Hamiltonian K corresponding to the Hamiltonian H given by (1.8) when the x coordinate is taken to be the independent variable. Assume that $\dot{x} > 0$ for the trajectories in question.

Answer:

$$K = -[(p_t + q\psi)^2/c^2 - m_o^2c^2 - (p_y - qA_y)^2 - (p_z - qA_z)^2]^{1/2} - qA_x. \tag{1.22}$$

Note that according to (1.16), p_t is usually negative.

Exercise 1.6: Find the Hamiltonian K corresponding to the Hamiltonian H given by (1.10) when the coordinate ϕ is taken to be the independent variable. Assume that $\dot{\phi} < 0$ for trajectories of interest. [This would be the case if the triad of unit vectors $\hat{e}$,$\hat{e}_\phi$, $\hat{e}_z$ is taken to form a right-handed coordinate system, if $\hat{e}_\rho$ points along the general direction of $\hat{e}_y$ of a rectangular triad $\hat{e}_x$, $\hat{e}_y$, $\hat{e}_z$, and if the two $\hat{e}_z$ vectors of the two triads agree. See Figs. (1.1) and (1.3)].

Answer:

$$K = +\rho[(p_t + q\psi)^2/c^2 - m_o^2c^2 - (p_\rho - qA_\rho)^2 - (p_z - qA_z)^2]^{1/2} - q\rho A_\phi. \tag{1.23}$$

Exercise 1.7: Show that a uniform electric field in the x direction can be derived from the scalar and vector potentials

$$\psi = 0$$

$$\vec{A} = -Et\hat{e}_x.$$

Exercise 1.8: Show that a uniform electric field in the x direction can be derived from the scalar and vector potentials

$$\psi = -Ex$$

$$\vec{A} = 0.$$

Exercise 1.9: Show that a uniform magnetic field in the z direction can be derived from the scalar and vector potentials

$$\psi = 0$$

$$\vec{A} = -By\hat{e}_x.$$

Exercise 1.10: Show that a magnetic quadrupole field with midplane symmetry can be derived from the scalar and vector potentials

$$\psi = 0$$

$$\vec{A} = (a_2/2)(z^2 - y^2)\hat{e}_x.$$

Exercise 1.11: Show that a magnetic sextupole field with midplane symmetry can be derived from the scalar and vector potentials

$$\psi = 0$$

$$\vec{A} = a_s(yz^2 - y^3/3)\ \hat{e}_x.$$

Exercise 1.12: Show that when cylindrical coordinates are used, a uniform magnetic field in the z direction can be derived from the scalar and vector potentials

$$\psi = 0$$

$$\vec{A} = (\rho/2)B\ \hat{e}_\phi.$$

Exercise 1.13: Suppose that the electric field $\vec{E}$ is zero and the magnetic field $\vec{B}$ is static. Show that in this case p_t has the constant value

$$p_t = -[m_o^2c^4 + c^2(\vec{p}^{\,mech})^2]^{1/2}.$$

Suppose further that the magnetic field can be derived from a vector potential having only an x component. Show that if one is only interested in determining trajectories and not in determining transit time, then one may use the Hamiltonian

$$K = -[(\vec{p}^{\,mech})^2 - p_y^2 - p_z^2]^{1/2} - qA_x$$

with x treated as the independent variable.

1.3 Hamiltonian formulation of light optics

Much of the remaining lectures will be devoted, at least indirectly, to the subject of charged particle beam optics. For this reason, it is useful to also consider the analogous topic of geometrical light optics. It too can be formulated in Hamiltonian terms.

Consider the optical system illustrated schematically in Fig. (1.4). A light ray originates at the general initial point P^i with spatial coordinates $\vec{r}^{\,i}$, and moves in an initial direction specified by the unit vector $\hat{s}^i$. After passing through an optical device, it arrives at the final point P^f with coordinates $\vec{r}^{\,f}$ in a direction

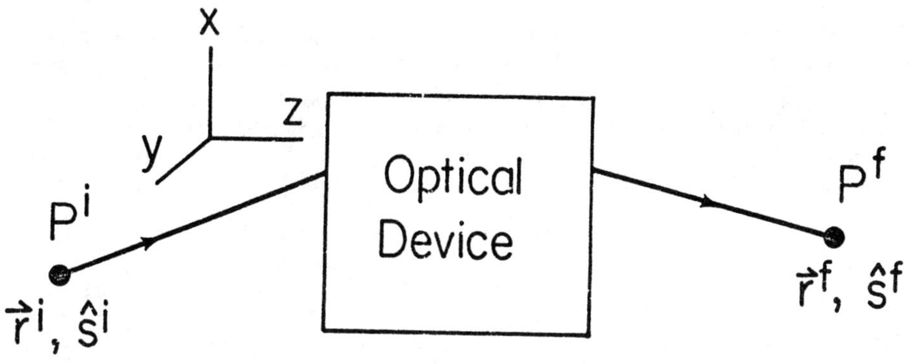

Fig. 1.4: An optical system consisting of an optical device preceded and followed by simple transit. A ray originates at P^i with location $\vec{r}^i$ and direction $\hat{s}^i$, and terminates at P^f with location $\vec{r}^f$ and direction $\hat{s}^f$.

specified by the unit vector $\hat{s}^f$. Given the initial quantities $(\vec{r}^i, \hat{s}^i)$, the fundamental problem of geometrical optics is to determine the final quantities $(\vec{r}^f, \hat{s}^f)$, and to design an optical device in such a way that the relation between the initial and final ray quantities has various desired properties.

Suppose the z coordinates of the initial and final points P^i and P^f as shown in Fig. (1.4) are held fixed. The planes $z=z^i$ and $z=z^f$ may be viewed as object and image planes respectively. Further, suppose the general light ray from P^i to P^f is parameterized using z as an independent variable. That is, the path of a general ray is described by specifying the two functions x(z) and y(z). Then the element ds of path length along a ray is given by the expression

$$\begin{aligned} ds &= [(dz)^2 + (dx)^2 + (dy)^2]^{1/2} \\ &= [1 + (x')^2 + (y')^2]^{1/2}\, dz. \end{aligned} \tag{1.24}$$

Here a prime denotes the differentiation d/dz. Consequently, the optical path length along a ray from P^i to P^f is given by the integral

$$A = \int_{z^i}^{z^f} n(z,y,z)\,[1 + (x')^2 + (y')^2]^{1/2}\, dz. \tag{1.25}$$

Here the function $n(x,y,z)=n(\vec{r})$ specifies the index of refraction at each point before and after the optical device and in the device itself.

Fermat's principle requires that A be an extremum for the path of an actual ray. Therefore, the ray path satisfies the Euler-Lagrange equations

$$\begin{aligned} d/dz(\partial L/\partial x') - \partial L/\partial x &= 0 \\ d/dz(\partial L/\partial y') - \partial L/\partial y &= 0 \end{aligned} \tag{1.26}$$

with a Lagrangian L given by the expressions

$$L = n(x,y,z)\,[1 + (x')^2 + (y')^2]^{1/2}. \tag{1.27}$$

<u>Exercise 1.14</u>: Calculate explicit expressions for the two "momenta" conjugate to the coordinates x and y defined by the standard relations

$$p_x = \partial L/\partial x' \quad , \quad p_y = \partial L/\partial y' \tag{1.28}$$

Find the Hamiltonian H corresponding to L.

Answer:

$$p_x = n(\vec{r})\, x'/[1 + (x')^2 + (y')^2]^{1/2} \qquad (1.29)$$

$$p_y = n(\vec{r})\, y'/[1 + (x')^2 + (y')^2]^{1/2},$$

$$H = -[n^2(\vec{r}) - p_x^2 - p_y^2]^{1/2}. \qquad (1.30)$$

2. SYMPLECTIC MATRICES

2.1 Definitions

The purpose of this section is to define symplectic matrices and explore some of their properties in preparation for future use.

To define symplectic matrices, it is first necessary to introduce a certain fundamental $2n \times 2n$ matrix J. It is defined by the equation

$$J = \begin{pmatrix} 0 & I \\ -I & 0 \end{pmatrix} . \qquad (2.1)$$

Here each entry in J is an $n \times n$ matrix, I denotes the $n \times n$ identity matrix, and all other entries are zero.

Exercise 2.1: Show that the matrix J has the following properties:

$$J^2 = -I \text{ or } J^{-1} = -J \qquad (2.2)$$

$$\det(J) = 1 \qquad (2.3)$$

$$\tilde{J} = -J \qquad (2.4)$$

$$\tilde{J}\, J = I. \qquad (2.5)$$

Here $\tilde{J}$ denotes the transpose of J.

With this background a $2n \times 2n$ matrix M is said to be symplectic if

$$\tilde{M}\, J\, M = J. \qquad (2.6)$$

Observe that symplectic matrices must be of even dimension by definition.

<u>Exercise 2.2</u>: Show that a symplectic matrix M has the following properties:

$$\det(M) = \pm 1 \tag{2.7}$$

$$M^{-1} = -J\,\tilde{M}\,J \text{ or } M^{-1} = J^{-1}\,\tilde{M}\,J \tag{2.8}$$

$$M\,J\,\tilde{M} = J. \tag{2.9}$$

<u>Comment</u>: It can be shown that det(M) actually always equals +1 for a symplectic matrix.[4] Also, as is easily checked, in the 2×2 case the necessary and sufficient condition for a matrix to be symplectic is that it have determinant +1.

<u>Exercise 2.3</u>: Show that the matrices I and J are symplectic.

<u>Exercise 2.4</u>: Supose M is a symplectic matrix. Show that M^{-1} is then also a symplectic matrix.

<u>Exercise 2.5</u>: Suppose M and N are symplectic matrices. Show that the produce MN is then also a symplectic matrix.

2.2 Group properties

A set of matrices forms a group G if it satisfies the following properties:

1. The identity matrix I is in G.
2. If M is in G, M^{-1} exists and is also in G.
3. If M and N are in G, so is the produce MN.

Evidently, according to exercise (2.3), Eq. (2.7), and exercises (2.4) and (2.5), the set of all $2n \times 2n$ symplectic matrices (for any particular value of n) form a group. This group is often denoted by the symbol Sp(2n).

2.3 Eigenvalue spectrum

The characteristic polynomial $P(\lambda)$ of any matrix M is defined by the equation

$$P(\lambda) = \det(M - \lambda I). \tag{2.10}$$

Evidently $P(\lambda)$ is a polynomial with real coefficients if the matrix M is real. Also, the eigenvalues of M are the roots of the equation

$$P(\lambda) = 0. \tag{2.11}$$

It follows that if M is a real matrix, then its eigenvalues must also be real or must occur in complex conjugate pairs $\lambda, \bar{\lambda}$.

Exercise 2.6: Show that a symplectic matrix cannot have $\lambda=0$ as an eigenvalue.

Suppose M is a symplectic matrix. Then it follows from (2.8) that

$$J^{-1}(\tilde{M}-\lambda I)J = M^{-1} - \lambda I = -\lambda M^{-1}(M - \lambda^{-1}I). \quad (2.12)$$

Now take the determinant of both sides of (2.12). The result is the relation

$$P(\lambda) = \lambda^{2n} P(1/\lambda). \quad (2.13)$$

It follows that if λ is an eigenvalue of a symplectic matrix, so is the reciprocal $1/\lambda$. Consequently, the eigenvalues of a symplectic matrix must form reciprocal pairs.

Exercise 2.7: Verify (2.13) starting with (2.12).

The symmetry between λ and $1/\lambda$ exhibited by (2.13) can be further displayed by rewriting the equation in the form

$$\lambda^{-n} P(\lambda) = \lambda^{n} P(1/\lambda). \quad (2.14)$$

Now define another function $Q(\lambda)$ by writing

$$Q(\lambda) = \lambda^{-n} P(\lambda). \quad (2.15)$$

The functions P and Q evidently have the same zeroes. Moreover, the condition (2.14) requires that Q have the symmetry property

$$Q(\lambda) = Q(1/\lambda). \quad (2.16)$$

Equation (2.16) shows not only that the eigenvalues of a symplectic matrix must occur in reciprocal pairs; it shows that they must also occur with the same multiplicity. That is, if the root λ_o has multiplicity k, so must the root $1/\lambda_o$.

Also, if either +1 or -1 is a root, then this root must have even multiplicity. To see this, suppose for example that $\lambda=1$ is a root. Introduce the variable μ by writing $\lambda=\exp \mu$. Then (2.16) shows that Q is an even function of the variable μ and hence must have an expansion of the form

$$Q = \sum_{m=0}^{\infty} c_m \mu^{2m}. \quad (2.17)$$

Moreover, when λ is near 1, λ and μ are related by the expansion

$$\mu = \log \lambda = \log[1 + (\lambda-1)] = (\lambda-1)[1 - (\lambda-1)/2 + \cdots]. \quad (2.18)$$

Comparison of (2.17) and (2.18) shows that $\lambda=1$ is not a root unless $c_o=0$. If $c_o=0$, then $\lambda=1$ is a root of multiplicity 2. If $c_1=0$ as well, then $\lambda=1$ is a root of multiplicity 4, etc. A similar argument holds near $\lambda=-1$ upon making the substitution $\lambda=-\exp\mu$.

In summary, it has been shown that the eigenvalues of a real symplectic matrix must satisfy the following properties:

1. They must be real or occur in complex conjugate pairs.
2. They must occur in reciprocal pairs, and each member of the pair must have the same multiplicity.
3. If either ± 1 is an eigenvalue, it must have even multiplicity.

When combined, the conditions just enumerated place strong restrictions on the possible eigenvalues of a real symplectic matrix. Consider first the simplest case of a 2×2 symplectic matrix (n=1). Call the eigenvalues λ_1 and λ_2. Then, by the reciprocal property, it follows that

$$\lambda_1 \lambda_2 = 1. \tag{2.19}$$

Suppose, now, that λ_1 is real, positive, and greater than 1. Then λ_2 is real, positive, and less than 1. Similarly, if λ_1 is real, negative, and less than -1, than λ_2 is real, negative, and greater than -1. On the other hand, if λ_1 is complex, then $\lambda_2 = \bar{\lambda}_1$. This condition, when combined with Eq. (2.19), shows that in this case λ_1 and λ_2 must lie on the unit circle in the complex plane. Finally, there are the two special cases $\lambda_1=\lambda_2=1$ and $\lambda_1= \lambda_2=-1$.

Altogether, there are five possible cases. They are listed below along with names and designations whose significance will become clear later on. See also Fig. (2.1).

1. Hyperbolic case (unstable): $\lambda_1 > 1$ and $0 < \lambda_2 < 1$.
2. Inversion hyperbolic case (unstable): $\lambda_1 < -1$ and $-1 < \lambda_2 < 0$.
3. Elliptic case (stable): $\lambda_1 = e^{i\phi}$, $\lambda_2 = e^{-i\phi}$.

   (Eigenvalues are complex conjugates and lie on unit circle).
4. Parabolic case (generally linearly unstable): $\lambda_1 = \lambda_2 = +1$.
5. Inversion parabolic case (generally linearly unstable): $\lambda_1 = \lambda_2 = -1$.

The next simplest case is that of a 4×4 symplectic matrix (n=2). In this case, one has to deal with four possible eigenvalues and then apply reasoning analogous to the 2×2 case. Figures (2.2)

Case 1.

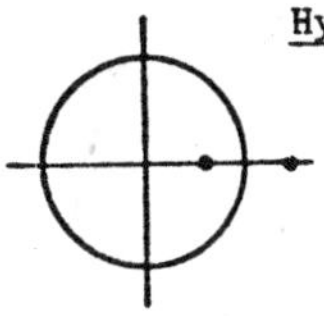

Hyperbolic (Unstable)

Case 2.

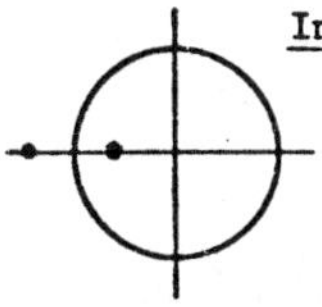

Inversion Hyperbolic (Unstable)

Case 3.

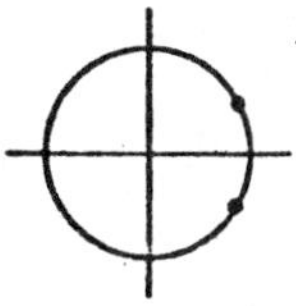

Elliptic (Stable)

Case 4.

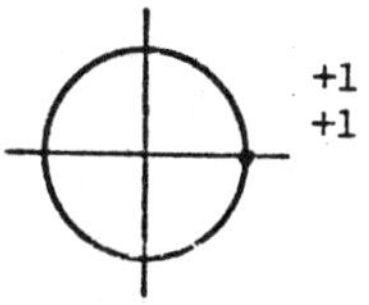

Parabolic

Transition between elliptic and hyperbolic cases can only occur by passage through this degenerate case. (Generally linearly unstable.)

Case 5.

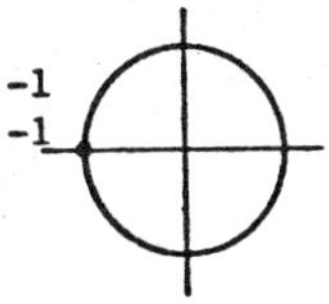

Inversion Parabolic

Transition between elliptic and inversion hyperbolic cases can only occur by passage through this degenerate case. (Generally linearly unstable.)

Fig. 2.1: Possible cases for the eigenvalues of a 2×2 real symplectic matrix.

Fig. 2.2: Possible eigenvalue configurations for a 4 × 4 symplectic matrix. The mirror image of each configuration is also a possible configuration, and therefore is not shown in order to save space.

A. Generic Configurations

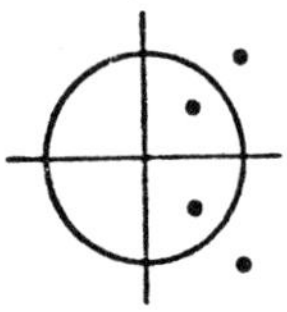

Case 1. All eigenvalues complex and off the unit circle. All eigenvalues can be obtained from a single one by the operations of complex conjugation and taking reciprocals. Unstable.

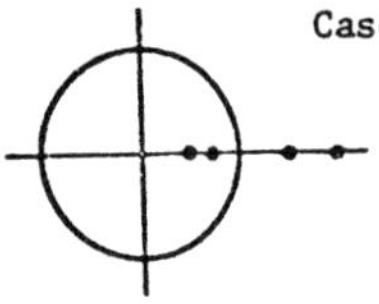

Case 2. All eigenvalues real, off the unit circle, and of same sign. Eigenvalues form reciprocal pairs. Unstable.

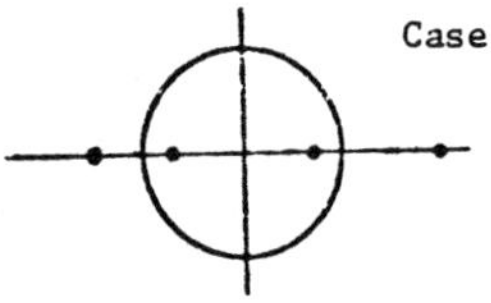

Case 3. All eigenvalues real, off the unit circle, and of differing signs. Eigenvalues form reciprocal pairs. Unstable.

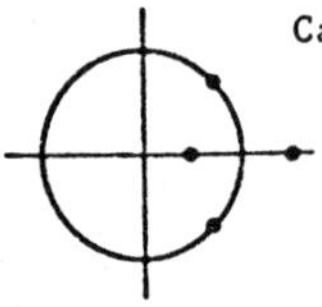

Case 4. Two eigenvalues complex and confined to unit circle. Two eigenvalues real. Eigenvalues form reciprocal pairs. Complex eigenvalues are also complex conjugate. Unstable.

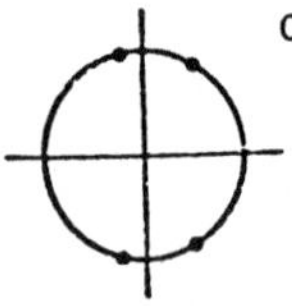

Case 5. All eigenvalues complex and confined to unit circle. Eigenvalues form reciprocal pairs which are also complex conjugate. Stable.

B. Degenerate Configurations. Transitions between generic configurations can only occur by passage through a degenerate configuration. Mirror image configurations are again possible, but not shown.

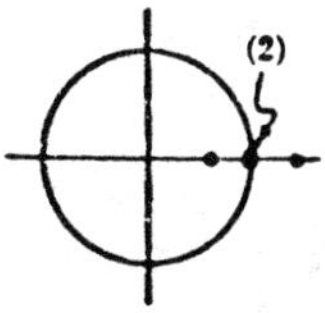

Case 1. Two eigenvalues equal, and two eigenvalues real. All of same sign. Occurs in transition between generic cases 2 and 4. Unstable.

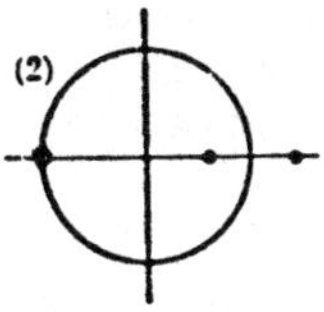

Case 2. Two eigenvalues equal, and two eigenvalues real. Signs differ. Occurs in transition between generic cases 3 and 4. Unstable.

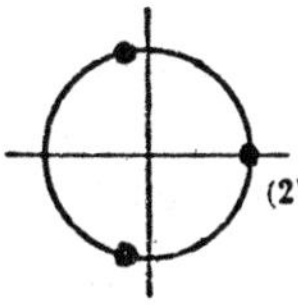

Case 3. Two eigenvalues equal, and two eigenvalues confined to unit circle. Occurs in transition between generic cases 4 and 5. Generally linearly unstable.

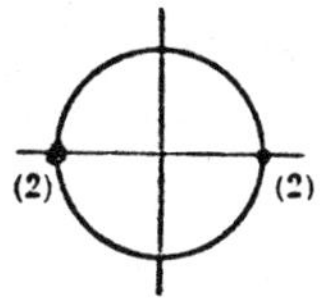

Case 4. Two eigenvalues equal +1 and two equal -1. Occurs in transition between generic cases 3 and 5, or 3 and 4, or 4 and 5. Also occurs in transition between degenerate cases 2 and 3. Generally linearly unstable.

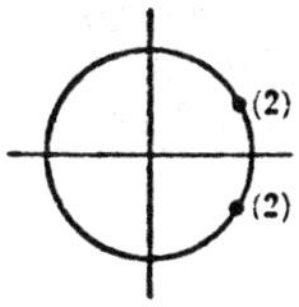

Case 5. Two pairs of eigenvalues equal, and confined to unit circle. Occurs in transition between generic cases 1 and 5. Not, however, a sufficient condition to guarantee that such a transition is possible. Stability also undetermined in absence of further conditions.

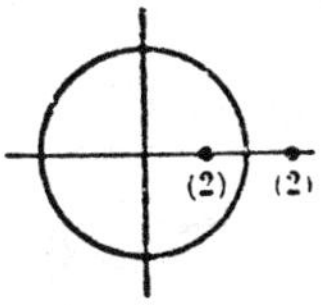

Case 6. Two pairs of eigenvalues equal and real. Occurs in transition between generic cases 1 and 2. Unstable.

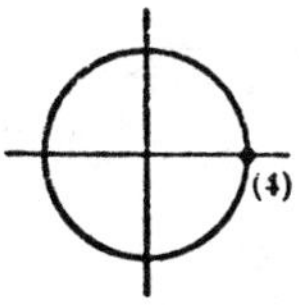

Case 7. All eigenvalues equal and have value ±1. Occurs in transitions between generic cases 1, 2, 4, and 5 and degenerate cases 1, 3, 5, and 6. Generally linearly unstable.

illustrate the various possibilities that can occur. Analysis of the possible spectrum of the 2n eigenvalues for the general $2n \times 2n$ symplectic matrix proceeds in a similar fashion.

Exercise 2.8: Show that the eigenvalues of a real symplectic matrix cannot all have absolute value less than 1.

Exercise 2.9: Show that the eigenvalues of a real symplectic matrix cannot all have absolute value greater than 1.

Exercise 2.10: Suppose that all the eigenvalues of a real symplectic matrix M lie on the unit circle and are distinct. Now suppose that M is changed slightly in such a way that it remains symplectic. Show that if the change in M is finite but small enough, then the eigenvalues must remain on the unit circle and must still be distinct. Show that all the generic configurations of Fig. (2.2) are unchanged by small perturbations. That is why they are called generic.

Exercise 2.11: In the case of a 4×4 symplectic matrix, show that $Q(\lambda)$ can be written in the form

$$Q(\lambda) = (\lambda+1/\lambda)^2 + 4b(\lambda+1/\lambda) + 4c, \tag{2.20}$$

where the quantities b and c are given by the relations

$$b = [P(1) - P(-1)]/[16] \tag{2.21}$$

$$c = [P(1) + P(-1)]/8 - 1. \tag{2.22}$$

Exercise 2.12: One might think that the determination of the four eigenvalues of a 4×4 symplectic matrix would in general require the solution of a quartic equation. However, using the results of exercise 2.11, show that thanks to the symplectic condition the eigenvalues can be found from the simple algebraic formulas

$$\lambda = w \pm (w^2-1)^{1/2} \tag{2.23}$$

where

$$w = -b \pm (b^2-c)^{1/2}. \tag{2.24}$$

Note that in general there are four choices of signs to be made corresponding to the four possible eigenvalues.

2.4 Lie algebraic properties

Let A be any matrix. The exponential of a matrix, written variously as e^A or exp(A), is defined by the exponential series

$$\exp(A) = \sum_{n=0}^{\infty} A^n/n! \qquad (2.25)$$

Exercise 2.13: Show that the series (2.25) converges for any matrix A.

Similarly, the logarithm of a matrix A is defined by the series

$$\log(A) = \log[I - (I-A)] = -\sum_{m=1}^{\infty} (I-A)^m/m. \qquad (2.26)$$

Exercise 2.14: Show that the series (2.26) converges for A sufficiently near the identity matrix I. Note that when A is near the identity, then log(A) is near the zero matrix.

As might be expected, the exponential and logarithmic functions are related. Specifically, if one has

$$B = \log(A), \qquad (2.27a)$$

then it follows that

$$A = \exp(B). \qquad (2.27b)$$

Exercise 2.15: Verify the relation given by Eqs. (2.27) in the case that A is sufficiently near the identity using the definitions (2.25) and (2.26).

With this background, suppose that M is a real symplectic matrix near the identity. Then M can be written in the form

$$M = \exp(\varepsilon B) = I + \varepsilon B + \varepsilon^2 B^2/2! + \cdots \qquad (2.28)$$

where ε is a small number and B is a real matrix. Next insert the expansion (2.28) into the symplectic condition (2.6). Then one finds the relation

$$(I + \varepsilon\tilde{B})\ J\ (I + \varepsilon B) = J + O(\varepsilon^2). \qquad (2.29)$$

Upon assuming that (2.29) holds term by term in powers of ε, it follows that B must satisfy the condition

$$\tilde{B}J + JB = 0. \qquad (2.30)$$

Exercise 2.16: Derive (2.30) directly from (2.28) using (2.8).

To understand the implications of the condition (2.30), suppose that B is written in the form

$$B = JS. \tag{2.31}$$

Exercise 2.17: Verify that it is always possible to find a real matrix S such that (2.31) is true.

Upon inserting (2.31) into (2.30), one finds the equivalent condition

$$\tilde{S}JJ + JJS = 0 \quad \text{or} \quad \tilde{S} = S. \tag{2.32}$$

That is, S must be a symmetric matrix.

Conversly, suppose that B is any matrix of the form (2.31) with S real and symmetric. Then the matrix M defined by (2.28) is symplectic. To see this, simply compute! One finds the results

$$\begin{aligned}
M &= \exp(\varepsilon JS), \\
\tilde{M} &= \exp(-\varepsilon SJ), \\
\tilde{M}JM &= \exp(-\varepsilon SJ)J\exp(\varepsilon JS) \\
&= JJ^{-1}\exp(-\varepsilon SJ)J\exp(JS) \\
&= J\exp(-\varepsilon J^{-1}SJ^{2})\exp(JS) \\
&= J\exp(-\varepsilon JS)\exp(JS) \\
&= J.
\end{aligned}$$

Exercise 2.18: Verify the details of the calculation above using the series definition of the exponential function as given by (2.23).

What has been shown is that any symplectic matrix sufficiently near the identity can be written in the form exp(JS) with S small and symmetric, and vice versa.

A set of matrices forms a Lie algebra if it satisfies the following properties:

1. If the matrix A is in the Lie algebra, then so is the matrix aA where a is any scalar.
2. If two matrices A and B are in the Lie algebra, then so is their sum.
3. If two matrices A and B are in the Lie algebra, then so is their commutator [A,B]. The commutator is defined by the relation

$$[A,B] = AB - BA. \tag{2.33}$$

The reason for introducing the concept of a Lie algebra at this stage is to point out that the set of matrices of the form JS with S symmetric is a Lie algebra.

> Exercise 2.19: Verify that the set of matrices of the form JS is indeed a Lie algebra by showing that conditions 1 through 3 above are satisfied.

It is a remarkable fact that there is a close connection between the concept of a Lie algebra and that of a group. The connection arises from a deep property of the exponential function which generally bears the names Campbell-Baker-Hausdorff. Their result may be stated as follows: Let A and B be any two matrices (of the same dimension). Form the matrices exp(sA) and exp(tB) where s and t are parameters. Then, for s and t sufficiently small, it is possible to write

$$\exp(sA)\,\exp(tB) = \exp(C), \qquad (2.34)$$

where C is some other matrix. The remarkable fact is that C is a member of the Lie algebra generated by A and B. That is, C is a sum of elements formed only from A and B and their multiple commutators. Specifically, one has the relation

$$C(s,t) = sA + tB + (st/2)[A,B] + (s^2t/12)[A,[A,B]]$$
$$+ (st^2/12)[B,[B,A]] + O(s^3t, s^2t^2, st^3). \qquad (2.35)$$

No terms of the form A^2, B^2, AB, A^2B, etc. occur! In general, the series for C contains an infinite number of terms and may only converge for sufficiently small s and t.

The proof of this theorem is difficult and is given elsewhere.[5] For present purposes, it shows that given any Lie algebra L of matrices, there exists a corresponding Lie group G. Furthermore, the rules for multiplying any two group elements are contained within the Lie algebra. To see the truth of this assertion, consider all matrices of the form $g(s) = \exp(s\ell)$ with ℓ contained in L. According to the previous result, one has

$$\exp(s\ell)\,\exp(t\ell') = \exp \ell''$$

for s, t sufficiently small. Also

$$g(0) = I \text{ and } g^{-1}(s) = g(-s).$$

Thus these matrices, at least those sufficiently near the identity, form a group. Once the group has been obtained near the identity, it can be extended to a global group by successively multiplying the different g's already obtained.

It has already been shown that symplectic matrices form a group. Furthermore, it has been shown that symplectic matrices near the identity can be written as the exponentials of elements of a Lie algebra. It follows that symplectic matrices form a Lie group.

Properties 1 and 2 of a Lie algebra indicate that the elements of a Lie algebra form a linear vector space. It is therefore natural to speak of the dimension of a Lie algebra and its associated Lie group. For the case of the symplectic group, elements of the Lie algebra are of the form (2.31) where S is any symmetric matrix. The dimension of the Lie algebra in this case, therefore, is just the dimensionality of the set of all $2n \times 2n$ symmetric matrices. This number is easily computed. There are 2n independent entries on the diagonal of a $2n \times 2n$ symmetric matrix, and $[(2n)^2 - 2n]/2$ independent entries above the diagonal. Finally, all the entries below the diagonal are given in terms of the entries above the diagonal by the symmetry condition. Therefore, the dimension of the symplectic group Lie algebra, which will be written as dim Sp(2n), is given by the relation

$$\dim Sp(2n) = 2n + [(2n)^2 - 2n]/2 = n(2n + 1). \quad (2.36)$$

2.5 Exponential representation

The discussion so far has shown that symplectic matrices sufficiently near the identity element can be written as exponentials of elements in the symplectic group Lie algebra. The next question to ask is what can be said about representing symplectic matrices in general.

To study this question, it is useful to employ polar decomposition. Let M be any real nonsingular matrix. Then M can be written uniquely in form

$$M = P\mathcal{O}. \quad (2.37)$$

Here P is a real positive definite symmetric matrix, and $\mathcal{O}$ is a real orthogonal matrix.[6] (A matrix $\mathcal{O}$ is orthogonal if $\mathcal{O}\tilde{\mathcal{O}} = \tilde{\mathcal{O}}\mathcal{O} = I$.) Now suppose that M is symplectic. Using (2.8), the symplectic condition can be written in the form

$$M = J^{-1} \tilde{M}^{-1} J. \quad (2.38)$$

Then, upon inserting the polar decomposition (2.37) into (2.38), one finds the relation

$$P\mathcal{O} = J^{-1} P^{-1} J J^{-1} \mathcal{O} J. \quad (2.39)$$

Exercise 2.20: Verify Eq. (2.39).

Next, observe that the matrix $J^{-1} P^{-1} J$ is real, symmetric, and positive definite, and observe that the matrix $J^{-1} \mathcal{O} J$ is real and orthogonal.

Exercise 2.21: Verify the two observations above.

Consequently, because polar decomposition is unique, Eq. (2.39) implies the relations

$$P = J^{-1} P^{-1} J \tag{2.40a}$$

$$\mathcal{O} = J^{-1} \mathcal{O} J. \tag{2.40b}$$

Using the fact that P is symmetric and $\mathcal{O}$ is orthogonal, Eqs. (2.40) can also be written in the form

$$P = J^{-1} \tilde{P}^{-1} J \tag{2.41a}$$

$$\mathcal{O} = J^{-1} \tilde{\mathcal{O}}^{-1} J. \tag{2.41b}$$

It follows that each of the matrices P and $\mathcal{O}$ are themselves symplectic.

Exercise 2.22: Verify Eqs. (2.41), and the claim that $\mathcal{O}$ and P are symplectic.

The next thing to do is to work with the matrices $\mathcal{O}$ and P. Consider first the matrix $\mathcal{O}$. Since $\mathcal{O}$ is real orthogonal and has determinant +1 ($\mathcal{O}$ is symplectic), it can be written in the form

$$\mathcal{O} = \exp(A), \tag{2.42}$$

where A is a unique real antisymmetric matrix,

$$\tilde{A} = -A. \tag{2.43}$$

Upon inserting the representation (2.42) into the condition (2.40b), one finds the condition

$$\mathcal{O} = \exp(A) = \exp(J^{-1}AJ). \tag{2.44}$$

Exercise 2.23: Verify Eq. (2.44) using (2.40b) and the definition (2.25).

Since the matrix $(J^{-1}AJ)$ is real antisymmetric if the matrix A is, and since A is unique, it follows from (2.44) that A has the property

$$J^{-1}AJ = A \text{ or } AJ = JA. \tag{2.45}$$

Using (2.43), the condition (2.45) can also be written in the form

$$\tilde{A}J + JA = 0. \tag{2.46}$$

Now compare (2.46) with Eq. (2.30). According to the argument applied earlier, the matrix A can be written in the form

$$A = JS^c, \tag{2.47}$$

where S^c is a real symmetric matrix. Further, since A commutes with J, see Eq. (2.45), it follows that S^c <u>commutes</u> with J.

$$S^cJ = JS^c. \tag{2.48}$$

In summary, it has been shown that $\mathcal{O}$ can be written in the form

$$\mathcal{O} = \exp(JS^c), \tag{2.49}$$

where S^c is a real symmetric matrix which commutes with J.

<u>Exercise 2.24</u>: Verify that $(J^{-1}AJ)$ is real antisymmetric if A is.

It remains to see what can be said about the matrix P. Since P is real, symmetric, and positive definite, it can be written in the form

$$P = \exp(B), \tag{2.50}$$

where B is unique, real, and symmetric,

$$\tilde{B} = B. \tag{2.51}$$

Now insert the representation (2.50) into the condition (2.40a) to obtain the result

$$P = \exp(B) = \exp(-J^{-1}BJ). \tag{2.52}$$

Since the matrix $(J^{-1}BJ)$ is real symmetric if the matrix B is, and since B is unique, it follows from (2.52) that B has the property

$$J^{-1}BJ = -B \quad \text{or} \quad BJ + JB = 0. \tag{2.53}$$

Using (2.51), the condition (2.53) can be re-expressed in the form

$$\tilde{B}J + JB = 0. \tag{2.54}$$

Consequently, B can also be written in the form

$$B = JS^a, \tag{2.55}$$

where S^a is a real symmetric matrix. However, in this case Eq. (2.51) implies the condition

$$JS^a + S^aJ = 0. \tag{2.56}$$

That is, S^a anticommutes with J. In summary, it has been shown that P can be written in the form

$$P = \exp(JS^a), \tag{2.57}$$

where S^a is a real symmetric matrix which anticommutes with J.

Now combine Eqs. (2.37), (2.49), and (2.57). The result is that any symplectic matrix can be written in the form

$$M = \exp(JS^a)\ \exp(JS^c). \tag{2.58}$$

It has been shown that the most general symplectic matrix can be written as the product of two exponentials of elements in the symplectic group Lie algebra, and each of the elements is of a special type.

It is interesting to examine the properties of commuting and anticommuting with J in a bit more detail. Let S be any symmetric matrix. Form the matrices S^a and S^c by the rules

$$S^a = (S - J^{-1}SJ)/2 \tag{2.59a}$$

$$S^c = (S + J^{-1}SJ)/2. \tag{2.59b}$$

It is easily verified that S^a and S^c are symmetric, and anticommute and commute respectively with J as the notation suggests.

Exercise 2.25: Verify the claims just made.

Also, it is obvious by construction that

$$S = S^a + S^c. \tag{2.60}$$

That is, any symmetric matrix can be uniquely decomposed into a sum of two symmetric matrices which anticommute and commute with J respectively.

2.6 Basis for Lie algebra

The last topic to be discussed is that of a suitable basis for the symplectic group Lie algebra. For simplicity, the discussion will be limited to the cases Sp(2) and Sp(4). These two cases, along with Sp(6), are those of primary interest for accelerator applications.

In selecting a basis for the Lie algebra of the symplectic group, it is convenient to begin with the observation that matrices of the form JS^c constitute a Lie algebra all by themselves. That is, the commutator of any two matrices of the form JS^c is again a matrix of the form JS^c. By contrast, the commutator of a matrix of the form JS^c with that of the form JS^a is again a matrix of the form JS^a. Finally, the commutator of two matrices of the form JS^a is a matrix of the form JS^c. It is therefore convenient to begin with the matrices of the form JS^c.

<u>Exercise 2.26</u>: Check the assertions just made about various commutators.

In the 2 × 2 case of Sp(2), the most general symmetric matrix S is of the form

$$S = \begin{pmatrix} \alpha & \beta \\ \beta & \gamma \end{pmatrix}, \tag{2.61}$$

and J is simply the matrix

$$J = \begin{pmatrix} 0 & 1 \\ -1 & 0 \end{pmatrix}. \tag{2.62}$$

Requiring that J commute with S gives the restrictions

$$\beta = 0, \quad \alpha = \gamma. \tag{2.63}$$

Consequently, the most general S^c in the 2 × 2 case is just a multiple of the identity,

$$S^c = \alpha I, \tag{2.64}$$

and JS^c is simply a multiple of J,

$$JS^c = \alpha J. \tag{2.65}$$

It is therefore convenient to select J itself as one of the basis elements of the Lie algebra.

<u>Exercise 2.27</u>: Verify that the requirement that J commute with S does indeed give the restrictions (2.63).

Next study the matrix S^a. Requiring that J anticommute with the S of (2.61) gives only the restriction

$$\gamma = -\alpha. \tag{2.66}$$

<u>Exercise 2.28</u>: Verify this fact.

Consequently, S^a is of the general form

$$S^a = \alpha \begin{pmatrix} 1 & 0 \\ 0 & -1 \end{pmatrix} + \beta \begin{pmatrix} 0 & 1 \\ 1 & 0 \end{pmatrix}, \qquad (2.67)$$

JS^a is of the general form

$$JS^a = \alpha \begin{pmatrix} 0 & -1 \\ -1 & 0 \end{pmatrix} + \beta \begin{pmatrix} 1 & 0 \\ 0 & -1 \end{pmatrix}. \qquad (2.68)$$

Upon combining the results of Eqs. (2.65) and (2.68), one sees that a convenient choice of basis elements for the Lie algebra of Sp(2) is given by the matrices

$$B_1 = J = \begin{pmatrix} 0 & 1 \\ -1 & 0 \end{pmatrix}, \qquad (2.69a)$$

$$B_2 = \begin{pmatrix} 0 & 1 \\ 1 & 0 \end{pmatrix}, \qquad (2.69b)$$

$$B_3 = \begin{pmatrix} 1 & 0 \\ 0 & -1 \end{pmatrix}, \qquad (2.69c)$$

They satisfy the commutation rules

$$[B_1, B_2] = 2B_3 \qquad (2.70a)$$

$$[B_2, B_3] = -2B_1 \qquad (2.70b)$$

$$[B_3, B_1] = 2B_2. \qquad (2.70c)$$

<u>Exercise 2.29</u>: Verify the commutation rules (2.70) for Sp(2).

Suppose the basis elements B_i are used to evaluate (2.58). Making this calculation, one finds that the most general 2×2 symplectic matrix can be written in the form

$$M = \exp(b_2B_2 + b_3B_3)\ \exp(b_1B_1), \tag{2.71}$$

where the b_i are arbitrary real coefficients. [Note that there are indeed three coefficients as predicted by Eq. (2.36) evaluated for n=1.] Thus, Eq. (2.71) gives a complete parameterization of the 2×2 symplectic group.

<u>Exercise 2.30</u>: Evaluate $\exp(b_2B_2 + b_3B_3)$ and $\exp(b_1B_1)$ using (2.25) to find the results

$$\exp(b_2B_2 + b_3B_3) = I \cosh[(b_2^2 + b_3^2)^{1/2}]$$

$$+ [(b_2B_2 + b_3B_3)/(b_2^2 + b_3^2)^{1/2}]\sinh[(b_2^2 + b_3^2)^{1/2}], \tag{2.72}$$

$$\exp(b_1B_1) = I \cos b_1 + B_1 \sin b_1. \tag{2.73}$$

It is interesting to note that, according to Eq. (2.72), $\exp(b_2B_2 + b_3B_3)$ has the topology of two dimensional Euclidean space E^2 since b_2 and b_3 can each range from $\pm\infty$ without any duplication of results. By contrast, $\exp(b_1B_1)$, according to (2.73), has the topology of a circle C since it is periodic in b_1 with period 2π. It follows that Sp(2) has the product topology $E^2 \times C$, and is therefore infinitely connected.

The case of Sp(4) is somewhat more complicated. The most general 4×4 real symmetric matrix S can be written in the block form

$$S = \begin{pmatrix} A & B \\ \tilde{B} & C \end{pmatrix}, \tag{2.74}$$

where the matrices A, B, and C are 2×2 and real, and the matrices A and C are themselves symmetric,

$$\tilde{A} = A\ ,\ \tilde{C} = C. \tag{2.75a,b}$$

Requiring that J commute with S gives the restrictions

$$\tilde{B} = -B \tag{2.76a}$$

$$C = A. \tag{2.76b}$$

<u>Exercise 2.31</u>: Verify the restrictions (2.76).

Thus, the most general S^c is of the form

$$S^c = \begin{pmatrix} A & B \\ -B & A \end{pmatrix} \tag{2.77}$$

with the restrictions (2.75a) and (2.76a).
Correspondingly, JS^c is of the form

$$JS^c = \begin{pmatrix} -B & A \\ -A & -B \end{pmatrix} . \tag{2.78}$$

The restriction (2.75a) requires that A be of the form

$$A = a\begin{pmatrix} 1 & 0 \\ 0 & 1 \end{pmatrix} +b\begin{pmatrix} 1 & 0 \\ 0 & -1 \end{pmatrix} + c\begin{pmatrix} 0 & 1 \\ 1 & 0 \end{pmatrix} \tag{2.79}$$

where a, b, and c are arbitrary coefficients. The restriction (2.76a) requires that B be of the form

$$B = d\begin{pmatrix} 0 & 1 \\ -1 & 0 \end{pmatrix} , \tag{2.80}$$

where d is also an arbitrary coefficient. Consequently, the Lie algebra spanned by matrices of the form JS^c is four dimensional. A convenient choice of basis elements is given by the matrices

$$B_0 = J = \begin{bmatrix} 0 & 0 & 1 & 0 \\ 0 & 0 & 0 & 1 \\ -1 & 0 & 0 & 0 \\ 0 & -1 & 0 & 0 \end{bmatrix} \tag{2.81a}$$

$$B_1 = \begin{bmatrix} 0 & 0 & 1 & 0 \\ 0 & 0 & 0 & -1 \\ -1 & 0 & 0 & 0 \\ 0 & 1 & 0 & 0 \end{bmatrix} \tag{2.81b}$$

$$B_2 = \begin{bmatrix} 0 & 0 & 0 & 1 \\ 0 & 0 & 1 & 0 \\ 0 & -1 & 0 & 0 \\ -1 & 0 & 0 & 0 \end{bmatrix} \qquad (2.81c)$$

$$B_3 = \begin{bmatrix} 0 & -1 & 0 & 0 \\ 1 & 0 & 0 & 0 \\ 0 & 0 & 0 & -1 \\ 0 & 0 & 1 & 0 \end{bmatrix} . \qquad (2.81d)$$

They satisfy the commutation rules

$$\begin{aligned} [B_0, B_i] &= 0 \quad , \; i = 0,\, 1,\, 2,\, 3 \\ [B_1, B_2] &= 2B_3 \\ [B_2, B_3] &= 2B_1 \\ [B_3, B_1] &= 2B_2 \, . \end{aligned} \qquad (2.82)$$

<u>Exercise 2.32</u>: Verify the commutation rules (2.82).

The commutation rules (2.82) will be recognized as a variant of the commutation rules of the unitary group U(2). Indeed, let V be the unitary transformation

$$V = \frac{1}{\sqrt{2}} \begin{pmatrix} I & iI \\ iI & I \end{pmatrix} . \qquad (2.83)$$

Then one finds from (2.78) the result

$$V^{-1}(JS^c)V = \begin{pmatrix} -B+iA & 0 \\ 0 & -B-iA \end{pmatrix} . \qquad (2.84)$$

Matrices of the form $-B+iA$ with A and B real and obeying (2.75a) and (2.76a) span the space of $n \times n$ anti-Hermitian matrices in the general $n \times n$ case. Consequently, upon exponentiation, the matrices of the form $-B+iA$ generate the unitary group U(n), and the matrices $-B-iA$ generate the complex conjugate representation $\bar{U}(n)$. Therefore, the

Lie algebra spanned by the matrices JS^c is reducible, and is a variant of the Lie algebra of U(n) in the general case.

To study the matrix S^a in the 4×4 case, require that J anti-commute with the S of (2.74). One now finds the restrictions

$$\tilde{B} = B \quad , \tag{2.85a}$$

$$C = -A \quad . \tag{2.85b}$$

<u>Exercise 2.33</u>: Verify the restrictions (2.85).

Thus, the most general S^a is of the form

$$S^a = \begin{pmatrix} A & B \\ B & -A \end{pmatrix} \tag{2.86}$$

with A and B real and subject to the restrictions (2.75a) and (2.85a). Correspondingly, JS^a is of the form

$$JS^a = \begin{pmatrix} B & -A \\ -A & -B \end{pmatrix} \quad . \tag{2.87}$$

As before, the most general matrix A satisfying (2.75a) can be written in the form

$$A = a\begin{pmatrix} 1 & 0 \\ 0 & 1 \end{pmatrix} + b\begin{pmatrix} 1 & 0 \\ 0 & -1 \end{pmatrix} + c\begin{pmatrix} 0 & 1 \\ 1 & 0 \end{pmatrix} \quad , \tag{2.88}$$

where a, b, and c are arbitrary coefficients. Similarly, the most general B satisfying (2.85a) can be written in the form

$$B = d\begin{pmatrix} 1 & 0 \\ 0 & 1 \end{pmatrix} + c\begin{pmatrix} 1 & 0 \\ 0 & -1 \end{pmatrix} + f\begin{pmatrix} 0 & 1 \\ 1 & 0 \end{pmatrix} \tag{2.89}$$

Consequently, the vector space spanned by matrices of the form JS^a is six dimensional. Since the Lie algebra generated by the matrices JS^c is four dimensional, the complete Lie algebra generated by both the matrices JS^c and JS^a is ten dimensional. This result is in accord with Eq. (2.36) evaluated for n=2. That is, the Lie algebra of Sp(4) is ten dimensional.

After some agony, one finds that a possible and perhaps convenient choice for the six matrices required to provide a basis for matrices of the form JS^a is as follows:

$$F_1 = \begin{bmatrix} 0 & -1 & 0 & -1 \\ -1 & 0 & -1 & 0 \\ 0 & -1 & 0 & 1 \\ -1 & 0 & 1 & 0 \end{bmatrix} \tag{2.90a}$$

$$F_2 = \begin{bmatrix} 1 & 0 & 1 & 0 \\ 0 & -1 & 0 & -1 \\ 1 & 0 & -1 & 0 \\ 0 & -1 & 0 & 1 \end{bmatrix} \tag{2.90b}$$

$$F_3 = \begin{bmatrix} 1 & 0 & -1 & 0 \\ 0 & 1 & 0 & -1 \\ -1 & 0 & -1 & 0 \\ 0 & -1 & 0 & -1 \end{bmatrix} \tag{2.90c}$$

$$G_1 = \begin{bmatrix} 0 & -1 & 0 & 1 \\ -1 & 0 & 1 & 0 \\ 0 & 1 & 0 & 1 \\ 1 & 0 & 1 & 0 \end{bmatrix} \tag{2.90d}$$

$$G_2 = \begin{bmatrix} 1 & 0 & -1 & 0 \\ 0 & -1 & 0 & 1 \\ -1 & 0 & -1 & 0 \\ 0 & 1 & 0 & 1 \end{bmatrix} \tag{2.90e}$$

$$G_3 = \begin{bmatrix} -1 & 0 & -1 & 0 \\ 0 & -1 & 0 & -1 \\ -1 & 0 & 1 & 0 \\ 0 & -1 & 0 & 1 \end{bmatrix} . \tag{2.90f}$$

With this choice of basis, and using (2.58), the most general 4 × 4 symplectic matrix can be written in the form

$$M = \exp\left(\sum_1^3 f_i F_i + g_i G_i\right) \exp\left(\sum_0^3 b_i B_i\right). \tag{2.91}$$

<u>Exercise 2.34</u>: Work out the commutation rules of the B's, F's, and G's for Sp(4).
<u>Answer</u>: See Table (2.1).

<u>Exercise 2.35</u>: Show that every matrix of the form $M = \exp(JS^a)$ is symplectic and has all its eigenvalues on the positive real axis. Show that every matrix of the from $M = \exp(JS^c)$ is symplectic and has all its eigenvalues on the unit circle.

<u>Exercise 2.36</u>: Show that the topology of Sp(2n) is that of $E^m \times U(n)$ with $m = n(n+1)$. Sp(2n) is therefore infinitely connected in general.
<u>Research problem 2.1</u>: Work out a suitable basis for the Lie algebra of Sp(6). According to the work done so far, elements of the form JS^c must generate a U(3). This requires 9 elements. Also, by (2.36), the Lie algebra of Sp(6) is 21 dimensional. Thus there must be 12 elements of the form JS^a.

	B_0	B_1	B_2	B_3	F_1	F_2	F_3	G_1	G_2	G_3
B_0		0	0	0	$2G_1$	$2G_2$	$2G_3$	$-2F_1$	$-2F_2$	$-2F_3$
B_1			$2B_3$	$-2B_2$	0	$2F_3$	$-2F_2$	0	$2G_3$	$-2G_2$
B_2				$2B_1$	$-2F_3$	0	$2F_1$	$-2G_3$	0	$2G_1$
B_3					$2F_2$	$-2F_1$	0	$2G_2$	$-2G_1$	0
F_1						$-4B_3$	$4B_2$	$-4B_0$	0	0
F_2							$-4B_1$	0	$-4B_0$	0
F_3								0	0	$-4B_0$
G_1									$-4B_3$	$4B_2$
G_2										$-4B_1$
G_3										

Table 2.1. Commutation rules for the Lie algebra Sp(4)

3. LIE ALGEBRAIC STRUCTURE OF CLASSICAL MECHANICS

3.1 Definition and properties of Poisson bracket

Let $H(q,p,t)$ be the Hamiltonian for some dynamical system and let f be any dynamical variable. That is, let $f(q,p,t)$ be any function of the phase space variables q,p and the time t. Consider the problem of computing the total time rate of change of f along a trajectory generated by H. According to the chain rule, this derivative is given by the expression

$$df/dt = \partial f/\partial t + \sum_i [(\partial f/\partial q_i)\, \dot{q}_i + (\partial f/\partial p_i)\, \dot{p}_i]. \qquad (3.1)$$

However, the $\dot{q}$'s and $\dot{p}$'s are given by Hamilton's equations of motion (1.11). Consequently, the expression for df/dt can also be written in the form

$$df/dt = \partial f/\partial t + \sum_i [(\partial f/\partial q_i)(\partial H/\partial p_i) - (\partial f/\partial p_i)(\partial H/\partial q_i)]. \qquad (3.2)$$

The second quantity appearing on the right of (3.2) occurs so often that it is given a special symbol and a special name in honor of Poisson. Let f and g be any two functions of the variables q,p,t. Then the Poisson bracket of f and g, denoted by the symbol $[f,g]$ is defined by the equation

$$[f,g] = \sum_i [(\partial f/\partial q_i)(\partial g/\partial p_i) - (\partial f/\partial p_i)(\partial g/\partial q_i)] . \qquad (3.3)$$

With this new notation, Eq. (3.2) can be written in the compact form

$$df/dt = \partial f/\partial t + [f,H]. \qquad (3.4)$$

Note that the Poisson bracket symbol [,] is the same as that used earlier for a commutator. This is somewhat awkward, but unfortunately there are not always enough convenient symbols to go around.

Exercise 3.1: Evaluate the so-called fundamental Poisson brackets $[q_i,q_j]$, $[p_i,p_j]$, $[q_i,p_j]$.

The Poisson bracket operation has several remarkable properties. Upon calculation one finds the relations

1. Distributive property

$$[(af+bg),h] = a[f,h] + b[g,h] \qquad (3.5)$$

for arbitrary constants a,b.

2. Antisymmetry condition

$$[f,g] = -[g,f]. \qquad (3.6)$$

3. Jacobi identify

$$[f,[g,h]] + [g,[h,f]] + [h,[f,g]] = 0 . \qquad (3.7)$$

4. Derivation with respect to ordinary multiplication

$$[f,gh] = [f,g]h + g[f,h] . \qquad (3.8)$$

Exercise 3.2: Verify Eqs. (3.5), (3.6), and (3.8). You may also try to verify (3.7). However, this is more easily done later on after the development of additional notation.

At this point it is useful to make some additional definitions of a mathematical nature: An algebra A over a field of numbers F is defined as a linear vector space supplemented by a rule for multiplying may two vectors to yield a third vector. This multiplication rule must satisfy certain properties. Indicating multiplication by a "o", we require that to every ordered pair of elements $x,y \in A$, there corresponds a third unique element of A, denote by x o y, and called the product of x and y. The product satisfies

1. $(cx) \circ y = x \circ (cy) = c(x \circ y)$

2. $(x + y) \circ z = x \circ z + y \circ z$

3. $x \circ (y + z) = x \circ y + x \circ z$

for any $x,y, z \in A$ and $c \in F$.

An example of an algebra is the set of all $N \times N$ matrices. The set of all $N \times N$ matrices forms an N^2 dimensional vector space. It also forms an algebra if we use for the "o" operation ordinary matrix multiplication. Note that in this case multiplication is associative, that is, the multiplication rule satisfies the property

$$(x \circ y) \circ z = x \circ (y \circ z).$$

A second example of an algebra is the set of all 3-vectors with the multiplication rule $\vec{a} \circ \vec{b} = \vec{a} \times \vec{b}$. Here "x" denotes the usual cross product. This algebra is not associative.

Exercise 3.3: Verify that this algebra is not associative.

A Lie algebra L is an algebra in which the multiplication rule (sometimes now called a Lie product) satisfies two further properties. For convenience, multiplication of x and y will now be denoted by the symbol [x,y],

$$[x,y] = x \circ y.$$

In using this customary notation, however, it should be understood that the bracket [,] does not necessarily refer to the Poisson bracket. Rather, in this context, it refers to the Lie product abstractly, and independently of any particular realization. The two additional properties for a Lie product are

4. $[x,y] = -[y,x]$ (antisymmetry)

5. $[x,[y,z]] + [y,[z,x]] + [z,[x,y]] = 0$ (Jacobi condition).

Exercise 3.4: Verify that the algebra of 3-vectors with multiplication defined by $[\vec{a},\vec{b}] = \vec{a} \times \vec{b}$ is a Lie algebra.

Exercise 3.5: Verify that the set of all $N \times N$ matrices with the multiplication rule defined by $[A,B] = AB - BA$ forms a Lie algebra. In particular, check the Jacobi condition.

The reader will recognize that the general definition of a Lie algebra as just given is a bit more complicated than that used previously for matrices. However, exercise (3.5) shows that the matrix case is a special case of the general definition.

Now the stage is set for a more subtle conclusion. Observe that the set of all functions of the variables q,p,t forms a linear vector space. That is, any linear combination of two such functions is again a function. Now define the Lie product of any two functions to be the Poisson bracket (3.3). Equations (3.5) and (3.6) show that conditions 1 through 4 of a Lie algebra are satisfied. And Eq. (3.7) shows that condition 5 is satisfied. Consequently, the set of functions of the variables q,p,t forms a Lie algebra! This Lie algebra will be called the Poisson bracket Lie algebra of dynamical variables.

Exercise 3.6: Determine the dimensionality of the Poisson bracket Lie algebra of dynamical variables.
Answer: The set of functions of q,p,t is an infinite dimensional vector space.

At this point it is convenient to introduce a more compact notation for the phase space variables $(q_1 \ldots q_n)$, $(p_1 \ldots p_n)$. To do this, introduce the 2n variables $(z_1 \ldots z_{2n})$ by this rule

$$z_i = q_i \quad , \quad i = 1, \quad n \tag{3.9a}$$

$$z_{n+i} = p_i \quad , \quad i = 1, \quad n. \tag{3.9b}$$

That is, the first n z's are the q's and the last n z's are the p's.

With the definition (3.9), it is easily verified that the fundamental Poisson brackets $[z_i,z_j]$ are given by the relation

$$[z_i,z_j] = J_{ij}, \tag{3.10}$$

where J is the fundamental 2n × 2n matrix given by (2.1) and used in defining symplectic matrices.

Exercise 3.7: Verify Eq. (3.10).

Also, suppose functions f and g of the variables q,p,t are written more compactly as f(z,t), g(z,t). Then the general Poisson bracket (3.3) can be written more compactly in the form

$$[f,g] = \sum_{i,j} (\partial f/\partial z_i)\, J_{ij}\, (\partial g/\partial z_j). \tag{3.11}$$

Suppose further that the 2n quantities $(\partial f/\partial z_i)$ are viewed as the components of a vector conveniently written as $\partial_z f$, etc. Then the right-hand side of (3.11) can be viewed as a combination of two vectors and a matrix which can be written even more compactly using matrix and scalar product notation,

$$[f,g] = (\partial_z f,\ J\, \partial_z g). \tag{3.12}$$

Exercise 3.8: Verify Eqs. (3.11) and (3.12) starting from the definition (3.3).

Exercise 3.9: Verify the Jacobi identity (3.7).

3.2 Equations, constants, and integrals of motion

It has already been shown that any dynamical variable f(z,t) of a dynamical system governed by a Hamiltonian H obeys the equation of motion

$$df/dt = \partial f/\partial t + [f,H]. \tag{3.13}$$

Exercise 3.10: Show that the dynamical variables z_i obey the equations of motion

$$z_i = (J\, \partial_z H)_i. \tag{3.14}$$

A dynamical variable f is called a constant of motion if its total time derivative vanishes. In view of (3.13), a constant of motion satisfies the equation

$$\partial f/\partial t + [f,H] = 0. \tag{3.15}$$

Exercise 3.11: Suppose that the dynamical variables f and g are constants of motion. Verify Poisson's theorem which states that the quantity [f,g] is then also a constant of motion.
Hint: Use the Jacobi identity.

It can be shown in general that any Hamiltonian dynamical system with n degrees of freedom has 2n functionally independent constants of motion.

Suppose that a constant of motion f does not explicitly depend on the time t,

$$\partial f/\partial t = 0. \tag{3.16}$$

A constant of motion which does not explicitly depend on the time will be called an integral of motion. Evidently, an integral of motion obeys the equation

$$[f,H] = 0. \tag{3.17}$$

Exercise 3.12: Suppose that f and g are integrals of motion. Show that [f,g] is then also an integral of motion.

Exercise 3.13: Suppose that the Hamiltonian H for a dynamical system does not depend explicitly on the time t. Show that then H is an integral of motion.

The question of the existence of integrals of motion is quite complicated. Observe that if f(z) is an integral of motion, then any given trajectory must remain for all time on a general hypersurface in phase space defined by an equation of the form

$$f(z) = \text{constant}.$$

If there are several functionally independent integrals of motion, then the general trajectory is further restricted to lie in the intersection of several hypersurfaces for all time. Thus, the greater the number of integrals, the more can be said about the behavior of a dynamical system.

Consider a time independent Hamiltonian H(z). A point z^o in phase space for which the vector $\partial_z H$ is zero is called a critical point. Evidently, according to Eq. (3.14), a critical point is some kind of equilibrium point. Now suppose some small region R of phase space contains no critical points. Then, it can be shown that provided R is small enough, the Hamiltonian H(z) has 2n-1 functionally independent integrals of motion in the region R. Furthermore, n of these integrals can be arranged to be in involution. (Two functions f and g are said to be in involution if their Poisson bracket [f,g] is zero.)

The result just stated is of limited use unless all trajectories starting in R happen to remain in R. In general, and contrary to the impression given by most textbooks, most dynamical Hamiltonian systems do not have global integrals of motion. If a time independent

Hamiltonian dynamical system with n degrees of freedom has n global integrals of motion in involution, the system is said to be completely integrable. In general, only the soluble problems found in textbooks fall into this category. Most Hamiltonian dynamical systems, including the majority encountered in real life, are not completely integrable and are therefore sufficiently complicated to be in some sense insoluble. In particular, the behavior of most Hamiltonian systems is sufficiently complicated that the trajectories are not generally confined to lie on hypersurfaces in phase space.[7-9]

3.3 Lie operators

Let $f(z,t)$ be some function, and let $g(z,t)$ be any other function. Associated with each f is a Lie operator which acts on general functions g. The Lie operator associated with the function f will be denoted by the symbol $:f:$ and it is defined in terms of Poisson brackets by the rule

$$:f:g = [f,g]. \tag{3.18}$$

In an analogous way, powers of $:f:$ are defined by taking repeated Poisson brackets. For example, $:f:^2$ is defined by the relation

$$:f:^2 g = [f,[f,g]]. \tag{3.19}$$

Finally, $:f:$ to the zero power is defined to be the identity operator,

$$:f:^0 g = g. \tag{3.20}$$

Evidently, a Lie operator, as well as its powers, is a linear operator because of Eq. (3.5). For the same reason, the sum of two Lie operators is again a Lie operator. Specifically, one finds the relation

$$a:f: + b:g: = :(af + bg): \tag{3.21}$$

for any two scalars a,b and any two functions f,g. Therefore, the set of Lie operators forms a linear vector space.

A Lie operator is also a derivation with respect to the operation or ordinary multiplication. That is, a Lie operator satisfies the product rule analogous to that for differentiation: Let g and h be any two functions. Then, according to Eq. (3.8), $:f:$ obeys the rule

$$:f:(gh) = (:f:g)h + g(:f:h). \tag{3.22}$$

Exercise 3.14: Starting from (3.22), show that $:f:^n$ obeys the Leibniz rule

$$:f:^n(gh) = \sum_{m=0}^{n} \binom{n}{m}(:f:^m g)(:f:^{n-m} h), \tag{3.23}$$

where $\binom{n}{m}$ is the binomial coefficient defined by

$$\binom{n}{m} = \frac{n!}{(m!)(n-m)!} \quad . \tag{3.24}$$

In addition to being a derivation with respect to ordinary multiplication, a Lie operator is also a derivation with respect to Poisson bracket multiplication. Suppose g and h are any two functions. Then the Jacobi identity (3.7) can be written in the form

$$[f,[g,h]] = [[f,g],h] + [g,[f,h]], \tag{3.25}$$

or equivalently, using Lie operator notation,

$$:f:[g,h] = [:f:g,h] + [g,:f:h] \ . \tag{3.26}$$

Exercise 3.15: Verify Eqs. (3.25) and (3.26).

Exercise 3.16: Write and verify the analog of the Leibniz rule of exercise (3.14) for the case of $:f:^n[g,h]$.

Since the set of Lie operators forms a linear vector space, it is of interest to inquire whether the vector space can be given a multiplication rule which will convert it into a Lie algebra. The answer is yes, as is nearly obvious, since Lie operators are linear operators and linear operators are quite similar to matrices. The Lie product of two Lie operators :f: and :g: is simply taken to be their commutator. Denoting the Lie product of two Lie operators by the symbol :f:, :g: , the Lie product is defined by the rule

$$\{:f:,:g:\} = :f::g: - :g::f:. \tag{3.27}$$

Note that there are now two Lie algebras which have to be kept in mind. First, there is the Lie algebra of functions of z,t with the Lie product defined to be the Poisson bracket. Second, there is the Lie algebra of Lie operators with the Lie product defined to be the commutator.

There is, however, one point which has been overloooked. Namely, is the right-hand side of (3.27) a Lie operator? To answer this question, it is useful to view the Jacobi identity (3.7) for Poisson brackets form yet another perspective. For any function h, the Jacobi identity can be written in the form

$$[f,[g,h]] - [g,[f,h]] = [[f,g],h]. \tag{3.28}$$

However, using Lie operator notation, this same equation can be written in the form

$$:f::g:h - :g::f:h = :[f,g]:h, \tag{3.29}$$

or more compactly, using (3.27),

$$\{:f:,:g:\}h = :[f,g]:h. \qquad (3.30)$$

But, since h is an arbitrary function, Eq. (3.29) can also be viewed as the operator identity

$$\{:f:,:g:\} = :[f,g]:. \qquad (3.31)$$

Evidently, the commutator of two Lie operators :f: and :g: is again a Lie operator, and is in fact the Lie operator associated with the function [f,g].

Put another way, Eq. (3.31) shows that there is a close connection between the Lie algebra of functions and the Lie algebra of Lie operators. Specifically, the Lie product (commutator) of two Lie operators is the Lie operator of the Lie product (Poisson bracket) of the two associated functions. Mathematicians have a word for such a situation. They would say that the two Lie algebras are homomorphic.

Exercise 3.17: Verify Eqs. (3.28), (3.29), and (3.30).

Exercise 3.18: Verify that the Lie product defined by (3.27) satisfies all the properties required to make the set of Lie operators into a Lie algebra. See exercise (3.5).

3.4 Lie Transformations

Since powers of :f: have been defined, it is also possible to deal with power series in :f:. Of particular importance is the power series exp(:f:). This particular object is called the Lie transformation associated with :f: or f. The Lie transformation is also a linear operator, and is formally defined as expected by the exponential series

$$\exp(:f:) = \sum_{n=0}^{\infty} :f:^n/n!. \qquad (3.32)$$

In particular, the action of exp(:f:) on any function g is given by the rule

$$\exp(:f:)\, g = g + [f,g] + [f,[f,g]]/2! + \cdots. \qquad (3.33)$$

Exercise 3.19: Let q and p be the phase-space coordinates for a system having one degree of freedom. Let f be the function

$$f = -\lambda p^2/2 .$$

Show that

$$\exp(:f:)\, p = p$$

$$\exp(:f:)\, q = q + \lambda p .$$

Here λ is an arbitrary parameter.
Hint: Observe that the series (3.33) terminates in this case.

Exercise 3.20: Repeat exercise (3.19) for the case $f = \lambda q^2$.

Exercise 3.21: Repeat exercise (3.19) for the case $f = \lambda q^3$.

Exercise 3.22: Repeat exercise (3.19) for the case $f = -\lambda pq$. Now you must sum an infinite series.

Answer:

$$\exp(:f:)\, q = (e^{\lambda})\ q$$

$$\exp(:f:)\, p = (e^{-\lambda})\ p.$$

Exercise 3.23: Repeat exercise (3.19) for the case $f = -\lambda(p^2 + q^2)/2$,
Answer:

$$\exp(:f:)\, q = q \cos \lambda + p \sin \lambda$$

$$\exp(:f:)\, p = -q \sin \lambda + p \cos \lambda .$$

The fact that :f: is a derivation with respect to ordinary multiplication, see Eq. (3.22), implies that the Lie transformation exp(:f:) is an isomorphism with respect to ordinary multiplication. This is another remarkable property of the exponential function! That is, suppose g and h are any two functions. Then the Lie transformation exp(:f:) has the property

$$\exp(:f:)\,(gh) = (\exp(:f:)g)\,(\exp(:f:)h). \qquad (3.34)$$

In words, Eq. (3.34) says that one can either let a Lie transformation act on the product of two functions, or act on each function separately and then take the product of the results. Both operations give the same net result.

Exercise 3.24: Verify Eq. (3.34) for the case $f = \lambda q^2$ and $g = h = p$.

Exercise 3.25: Prove Eq. (3.34), using the definition (3.32) and the Leibniz rule (3.23), by expansion and resummation of various series.

The isomorphism property of exp(:f:) described by (3.34) often facilitates computations involving Lie transformations. Let the symbol z stand, as usual, for the collection of quantities $z_1 \ldots z_{2n}$. Similarly, let the symbol exp(:f:)z stand for the collection of quantities $\exp(:f:)z_1 \ldots \exp(:f:)z_{2n}$. Now let g(z) be any function. Then it follows from (3.34) that

$$\exp(:f:)\ g(z) = g[\exp(:f:)z]. \qquad (3.35)$$

That is, the action of a Lie transformation on a function is to perform a Lie transformation on its arguments.

To see the truth of (3.35), suppose first that g were a polynomial in the quantities $z_1 \dots z_{2n}$. But a polynomial is just a sum of monomials of the form

$$z_1^{m_1}\ z_2^{m_2} \dots z_{2n}^{m_{2n}} \quad .$$

It follows from (3.34) that

$$\exp(:f:)\ z_1^{m_1} \dots z_{2n}^{m_{2n}} = [\exp(:f:)z_1]^{m_1} \cdots [\exp(:f:)z_{2n}]^{m_{2n}} \quad . \qquad (3.36)$$

Also, as commented earlier, exp(:f:) is a linear operator. Therefore a Lie transformation has the advertised property (3.35) when acting on polynomials. But the set of polynomials is dense in the complete set of functions on any bounded domain. Consequently, (3.35) holds in general by continuity.

The last observation to be made is that since :f: is also a derivation with respect to Poisson bracket multiplication, the Lie transformation exp(:f:) must also be an isomorphism with respect to Poisson bracket multiplication. That is, suppose g and h are any two functions. Then the Lie transformation exp(:f:) has the property

$$\exp(:f:)\ [g,h] = [\exp(:f:)\ g,\ \exp(:f:)\ h]. \qquad (3.37)$$

This property will be essential for subsequent discussions of symplectic maps and charged particle beam transport.

> Exercise 3.26: Derive (3.37) from the definition (3.32) and the results of exercise (3.16). Mutatis mutandis, what is required is a repeat of exercise (3.25).

4. SYMPLECTIC MAPS

4.1 Definitions

Let $z_1 \dots z_{2n}$ be a set of canonical variables for some Hamiltonian dynamical system. Suppose a transformation is made to some new set of variables $\bar{z}_1(z,t) \dots \bar{z}_{2n}(z,t)$.[10] Such a transformation will be called a mapping, and will be denoted by the symbol $\mathcal{M}$,

$$\mathcal{M}:\ z \to \bar{z}(z,t). \qquad (4.1)$$

Also, let M(z,t) be the Jacobian matrix of the map $\mathcal{M}$. It is defined by the equation

$$M_{ab}(z,t) = \partial \bar{z}_a / \partial z_b. \tag{4.2}$$

The map $\mathcal{M}$ is said to be <u>symplectic</u> if its Jacobian matrix M is a symplectic matrix for all values of z and t,

$$\tilde{M}\ J\ M = J, \quad \text{or} \quad M\ J\ \tilde{M} = J. \tag{4.3}$$

Note that in general M depends on z and t. However, the particular combinations $\tilde{M}\ J\ M$ or $M\ J\ \tilde{M}$ must be z and t independent. Therefore, a symplectic map must have very special properties.

To appreciate the significance of a symplectic mapping, consider the Poisson brackets of the various $\bar{z}$'s with each other. Using Eq. (3.11), one finds the result

$$[\bar{z}_a, \bar{z}_b] = \sum_{c,d} (\partial \bar{z}_a / \partial z_c)\ J_{cd}\ (\partial \bar{z}_b / \partial z_d). \tag{4.4}$$

By using the definition (4.2) of the Jacobian matrix M, Eq. (4.4) can also be written in the form

$$\begin{aligned} [\bar{z}_a, \bar{z}_b] &= \sum_{c,d} M_{ac}\ J_{cd}\ M_{bd} \\ &= \sum_{c,d} M_{ac}\ J_{cd}\ \tilde{M}_{db} \\ &= (M\ J\ \tilde{M})_{ab}\ . \end{aligned} \tag{4.5}$$

Finally, upon using the symplectic condition (4.3), one finds the result

$$[\bar{z}_a, \bar{z}_b] = (M\ J\ \tilde{M})_{ab} = J_{ab} = [z_a, z_b]. \tag{4.6}$$

Consequently, the necessary and sufficient condition for a map $\mathcal{M}$ to be symplectic is that it preserve the fundamental Poisson brackets (3.10). This statement is equivalent, in turn, to the condition that the map $\mathcal{M}$ must preserve the Poisson bracket Lie algebra of all dynamical variables.

Symplectic mappings also have a geometrical apsect. Let z^o be some point in phase space, and suppose it is sent to the point $\bar{z}^o$ under the action of a symplectic map $\mathcal{M}$. Also, let dz and δz be two small vectors originating at z^o. Under the action of $\mathcal{M}$, they are sent to two vectors $d\bar{z}$ and $\delta \bar{z}$ originating at the point $\bar{z}^o$. See Fig. 4.1. According to the chain rule, one has the relation

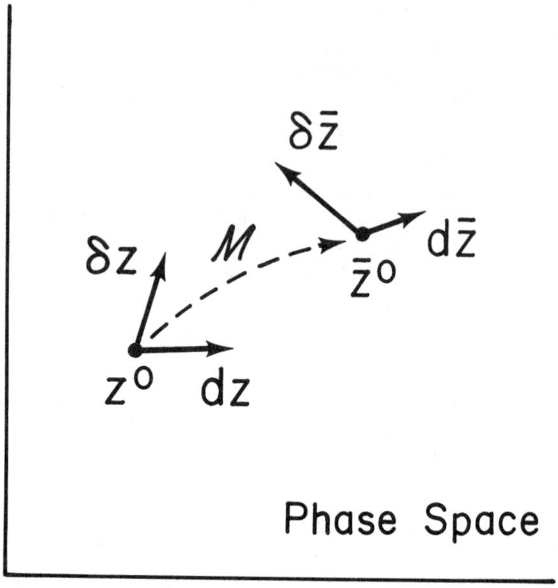

Fig. 4.1: The action of a symplectic map M on phase space. The general point z^o is mapped to the point $\bar{z}^o$, and the small vectors dz and δz are mapped to the small vectors $d\bar{z}$ and $\delta\bar{z}$. The figure is only schematic since in general phase space has a large number of dimensions.

$$d\bar{z}_a = \sum_b (\partial \bar{z}_a/\partial z_b)\, dz_b, \tag{4.7}$$

or more compactly, using (4.2),

$$d\bar{z} = M\, dz. \tag{4.8}$$

Similarly, the vectors δz and $\delta \bar{z}$ are related by the equation

$$\delta \bar{z} = M\, \delta z. \tag{4.9}$$

Now use the two vectors $\delta \bar{z}$, $d\bar{z}$ and the matrix J to form the quantity $(\delta \bar{z}, J d\bar{z})$. This quantity is called the <u>fundamental symplectic two-form</u>. (Given any two vectors, a two-form is a rule for computing an associated number. The rule must be such that the answer depends linearly on each vector.) Suppose the relations (4.8) and (4.9) are inserted into the two-form $(\delta \bar{z}, J d\bar{z})$. Then, using matrix manipulation and the symplectic condition (4.3), one finds the relation

$$(\delta \bar{z},\ J\, d\bar{z}) = (M\, \delta z,\ J\, M\, dz) = (\delta z,\ \tilde{M}\, J\, M\, dz) = (\delta z,\ J\, dz). \tag{4.10}$$

That is, the value of the fundamental symplectic two-form is unchanged by a symplectic map. Evidently, a necessary and sufficient condition for a map to be symplectic is that it preserve the fundamental symplectic two-form at all points of phase space and for all time.

> <u>Exercise 4.1</u>: Consider a two-dimensional phase space consisting of the variables q,p. Evaluate the quantity $(\delta z,\ J\, dz)$ and show that it is related to the area formed by the small parallelogram with sides δz and dz.

There is a third aspect of symplectic mappings which should already be familiar. In the usual treatments of Classical Mechanics, an important topic is that of canonical transformations. Canonical transformations are usually defined as those transformations which either

a. preserve the Hamiltonian form of the equations of motion for all Hamiltonian dynamical systems, or

b. preserve the fundamental Poisson brackets.

In case b, according to the previous discussion, canonical transformations and symplectic maps are the same thing. In case a, it can be shown that the most general canonical transformation is a map $\mathcal{M}$ whose Jacobian matrix satisfies the condition

$$\tilde{M}\, J\, M = \lambda J, \tag{4.11}$$

where λ is some real nonzero constant independent of z and t.[11] Furthermore, it can be shown that $\mathcal{M}$ in this case consists of a symplectic map followed or preceded by a simple scaling of phase-space variables. Therefore, in either case, the central object of interest is a symplectic map.

4.2 Group Properties

Let $\mathcal{M}$ be a symplectic mapping of z to $\bar{z}$, and suppose it has an inverse $\mathcal{M}^{-1}$,

$$\mathcal{M} : z \to \bar{z} \tag{4.12a}$$

$$\mathcal{M}^{-1} : \bar{z} \to z. \tag{4.12b}$$

According to (4.8), the relation between a small change dz in z, and the associated small change $d\bar{z}$ in $\bar{z}$ is given by the Jacobian matrix M of $\mathcal{M}$. Since M is symplectic, it has an inverse M^{-1}. Therefore, Eq. (4.8) can be inverted to give the relation

$$dz = M^{-1}\, d\bar{z}. \tag{4.13}$$

But now, comparison of (4.12b) and (4.13) shows that the Jacobian matrix of $\mathcal{M}^{-1}$ is M^{-1}. Note also that the local existence of $\mathcal{M}^{-1}$ did not really have to be assumed, but follows instead from the implicit function theorem since M^{-1} is known to exist from the symplectic condition. Finally, the matrix M^{-1} is symplectic since the inverse of a symplectic matrix is also a symplectic matrix. It follows that $\mathcal{M}^{-1}$ is a symplectic map. What has been shown is that if $\mathcal{M}$ is a symplectic map, then $\mathcal{M}^{-1}$ exists (at least locally) and is also a symplectic map.

Next suppose that $\mathcal{M}^{(1)}$ is a symplectic mapping of z to $\bar{z}$ and $\mathcal{M}^{(2)}$ is a symplectic mapping of $\bar{z}$ to another set of variables $\bar{\bar{z}}$. Now consider the composite mapping $\mathcal{M} = \mathcal{M}^{(2)}\, \mathcal{M}^{(1)}$ which sends z to $\bar{\bar{z}}$.

$$\mathcal{M} = \mathcal{M}^{(2)}\, \mathcal{M}^{(1)} \tag{4.14}$$

$$\mathcal{M}^{(1)} : z \to \bar{z} \tag{4.15a}$$

$$\mathcal{M}^{(2)} : \bar{z} \to \bar{\bar{z}} \tag{4.15b}$$

$$\mathcal{M}^{(2)}\mathcal{M}^{(1)} : z \to \bar{\bar{z}}. \tag{4.15c}$$

According to the chain rule, the Jacobian matrix M of the composite mapping $\mathcal{M}$ is the product of the Jacobian matrices of $\mathcal{M}^{(2)}$ and $\mathcal{M}^{(1)}$,

$$M = M^{(2)}\, M^{(1)}. \tag{4.16}$$

However, the matrices $M^{(2)}$ and $M^{(1)}$ are symplectic since they are the Jacobian matrices of symplectic maps. It follows from (4.16) and the group property for symplectic matrices that M is also a symplectic matrix. Consequently, the composite mapping $\mathcal{M}$ is also a symplectic map. What has been shown is that if $\mathcal{M}^{(1)}$ and $\mathcal{M}^{(2)}$ are symplectic maps, so is their product $\mathcal{M}^{(2)}\,\mathcal{M}^{(1)}$.

It is also obvious that the identity mapping which sends each z into itself is a symplectic map, for the Jacobian matrix of this map is evidently the identity matrix, and the identity matrix is symplectic.

The previous discussion has shown that the set of symplectic maps has properties very analogous to the group properties of the group of symplectic matrices. As defined earlier, the concept of a group applied only to matrices. However, it is clear that the concept of a group can be enlarged to include the possibility of general mappings. When this is done, the set of all symplectic maps is entitled to be called a group.

4.3 Relation to Hamiltonian Flows

Let H(z,t) be the Hamiltonian for some dynamical system. Consider a large Euclidean space with 2n+1 axes labeled by the variables $z_1 \ldots z_{2n}$ and the time t. This construction will be called state space. See Fig. 4.2. Suppose the 2n quanties $z_1(t^o) \ldots z_{2n}(t^o)$ are specified at some initial time t^o. Then the quantities $z_1(t) \ldots z_{2n}(t)$ at some other time t are uniquely determined by the initial conditions $z_1(t^o) \ldots z_{2n}(t^o)$ and Hamilton's equations of motion (3.14). The set of all trajectories in state space for all possible initial conditions will be called a Hamiltonian flow.

Let t^o be some initial time, and let t^f be some other final time. Also, let $z^{(0)}$ denote the set of quantities $z_1(t^o) \ldots z_{2n}(t^o)$, and let $z^{(f)}$ denote the corresponding set $z_1(t^f) \ldots z_{2n}(t^f)$. Then it is a remarkable fact that the relation between the quantities $z^{(0)}$ and $z^{(f)}$ is a symplectic map.

Theorem 4.1: Let H(z,t) be the Hamiltonian for some dynamical system, and let $z^{(0)}$ denote a set of initial conditions at some initial time t^o. Also, let $z^{(f)}$ denote the coordinates at some time t^f of the trajectory with initial conditions $z^{(0)}$. Finally, let $\mathcal{M}$ denote the mapping from $z^{(0)}$ to $z^{(f)}$ obtained by following the Hamiltonian flow specified by H,

$$\mathcal{M}: \; z^{(0)} \rightarrow z^{(f)}. \tag{4.17}$$

Then the mapping $\mathcal{M}$ is symplectic.

Proof: Suppose the flow takes place for a time interval of duration T so that t^o and t^f are related by the equation

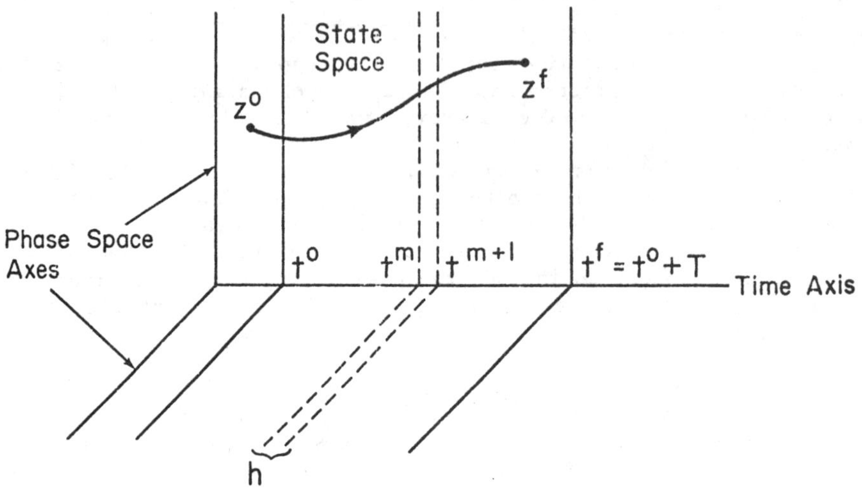

Fig. 4.2: A trajectory in state space. Under the Hamiltonian flow specified by a Hamiltonian H, the general phase-space point z^o is mapped into the phase-space point z^f. The mapping is symplectic for any Hamiltonian.

$$t^f = t^o + T. \tag{4.18}$$

Divide the interval T into N small steps each of duration h. Evidently, T, N, and h are related by the equation

$$T = Nh. \tag{4.19}$$

Also, define intermediate times t^m at each step by the rule

$$t^m = t^o + mh, \quad m = 0,1, \cdots N. \tag{4.20}$$

Suppose that the mapping $\mathcal{M}$ is viewed as a composite of mappings betwen adjacent times t^m and t^{m+1}. That is, $\mathcal{M}$ is written in the form

$$\mathcal{M} = \mathcal{M}^{t^f \leftarrow t^{N-1}} \ldots \mathcal{M}^{t^{m+1} \leftarrow t^m} \ldots \mathcal{M}^{t^1 \leftarrow t^o} \tag{4.21}$$

with the notation that $\mathcal{M}^{t^{m+1} \leftarrow t^m}$ denotes the mapping between the quantities

$$z^{(m)} = z_1(t^m) \cdots z_{2n}(t^m)$$

and

$$z^{(m+1)} = z_1(t^{m+1}) \cdots z_{2n}(t^{m+1}). \tag{4.22}$$

Corresponding to the relation (4.21), the Jacobian matrix M of the mapping $\mathcal{M}$ can be written using the chain rule in the product form

$$M = M^{t^f \leftarrow t^{N-1}} \ldots M^{t^{m+1} \leftarrow t^m} \ldots M^{t^1 \leftarrow t^0}, \tag{4.23}$$

where, as the notation is meant to indicate, $M^{t^{m+1} \leftarrow t^m}$ is the Jacobian matrix for the map $\mathcal{M}^{t^{m+1} \leftarrow t^m}$.

Next it will be shown that each matrix in the product (4.23) is symplectic at least through terms of order h. According to Taylor's series, the relation between $z^{(m+1)}$ and $z^{(m)}$ can be written in the form

$$\begin{aligned} z_a^{(m+1)} &= z_a(t^{m+1}) = z_a(t^m+h) \\ &= z_a(t^m) + h\, \dot{z}_a(t^m) + O(h^2) \\ &= z_a^{(m)} + h(J\, \partial_z H)_a + O(h^2). \end{aligned} \tag{4.24}$$

Here use has also been made of the equations of motion (3.14). Suppose Eq. (4.24) is used to compute the associated Jacobian matrix. The result of this computation is the relation

$$M_{ab}^{t^{m+1} \leftarrow t^m} = \partial z_a^{(m+1)} / \partial z_b^{(m)}$$

$$= \delta_{ab} + h \sum_c J_{ac} \, \partial^2 H / \partial z_c \partial z_b + O(h^2). \quad (4.25)$$

Using matrix notation, Eq. (4.25) can be written more compactly in the form

$$M^{t^{m+1} \leftarrow t^m} = I + h\, J\, S + O(h^2), \quad (4.26)$$

where S is the <u>symmetric</u> matrix

$$S_{cb} = \partial^2 H / \partial z_c \partial z_b . \quad (4.27)$$

Now compare Eq. (4.26) with Eqs. (2.28) and (2.31). Evidently, the Jacobian matrix (4.26) is a symplectic matrix at least through terms of order h.

The desired proof is almost complete. Since symplectic matrices form a group, the product matrix M given by (4.23) differs from a symplectic matrix by terms at most of order Nh^2 because each of the N terms in the product differs from a symplectic matrix by terms at most of order h^2. Now take the limit $h \to 0$ and $N \to \infty$. In this limit terms proportional to Nh^2 vanish since, using (4.19),

$$Nh^2 = (T/h)h^2 = Th, \quad (4.28)$$

and the quantity Th vanishes as h goes to zero. It follows that M is a symplectic matrix, and $\mathcal{M}$ is a symplectic map.

What has been shown is that the problem of describing and following Hamiltonian flows, which is one of the fundamental aspects of classical mechanics, is equivalent to the problem of calculating and representing symplectic maps.

<u>Exercise 4.2</u>: Let H be the harmonic oscillator Hamiltonian given by

$$H = (z_2^2 + \omega^2 z_1^2)/2. \quad (4.29)$$

Find the map $\mathcal{M}$ produced by the Hamiltonian flow governed by H. Verify that $\mathcal{M}$ is a symplectic map.

Answer: The mapping $\mathcal{M}$ is given by

$$\mathcal{M}: \quad z^o \rightarrow z^f$$

with

$$z_1^f = z_1^o \cos\phi + (z_2^o/\omega)\sin\phi$$

$$z_2^f = -\omega z_1^o \sin\phi + z_2^o \cos\phi \tag{4.30}$$

where

$$\phi = \omega(t^f - t^o). \tag{4.31}$$

The reader should find the associated Jacobian matrix M and verify that it is symplectic.

Exercise 4.3: Consider the behavior of light rays in an optical system as shown in Fig. (1.4) and described in exercise (1.14). Let $\vec{p}$ be a two-component vector with entries p_x and p_y, and let $\vec{q}$ be a two-component vector with entries q_x=x and q_y=y. Evidently, a ray leaving the initial point P^i is characterized by the initial quantities $\vec{q}^{\,i}$ and $\vec{p}^{\,i}$. The quantity $\vec{q}^{\,i}$ specifies the initial point of origin of the ray in the object plane, and, according to (1.29), $\vec{p}^{\,i}$ describes the initial direction of the ray. Similarly, $\vec{q}^{\,f}$ and $\vec{p}^{\,f}$ characterize the ray as it arrives at the final point P^f in the image plane. Show that the relationship between the initial quantities $\vec{q}^{\,i}$, $\vec{p}^{\,i}$ and the final quantities $\vec{q}^{\,f}$, $\vec{p}^{\,f}$ is given by a symplectic map.

4.4 Liouville's theorem and the Poincare invariants

It has been seen that Hamiltonian flows generate symplectic maps, and that symplectic maps have special properties. It follows that Hamiltonian flows must have special properties. One of these properties is described by Liouville's theorem.

Consider an ensemble of noninteracting systems with each member of the ensemble governed by the same Hamiltonian H(z,t). At some initial instant t^i, let each member of the ensemble be characterized by a point in phase space corresponding to its initial conditions. Suppose, further that all the points of the ensemble at the initial instant t^i occupy a certain region R^i of phase space. The volume V^i initially occupied by the ensemble in phase space is given by the integral

$$V^i = \int_{R^i} dz_1^i \dots dz_{2n}^i. \tag{4.32}$$

Now follow all the trajectories of the members of the ensemble through state space to some later instant t^f. The members of the ensemble will then occupy some final region R^f of phase space, and the volume of this region will be given by the integral

$$V^f = \int_{R^f} dz_1^f \dots dz_{2n}^f . \tag{4.33}$$

Evidently, the relation between the quantities z^i and z^f for each member of the ensemble is given by a common symplectic map $\mathcal{M}$. Also, the relation between the differential quantities dz^i and dz^f is given by the Jacobian matrix M belonging to $\mathcal{M}$,

$$dz^f = M\, dz^i. \tag{4.34}$$

It follows from the standard rules for changing variables of integration that the volume V^f is also given by the relation

$$V^f = \int_{R^i} |\det M|\, dz_1^i \dots dz_{2n}^i . \tag{4.35}$$

But, since M is a symplectic matrix, it must have determinant +1. Therefore, comparison of (4.35) and (4.32) shows that the two volumes V^f and V^i are the same,

$$V^f = V^i . \tag{4.36}$$

This is Liouville's theorem.

There is a slighly different phrasing of Liouville's theorem which is also worth mentioning. By construction, the number of ensemble points in V^f and V^i is the same. Therefore, since V^f equals V^i, one may also say that the <u>density</u> of points in phase space is preserved by Hamiltonian flows.

The volume invariant of Liouville's theorem is actually the last in a hierarchy of invariants called the Poincare invariants. The first invariant in the series consists of a certain 2-dimensional integral over a 2-dimensional hypersurface in phase space. The next consists of a 4-dimensional integral over a 4-dimensional hypersurface, etc. The last consists of a 2n dimensional integral which is just the volume of Liouville's theorem.

A complete and proper discussion of all the Poincare invariants requires the use of the calculus of differential forms. However, the first in the series of invariants is easily discussed using the fundamental symplectic two-form (δz, Jdz) introduced earlier.

Suppose the points of an ensemble at the instant t^i lie on a 2-dimensional hypersurface R_2^i. Let this surface be parameterized by the two quantities α, β. That is, there are 2n relations of the form

$$z_j = f_j(\alpha, \beta) \tag{4.37}$$

which describe the surface. Next consider the integral over the surface R_2^i given by

$$I_2^i = \int_{R_2^i} (\delta z^i, Jdz^i). \tag{4.38}$$

The integral (4.38) is to be understood as follows: Let dz^i be the vector formed using (4.37) when only α is allowed to vary

$$dz_j^i = (\partial f_j/\partial\alpha)\, d\alpha \tag{4.39a}$$

or, in vector notation,

$$dz^i = \partial_\alpha f\, d\alpha. \tag{4.39b}$$

Similarly, let δz^i be the vector formed when only β is allowed to vary,

$$\delta z^i = \partial_\beta f\, d\beta\,. \tag{4.39c}$$

Then the integral (4.38) can also be written in the form

$$I_2^i = \int_{R_2^i} (\partial_\beta f,\ J\, \partial_\alpha f)\, d\alpha d\beta. \tag{4.40}$$

Now follow, as before, the trajectories of the members of the ensemble through state space to some later instant t^f. Then the points will lie on some other 2-dimensional hypersurface R_2^f, and one can form the associated integral

$$I_2^f = \int_{R_2^f} (\delta z^f,\ J\, dz^f). \tag{4.41}$$

However, by a change of variables using (4.34) and its counterpart for δz, the integral (4.41) can also be written in the form

$$I_2^f = \int_{R_2^i} (M\delta z^i, J M dz^i). \quad (4.42)$$

But, as found earlier in Eq. (4.10) using the symplectic condition, one has the relation

$$(M \delta z^i, J M dz^i) = (\delta z^i, J dz^i). \quad (4.43)$$

It follows that the two-dimensional integral based on the fundamental symplectic two-form is conserved,

$$I_2^f = I_2^i. \quad (4.44)$$

One final point is worth mentioning. In constructing the general higher order Poincare invariants, the calculus of differential forms is used to make general 2m-forms for $m = 2, 3, \ldots, n$ from the fundamental symplectic two-form. The invariance of all these forms, including the last of the hierarchy ($m=n$) which is just the volume element, follows from the invariance (4.10) of the fundamental symplectic two-form. This invariance is in turn equivalent to the symplectic condition (4.3). Thus, the symplectic condition is really the fundamental condition from which everything else follows.

Exercise 4.4: Suppose a "burst" of protons is injected into a uniform electric field $\vec{E} = E_o\hat{e}_z$. Assume the burst is initially concentrated at x and $y = 0$ and v_x and $v_y = 0$, but is uniformly spread in z and v_z about the values $z = 0$ and $v_z = v_z^o$ within intervals $\pm\Delta z$ and $\pm\Delta v_z$. Thus the problem is essentially that of one-dimensional motion along the z axis. The initial distribution is shown schematically in figure 4.3.

Find the distribution at later times, and verify Liouville's theorem. Do not assume Δz and Δv_z are infinitesimal. Neglect Coulomb interactions between particles.

5. LIE ALGEBRAS AND SYMPLECTIC MAPS

5.1 Lie transformations as symplectic maps

Let $f(z,t)$ be any dynamical variable, and let $\exp(:f:)$ be the Lie transformation associated with f. This Lie transformation can be used to define a map $\mathcal{M}$ which produces new variables $\bar{z}(z,t)$ by the rule

$$\bar{z}_a(z,t) = \exp(:f:)\, z_a, \quad a = 1, 2, \ldots, 2n. \quad (5.1)$$

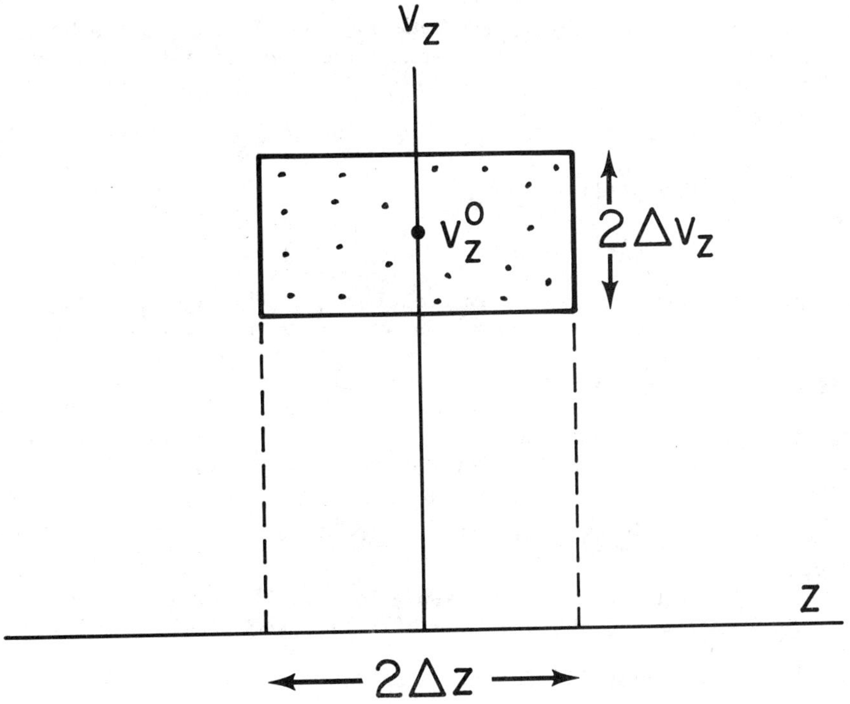

Figure (4.3). Initial phase-space distribution for exercise 4.4.

The relations (5.1) can also be expressed more compactly by writing

$$\bar{z} = \mathcal{M} z \tag{5.2}$$

with

$$\mathcal{M} = \exp(:f:). \tag{5.3}$$

Consider the Poisson brackets of the various $\bar{z}$'s with each other. Using the definition (5.1) and the ismorphism condition (3.37), one finds the result

$$\begin{aligned}[\bar{z}_a, \bar{z}_b] &= [\exp(:f:)\ z_a,\ \exp(:f:)\ z_b] \\ &= \exp(:f:)\ [z_a, z_b] \\ &= \exp(:f:)\ J_{ab} = J_{ab}.\end{aligned} \tag{5.4}$$

It follows from (5.4) that $\mathcal{M}$ is a symplectic map! What has been shown is that every Lie transformation may be viewed as a symplectic map.

<u>Exercise 5.1</u>: Verify Eq. (5.4).

The symplectic map defined by (5.3) has the particular property that f is an <u>invariant function</u> for the map. That is, one has the relation

$$f(\bar{z},t) = f(z,t). \tag{5.5}$$

To see the truth of this assertion, apply Eq. (3.35) to the case where $g = f$. One finds, using the notation of (5.1), the result

$$\exp(:f:)\ f(z,t) = f(\bar{z},t). \tag{5.6}$$

However, using the expression (3.33), one also obtains the result

$$\exp(:f:)\ f(z,t) = f(z,t) \tag{5.7}$$

since the Poisson bracket $[f,f]$ is zero by the antisymmetry condition. Comparison of (5.6) and (5.7) shows that Eq. (5.5) is indeed correct. Note that in all these calculations, the time t plays no essential role and may be regarded simply as a parameter.

Suppose the symplectic map exp(-:f:) is applied to both sides of (5.1). One finds the result

$$\exp(-:f:)\ \bar{z}_a = \exp(-:f:)\ \exp(:f:)\ z_a. \tag{5.8}$$

Consider first the problem of evaluating the right-hand side of (5.8). Observe that the Lie operators :f: and -:f: commute. Indeed, using Eq. (3.31), one finds the result

$$\{:f:, -:f:\} = :[f,-f]: = :0: = 0. \tag{5.9}$$

One might therefore imagine that the two operators in the product exp(-:f:) exp(:f:) cancel each other to give a net identity operation. This is indeed the case. Consequently, when read from right to left, Eq. (5.8) may be rewritten in the form

$$z_a = \exp(-:f:)\ \bar{z}_a \ . \qquad (5.10)$$

What has been shown is that if $\mathcal{M}$ is given by the Lie transformation relation (5.3) then $\mathcal{M}^{-1}$ is given by the relation

$$\mathcal{M}^{-1} = \exp(-:f:). \qquad (5.11)$$

<u>Exercise 5.2</u>: Suppose f and g are in <u>involution</u>. That is,

$$[f,g] = 0. \qquad (5.12)$$

Show from the power series definition (3.32) that in this case

$$\exp(:f:)\ \exp(:g:) = \exp(:f+g:)\ . \qquad (5.13)$$

In attempting to evaluate (5.10) explicitly, the reader may be at somewhat of a loss as to exactly how to proceed since what appears to be called for is a set of Poisson brackets with respect to the variables z. However, in view of the invariance property (5.5), one may equally well compute all Poisson brackets with respect to the variables $\bar{z}$.

5.2 Factorization theorem

Note what has been accomplished so far. Section 2.4 showed that matrices of the form JS with S symmetric produce a Lie algebra. It also showed that any symplectic matrix sufficiently near the identity can be written in the form exp(JS). Similarly, section 3.3 showed that the set of Lie operators :f: forms a Lie algebra. And section 5.1 has just shown that Lie transformations exp(:f:) are symplectic maps. What remains to be studied is the question of whether any symplectic map $\mathcal{M}$ can be written in exponential form.

The answer to this question is given by the factorization theorem:

<u>Theorem 5.1</u>: Let $\mathcal{M}$ be an analytic symplectic mapping which maps the origin into itself. That is, the relation between the $\bar{z}$'s and z's is assumed to be expressible in a power series of the form

$$\bar{z}_a = \sum_b L_{ab}\, z_b + \text{higher order terms in } z. \qquad (5.14)$$

Under these conditions the map $\mathcal{M}$ can be written as a product of Lie transformations,

$$\mathcal{M} = \exp(:f_2^c:)\ \exp(:f_2^a:)\ \exp(:f_3:)\ \exp(:f_4:)\ \ldots \qquad (5.15)$$

Moreover, the functions f_m are homogeneous polynomials of degree m in the variable $z_1 \ldots z_{2n}$. Finally, the polynomials f_2^c and f_2^a are quadratic polynomials of the form

$$f_2^c = -(1/2) \sum_{ij} S_{ij}^c z_i z_j \tag{5.16a}$$

$$f_2^a = -(1/2) \sum_{ij} S_{ij}^a z_i z_j \tag{5.16b}$$

where S^c and S^a are real symmetric matrices which commute and anti-commute with J respectively.

Partial Proof: Let M(z) be the Jacobian matrix of the map $\mathcal{M}$. Then from Eq. (5.14), it is evident that

$$M(0) = L. \tag{5.17}$$

Also, since M is symplectic for all values of z, see Eq. (4.3), it follows that L is a symplectic matrix. Consequently, using the results summarized in Eqs. (2.37) and (2.58), the matrix L can be written in the form

$$L = P\mathcal{O} = \exp(JS^a)\exp(JS^c). \tag{5.18}$$

As indicated in Eq. (5.16a), let f_2^c be a quadratic polynomial defined in terms of the matrix S^c appearing in the decomposition (5.18). Now consider the Lie operator $:f_2^c:$. Suppose this Lie operator acts on the various z's. One finds the result

$$\begin{aligned} :f_2^c: z_k &= -(1/2) \sum_{ij} S_{ij}^c [z_i z_j, z_k] \\ &= -(1/2) \sum_{ij} S_{ij}^c \{[z_i, z_k] z_j + [z_j, z_k] z_i\} \\ &= -(1/2) \sum_{ij} S_{ij}^c \{J_{ik} z_j + J_{jk} z_i\} \\ &= \sum_i (JS^c)_{ki} z_i . \end{aligned} \tag{5.19}$$

Here use has been made of the antisymmetry of J and the symmetry of S^c. Using matrix and vector notation, Eq. (5.19) can also be written in the more compact form

$$:f_2^c: z = (JS^c) z. \tag{5.20}$$

From this form it is easy to see that one has the general relation

$$:f_2^c:^m \; z = (JS^c)^m \; z. \tag{5.21}$$

Finally, it follows from (5.21) that one also has the relation

$$\exp(:f_2^c:) \; z = \exp(JS^c) \; z = \mathcal{O}z. \tag{5.22}$$

Exercise 5.3: Verify Eqs. (5.19) through (5.22).

Exercise 5.4: Verify the analogous relation

$$\exp(:f_2^a:) \; z = \exp(JS^a) \; z = Pz. \tag{5.23}$$

By using matrix and vector notation, and the representation (5.18), Eq. (5.14) can be written in the more compact form

$$\bar{z} = P \, \mathcal{O} \, z + \text{higher order terms in } z. \tag{5.24}$$

Now apply the Lie transformation $\exp(-:f_2^c:)$ to both sides of (5.24). One finds the result

$$\exp(-:f_2^c:) \; \bar{z} = P \, \mathcal{O} \, \exp(-:f_2^c:) \; z + \text{higher order terms in } z. \tag{5.25}$$

However, it follows from (5.22) that one also has the relation

$$\exp(-:f_2^c:) \; z = \exp(-JS^c) \; z = \mathcal{O}^{-1} \; z. \tag{5.26}$$

Consequently, Eq. (5.25) can be rewritten in the form

$$\exp(-:f_2^c:) \; \bar{z} = P \; z + \text{higher order terms in } z. \tag{5.27}$$

Next apply the Lie transformation $\exp(-:f_2^a:)$ to both sides of (5.27). By making arguments analogous to those of the previous paragraph, one finds the result

$$\exp(-:f_2^a:) \; \exp(-:f_2^c:) \; \bar{z} = z + \text{higher order terms in } z. \tag{5.28}$$

Exercise 5.5: Verify Eqs. (5.24) through (5.28).

Suppose f_3 is some cubic polynomial in the z's. Then one finds the result

$$\exp(-:f_3:) \; z_b = z_b + \underbrace{:-f_3: \; z_b}_{\substack{\text{quadratic}\\ \text{terms}}} + \underbrace{(1/2!):-f_3:^2 \; z_b}_{\substack{\text{cubic}\\ \text{terms}}} \; \ldots \, , \tag{5.29}$$

where the degrees of the various terms have been indicated. Consider now the effect of multiplying both sides of (5.28) by the Lie transformation $\exp(-:f_3:)$. Using (5.29), one finds the result

$$\exp(-:f_3:)\ \exp(-:f_2^a:)\ \exp(-:f_2^c:)\ \bar{z} = z$$
$$- :f_3:\ z + \text{higher order terms of } z. \qquad (5.30)$$

At this point, one might hope that by a suitable choice of f_3, all quadratic terms from the right-hand side of (5.30) could be eliminated. This can indeed be shown to be the case.[12] Consequently, there exists an f_3 (which is in fact unique) such that (5.30) takes the form

$$\exp(-:f_3:)\ \exp(-:f_2^a:)\ \exp(-:f_2^c:)\ \bar{z} = z$$
$$+ \text{ cubic and higher order terms in } z. \qquad (5.31)$$

Similarly, there exist unique higher order polynomials f_4, f_5, etc. which can be used successively to remove degree by degree each of the higher order terms on the right-hand side of (5.31). Consequently, one has the final relation

$$\exp(-:f_4:)\ \exp(-:f_3:)\ \exp(-:f_2^a:)\ \exp(-:f_2^c:)\ \bar{z} = z. \qquad (5.32)$$

Now multiply both sides of (5.32) by the $\mathcal{M}$ of Eq. (5.15). One finds the relation

$$\bar{z} = \mathcal{M}z \qquad (5.33)$$

which is the desired result.

Exercise 5.6: Suppose f_m and g_n are homogeneous polynomials of degrees m and n respectively. Show that $[f_m,g_n]$ is then also a homogeneous polynomial, and determine its degree.
Answer: Degree of $[f_m,g_n] = (m+n-2)$.

Exercise 5.7: Verify Eq. (5.29) and the assertion about the degree of the various terms. Verify the derivation of (5.33) from (5.32).

Theorem (5.1) is a key result. Recall that in theorem (4.1) it was shown that Hamiltonian flows produce symplectic maps. Now, thanks to theorem (5.1), it is possible to describe the most general analytic symplectic map (which sends the origin into itself) simply in terms of various homogeneous polynomials. Finally, it can be shown that the restriction of preserving the origin can be removed by including Lie transformations of the form $\exp(:f_1:)$ where f_1 is a suitably chosen polynomial linear in the z's. Consequently, any analytic symplectic map can be represented as a product of Lie transformations generated by polynomials.

At this point, two comments are appropriate. First, suppose the factored product representation (5.15) is truncated at any point. Then the resulting expression is still a symplectic map because each term in the product is a symplectic map. Also, if the truncation consists of dropping all terms in the product (5.15) beyond $\exp(:f_m:)$ for some m, then according to exercise 5.6 the power series expansion (5.14) for the truncated map agrees with that of the original map through terms of degree (m-1). Consequently, a truncated map provides a symplectic approximation to the exact map.

Second, suppose Eq. (5.15) is decomposed, as shown below, into those factors involving only quadratic polynomials, and the remaining factors involving cubic and higher degree polynomials.

$$\mathcal{M} = \overbrace{\exp(:f_2^c:)\exp(:f_2^a:)}^{\text{"Gaussian optics"}} \quad \overbrace{\underbrace{\exp(:f_3:)}_{\substack{\text{second order}\\ \text{effects or}\\ \text{aberrations}\\ \text{due to}\\ \text{sextupoles}}}\underbrace{\exp(:f_4:)\ \ldots}_{\substack{\text{third order effects}\\ \text{or aberrations due to}\\ \text{octupoles or iterated}\\ \text{sextupoles}}}}^{\text{Aberrations or nonlinear corrections}}$$

It will be demonstrated in subsequent sections that dropping all terms beyond those involving the quadratic polynomials leads to a lowest order approximation for $\mathcal{M}$ which is equivalent to the paraxial Gaussian optics approximation in the case of light optics, and the usual linear matrix approximation in the case of charged particle beam optics. Moreover, the remaining factors $\exp(:f_3:)$ $\exp(:f_4:)\ldots$represent aberrations or nonlinear corrections to the lowest order approximation. In particular, in the case of charged particle beam optics, the factor $\exp(:f_3:)$ describes various chromatic effects and the effects due to sextupoles. Similarly, the factor $\exp(:f_4:)$ describes higher order chromatic effects, the effects due to iterated sextupoles, and the effects due to octupoles. Finally in some cases, f_3, f_4, etc. also describe what may be called "kinematic" nonlinearities in the equations of motion. They arise, for example, from the fact that the Hamiltonians (1.22) and (1.23) are intrinsically nonlinear even in the absence of electric and magnetic fields.

5.3 Symplectic map for time independent Hamiltonian

Suppose the Hamiltonian of theorem (4.1) does not explicitly depend on the time. Then the symplectic map (4.17) obtained by following the Hamiltonian flow specified by H can be written immediately in the form

$$\mathcal{M} = \exp\{-(t^f-t^o)\ :H:\}. \qquad (5.34)$$

To see the truth of (5.34), let $\mathcal{M}$ act on z to give the result

$$\bar{z} = z = \sum_{m=0}^{\infty} (1/m!)\ (t^f-t^o)^m :-H:^m z. \qquad (5.35)$$

However, Taylor's theorem gives the result

$$\bar{z} = z(t^f) = z(t^o) + \sum_{m=1}^{\infty} (1/m!)(t^f - t^o)^m \ (d/dt)^m z(t)\Big|_{t^o} \ . \quad (5.36)$$

Also, Hamilton's equations of motion for the z's can be written in the form

$$\dot{z} = [z,H] = [-H,z] = :-H:z$$

$$\ddot{z} = [-H,\dot{z}] = :-H:\dot{z} = :-H:^2 z$$

$$(d^3z)/(dt)^3 = :-H:^3 z, \text{ etc.} \quad (5.37)$$

Upon inserting the results of (5.37) into (5.36), one obtains the desired result (5.35).

Exercise 5.8: Evaluate $\mathcal{M}z$ with $\mathcal{M}$ given by (5.34) for the H of Eq. (4.29). See exercise (3.23).

Exercise 5.9: Suppose that the Hamiltonian H is not necessarily time independent, but does have the property that the Lie operators :H(z,t): for various times all commute. That is, one has the relation

$$\{:H(z,t):, :H(z,t'):\} = 0 \text{ for all } t,t'. \quad (5.38)$$

Show that in this case the symplectic map obtained by following the Hamiltonian flow specified by H can be written in the form

$$\mathcal{M} = \exp(-\int_{t^o}^{t^f} :H: dt). \quad (5.39)$$

5.4 A calculus for Lie transformations and noncommuting operators

A. Introductory Background

Section 4.3 showed that Hamiltonian flows produce symplectic maps, and section 5.2 showed that symplectic maps which send the origin into itself can be written in the factored product form (5.15). In addition, Eq. (5.34) gives an explicit representation of the symplectic map in the case of a time independent Hamiltonian. In subsequent sections these results will be applied to charged particle beam transport, light optics, and orbits in circular machines. The purpose of this section is to provide a collection of formulas for the manipulation of Lie transformations and noncommuting operators in general. These formulas have been derived elsewhere, and some will be stated without proof.[12] Some formulas will be used to compute the product of two symplectic maps when each is written in factored product form. Others will be used to combine various exponents in a factored product decomposition into a single exponent. Still others will be used to produce factored product decompositions.

Work with noncommuting quantities is often facilitated by the concept of an adjoint Lie operator. Let :f: be some Lie operator, and let :g: be any other Lie operator. The adjoint of the Lie operator :f:, which will be denoted by the symbol #:f:#, is a kind of super operator which acts on other Lie operators according to the rule[13]

$$\#{:}f{:}\#{:}g{:} = \{{:}f{:},{:}g{:}\}. \tag{5.40}$$

Here, the right-hand side of (5.40) denotes the commutator as in Eq. (3.27). Powers of #:f:# can also be defined by repeated applications of (5.40). For example, $\#{:}f{:}\#^2$ is defined by the relation

$$\#{:}f{:}\#^2 \;{:}g{:} = \{{:}f{:},\{{:}f{:},{:}g{:}\}\}. \tag{5.41}$$

Finally, #:f:# to the zero power is defined to be the identity operator,

$$\#{:}f{:}\#^0 \;{:}g{:} = {:}g{:}\ . \tag{5.42}$$

Exercise 5.10: As usual, let the symbols { , } denote the operation of commutation,

$$\{\#{:}f{:}\#,\#{:}g{:}\#\} = \#{:}f{:}\#\#{:}g{:}\# - \#{:}g{:}\#\#{:}f{:}\#. \tag{5.43}$$

As a test of your grasp of what is going on, prove the relations

$$\{\#{:}f{:}\#,\#{:}g{:}\#\} = \#\{{:}f{:},{:}g{:}\}\# = \#{:}[f,g\,]{:}\#. \tag{5.44}$$

Hint: Use the Jacobi condition for commutators.

To simplify notation in some cases where no confusion can arise, the set of colons in the symbols #:f:# for the adjoint of the Lie operator :f: will be omitted. That is, the abbreviated symbol #f# will be used to serve for the complete symbol #:f:#.

Exercise 5.11: Prove the relation

$$\#f\#^n{:}g{:} = {:}({:}f{:}^n g){:}\ . \tag{5.45}$$

Exercise 5.12: Prove the relation

$$\#f\#({:}g{:}{:}h{:}) = (\#f\#{:}g{:}){:}h{:} + {:}g{:}(\#f\#{:}h{:})\ . \tag{5.46}$$

That is, #f# is also a derivation with respect to the multiplication of Lie operators.

Armed with this new notation, it is now possible to state some useful results. Suppose :f: and :g: are two Lie operators. Then, it can be shown that

$$\exp(:f:)\ :g:\ \exp(-:f:) = \exp(\#f\#)\ :g: \quad . \tag{5.47}$$

Here, as the notation suggests,

$$\exp(\#f\#) = \sum_{m=0}^{\infty} \#f\#^m/m! \quad . \tag{5.48}$$

<u>Exercise 5.13</u>: Verify (5.47) term by term for at least the first few terms by comparing power series expansions.

Now, using (5.47), it follows that

$$\exp(\#f\#)\ :g: = :\exp(:f:)g: \quad . \tag{5.49}$$

Consequently, Eq. (5.47) can also be written in the form

$$\exp(:f:)\ :g:\ \exp(-:f:) = :\exp(:f:)g: \quad . \tag{5.50}$$

Next observe that because #f# is a derivation, see Eq. (5.46), it follows from (5.50) that a similar relation holds for powers of :g:,

$$\exp(:f:):g:^m\ \exp(-:f:) = :\exp(:f:)g:^m \quad . \tag{5.51}$$

Moreover, Eq. (5.51) in turn implies the relation

$$\exp(:f:)\ \exp(:g:)\exp(-:f:) = \exp(:\exp(:f:)g:). \tag{5.52}$$

<u>Exercise 5.14</u>: Verify Eqs. (5.49) through (5.52) starting with Eq. (5.47).

A second useful result is the analog of formulas (2.34) and (2.35) for Lie operators. Suppose :f: and :g: are any two Lie operators. Then one has the Campbell-Baker-Hausdorff formula

$$\exp(s:f:)\ \exp(t:g:) = \exp(s:f: + t:g:$$
$$+ (st/2)\{:f:,:g:\} + (s^2t/12)\{:f:,\{:f:,:g:\}\}$$
$$+ (st^2/12)\{:g:,\{:g:,:f:\}\} + \ldots). \tag{5.53}$$

Moreover, using Eqs. (3.31) and (5.45), Eq. (5.53) can also be written in the form

$$\exp(s:f:)\ \exp(t:g:) = \exp(:h:) \tag{5.54a}$$

with

$$h = sf + tg + (st/2)[f,g]$$
$$+ (s^2t/12):f:^2g + (st^2/12):g:^2f + \ldots \quad . \tag{5.54b}$$

B. Concatenation Formulas

As an application of the formulas developed so far, consider the problem of computing the product of two symplectic maps when each is expressed in factored product form. This problem arises, for example, in the case that one knows the effect of two beam elements separately, and one wants to know the net effect when one beam element is followed by another.

Let $\mathcal{M}^f$ and $\mathcal{M}^g$ denote the symplectic maps given by the expressions

$$\mathcal{M}^f = \exp(:f_2:)\ \exp(:f_3:)\ \exp(:f_4:), \tag{5.55a}$$

$$\mathcal{M}^g = \exp(:g_2:)\ \exp(:g_3:)\ \exp(:g_4:). \tag{5.55b}$$

Also, let $\mathcal{M}^h$ be the product of $\mathcal{M}^f$ and $\mathcal{M}^g$,

$$\mathcal{M}^h = \mathcal{M}^f\ \mathcal{M}^g. \tag{5.56}$$

The problem is to find polynomials h_2, h_3, etc., such that

$$\mathcal{M}^h = \exp(:h_2:)\ \exp(:h_3:)\ \exp(:h_4:)\ \dots\ . \tag{5.57}$$

That is, the problem is to express $\mathcal{M}^h$ as given by (5.56) in the factored product form (5.57). For simplicity, only expressions for h_2, h_3, and h_4 will be found explicitly.

Before proceeding further, it is necessary to establish a few simple facts.

Exercise 5.15: Suppose g_2 is a quadratic polynomial written in the form

$$g_2 = -(1/2) \sum_{ij} S_{ij}\ z_i z_j\ , \tag{5.58}$$

where S is some symmetric matrix. Suppose further that f_m is some homogeneous polynomial of degree m. Show that then $\exp(:g_2:)\ f_m$ is also a homogeneous polynomial of degree m.

Solution:

$$\exp(:g_2:)\ f_m(z) = f_m(M^g\ z) \tag{5.59}$$

where M^g is the linear transformation defined by the equation

$$M^g = \exp(JS). \tag{5.60}$$

See Eqs. (3.35), (5.16), and (5.22).

Simply from its definition, $\mathcal{M}^h$ can be written in the form

$$\mathcal{M}^h = \exp(:f_2:)\exp(:f_3:)\exp(:f_4:)\times$$
$$\exp(:g_2:)\exp(:g_3:)\exp(:g_4:). \qquad (5.61)$$

Next, by suitable insertions of various Lie transformations and their inverses, Eq. (5.61) can be rewritten in the form

$$\mathcal{M}^h = \exp(:f_2:)\exp(:g_2:)\exp(-:g_2:)\exp(:f_3:)\exp(:g_2:)\times$$
$$\exp(-:g_2:)\exp(:f_4:)\exp(:g_2:)\exp(:g_3:)\exp(:g_4:). \qquad (5.62)$$

<u>Exercise 5.16</u>: Verify Eq. (5.62) starting from (5.61).

Evidently, comparison of (5.62) and (5.57) shows that h_2 is determined by the equation

$$\exp(:h_2:) = \exp(:f_2:)\exp(:g_2:) \quad . \qquad (5.63)$$

Indeed, using the notation of Eq. (5.61) of exercise (5.15), one has the result[14]

$$M^h = M^g M^f. \qquad (5.64)$$

<u>Exercise 5.17</u>: Verify Eq. (5.64).

Next observe that Eq. (5.63) contains the factors $\exp(-:g_2:)$ $\exp(:f_3:)\exp(:g_2:)$. Here is where Eq. (5.52) comes into play. Employing Eq. (5.52), one finds the result

$$\exp(-:g_2:)\exp(:f_3:)\exp(:g_2:) = \exp(:\exp(-:g_2:)f_3:). \qquad (5.65)$$

Also, according to exercise (5.15), one has the relation

$$\exp(-:g_2:)\, f_3(z) = f_3\,[\,(M^g)^{-1}\, z\,] \qquad (5.66)$$

where $(M^g)^{-1}$ is the matrix

$$(M^g)^{-1} = \exp(-JS) = -J\,\tilde{M}^g\,J. \qquad (5.67)$$

Here use has also been made of (2.8).

<u>Exercise 5.18</u>: Verify Eqs. (5.65) through (5.67).

In order to simplify further expressions, introduce the notation

$$f_m^t(z) = f_m[(M^g)^{-1} z] \qquad (5.68)$$

which indicates that the homogeneous polynomial $f_m(z)$ of degree m has been <u>transformed</u> to the new homogeneous polynomial $f_m[(M^g)^{-1} z]$. With this notation, Eq. (5.65) can be written in the more compact form

$$\exp(-:g_2:)\ \exp(:f_3:)\ \exp(:g_2:) = \exp(:f_3^t:) \quad . \qquad (5.69)$$

Similarly, one finds for the factor $\exp(-:g_2:)\ \exp(:f_4:)\ \exp(:g_2:)$, which also occurs in (5.62), the result

$$\exp(-:g_2:)\ \exp(:f_4:)\ \exp(:g_2:) = \exp(:f_4^t:). \qquad (5.70)$$

<u>Exercise 5.19</u>: Verify Eq. (5.70).

Putting together the work done so far, one finds that Eq. (5.62) can also be written in the form

$$\mathcal{M}^h = \exp(:h_2:)\ \exp(:f_3^t:)\ \exp(:f_4^t:)\ \exp(:g_3:)\ \exp(:g_4:). \qquad (5.71)$$

Again, by a suitable insertion of a Lie transformation and its inverse, Eq. (5.71) can be written in the form

$$\mathcal{M}^h = \exp(:h_2:)\ \exp(:f_3^t:)\ \exp(:g_3:)\ \exp(-:g_3:)\times$$
$$\exp(:f_4^t:)\ \exp(:g_3:)\ \exp(:g_4:). \qquad (5.72)$$

Now consider the factor $\exp(-:g_3:)\ \exp(:f_4^t:)\ \exp(:g_3:)$. Again using Eq. (5.52), this factor can be written in the form

$$\exp(-:g_3:)\ \exp(:f_4^t:)\ \exp(:g_3:) = \exp(:\exp(-:g_3:)f_4^t:). \qquad (5.73)$$

However, using the results of exercise (5.6), one finds the relation

$$\exp(-:g_3:)\ f_4^t = f_4^t + \text{polynomials of degree 5 and greater.} \qquad (5.74)$$

Therefore, if one is not interested in computing the h_m of degree 5 and higher, Eq. (5.72) can also be written in the form

$$\mathcal{M}^h = \exp(:h_2:)\ \exp(:f_3^t:)\ \exp(:g_3:)\ \exp(:f_4^t:)\ \exp(:g_4):\ \cdots\ . \qquad (5.75)$$

At this point Eqs. (5.53) and (5.54) can be brought into play. Using these equations, one finds for the product $\exp(:f_3^t:)\ \exp(:g_3:)$ occurring in (5.75) the result

$$\exp(:f_3^t:)\ \exp(:g_3:) = \exp(:f_3^t: + g_3: + (1/2)\ :[f_3^t,g_3]: + \cdots). \tag{5.76}$$

Note that $[f_3^t,g_3]$ is a homogeneous polynomial of degree 4, and that the remaining terms not shown on the right-hand side of (5.76) are polynomials of still higher degree. Consequently, Eq. (5.76) can also be written in the form

$$\exp(:f_3^t:)\ \exp(:g_3:) = \exp(:f_3^t: + :g_3:)\ \exp(:[f_3^t,g_3]/2: \cdots). \tag{5.77}$$

Finally, again using (5.53) and (5.54), it is clear that Eq. (5.72) for $\mathcal{M}^h$ can be written in the form

$$\mathcal{M}^h = \exp(:h_2:)\ \exp(:f_3^t: + :g_3:)\ \exp(:[f_3^t,g_3]/2: + :f_4^t: + :g_4:)\ \ldots . \tag{5.78}$$

Comparison of Eqs. (5.57) and (5.78) shows that the polynomials h_3 and h_4 are given by the expressions

$$h_3 = f_3^t + g_3 \tag{5.79}$$

$$h_4 = [f_3^t,g_3]/2 + f_4^t + g_4. \tag{5.80}$$

With further work, it is possible to find the polynomials h_5, h_6, etc.

C. Formulas for Combining Exponents

Sometimes, as will be shown later, it is useful to be able to write the product of two Lie transformations as a single Lie transformation. This is what the Campbell-Baker-Hausdorff formula (5.53) attempts to do. In general, there are no known convenient expressions for all the terms on the right-hand side of (5.54b). However, it is possible to sum the series completely with respect to s and the first few powers in t.[12] One such result can be written in the form

$$h = sf + s:f:[1 - \exp(-s:f:)]^{-1}\ (tg) + O(t^2). \tag{5.81}$$

Here the operator expression involving :f: is to be interpreted as the infinite series

$$\begin{aligned} s:f:[1 - \exp(-s:f:)]^{-1} &= s:f:[1 - \sum_{m=0}^{\infty} (-s:f:)^m/m!]^{-1} \\ &= s:f:[- \sum_{m=1}^{\infty} (-s:f:)^m/m!]^{-1} \qquad (5.82) \\ &= 1 + (s/2):f: + (s^2/12):f:^2 + \cdots \end{aligned}$$

Equations (5.54a) and (5.81) may be combined to give the result

$$\exp(s{:}f{:})\exp(t{:}g{:})=\exp[s{:}f{:}+{:}\{s{:}f{:}[1-\exp(-s{:}f{:})]^{-1}(tg)\}{:}+O(t^2)]. \tag{5.83}$$

D. Factorization Formulas

Equation (5.83) gives a formula for combining two exponentials into one grand exponential. Sometimes, as in, for example the construction of a factored product decomposition, it is useful to be able to turn the process around. Define a quantity h by writing

$$s{:}f{:}[1 - \exp(-s{:}f{:})]^{-1}\, g = h. \tag{5.84}$$

Observe that Eq. (5.84) may be solved for the quantity g to give the relation

$$g = \{[1 - \exp(-s{:}f{:})]/[s{:}f{:}]\}\, h. \tag{5.85}$$

Here the operator expression appearing on the right of (5.85) is interpreted to be the series

$$[1 - \exp(-s{:}f{:})]/[s{:}f{:}] = -\sum_{m=1}^{\infty} (-s)^m{:}f{:}^m/[m!\,s{:}f{:}]$$

$$= -\sum_{m=1}^{\infty} (-s)^{m-1}{:}f{:}^{m-1}/m!. \tag{5.86}$$

Now insert Eq. (5.84) into Eq. (5.83). One finds, upon reading right to left, the result

$$\exp[s{:}f{:} + t{:}h{:} + O(t^2)] = \exp(s{:}f{:})\exp(t{:}g{:}). \tag{5.87}$$

Finally, the term of $O(t^2)$ can be taken from the left to the right-hand side of (5.87) to produce the relation

$$\exp[s{:}f{:} + t{:}h{:}] = \exp(s{:}f{:})\exp(t{:}g{:})\exp[{:}O(t^2){:}]. \tag{5.88}$$

Equation (5.88) gives a formula for writing the sum of two exponentials as a product of two exponentials.

It is worth remarking, in closing, that the operation described by Eq. (5.85), which is required for evaluating (5.88), can be written in a more compact form. First, observe the formal integral identity

$$[1 - \exp(-s{:}f{:})]/[s{:}f{:}] = \int_0^1 d\tau\, \exp(-\tau s{:}f{:}). \tag{5.89}$$

Using this identity, Eq. (5.85) can be written in the form

$$g = \int_0^1 d\tau \exp(-\tau s{:}f{:})\, h. \qquad (5.90)$$

But now Eq. (3.35) can be employed to give the final integral formula

$$g(2) = \int_0^1 d\tau \;\; h \exp(-\tau s{:}f{:}) z \, . \qquad (5.91)$$

Exercise 5.20: Verify the expansions (5.82) and (5.86).

Exercise 5.21: Verify the integral identity (5.89).

6. APPLICATIONS TO CHARGED PARTICLE BEAM TRANSPORT AND LIGHT OPTICS

Section 4.3 showed that Hamiltonian flows produce symplectic maps, and section 5.2 showed that symplectic maps which send the origin into itself can be written in the form

$$\mathcal{M} = \exp(:f_2^c:) \exp(:f_2^a:) \exp(:f_3:) \cdots . \qquad (6.1)$$

Thus, the result of a Hamiltonian flow which sends the origin into itself can be completely described by a set of polynomials f_2, f_3, etc. The purpose of this section is to give examples of the computation of the first few polynomials for quadrupole, sextupole, dipole, and drift beam elements. These calculations serve as the beginning of a complete catalog of Lie operators for the commonly used beam elements. Further results may be found elsewhere, and additional work is still in progress.[15] A brief discussion will also be given of Lie operators corresponding to various simple optical elements in the case of light optics.

6.1 The perfect quadrupole

Consider the case of a perfect magnetic quadrupole of length ℓ. By "perfect" it is meant that the magnetic field in the body of the magnet is assumed to have only a quadrupole component, and end effects and fringe fields are neglected. Then, employing the coordinates shown in Fig. (1.1) and utilizing the results of exercises 1.5 and 1.10, one finds for the Hamiltonian K the result

$$K = -[(p_t^2/c^2) - m_o^2c^2 - p_y^2 - p_z^2]^{1/2} - (qa_2/2)(z^2 - y^2). \qquad (6.2)$$

The design orbit for a quadrupole passes through the center of the quadrupole, and has a certain design energy. Consequently, the design orbit can be characterized by writing the equations

$$p_y = 0 \quad , \; y = 0 \tag{6.3a,b}$$

$$p_z = 0 \quad , \; z = 0 \tag{6.4a,b}$$

$$p_t = p_t^o \quad , \; t(x) = x(v_x^o)^{-1} + \text{constant.} \tag{6.5a,b}$$

The meaning of Eqs. (6.3) and (6.4) should be evident. Equations (6.5) may require some explanation. Applying Hamilton's equations, one has the relations

$$p_t' = dp_t/dx = -\partial K/\partial t = 0, \tag{6.6a}$$

$$t' = dt/dx = \partial K/\partial p_t = -(p_t/c^2)[(p_t^2/c^2) - m_o^2c^2 - p_y^2 - p_z^2]^{-1/2}. \tag{6.6b}$$

It follows from Eq. (6.6a) that p_t has a constant value. Hence, Eq. (6.5a) is correct. Inserting this constant value p_t^o (corresponding to a particular value for the design energy) and Eqs. (6.3) and (6.4) into Eq. (6.6b) gives the result that on the design orbit

$$t' = dt/dx = (v_x^o)^{-1}, \tag{6.7a}$$

where v_x^o is the velocity of the design orbit given by the relation

$$v_x^o = [(p_t^o/c)^2 - m_o^2c^2]^{1/2}[-p_t^o/c^2]^{-1} \quad . \tag{6.7b}$$

Integration of (6.7a) gives the advertized result (6.5b).

According to Eq. (6.5), following the Hamiltonian flow generated by K along the design orbit does not lead to a mapping of the origin of phase space into itself since the time along the orbit changes, and p_t usually differs from zero. This situation can be remedied by working with a different set of canonical variables. Define "new" variables T, Y, Z, P_T, P_Y, P_Z by the relations

$$t = T + x(v_x^o)^{-1} \tag{6.8a}$$

$$y = Y \tag{6.8b}$$

$$z = Z \tag{6.8c}$$

$$p_t = p_t^o + P_T \tag{6.8d}$$

$$p_y = P_Y \tag{6.8e}$$

$$p_z = P_Z . \tag{6.8f}$$

It is readily verified that this change of variables is a canonical transformation and arises from the transformation function

$$F_2(q,P) = yP_Y + zP_Z + [t - x(v_x^o)^{-1}][P_T + p_t^o], \tag{6.9}$$

where v_x^o and p_t^o are to be regarded as constant numbers. For example, one finds the relations[16]

$$T = \partial F_2/\partial P_T = t - x(v_x^o)^{-1} \tag{6.10a}$$

$$p_t = \partial F_2/\partial t = P_T + p_t^o \tag{6.10b}$$

which are in agreement with Eqs. (6.8a) and (6.8d).

In terms of these new variables, it is evident that the design orbit can be taken to be given by the equations

$$T = X = Y = P_T = P_X = P_Y = 0. \tag{6.11}$$

Note that the variable T measures deviations in transit time from that of the design orbit, and its conjugate momentum P_T measures deviations in p_t from the design orbit value p_t^o.

Let K be regarded as the "old" Hamiltonian, and let H denote the "new" Hamiltonian for the new variables. [Note that in this context H is not to be confused with the Hamiltonians of exercises (1.3) or (1.4).] Then one has the relation[16]

$$H = K + \partial F_2/\partial x \tag{6.12}$$

since in the current procedure x plays the role of the independent "time-like" variable. Carrying out the prescription (6.12), one finds the result

$$H = -[(p_t^o + P_T)^2/c^2 - m_o^2c^2 - p_y^2 - p_z^2]^{1/2} - (qa_2/2)(z^2 - y^2)$$
$$-(v_x^o)^{-1}(P_T + p_t^o). \tag{6.13}$$

Here the quantities Y, Z, P_Y, P_Z have been replaced by their lower case counterparts in view of Eqs. (6.8b), (6.8c), (6.8e), and (6.8f) in order to simplify notation. From Eqs. (6.11) it follows that the mapping $\mathcal{M}$ produced by following the Hamiltonian flow generated by H does indeed send the origin of phase space into itself. Moreover, according to section (5.3), the mapping $\mathcal{M}$ is given explicitly by the expression

$$\mathcal{M} = \exp(-\ell :H:), \tag{6.14}$$

where, as assumed earlier, ℓ is the length of the quadrupole.

The problem at this point is to take the $\mathcal{M}$ given by (6.14) and re-express it in the factored product form (6.1). The first step in solving this problem is to expand H in a power series about the design orbit to give an expression of the form

$$H = H_0 + H_1 + H_2 + H_3 + \cdots \tag{6.15}$$

where H_n denotes a homogeneous polynomial of order n. One finds for the first few polynomials the results

$$H_0 = m_0^2 c^2/p^o \tag{6.16a}$$

$$H_1 = 0 \tag{6.16b}$$

$$H_2 = (P_T^2/2)(m_0^2/p^{o3}) + (p_y^2 + p_z^2)/(2p^o) - (qa_2/2)(z^2 - y^2) \tag{6.16c}$$

$$H_3 = [P_T^3/2]\,[(m_0^2)/(p^{o4}v_x^o)] + [P_T(p_y^2 + p_z^2)/(2p^{o2}v_x^o)] \;. \tag{6.16d}$$

Here p^o denotes the magnitude of the design mechanical momentum, and is related to p_t^o by the equation

$$p_t^o = -[m_0^2c^4 + c^2p^{o2}]^{1/2} \;. \tag{6.17}$$

See exercise (1.13).

<u>Exercise 6.1</u>: Verify Eqs. (6.16). Note that the vanishing of H_1 is equivalent to the condition that $\mathcal{M}$ send the origin of phase space into itself.

Next insert the expansion (6.15) into the expression (6.14) for $\mathcal{M}$ and imagine that the result is written in factored product form. Doing so gives the relation

$$\begin{aligned} \mathcal{M} &= \exp\{-\ell(:H_2: + :H_3: + \cdots)\} \\ &= \exp(:f_2:)\exp(:f_3:) \cdots \;. \end{aligned} \tag{6.18}$$

Note that as expected the constant part of the Hamiltonian plays no role since $:H_0: = 0$.

The calculation of f_2 and f_3 can now be carried out with the aid of Eqs. (5.88) and (5.90). One finds the results

$$f_2 = -\ell H_2 \tag{6.19}$$

$$f_3 = -\ell\int_0^1 d\tau \exp(\tau\ell :H_2:)H_3 \;. \tag{6.20}$$

Note that Eq. (5.88) involves uncalculated terms of order t^2 where t is an expansion parameter. In the present context, these terms are those in the expansion which involve multiple Poisson brackets containing any number of H_2's and two H_3's. However, according to exercise (5.6), these terms are all polynomials of order 4. Therefore, they contribute to f_4, but not to f_2 or f_3. Similarly, the terms H_4 and higher in H do not contribute to f_2 and f_3.

Exercise 6.2: Verify Eqs. (6.19) and (6.20) and the assertions just made.

To evaluate f_3 as given by Eq. (6.20), and to gain additional insight, it is useful to study in detail the first factor in M given by $\exp(:f_2:) = \exp(-\ell:H_2:)$. Observe that H_2 can be written as the sum of three terms which are mutually in involution (have vanishing Poisson brackets),

$$H_2 = H_2^T + H_2^y + H_2^z \tag{6.21}$$

where

$$H_2^T = (P_T^2/2)(m_0^2/p^{o^3}) \tag{6.22a}$$

$$H_2^y = p_y^2/(2p^o) + (qa_2/2)y^2 \tag{6.22b}$$

$$H_2^z = p_z^2/(2p^o) - (qa_2/2)z^2 \quad . \tag{6.22c}$$

Consequently, using exercise 5.2, the quantity of interest can also be written in the form

$$\exp(-\ell:H_2:) = \exp(-\ell:H_2^T:)\exp(-\ell:H_2^y:)\exp(-\ell:H_2^z:). \tag{6.23}$$

Each of the three factors appearing in (6.23) can now be evaluated individually.

Consider first the factor $\exp(-\ell:H_2^T:)$. Following exercise (3.19), it is easily verified that $\exp(-\ell:H_2^T:)$ leaves the variables y, p_y, z, p_z in peace, and transforms the variables P_T and T according to the rules

$$\exp(-\ell:H_2^T:)T = T + (\ell m_0^2/p^{o^3})P_T \tag{6.24a}$$

$$\exp(-\ell:H_2^T:)P_T = P_T \quad . \tag{6.24b}$$

The above relations can be written more compactly in the form

$$\exp(-\ell:H_2^T:)\begin{pmatrix} T \\ P_T \end{pmatrix} = \begin{pmatrix} 1 & \ell m_0^2/P^{o^3} \\ 0 & 1 \end{pmatrix}\begin{pmatrix} T \\ P_T \end{pmatrix}. \tag{6.25}$$

Physically, they describe the dependence of transit time on particle energy.

Next consider the factor $\exp(-\ell:H^y:)$. It affects only the coordinates y and p_y. Furthermore, the action of powers of $(-\ell:H_2^y:)$ on y and p_y is easily evaluated. One finds the result

$$(-\ell:H_2^y:)y = -\ell(2p^o)^{-1}[p_y^2,y] = (\ell/p^o)p_y \tag{6.26a}$$

$$(-\ell:H_2^y:)p_y = -(\ell qa_2/2)[y^2,p_y] = -q\ell a_2 y \tag{6.26b}$$

$$\begin{aligned}(-\ell:H_2^y:)^2 y &= (-\ell:H_2^y:)(-\ell:H_2^y:)y \\ &= -(\ell:H_2^y:)(\ell/p^o)p_y = -(q\ell^2 a_2/p^o)y\end{aligned} \tag{6.26c}$$

$$\begin{aligned}(-\ell:H_2^y:)^2 p_y &= (-\ell:H_2^y:)(-\ell:H_2^y:)p_y \\ &= (-\ell:H_2^y:)(-q\ell a_2 y) = -(q\ell^2 a_2/p^o)p_y, \text{ etc.}\end{aligned} \tag{6.26d}$$

Thus, as in exercise 3.23, the infinite series describing the effect of $\exp(-\ell:H_2^y:)$ on y and p_y can be summed. Writing the result in matrix form as before, one finds the relation

$$\exp(-\ell:H_2^y:)\begin{pmatrix} y \\ p_y \end{pmatrix} = \begin{pmatrix} \cos k\ell & (kp^o)^{-1}\sin k\ell \\ -kp^o\sin k\ell & \cos k\ell \end{pmatrix}\begin{pmatrix} y \\ p_y \end{pmatrix}, \tag{6.27}$$

where k denotes the quantity

$$k = (qa_2/p^o)^{1/2}. \tag{6.28}$$

Observe that Eq. (6.27) gives just the result expected in y and p_y space for the effect of a horizontally focussing quadrupole described in the linear matrix approximation.[17]

Exercise 6.3: Verify Eq. (6.27). Summation of the series is facilitated by the identity

$$\exp(w) = \cosh w + \sinh w$$

which decomposes the exponential series into even and odd parts.

Finally, consider the factor $\exp(-\ell{:}H_2^z{:})$. It affects only z and p_z. According to Eqs. (6.22b) and (6.22c), the forms of H_2^y and H_2^z differ only in the sign of the quantity a_2. Consequently, the effect of $\exp(-\ell{:}H_2^z{:})$ on z and p_z can be inferred by analogy with Eq. (6.27). One finds the result

$$\exp(-\ell{:}H_2^z{:})\begin{pmatrix} z \\ p_z \end{pmatrix} = \begin{pmatrix} \cosh k\ell & (kp^o)^{-1}\sinh k\ell \\ kp^o \sinh k\ell & \cosh k\ell \end{pmatrix}\begin{pmatrix} z \\ p_z \end{pmatrix}. \quad (6.29)$$

Note that Eq. (6.29) gives just the vertical defocussing action expected in linear matrix approximation for the quadrupole in question.[17]

Exercise 6.4: Verify Eq. (6.29).
Hint: Just write the expression analogous to (6.27) and make the analytic continuation $k \to ik$ (or, if you wish, $k \to -ik$).

So far, it has been shown that the factor $\exp({:}f_2{:})$ reproduces the usual linear matrix approximation to charged particle beam optics. The next term to compute is f_3.

Observe that H_3 as given by (6.16d) can be written as a sum of terms in the form

$$H_3 = H_3^T + H_3^{Ty} + H_3^{Tz} \quad (6.30)$$

where

$$H_3^T = [P_T^3/2][(m_o^2)/(p^{o^4} v_x^o)] \quad (6.31a)$$

$$H_3^{Ty} = P_T\, p_y^2/(2p^{o^2} v_x^o) \quad (6.31b)$$

$$H_3^{Tz} = P_T\, p_z^2/(2p^{o^2} v_x^o). \quad (6.31c)$$

Moreover, using the fact that some of the terms in (6.31) are in involution with some of the terms in (6.22), it can easily be verified that the quantity exp $-\ell(:H_2: + :H_3)$ can be written in the form

$$\exp\{-\ell(:H_2: + :H_3:)\} =$$

$$\exp(-\ell:H_2^T:)\exp(-\ell:H_3^T:)\exp\{-\ell(:H_2^y:+:H_3^{Ty}:)\} \times$$

$$\exp\{-\ell(:H_2^z:+:H_3^{Tz}:)\}. \qquad (6.32)$$

This facilitates the computation of f_3 since the only terms that need to be factored now are $\exp\{-\ell(:H_2^y:+:H_3^{Ty}:)\}$ and its z counterpart.

Exercise 6.5: Verify Eq. (6.32).

Suppose the factorization of $\exp\{-\ell(:H_2^y:+:H_3^{Ty}:)\}$ is written in the form

$$\exp\{-\ell(:H_2^y: + :H_3^{Ty}:)\} = \exp(:f_2^y:)\exp(:f_3^{Ty}:)\exp(:f_4:)\ldots . \qquad (6.33)$$

Then it follows from (5.88) and (5.90) that f_2^y and f_3^{Ty} are given by the expressions

$$f_2^y = -\ell H_2^y \qquad (6.34)$$

$$f_3^{Ty} = -\ell \int_0^1 d\tau \, \exp(\tau\ell:H_2^y:) \, H_3^{Ty} . \qquad (6.35)$$

Moreover, there is the simplification

$$\exp(\tau\ell:H_2^y:)H_3^{Ty} = P_T/(2p^{o^2} v_x^o)\{\exp(\tau\ell:H_2^y:)p_y\}^2 . \qquad (6.36)$$

Here use has been made of the isomorphism property (3.34) and (3.35). It is therefore only necessary to evaluate the quantity $\exp(\tau\ell:H_2^y:)p_y$. This can be done easily by reference to (6.27). One finds the result

$$\exp(\tau\ell:H_2^y:)p_y = ykp^o \sin(k\tau\ell) + p_y \cos(k\tau\ell). \qquad (6.37)$$

Consequently, f_3^{Ty} is given by the relation

$$f_3^{Ty} = -\ell P_T/(2p^{o^2} v_x^o) \int_0^1 d\tau\{ykp^o \sin(k\tau\ell) + p_y \cos(k\tau\ell)\}^2 . \qquad (6.38)$$

The integration can easily be carried out to give the explicit result

$$f_3^{Ty} = -\ell P_T/(4p^{o^2} v_x^o)\{p_y^2[1 + (2k\ell)^{-1} \sin(2k\ell)]$$
$$+ y^2(kp^o)^2[1 - (2k\ell)^{-1} \sin(2k\ell)]$$
$$+ yp_y(p^o/\ell)[1 - \cos 2k\ell]\} . \qquad (6.39)$$

<u>Exercise 6.6</u>: Verify Eqs. (6.37) through (6.39).

In a similar fashion, the factor $\exp\{-\ell(:H_2^z: + :H_3^{Tz}:)\}$ can be handled by writing

$$\exp\{-\ell(:H_2^z: + :H_3^{Tz}:)\} = \exp(:f_2^z:)\exp(:f_3^{Tz}:)\exp(:f_4:)\cdots \quad . \qquad (6.40)$$

As observed before, H_2^y and H_2^z are of the same form except for the sign of a_2. Also, inspection of Eqs. (6.31b) and (6.31c) show that H_3^{Ty} and H_3^{Tz} are of the same form. Consequently, the expression for f_3^{Tz} can be inferred by analogy from that for f_3^{Ty} as given by (6.39). One finds the result

$$f_3^{Tz} = -\ell P_T/(4p^{o^2} v_x^o)\{p_y^2[1 + (2k\ell)^{-1} \sinh(2k\ell)]$$
$$- y^2(kp^o)^2[1 - (2k\ell)^{-1} \sinh(2k\ell)]$$
$$+ yp_y(p^o/\ell)[1 - \cosh(2k\ell)]\}. \qquad (6.41)$$

<u>Exercise 6.7</u>: Verify (6.41) using the method of exercise (6.4).

Let us now summarize what has been accomplished. Upon combining Eqs. (6.18), (6.32), (6.33), and (6.40), one finds for $\mathcal{M}$ the result

$$\mathcal{M} = \exp(-\ell:H_2^T:)\exp(-\ell:H_3^T:)\exp(:f_2^y:)\exp(:f_3^{Ty}:)\times$$
$$\exp(:f_2^z:)\exp(:f_3^{Tz}:). \qquad (6.42a)$$

It is easily verified that the various factors in (6.42a) can be rearranged to give the result

$$\mathcal{M} = \exp(-\ell:H_2:)\exp\{:(-\ell H_3^T + f_3^{Ty} + f_3^{Tz}):\}. \qquad (6.42b)$$

Consequently, the quantity f_3 appearing in (6.18) is given by the relation

$$f_3 = -\ell H_3^T + f_3^{Ty} + f_3^{Tz}. \tag{6.43}$$

This completes the calculation of f_3.

Exercise 6.8: Verify that (6.42a) can be rearranged to give (6.42b). That is, show that all rearrangements involve commuting Lie operators of functions which are in involution.

Observe that all the terms in $f_2 = - H_2$ and f_3 are independent of the time. This means that the map $\mathcal{M}$ leaves P_T unchanged. This is to be expected, of course, since a static quadrupole field does not change the energy of a particle. Observe also that all terms in f_3 contain at least one factor of P_T. This means that f_3 is zero at the design energy (or, equivalently, at the design momentum). Consequently f_3, in the case of a perfect quadrupole, describes purely chromatic effects.

6.2 The perfect sextupole

Consider next the case of a perfect sextupole of length ℓ. Again employing the coordinates shown in Fig. (1.1), and using the results of exercises 1.5 and 1.11, one finds in this case for the Hamiltonian K the result

$$K = -[(p_t^2/c^2) - m_0^2c^2 - p_y^2 - p_z^2]^{1/2} - qa_s(yz^2 - y^3/3). \tag{6.44a}$$

A moment's thought shows that the design orbit for a sextupole is the same as that for a quadrupole, and hence it is advantageous to make the same canonical transformation (6.8). When this is done, the Hamiltonian K will be replaced by the new Hamiltonian H given by the expression

$$\begin{aligned} H = &-[p_t^o + P_T)^2/c^2 - m_o^2c^2 - p_y^2 - p_z^2]^{1/2} \\ &-qa_s(yz^2 - y^3/3) - (v_x^o)^{-1}(P_T + p_t^o). \end{aligned} \tag{6.44b}$$

Finally, H can be expanded about the design orbit as in (6.15) to give for the first few homogeneous polynomials H_n the result

$$H_0 = m_o^2c^2/p^o \tag{6.45a}$$

$$H_1 = 0 \tag{6.45b}$$

$$H_2 = (P_T^2/2)(m_o^2/p^{o^3}) + (p_y^2 + p_z^2)/(2p^o) \tag{6.45c}$$

$$H_3 = [P_T^2/2][(m_o^2/p^{o^4} v_x^o)]$$

$$+ P_T(p_y^2 + p_z^2)/(2p^{o^2} v^o{}_x) \qquad (6.45d)$$

$$- qa_s(yz^2 - y^3/3).$$

As before, the symplectic map M produced by following through the sextupole the Hamiltonian flow generated by H is given by the expression

$$M = \exp\{-\ell(:H_2: + :H_3: + \ldots\} \; . \qquad (6.46)$$

Eventually, we will want to factor M. However, before doing so, it is interesting to examine the expression for M in the limiting case of a very short but very strong sextupole. That is, consider the limit $\ell \to 0$, $a_s \to \infty$ in such a way that $\lim \ell a_s = \alpha$. In this limit, one finds

$$\lim(\ell H) = -q\alpha(yz^2 - y^3/3). \qquad (6.47)$$

Therefore, the mapping M in this case is given by the simple expression

$$M = \exp\{q\alpha:(yz^2 - y^3/3):\} \; . \qquad (6.48)$$

It is evident that the M of (6.48) leaves the quantities P_T, T, y, and z unchanged. Moreover, it is easily checked that

$$q\alpha:(yz^2 - y^3/3):p_y = -q\alpha y^2 \qquad (6.49a)$$

$$\{q\alpha:(yz^2 - y^3/3):\}^2 p_z = 0, \text{ etc.} \qquad (6.49b)$$

$$q\alpha:(yz^2 - y^3/3):p_z = 2q\alpha yz \qquad (6.50a)$$

$$\{q\alpha:(yz^2 - y^3/3):\}^2 p_z = 0, \text{ etc.} \qquad (6.50b)$$

Consequently, one finds the relations

$$M p_y = p_y - q\alpha y^2 \qquad (6.51a)$$

$$M p_z = p_z + 2q\alpha yz. \qquad (6.51b)$$

This is just the usual thin lens or impulsive "kick" approximation for the action of a short sextupole. What this brief calculation has demonstrated is that the kick approximation does indeed produce a symplectic map. Therefore its use in many applications is a very reasonable procedure.

Exercise 6.9: Evaluate $\mathcal{M}$ for a quadrupole in the kick approximation.

We turn now to the task of factoring $\mathcal{M}$. The required calculations are similar to those already performed for the case of the quadrupole. First, one finds that H_2 can be decomposed as in (6.21) with the result

$$H_2^T = (P_T^2/2)(m_o^2/p^{o^3}) \tag{6.52a}$$

$$H_2^y = p_y^2/(2p^o) \tag{6.52b}$$

$$H_2^z = p_z^2/(2p^o). \tag{6.52c}$$

Next, the actions of $\exp(-\ell{:}H_2^T{:})$, $\exp(-\ell{:}H_2^y{:})$, and $\exp(-\ell{:}H_2^z{:})$ are easily found. The action of $\exp(-\ell{:}H_2^T{:})$ is the same as in (6.25), and those of $\exp(-\ell{:}H_2^y{:})$ and $\exp(-\ell{:}H_2^z{:})$ are the $a_2 \to 0$ limits of (6.27) and (6.29) respectively. One finds, for example, the result

$$\exp(-\ell{:}H_2^y{:})\begin{pmatrix} y \\ p_y \end{pmatrix} = \begin{pmatrix} 1 & (\ell/p^o) \\ 0 & 1 \end{pmatrix}\begin{pmatrix} y \\ p_y \end{pmatrix} \tag{6.53}$$

which is just the expression expected for the effect of a drift of length ℓ in the linear matrix approximation. Similarly, $\exp(-\ell{:}H_2^z{:})$ gives the same transformation in z,p_z space.

The cubic portion H_3 of the Hamiltonian can be decomposed in the form

$$H_3 = H_3^T + H_3^{Ty} + H_3^{Tz} + H_3^{yz} \quad . \tag{6.54}$$

The first three components in (6.54) are the same as those given in (6.31); and the fourth, H_3^{yz}, is given by the expression

$$H_3^{yz} = -qa_s(yz^2 - y^3/3). \tag{6.55}$$

Now observe that many of the components of H_2 as given by (6.52) are in involution with the components of H_3. It follows that $\mathcal{M}$ as given by (6.46) can also be written in the form

$$\mathcal{M} = \exp(-\ell{:}H_2^T{:})\exp\{-\ell{:}(H_2^y + H_2^z + H_3^{yz}){:}\}\times$$
$$\exp\{-\ell{:}(H_3^T + H_3^{Ty} + H_3^{Tz}){:}\} \quad . \tag{6.56}$$

Therefore, the only factorization that needs to be carried out is that for the quantity $\exp\{-\ell:(H_2^y + H_2^z + H_3^{yz}):\}$.

Exercise 6.10: Verify the factorization (6.56) starting from (6.46).

Suppose the desired factorization is written in the form

$$\exp\{-\ell:(H_2^y + H_2^z + H_3^{yz}):\} = \exp(-\ell:H_2^y:)\exp(-\ell:H_2^z:)\times \exp(:f_3^{yz}:)\exp(:f_4:)\cdots . \quad (6.57)$$

Then, following the usual procedure, the quantity f_3^{yz} is given by the expression

$$f_3^{yz} = -\ell \int_0^1 d\tau \exp(\tau\ell:H_2^y:)\exp(\tau\ell:H_2^z:)H_3^{yz}. \quad (6.58)$$

The integrand may be simplified using the isomorphism property (3.34) and (3.35) to give the result

$$\exp(\tau\ell:H_2^y:)\exp(\tau\ell:H_2^z:)H_3^{yz} =$$

$$-qa_s\{[\exp(\tau\ell:H_2^y:)y][\exp(\tau:H_2^z:)z]^2 - [\exp(\tau\ell:H_2^y:)y]^3/3\}. \quad (6.59)$$

Moreover, from (6.53), and its z counterpart, it follows that

$$\exp(\tau\ell:H_2^y:)y = y - (\tau\ell/p^o)p_y \quad (6.60a)$$

$$\exp(\tau\ell:H_2^z:)z = z - (\tau\ell/p^o)p_z. \quad (6.60b)$$

Consequently, the expression for f_3^{yz} can also be written in the form

$$f_3^{yz} = \ell qa_s \int_0^1 d\tau\{[y-(\tau\ell/p^o)p_y][z-(\tau\ell/p^o)p_z]^2-[y-(\tau\ell/p^o)p_y]^3/3\}. \quad (6.61)$$

The integration can now easily be carried out to give the explicit result

$$f_3^{yz} = \ell qa_s\{(yz^2 - y^3/3) - (\ell/p_o)(z^2p_y - y^2p_y + 2yzp_z)/2 + (\ell/p^o)^2(yp_z^2 - yp_y^2 + 2zp_yp_z)/3 - (\ell/p^o)^3(p_yp_z^2 - p_y^3/3)/4\}. \quad (6.62)$$

Observe that all terms in (6.62) except the first vanish in the short and strong sextupole limit as expected.

Combining (6.57) with (6.56) gives the final result

$$f_3 = -\ell H_3^T - \ell H_3^{Ty} - \ell H_3^{Tz} + f_3^{yz} . \quad (6.63)$$

This completes the calculation of f_3 for the sextupole. Note that the first three terms in (6.63) describe chromatic effects since, according to (6.31), they all depend on P_T. However, the term f_3^{yz} is independent of P_T, and describes nonlinear effects due to the sextupole magnetic field.

<u>Exercise 6.11:</u> Verify (6.62) starting with (6.58).

6.3 The normal entry and exit dipole

For the sake of simplicity, the discussion of dipole magnets will be limited to the case of normal entry and exit as illustrated in Fig. (1.3). Also, the discussion of fringe field effects will be omitted although they are important.

The treatment of the normal entry and exit dipole is most easily carried out in cylindrical coordinates. It begins with the Hamiltonian (1.23) and the potentials of exercise 1.12. That is, the Hamiltonian with ϕ as an independent variable is given by the expression

$$K = \rho[p_t^2/c^2 - m_o^2c^2 - p_\rho^2 - p_z^2]^{1/2} - q\rho^2B/2. \quad (6.64)$$

Let p^o be the momentum of the design orbit. Then the design orbit in the dipole is a circular arc with radius ρ_o given by the familiar relation

$$\rho_o = p^o/(qB). \quad (6.65)$$

In terms of these quantities, the design orbit is given by the equations

$$p_\rho = 0 \quad , \quad \rho = \rho_o \quad (6.66a,b)$$

$$p_z = 0 \quad , \quad z = 0 \quad (6.67a,b)$$

$$p_t = p_t^o \quad , \quad t(\phi) = \phi(v_\phi^o)^{-1} + \text{constant}. \quad (6.68a,b)$$

Here p_t^o and p^o are related by Eq. (6.17) as before; and the quantity v_ϕ^o is specified by the condition

$$dt/d\phi \Big|_{\substack{\text{design}\\ \text{orbit}}} = \partial K/\partial p_t \Big|_{\substack{\text{design}\\ \text{orbit}}} = (v_\phi^o)^{-1}. \qquad (6.69)$$

Upon evaluation (6.69), one finds the result

$$v_\phi^o = [(p_t^o/c)^2 - m_o^2c^2]^{1/2}[\rho_o p_t^o/c^2]^{-1} \quad . \qquad (6.70)$$

As was the case with the quadrupole and sextupole, it is convenient to introduce new coordinates which facilitate expansion about the design orbit. They are defined in this instance by the relations

$$t = T + \phi(v_\phi^o)^{-1} \qquad (6.71a)$$

$$\rho = \rho_o + Y \qquad (6.71b)$$

$$z = Z \qquad (6.71c)$$

$$p_t = p_t^o + P_T \qquad (6.71d)$$

$$p_\rho = P_Y \qquad (6.71e)$$

$$p_z = P_Z \quad . \qquad (6.71f)$$

Evidently the design orbit corresponds to all the new variables having zero values. Simple calculation shows that the indicated change of variables can be obtained from the transformation function

$$F_2 = (\rho - \rho_o)P_Y + zP_Z + [t - \phi(v_\phi^o)]^{-1}[P_T + p_t^o] . \qquad (6.72)$$

Now let H be the Hamiltonian for the new variables. It is given by the relation

$$H = K + \partial F_2/\partial\phi \qquad (6.73)$$

or, more explicitly, by the relation

$$H = (\rho_o + y)[(p_t^o + P_T)^2/c^2 - m_o^2c^2 - p_y^2 - p_z^2]^{1/2}$$

$$- q(\rho_o + y)^2B/2 - (v_\phi^o)^{-1}(P_T + p_t^o) \quad . \qquad (6.74)$$

Here, as before, lower case letters have again been introduced for simplicity. Note, however, that in this application y is not a Cartesian coordinate but instead describes deviations in ρ from the design value ρ_o.

The next step is to expand H in a power series about the design orbit as in (6.15). Upon expanding (6.74), one finds for the first few terms the result

$$H_0 = -\rho_o [(p^o/2) + (m_o^2 c^2/p^o)] \tag{6.75a}$$

$$H_1 = 0 \tag{6.75b}$$

$$H_2 = -\rho_o p_y^2/(2p^o) - qy^2 B/2 - \rho_o p_z^2/(2p^o)$$

$$+ yP_T p_t^o/(p^o c^2) - P_T^2 \rho_o m_o^2/(2p^{o^3}) \tag{6.75c}$$

$$H_3 = P_T^3 \rho_o p_t^o m_o^2/(2p^{o^5} c^2) + P_T(p_y^2 + p_z^2) p_t{}^o/(2p^{o^3} c^2)$$

$$- yP_T^2 m_o^2/(2p^{o^3}) - y(p_y^2 + p_z^2)/(2p^o). \tag{6.75d}$$

Here use has been made of (6.65).

Note that H_1 is zero as expected. Observe also that the last term in H_3 is independent of P_T. It therefore produces effects similar to that of a sextupole. Indeed, terms of this sort from the various dipoles in a lattice make important contributions to the chromaticity in the case of a small ring.[19] A remark is also in order about the quadratic terms. The first two terms in H_2 are of the same sign and similar to those in (6.22b). They produce the horizontal focussing associated with a normal entry and exit magnet. The third term shows there is no focussing in the vertical direction. The fourth term involving the product yP_T describes the dispersion produced by a bend. Finally, the last term describes the dependence of transit time on energy.

<u>Exercise 6.12</u>: Verify Eqs. (6.75).

We are ready to consider the symplectic map $\mathcal{M}$ produced by following the Hamiltonian flow specified by H. Since ϕ now plays the role of an independent variable, the map $\mathcal{M}$, in analogy to (5.34), can be written in the form

$$\mathcal{M} = \exp\{-(\phi^f - \phi^i):H:\}. \tag{6.76}$$

Here ϕ^i is the initial value of ϕ at the entry to the dipole, and ϕ^f is the final value upon exit. But according to Figs. (1.1) and (1.3), ϕ decreases as a particle moves through a dipole. Therefore, the change in ϕ can be written in the form

$$\phi^f - \phi^i = -\theta,$$

where θ is a positive quantity equal to the bend angle as illustrated in Fig. (1.3). With this explanation, the symplectic map $\mathcal{M}$ can be written in the form

$$\mathcal{M} = \exp\{\theta(:H_2: + :H_3: + \ldots)\} \quad . \tag{6.77}$$

All that remains to be done is to factor $\mathcal{M}$. As a first step in this process, suppose H_2 as given by (6.75c) is written as a sum of three terms mutually in involution,

$$H_2 = H_2^T + H_2^z + H_2^{yT} \tag{6.78}$$

$$H_2^T = -P_T^2 \rho_o m_o^2/(2p^{o^3}) \tag{6.79a}$$

$$H_2^z = -\rho_o p_z^2/(2p^o) \tag{6.79b}$$

$$H^{yT} = -\rho_o p_y^2/(2p^o) + yP_T/(\rho_o v_\phi^o) - qy^2 B/2 \quad . \tag{6.79c}$$

Here, in simplifying (6.79c), use has been made of the relation

$$(v_\phi^o)^{-1} = (\rho_o p_t^o/c^2)[(p_t^o/c)^2 - m_o^2 c^2]^{-1/2} = (\rho_o p_t^o)/(p^o c^2). \tag{6.80}$$

The evaluation of $\exp(\theta:H_2^T:)$ and $\exp(\theta:H_2^z:)$ is straight forward, and introduces no new features beyond those already encountered in earlier sections. However, the calculation of $\exp(\theta:H_2^{yT}:)$ requires a bit more work since H_2^{yT} depends on the three variables y, p_y, and P_T. Evidently, $\exp(\theta:H_2^{yT}:)$ leaves z, p_z, and P_T unaffected. Moreover, the action of powers of $:H_2^{yT}:$ on the variables y, p_y, and T is easily calculated. One finds the result

$$:H_2^{yT}: \begin{bmatrix} y \\ p_y \\ T \\ P_T \end{bmatrix} = G \begin{bmatrix} y \\ p_y \\ T \\ P_T \end{bmatrix} \tag{6.81}$$

where G denotes the matrix

$$G = \begin{bmatrix} 0 & \rho_o/P^o & 0 & 0 \\ -qB & 0 & 0 & (\rho_o v_\phi^o)^{-1} \\ -(\rho_o v_\phi^o)^{-1} & 0 & 0 & 0 \\ 0 & 0 & 0 & 0 \end{bmatrix} . \qquad (6.82)$$

It follows that the action of $\exp(:H_2^{yT}:)$ is given by the relation

$$\exp(\theta :H^{yT}:) \begin{bmatrix} y \\ P_y \\ T \\ P_T \end{bmatrix} = \exp(\theta G) \begin{bmatrix} y \\ P_y \\ T \\ P_T \end{bmatrix} . \qquad (6.83)$$

The determination of $\exp(\theta G)$ is facilitated by an algebraic property of the matrix G. A short calculation shows that G has the characteristic polynomial $P(\lambda)$ given by the equation

$$P(\lambda) = \det(G - \lambda I) = \lambda^2(\lambda^2 + 1). \qquad (6.84)$$

Here use has been made of (6.65). Consequently, by the Cayley-Hamilton theorem, the matrix G has the special property[18]

$$G^4 = -G^2. \qquad (6.85)$$

The property (6.85) makes it easy to sum the exponential series for exp(G). One finds the result

$$\begin{aligned} \exp(\theta G) &= \cosh(\theta G) + \sinh(\theta G) \\ &= [I + (\theta G)^2/2! + (\theta G)^4/4! + (\theta G)^6/6! + \cdots] \\ &+ [(\theta G) + (\theta G)^3/3! + (\theta G)^5/5! + (\theta G)^7/7! + \cdots] \\ &= [I + G^2(\theta^2/2! - \theta^4/4! + \theta^6/6! + \cdots)] \\ &+ [\theta G + G^3(\theta^3/3! - \theta^5/5! + \theta^7/7! + \cdots)] \\ &= I + \theta G + G^2(1 - \cos\theta) + G^3(\theta - \sin\theta). \end{aligned} \qquad (6.86)$$

Upon inserting the explicit form (6.82) for G into (6.86), one finds the desired expression

$$\exp(\theta G) = \begin{bmatrix} \cos\theta & (\rho_o/p^o)\sin\theta & 0 & (p^o v^o_\phi)^{-1}(1-\cos\theta) \\ -(p^o/\rho_o)\sin\theta & \cos\theta & 0 & (\rho_o v^o_\phi)^{-1}\sin\theta \\ -(\rho_o v^o_\phi)^{-1}\sin\theta & (1-\cos\theta)/(p^o v^o_\phi) & 1 & (p^o\rho_o)^{-1}(v^o_\phi)^{-2}(\sin\theta-\theta) \\ 0 & 0 & 0 & 1 \end{bmatrix} \tag{6.87}$$

Note that (6.87) describes the expected effect of dispersion in the median plane of the bend.

All tools have now been assembled for the calculation f_3. Because no new techniques are involved, the actual work will be left as an exercise for the reader.

Exercise 6.13 Verify Eqs. (6.84) through (6.87).

Exercise 6.14: Calculate f_3 for the normal entry and exit dipole.

6.4 The straight section drift

The calculation of results for a drift have been saved for last because they are particularly easy. Consider the quadrupole Hamiltonian (6.13) in the case of zero field strength. Then one finds for H an expansion of the form (6.15) with the homogeneous polynomials H_0 through H_4 given explicitly by the expressions

$$H_0 = m_0^2c^2/p^0 \tag{6.88a}$$

$$H_1 = 0 \tag{6.88b}$$

$$H_2 = (P_T^2/2)(m_0^2/p^{0^3}) + (p_y^2 + p_z^2)/(2p^0) \tag{6.88c}$$

$$H_3 = [P_T^3/2][(m_0^2)/(p^{0^4} v_x^0)] + [P_T(p_y^2 + p_z^2)/(2p^{0^2} v_x^0)] \tag{6.88d}$$

$$H_4 = P_T^4 m_0^2 c^2(5m_0^2c^2 + 4p^{0^2})/(8p^{0^7} c^4) \tag{6.88e}$$

$$+P_T^2(p_y^2 + p_z^2)(3m_0^2c^2 + 2p^{0^2})/(4p^{0^5} c^2) + (p_y^2 + p_z^2)^2/8p^{0^3}).$$

Observe that all the terms in Eqs. (6.88) are mutually in involution.

It follows that the symplectic map $\mathcal{M}$ for a drift can be factored immediately to give the result

$$\mathcal{M} = \exp\{-\ell:(H_2 + H_3 + H_4 + \cdots):\}$$
$$= \exp(-\ell:H_2:)\exp(-\ell:H_3:)\exp(-\ell:H_4:) \cdots . \qquad (6.89)$$

Note that H_4 contains the term $(p_y^2 + p_z^2)^2$. This is an example of a "kinematic" nonlinearity which occurs even apart from chromatic effects and the effects of electric and magnetic fields.

Exercise 6.15: Verify Eqs. (6.88) and (6.89).

6.5 Application to light optics

According to section 1.3 and exercise 4.3, the passage of light rays through an optical system can be described by a symplectic map. When applied to optics, the factorization theorem indicates that the effect of any collection of lenses, prisms, and mirrors can be characterized by a set of homogeneous polynomials. It is easy to verify that the polynomials f_2 reproduce Gaussian optics, and the higher order polynomials f_3, f_4, etc. describe departures from Gaussian optics, and are related to aberrations in the case of an imaging system. Thus, from a Lie algebraic perspective, the fundamental problem of geometrical optics is to study what polynomials correspond to various optical elements, and to study what polynomials correspond to various desired optical properties.

Exercise 6.16: Show that $\mathcal{M}$ given by

$$\mathcal{M} = \exp\{-\ell/(2n):(\vec{p}^{\,i})^2:\}$$

corresponds to transit by a distance ℓ through a medium of refractive index n in the Gaussian approximation.

Answer:

$$p_\alpha^f = \mathcal{M}\, p_\alpha^i = p_\alpha^i$$

$$q_\alpha^f = \mathcal{M}\, q_\alpha^i = q_\alpha^i + (\ell/n)p^i .$$

Exercise 6.17: Show that $\mathcal{M}$ given by

$$\mathcal{M} = \exp\{(n_2-n_1)/(2r):(\vec{q}^{\,i})^2:\}$$

corresponds to refraction in the Gaussian approximation for rays passing through a spherical interface of radius r from a medium having index of refraction n_1 to a second medium having index n_2.

Answer:

$$q_\alpha^f = \mathcal{M} q_\alpha^i = q_\alpha^i$$

$$p_\alpha^f = \mathcal{M} p_\alpha^i = p_\alpha^i + \{(n_2-n_1)/r\} q_\alpha^i.$$

To simplify discussion, suppose, as is often the case, that the optical system under consideration is axially symmetric about some axis and is also symmetric with respect to reflections through some plane containing the axis of symmetry. Then axial symmetry requires that the various f_n be functions only of the variables $\vec{p}^2$, $\vec{q}^2$, $\vec{p}\cdot\vec{q}$, and $\vec{p}\times\vec{q}$; and reflection symmetry rules out the variable $\vec{p}\times\vec{q}$. It follows that all the f_n with odd n must vanish since it is impossible to construct an odd order homogeneous polynomial using only the variables $\vec{p}^2$, $\vec{p}\cdot\vec{q}$, and $\vec{q}^2$. Consequently, in any case having the assumed symmetries, the optical symplectic map $\mathcal{M}$ must be of the general form

$$\mathcal{M} = \exp(:f_2:)\exp(:f_4:)\exp(:f_6:) \cdots . \tag{6.90}$$

Of course in the general case of no particular symmetries, the odd degree polynomials f_3, f_5, etc., can also occur, and all the polynomials f_n can in principle depend on the components of the vectors $\vec{p}$ and $\vec{q}$ in an unrestricted fashion.

It can be shown that the polynomials f_4, f_6, etc. in the case of an imaging system are related to third-order, fifth-order, etc. aberrations respectively. Consider the case of f_4. According to the previous discussion of symmetry, f_4 in the cases of present interest can depend only on the variables $\vec{p}^2$, $\vec{p}\cdot\vec{q}$, and $\vec{q}^2$. Consequently, f_4 must be of the general form

$$f_4 = A(\vec{p}^2)^2 + B\,\vec{p}^2(\vec{p}\cdot\vec{q}) + C(\vec{p}\cdot\vec{q})^2$$
$$+ D\,\vec{p}^2\vec{q}^2 + E(\vec{p}\cdot\vec{q})\vec{q}^2 + F(\vec{q}^2)^2. \tag{6.91}$$

Here the quantities A through F are arbitrary coefficients whose values depend on the particular optical system under consideration. When employed in (6.90), the last term $F(\vec{q}^2)^2$ has no effect on the quality of an image since $[F(\vec{q}^2)^2, q_\alpha]=0$. (It does, however affect the arrival direction $\vec{p}$ of a ray at the image plane, and therefore may be important if the optical system under study is to be used for some other purpose as part of a larger optical system.) The remaining terms do affect the image. Specifically, one finds in the imaging case the following one to one correspondence between the terms in the expansion (6.91) and the classical Seidel third-order monochromatic aberrations:

TABLE OF ABBERATIONS

Term	Seidel Aberration
$A(\vec{p}^2)^2$	Spherical Aberration
$B\vec{p}^2(\vec{p}\cdot\vec{q})$	Coma
$C(\vec{p}\cdot\vec{q})^2$	Astigmatism
$D\ \vec{p}^2\vec{q}^2$	Curvature of Field
$E(\vec{p}\cdot\vec{q})\vec{q}^2$	Distortion

In reflecting upon what has been accomplished so far, it is evident that all that has been assumed is that the optical map $\mathcal{M}$ is symplectic (a consequence of Fermat's principle) and has certain symmetries. As a consequence of these assumptions, Gaussian optics was obtained as a first approximation, and it was found that for an imaging system only five well-defined kinds of monochromatic aberrations can occur in third order. The use of Lie algebraic methods seems to be an optimal way of arriving at and understanding these basic results.

To proceed further, it is necessary to have a catalog of Lie operators corresponding to various simple optical elements. The first simple optical element to be considered is transit through a slab of thickness ℓ composed of a homogeneous medium having constant refractive index n. In this case, the associated symplectic map $\mathcal{M}$ is found to be given by the expression

$$\mathcal{M} = \exp\{\ell:(n^2 - \vec{p}^2)^{1/2}:\} = \exp\{(-\ell)/(2n):\vec{p}^2:\}\times$$
$$\exp(-\ell)/(8n^3):(\vec{p}^2)^2:\}\cdots \quad . \qquad (6.92)$$

<u>Exercise 6.18</u>: Verify (6.92) starting with (1.30) and (6.14) and making a power series expansion.

Next consider a lens. Figure 6.1 shows a lens with planar entrance and exit faces. It is composed of two media having indices of refraction n_1 and n_2 separated by a curved interface. Suppose a cartesian coordinate system is located in the exit face (right face) of the lens with the z axis along the optical axis and the x and y axes lying in the face of the lens. Then the shape of the curved interface is taken to be given by the equation

$$z = -\alpha(x^2 + y^2) + \beta(x^2 + y^2)^2 + \gamma(x^2 + y^2)^3 + \cdots . \qquad (6.93)$$

For the case of a spherical lens with radius of curvature r, the shape of the surface is described by the relation

$$z = -r + (r^2 - x^2 - y^2)^{1/2} \quad . \qquad (6.94)$$

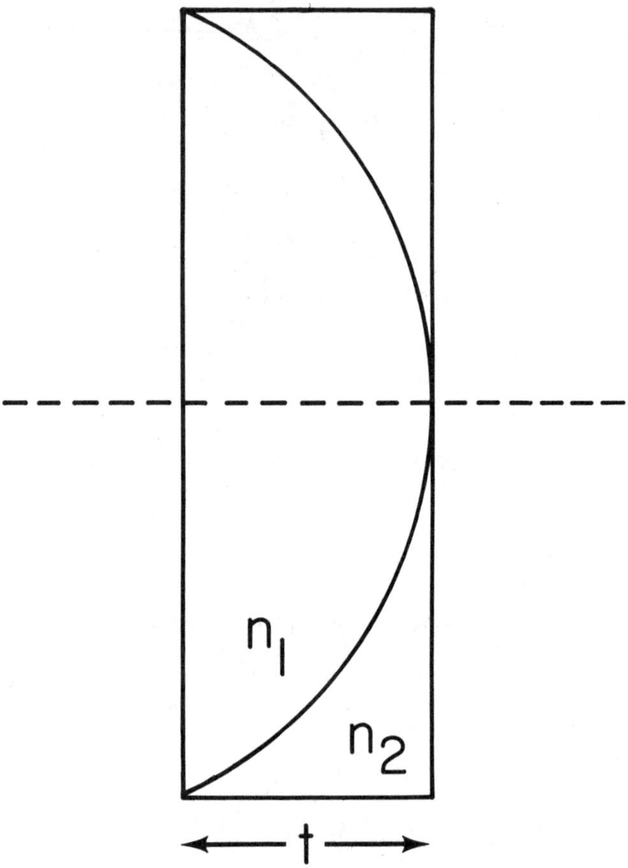

Fig. 6.1: A lens with planar entrance and exit faces. It is composed of two media having indices of refraction n_1 and n_2, and separated by a curved interface.

In this case the quantities α and β have the explicit values

$$\alpha = 1/(2r)$$

$$\beta = -\alpha^3 = -1/(8r^3). \qquad (6.95)$$

Evidently, the most general lens with axial symmetry (including the case of a corrector plate) can be obtained by joining together lenses of the type shown in Fig. 6.1 and their reversed counterparts, perhaps with slabs of constant index material in between.

Calculations show that the symplectic map $\mathcal{M}$ for this lens, when light passes from left to right, is given through fourth order by the expression

$$\mathcal{M} = \exp\{(-t)/(2n_1):\vec{p}^2:\}\exp\{(-t)/(8n_1^3):(\vec{p}^2)^2:\}\times$$
$$\exp\{\alpha(n_2-n_1):\vec{q}^2:\}\exp(:f_4:). \qquad (6.96)$$

Each of the terms appearing in (6.96) has a simple interpretation. Proceeding from left to right, comparison with (6.92) shows that the first two terms in (6.96) simply correspond to propagation through a slab of thickness t and refractive index n_1. Here, as shown in Fig. 6.1, t is the thickness of the lens. The third term, $\exp\{\alpha(n_2-n_1):\vec{q}^2:\}$, simply produces the refraction expected for a thin lens in Gaussian approximation. See exercise 6.17 . Finally, the last term, $\exp(:f_4:)$, describes third order departures from Gaussian optics produced at the refracting interface.

In general, f_4 has an expansion of the form (6.91) because the lens shown in Fig. 6.1 and described by (6.93) has the required symmetry properties. Moreover, calculation shows that the coefficients A through F have the explicit values

$$A = 0 \quad , \quad B = 0 \quad , \quad C = 0$$

$$D = \alpha(n_2-n_1)/(2n_1n_2)$$

$$E = 2\alpha^2(n_1-n_2)/n_1 \qquad (6.97)$$

$$F = \alpha^3(n_1-n_2)\{n_1[(\beta/\alpha^3)+2] - 2n_2\}/n_1.$$

Observe that three of the six possibly nonzero terms are in fact zero. This fact may be interpreted as a good omen indicating that the Lie algebraic description is ideally suited in some sense to the task of characterizing the optical properties of a lens. However, Eqs. (6.97) should not be interpreted to mean that the lens of Fig. 6.1 has no spherical aberration, coma, or astigmatism. Strictly speaking, as described earlier, aberrations are not properties of a lens by itself, but rather are properties of an optical system which is imaging in the

Gaussian approximation. As will be seen shortly, when the lens in question is preceeded and followed by transit in order to produce an imaging system, then the complete system generally does have these and other aberrations.

To see how Lie transformations may be concatenated to describe a compound optical system, consider the simple case of an imaging system consisting of transit in air (n=1) over a distance d_1, refraction by the lens of Fig. 6.1 with $n_1 = n$ and $n_2 = 1$, and final transit in air over a distance d_2. Using the results already obtained, the symplectic map $\mathcal{M}$ for this sytem is given by the product of transit, refraction, and transit:

$$\begin{aligned}\mathcal{M} = {} & \exp\{(-d_1/2):\vec{p}^2:\}\exp\{(-d_1/8):(\vec{p}^2)^2:\}\times \\ & \exp\{(-t)/(2n):\vec{p}^2:\}\exp\{(-t)/(8n^3):(\vec{p}^2)^2:\}\times \\ & \exp\{\alpha(1-n):\vec{q}^2:\}\exp(:f_4:)\times \\ & \exp\{(-d_2/2):\vec{p}^2:\}\exp\{(-d_2/8):(\vec{p}^2)^2:\} \quad .\end{aligned} \tag{6.98}$$

Because the optical system is assumed to be imaging in the Gaussian approximation, the quantities d_1, d_2, t, α, and n are taken to be related by the familiar imaging condition

$$1/(d_1 + t/n) + 1/d_2 = 1/f. \tag{6.99}$$

Here f is the focal length of the lens defined by the relation

$$1/f = (n-1)/r = 2\alpha(n-1). \tag{6.100}$$

Using the tools developed previously, the expression (6.98) can be re-expressed in the factored product form

$$\mathcal{M} = \mathcal{M}_G \exp(:f_4^*:) \cdots . \tag{6.101}$$

Here the quantity $\mathcal{M}_G$ denotes the Gaussian portion of the map, and is given by the relation

$$\begin{aligned}\mathcal{M}_G = {} & \exp\{(d_1/2):\vec{p}^2:\}\times \\ & \exp\{(-t)/(2n):\vec{p}^2:\}\exp\{\alpha(1-n):\vec{q}^2:\}\times \\ & \exp\{(-d_2/2):\vec{p}^2:\}\end{aligned} \tag{6.102}$$

as would be expected from Gaussian optics. The determination of the function f_4^* requires somewhat more effort. In general, f_4^* has an expansion of the same form as that in (6.91). The explicit determination of the coefficient can be carried out using Eqs. (5.70) and (5.80).

Denoting the coefficients occurring in f_4^* by the symbols A^* through F^*, one finds the results

$$\begin{aligned}
A^* &= -m^4(d_1 + tn^{-3})/8 + (d_2^2 D - d_2^3 E + d_2^4 F) - d_2/8 \\
B^* &= -m^3(d_1 + tn^{-3})/(2f) + (-2d_2 D + 3d_2^2 E - 4d_2^3 F) \\
C^* &= -m^2(d_1 + tn^{-3})/(2f^2) + (-2d_2 E + 4d_2^2 F) \qquad (6.103) \\
D^* &= -m^2(d_1 + tn^{-3})/(4f^2) + (D - d_2 E + 2d_2^2 F) \\
E^* &= -m(d_1 + tn^{-3})/(2f^3) + (E - 4d_2 F) \\
F^* &= -(d_1 + tn^{-3})/(8f^4) + F.
\end{aligned}$$

Here the quantities D, E,and F are those given by (6.97).

Evidently all the coefficients A^* through F^* are nonzero in the general case. Thus an imaging system made with a single lens having one flat face generally suffers form all five Seidel aberrations.

By looking at the terms appearing in the coefficients A^* through F^*, it is possible to determine the various sources of aberration. For example, consider A^* which describes spherical aberration. The first term in A^* arises from transit in air over the distance d_1 and transit through the thickness of the lens, the second set of terms is produced by the lens itself, and the last term is caused by the final transit over the distance d_2. From this result it is clear that simple transit, when combined with even perfect Gaussian refraction, is itself a source of aberration.

Further discussion of the use of Lie algebraic methods in light optics is outside the scope of this section. Additional results may be found elsewhere.[20]

<u>Exercise 6.19</u>: Show that $\mathcal{M}_G$ as given by (6.102) has the action

$$\mathcal{M}_G \begin{bmatrix} \vec{q} \\ \vec{p} \end{bmatrix} = \begin{bmatrix} m & 0 \\ -1/f & 1/m \end{bmatrix} \begin{bmatrix} \vec{q} \\ \vec{p} \end{bmatrix}$$

where m is the magnification defined by the ratio

$$m = -d_2/(d_1 + t/n).$$

<u>Exercise 6.20</u>: Verify the expressions for A^* through F^*.

6.6 Work in progress

The preparation of a catalog of Lie operators as outlined in sections 6.1 through 6.4 is being continued to embrace all common beam elements including active elements such as accelerating cavities and bunchers. In addition, the catalog will contain all fourth order polynomial terms f_4 as well as the f_2 and f_3 terms previously described. Consequently the catalog will include octupoles as well as sextupoles.

Second, this catalog is being incorporated in a computer program which is being written to concatenate various beam elements, using the calculus for Lie transformations described in Section 5.1, in order to find the complete behavior for any collection of beam elements. The goal of this effort is to produce a charged particle beam transport code whose function is similar to the widely used program TRANSPORT,[21] but which will routinely work to one higher order in nonlinear effects so as to treat octupole elements, iterated sextupoles, and all other sources of third-order nonlinearities. Because the Lie algebraic methods automatically take into account the symplectic nature of $\mathcal{M}$, it is expected that storage requirements for the program under development will be approximatley 34 times less than what would be required if one were to try to extend the methods of TRANSPORT to one order higher. The methods of TRANSPORT, when extended to the next higher order, would require the use of 83×83 matrices in the general case. This requires $(83)^2 = 6889$ storage locations for each matrix. By contrast, the Lie algebraic methods require only the storage of the coefficients in the polynomials f_2, f_3, and f_4. This requires only 203 storage locations for each symplectic map in the general case. It is also hoped that the program will be fast enough so that it can be run iteratively to produce self-consistent space-charge forces in the case of intense beams. Finally, it is hoped that the description of nonlinearities in terms of polynomials rather than in terms of large matrices will provide additional insight into methods for both controlling and exploiting nonlinear behavior.

7. APPLICATIONS TO ORBITS IN CIRCULAR MACHINES AND COLLIDING BEAMS

The purpose of this chapter is to apply the results developed in previous chapters to the case of orbits in circular machines and colliding beam machines. By "circular" it is meant that each orbit passes through the same set of beam elements over and over again. Attention will be devoted to the existence of closed orbits and their stability. Stability will be discussed both in the linear approximation and in the general case where nonlinear structure resonances must be considered. Finally, the beam-beam interaction will be treated using methods analogous to those developed for structure resonances.

7.1 Existence of closed orbits

A perfectly constructed lattice for a "circular" machine has at least one closed orbit, namely the design orbit associated with the design momentum. In this section, it is shown that, under very general conditions, a perfect lattice also has a unique closed orbit for every other momentum value. Moreover, an efficient method is presented for its computation. Finally, as a byproduct of the discussion, it is shown that the same conclusions hold for a perturbed lattice. This is comforting, because lattices as actually constructed are, of course, unavoidably imperfect.

The proof of the existence and the computation of the off-momentum closed orbits are facilitated by topological methods that have their basis in the work of Poincare. These methods are both elegant and powerful, and therefore worthy of a brief explanation. Basically, they consist of converting the problem of studying orbits into an equivalent mathematical problem of studying a related symplectic map. This symplectic map is in turn studied by fixed point methods.

The first step is the construction of a Poincare surface of section. Consider an imaginary plane which intersects the design orbit at right angles somewhere in some straight section. Then it is obvious that every other orbit will also intersect this plane. Indeed, every orbit intersects this plane each time it goes around the lattice.

In our subsequent discussion, we will assume that the total momentum p remains a constant along an orbit. This is equivalent to assuming that all accelerating sections and bunchers have been turned off, and the effect of synchrotron radiation, if any, is ignored. Then every orbit with a given fixed total momentum p is completely determined by the values of its two transverse coordinates and two transverse momenta at the point of intersection. Indeed, the orbit equations can be written in the form

$$u_i' = g_i(\vec{u},\theta) \qquad i = 1,\ldots 4. \qquad (7.1)$$

Here $\vec{u}$ denotes a four-component vector whose entries are the two coordinates and two momenta transverse to the beam axis. The quantity θ denotes some generalized angle which increases by 2π in going around the lattice, and a prime denotes differentiation with respect to θ. Consequently, the values of the four quantities u_i at the point of intersection may be viewed as a complete set of initial conditions for the four first-order orbit differential equations. Finally, it is only necessary to recall that the solution of a set of first-order differential equations is uniquely and completely specified by initial conditions. The whole situation may be summarized by saying that there is a certain four-dimensional hypersurface in phase space which cuts across every phase-space trajectory for the orbits under study, and therefore is appropriately called a surface of section. In addition, each orbit is uniquely specified in terms of any point (four coordinates) at which it crosses the surface.

> Exercise 7.1: Supply the necessary reasoning to show that the equations of motion for the transverse coordinates and momenta can be written in the form (7.1). Show that in fact the equations of motion come from a Hamiltonian.
>
> Hint: Imagine applying theorem (1.1) to each beam element in the lattice. In each element, select θ to be proportional to q_1, perhaps with a proportionality constant which varies from element to element. Observe that, according to (1.21c), p_t (or equivalently the total momentum p) is a constant since under the assumptions made, K is time independent for each beam element. Therefore, as far as the equations of motion for the remaining variables are concerned, p_t may be treated as a constant parameter. Also, Eq. (1.21a) may be ignored since it is unnecessary to know the transit time in order to determine the orbit. The remaining Eqs. (1.21b) and (1.21d), with suitable scaling from element to element to reflect the varying proportionality between q_1 and θ, then lead to Eqs. (7.1).

The next step is to observe that orbits in the lattice generate a symplectic mapping $\mathcal{M}$ of the surface of section into itself. Consider a point on the surface of section. Since any such point requires four numbers for its specification, it is convenient to denote these four numbers collectively by a four-component vector $\vec{a}$. Now use the coordinates of $\vec{a}$ as initial conditions, and follow the orbit with these initial conditions once around the lattice until it again crosses the surface of section at some point $\vec{b}$. The mapping $\mathcal{M}$, called the Poincare map, is simply defined by the relation

$$\vec{b} = \mathcal{M}\vec{a}. \tag{7.2}$$

That is, $\mathcal{M}$ describes the effect of one circuit around the lattice.

Note that since the equations of motion are in general nonlinear, the relation between $\vec{b}$ and $\vec{a}$ defined by the Poincare map $\mathcal{M}$ is also nonlinear. Finally, the symplectic nature of the map follows from theorem (4.1) with the independent variable θ playing the role of the time variable. That is, the Poincare map $\mathcal{M}$ is the result of following a Hamiltonian flow. (See exercise 7.1).

Much of what one wants to know about orbits is equivalent to a knowledge of $\mathcal{M}$. For example, the key question of the long term behavior of orbits for a large number of turns is equivalent to a knowledge of $\mathcal{M}^n$ for large n. More particularly of interest for the current discussion, the determination of a <u>closed orbit</u> is equivalent to the discovery of a point $\vec{f}$, called a <u>fixed point</u>, which is sent into <u>itself</u> under the action of $\mathcal{M}$,

$$\mathcal{M}\vec{f} = \vec{f}. \tag{7.3}$$

That is, the set of initial conditions $\vec{f}$ for a closed orbit must, by definition, be mapped into itself by one circuit around the lattice.

Another concept needed is that of the <u>linear part</u> of a map. Let $\vec{a}$ be an arbitrary point in the surface of section and let $\vec{a} + \vec{\varepsilon}$, where $\vec{\varepsilon}$ is a small vector, be a point near $\vec{a}$. Now consider the point $\mathcal{M}(\vec{a} + \vec{\varepsilon})$. According to Eq. (7.2) this point should be near $\vec{b}$ since $\vec{a} + \vec{\varepsilon}$ is near $\vec{a}$. In fact, there is a power series expansion in $\vec{\varepsilon}$ of the form

$$\mathcal{M}(\vec{a} + \vec{\varepsilon}) = \vec{f} + M_a\vec{\varepsilon} + O(\vec{\varepsilon}^2) \tag{7.4}$$

where M_a is the 4×4 Jacobian matrix of $\mathcal{M}$ at the point $\vec{a}$. (See section 4.1.) Evidently, M_a describes the linearized behavior of $\mathcal{M}$ in the vicinity of $\vec{a}$. For this reason, the Jacobian matrix M_a can also be viewed as the linear part of $\mathcal{M}$ at $\vec{a}$.

The linear part of $\mathcal{M}$ can be conveniently calculated from the <u>variational equations</u> associated with the main orbit Eqs. (7.1). Variational equations describe all orbits near a particular orbit. In contrast to the main orbit equations, the variational equations are linear by construction. Let $\vec{u}_*(\theta)$ denote a particular orbit of interest. Then orbits near this orbit can be written in the form

$$\vec{u} = \vec{u}_*(\theta) + \varepsilon\vec{w} \tag{7.5}$$

where ε is a small quantity. By definition, $\vec{u}_*(\theta)$ satisfies the Eq. (7.1). Inserting the prescription (7.5) into the equation of motion (7.1), and retaining terms of lowest order in ε, shows that $\vec{w}$ must satisfy the variational equation

$$\vec{w}' = A_*(\theta)\vec{w}. \tag{7.6}$$

Here A_* is a 4 × 4 theta-dependent matrix defined by

$$A_{*ij}(\theta) = \partial g_i(\vec{u},\theta)/\partial u_j \Big|_{\vec{u} = \vec{u}_*(\theta)}. \tag{7.7}$$

It is evident that the variational Eq. (7.6) is linear in $\vec{w}$, and that the θ dependence of A_*, as indicated by the subscripted star, depends in general on the orbit $\vec{u}_*(\theta)$ about which variations are being studied.

Because the variational equations are linear, their solution for all initial conditions can be obtained by a finite amount of computation: Let $\theta = 0$ denote some arbitrary point in the lattice. Consider the first-order linear <u>matrix</u> variational equation defined by

$$B_*'(\theta) = A_*(\theta)\, B_*(\theta) \tag{7.8}$$

with the initial condition

$$B_*(0) = I. \tag{7.9}$$

Here B_* is a 4 × 4 matrix, I denotes the 4 × 4 identity matrix, and A_* is the same matrix as defined in (7.7). The solution to (7.8), with the initial condition (7.9), is uniquely defined and can be obtained if desired by a finite amount of numerical integration. Indeed, the integration of (7.8) is equivalent to the integration of 16 linear first-order equations (since B_* is 4 × 4) with the single set of initial conditions specified by (7.9). Now let $\vec{w}^0$ be an arbitrary four-component vector. Consider $\vec{w}(\theta)$ defined by the equation

$$\vec{w}(\theta) = B_*(\theta)\vec{w}^0. \tag{7.10}$$

One easily checks that $\vec{w}(\theta)$ is a solution to the variational equations (7.6) and satisfies the arbitrarily prescribed initial condition

$$\vec{w}(0) = \vec{w}^0. \tag{7.11}$$

It is now easy to see that the linear part of $\mathcal{M}$ is available from the variational equations. Suppose the surface of section is located at $\theta = 0$. Let $\vec{u}_a(\theta)$ be the orbit with initial conditions $\vec{a}$. That is,

$$\vec{u}_a(0) = \vec{a}. \tag{7.12}$$

Then this trajectory must also satisfy the equation

$$\vec{u}_a(2\pi) = \mathcal{M}\vec{a} = \vec{b}. \tag{7.13}$$

Next, with the aid of the variational equations, the nearby trajectory $\vec{u}_{a+\varepsilon}(\theta)$ with initial conditions $\vec{a} + \vec{\varepsilon}$ is expressible in the form

$$\vec{u}_{a+\varepsilon}(\theta) = \vec{u}_a(\theta) + B_a(\theta)\vec{\varepsilon} + O(\vec{\varepsilon}^2). \qquad (7.14)$$

[See Eqs. (7.5), (7.10), and (7.11).]

Now put $\theta = 2\pi$ in Eq. (7.14). The result is the relation

$$\mathcal{M}(\vec{a} + \vec{\varepsilon}) = \vec{u}_{a+\varepsilon}(2\pi) = \vec{u}_a(2\pi) + B_a(2\pi)\vec{\varepsilon} + O(\vec{\varepsilon}^2). \qquad (7.15)$$

Comparison of (7.15) and (7.4), with the aid of (7.13), gives the result

$$M_a = B_a(2\pi). \qquad (7.16)$$

The stage has now been set for the determination of the fixed points of $\mathcal{M}$. This will be done with the aid of another map $\mathcal{C}$, called a contraction map, defined in terms of $\mathcal{M}$. Let $\vec{e}$ be an arbitrary point in the vicinity of a fixed point $\vec{f}$. The contraction map $\mathcal{C}$ will be shown to have the remarkable property

$$\vec{f} = \lim_{n\to\infty} \mathcal{C}^n \vec{e}. \qquad (7.17)$$

That is, a guess as to the whereabouts of a closed orbit is sufficient starting information to contract in on it exactly. In practice, the starting guess can be taken to be the initial conditions for the on-momentum design orbit.

The construction of the contraction mapping $\mathcal{C}$ is a generalization of Newton's method to the case of several variables.[22] The map $\mathcal{C}$ is defined by requiring that its action on the arbitrary point $\vec{a}$ be given by the rule

$$\mathcal{C}\vec{a} = \vec{a} - (I - M_a)^{-1}(\vec{a} - \mathcal{M}\vec{a}). \qquad (7.18)$$

It is easily verified that the map $\mathcal{C}$ defined by (3.10) has the advertised property (7.17). First, suppose that $\vec{f}$ is a fixed point of $\mathcal{M}$. Then, it is easily verified that $\vec{f}$ is also a fixed point of $\mathcal{C}$,

$$\mathcal{C}\vec{f} = \vec{f}. \qquad (7.19)$$

Next, suppose that $\vec{e}$ is some point in the vicinity of $\vec{f}$. Then $\vec{e}$ is of the form $\vec{f} + \vec{\varepsilon}$ where $\vec{\varepsilon}$ is some small vector. Inserting $\vec{e}$ in Eq. (7.19), one finds after a short calculation that

$$\begin{aligned}\mathcal{C}\vec{e} = \mathcal{C}(\vec{f} + \vec{\varepsilon}) &= \vec{f} + \vec{\varepsilon} - (I - M_{f+\varepsilon})^{-1}[(\vec{f} + \vec{\varepsilon}) - \mathcal{M}(\vec{f} + \vec{e})] \\ &= \vec{f} + \vec{e} - (I - M_{f+\varepsilon})^{-1}[(I - M_f)\vec{\varepsilon} + O(\vec{\varepsilon}^2)] \\ &= \vec{f} + O(\vec{\varepsilon}^2). \qquad (7.20)\end{aligned}$$

Here use has been made of the relation

$$\mathcal{M}(\vec{f} + \vec{\varepsilon}) = \vec{f} + M_f\vec{\varepsilon} + O(\vec{\varepsilon}^2) \qquad (7.21)$$

and the observation that

$$M_{f+\varepsilon} = M_f + O(\varepsilon). \qquad (7.22)$$

Thus, according to (7.20), although the initial point $\vec{e}$ differs from the desired fixed point $\vec{f}$ by an amount $\vec{\varepsilon}$, the point $\mathcal{C}\,\vec{e}$ differs from the point $\vec{f}$ only by an amount of order $\vec{\varepsilon}^2$. Similarly, the point $\mathcal{C}^2\vec{e}$ differs from $\vec{f}$ only by an amount of order $(\vec{\varepsilon}^2)^2$, and $\mathcal{C}^n\vec{e}$ differs from $\vec{f}$ only by an amount of order $|\vec{\varepsilon}|^{2n}$. Consequently the convergence of the limit (7.17) to $\vec{f}$ is extremely fast. When applied to a typical problem in actual practice, one finds that the fixed point corresponding to an off-momentum closed orbit is given accurately to 1 part in 10^{10} by the time $n = 4$.[23]

Let p^o denote the design momentum, and p the momentum value of interest. Write

$$p = p^o(1 + \delta) \qquad (7.23)$$

and give δ a small value. Also, let $\vec{f}(\delta)$ denote the fixed point (set of closed orbit initial conditions) corresponding to the momentum given by the relation (7.23). Examination of (7.18) and (7.20) shows that Newton's method succeeds as long as the matrix $I - M_{f(\delta)}$ has an inverse. Thus, as one gradually changes the momentum away from the design value, $\delta = 0$, the closed design orbit will continuously deform into the off-momentum closed orbit.

In order to complete the discussion, it is necessary to check whether the matrix $[I - M_{f(\delta)}]$ has an inverse. Evidently the inverse exists unless the related determinant satisfies the condition

$$\det[M_{f(\delta)} - I] = 0. \qquad (7.24)$$

Equation (7.24) is equivalent to the condition that the Jacobian matrix $M_{f(\delta)}$ has eigenvalue +1.

Since $\mathcal{M}$ is a symplectic map, the Jacobian matrix $M_{f(\delta)}$ must be symplectic. However, according to section 2.3 and Fig. 2.2, all the eigenvalues of a symplectic matrix generally differ from one. Moreover, exercise 2.10 shows that if the eigenvalues of $M_{f(\delta)}$ differ from 1 for some value of δ, then the same will be true for nearby values of δ. Thus, if the eigenvalues of $M_{f(0)}$ for the design orbit are far away from +1, then the eigenvalues of $M_{f(\delta)}$ for the closed orbit associated with the initial condition $\vec{f}(\delta)$ will also differ from 1 for a range of δ values, and there will be one closed orbit for each value of δ. Finally, it will be shown in the next section that the eigenvalues of $M_{f(\delta)}$ for a closed orbit are related to the tunes for small betatron oscillations about this orbit. In particular, $M_{f(\delta)}$ can have eigenvalue +1 only if some tune has an integer value. Thus, the existence of closed orbits is assured for a range of momentum values provided the tunes of the design orbit are far from integer values.

A moment's reflection on the arguments just made for the general existence of an off-momentum closed orbit shows that similar arguments are applicable to other situations. The quantity δ simply appears as a parameter in the equations of motion, and what is being studied is the continued existence of a closed orbit as a parameter is continuously varied to a new value. Now, the difference between a perfect lattice and the perturbed lattice realized in actual construction may also be regarded as the result of a variation in certain parameters. Consequently, if the tunes for the closed design orbit in the perfect lattice have noninteger values, then the imperfect lattice will also have a closed orbit at the design momentum (and other nearby momenta as well) providing the perturbations in the lattice are not so large as to drive some tune to an integer value. In particular, small imperfections in a lattice, such as arise from magnet misalignment and misplacement, magnet under or over powering, magnetic fringe fields and general magnetic field inhomogenities, etc., do not destroy the existence of closed orbits but merely cause them to be slightly distorted.

Exercise 7.2: Verify Eq. (7.20).

7.2 Stability of closed orbits in the linear approximation

Let $\vec{f}(\delta)$ be the initial conditions in the Poincare surface of section for the closed orbit corresponding to the momentum deviation δ, and consider a nearby orbit having the same total momentum and the initial conditions $\vec{f} + \vec{\varepsilon}$ where $\vec{\varepsilon}$ is a small vector. Consider the action of $\mathcal{M}$ on the nearby orbit with initial conditions $\vec{f} + \vec{\varepsilon}$. Using (7.2), (7.3), and (7.4) one finds the result

$$\mathcal{M}(\vec{f} + \vec{\varepsilon}) = \vec{f} + M_f\vec{\varepsilon} + O(\vec{\varepsilon}^2). \qquad (7.25)$$

This relation was already employed in verifying (7.20).

Similarly, the action of $\mathcal{M}^n$ is given by the relation

$$\mathcal{M}^n(\vec{f} + \vec{\varepsilon}) = \vec{f} + M_f^n\vec{\varepsilon} + O(\vec{\varepsilon}^2). \qquad (7.26)$$

Thus, for nearby orbits, the long term behavior resulting from many lattice circuits is governed, in lowest approximation, by the matrix powers M_f^n.

But the behavior of large powers of a matrix is controlled, in turn, by the eigenvalue spectrum of the matrix. Moreover, as already seen earlier, M_f is a 4×4 symplectic matrix. Therefore, the various possibilities for the spectrum must be as illustrated in Fig. 2.2.

The behavior of M_f^n for large n is easily analyzed in each case. Suppose that some eigenvalue λ of M_f has a magnitude which exceeds one as in cases 1 through 4 of the generic configurations and cases 1, 2, and 6 of the degenerate configurations of Fig. 2.2. Then in these cases M_f^n grows exponentially without bound as n increases [essentially as $|\lambda|^n = \exp(n \log|\lambda|)$]. Correspondingly, the closed orbit associated with the fixed point $\vec{f}$ is unstable. That is, orbits with the same total momentum and initially nearby will, according to (7.26), deviate exponentially away from the closed orbit over the course of successive circuits around the lattice.

Next suppose that some eigenvalue λ of M_f has the value ± 1 while all others have absolute value one as in cases 3, 4, and 7 of the degenerate configurations. Then, as shown earlier in section 2.3, this eigenvalue must have even multiplicity. Consider, to be concrete, the case $\lambda = +1$, and suppose that M_f is brought to Jordan normal form by a similarity transformation S.[24] Then, in general the result will be a relation of the form

$$S^{-1} M_f S = \left(\begin{array}{cc|c} 1 & a & 0 \\ 0 & 1 & \\ \hline 0 & & \end{array}\right) . \tag{7.27}$$

Here only the upper left-hand 2 × 2 block of (7.27) has been specified, and the quantity a denotes a number which is generally nonzero. See, for example, Eq. (6.53) for a 2 × 2 matrix of this type. Then when this 2 × 2 matrix is raised to the n'th power, as occurs in the calculation of M_f^n, one finds the result

$$\begin{pmatrix} 1 & a \\ 0 & 1 \end{pmatrix}^n = \begin{pmatrix} 1 & na \\ 0 & 1 \end{pmatrix} . \tag{7.28}$$

Thus, unless the quantity a vanishes, which is generally not the case, M_f^n grows linearly with n. Correspondingly, the closed orbit associated with the fixed point $\vec{f}$ is again unstable, although not as unstable as the exponential growth case of the previous paragraph. Note also that, under a small perturbation, an eigenvalue in this case can leave the unit circle through the point +1, and then the closed orbit becomes exponentially unstable.

Exercise 7.3: Carry out similar reasoning for the case $\lambda = -1$ to show that M_f^n again grows linearly with n, and hence the closed orbit associated with $\vec{f}$ is again linearly unstable. Also show that under a small perturbation an eigenvalue can leave the unit circle through the point -1, and the closed orbit then comes exponentially unstable.

Examination of Fig. 2.2 shows that there are still two possibilities which have not been discussed, namely case 5 of the generic configurations, and case 5 of the degenerate configurations. In the generic configuration of case 5, the matrix M_f can be diagonalized by a similarity transformation since all its eigenvalues are distinct. For this possibility, let λ_1 and λ_2 be two of the distinct eigenvalues which are not complex conjugates. Since they lie on the unit circle, they can be written in the form

$$\lambda_1 = e^{i\psi_1}, \ \lambda_2 = e^{i\psi_2} \tag{7.29}$$

where ψ_1 and ψ_2 are two real numbers. The remaining two distinct eigenvalues are then just the numbers $e^{-i\psi_1}$ and $e^{-i\psi_2}$. It follows that in this situation M_f^n merely oscillates with increasing n. That is, all matrix elements remain bounded. Correspondingly, the closed orbit associated with $\vec{f}$ must be <u>stable</u>. That is, to the accuracy of the linear approximation (7.26), nearby orbits exhibit bounded betatron oscillations about the closed orbit.

At this point it is possible to make another fundamental observation about imperfect lattices. Suppose a perfect lattice has been designed, all its closed orbits for a range of off-momentum values have been found, and all these closed orbits prove to be stable. What happens if this lattice is slightly perturbed? It has already been argued that the various closed orbits generally continue to exist and are only slightly deformed under perturbation. But do they remain stable? Fortunately, because not even the Swiss can build perfect lattices, the answer is yes thanks to the symplectic condition.

To understand this result, suppose the eigenvalues of M_f were originally all complex, of absolute value 1, and distinct. Then, according to exercise 2.10, this situation persists under a small perturbation. Indeed, for an eigenvalue to leave the unit circle, the eigenvalue spectrum must first pass through one of the degenerate configurations shown in part B of Fig. 2.2. Thus, from this perspective, the symplectic condition is really the key reason why it is possible to build workable accelerators and storage rings.

What about the last possibility, case 5 of the degenerate configurations? In this situation more needs to be known about the matrix M_f. If it can be diagonalized despite the fact that its eigenvalues are not distinct, then the same reasoning can be used as in the generic case, and the closed orbit is stable. However, suppose M_f cannot be diagonalized, but can only be brought to the Jordan form. Then, as in the other degenerate cases, the closed orbit is linearly unstable. To discover which of these possibilities holds, and the effect of small perturbations, requires considerably more discussion which is beyond the scope of these lectures.[25]

<u>Exercise 7.4</u>: Suppose that $\vec{f}$ is a fixed point of $\mathcal{M}$ and that M_f has a spectrum corresponding to one of the five <u>generic</u> configurations of Fig. 2.2. Show that in this case $\vec{f}$ must be an <u>isolated</u> fixed point. That is, there are no other fixed points near $\vec{f}$.

<u>Hint</u>: Assume that $\vec{g}$ is a nearby fixed point of $\mathcal{M}$. Then $\vec{g} = \vec{f} + \vec{\varepsilon}$ where $\vec{\varepsilon}$ is small. Now show that $\vec{\varepsilon}$ must satisfy $M_f\vec{\varepsilon} = \vec{\varepsilon}$, and that this is impossible unless $\vec{\varepsilon} = 0$. For extra credit, show that if two or more fixed points are to coalesce as some parameter is varied, then the spectrum of M_f when they meet must be one of the degenerate configurations of Fig. 2.2 with an eigenvalue +1.

As intimated earlier, the eigenvalues λ_1 and λ_2 in the case of a stable closed orbit are related to the betatron oscillation tunes. Roughly speaking, the tunes of a closed stable orbit can be defined to be the number of vertical and horizontal oscillations made about the closed orbit by a nearby orbit during one lattice circuit. More precision in definition is possible for the case of the design orbit, if the vertical and horizontal degrees of freedom are uncoupled owing to the lattice having midplane symmetry, by the method of transfer matrices. In that case, tunes are related to the eigenvalues of the transfer matrix.

But a moment's reflection shows that the transfer matrix for the design orbit from the point $\theta=0$ to the general point θ coincides with $B_{f(0)}(\theta)$. [Here $\vec{f}(0)$ denotes $\vec{f}(\delta)$ with $\delta=0$.] Now consider the general off-momentum closed orbit. Assume that $\mathcal{M}$ is the Poincare map for a complete circuit about the lattice so that M_f and B_f for the general fixed point $\vec{f}(\delta)$ are related by the equation

$$M_f = B_f(2\pi). \tag{7.30}$$

It follows from a general Floquet analysis, in analogy with the method of transfer matrices, that the tunes T_1, T_2 of a general closed orbit (even in the absence of midplane symmetry for the lattice under consideration) can be defined to be the numbers

$$\begin{aligned} T_1 &= \pm(\psi_1/2\pi) + \text{some integer} \\ T_2 &= \pm(\psi_2/2\pi) + \text{some integer.} \end{aligned} \tag{7.31}$$

Here the quantities ψ_i are related to the eigenvalues of M_f by (7.29).

Note that if +1 is an eigenvalue of M_f, then according to (7.29) and (7.31) at least one tune must be an integer. As indicated earlier, the proof of the existence and uniqueness of off-momentum closed orbits breaks down under these conditions. Moreover, according to the previous orbit stability discussion, if a closed orbit does exist when $\lambda = 1$, it is generally linearly unstable and subject to becoming exponentially unstable under small perturbation. Note also that when a tune is half integer, some eigenvalue of M_f has the value -1. Then the orbit is again generally linearly unstable and again subject to becoming exponentially unstable under a small perturbation. Consequently, integer and half integer tunes are generally to be avoided unless, perhaps, one is trying to exploit orbit instability combined with suitable nonlinear effects to achieve extraction.

Equation (7.31) states that tunes, in the general case, are only defined modulo an integer. This should not be surprising. It is not possible to detect how many whole oscillations are made simply by observing the orbit at one surface of section. Moreover, if the horizontal and vertical degrees of freedom are strongly coupled, it is not always possible to define what is meant by a whole oscillation even if the orbit is observed everywhere.

In the case that the lattice has N identical periods, the integer ambiguity can be resolved providing it is assumed, as is usually the case, that the number of betatron oscillations made in the passage through a single lattice period is less than 1. (This is equivalent to assuming that the phase advance per lattice period is less than 2π in the case that horizontal and vertical oscillations are uncoupled.) Suppose that we now take $\mathcal{M}$ to be the Poincare map for <u>one lattice period</u> so that M_f and B_f are related by the equation[26]

$$M_f = B_f(2\pi/N). \tag{7.32}$$

Again, as before, let λ_1 and λ_2 be two eigenvalues of M_f written in the form (7.29) with ψ_1, ψ_2 distinct and lying in the interval $(0,2\pi)$. In this case the numbers ψ_1, ψ_2 can be regarded as the phase advances per period, and the overall lattice tunes can be defined to be the numbers

$$\begin{aligned} T_1 &= N(\psi_1/2\pi) \\ T_2 &= N(\psi_2/2\pi). \end{aligned} \tag{7.33}$$

By contrast, the ψ values appearing in (7.31) are essentially phase advances, modulo signs and an unspecified number of 2π's, for an entire circuit of the whole lattice. Bearing this distinction in mind, it is easily verified that the definition (7.33) coincides with one of the set of values given by (7.31).

Exercise 7.5: In the case that the horizontal and vertical degrees of freedom of a lattice are uncoupled, the matrix M_f takes the reduced block form

$$M_f = \begin{pmatrix} \text{hor} & 0 \\ 0 & \text{ver} \end{pmatrix} \tag{7.34}$$

where the abbreviations "hor" and "ver" refer to 2 × 2 matrices which separately describe the horizontal and vertical motions. In this case, show that the phase advances ψ are given by the simple formulas

$$\psi_1 = \cos^{-1} [\,(1/2)\ \mathrm{tr}\ (\text{hor})\,] \tag{7.35a}$$

$$\psi_2 = \cos^{-1} [\,(1/2)\ \mathrm{tr}\ (\text{ver})\,]. \tag{7.35b}$$

In the more general case in which the horizontal and vertical degrees of freedom are coupled, it might be thought that a general diagonalization procedure would be required to find the eigenvalues of M_f. However show, using the results of exercise 2.11, that the two phase advances, now denoted by $\psi_\pm$, can be found directly from the relations

$$\psi_\pm = \cos^{-1} [-b \pm (b^2 - c)^{1/2}\,]. \tag{7.36}$$

In the discussion so far, it has been assumed that the total momentum p remains constant along an orbit. It is also possible to treat more general situations. Consider, for example, the case of a proton storage ring with one or several radio frequency bunching elements. Then in this case there is one design orbit with constant total momentum, and other orbits which exhibit synchrotron oscillations. To describe these various orbits, it is necessary to introduce the additional variables t and p_t as well as the previously used two coordinates and two momenta transverse to the beam. Indeed, it is useful to employ, as in Section 6, the time and conjugate momentum deviation variables T and P_T. Then the Poincare surface of section becomes a full six dimensional phase space, and the Poincare mapping $\mathcal{M}$ maps this phase space into itself.

As before, a fixed point $\vec{f}$ of $\mathcal{M}$ corresponds to a closed orbit. In particular, for a perfect machine there is a fixed point corresponding to the synchronous on-momentum design orbit, and the T and P_T components of $\vec{f}$ satisfy the conditions $T = 0$ and $P_T = 0$.

Now consider M_f, the linear part of $\mathcal{M}$ at $\vec{f}$. The matrix M_f will be 6 × 6 and symplectic. Suppose all its eigenvalues are distinct and lie on the unit circle. This is the 6 × 6 analog of case 5 of the generic configurations. (See Fig. 2.2). Then in this case the design orbit will be stable, and all other nearby orbits, including off-momentum and off-synchronous orbits, will make synchro-betatron oscillations (possibly coupled) about the design orbit. Of course, now there will be three tunes, and one of them will be quite small because synchrotron oscillation frequencies are usually much lower than betatron oscillation frequencies. Finally one can show, by arguments similar to those made before, that the fixed point $\vec{f}$ of $\mathcal{M}$ is isolated, and continues to exist and to remain stable under small perturbations of the lattice including variations in the strength and frequency of the bunching fields. These are all again benefits of the symplectic condition.

The preceeding discussion has shown that a fixed point is stable if the eigenvalues of M_f are distinct and lie on the unit circle. Otherwise, the fixed point is generally unstable. Still more can be said about the fixed points of two dimensional symplectic mappings. This restricted case is of interest in its own right, and is directly applicable to accelerators in the approximation that the various degrees of freedom are uncoupled. Moreover, the restricted case suggests results that may have useful analogs in the general case of four and six dimensional symplectic mappings.[27] The series of exercises below develops what is known about the two dimensional case, and explains the significance of the names hyperbolic, elliptic, etc. that occur in Fig. (2.1).

<u>Exercise 7.6</u>: Suppose $\mathcal{M}$ is a symplectic map in two dimensions with a fixed point f. Without loss of generality, the fixed point may be taken to be located at the origin. Let M denote the linear part of $\mathcal{M}$ at the fixed point. Suppose further that the spectrum of M corresponds to case 3 of Fig. (2.1). Show that M then sends a certain set of nested ellipses around the origin into themselves. Its only action is to advance or "rotate" points along the ellipses. (The rate of advance is related to the tune of the associated closed orbit.) For this reason, if the spectrum of M is as in case 3, the fixed point is said to be <u>elliptic</u>. It follows from (7.26) that, in the linear approximation, points initially very near the origin remain near the origin under repeated application of $\mathcal{M}$. (Each point must remain on its particular ellipse.) For this reason, an elliptic fixed point is said to be <u>stable</u>.

<u>Solution</u>: Suppose the eigenvalues of M are $\exp(\pm iw)$. Then tr M = 2 cos w. Introduce Twiss parameters α, β, γ by writing without loss of generality

$$M = \begin{pmatrix} \cos w + \alpha \sin w & \beta \sin w \\ -\gamma \sin w & \cos w - \alpha \sin w \end{pmatrix}. \tag{7.37}$$

The symplectic condition that det M=1 requires the relation

$$\beta\gamma = 1 + \alpha^2. \tag{7.38}$$

Next observe that M can be expressed in the form

$$M = I \cos w + K \sin w \tag{7.39}$$

where K is the matrix

$$K = \begin{pmatrix} \alpha & \beta \\ -\gamma & -\alpha \end{pmatrix} . \tag{7.40}$$

Moreover, thanks to (7.38), K has the property

$$K^2 = -I. \tag{7.41}$$

Consequently, M can also be written in the exponential form

$$M = \exp(wK). \tag{7.42}$$

Finally, K can be written in the form

$$K = J\ S \tag{7.43}$$

where S is the symmetric matrix

$$S = \begin{pmatrix} \gamma & \alpha \\ \alpha & \beta \end{pmatrix} . \tag{7.44}$$

Now introduce a polynomial g_2 by the definition

$$\begin{aligned} g_2(z) &= -(1/2) \sum_{ij} S_{ij}\, z_i\, z_j \\ &= -(1/2)(\gamma q^2 + 2\alpha qp + \beta p^2). \end{aligned} \tag{7.45}$$

It follows from section 5.1 that M satisfies the relation

$$\exp(w{:}g_2{:})z = Mz. \tag{7.46}$$

Therefore g_2 is an invariant function. That is, g_2 has the property

$$g_2(Mz) = g_2(z). \tag{7.47}$$

Here, as usual, z stands for the complete collection of phase-space variables. See Eqs. (5.5) through (5.7) for a review of why g_2 should have the invariance property (7.47). Indeed, apart from the factor (-1/2), g_2 is the Courant-Snyder invariant.[28] Note that the discrimanant of the quadratic form (7.45), namely the quantity $(2\alpha)^2 - 4\beta\gamma$, has the negative value -4 thanks to Eq. (7.38). Consequently, the level lines of g_2 are indeed nested ellipses.

Exercise 7.7: Repeat exercise 7.6 under the supposition that the spectrum of M corresponds to case 1 of Fig. (2.1). Show that M then sends a certain set of hyperbolae into themselves. For this reason, if the spectrum of M is as in case 1, the fixed point is said to be hyperbolic. Show also that if a point is initially very near the fixed point, then in the linear approximation it will in general eventually be moved away from the fixed point under repeated action of M. Therefore, a hyperbolic fixed point is said to be unstable.

Hint: Suppose the eigenvalues of M are written in the form $\exp(\pm w)$. Then tr M = 2 cosh w. In this case, introduce parameters α, β, γ by writing

$$M = \begin{pmatrix} \cosh w + \alpha \sinh w & \beta \sinh w \\ -\gamma \sinh w & \cosh w - \alpha \sinh w \end{pmatrix} . \tag{7.48}$$

Show that in this case

$$\beta\gamma = \alpha^2 - 1. \tag{7.49}$$

Also show that M can be expressed in the form

$$M = I \cosh w + K \sinh w \tag{7.50}$$

with K as before. Next show that K now satisfies the relation

$$K^2 = +I. \tag{7.51}$$

Now show that the rest of the exercise goes through as before except that the discrimant of g_2 now has the positive value +4, and therefore level lines of g_2 are hyperbolae. Note that the level line $g_2 = 0$ corresponds to the fixed point itself and the asymptotes of the hyperbolae. Check that the asymptotes are given by the equations

$$q = -p\beta/(\alpha \pm 1). \tag{7.52}$$

Also verify that the eigenvectors of M lie along the asymptotes. Consider the asymptote corresponding to the eigenvector with eigenvalue greater than 1. Show that the action of M is to move points on this asymptote away from the origin, and to move points on the other asymptote toward the origin. For this reason, these two asymptotes are called the <u>unstable</u> and <u>stable manifolds</u> respectively. Show that the action of M on any given hyperbola is to move points along the hyperbola. See Fig. (7.1) which illustrates the action of M on a general point in the hyperbolic case.

<u>Exercise 7.8</u>: Repeat exercise 7.6 under the supposition that the spectrum of M corresponds to case 4 of Fig. (2.1). Show that M then sends a family of straight lines into themselves. In this case the fixed point is said to be parabolic, although the invariant curves are straight lines and <u>not</u> parabolas. Note that the parabolic case is a transitional case between the elliptic and hyperbolic cases. It is called parabolic out of deference to the convention that the word "parabolic" is used in other contexts to describe a circumstance intermediate to an elliptic or hyperbolic case. Show that the invariant line through the origin consists <u>entirely</u> of fixed points. (A parabolic fixed point is <u>not</u> isolated in the linear approximation. See exercise 7.4.) Also show that the action of M on each line is that of a linear displacement. For this reason, a parabolic fixed point is said to be <u>linearly unstable</u>.

<u>Hint</u>: Since M has +1 as an eigenvalue with multiplicity two, it must satisfy tr M = 2. Suppose M is parameterized by writing it in the form

$$M = \begin{pmatrix} 1+\alpha & \beta \\ -\gamma & 1-\alpha \end{pmatrix}. \tag{7.53}$$

Then the condition det M = 1 entails the restriction

$$\alpha^2 = \beta\gamma. \tag{7.54}$$

Next define a matrix K as in (7.46). Show that K satisfies the relation

$$K^2 = 0, \tag{7.55}$$

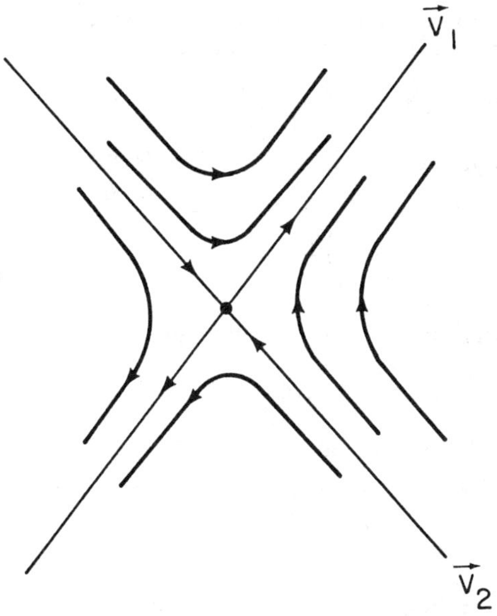

Figure (7.1): The action of M on points near a hyperbolic fixed point. Suppose that $\vec{v}_1$ and $\vec{v}_2$ are eigenvectors of M having eigenvalues greater and less than one respectively. Then points on $\vec{v}_1$ are moved outward, those on $\vec{v}_2$ are moved inward, and others are moved on hyperbolas.

and that M can be written in the form

$$M = \exp(K)\ . \tag{7.56}$$

Now proceed as before. Then show that (7.54) implies that g_2 can be written as a perfect square in the form

$$g_2(z) = -(1/2)(\gamma^{1/2}q + \beta^{1/2}p)^2 . \tag{7.57}$$

Let $\vec{u}$ be the vector with entries

$$\vec{u} = \begin{pmatrix} \beta \\ -\alpha \end{pmatrix}.$$

Show that $\vec{u}$ has eigenvalue +1. Next show that the line

$$g_2(z) = g_2(0) = 0 \tag{7.58}$$

is along $\vec{u}$. Therefore, all points on this line will remain fixed under the action of M. Let $\vec{v}$ be the vector with entries

$$\vec{v} = \begin{pmatrix} \alpha \\ \beta \end{pmatrix}.$$

Verify that $\vec{v}$ is orthogonal to $\vec{u}$. Suppose $\vec{w}$ is a general point in phase space. Since $\vec{u}$ and $\vec{v}$ are orthogonal, $\vec{w}$ can be expanded in terms of $\vec{u}$ and $\vec{v}$ to give a relation of the form

$$\vec{w} = a\vec{u} + b\vec{v}$$

where <u>a</u> and <u>b</u> are expansion coefficients. Verify that the action of M on the general point $\vec{w}$ is given by the relation

$$M\vec{w} = [\, a + b(\beta+\gamma)]\ \vec{u} + b\vec{v} = \vec{w} + b(\beta+\gamma)\ \vec{u}.$$

Thus, the action of M on a general point, in the parabolic case, is as illustrated in Fig. (7.2).

<u>Exercise 7.9</u>: Suppose the spectrum of M corresponds to the inversion hyperbolic or inversion parabolic cases of Fig. (3.1). Show that M can then be written as a product in the form

$$M = (-I)\ N \tag{7.59}$$

where N is symplectic and either hyperbolic or parabolic, respectively. Thus, M is a product of a hyperbolic or parabolic symplectric matrix followed by <u>inversion</u> through the origin. Let g_2 be the invariant function for N as constructed in exercises 7.7 or 7.8. For a fixed value, g_2 generally has two branches since it is an even function. The action of N is to

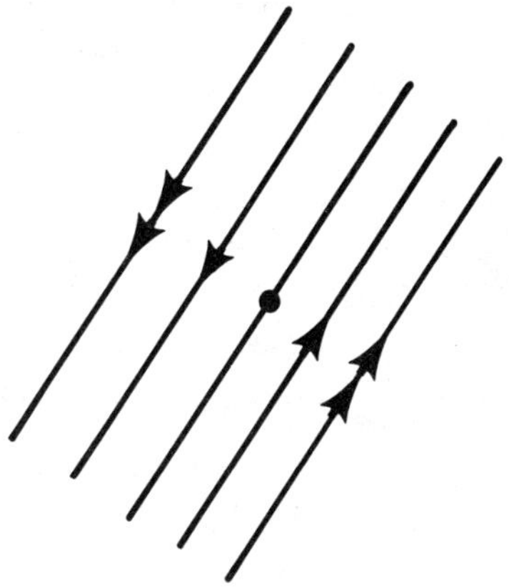

Figure (7.2): The action of M on points near a parabolic fixed point. The action is such as to produce a "sheared" flow along straight lines. The displacement along a line is proportional to the distance of that line from the fixed point.

move points along on a given branch, and the action of (-I) is to jump between corresponding branches. Consequently, the action of M^n for increasing n is to send a point jumping back and forth between two branches. Observe that M^2 is either hyperbolic or parabolic, respectively. Therefore, after two jumps, a point is always back on the same branch, and the net effect is motion along that branch just as in the hyperbolic or parabolic cases.

7.3 Stability of closed orbits including nonlinear effects

Suppose $\mathcal{M}$ is a symplectic map with fixed point $\vec{f}$. Without loss of generality, one can select a new canonical coordinate system in such a way that $\vec{f}$ is located at the origin. With this choice of coordinates, $\mathcal{M}$ becomes a symplectic map which sends the origin into itself.

Since $\mathcal{M}$ is a symplectic map which sends the origin into itself, it follows from the factorization theorem 5.1 that there exist homogeneous polynomials g_2, g_3, g_4, etc. such that $\mathcal{M}$ can be written in the form

$$\mathcal{M} = \exp(:g_2:) \exp(:g_3:) \exp(:g_4:) \cdots . \qquad (7.60)$$

(Here, in order to avoid confusion with the symbol f employed to refer to the fixed point $\vec{f}$, the letters g_n rather than f_n will be used to denote polynomials.)

From this perspective, the linear stability analysis made for closed orbits in the previous section is equivalent to a study of the "Gaussian" piece of $\mathcal{M}$, namely $\exp(:g_2:)$. Indeed, if the variables in the surface of section on which $\mathcal{M}$ acts are denoted by the letters z_1 z_{2n}, then one has the relation

$$\exp(:g_2:)\, z_i = \sum_j (M_f)_{ij}\, z_j . \qquad (7.61)$$

The linear stability analysis neglected the effect of the factors $\exp(:g_3:) \exp(:g_4:)\cdots$. It showed that closed orbits are stable, in lowest approximation, if the eigenvalues of M_f are distinct and all lie on the unit circle. The purpose of this section is to examine the effect of the higher order terms.

As stated earlier, the long-term behavior of orbits for a large number of turns is equivalent to a knowledge of $\mathcal{M}^n$ for large n. Suppose that $\mathcal{M}$ could be written as the exponential of a single Lie operator :h: in the form

$$\mathcal{M} = \exp(:h:). \qquad (7.62)$$

Then the evaluation of $\mathcal{M}^n$ would be simple, for one would simply have the result

$$\mathcal{M}^n = \exp(n:h:). \qquad (7.63)$$

But, thanks to remarkable pedagogical foresight, formulas for combining exponents into a single exponent were developed in section 5.4. To see how they may be applied, consider as a simple example the case where all terms in (7.60) beyond $\exp(:g_3:)$ are neglected. That is, suppose $\mathcal{M}$ is approximated by the truncated symplectic map

$$\mathcal{M} = \exp(:g_2:)\ \exp(:g_3:). \tag{7.64}$$

Then $:g_2:$ and $:g_3:$ can be combined using Eqs. (5.53), (5.81), and (5.83). One finds the result

$$h = g_2 + :g_2:[1 - \exp(-:g_2:)]^{-1}\ g_3$$
$$+ \text{ terms of degree 4 and higher.} \tag{7.65}$$

Note that Eqs. (5.81) and (5.83) omit terms of degree t^2 and higher. In the present context, this omission is equivalent to neglecting all commutators in an expression of the form (5.53) which contain two or more g_3's and any number of g_2's. However, according to exercise 5.6, these neglected terms are of degree 4 and higher as indicated in (7.65). It should also be remarked that these omitted terms can in principle be computed if needed. In particular, the term of degree t^2 is known.[29] Thus, h can be computed through degree 4 with existing tools. If such a calculation were to be made, then for consistency the contribution of degree four from the g_4 term in (7.60), which is easily computed, should be included as well.

In order to continue with the calculation for h as given by (7.65), it is necessary to specify more about g_2 or, equivalently, M_f. Consider for simplicity the case of a coasting proton beam. That is, all accelerating sections and bunchers have been turned off. Suppose further that all particles in the beam have the design momentum. Then attention can be restricted to the two transverse positions and momenta. Moreover, if the Poincare surface is indeed located in some straight section as has been assumed, then these transverse coordinates can be conveniently taken to be the Cartesian quantities y, z, p_y, p_z at some fixed location. Finally, assume that the horizontal and vertical betatron oscillations are uncoupled in the linear approximation, and that the design orbit is stable with tunes $T_y = (w_y)/(2\pi)$ and $T_z = (w_z)/(2\pi)$.[30] Then, without loss of generality, the polynomial g_2 can be taken to be the expression[31]

$$g_2 = -w_y(p_y^2 + y^2)/2 - w_z(p_z^2 + z^2)/2. \tag{7.66}$$

It is easily verified that $\exp(:g_2:)$ has the desired action

$$\exp(:g_2:)\begin{bmatrix} y \\ p_y \\ z \\ p_z \end{bmatrix} = \begin{bmatrix} \cos(w_y) & \sin(w_y) & 0 & 0 \\ -\sin(w_y) & \cos(w_y) & 0 & 0 \\ 0 & 0 & \cos(w_z) & \sin(w_z) \\ 0 & 0 & -\sin(w_z) & \cos(w_z) \end{bmatrix} \begin{bmatrix} y \\ p_y \\ z \\ p_z \end{bmatrix} . \tag{7.67}$$

Exercise 7.10: Verify (7.67) and convince yourself that the quantities $T_y = (w_y)/(2\pi)$ and $T_z = (w_z)/(2\pi)$ are tunes.

As indicated in (5.82), the evaluation of the operator expression in (7.65) requires the summation of an infinite series. In order to carry out the summation, it is convenient to expand the quantity g_3 in terms of eigenfunctions of the operator $:g_2:$. At first sight, it may not be obvious that $:g_2:$ should have eigenfunctions. However, suppose that g_n is a homogeneous polynomial of degree n. Then, according to exercise 5.6, the quantity $:g_2:g_n$ is also a homogeneous polynomial of degree n. Therefore, $:g_2:$ may be viewed as a linear operator which maps the vector space of homogeneous polynomials of degree n into itself. It follows from simple matrix theory that $:g_2:$ must have at least one eigenvector.

Indeed, it is easy by explicit construction to see that $:g_2:$ has a complete set of eigenfunctions. For ease of exposition, consider first the simpler case of one degree of freedom with canonical coordinates q,p. Also, for the moment, let g_2 denote the polynomial

$$g_2 = -w(p^2 + q^2)/2. \tag{7.68}$$

Now replace the variables q,p by the action-angle variables a,ϕ defined by the relation

$$q = (2a)^{1/2} \sin \phi \tag{7.69a}$$

$$p = (2a)^{1/2} \cos \phi. \tag{7.69b}$$

Then it is easily verified that

$$2a = q^2 + p^2 = -2g_2/w . \tag{7.70}$$

It follows that

$$:g_2:a = 0. \tag{7.71}$$

Consider the quantity $(p + iq)^m$. One finds the relation

$$(p + iq)^m = [(2a)^{1/2} (\cos\phi + i \sin\phi)]^m$$
$$= (2a)^{m/2} e^{im\phi} . \qquad (7.72)$$

Also compute the quantity $[\exp(\tau{:}g_2{:})](p + iq)$ where τ is some parameter. The result is easily found to be the relation

$$[\exp(\tau{:}g_2{:})](p + iq) = e^{i\tau w} (p + iq). \qquad (7.73)$$

Next, using the isomorphism property (3.35), one finds the general relation

$$[\exp(\tau{:}g_2{:})](p + iq)^m = \{[\exp(\tau{:}g_2{:})](p + iq)\}^m$$
$$= e^{im\tau w}(p + iq)^m. \qquad (7.74)$$

Now differentiate both sides of (7.74) with respect to the parameter τ, and then set $\tau = 0$. The result is the relation

$$:g_2:(p + iq)^m = imw(p + iq)^m$$

or

$$:g_2:[(2a)^{m/2}e^{im\phi}] = imw(2a)^{m/2}e^{im\phi} . \qquad (7.75)$$

Finally, in view of (3.34) and (7.71), the factor $(2a)^{m/2}$ may be removed from both sides of (7.75) to give the result

$$:g_2:e^{im\phi} = imwe^{im\phi} . \qquad (7.76)$$

It follows that the eigenfunctions of $:g_2:$ are the functions $e^{im\phi}$ with the eigenvalues imw.

Evidently, an equivalent result holds in the two-dimensional case. Introduce two pairs of action-angle variables by the relations

$$y = (2a_y)^{1/2} \sin\phi_y \qquad (7.77a)$$

$$p_y = (2a_y)^{1/2} \cos\phi_y \qquad (7.77b)$$

$$z = (2a_z)^{1/2} \sin\phi_z \qquad (7.77c)$$

$$p_z = (2a_z)^{1/2} \cos\phi_z. \qquad (7.77d)$$

Then it follows that the eigenfunctions of the full g_2, now again given by (7.66), are the functions $e^{im\phi_y} e^{in\phi_z}$ with the eigenvalue relations

$$:g_2: e^{im\phi_y} e^{in\phi_z} = i(mw_y + nw_z) e^{im\phi_y} e^{in\phi_z} \quad . \qquad (7.78)$$

Exercise 7.11: Verify Eqs. (7.73), (7.75), (7.76), and (7.78).

With this background, we are ready to proceed. From (7.77) it follows that g_3 has an expansion of the form

$$g_3 = \sum_{m,n} c_{mn}(a_y, a_z)\, e^{im\phi_y} e^{in\phi_z} \qquad (7.79)$$

where, as indicated, the coefficients c_{mn} depend on the variables a_y, a_z. Using this expansion, it is easy to evaluate the operator expression in (7.65). One finds the result

$$:g_2:[1 - \exp(-:g_2:)]^{-1} g_3 = \sum_{m,n} c_{mn}\, e^{im\phi_y} e^{in\phi_z} \times$$

$$i(mw_y + nw_z)/[1 - \exp(-imw_y - inw_z)]. \qquad (7.80)$$

Consequently, after a little trigonometric tidying up and neglecting the terms of degree 4 and higher, the expression (7.65) can be written in the explicit form

$$h = g_2 + \sum_{m,n} c_{mn}\, e^{im(\phi_y + w_y/2)} e^{in(\phi_z + w_z/2)} \times$$

$$(mw_y/2 + nw_z/2)/[\sin(mw_y/2 + nw_z/2)]. \qquad (7.81)$$

Exercise 7.12: Verify Eqs. (7.80) and (7.81).

Examination of (7.81) shows that the denominator in the summation vanishes if the argument of the sine function vanishes or equals an integral multiple of π. That is, there are possible difficulties in computing h when there is a relation involving integers of the form

$$mw_y/2 + nw_z/2 = \ell\pi$$

or

$$m(w_y)/(2\pi) + n(w_z)/(2\pi) = \ell \quad . \qquad (7.82)$$

Here the quantities ℓ, m, and n can in principle take all possible positive and negative integer values including zero. Equation (7.82) is the usual condition for nonlinear structure resonances.[32] It indicates that, in principle, difficulties can occur whenever a tune has a rational value, or two tunes are relatively rational. In actual practice, the summation range of m and n in (7.81) is limited by the degree of g. For example, in (7.79) and hence in (7.81), the summations over m and n are restricted to lie between +3 and −3. Similarly, for a g_4, they would be restricted to lie between +4 and −4. Thus, if one assumes that the higher degree g's are much smaller or less important than the lower order g's, then the lower order resonances dominate.

The complete details of nonlinear structure resonances have yet to be worked out from a Lie algebraic perspective. However, some work has been done on the simpler one-dimensional case in which g_2 is given by (7.68) and g_3 has the form

$$g_3 = sq^3/3. \tag{7.83}$$

This may be viewed as the case of a perfect ring with a short strong sextupole insertion, provided attention is restricted to horizontal betatron oscillations and all orbits are assumed to lie in the midplane.[33]

Indeed, Eq. (6.47) takes the form (7.83) in the midplane z = 0 when due account is taken of the change of notation. Here s is a measure of the sextupole strength integrated over its length.

Exercise 7.13: Suppose $\mathcal{M}$ is given by (7.64) with g_2 and g_3 given by (7.68) and (7.83) respectively. Find the action of $\mathcal{M}$ on the general point (q,p).

Answer:

$$\begin{aligned}\bar{q} = \mathcal{M}q &= q \cos w + p \sin w \\ \bar{p} = \mathcal{M}p &= -q \sin w + p \cos w + s(q \cos w + p \sin w)^2.\end{aligned} \tag{7.84}$$

When (7.69) is inserted into (7.83), one finds the expansion

$$\begin{aligned}g_3 &= (s/3)(2a)^{3/2}(\sin\phi)^3 = \sum_m c_m(a)e^{im\phi} \\ &= (s/3)(2a)^{3/2}(e^{3i\phi} - 3e^{i\phi} + 3e^{-i\phi} - e^{-3i\phi})/(2i)^3.\end{aligned} \tag{7.85}$$

Thus, as expected, only a few of the coefficients c_m are nonzero. In analogy to (7.81), one now has the relation

$$h = g_2 + \sum_m c_m e^{im(\phi+w/2)}(mw/2)/[\sin(mw/2)]. \qquad (7.86)$$

Finally, when the explicit coefficients as given by (7.85) are inserted into (7.86) and use is made of (7.70), one finds the result

$$h = -wa - (ws/8)\,(2a)^{3/2}\{[\sin(3\phi+3w/2)]/[\sin(3w/2)]$$
$$- [\sin(\phi+w/2)]/[\sin(w/2)]\}. \qquad (7.87)$$

<u>Exercise 7.14</u>: Verify Eqs. (7.85), (7.86), and (7.87).

Inspection of (7.87) shows that resonances occur whenever the quantities (3w/2) or (w/2) are multiples of π. That is, the resonance conditions are

$$T = w/(2\pi) = \ell/3 \text{ or } T = w/(2\pi) = \ell. \qquad (7.88)$$

Consequently, for a short sextupole insertion, there are resonances when the tune T is an integer multiple of (1/3), and at an integer tune values.

The behavior of the symplectic map $\mathcal{M}$ at and near these resonant values will be examined shortly. Suppose for the moment that w is far from these resonant values. Then the expression for h is well behaved. Moreover, since $\mathcal{M}$ can be written in the form (7.62), it follows that h is an <u>invariant function</u> for the mapping $\mathcal{M}$. That is, denoting the general action of $\mathcal{M}$ as in (5.2), one has the relation

$$h(\mathcal{M}z) = h(z). \qquad (7.89)$$

Here, as usual, z stands for the complete collection of phase-space variables. See Eqs. (5.5) through (5.7) for a review of why h should have the invariance property (7.89).

Equation (7.89) says that the value of h should be the same at all those points in the surface of section which are the image of a given point under successive applications of the map $\mathcal{M}$. Or, put another way, the mapping $\mathcal{M}$ must send each level line of the function h into itself. This condition may place great restrictions on the action of $\mathcal{M}$. For example, if h has a level line which is a simple closed curve, then $\mathcal{M}$ can never map the interior of this curve to the exterior because a continuous mapping, which $\mathcal{M}$ is, must preserve topological properties. Thus, if there is indeed an invariant function with a closed curve level line, then orbits having initial conditions in the surface of section which are initially inside this curve must always intersect the surface of section at points inside the curve for all future times. Therefore such orbits are <u>stable</u> because they are bounded for all time.

Indeed, even more can be said. Comparison of (7.62) and (5.34) shows that the function (-h) can be viewed as playing the role of an "effective Hamiltonian" for the Poincare map $\mathcal{M}$. That is, suppose (-h) is viewed as a Hamiltonian defined on the surface of section. Then following the Hamiltonian flow specified by (-h) for n units of time from some initial point is equivalent to computing the action of $\mathcal{M}^n$ on the initial point.

As an example of how this works out in practice, look at Fig. 7.3. The points on the inner curve are the images of the point

$$(q_o,p_o) = (0.5,0.0) \qquad (7.90)$$

under successive applications of $\mathcal{M}$.

That is, they are the points

$$(q_n,p_n) = \mathcal{M}^n(q_o,p_o) \qquad (7.91)$$

where $\mathcal{M}$ is the symplectic map given explicitly by (7.84). The tune angle is set at the nonresonant value w=5 degrees, which corresponds to the tune T having the value T = .0139, and the sextupole has the value s = .05 for its strength.

Similarly, the points on the outer curve are the images of an initial point on the q axis with a larger radius.

For comparison, the circles on the inner curve are centered on the points obtained by following (by numerical integration) the Hamiltonian flow specified by (-h) for successive units of time starting with the initial conditions (7.90). Here h is the quantity given by (7.87). The actual points themselves are not shown, but only the open circles surrounding them. Similarly, the circles on the outer curve are centered about points, again not shown, which are the result of following the flow specified by (-h) starting with, as initial conditions, the point on the q axis having a larger radius.

Evidently, the circles on the inner curve appear to surround the points (7.91). Similarly, the circles on the outer curve appear to surround the points on the outer curve. The close agreement of the points and circles illustrates the close agreement between $\mathcal{M}^n$ and exp(n:h:) for the case of a tune far away from resonant values.

In the absence of the sextupole term (i.e. when s=0), the mapping $\mathcal{M}$ would map the nested circles $q^2 + p^2 = 2a$ = constant into themselves. This fact can easily be seen from (7.84). That is, the quantity g_2 given by (7.70) is an invariant of $\mathcal{M}$ in lowest approximation. According to exercise (7.6), this quantity is the Courant-Snyder invariant for betatron oscillations.

Apparently, in the nonresonant case, the only effect of a sextupole of moderate strength is to produce an "egg shaped" distortion of the invariant curves. Moreover, the shape of the distorted curves is well described by the level lines h = constant. Consequently, h provides a generalization of the Courant-Snyder invariant to the case of nonlinear motion.

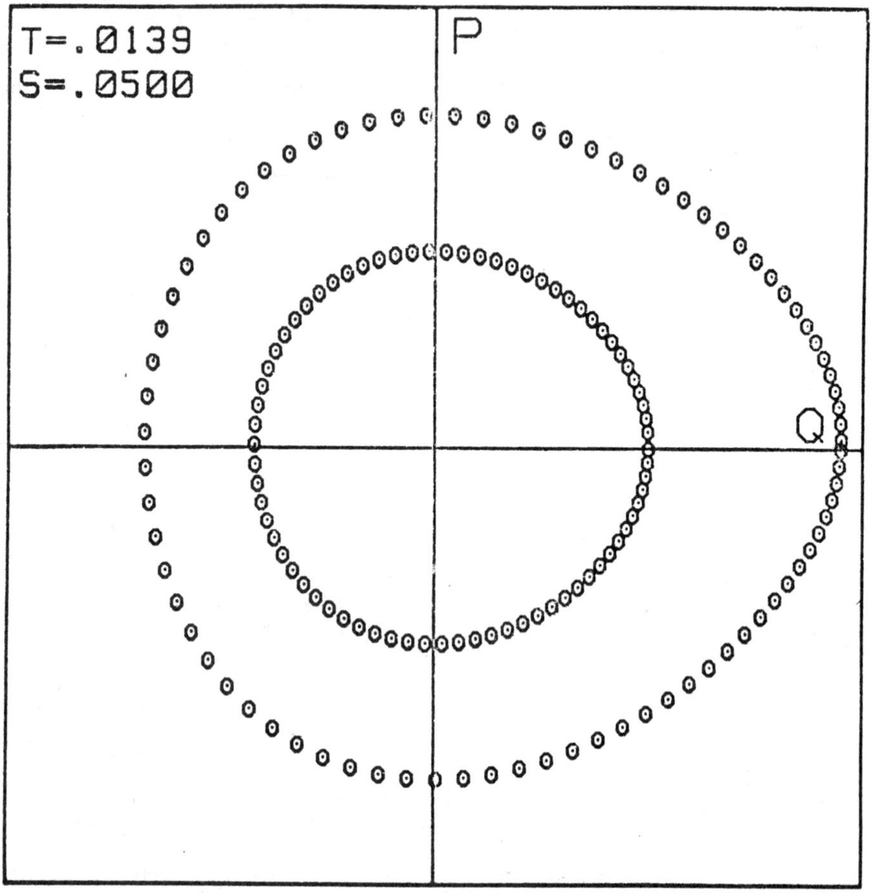

Figure (7.3): Typical behavior when the tune is far from resonance. Orbits are stable, and iterates of appear to agree perfectly with trajectories derived from h. The tune is T = .0139, and the sextupole strength is s=.05. In the viewing area the coordinates q and p range from -1 to 1.

The fact that the invariant curves are now "egg shaped" rather than circular means that the amplitude of betatron oscillation is not constant, but instead varies between certain minimum and maximum values. The extent of this variation is sometimes referred to as the beating range.[34] Also, it can be shown that the rate of progress of successive points along a given curve, which is a measure of the tune of an orbit, depends on which curve is considered. This circumstance may be described roughly by saying that, in the nonlinear case, the tune of an orbit depends on the betatron amplitude. Evidently, complete information about both these nonlinear phenomena is given by h.

Figure 7.4 shows another example. In this case, the sextupole strength is as before, but the tune angle has the almost-zero value w=1 degree. Again, there is good agreement between points, which are the result of computing the action of $\mathcal{M}^n$ on various initial points, and circles which are the result of following the Hamiltonian flow generated by (-h). Only a slight departure of the circles from the points is visible in the far upper right-hand corner of the figure. Presumably this departure arises from the omitted terms in (7.65). Note that g_3 as given by (7.83) is proportional to the quantity s which is small, and that multiple commutators involving several factors of g_3 will contain higher powers of s. It follows that the omitted terms are not only of degree 4 and higher as indicated, but also in this case are proportional to s^2 and higher powers of s. Therefore, one should expect very good agreement near the origin where the higher degree terms are small, and moderately good agreement away from the origin where the contributions of the higher degree terms are still diminished by the higher powers of s.

Observe that in this case the map has two fixed points and therefore there are two closed orbits. One of the fixed points is at the origin, as expected, and the second is located somewhere near the q axis as indicated in Fig. 7.4.

To the extent that (-h) acts as an effective Hamiltonian for $\mathcal{M}$, then fixed points of $\mathcal{M}$ should correspond to equilibrium (critical) points of h. That is, fixed points should be solutions to the equations

$$\partial h/\partial q = 0 \qquad (7.92a)$$

$$\partial h/\partial p = 0. \qquad (7.92b)$$

Evidently, Eqs. (7.92) are satisfied at the origin. Indeed, at the origin h has an expansion of the form

$$h = -w(p^2 + q^2)/2 + \text{terms of degree 3 and higher}\,. \qquad (7.93)$$

Consequently, the quadratic terms dominate at the origin. It follows that the linear part of the map at the origin is given by the matrix

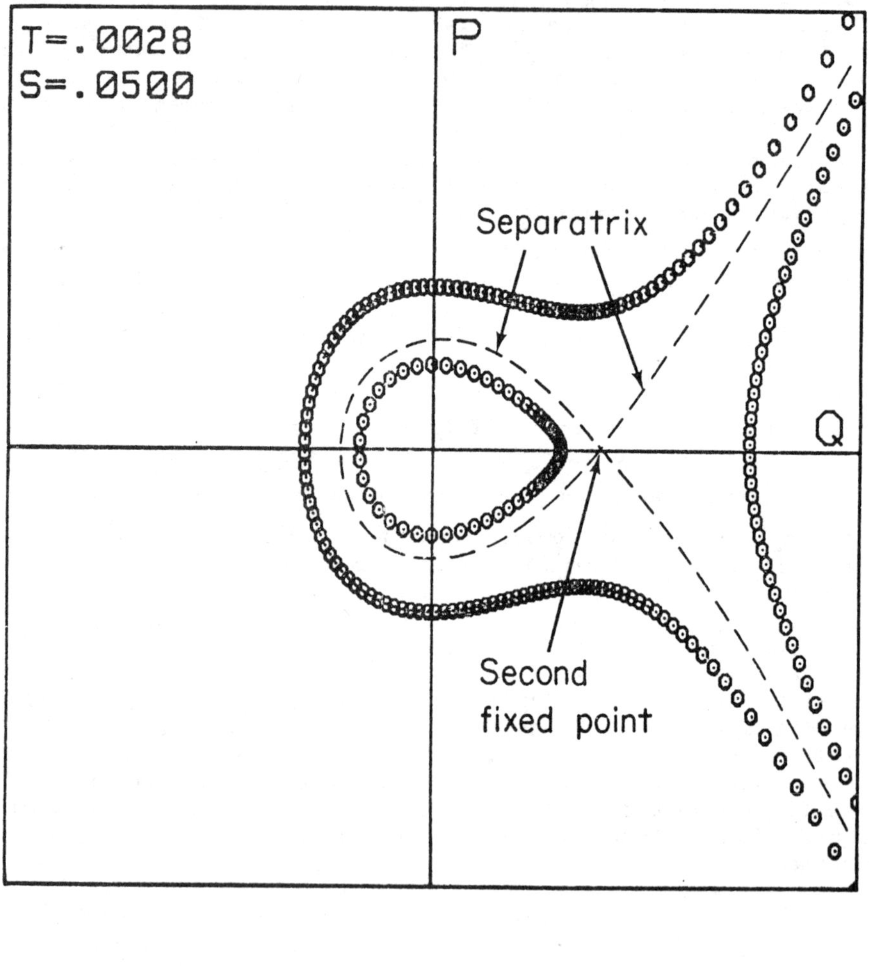

Figure (7.4): Behavior near an integer tune showing stable and unstable regions and their separatrix. Note that the regions of stability and instability are well predicted by h. In the viewing area the coordinates q and p range from -1 to 1.

$$M = \begin{pmatrix} \cos w & \sin w \\ -\sin w & \cos w \end{pmatrix} \quad . \tag{7.94}$$

This result is also obvious from Eq. (7.84). Note that the eigenvalues of M are $\exp(\pm iw)$. Therefore, according to exercise 7.6, the fixed point at the origin is elliptic. Consequently, the fixed point at the origin is stable in the sense that points initially very near the origin remain near the origin (in the linear approximation) under repeated applications of $\mathcal{M}$.

Equations (7.92) are also satisfied at a second point. This point is most easily found using action-angle variables. Since the transformation (7.49) is canonical, equations (7.92) can be written in the equivalent form

$$\partial h/\partial\phi = 0 \tag{7.95a}$$

$$\partial h/\partial a = 0. \tag{7.95b}$$

Upon carrying out the indicated differentiations and solving Eqs. (7.95) using (7.87), one finds the results

$$\phi = \pi/2 - w/2 \tag{7.96a}$$

$$(2a)^{1/2} = [(8)/(3s)][\sin(3w/2)\sin(w/2)]/[\sin(3w/2) + \sin(w/2)]. \tag{7.96b}$$

These results can then be substituted in Eqs. (7.69) to find, in the approximation that the terms of degree 4 and higher in (7.65) can be neglected, that the fixed point should have coordinates q,p satisfying the relations

$$q = 2s^{-1} (\sin w)(1 + \cos w)^{-1} (1 + 2 \cos w)/3 \tag{7.97a}$$

$$p/q = (\sin w)/(1 + \cos w). \tag{7.97b}$$

Because the mapping under study is rather simple, its fixed points can also be found directly and exactly. Suppose one requires that $\bar{q} = q$ and $\bar{p} = p$ in Eqs. (7.84). One finds, in addition to the fixed point at the origin, the result

$$q = 2s^{-1}(\sin w)(1 + \cos w)^{-1} \tag{7.98a}$$

$$p/q = (\sin w)/(1 + \cos w). \tag{7.98b}$$

Evidently, Eqs. (7.97b) and (7.98b) are identical, and hence Eq. (7.96a), although derived from an expression for h which neglects terms of degree 4 and higher, is actually exact. The exactness of this result is related to the observation that, as a result of time reversal invariance, the function h must be symmetric to all orders about the line $\phi = \pi/2 - w/2$.

Comparison of Eqs. (7.97a) and (7.98a) shows that they differ by the factor $(1 + 2\cos w)/3 = 1 - w^2/3 + \cdots$. Thus, the approximate and exact expressions for q agree in the small w limit, and differ only by terms of order w^2. This is what should be expected. Note that, according to (7.98a), when w is of order 1, then the quantity (qs) is also of order 1. But, in this case, the neglected terms of higher degree in the expansion for h can no longer be ignored.

At this point it is interesting to recall that section 7.1 and exercisé 7.4 showed the existence and local uniqueness of closed orbits as long as tunes are away from integer values. Inspection of Eqs. (7.97) or (7.98) shows that the fixed point away from the origin coalesces with the origin as $w \to 0$. That is, this simple example illustrates that two distinct closed orbits can merge as a tune goes through an integer value.

<u>Exercise 7.15</u>: Verify Eqs. (7.96), (7.97), and (7.98).

The stability of the fixed point away from the origin can be examined by expanding h about the fixed point. Since the fixed point is an equilibrium point, the first nonvanishing derivatives are given by the expressions

$$\begin{aligned}\partial^2 h/\partial\phi^2 &= (ws/8)(2a)^{3/2}\{[9\sin(3\phi + 3w/2)]/[\sin(3w/2)] \\ &\quad - [\sin(\phi + w/2)]/[\sin(w/2)]\} \\ &= -(ws/8)(2a)^{3/2}\{9[\sin(3w/s)]^{-1} + [\sin(w/2)]^{-1}\},\end{aligned} \tag{7.99a}$$

$$\partial^2 h/\partial\phi\partial a = 0\,, \tag{7.99b}$$

$$\begin{aligned}\partial^2 h/\partial a^2 &= -(3/4)(ws/8)(2)^{3/2}\, a^{-1/2}\{[\sin(3\phi + 3w/2]/[\sin(3w/2)] \\ &\quad - \sin(\phi + w/2)]/[\sin(w/2)]\} \\ &= (3/4)(ws/8)(2)^{3/2}\, a^{-1/2}\{[\sin(3w/2)]^{-1} + [\sin(w/2)]^{-1}\}.\end{aligned} \tag{7.99c}$$

Here use has been made of (7.96a). Observe that the two terms (7.99a) and (7.99c) are of opposite signs. Consequently, in the neighborhood of the equilibrium point, h has an expansion which looks like that of an harmonic oscillator with a negative spring constant. Therefore, the equilibrium is unstable, and correspondingly the fixed point is hyperbolic.

Let z^0 be the coordinates of the hyperbolic fixed point as given by Eq. (7.97) or (7.98). Near z^0, $\mathcal{M}$ can be approximated by its linear part. Therefore, there is a local hyperbolic structure as described in exercise 7.7. In particular, there are the asymptotes which were referred to as the stable and unstable manifolds. The concept of stable and unstable manifolds can be generalized to the full map. Let z denote a general point. Then, the stable manifold, denoted by W_s, is defined to be the set of all points z which are sent into z^0 by $\mathcal{M}^n$ in the limit of large n. In set theoretic notation, one has the definition

$$W_s = \{z \mid \lim_{n\to\infty} \mathcal{M}^n z = z^0\}. \tag{7.100}$$

Evidently, the stable manifold is the analog of the asymptote corresponding to that eigenvector of the linear part having eigenvalue less than one. Similarly, the unstable manifold is the analog of the asymptote corresponding to the eigenvector having eigenvalue greater than one. Note that points on the unstable manifold, W_u, should move away from z^0 under the action of $\mathcal{M}$, and therefore they should approach z^0 under the action of $\mathcal{M}^{-1}$. Consequently, the unstable manifold is cleverly defined as the set of all points which are sent into z^0 by $\mathcal{M}^{-n}$ in the large n limit,

$$W_u = \{z \mid \lim_{n\to\infty} \mathcal{M}^{-n} z = z^0\}. \tag{7.101}$$

To the extent that h is an invariant function, the stable and unstable manifolds are given by the level lines of h that pass through z^0. That is, the stable and unstable manifolds are solutions of the equation

$$h(z) = h(z^0). \tag{7.102}$$

These curves are indicated schematically in Fig. 7.4. Note that two branches of the stable and unstable manifold join, and two branches go off to infinity. Note also that the behavior of points under the action of $\mathcal{M}$ is different on different sides of the stable and unstable manifolds. For this reason, the stable and unstable manifolds are also sometimes referred to together as the separatrix. That is, the stable and unstable manifolds separate phase space into regions of qualitatively different behavior. In particular, points inside that portion of the separatix surrounding the origin remain inside, and points outside are eventually sent to infinity. In the content of a circulating charged particle beam, points inside the separatrix loop correspond to trajectories which remain within the machine, and points outside correspond to trajectories which eventually escape.

It can be shown that for most symplectic maps the stable and unstable manifolds of a hyperbolic fixed point do not join smoothly as Fig. 7.4 would suggest, but rather they intersect at some point at some finite angle. This phenomena is called separatrix splitting. The point of intersection is called a homoclinic point, and the angle is referred to as a homoclinic angle.[35]

When separatrix splitting occurs, the symplectic nature of the map causes the stable and unstable manifolds to oscillate about each other. Correspondingly, the concept of a separatrix loses its meaning. Separatrix splitting occurs for the simple map we have been studying.[33,36] Figure 7.5 shows the behavior of the stable and unstable manifolds for the map (7.84) for a tune angle of $w \simeq 70$ degrees. For convenience of plotting, the origin has been shifted to the hyperbolic fixed point, and the axes, now called x and y, have been lined up along the asymptotes. The existence of a homoclinic point is evident. Figures 7.6 and 7.7 show how the stable and unstable manifolds oscillate about each other as they return to the neighborhood of the hyperbolic fixed point.

In many cases, the angle of intersection between the stable and unstable manifolds at the homoclinic point is too small to be readily visible. For example, when the tune angle has the value of 5 degrees corresponding to Fig. (7.3), the homoclinic angle is less than 10^{-13} degrees. However, the existence of homoclinic angles, no matter how small, rules out the true existence of invariant functions.[36,37] Consequently, the series (7.65) is generally divergent when all terms are included. Nevertheless, the truncated series is still useful, as has been seen, provided the homoclinic angle is small or one does not inquire about properties of the map that are too detailed.

A second and related consequence of the intersection of stable and unstable manifolds is the appearance of chaotic behavior. Figure 7.8 displays the result of iterating the map (7.84) for a variety of initial conditions when the tune angle has the value $w \simeq 76$ degrees. The scale is such as to show only the region in the neighborhood of the elliptic fixed point at the origin. The hyperbolic fixed point given by (7.98) is outside the viewing area of the drawing. Inspection shows that there appear to be five other elliptic fixed points and five other hyperbolic fixed points. These points are actually fixed points of $\mathcal{M}^5$, the fifth power of the map. Figure 7.9 shows a magnification of the behavior near one of these five hyperbolic fixed points.[38] Many of the points in the figure appear to be distributed in a random or chaotic fashion. This behavior arises from the fact that the stable manifolds from one hyperbolic fixed point intersect (at a finite angle) the unstable manifolds from another fixed point. Such points of intersection of manifolds from different fixed points are called heteroclinic points. As is the case with homoclinic points, the manifolds then go into wild oscillation about each other thus producing the apparently chaotic behavior.

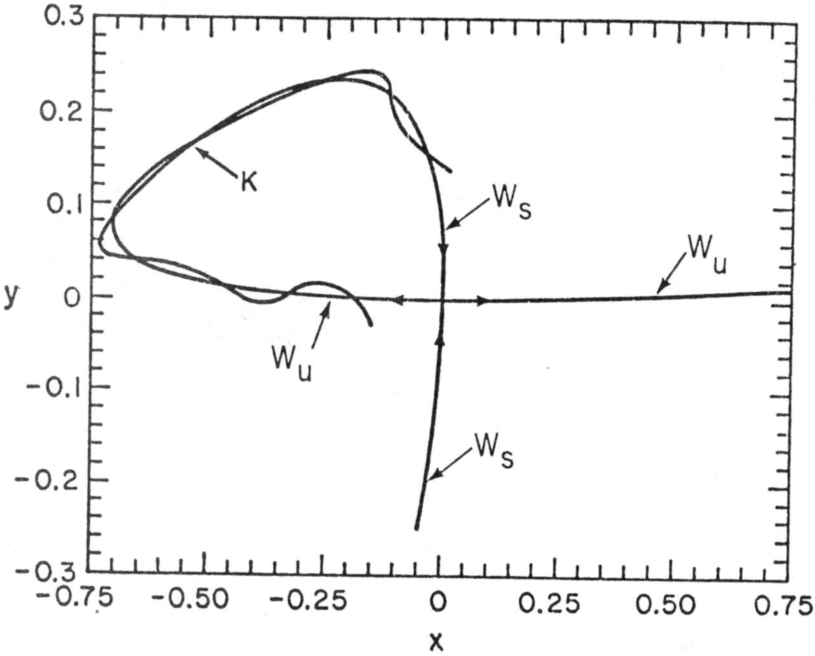

Figure (7.5): The intersection of stable and unstable manifolds emanating from a hyperbolic fixed point resulting in a homoclinic point K.

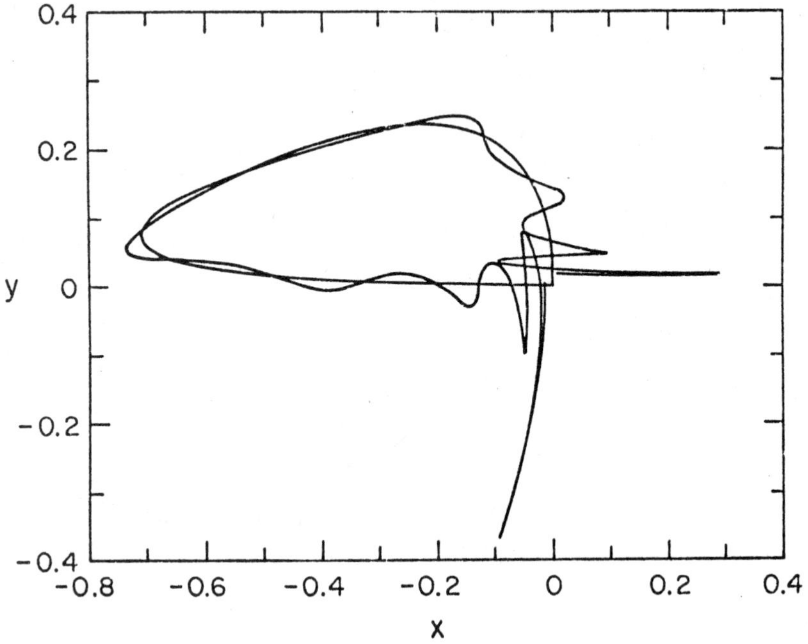

Figure (7.6): Successive homoclinic intersections and oscillations of the stable and unstable manifolds. The other halves of W_u and W_s, those pieces that go off to infinity, are not shown.

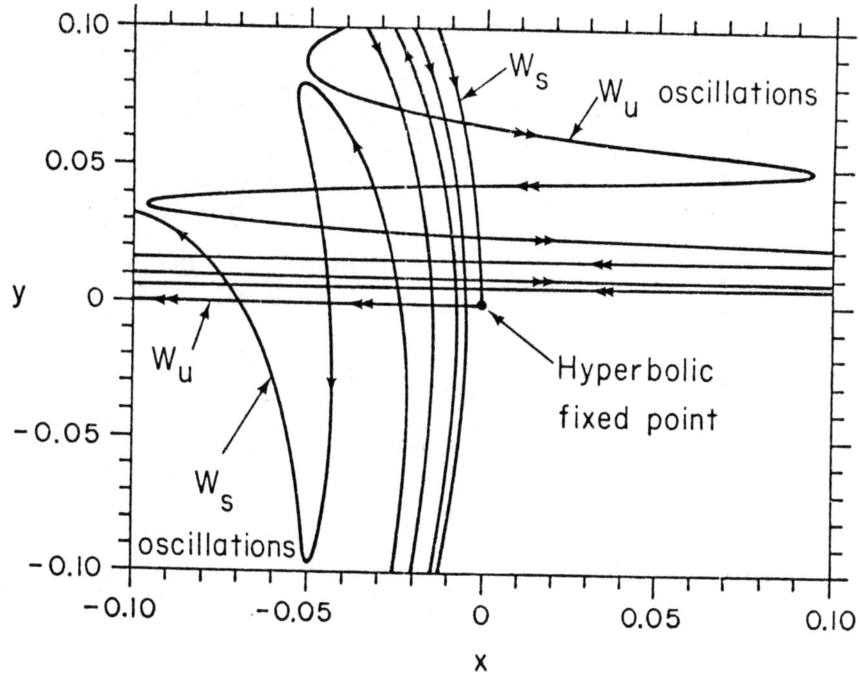

Figure (7.7): A continuation of Fig. (7.6) near the origin showing the formation of a grid of intersecting lines. The spacing of the grid becomes finer and finer as it approaches the hyperbolic fixed point. Each grid intersection is a homoclinic point. The result of all these intersections is an ever denser cloud of homoclinic points which has the hyperbolic fixed point as a limit point.

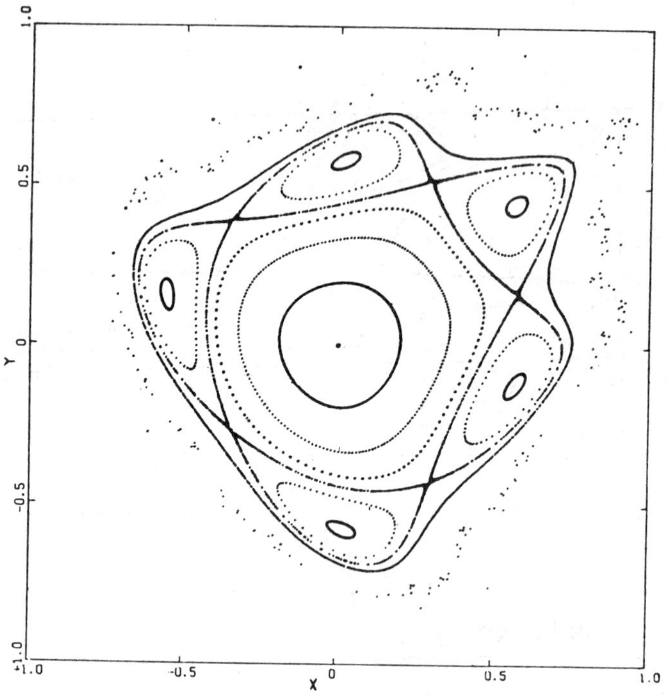

Figure (7.8): Behavior of M near a tune of 1/5. There are five elliptic and five hyperbolic fixed points of M^5. The variables are selected so that the hyperbolic fixed point given by equation (7.98) has coordinates $x \simeq 1.6$, $y \simeq 1.2$.

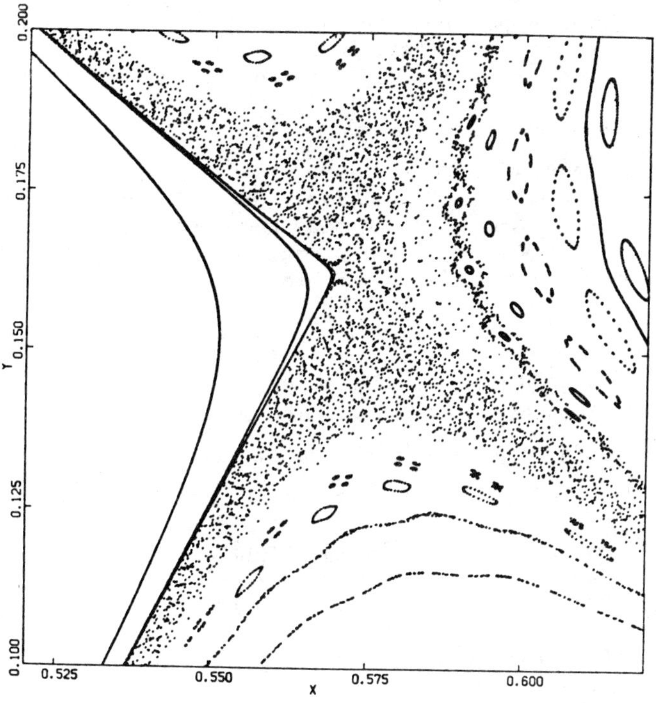

Figure (7.9): A magnification of Fig. (7.8) near one of the hyperbolic fixed points. The chaotic behavior arises from the heteroclinic intersections of the stable and unstable manifolds emanating from the hyperbolic fixed points. Note also the presence of many small islands corresponding to very high order resonances. Indeed, it can be shown that in the case of homoclinic or heteroclinic seperatrix splitting for a hyperbolic fixed point, there must be fixed points of high powers of M in every neighborhood of the "parent" hyperbolic point.

Evidently, even the simplest of nonlinear symplectic maps can exhibit an extremely complicated behavior.

Our discussion of the example of a perfect ring with a short sextupole insertion has so far been restricted to the nonresonant case. Suppose, now, that the tune is at or near one of the resonant values 1/3 or 2/3. In these cases the expansion (7.87) is no longer valid, and the symplectic map $\mathcal{M}$ cannot be written in the exponential form (7.62).

However, we will see that although $\mathcal{M}$ cannot be written in exponential form at a third integer resonance, the cube of the map, $\mathcal{M}^3$, can still be written in exponential form. For brevity, attention will be restricted to tunes at and near the resonant value 1/3. The case of the resonant value 2/3 can be handled analogously.

Suppose the tune is written in the form

$$T = w/(2\pi) = (1/3) + \delta \tag{7.103}$$

where the quantity δ measures departure from exact resonance. Now watch closely a bit of algebraic sleight of hand! If $\mathcal{M}$ is written in the form (7.64), then $\mathcal{M}^3$ is obviously the product

$$\begin{aligned} \mathcal{M}^3 = {} & \exp(:g_2:)\exp(:g_3:) \times \\ & \exp(:g_2:)\exp(:g_3:) \times \\ & \exp(:g_2:)\exp(:g_3:). \end{aligned} \tag{7.104}$$

However, this product can be rewritten in the form

$$\begin{aligned} \mathcal{M}^3 = {} & \exp(3:g_2:)\exp(-2:g_2:)\exp(:g_3:)\exp(2:g_2:) \times \\ & \exp(-:g_2:)\exp(:g_3:)\exp(:g_2:) \times \\ & \exp(:g_3:). \end{aligned} \tag{7.105}$$

Several simplications are now possible. The term $\exp(3:g_2:)$ can be rewritten with the aid of (7.103) and (7.68). One finds the result

$$\exp(3:g_2:) = \exp(-\pi:p^2 + q^2:)\exp(:f_2:) \tag{7.106}$$

where f_2 denotes the function

$$f_2 = -3\delta\pi(p^2 + q^2). \tag{7.107}$$

However, the first Lie transformation in (7.106) corresponds to rotation by 2π in the q,p phase space, and is therefore just the identity operator. See exercise 3.23. Therefore, one also has the relation

$$\exp(3{:}g_2{:}) = \exp({:}f_2{:}). \tag{7.108}$$

The remaining terms in (7.105) can be simplified using the calculus developed in section 5.4. Employing Eq. (5.52), one finds the results

$$\exp(-2{:}g_2{:})\exp({:}g_3{:})\exp(2{:}g_2{:}) = \exp({:}\exp(-2{:}g_2{:})g_3{:}), \tag{7.109a}$$

and

$$\exp(-{:}g_2{:})\exp({:}g_3{:})\exp({:}g_2{:}) = \exp({:}\exp(-{:}g_2{:})g_3{:}). \tag{7.109b}$$

It follows from exercise 5.6 that the quantities $\exp(-2{:}g_2{:})g_3$ and $\exp(-{:}g_2{:})g_3$ are both homogeneous polynomials of degree three. Therefore, using the Campbell-Baker-Hausdorff formula to combine exponents and neglecting polynomials of degree four and higher, which is in keeping with the spirit of our calculations, one finds that the remaining terms in (7.105) can be rewritten in the form

$$\begin{aligned} &\exp(-2{:}g_2{:})\exp({:}g_3{:})\exp(2{:}g_2{:}) \times \\ &\exp(-{:}g_2{:})\exp({:}g_3{:})\exp({:}g_2{:}) \times \\ &\exp({:}g_3{:}) = \\ &\exp({:}[\exp(-2{:}g_2{:}) + \exp(-{:}g_2{:}) + 1]g_3{:}). \end{aligned} \tag{7.110}$$

Combining (7.108) and (7.110) gives the result

$$\mathcal{M}^3 = \exp({:}f_2{:})\exp({:}[\exp(-2{:}g_2{:}) + \exp(-{:}g_2{:}) + 1]g_3{:}). \tag{7.111}$$

Suppose the Campbell-Baker-Hausdorff formula is now used to combine the two exponents in (7.111) to obtain a single exponent. For later convenience, call this exponent $3h_r$. Then one finds the result

$$\mathcal{M}^3 = \exp(3{:}h_r{:}) \tag{7.112}$$

with h_r given by the relation

$$3h_r = f_2 + :f_2:[1 - \exp(-:f_2:)]^{-1} \times$$
$$[\exp(-2:g_2:) + \exp(-:g_2:) + 1]\, g_3. \qquad (7.113)$$

See Eq. (5.81). However, from the algebraic identity

$$a^2 + a + 1 = (1 - a^3)/(1 - a), \qquad (7.114)$$

there follows the operator identity

$$\exp(-2:g_2:) + \exp(-:g_2:) + 1$$
$$= [1 - \exp(-3:g_2:)]/[1 - \exp(-:g_2:)]. \qquad (7.115)$$

Also, from (7.108), it follows that

$$1 - \exp(-3:g_2:) = 1 - \exp(-:f_2:). \qquad (7.116)$$

Therefore, using (7.115) and (7.116), the expression (7.113) for h_r can also be written in the form

$$3h_r = f_2 + :f_2:[1 - \exp(-:g_2:)]^{-1} g_3. \qquad (7.117)$$

Inspection of Eqs. (7.65) and (7.117) shows that the expressions for h and h_r are rather similar. Consequently, for g_3 given by (7.83), we can use our previous calculations to find the specific form of h_r. One finds the result

$$h_r = -2\pi\delta a - (2\pi\delta s/8)(2a)^{3/2} \times$$
$$\{[\sin(3\phi+3w/2)]/[\sin(3w/2)] - [\sin(\phi+w/2)]/[\sin(w/2)]\}.$$
$$(7.118)$$

Suppose that the tune $(w/2\pi)$ approaches the resonant value (1/3). Then one has the pleasure of evaluating the limiting ratio

$$\delta/[\sin(3w/2)] = \delta/\sin(\pi+3\pi\delta) = -1/(3\pi) + O(\delta). \qquad (7.119)$$

Evidently the function h_r, unlike h, is well behaved both at and near resonant tunes.

Exercise 7.16: Verify Eqs. (7.105), (7.106), (7.113), (7.117), and (7.118).

How well does all this work in practice? Figure 7.10 shows a phase-space plot in the Poincare surface of section for the near-resonant case w = 119 degrees. The points on the various curves are the images, under successive applications of $\mathcal{M}^3$, of a given initial point on each curve. For comparison, the circles are centered on points (again not shown) obtained by taking the given initial point as an initial condition, and following the Hamiltonian flow generated by $-h_r$ for 3n units of time. Evidently, there are two kinds of curves. Those that are far from the origin lead to infinity, and correspond to orbits on which particles are eventually lost. Those that are sufficiently close to the origin (corresponding to sufficiently small betatron oscillation amplitudes) encircle the origin, and hence describe stable orbits. In the limit of exact resonance this stable region shrinks to zero, and all of phase space becomes unstable.

Observe also that there is good agreement between the points and the circles. That is, Eq. (7.112) is well satisfied with h_r given by (7.118). Consequently, in the resonant or near resonant case, h_r can be used to predict the stable and unstable regions of phase space, and the behavior of orbits in each region. In particular, h_r can be used to predict regions of stability and instability, resonance widths, and the betatron amplitude growth rates of unstable orbits.

The examples just described have all been limited to the one-dimensional case. It has been seen that the functions h and h_r can be used to compute all relevant linear and nonlinear features of orbits (apart from the consequences of possible homoclinic and heteroclinic behavior which are assumed to be visible only on a fine scale) in both the nonresonant and resonant cases. Preliminary calculations indicate that similar results can be obtained in the full two-dimensional case. In this case, the function h is given by (7.81), and there is a related formula for h_r to be used under resonance and near resonance conditions. As before, h and h_r will be invariant functions, and their negatives will also act as effective Hamiltonians. However, since the Poincare surface of section is now a four-dimensional phase space, a knowledge of the "level surfaces" of h or h_r may not be sufficient to give complete qualitative information about orbits. Indeed, one is now in principle faced with the full general problem of motion in four-dimensional phase space, and such motion can be extremely complicated. For this reason, it may be useful to look for additional invariant functions, perhaps by the method of normal forms.[29,39] These additional invariant functions would be formal integrals of motion for the Hamiltonians -h or $-h_r$.

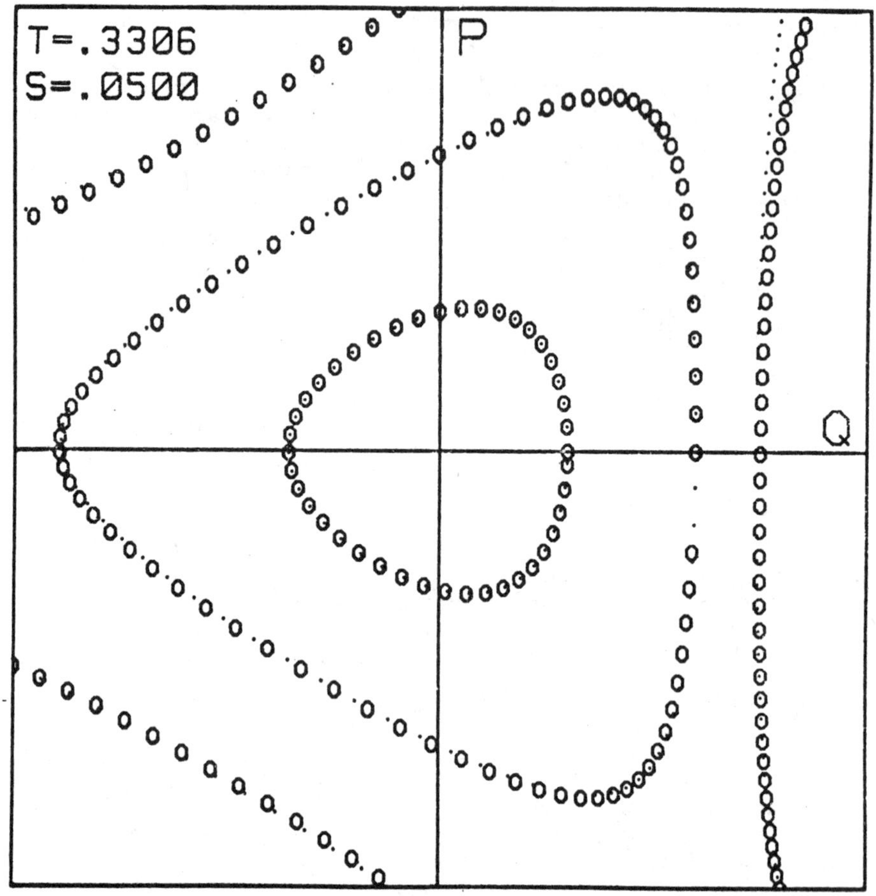

Figure (7.10): The behavior of M^3 near a third-integer tune. Observe that the regions of stability and instability are well predicted by h_r. In the viewing area the coordinates q and p range from -1 to 1.

7.4 The beam-beam interaction

Consider two charged particle beams, cirulating in the same or different rings, which intersect in a certain collision region. Suppose further that attention is restricted to the "weak-strong" limit or approximation. In this case, one beam, the strong beam, is taken to be fixed and unaffected by the second beam. The other beam, the weak beam, is treated as a collection of particles that are affected by their passage through the strong beam, but not by each other.

In the weak-strong limit, the net motion of a particle in the weak beam can be viewed as the continual repetition of two sequential motions: passage through the storage ring followed by passage through the strong beam. See Fig. 7.11. The equations of motion for each of these two passages (through the ring and through the strong beam) are derivable from Hamiltonians, and therefore each passage is described by a symplectic map.

Suppose the passages through the ring and the strong beam are described by the symplectic maps $\mathcal{M}_R$ and $\mathcal{M}_B$ respectively. By design, the passage through the storage ring is well described by a linear map. Thus, in lowest approximation, $\mathcal{M}_R$ can be written in the form

$$\mathcal{M}_R = \exp(:g_2:). \tag{7.120}$$

Higher order nonlinear corrections, as described in the previous sections, can be included if desired; but they will be omitted for the time being for simplicity.

What can be said about the map $\mathcal{M}_B$ that describes passage through the strong beam? According to the factorization theorem, $\mathcal{M}_B$ can be written in the factored product form (5.15). However, in this case the factored product representation, if truncated after a small number of terms, is not a particularly good representation of $\mathcal{M}_B$. This is because by design the charge and current density of the strong beam vary rapidly over a small spatial region, and hence $\mathcal{M}_B$ is very nonlinear. This apparent difficulty can be overcome by observing that, for $\mathcal{M}_B$, each exponent occurring in the factored product representation must be rather small. (This is because the beam-beam interaction is quite weak in the sense that not very much happens in a single passage through the strong beam.) Consequently, the different exponents can in principle be combined into one grand exponent using the Campbell-Baker-Hausdorff formula. Thus there is some function f_b, expected to be small but quite nonlinear, such that $\mathcal{M}_B$ can be written in the form

$$\mathcal{M}_B = \exp(:f_b:). \tag{7.121}$$

Upon combining the results of the previous two paragraphs, it follows that the grand symplectic map $\mathcal{M}$ that describes the result of passage through the ring and then the strong beam can be written in the form

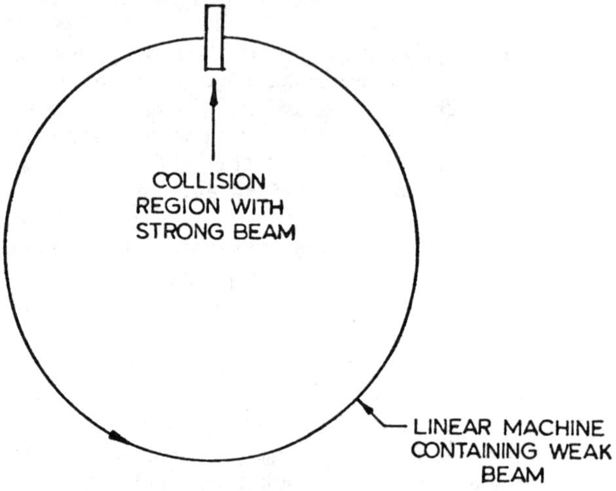

Figure (7.11): Schematic representation of particle motion in a storage ring and a colliding beam region.

$$\mathcal{M} = \exp(:g_2:)\exp(:f_b:). \tag{7.122}$$

In order to compute the long-time behavior of particles in the weak beam, the quantity of interest is again $\mathcal{M}^n$ for large n. As before, we try to find a function h such that $\mathcal{M}$ can be written in the form (7.62).

From this point on, the discussion of the beam-beam interaction in the weak-strong limit parallels the discussion of the stability of closed orbits in the presence of nonlinearities. In particular, formulas (7.65), (7.79), and (7.81) may be taken over directly simply by replacing g_3 by the function f_b. Similarly, there are related expressions for h_r in resonant and near resonant cases. Only two important differences must be kept in mind. First, the remainder terms in (7.65) are no longer of degree 4 and higher, but rather are of second order and higher in the beam-beam interaction strength.[40] Second, because of the possibility of nonlinearities of arbitrarily high degree in f_b, the summation ranges in (7.79) and (7.81) are generally infinite. Thus, in principle, nonlinear resonances of arbitrarily high order can occur.

As is the case with nonlinear structure resonances, the complete details of the effect of the beam-beam interaction in the full two-dimensional problem have not yet been worked out from a Lie algebraic perspective. Some work has been done on a one-dimensional model, and a discussion for a simple example will be given to close this section.[41]

Suppose the strong beam is assumed to be an unbunched ribbon in the horizontal plane. That is, the horizontal dimension of the strong beam is much larger than the vertical dimension. Assume also that the vertical charge distribution of the strong beam is well described by a Gaussian shape. Finally, assume that the weak beam lies in the same horizontal plane and crosses the strong beam at a fixed angle. Then the primary effect of the strong beam on particles in the weak beam is to cause vertical deflections.[42] Consequently, attention will be restricted to the vertical degree of freedom of the weak beam.

To find the map $\mathcal{M}_B$ exactly, it is necessary to integrate the nonlinear equations of motion for a particle passing through the strong beam. However, a good approximation to this map is given by assuming that the particle suffers a vertical momentum change depending only upon its initial vertical position, and that the vertical position itself remains unaffected. Thus, there is some "deflection function" u(q) in terms of which $\mathcal{M}_B$ has the effect

$$\begin{aligned} \mathcal{M}_B q &= q \\ \mathcal{M}_B p &= p + u(q). \end{aligned} \tag{7.123}$$

This impulse approximation becomes exact in the limit that the interaction region becomes a point and/or the transit time through the region approaches zero. In any case, the mapping (7.123) is symplectic, and therefore its use will produce no qualitative error.

The function u is proportional to the electrostatic force exerted by the strong beam. In the Gaussian model employed and with a suitable choice of coordinates, u is given by the relation[43]

$$u(q) = 4\pi D/\sqrt{3} \int_0^{q\sqrt{3}} dt\, e^{-t^2}. \qquad (7.124)$$

Here D is the beam-beam strength parameter that typically has values ranging from 10^{-3} to 10^{-2}. It is normalized in such a way that the beam-beam interaction depresses the tune for infinitesimal betatron oscillations by an amount D when D is small. Now let f_b be the function defined by

$$f_b(q) = \int_0^{q} u(q')dq'. \qquad (7.125)$$

Then it is readily verified that $\mathcal{M}_B$ as given by the Lie transformation (7.121) produces the desired action (7.123).

<u>Exercise 7.17</u>: Verify that (7.121) with f_b given by (7.125) satisfies (7.123).

Explicit calculation shows that f_b has an expansion of the form

$$f_b = \sum_{-\infty}^{\infty} c_n(a) \exp(2in\phi) \qquad (7.126)$$

where the coefficients $c_n(a)$ are given by the formulas[41]

$$c_0(a) = 2\pi Da \sum_{m=0}^{\infty} (-3a/2)^m (2m)!/\{(m!)[(m+1)!]^2\} \qquad (7.127a)$$

and

$$c_n(a) = -(4\pi/3)D(3a/2)^n \times$$

$$\sum_{m=0}^{\infty}(-3a/2)^m(2n+2m-2)!/[m!(n+m-1)!(m+2n)!] \qquad (7.127b)$$

when $n > 0$.

When use is made of this expansion, and g_2 is taken to be given by (7.68) as before, then h is found to be given by the relation

$$h = -wa + c_0(a) + 2\sum_{1}^{\infty} c_n(a)[nw/\sin(nw)]\cos[2n(\phi+w/2)]$$

$$+ \text{ terms of order } D^2 \text{ and higher.} \qquad (7.128)$$

Inspection of (7.128) indicates that the beam-beam interaction should produce nonlinear resonances whenever

$$nw = \ell\pi \text{ or } T = w/(2\pi) = \ell/(2n). \qquad (7.129)$$

Thus, there should be resonances at half-integer, quarter-integer, sixth-integer tunes, etc. The strength of these resonances should be proportional to $nc_n(a)$ for n=1,2,3, etc. It can be shown that the quantity (nc_n) falls off faster than exponentially in n with increasing n.[41] Moreover, according to (7.127b), c_n for large n falls off rapidly as $a \to 0$. Consequently, the sizes of various resonance features in phase space should decrease rapidly with the order of the resonance. In addition, these features should decrease in size according to their proximity to the origin in phase space.

How does this work out in practice? Figures (7.12) through (7.15) show phase-space plots generated numerically by applying successive iterates of $\mathcal{M}$ to various initial conditions, and using various tune values. The phase-space coordinates range over (-2,2), and the scale is chosen so that the strong beam lies within $-1 \leq q \leq 1$. The tunes are near the resonant values 1/2, 1/4, 1/6, and 1/8 respectively, and the beam-beam interaction strength is 10^{-2}. Observe that the expected resonances are all present, and that the size of resonance features, e.g. island dimensions, do indeed decrease with increasing order of the resonance.

Figure (7.16) shows a tenth-order resonance obtained by running near a tune of 1/10. It was not shown as part of the previous sequence because the island structure becomes too small to see when (by adjusting the tune) it is located closer to the origin. This example illustrates that the sizes of resonant features do indeed decrease with proximity to the origin, and in fact the higher the order of the resonance, the more rapid is the decrease.

Figure (7.17) shows the case of running with a nonresonant tune of 77/100. On the scale shown, and for the number of iterations made, there seems to be no evidence that any points will escape from the beam. Just as in the treatment given in the previous section of the effect of nonlinearities, the quantity h should again act as a generalization of the Courant-Snyder invariant in the non-resonant case. Figure (7.18) illustrates this proposition. It shows plots of the two quantities (-wa) and h as a function of $[\phi/(2\pi)]$ for each of the curves in Fig. (7.17). It is evident that the quantity (-wa), which is proportional to the ordinary Courant-Snyder invariant, can, in some cases, show substantial variations. By contrast, the quantity h is more nearly constant in all cases.

In the situation of tunes near and at resonant values, it is again useful to work with powers of $\mathcal{M}$. Suppose the tune is written in the form

$$T = w/(2\pi) = k/m + \delta \qquad (7.130)$$

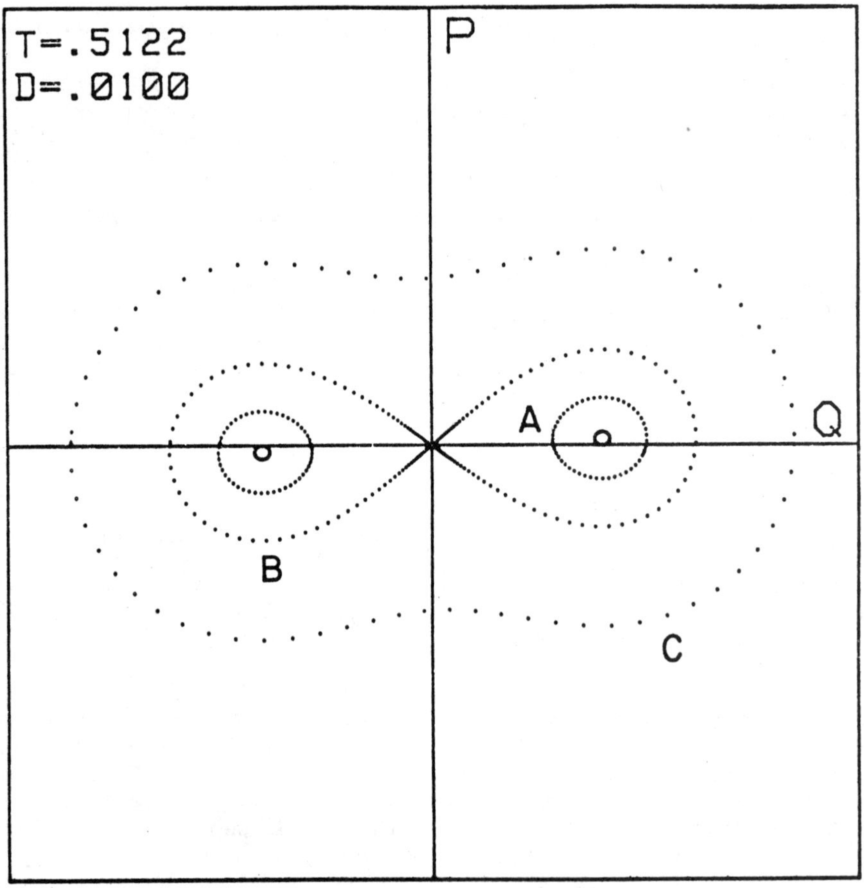

Figure (7.12): Phase-space plot generated by successive iterations of M for various initial conditions. The tune is near one half, and the beam-beam interaction strength is 10^{-2}. The coordinates extend from −2 to 2, and are normalized in such a way that the strong beam has fallen off by $1/e^3$ at $q = \pm 1$.

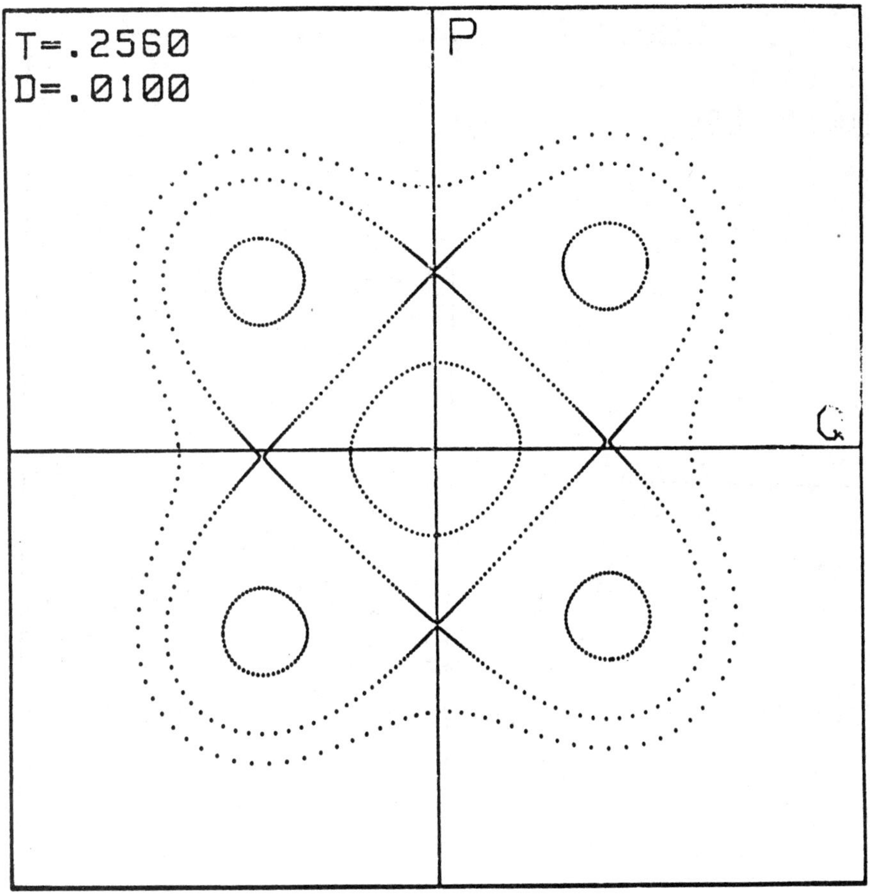

Figure (7.13): Phase-space plot when the tune is near one fourth.

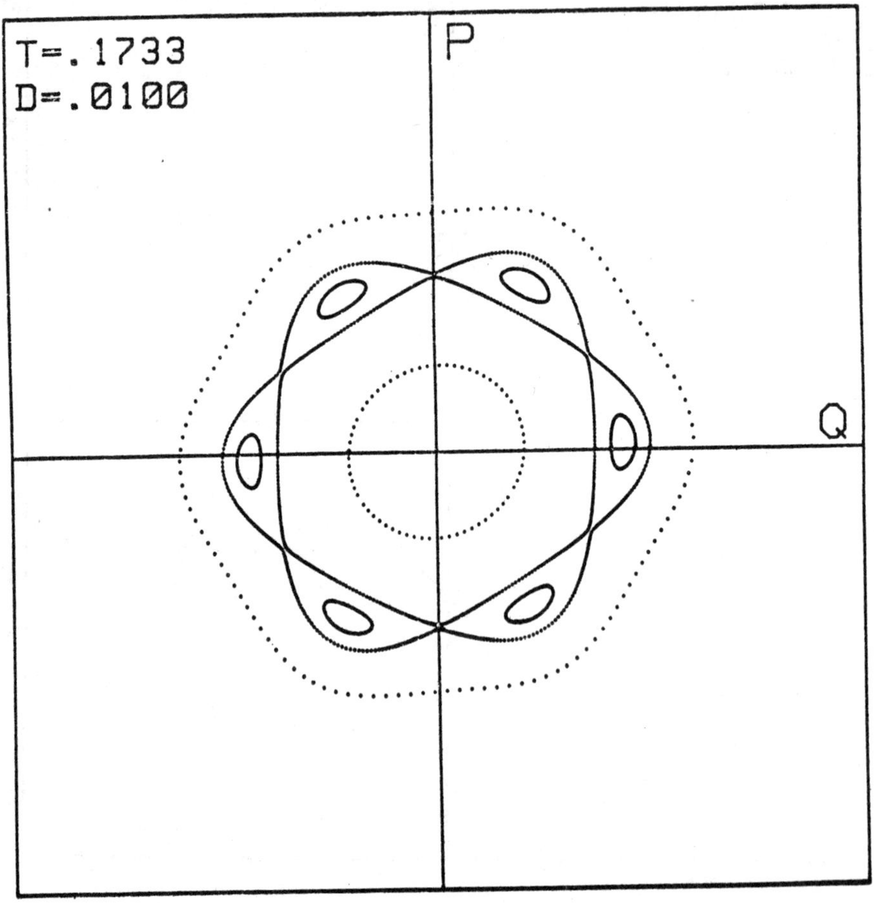

Figure (7.14): Phase-space plot when the tune is near one sixth.

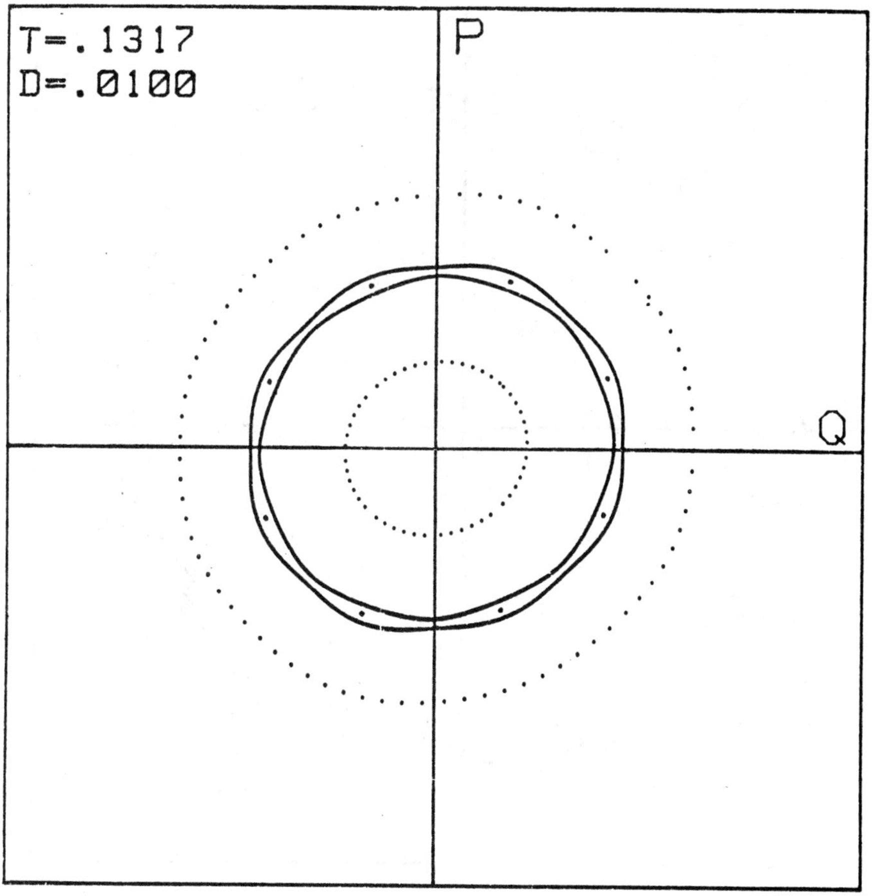

Figure (7.15): Phase-space plot when the tune is near one eigth.

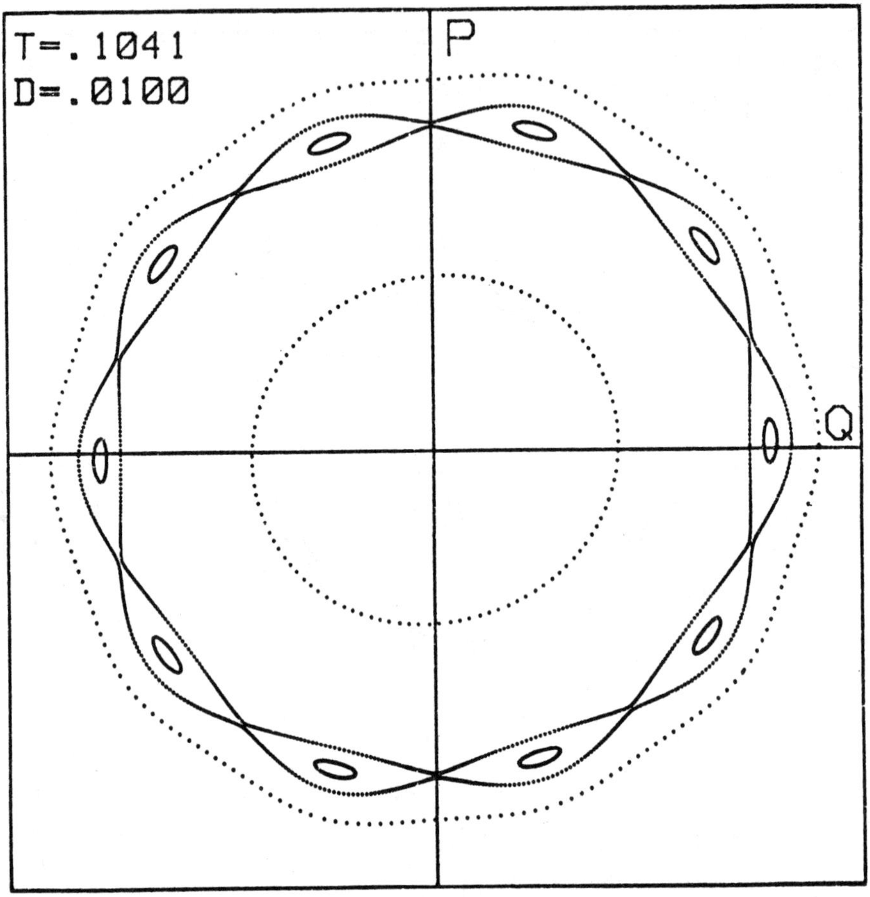

Figure (7.16): Phase-space plot near a tune of one tenth.

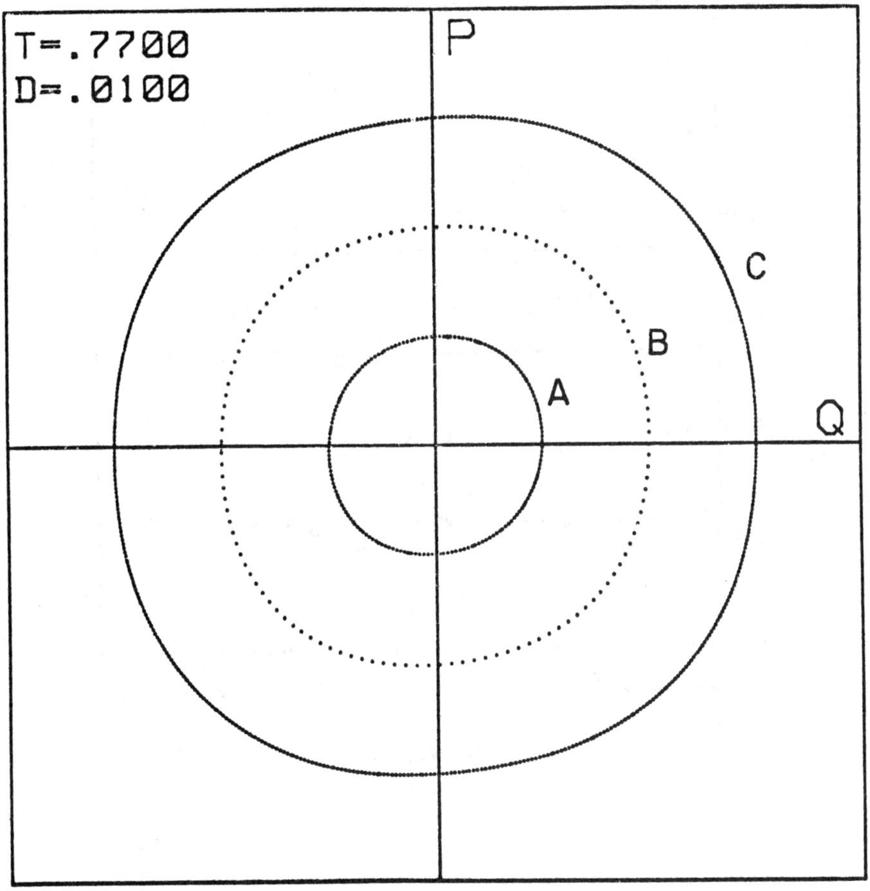

Figure (7.17): Phase-space plot for a nonresonant tune.

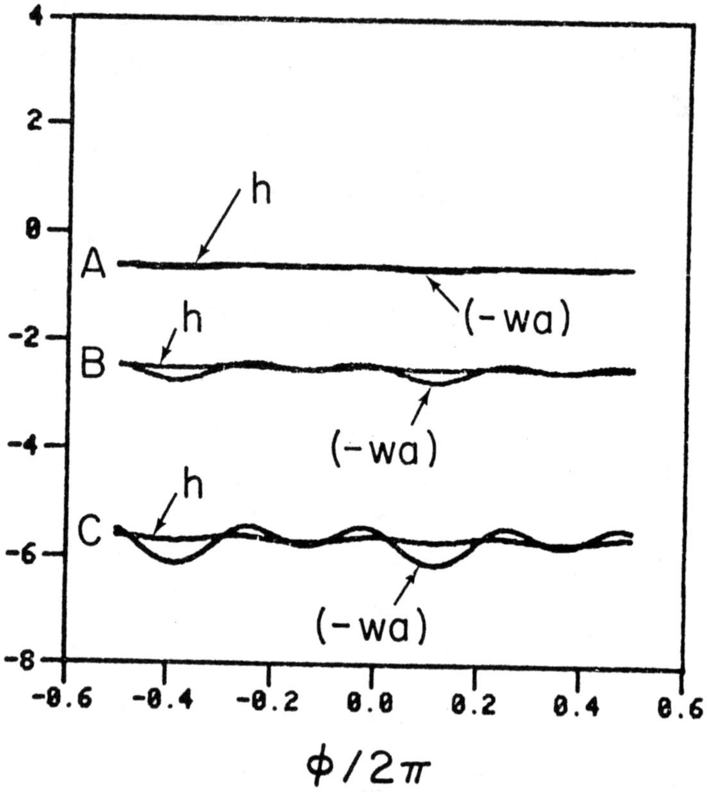

Figure (7.18): Plots of the quantities (-wa) and h for each of the three cases of Fig. (7.17). Observe that h is more nearly constant than the ordinary Courant-Snyder invariant (-wa). Observe also that the larger variations in the ordinary Courant-Snyder invariant occur in the cases with the larger betatron amplitudes, i.e., the cases where nonlinearities are more important.

where the quantity δ measures departures from the exact resonance values k/m. Now consider the m'th power of the map $\mathcal{M}$. By proceeding in a manner analogous to that used for the special case m=3 in the last section, one finds that the m'th power of the map $\mathcal{M}$ can be written in the form

$$\mathcal{M}^m = \exp(m{:}h_r{:}) \tag{7.131}$$

with h_r given by the formula

$$h_r = -\delta a + c_o(a) + 2\sum_1^\infty c^n(a)[n\delta/\sin(nw)]\cos[2n(\phi+w/2)]$$

$$+ \text{ terms of order } D^2 \text{ and higher.} \tag{7.132}$$

Again, h_r is well defined both near and at resonant tune values.

Exercise 7.18: Provide the derivation of Eq. (7.132).

The quantity h_r should be an invariant function in the resonant case. As an illustration of how this works out in the particular example of a near half-integer tune, Fig. (7.19) shows values of h_r plotted as functions of q for each of the three labeled curves of Fig. (7.12). Observe that the quantity h_r remains remarkably constant. Thus, the curves of Fig. (7.12) are very nearly level lines of h_r, and the major features of Fig. (7.12) can be predicted using $(-h_r)$ as an effective Hamiltonian.

We have again seen that the functions h and h_r can be used to compute all relevant nonlinear features of orbits in both the nonresonant and resonant regimes for the one-dimensional case. Similar results should be within reach for the two-dimensional case. This general problem should be examined after the related but simpler two-dimensional problem of the previous section has been thoroughly studied.

We close this section with an illustration of separatrix splitting in the case of the beam-beam interaction. Study of the fixed point at the origin of Fig. (7.12) shows that it is inversion hyperbolic. Consequently, the origin is a hyperbolic fixed point of $\mathcal{M}^2$. Next observe that the stable and unstable manifolds of $\mathcal{M}^2$ emanating from the origin appear to join smoothly to form the figure eight curve labeled B. On general grounds, according to the discussion of the previous section, one would expect these manifolds to intersect and then go into oscillation about each other. Apparently, the angle of homoclinic intersection in Fig. (7.12), if it exists at all, must be very small.

By contrast, Fig. (7.20) shows the case of Fig. (7.12) extended to larger values of the beam-beam interaction with the value of the tune suitably adjusted to maintain a near resonance condition. All

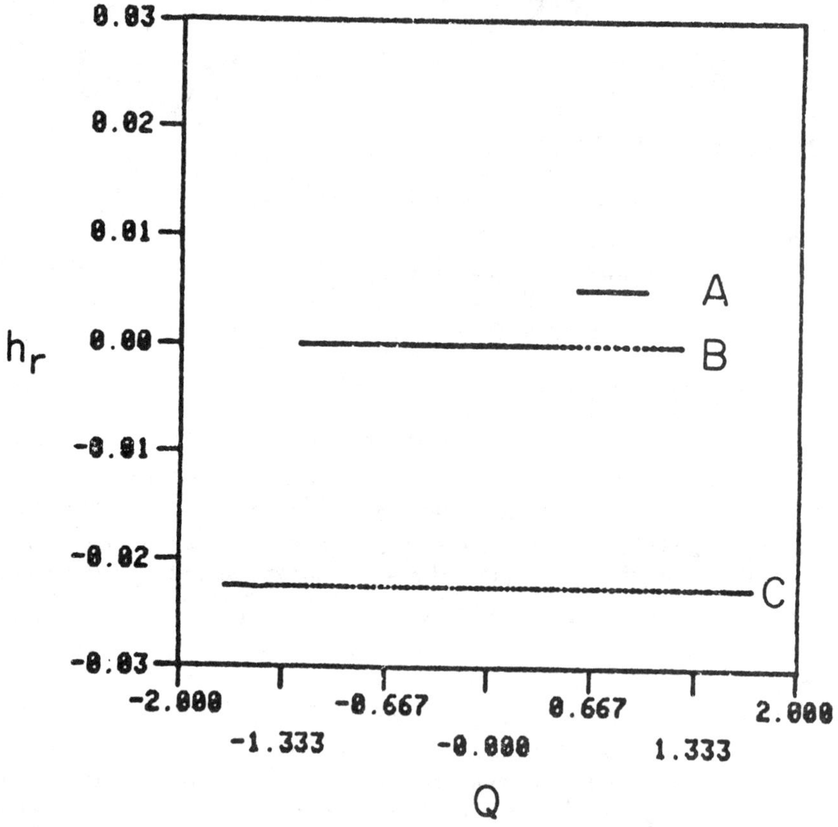

Figure (7.19): Values of h_r, the generalized Courant-Snyder invariant in the resonant case, plotted for each of the three cases of Fig. (7.12). Note that h_r is remarkably constant despite the large distortions from circular behavior evident in Fig. (7.12).

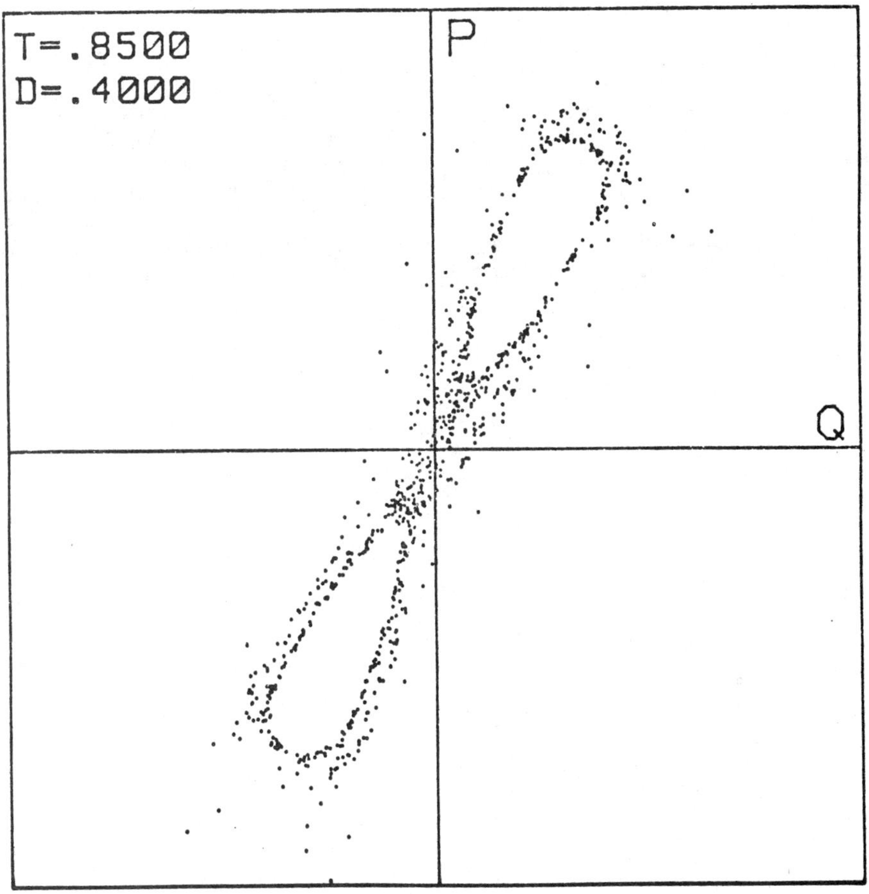

Figure (7.20): A phase-space plot showing chaotic behavior for a large value of the beam-beam interaction strength. All the points displayed are the images of a single point under repeated application of M.

the points shown are the image of a single point under the action of successive powers of . Chaotic behavior is clearly now in evidence. This behavior arises because the angle of homoclinic intersection, as illustrated schematically in Fig. (7.21), is now quite large.

ACKNOWLEDGMENTS

Much of this work was carried out with the collaboration of Mr. David R. Douglas. The author is also grateful to the Los Alamos National Laboratory for the fine hospitality provided during a sabbatical year in which many of the ideas for this work were conceived.

This work was supported in part by U.S. Department of Energy contract number DE-AS05-80ER10666.A000.

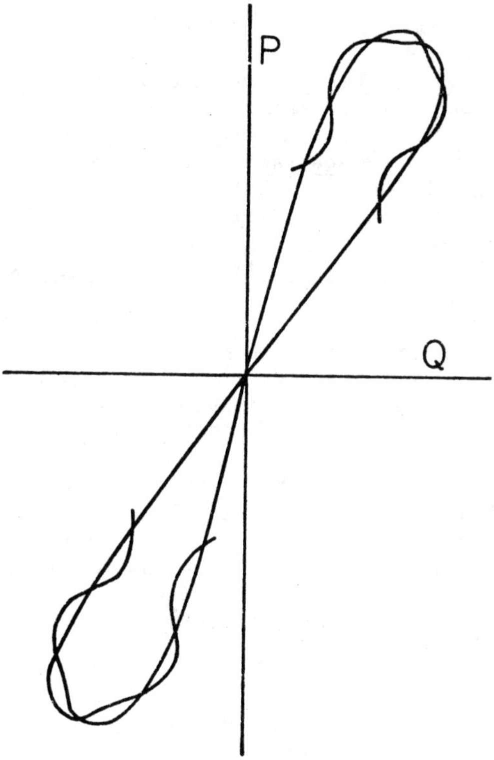

Figure (7.21): Schematic presentation of the intersections of the stable and unstable manifolds in the case of Fig. (7.20) leading to chaotic behavior.

REFERENCES

[1] H. Goldstein, Classical Mechanics 2nd ed. (Addison-Wesley, Reading, Massachusetts, 1980).

[2] J. Moser, M.V. Berry, J. Ford, Y.M. Treve et. al., Topics in Nonlinear Dynamics, AIP Conference Proceedings #46, Siebe Jorna, Ed. (1978).

[3] V. I. Arnold, Mathematical Methods of Classical Mechanics (Springer-Verlag, 1978).

[4] M. Hammermesh, Group Theory and its Applications to Physical Problems (Addison-Wesley, Reading, Massachusetts, 1962), p. 406.

[5] A.J. Dragt and J.M. Finn, J. Math. Phys. 17, 2215 (1976).

[6] N. Jacobson, Lectures in Abstract Algebra, Vol. II - Linear Algebra (D. Van Nostrand, Princeton, 1953), p. 188.

[7] A.J. Dragt and J.M. Finn. J. Geophys. Res. 81, 2327 (1976).

[8] A.J. Dragt and J.M. Finn, J. Math. Phys. 20, 2649 (1979).

[9] J. Moser, Stable and Random Motion in Dynamical Systems (Princeton University Press, 1973).

[10] In this discussion, the time t simply plays the role of a parameter. It is included in the notation to indicate that the transformation may depend on the time. That is, the transformation may be different at different times.

[11] C.L. Siegel and J.K. Moser, Lectures on Celestial Mechanics (Springer, New York, 1971).

[12] A.J. Dragt and J.M. Finn, J. Math. Phys. 17, 2215 (1976).

[13] The concept of an "adjoint" operator in the Lie algebraic context should not be confused with the concept of Hermitian adjoint often employed in the theory of linear operators on Hilbert spaces.

[14] In the general case, one may require a product of two exponentials. That is, Eq. (5.63) has the general form

$$\exp(:h_2^c:)\ \exp(:h_2^a:) = \exp(:f_2^c:) \times$$
$$\exp(:f_2^a:)\ \exp(:g_2^c:)\ \exp(:g_2^a:). \tag{5.92}$$

Similarly, Eq. (5.60) then takes the more general form

$$M^g = \exp(JS^a)\ \exp(JS^c). \tag{5.93}$$

With this understanding, Eq. (5.64) and all subsequent equations hold in the general case. It is also worth remarking that in many cases of practical interest a single exponential, which may not have either the "a" or "c" symmetry, is sufficient. See Ref. 12.

[15] D.R. Douglas and A.J. Dragt, IEEE Trans. Nucl. Sci. NS-28, 2522 (1981).

[16] H. Goldstein, Classical Mechanics, 2nd ed. (Addison-Wesley, 1980).

[17] H. Bruck, Circular Particle Accelerators, Los Alamos National Laboratory Translation of Accelerateurs Circulaires de Particules, Presses Universitaires de France (1966).

[18] B. Friedman, Principles and Techniques of Applied Mathematics, (John Wiley, New York, 1956), p. 92.

[19]A.J. Dragt, Exact Numerical Calculation of Chromaticity in Small Rings, Univ. of Maryland Physics Publication No. 81-115 (1981).
[20]A.J. Dragt, A Lie Algebraic Theory of Geometrical Optics and Optical Aberrations, J. Opt. Soc. Am. 72, 372 (1982).
[21]K.L. Brown, SLAC-75, Rev. 3 (1975).
[22]P. Henrici, Elements of Numerical Analysis, John Wiley, New York (1964).
[23]A. J. Dragt, Exact Numerical Calculation of Chromaticity in Small Rings, to appear in Particle Accelerators (1982).
[24]B. Friedman, Principles and Techniques of Applied Mathematics, John Wiley, New York (1956).
[25]For a discussion of some of the relevant mathematical background required for a proper treatment of this subject, see V. I. Arnold, Mathematical Methods of Classical Mechanics, page 228, Springer-Verlag, New York (1978); I. M. Gelfand, and V. B. Lidskii, Amer. Math. Soc. Translations, Series 2, Volume 8, page 143 (1958); V. A. Yakubovic, Amer. Math. Soc. Translations, Series 2, Volume 10, page 125 (1958).
[26]In this case, it is necessary to set up two surfaces of section. One is located at the beginning of the lattice period, and the other at its end. The Poincare map $\mathcal{M}$ now sends points in the initial surface of section at the beginning of the lattice period to image points in the final surface of section.
[27]A preliminary study of the four-dimensional case has been made by D. A. Edwards, and L. C. Teng, IEEE Trans. on Nucl. Sci. Vol. 20, No. 3, 885 (1973). See also A. J. Dragt, Symplectic Floquet Theory, manuscript in preparation.
[28]E. D. Courant, and H. S. Snyder, Annals of Physics 3, 1 (1958); H. Bruck, Circular Particle Accelerators, Chapters 4-8, Los Alamos National Laboratory technical report LA-TR-72-10 Rev., an English translation of Accelerateurs Circulaires de Particules, Presses Universitaires de France (1966).
[29]A. J. Dragt, and J. M. Finn, J. Math. Phys. 17, 2215 (1976).
[30]This assumption is made only for simplicity. In point of fact, the betatron oscillations can always be uncoupled by a suitable choice of coordinates.
[31]C. Bovet, R. Gouiran, I. Gumowski, and K. H. Reich, A Selection of Formulae and Data Useful for the Design of A. G. Synchrotrons, CERN/MPS-SI/Int. DL/70/4, page 17 (25 April 1970); J. C. Herrera, M. Month, and R. F. Peierls, Nonlinear Dynamics and the Beam-Beam Interaction, page 202, M. Month and J. C. Herrera, Edits., AIP Conf. Proc. No. 57, Amer. Inst. Phys., N.Y. (1979).
[32]G. Guignard, A General Treatment of Resonances in Accelerators, Report CERN 78-11 (1978).
[33]A. J. Dragt, IEEE Trans. Nuc. Sci. Vol. NS-26, No. 3, page 3601 (1979).
[34]The term beating range is also used to refer to a phenomenon which occurs in the more general problem of two degrees of freedom (with a four-dimensional phase space). In this case a resonance can cause a bounded periodic exchange of amplitude between the betatron oscillations in the two degrees of freedom.

[35]V. I. Arnold, Mathematical Methods of Classical Mechanics, page 394, Springer-Verlag, New York (1978); J. Moser, Stable and Random Motions in Dynamical Systems, Princeton University Press, Princeton, N.J. (1973).

[36]A. J. Dragt, and J. M. Finn, J. Geophys. Res. 81, No. 13, page 2327 (1976).

[37]In the absence of the existence of invariant functions, it is generally not possible within our present knowledge to make positive statements about long-term stability. If the use of the first few terms of h predicts instability, as in the unstable region of Fig. 7.4, then we may be sure that orbits in this region are in fact unstable. However, if stability is predicted by the first few terms of h, then we are in fact assured of stability only for moderately long periods of time. That is, given enough time, there is the possibility that homoclinic behavior will bring a point from what was judged to be the stable region into the unstable region. Thus, when nonlinear effects are taken into account, we are in the uncomfortable position of being generally unable to make any rigorous statements about very long term behavior. From this perspective, the fact that storage rings work so well must be viewed, at the very least, as a minor miracle.
Under certain conditions it can be shown in the one degree of freedom case (two-dimensional phase space) that there are still isolated closed invariant curves despite the nonexistence of invariant functions. The existence of these curves was first shown by Komolgorov, Arnold, and Moser, and consequently they are often referred to as KAM curves. See, for example, J. Moser, Stable and Random Motions in Dynamical Systems, Princeton University Press, Princeton, N.J. (1973). The existence of closed invariant curves would guarantee the stability of all orbits corresponding to initial conditions inside these curves. There is some optimism that such curves exist for regions of phase space of physical interest. However, the present mathematical proof of their existence holds only for regions of phase space much too small to be of physical significance.

[38]The map (7.84) was first studied by Henon in a different context. Figures 7.8 and 7.9 are taken from his work. See, M. Henon, Quarterly of Applied Mathematics 27, 291 (1969). For remarkably beautiful color pictures of phase-space plots of this map, see the work of M. Feigenbaum in Los Alamos Science, Volume 1, Number 1 (1980), published by the Los Alamos National Laboratory, Los Alamos, New Mexico.

[39]A. J. Dragt, and J. M. Finn, J. Math. Phys. 20, p. 2649 (1979).

[40]The remainder terms are all commutators which contain two or more f_b's and any number of g_2's. Therefore, these terms are of second order and higher in the beam-beam interaction strength. The terms of order two can be computed if needed. See Ref. 29.

[41]A. J. Dragt, Nonlinear Dynamics and the Beam-Beam Interaction, p. 143, M. Month and J. Herrera, Edits., Am. Inst. of Phys. Proceedings

No. 57 (1979); A. J. Dragt, and O. Jakubowicz, Analysis of the Beam-Beam Interaction Using Transfer Maps, Proceedings of the SLAC Symposium on the Beam-Beam Interaction, SLAC-Pub-2624, Stanford University, Stanford, California (1980).

[42] J. C. Herrera, Nonlinear Dynamics and the Beam-Beam Interaction, p. 29, M. Month and J. C. Herrera, Edits., Am. Inst. Phys. Proceedings No. 57 (1979).

[43] J. C. Herrera, M. Month, and R. F. Peierls, Nonlinear Dynamics and the Beam-Beam Interaction, p. 202, M. Month and J. C. Herrera, Edits., Amer. Inst. Phys. Proc. No. 57 (1979).

BEAM-BEAM INTERACTION

Part One. Luminosity in Electron-Positron Colliding-Beam Storage Rings: Introduction to the Problem

Jonathan F. Schonfeld
Fermi National Accelerator Laboratory, Batavia, Illinois 60510

TABLE OF CONTENTS

Part Two. The Dynamics of the Beam-Beam Interaction

J. L. Tennyson
Electronics Research Laboratory
University of California, Berkeley, California 94720

00940243X/82/870314-32$3.00 Copyright 1982 American Institute of Physics

The goal of the colliding-beam storage ring designer and operator is optimization of luminosity, which is defined as the number of elementary-particle events of unit cross-section that take place at a single beam-beam encounter point, per unit time.

At the newest electron-positron storage rings-CESR, PEP, and PETRA - the luminosities observed have so far been at best about an order of magnitude below the design values[1] of $\sim 10^{32}$ cm^{-2} sec^{-1}. This means that high energy physics experiments, so dependent on good statistics for analyzing rare processes, must run much longer than had been originally expected for acceptable results.

In the first two sections of this pedogogical report, I shall discuss two assumptions on which these design expectations were based, and specific ways in which subsequent operating experience has shown these assumptions to be naive. One assumption led to an overestimate of luminosity at a given current, while the other led to an overestimate of the largest current that could be stored. (I shall consider in detail only phenomena related directly to beam-beam collisions; single-beam effects will be mentioned only in passing.) In the third and final section I shall describe some recent theoretical attempts to go beyond these assumptions.

I have chosen to concentrate here exclusively on e^+e^- machines, because there is published data on high-energy e^+e^- storage from at least seven separate laboratories, collected over a period of, by now, about twenty years. By contrast, there is as yet no significant data on $\bar{p}p$ colliders; and all the pp data comes from only one facility - the ISR - whose operating conditions differ from those of electron-positron rings.

I. HOW CONVENTIONAL DESIGN PROCEDURE OVERESTIMATES LUMINOSITY AT A GIVEN CURRENT

Mathematically, luminosity is defined by the formula

$$L = \left(\frac{I_1}{A_1}\right)\left(\frac{I_2}{A_2}\right) \frac{A}{e^2 Bf} , \tag{1}$$

where f is the revolution frequency of a stored particle (i.e., speed of light divided by circumference of ring); e is the positron charge; B is the number of bunches per beam; I_1 and I_2 are the two beam currents; A_1 and A_2 are the two beam areas, transverse to the collision axis; and A is the transverse area in which the two beams overlap upon intersection. (When the beam particles are not distributed uniformly in the transverse plane, these A's are effective areas, defined by appropriately weighted averages. See Ref. 35.)

It has been standard practice, when computing anticipated luminosity, to ignore the effect that one beam has on the size and shape of the other.[2] This is the nonperturbation assumption. It means, in particular, that A_1, A_2 and A are all equal and independent of the number of particles in the storage ring. For equal currents ($I_1 = I_2 \equiv I$) we may then write

$$L = I^2/e^2\ BfA. \qquad (2)$$

In particular, luminosity in this configuration should increase with current as (constant)xI^2, where the proportionality constant is computable from first principles.

It turns out that this is not consistent with experiment: Direct measurements of luminosity show that L always grows much more slowly than (constant) xI^2 at large current. Direct observations of beam size show that the cross-sectional area of at least one of two colliding beams always increases significantly with I at large current. (When both beams blow up equally, Eq. (2) is still valid, but one may no longer take A to be independent of I.) These trends are illustrated in Figs. 1-12, to which we now turn.

(It should be noted that by unperturbed A I mean the single-beam cross-sectional area as computed from the beta-functions and emittances of the real storage ring. Because of single-beam instabilities,[33] (among other things), these β's may have to be held at larger values than anticipated by the designers. Thus one should really distinguish two principal reasons for low luminosity, relative to design: colliding-beam effects, which lower the luminosity vs. current curve relative to some quadratic; and single-beam effects, which lower the normalization of the reference quadratic itself.)

Figures 1-5 show data on luminosity as a function of current, from several laboratories. Figure 1 represents data taken at SPEAR.[3]

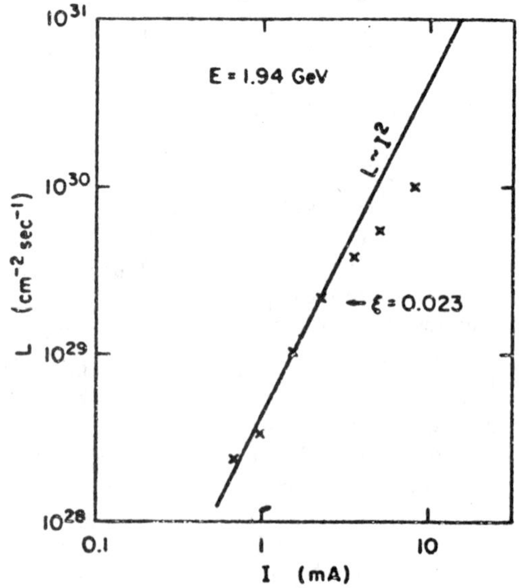

Figure 1

Luminosity vs. current per beam, at SPEAR, from Ref. 3

The data point with the largest current (about 8 mA) falls short of the quadratic curve (extrapolated from small current) by a factor of about three. Figure 2 represents data taken at CESR.[4] For $\beta_y^* = 29$ cm, the data point with the largest current falls short of the

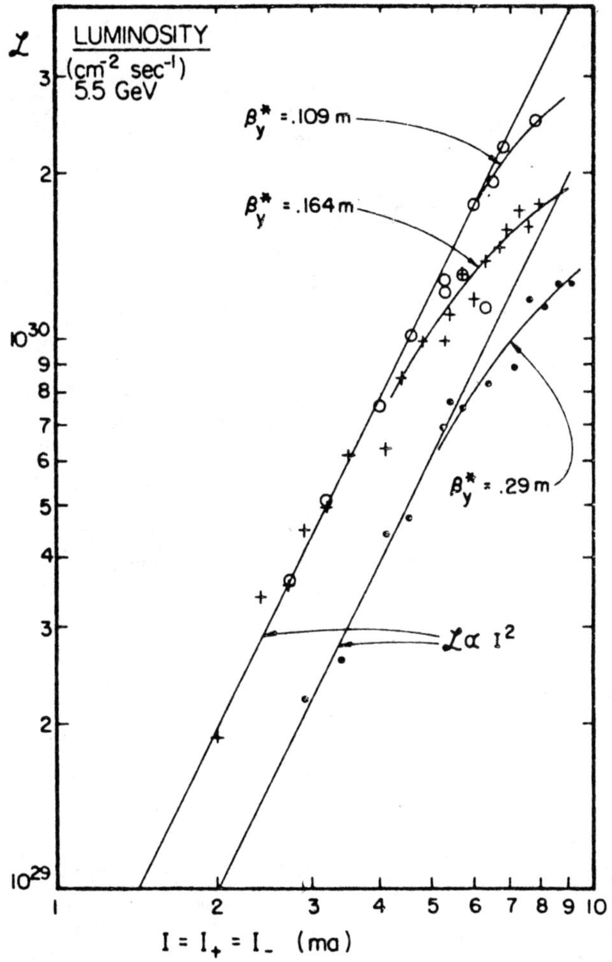

Figure 2

Luminosity vs. current per beam, at CESR, from Ref. 4.

corresponding quadratic extrapolation by a factor of about 1.6. Figure 3 shows data taken at PETRA.[5] For these graphs, PETRA was operated with two bunches per beam. (The SPEAR and CESR data described above were taken with one bunch per beam.) In each graph, the abscissa corresponds to the single-beam current per bunch, and the ordinate corresponds to the quotient of luminosity and the square of the single-bunch current. If L were proportional to I^2, then the

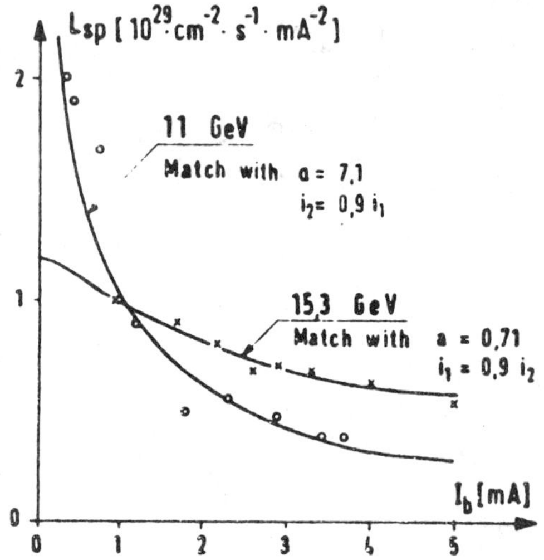

Figure 3
Specific luminosity vs. single-bunch current
at PETRA, from Ref. 5.

curves in Fig. 3 would be straight horizontal lines. Instead, the 15.3 GeV curve is lower at 5 mA, by a factor of about two, than it is at very small current. The effect is clearly much worse at the lower energy of 11 GeV. Figures 4 and 5 represent data taken at PEP,[6] with two different sets of magnetic lattice parameters. The two curves in each figure correspond to operation with one and three bunches per beam. In Fig. 4, the one-bunch curve is quadratic at small current, but falls below the extrapolation of the quadratic curve by a factor of about two at the largest current shown; in the same figure, the curve interpolated between the three-bunch data points grows more slowly than quadratic (more like $I^{1.6}$) even at small current. At the

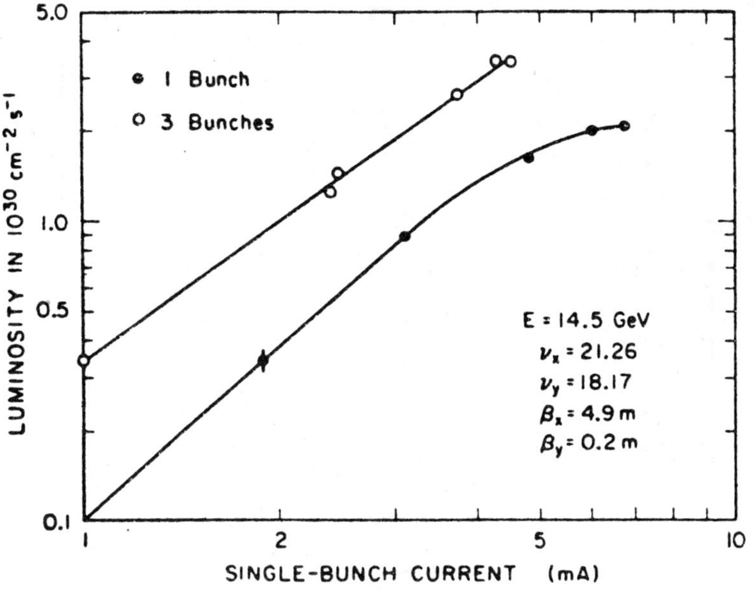

Figure 4

Luminosity vs. single-bunch current at PEP, from Ref. 6.

highest current for which data with both bunch numbers is shown, the three-bunch curve exceeds the two-bunch curve by a factor of two, not the naively expected three. (Three would arise in formula (2) from combining a factor of nine, corresponding to the ratio of I^2 in the two cases, with a factor of one-third, corresponding to the ratio of the two values of B^{-1}.) In Fig. 5, the single-bunch curve at the largest current falls short of the quadratic extrapolation from small current by a factor of about 5.4; the ratio of three-bunch to single-bunch luminosity falls to 2.2 at the largest current for which the comparison can be made.

(A number of workers have attempted to fit the high-current luminosity of different high-energy e^+e^- storage rings to a single phenomenological scaling law. See Refs. 3 and 34 for details.)

Figures 6-12 show data on the increase of transverse beam size with current. Let me first make a few comments before discussing this data in detail.

When the currents in two colliding beams are not equal ("weak-strong" case), the more tenuous ("weak") beam is always the one to be more enlarged. When the two currents are equal ("strong-strong" case) there is no general rule for predicting which beam (if either) will expand more than the other, and a perturbation (for example momentary

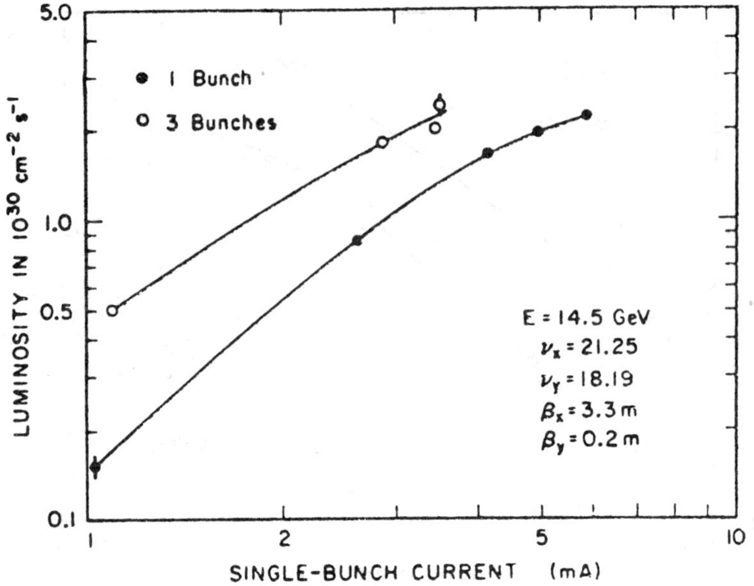

Figure 5

Luminosity vs. single-bunch current at PEP, from Ref. 6

excitation at a betatron frequency) can turn the thicker beam thinner and vice versa. At SPEAR[7] this kind of interchange can be induced in a gradual and controlled manner by slowly varying the phases between RF cavities. In particular, for any energy there is a setting of the relative phases for which the two beams are blown up equally, and this setting turns out to maximize luminosity. (There are indications that the same thing might be possible at PEP.[6]) Unless otherwise noted, all the SPEAR data discussed in this report were obtained with the beams deliberately matched in this way. (Data from other machines may also have been taken under conditions of equal blowup, but only at SPEAR is this situation claimed to be a result of conscious manipulation.)

This suggests the first of a number of exercises that will be proposed in the course of this paper.

<u>Problem 1</u>

How should one place two identical RF cavities in a given storage ring so that any variation of their relative phase will not affect the locations of the points at which the electron and positron synchronous design orbits cross?

While you're at it, a second exercise:

Problem 2

Even without colliding-beam effects, the shape of an electron bunch in a storage ring is not an entirely trivial matter. For example: A common practice is the deliberate rotation of a number of quadrupole magnets in order to couple horizontal and vertical betatron motion and thus enhance the vertical extent of the beam. (Even with such enhancement, electron and positron beams are much wider than they are high.) This linear xy coupling also has the effect of tilting the transverse profiles of the bunches. Look up the theory of equilibrium beam distributions in linearly focused storage rings (Ref. 8), and then use the figures in Table I to estimate the magnitude of the transverse tilt (it should be small) at an interaction point in PEP. Show that counter-rotating electron and positron bunches tilt in the same sense (i.e. both inclined toward the same general area above or below the center of the ring), when they pass the same azimuth. If they did not tilt in the same sense, then the two beams would not overlap totally at the interaction point even at low currents, and the incompleteness of this overlap would be one reason for low luminosity.

We now return to experimental data.

Figures 6-11 are obtained from observations of the synchrotron light radiated by beam particles. The radiation due to an ultra-relativistic charged particle is (to a good approximation) emitted tangent to the particle's path, so that (when the detector is sensitive enough) a picture of this light can be interpreted as a graph of the transverse beam distribution, with light intensity serving as a measure of particle density. Such radiation profiles provide the clearest indications (essentially self-evident) of the extent and shape of beam enlargement.

Figure 6 shows six television photographs of electron and positron synchrotron images taken under various conditions at SPEAR.[9] (Current and energy are not specified.) The indication "with flip-flop effect" means that the RF relative phases have not been adjusted so as to deliberately match the e^+ and e^- profiles. "Flip-flop balanced" means that such an adjustment has been made. These photos are not fine enough to reveal details of light intensity, for which one is referred to Fig. 7, depicting the results of photometric measurements[9] of bunch dimensions at one and four mA of current per colliding beam. The number under each peak in Fig. 7 is its full width at half-maximum, in millimeters.

The T.V. photographs shown in Fig. 8 were taken at PETRA[11] (the beam energy is not indicated, but is probably around 7 GeV), as was the data represented in the graph of bunch heights vs. time[5] shown here as Fig. 9. Figure 10, similar to Fig. 7, shows synchrotron light profiles measured at ADONE.[12] The energy is not indicated, but is in the vicinity of 1 GeV. ("r" and "v" stand for "radial" and "vertical", equivalent to "x" and "y" in Fig. 7.) Figure 11 shows strong and weak beam profiles for various values of vertical tune, as observed by a television monitor at ACO.[13] (Energy is once again

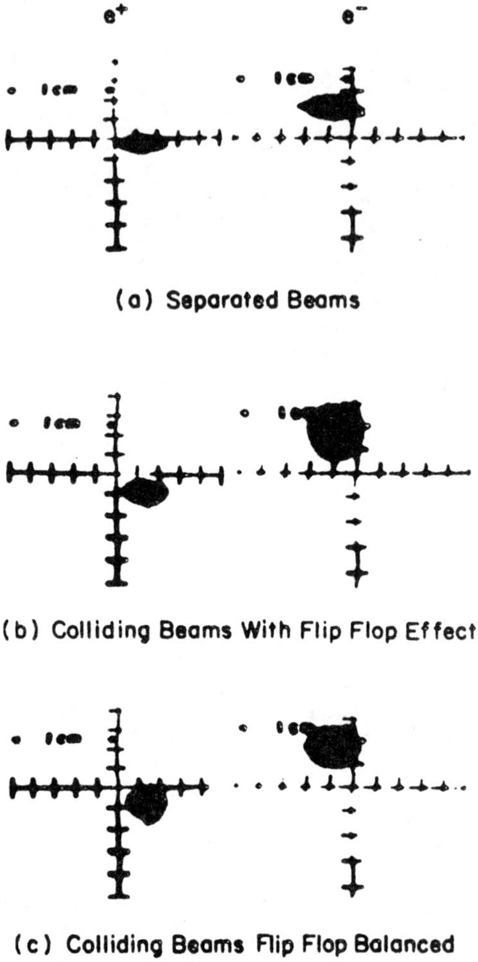

Figure 6

Television photographs of beam cross-sections at SPEAR, from Ref. 9.

around 1 GeV; current not specified.) Similar blow-up effects have been described in reports from CESR [4] and from PEP.[6]

The results of some less direct SPEAR measurements of the growth of beam height with current are shown in Fig. 12. (As is clear upon inspection of Figs. 6-11, beam width varies very little, if at all, with current.) I introduce this figure more to illustrate the range of experimental and interpretative techniques that have been applied to this phenomenon, than to add to the evidence of beam blow-up already indicated in the preceeding six figures.

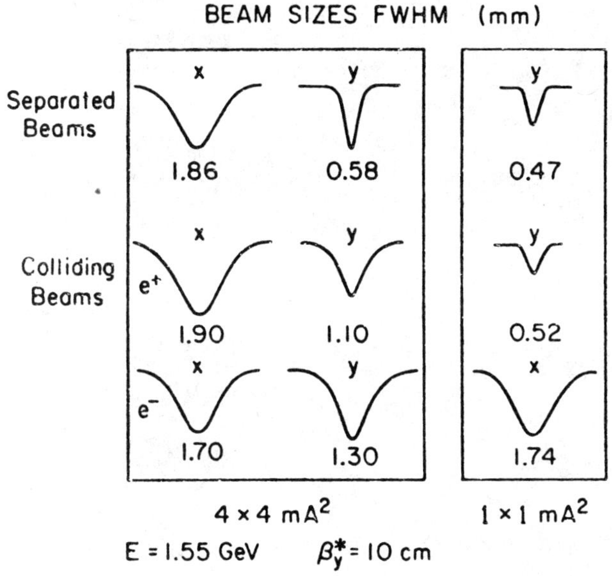

Figure 7

Transverse beam distributions at SPEAR, from Ref. 9.

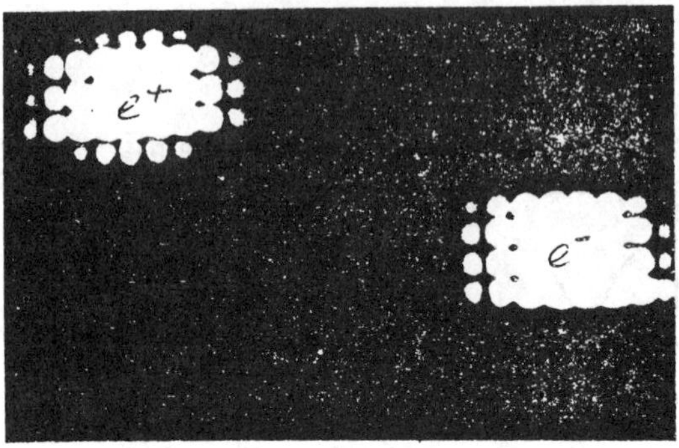

Beam dimensions at $I^+ = I^- = 0.3$ mA.

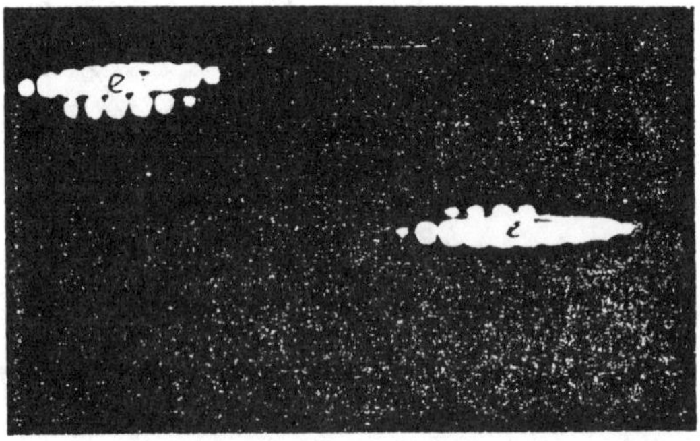

Beam dimensions at $I^+ = I^- = 0.15$ mA.

Figure 8

Television photographs of beam cross-sections at PETRA, from Ref. 11.

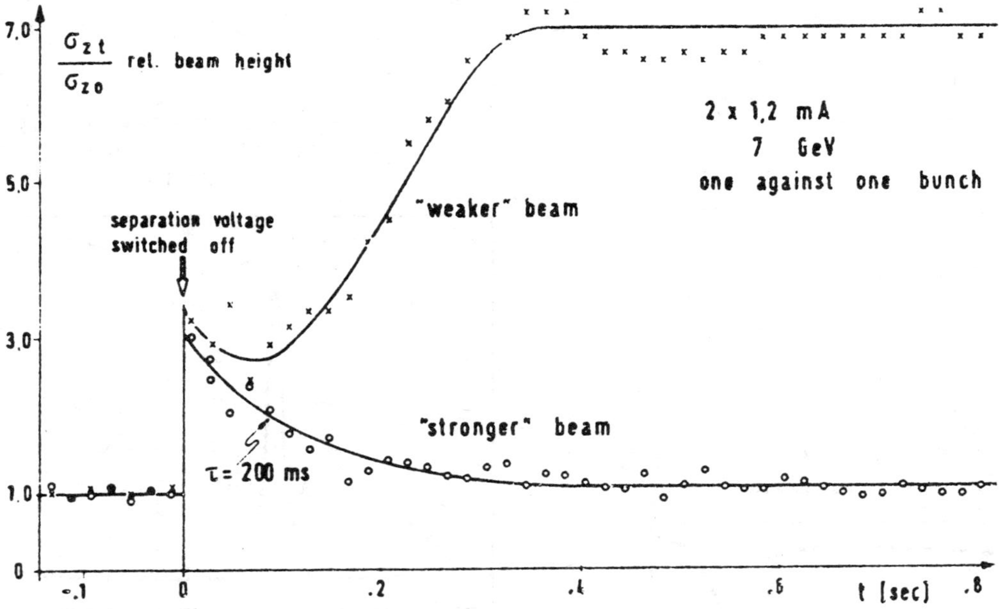

Figure 9

Beam heights vs. time at PETRA, from Ref. 5.

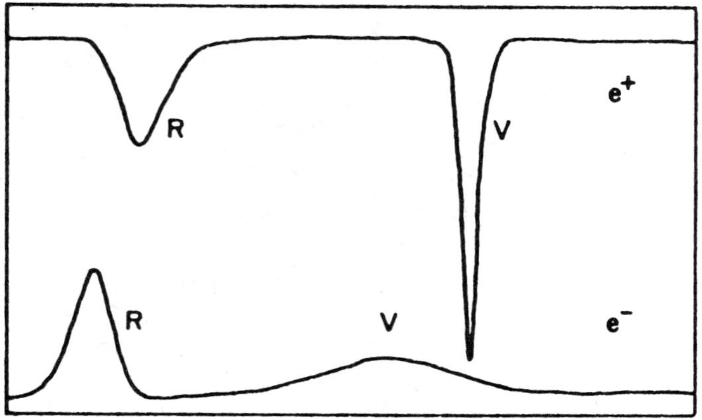

Weak beam is "flipped" due to excessive current in the strong beam. The strong beam has its unperturbed vertical dimension.

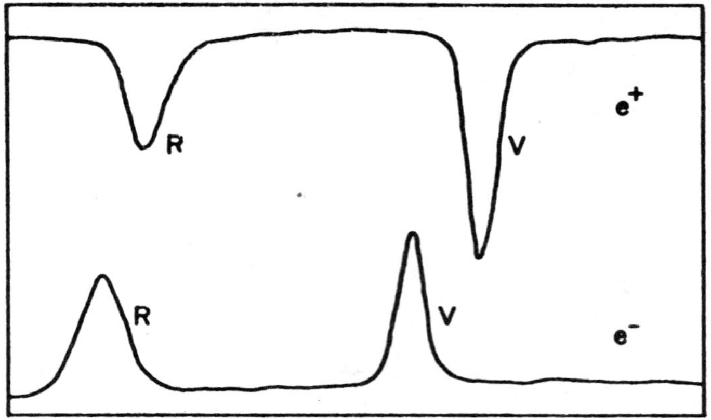

Interacting strong beams at or near the space limit.

Figure 10

Transverse beam distributions at ADONE, from Ref. 12.

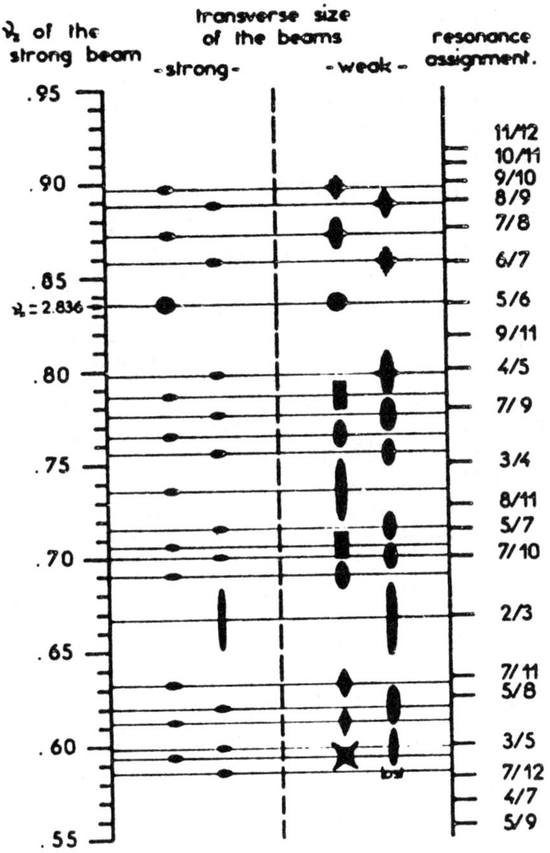

Figure 11

Transverse beam shapes vs. fractional part of vertical tune at ACO, from Fig. 13.

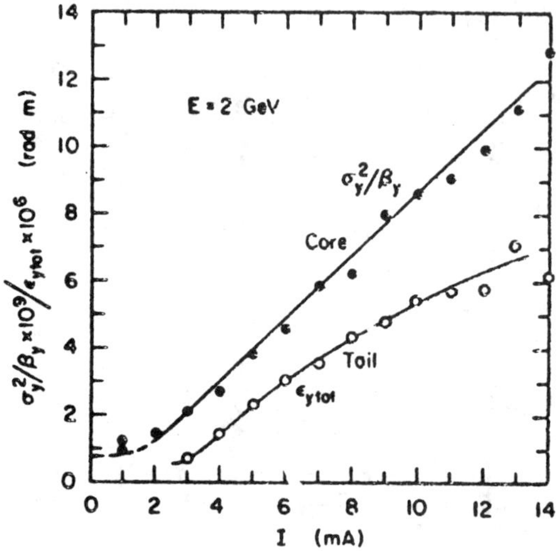

Figure 12

Square of beam height, normalized to β_y, vs. current per beam at SPEAR. Solid points derived from luminosity data; open points derived from lifetime measurements. From Ref. 3.

The two curves in Fig. 12 were obtained in two entirely different types of measurements, so there is nothing a priori problematical in their disagreement.

The black dots in Fig. 12 represent a reassembly of data that we've already examined, rather than a direct and independent measurement of beam size. They were obtained by inserting luminosity data (like that shown in Fig. 1) and the unperturbed value of σ_x^* (the rms bunch width) into the right-hand side of the following formula for σ_y^* (the rms bunch height)

$$\sigma_y^* = I^2/L\sigma_x^* e^2Bf. \tag{3}$$

Equation (3) is derived from (2) by writing the effective area A as $4\pi\sigma_x^*\sigma_y^*$, where the assumption of a gaussian distribution accounts for the exact number 4π as a correction for the presence of particles beyond the rms distances.

The white dots in Fig. 12 represent calculations based on beam-lifetime measurements, which are independent of luminosity measurements, and are more directly related to beam size, being immediately sensitive to the flux of particles at large y. The calculations were based on the following formula (derived again assuming a gaussian bunch distribution)[2]

$$\tau = \tau_\beta\left(\frac{\sigma_y}{h}\right)^2 \exp\frac{1}{2}\left(\frac{h}{\sigma_y}\right)^2 . \tag{4}$$

Equation (4) expresses the beam lifetime τ in terms of: the vertical betatron damping time τ_β; the half-height, h, of the beam pipe at the location of a scraper (assuming a pipe much wider than high); and the rms beam height, σ_y, at that point. The data for τ at different beam currents is substituted into the left-hand-side of (4), the known values of τ_β and h are substituted into the right-hand side, and the resulting equation is solved for σ_y to produce the open dots in Fig. 12.

(Note that the values of $(\sigma_y)^2$ shown in Fig. 6 are normalized to β_y. According to the theory of quantum (photon) noise fluctuations, σ_y and σ_x are proportional, as functions of storage-ring azimuth, to $\sqrt{\beta_y}$ and $\sqrt{\beta_x}$ (at least for vanishing dispersion). Thus, although the open and closed dots correspond to data taken at two different points (the interaction point and the scraper location), one may interpret the figure as if all the data were extracted at a single location.)

One should be aware that both curves are intrinsically inaccurate, in that both presuppose gaussian charge distributions, while other, independent experiments have shown nongaussian behavior at large distances from the bunch center.[9,10]

II. HOW CONVENTIONAL DESIGN PROCEDURE OVERESTIMATES MAXIMUM ACHIEVABLE CURRENT

To explain the method by which maximum storable colliding currents have conventionally been predicted, it is first necessary to introduce the concept of "beam-beam linear tuneshifts," which are measures of the effect that the whole of one beam has on a single particle of the other beam. (Thus, as there are two beams, there are, strictly speaking, two sets of beam-beam tuneshifts.) The conventional prediction of maximum current is based on a picture of instability that is most immediately concerned with maximum achievable tuneshifts. This point of view is becoming obsolete today, but has in the past enjoyed considerable influence, despite a very poor predictive record.

The tuneshift is a notion most naturally associated with beams whose distributions have reached a steady state, and whose bunches are not significantly tilted relative to the usual comoving axes. Because of the steady state, a single particle in one beam sees the perturbation due to the whole of the opposing beam as periodic in the usual path length parameter s. (The period is the path length between interaction points, which is the same as the circumference of the ring divided by twice the number (B) of identical bunches in a beam.) Because the tilt angles are negligible, the linear part of this perturbation does not couple x and y; so if xy coupling in the ring magnets is also neglected, then the motion of a test particle around the magnetic lattice and through the perturbation is described in linear approximation by two uncoupled linear 2nd order ordinary differential equations with periodic coefficients. Such a system can be described, as usual, in terms of (horizontal and vertical) tunes per period. The differences, ξ_x and ξ_y, between these tunes and the unperturbed ones are the horizontal and vertical linear beam-beam tune-shifts (per bunch per collision - the total tuneshifts would be these numbers times 2B) due to the beam in question.

Twenty years ago, it was expected that colliding-beam performance would be limited primarily by linear instability,[14] so that the maximum achievable tuneshift (x or y) would be the smallest number that would give an integer or half-integer when added to the corresponding unperturbed tune. Storage rings before SPEAR encountered instabilities at tuneshifts much smaller than those predicted by the linear theory.[15] The obvious next guess - that <u>non</u>linear resonances are responsible for upper limits on the ξ's - never developed into a body of unambiguous criteria. Instead, the idea that the maximum ξ's could be easily computed for each machine from first principles was supplanted by the idea that - for whatever reason - all e^+e^- storage rings are characterized by the same maximum achievable ξ_y. (As will be explained later, rules of thumb for ξ_x are not commonly employed in estimates of the maximum luminosity of e^+e^- machines.) Accordingly, the designers of SPEAR based their maximum current predictions on the maximum vertical tuneshift - .025 - reached at any of the earlier e^+e^- colliders.[16] SPEAR II subsequently performed up to $\xi_y \sim .06$ (although not at all energies - see Fig. 13) so the designers of CESR,[17] PEP,[18] and PETRA[5] based their optimum performance predictions on the supposition that .06 would be the largest ξ_y achievable. At present, none of these machines have exceeded $\xi_y \sim .03$.

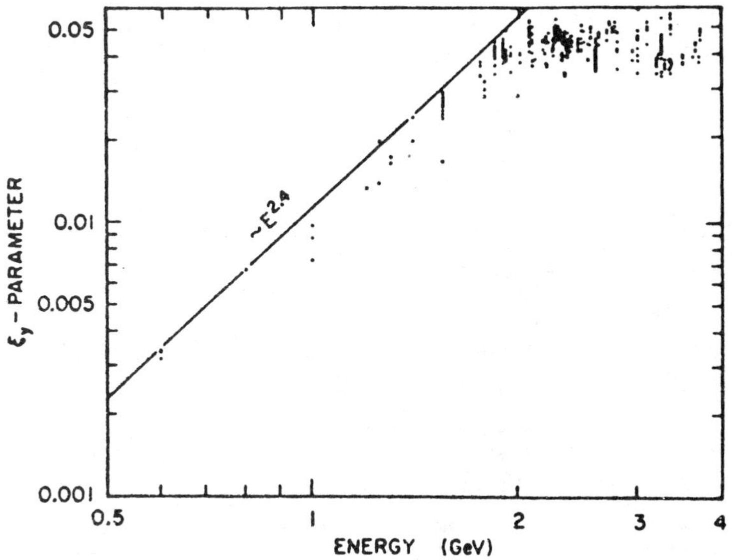

Figure 13

Maximum vertical tuneshift vs. energy at SPEAR, from Ref. 3.

In view of this record, how one should predict (short of large-scale computer simulation) the limiting parameters of the next generation of storage rings is not clear.

We now derive expressions for tuneshifts in terms of beam current and beam dimensions, so that we can see explicitly how an overestimate of maximum vertical tuneshift contributes to an overestimate of maximum current and luminosity. For convenience, we shall assume that the beams are Gaussian. The first two steps in the derivation are proposed as problem 3.

Problem 3

a) Consider a collision between a positron and an electron, initially very far apart, with initial velocities $\vec{v}$ and $-\vec{v}$, and initial impact parameter $\vec{\rho}$ (perpendicular to $\vec{v}$, pointing from positron to electron). Show that in the limit of very high (relativistic) energy (i.e. $v \sim c$), the total change (accumulated from very large negative to very large positive times) in electron momentum due to this collision is

$$\Delta\vec{p} \; -\vec{\rho}\,\frac{2e^2}{c} \tag{5}$$

($\Delta\vec{\rho}$ = 0 in the same approximation.)

b) Now replace the positron by a bunch of N positrons whose density function in the spatial plane perpendicular to $\vec{v}$ is a Gaussian of rms widths σ_x and σ_y. Show that when (5) is integrated over the entire positron distribution (beware of singular integrals), the result is, in linear approximation,

$$\Delta p_x = \frac{-2e^2 xN}{c\sigma_x(\sigma_x+\sigma_y)}$$

$$\Delta p_y = \frac{-2e^2 yN}{c\sigma_y(\sigma_x+\sigma_y)}, \tag{6}$$

where x and y are the horizontal and vertical transverse displacements of the electron relative to the positron bunch center.

Formula (6) is applicable to the typical collision in a storage ring when (as is usually the case) the bunch lengths (along $\vec{v}$) are small compared to the scale on which β_x and β_y vary in the interaction region. In such cases, one commonly regards the collisions between single particles and whole bunches as instantaneous; so that one may relate the momentum changes to the more conventional variables of beam dynamics according to $\Delta p_x = mc\gamma\Delta x'$ (and similarly for y), where m is the electron mass and mc^2 its energy; and one may replace the width and height, σ_x and σ_y, by σ_x^* and σ_y^*, their values at the point where the e^+ and e^- synchronous design orbits cross. In toto, for the linear approximation

$$\Delta x' = -\left(\frac{2r_o N}{\gamma\sigma_x^*(\sigma_x^* + \sigma_y^*)}\right) x$$

$$\Delta y' = -\left(\frac{2r_o N}{\gamma\sigma_y^*(\sigma_x^* + \sigma_y^*)}\right) y, \tag{7}$$

where $r_o = e^2/mc^2 = 2.82 \times 10^{-13}$ cm. Of course the words "electron" and "positron" in the foregoing may be freely interchanged.

Before we continue, you might want, as proposed below in problems 4 and 5, to estimate a couple of orders of magnitude that might give you a feel for the scale of the beam-beam interaction.

Problem 4

How does the typical change in x' or y' due to a beam-beam encounter in PEP compare with the components of transverse velocity of the typical stored electron? To estimate the typical changes in x' and y', use formula (7): γ and N can be computed from Table I and Eq. (12); x and σ_x^* may be estimated from the table by setting $x \sim \sigma_x^* \sim$ unperturbed σ_x^*, and similarly for y and σ_y^*. To estimate the typical velocity components, use

$$\sqrt{\langle (y')^2 \rangle^*} \sim \sqrt{\langle y^2 \rangle^*}/\beta_y^* \sim \sigma_y^* \Big|_{\text{unperturbed}} /\beta_y^* \qquad (8)$$

(and similarly for x), suggested (at least when β' can be neglected) by the form

$$W_y \equiv \frac{y^2}{\beta_y} + \beta_y \left(y' - \frac{\beta_y'}{2\beta_y}\, y\right)^2, \qquad (9)$$

of the Courant-Snyder invariant.[19]

Problem 5

How does the electromagnetic energy radiated as a result of the abrupt kick (7) compare with the energy lost to synchrotron radiation during a single turn through the storage ring? To estimate the energy radiated because of the collision, use, for example, formulae in Ref. 20, assuming that the time elapsed during the kick is of the order of the unperturbed bunch length divided by the speed of light. (For consequences of this kind of radiation - known as "beam-strahlung" - see Ref. 36.) To estimate the energy radiated during one machine revolution, use the approximation (energy radiated per turn) $\sim E/\tau_e f$, where τ_e is the energy damping time.

We continue toward the tuneshifts:

Following (7), transit of an electron (in linear approximation) through one machine period can be represented by the transformation

$$\begin{pmatrix} y \\ y' \end{pmatrix} \rightarrow \begin{pmatrix} \cos 2\pi\mu_y & \sigma_y^* \sin 2\pi\mu_y \\ -\frac{1}{\beta_y^*} \sin 2\pi\mu_y & \cos 2\pi\mu_y \end{pmatrix} \begin{pmatrix} 1 & 0 \\ \frac{-2r_0 N}{\gamma\sigma_y^*(\sigma_y^* + \sigma_x^*)} & 1 \end{pmatrix} \begin{pmatrix} y \\ y' \end{pmatrix} \tag{10}$$

(and similarly for x). The first matrix factor corresponds to linear transport between interaction points (μ_y is the unperturbed tune <u>per period</u>; the total unperturbed tune is $\nu_y = 2B\mu_y$), and the second factor corresponds, according to (7), to transport through an interaction region. The last step in the calculation of the ξ's proceeds from (10), and is proposed as problem 6:

<u>Problem 6</u>

Show that the eigenvalues of the transformation in (10) are $\exp\ 2\pi i\,(\mu_y + \xi_y)$, where

$$\cos 2\pi(\mu_y + \xi_y) = \cos 2\pi\mu_y - \left(\frac{r_0 N}{\gamma\sigma_y^*(\sigma_y^* + \sigma_x^*)}\right)^* \sin 2\pi\mu_y . \tag{11}$$

For small current (i.e., small N), this reduces to

$$\xi_y \sim \frac{r_0 N \beta_y^*}{2\pi\gamma\sigma_y^*(\sigma_y^* + \sigma_x^*)} \tag{12}$$

and similarly for x.

Equation (12) (together with $N = I/eBf$ and $\sigma_y^* \ll \sigma_x^*$, which implies that $\sigma_y^*(\sigma_y^* + \sigma_x^*) \sim \sigma_y^*\sigma_x^* = A/4\pi$) leads directly to

$$I \simeq \left(\frac{\gamma\, eBf}{2r_0\beta_y^*}\right) \xi_y A, \tag{13}$$

which is the basis for the standard prediction of maximum current (and, through (2), maximum luminosity). The maximum current for each of CESR, PEP, and PETRA was predicted by substituting $\xi_{y\ max} = .06$, and the unperturbed A, into the right-hand-side of (13).

How accurate are these predictions? As an example, let us compare prediction with outcome for three-bunch operation of PEP, according to the data shown in Fig. 4. Prediction gives $I_{max} \sim 29$ mA, when $\xi_{y\ max} = .06$ and $A_{unpert} \sim .0029\ cm^2$ are substituted into (13) for $B = 3$ (A_{unpert} is determined by comparing the quadratic part of the one-bunch curve with formula (2)), while the highest three-bunch current shown in Fig. 4 is $3 \times 5 = 15$ mA, i.e. a shortfall of about 50% (actually 50% over and above single-beam effects, as explained earlier).

In accordance with (13) (despite misgivings one might have about our initial gaussian approximation) we can identify the one-half with a product of one-third from the ratio of $\xi_{y\ max}$ observed at PEP to that predicted, and three-halves from the ratio of A observed (at maximum current) to that predicted. (At least for a rough calculation like this, the observed area can be determined from the three-bunch curve in Fig. 4 by applying formula (2), since for this data the two beams in PEP turned out to be about equally enlarged.[33]) According to (2), a maximum current 50% of that predicted, and a cross-sectional area 150% of that predicted, together mean an 83% shortfall in luminosity (not counting single-beam effects).

At this point a comment on experimental methodology is in order: Colliding beam facilities do not all employ the same criterion in determining the current beyond which the machine should not be operated. At SPEAR,[21] PEP,[21] and PETRA,[5] one considers the threshold to be crossed when background is excessive in high-energy-physics experimental detectors. The rise in background near maximum current is sharp enough that appreciable changes in the cutoff on background would result in only small changes in the measured maximum beam current.[21] A number of reports[11,12,13] from other labs have mentioned beam lifetime as the quantity used to operationally define the threshold.

The skeptical reader may wonder if there is in fact much difference between the maximum currents achievable in single- and colliding-beam modes. In this connection we note that at PETRA,[5] as of 1980, 20 mA of electrons could be stored in single-beam operation at about 7 GeV, but only up to about 5 mA per beam could be maintained at the same energy in colliding mode.

III. THEORETICAL TRENDS

This concludes our discussion of the experimental situation. I now describe some of the main themes in current theoretical work. This is not meant to be a comprehensive review (see References I, II, and 34). I want simply to give you a feeling for what seem to me to be the more ambitious recent efforts. (For a more detailed account of some of the topics mentioned here, see J. Tennyson's contribution to these Proceedings.)

The focus will be on three general (and overlapping) areas: computer simulation, models with reduced numbers of degrees of freedom, and instabilities of single-particle motion.

COMPUTER SIMULATIONS

Three computer analyses, numeral simulations of CESR, PETRA, and SPEAR, have attracted particular attention in the last year.

The CESR analysis simulated strong-strong operation. Motion was fully three-dimensional. Radiation damping and quantum noise were taken into account. The published report[22] includes a figure showing a computer-generated curve of luminosity vs. current (at an unspecified tune and energy), superimposed on a scatter of data points from real machine operation. This graph is reproduced here as Fig. 14; the agreement between simulation and experiment is encouraging. The simulation also produced a value for maximum storable colliding current that comes within about 10% of an experimentally observed value.

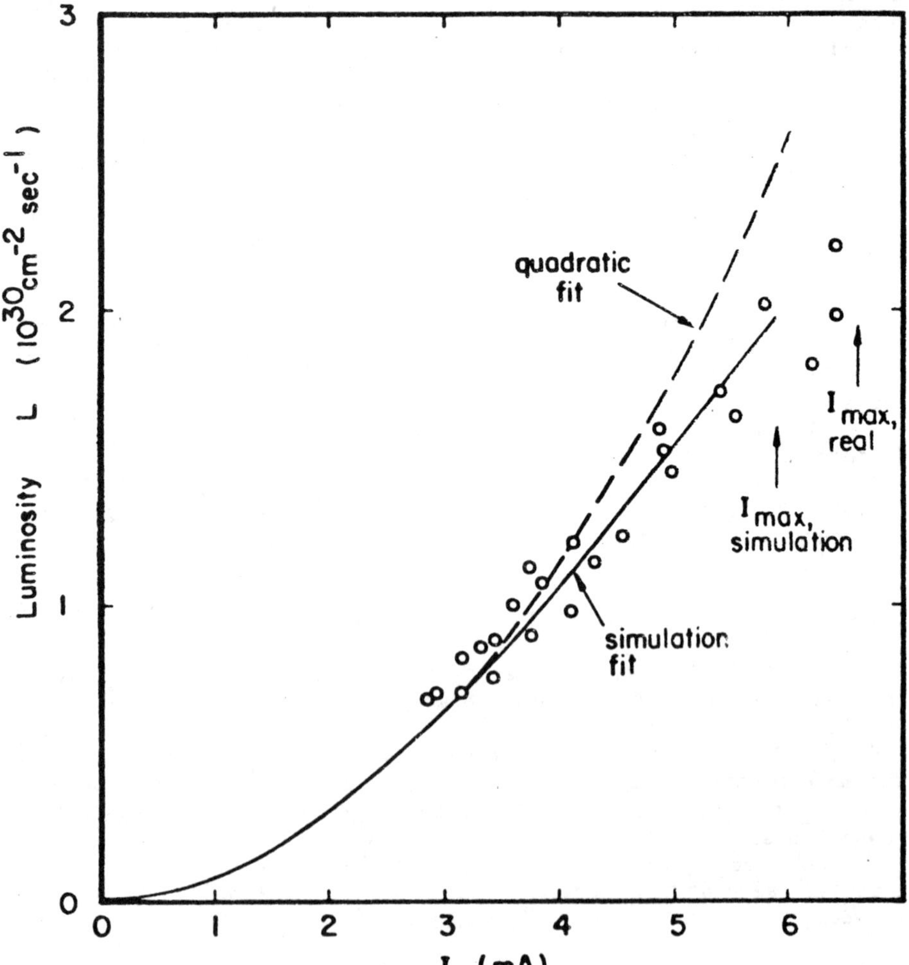

Figure 14

Luminosity vs. current at CESR, according to numerical simulation (solid curve) and experimental observation (open) dots), from Ref. 22.

The PETRA analysis simulated both strong-strong and weak-strong operation. As in the Cornell study, three degrees of freedom, radiation damping, and quantum noise were included. In addition, some effects of machine imperfection and finite bunch length were taken into account. In the published reports,[23] most of the simulation results are presented in graphs of beam heights vs. time (t=0 is the onset of collisions), and of large-time steady-state beam heights vs. betatron tunes, for different values of energy, tuneshifts, machine imperfection parameters, etc. The accompanying texts contain the suggestion that machine imperfection is the major cause of beam blow-up. These reports claim good agreement between some of the computer results and measurements made directly on PETRA. Unfortunately, very little real storage ring data is actually shown, so a reader cannot evaluate the agreement for himself.

The computer analysis of SPEAR[24] simulated weak-strong operation, and was simplified in that longitudinal oscillations and energy oscillations were neglected. The novel aspect of the work was the use of the x and y damping and fluctuation strengths as variable parameters. The results appear to indicate that vertical beam blow-up is driven primarily by horizontal damping and quantum noise, and is largely insensitive to vertical noise. Spurred by this observation, Tennyson[25] has attempted to formulate an intuitive picture of beam blow-up based on the notion of "resonance streaming." (No comparable heuristic ideas have yet developed from the other storage ring computer simulations.) I leave the description of resonance streaming to Tennyson's contribution to this summer school.

When personally assessing the importance of computer simulation - as compared with analytical work - the reader ought to bear in mind first that those who actually program and interpret simulations do not all seem to agree on what is "at the bottom" of colliding-beam effects in real machines. Second, successes of electron-positron computer simulations do not ipso facto ensure comparable successes in the proton-antiproton or proton-proton cases. The amount of computer time required for a simulation grows with the number of storage-ring turns to be simulated; a meaningful study models at least the number of revolutions in a radiation damping time. For electrons in CESR, this number is about 10^3. For protons in existing or foreseeable high-energy machines, this number is larger by several orders of magnitude.

Problem 7

Compute the number of machine revolutions in the radiation damping time of a proton in the CERN SPS; in the Fermilab Main Ring; in the Tevatron, presently under construction at Fermilab. (See Ref. 32 for relevant specifications.)

MODELS WITH FEW DEGREES OF FREEDOM

Many analytical investigations of colliding beam effects have concentrated on simplified model systems having fewer than three degrees of freedom per particle.

The one-dimensional model involves motion only in the vertical coordinate. Such a simplification is justified to the extent that one can: neglect finite bunch lengths; neglect xy coupling in the magnetic lattice; and take σ_x^*, for either beam, to be so much larger than σ_y^* that in the integral of the impulse (5) over the beam distribution, one need retain only the leading order in $1/\sigma_x^*$, i.e.

$$y' \simeq - \frac{2r_0 N}{\sigma_x^* \gamma} \int_0^{y/\sigma_y^*} e^{-u^2/2}\, du, \quad \Delta y \simeq 0 \tag{14a}$$

$$\Delta x' \simeq 0 \simeq \Delta x, \tag{14b}$$

which clearly does not couple x and y. In the one-dimensional model, the development of y and y' upon transit of an electron through one machine period is described by the functional composition of (14a) with the linear transformation corresponding to the first of the two matrix factors in (10). The importance of ζ_y, compared to ζ_x, in the conventional approach to estimates of anticipated maximum luminosity is based on the approximations underlying the one-dimensional model.

The one-and-one-half dimensional models are refinements of the one-dimensional model. They were introduced in order to study the effect of synchrotron oscillations on the stability of motion in the vertical coordinate. They differ from the one-dimensional case in the appearance of explicit time-dependence in the transformation that describes the development of y and y' upon transit through a machine period. The explicit time dependence is typically periodic, at the synchrotron frequency. (For a discussion of modulation at the horizontal betatron frequency, see Ref. 22.) The time parameter must of course be limited to those discrete values at which the beam-beam collisions actually take place; i.e. to $nt_0 + \delta$, where n is an unrestricted integer, t_0 is the time required for a beam particle to travel between interaction points, and δ is an arbitrary but fixed offset.

Problem 8

In positing the kind of explicit, steady time-dependence that is customary in such models, one tacitly assumes that all particles in a given bunch undergo uninterrupted harmonic synchrotron oscillations, all with the same amplitude, all in phase. This can be, at best, a tentative working hypothesis because there are too many particles in a bunch to be correlated in this way, because synchrotron oscillation can have an harmonic component, and because the synchrotron oscillation of even a single particle is subject to frequent random disturbance by photon emission. Estimate, for a typical particle in single-beam operation of PEP, the times required for random photon emission to substantially change the amplitude and phase of linear synchrotron oscillation; do the same for horizontal and vertical betatron oscillation. (Phase and amplitude decorrelations due to nonlinearities are harder to estimate, although they may be more important.[27]) Presumably, any one-and-one-half-dimensional instability that develops slowly compared with such a decorrelation time is of *a priori* questionable importance in a real storage ring.

Three forms of time-dependence are most frequently discussed: 1) Energy oscillations of the transverse coordinates are modeled by making the replacement

$$y \to y + a\sin\Omega t \tag{15}$$

in the right-hand-side of (14a). Ω is the synchrotron frequency; a is proportional to the vertical dispersion (η_y^*) at the interaction point. The relevence of this substitution is debatable, as reports often claim $\eta_y \equiv 0$.

2) Energy oscillations of the tune are modeled by making the replacement

$$\mu_y \to \bar{\mu}_y + m_y \sin\Omega t \tag{16}$$

in the factor that corresponds to magnet transport in the right-hand-side of (10). The amplitude m_y is proportional to the chromaticity.

3) Finite bunch lengths are modeled by making the decompositions

$$\begin{aligned} \beta_y^* &= \bar{\beta}_y^* + b_y \sin\Omega t \\ \beta_x^* &= \bar{\beta}_x^* + b_x \sin\Omega t \end{aligned} \tag{17a}$$

(and therefore)

$$\begin{aligned} \sigma_y^* &\propto \sqrt{\bar{\beta}_y^* + b_y \sin\Omega t} \\ \sigma_x^* &\propto \sqrt{\bar{\beta}_x^* + b_x \sin\Omega t} \,, \end{aligned} \tag{17b}$$

in the right-hand-side of (14a), and in the first matrix factor in (10). The amplitudes b are determined by the extent to which the β-functions near the interaction points vary over a scale set by the bunch length. At SPEAR, such variations can reportedly be appreciable.[27]

Problem 9

As is well-known,[2] the beta functions in an interaction region obey

$$\beta_{x,y} = \beta^*_{x,y} \, (1 + (s/\beta^*_{x,y})^2), \tag{18}$$

where s is distance, measured along the beam pipe, from the intersection point of the e^+ and e^- synchronous design orbits. Using (18) and the parameter values in Table I, compare β_x and β_y at s = 0 with β_x and β_y at s = one bunch length, for PEP.

Problem 10

For certain parameter values, a one-and-one-half dimensional model can be reduced to a one dimensional model by a change of variables. Since the typical one dimensional model is thought to be more stable than the typical one-and-one-half-dimensional model (see below), it may be desirable to build machines that can be operated in such exceptional configurations.[28] Consider, for example, the type-3 one-and-one-half-dimensional model of an imaginary storage ring whose horizontal and vertical β-functions are identically equal. Show that the transformation describing transit of an electron through one period of this machine, in this model, loses its explicit time dependence when expressed as a transformation on $\tilde{y} \equiv y/\sqrt{\beta_y^*}$ and $\tilde{\tilde{y}} \equiv y'\sqrt{\beta_y^*}$, rather than on y and y'.

INSTABILITY STUDIES

A number of authors[26,29,30,31] have performed numerical and analytical studies of the long-time stability of single-particle motion in one- and one-half-dimensional models. These analyses all neglect damping and noise. In mathematical terms, they are concerned only with maps of (y,y')-space formed by iterating, over and over, the single-machine-period transformations. For a number of years, efforts have been underway to determine what relations, if any, exist between the results of such stability analyses and the observed behavior of real e^+e^- storage rings.

Studies of the one-dimensional model have revealed that for ξ_y [$\simeq (r_0 \beta_y^* N)/(2\pi\gamma\sigma_x^* \sigma_y^*)$] less than a critical value near .25, all orbits in vertical phase space are bounded. An orbit is a sequence of points in phase space, each related to the one before it by an application of the single-period transformation. (In the one-and-one-half-dimensional case, the time parameter in each application of the transformation exceeds that in the preceeding application by the time required for a beam particle to travel between interaction points.) For ξ_y significantly less than the critical value, the typical orbit appears to the eye to lie on a closed curve, approximated well by an ellipse

$$\frac{y^2}{\beta_y^*} + (y')^2\beta_y^* = \text{constant.} \tag{19}$$

(The left-hand-side of (19) is just the Courant-Snyder invariant for the linear system obtained by setting N=0.) For ξ_y just below the critical value, there are orbits that appear to the eye to densely fill two-dimensional regions of the phase plane. The dynamics in such a region are highly unstable, in the sense that no two orbits that momentarily approach one another there can stay close indefinitely. For ξ_y greater than the critical value, one of these regions contains both the origin and points with arbitrarily large y or y' - so that most orbits that begin near the origin are unbounded.

One would have liked to identify this critical ξ_y with the maximum tuneshifts observed in electron-positron colliders, although as yet there is no solid theoretical justification for doing so; in any case, .25 is too large. In this regard, the one-and-one-half-dimensional models are more appealing. There are one-and-one-half-dimensional models exhibiting the same type of stability-to-instability transition, but at values of $< \xi_y >$ (i.e. ξ_y averaged over a modulation period) considerably smaller than the one-dimensional critical value.

OUTLOOK

At present, it is hard to designate any one theoretical program as the most promising. The criterion should be agreement with experiment. So far, only the largest numerical stimulations have produced results that can be meaningfully compared with experimental observations. Analytical theory remains primitive.

ACKNOWLEDGEMENTS

I am grateful to J. D. Bjorken, M. Month, and C. Quigg for encouraging my interest in this problem, and to J. Rees and J. Tennyson for helpful conversations.

TABLE I. Selected PEP Design Parameters

(for use in doing the exercises, from Ref. 18)

f= 136 kHz	
E	$= mc^2 = 15$ GeV
B	= 1 or 3
β_x^*	= 3.7 m
β_y^*	= 0.2 m
σ_x^*	= 1.1 mm
σ_y^*	= .06 mm
ν_x	= 18.77
ν_y	= 19.26
τ_β	= 8.2 msec
τ_e	= 4.1 msec
bunch length = 4 cm	
Ω	= .4 radian/revolution
$\xi_{y\text{-max}}$	= .06 (but use more realistic .02 in the exercises)
κ	= xy linear coupling = .27
α	= momentum compaction factor = .004

REFERENCES

I. M. Month and J. C. Herrera, eds., Nonlinear Dynamics and the Beam-Beam Interaction, AIP Conference Proceedings no. 57 (1979).

II. Proc. of the Beam-Beam Interaction Seminar, Stanford Linear Accelerator Center, May 22-23, 1980, SLAC-PUB-2624.

III. Proc. 11th Int. Conf. High Energy Accelerators, Geneva, 1980 (Birkhauser, Basel, 1980).

IV. Proc. of the Workshop on Long-Time Prediction in Nonlinear Conservative Dynamical Systems, University of Texas, Austin, March 16-19, 1981 (to be published).

[1] CERN Courier, May 1981, pp. 143-144.

[2] M. Sands, in Physics With Intersecting Storage Rings, B. Touschek, ed (Academic Press, New York, 1971) pp. 257-411.

[3] H. Wiedemann, in II, pp. 33-39.

[4] CESR Operations Group, in III, pp. 26-37.

[5] PETRA Storage Ring Group, in III, pp. 16-25.

[6] J. Rees, invited paper presented at 1981 Particle Accelerator Conference, Washington, D.C., March 11-13, 1981, SLAC-PUB-2684, PEP-NOTE-347.

[7] M. H. R. Donald and J. M. Paterson, IEEE Trans. Nucl. Sci. NS-26, 3580 (1979).

[8] A. W. Chao and M. J. Lee, J. Appl. Phys., 47, 4453 (1976); IEEE Trans. Nucl. Sci., NS-24, 1203 (1977).

[9] H. Wiedemann, in I, pp. 84-98.

[10] M. Cornacchia, in I, pp. 99-114.

[11] A. Piwinski, in I, pp. 115-127.

[12] S. Tazzari, in I, pp. 128-135.

[13] H. Zyngier, in I, pp. 136-142.

[14] F. Amman and D. Ritson, in Proc. Int. Conf. High Energy Accelerators, Brookhaven, 1961, pp. 471-475.

[15] E. D. Courant, IEEE Trans. Nucl. Sci. NS-12, 550 (1965).

[16] B. Richter, in Proc. Int. Symp. Electron and Positron Storage Rings, Saclay, 1966, pp. (I-1-1)-(I-1-24).

[17] CESR Design Report, Cornell Lab. for Nuclear Studies, Ithaca (1977).

[18] PEP Conceptual Design Report, LBL-4288, SLAC-189 (T/E/A) UC-28 (1976).

[19] E. D. Courant and H. S. Snyder, Ann. Phys. 3, 1 (1958).

[20] J. D. Jackson, Classical Electrodynamics (Wiley, New York, 1962).

[21] H. Wiedemann, private communication.

[22] S. Peggs and R. Talman, in II, pp. 21-32.

[23] A. Piwinski, in III, pp. 751-75; DESY preprint 80/131; DESY internal report M 81 G.

[24] J. Tennyson, in II, pp. 1-20.

[25] J. Tennyson, in IV.

[26]F. M. Izraelev, Physica 1D, 243 (1980).
[27]J. Tennyson, private communication.
[28]J. B. Vasserman, F. M. Izraelev, G. M. Tumaikin, preprint I. Ya. F. 79-74 (in Russian), Institute for Nuclear Physics, Novosibirsk.
[29]B. V. Chirikov, Phys. Repts. 52, 263 (1979).
[30]J. Tennyson, in I, pp. 158-193.
[31]A. Ruggiero, in Proc. 9th Int. Conf. High Energy Accelerators, Stanford, 1974, pp. 419-423.
[32]J. C. Herrera, in I, pp. 29-41.
[33]J. Rees, private communication.
[34]S. Kheifets, in IV (SLAC-PUB-2700, PEP-NOTE-346).
[35]A. Ruggiero, Fermilab internal report FN271 (1974).
[36]J.-E. Augustin et.al., in Proc. Workshop on Possibilities and Limitations of Accelerators and Detectors, Fermilab, 1979, pp. 87-105.

THE DYNAMICS OF THE BEAM-BEAM INTERACTION

J. L. Tennyson
Electronics Research Laboratory,
University of California, Berkeley CA 94720

1. INTRODUCTION

The beam-beam interaction is an important phenomenon from a practical (or, perhaps more correctly, a financial) point of view, since it has resulted in disappointingly low luminosities at virtually every colliding beam facility built to date. But there is another aspect of the beam-beam system that makes it interesting from a purely theoretical point of view. This is its lack of both dynamical and statistical invariants. The beam-beam system is a "near-integrable" system. [1] The first integrals of the linear magnet lattice are destroyed by the beam-beam force, and yet the motion remains nonergodic. As a result, the long term behavior of this system cannot be understood via treatments that are either purely dynamical or purely statistical; a hybrid theory is necessary. Although such a theory has been constructed, it is something of patchwork since its various pieces do not fit together in a natural way. This has made it difficult to derive precise values for such quantities as the maximal tune shift. Computer simulations have thus been used extensively both to guide the development of the theory and to confirm its results.

The analysis reviewed here describes only the behavior of the simplest model of the beam-beam interaction, the so-called "weak-strong" beam model. It should be stressed that most of the ideas here have not yet been substantiated by experimental observations. Although some observations seem to agree with the theory to within the theoretical uncertainties, these uncertainties are still large and it would be hard to consider this agreement a verification.

In studying the beam-beam interaction, we are primarily concerned with the long term evolution of the amplitudes of the betatron oscillations. It is therefore convenient to study the motion of particles in the euclidean amplitude space (whose coordinates a_x and a_y are the amplitudes of the radial and vertical oscillations as measured at an interaction point). The most important feature of the beam-beam system (and of all such "near-integrable" systems), is the web of nonlinear resonances which appears in the amplitude space (see fig. 17 for a preview). These resonances can contribute to the growth of the oscillation amplitudes via several different mechanisms. The influence of each particular mechanism depends, as we shall see, on certain parameters such as the tunes, the tune shifts, the beam shape, and the level of background noise.

In the discussion that follows, we shall adopt the following program:

1) We shall describe first the linear system; the single beam case and the linear beam-beam interaction.
2) We shall then examine the nonlinear beam-beam interaction for the one degree of freedom model. This will include calculations for resonance overlap, modulation-induced stochasticity, and parametric fluctuations.
3) Finally, we shall investigate the full two degrees of freedom system: the libration angle, coupling resonances, and direct diffusion enhancement along the resonances. A brief discussion of Arnold diffusion and modulational diffusion is included at the end.

1.1 The Linear System

When particles circulate in only one direction in a colliding beam machine, (the "single beam case"), the transverse oscillations about the design orbit are approximately linear. Traditionally,[2] the transverse displacement from the equilibrium orbit is described as an oscillation with varying amplitude and frequency,

0094-243X/82/870345-50$3.00 Copyright 1982 American Institute of Physics

$$x = A\beta^{1/2}(s)\cos[\phi(s)]. \tag{1.1}$$

Here, s is displacement along the equilibrium orbit, A is a constant, and $\beta(s)$ is a periodic function of s with period L, where L is the circumference of the equilibrium orbit. The function $\phi(s)$ is determined, to within a constant, by $\beta(s)$;

$$\phi(s) = \int_0^s \frac{ds'}{\beta(s')} + \phi(0), \tag{1.2}$$

or

$$\frac{d\phi(s)}{ds} = \beta^{-1}(s). \tag{1.3}$$

Thus, both the amplitude and frequency of the transverse oscillation are periodic with period L. The change in phase after one revolution is

$$\phi(s+L) - \phi(s) = \int_s^{s+L} \frac{ds'}{\beta(s')}. \tag{1.4}$$

The number of oscillations ν_0 per revolution is called the "tune"

$$\nu_0 = \frac{1}{2\pi}\int_s^{s+L} \frac{ds'}{\beta(s')}. \tag{1.5}$$

The tune is independent of both s and A.

To illustrate this linear motion, it is convenient to introduce a conjugate variable x which varies according to

$$x' = -A\beta^{1/2}(s)\sin[\phi(s)]. \tag{1.6}$$

This new variable is related to $\dot{x} \equiv \dfrac{\partial x}{\partial s}$ by

$$x' = \beta(s)\dot{x} - \frac{\dot{\beta}(s)}{2}x. \tag{1.7}$$

Using x and x' as coordinates, it is possible to construct a two dimensional euclidian "phase plane". If an observer who is situated at a particular point s on the equilibrium orbit records a particle's transverse position x and conjugate velocity x' every time the particle passes, the resulting set of points will lie on a circle in the phase plane, fig. 1. These points represent a section of the three dimensional phase space trajectory. The angle between two consecutively plotted points is $\Delta\phi = 2\pi\nu_0$. If ν_0 is a rational number, then the system is in linear resonance, and after some number of revolutions, the points will begin to repeat themselves.

The vertical betatron oscillation is similar to the radial, but has a different beta function $\beta(s)$ and a different tune ν_0. In this, and in the following section, we shall discuss only one transverse degree of freedom. The treatment, however, is equally applicable to both radial and vertical motion.

When particles circulate in both directions around the ring (the colliding beams case), the two countercirculating beams are each divided into one or more short bunches. These bunches collide with each other (actually pass through each other) at fixed "interaction" points on the equilibrium orbit. When a particle of one bunch passes through an opposing bunch at an interaction point, it receives a transverse kick from the relativistically enhanced collective electromagnetic field of the opposing bunch. As a result, the particle's transverse velocities $\dot{x}$ and $\dot{y}$ change very suddenly. The magnitudes of these changes, $\Delta\dot{x}$ and $\Delta\dot{y}$, depend only on the transverse positions x and y at the moment of passing.

To examine this process more exactly, it is necessary to adopt a model for the functions $\Delta\dot{x}(x,y)$ and $\Delta\dot{y}(x,y)$. To keep things simple (but hopefully not too simple), we make

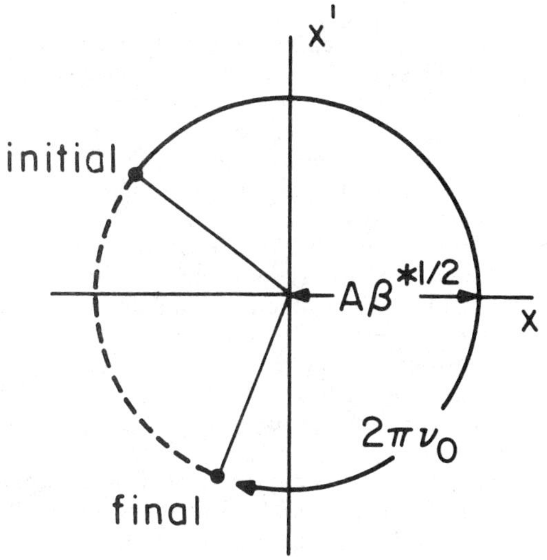

Fig. 1. Phase space mapping for the single beam system. The trajectory lies on a circle. The anglular velocity is $2\pi\nu_0$ per revolution.

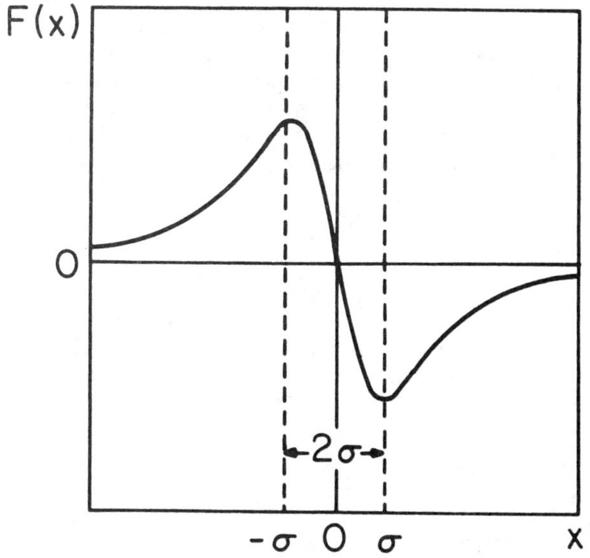

Fig. 2. A typical radial beam-beam force. The force is roughly linear at small values of x, and tends to zero with x^{-1} at large values of x.

the following assumptions: 1) The source distribution for the beam-beam interaction is continuous, approximately bi-Gaussian and static, 2) its center corresponds to the centers of the transverse oscillation, 3) the associated restoring force is antisymmetric with respect to both x and y, and 4) the force may be faithfully represented by a delta function impulse in s at the moment of passing. An example of such a force is shown in fig. 2.

Let us suppose that there is only one interaction point on the ring. To investigate the way in which the periodic jumps $\Delta \dot{x}$ and $\Delta \dot{y}$ affect the linear motion of the single beam case, we again call on our observer and station him at the interaction point. His instructions are to record a particle's x and x' coordinates immediately preceding, and then again immediately following, each collision. After three revolutions, the results might look something like fig. 3. The jump sizes in fig. 3 correspond approximately to those in a real system such as PEP or PETRA. The beam-beam force destroys, of course, the nice invariant circles of the single beam case. In fact, each jump changes the amplitude of oscillation by several percent on the average. If the change in phase between consecutive jumps were random, rather than uniform, the beam would clearly be lost very quickly.

The fact that the phase change between interaction points is fixed does not, however, imply an absence of diffusion; it simply makes it much harder to calculate diffusion rates. In this respect, the early pioneers of colliding beams were rather lucky. Beam-beam effects might have been serious enough to preclude the practical operation of present day colliding beam machines.

1.2 Linear Beam-Beam Interaction

When the amplitudes of the horizontal and vertical oscillations are much smaller than the beam width σ_x and beam height σ_y respectively, the jumps $\Delta \dot{x}$ and $\Delta \dot{y}$ are approximately proportional to x and y; so the system remains linear and thus, also integrable. In this case, the observer at the interaction point accumulates (after many thousands of collisions) a set of points falling on two symmetrically situated ellipses (fig. 4). The left leaning ellipse contains the points plotted immediately following the interaction. It is interesting to consider the way in which this ellipse evolves during the course of one revolution. As it drifts around the ring (between interactions), the ellipse simply rotates clockwise through an angle $2\pi\nu_0$. Then, when it passes the interaction point, it receives a kick which brings it back to its original position. If the beam-beam kicks are made stronger, the eccentricity of the ellipse increases until, at some critical value, the ellipse becomes a line, and the system goes unstable.

From fig. 4, it is clear that the beam-beam interaction always advances the phase ϕ. If $\Delta \dot{x} = -bx$, the average phase advance for $b<<1$ is (from (1.1) and (1.7)),

$$2\pi\xi = -< \Delta x'(A\beta^{*1/2})^{-1} \cos(\phi) >_\phi = \frac{\beta^* b}{2}, \tag{1.8}$$

where β^* is the value of $\beta(s)$ at the interaction point. The parameter ξ is called the "linear tune shift" and is often used to indicate the strength of the linear beam-beam interaction. The total betatron frequency, again for $b<<1$, is

$$\omega = 2\pi(\nu_0 + \xi). \tag{1.9}$$

1.3 Nonlinear Beam-Beam Interaction

In reality, the beam-beam force is only linear in a small central region of the beam $x << \sigma$ where the charge distribution is roughly uniform. As x becomes larger than σ, the force first levels off, and then begins to decline (see fig. 2). At large x ($x >> \sigma$), the force falls approximately[3] in proportion to x^{-1}. If we define the normalized amplitude of the

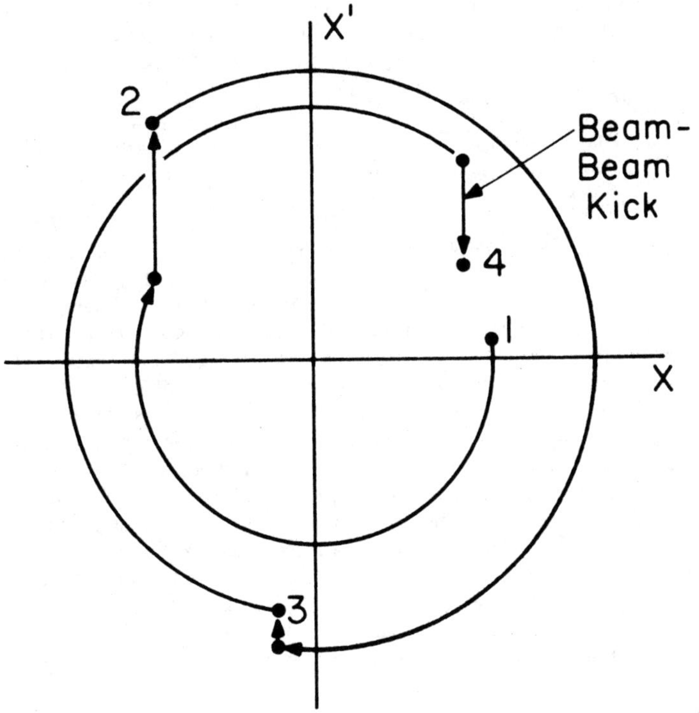

Fig. 3. Phase plane mapping for the colliding beam system. The trajectory is composed of circular drifts and vertical kicks.

oscillation as

$$a = \frac{A\beta^{*1/2}}{\sigma}, \tag{1.10}$$

we see that the particles with the smallest oscillation amplitudes will have the largest changes in tune, ($\Delta\nu \rightarrow \xi$ as $a \rightarrow 0$) and those with very large amplitudes will have almost no change in tune, ($\Delta\nu \rightarrow 0$ as $a \rightarrow \infty$). From now on, since the tune varies with amplitude,the term "linear tune shift" and its corresponding symbol "ξ", will be reserved for the maximal value of $\Delta\nu$. The total amplitude-dependent frequency of the betatron oscillation is now

$$\omega(a) = 2\pi[\nu_o + \Delta\nu(a)]. \tag{1.11}$$

Because the frequency changes with amplitude, there are certain amplitudes a_{mn} at which the system becomes resonant, ie. satisfes the condition

$$m\omega(a_{mn}) = 2\pi n, \qquad (n, m \text{ integers}). \tag{1.12}$$

These nonlinear "parametric" resonances are unavoidable since they may be found in any finite amplitude interval Δa. They are potentially dangerous because they tend to facilitate the exchange of energy between the longitudinal and transverse motions. Examples of nonlinear resonance are shown in fig.s 5 and 6. In fig.5, the total tune is shown as a function of amplitude for different tune shifts ξ. The tune shift here is negative because the system represented[4] is an intersecting beam device (both beams have the same charge). The dashed line represents the $m = 4$ $n = 15$ resonance. In fig. 6, phase plane plots are shown for two values of ξ, one such that $\nu_o + \xi$ is below the resonance, and the other such that it is above. In each case, several thousand points from each of ten trajectories were recorded. Although these points no longer fall on ellipses, as in the linear case, they do seem to remain on smooth curves. Furthermore, the resonant exchange of energy is strictly bounded by the nonlinearity, as seen in fig. 6b. The nonlinear resonance itself is characterized by a set of "islands", or closed curves that do not enclose the origin of the phase plane. We shall call the motion associated with these islands "resonant libration", in contrast with the ordinary rotation that characterizes the non-resonant orbits. Note that so far, there is no significant loss of correlations in the beam-beam kicks.

1.4 Deterministic Stochasticity

It turns out that not all trajectory sections are confined to closed curves on the phase plane. Some trajectories (a finite portion of them), actually diffuse slightly in amplitude, but only over a limited range. They fill in thin bands of finite thickness which may be too thin to see in plots such as fig. 6. These stochastic bands always form in the vicinities of the separatrices of nonlinear resonances, fig. 7a. They are said to result from the splitting of the separatrices of one resonance due to a slight interaction with the other resonances. [5,6] Smooth separatrix curves do not exist in this system.

When the stochastic bands are very thin, they are of no practical importance for the one degree of freedom system. But as the strength ξ of the beam-beam interaction increases, the widths of these bands also increase and neighboring resonances move closer together. When adjacent resonances begin to "overlap" one another, the stochasticity expands to fill almost the entire region (see fig. 7b). This results in the randomization of the beam-beam kicks and a subsequent rapid loss of particles from the beam. The rate at which particles diffuse in amplitude depends on the strengths and libration frequencies of the particular resonances which are overlapping. Although the resonance overlap criterion gives us an absolute upper limit on the tune shift ξ, the criterion is sometimes difficult to apply since neighboring resonances may vary greatly in size and spacing.

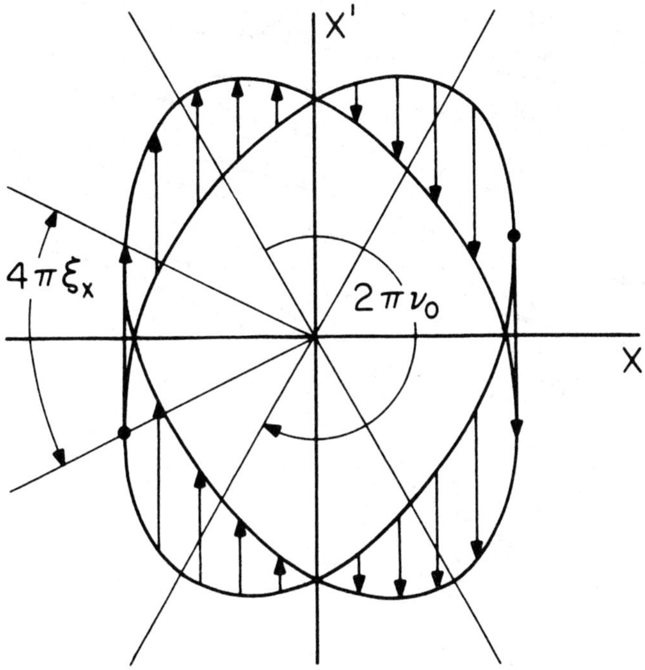

Fig. 4. The invariant ellipse of the linear colliding beam system. The ellipse rotates during the drift, then jumps back to its original position when it receives the kick.

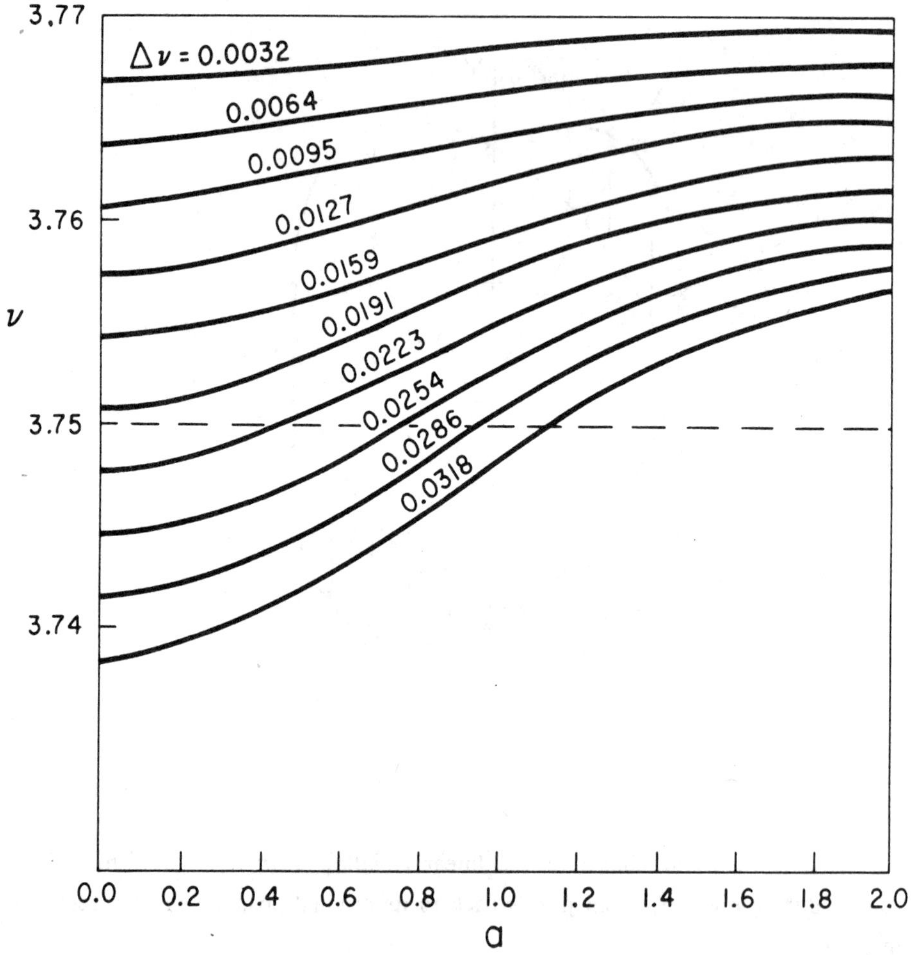

Fig. 5. The vertical tune versus the vertical amplitude (for ISABELLE, ref. 4). The dashed line shows the position of the 15/4 resonance.

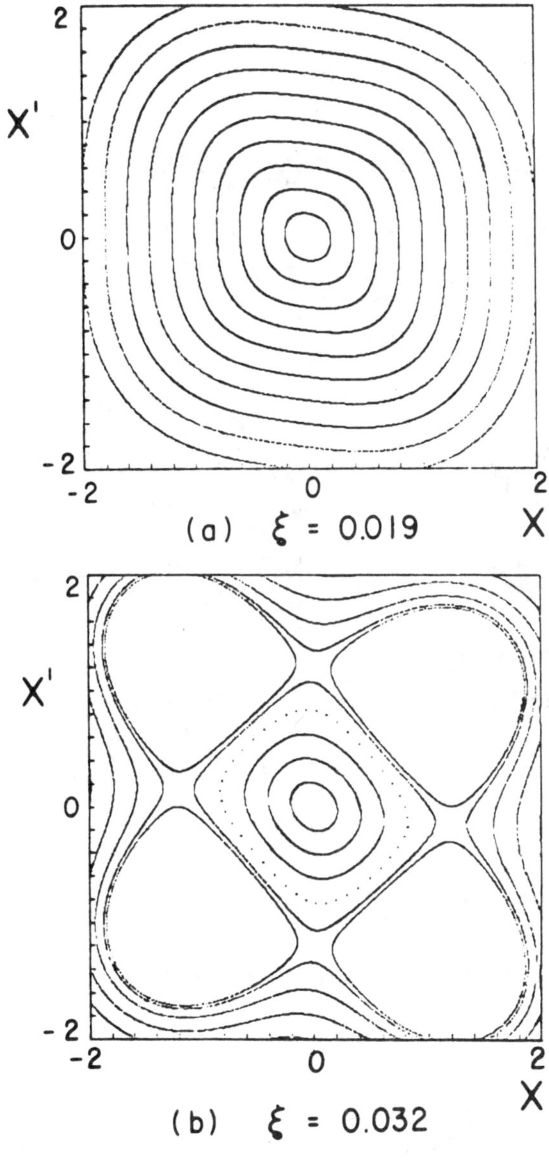

Fig. 6. Phase plane plots for the vertical motion at two values of the tune shift ξ. In a), ξ is too small to bring the resonance out of the origin into the phase plane. In b), ξ is larger and the powerful 15/4 resonance becomes the dominant feature. The tune here is $\nu_0 = 3.770$, the same as in fig. 5 (this figure is also from ref. 4).

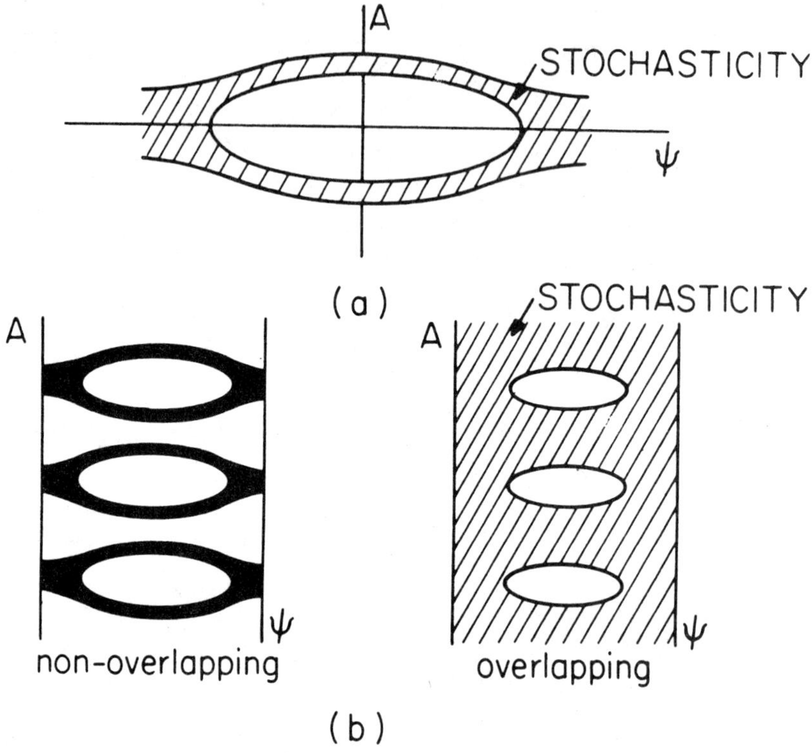

Fig. 7. Stochastic regions of the phase space. a) around the separatrix of a nonlinear resonance, and b) due to the overlap of neighboring resonances.

We now turn our attention to the basic time and length scales that characterize each nonlinear resonance.

2. THE ONE DEGREE OF FREEDOM SYSTEM

2.1 The Hamiltonian

To calculate the resonance overlap threshold (and other quantities yet to be discussed), it is necessary to know the resonance widths, spacings and libration frequencies. Providing it is possible to estimate a potential $V(x)$ for the beam-beam force, these quantities can be easily derived from the Hamiltonian for the system. The Hamiltonian consists of a linear oscillator (with a time-dependent frequency), and a delta function perturbation representing the beam-beam interaction,

$$H = \frac{\dot{x}^2}{2} + K(s)\frac{x^2}{2} + \sum_{n=-\infty}^{\infty} \delta(n-s)\epsilon V(x), \tag{2.1}$$

where $K(s)$ is related to $\beta(s)$ by

$$\frac{\beta\ddot{\beta}}{2} - \dot{\beta}^2 + K(s)\beta^2 = 1, \tag{2.2}$$

and the unit of s is the circumference of the ring $L = 1$. The parameter ϵ represents the strength of the beam-beam force (the beam current). The action-angle variables for this system may be obtained from the generating function,[7]

$$F = -\frac{x^2}{2\beta}\tan(\psi - \omega_o s + \int_0^s \frac{ds'}{\beta(s')}) + \frac{x^2}{4\beta}\dot{\beta}, \tag{2.3}$$

where $\omega_o = 2\pi\nu_o$ and the dependence of β on s is implied. The new variables replace the old according to

$$x = \sqrt{2I\beta}\cos(\psi - \omega_o s + \int_0^s \frac{ds'}{\beta(s')}) \tag{2.4}$$

$$\dot{x} = -\left(\frac{2I}{\beta}\right)^{1/2} \sin(\psi - \omega_o s + \int_0^s \frac{ds'}{\beta(s')}) + \frac{\dot{\beta}}{2}\left(\frac{2I}{\beta}\right)^{1/2} \times \cos(\psi - \omega_o s + \int_0^s \frac{ds'}{\beta(s')}). \tag{2.5}$$

The new Hamiltonian is

$$H = I\omega_o + \delta_s \epsilon f(I,\psi), \tag{2.6}$$

where

$$\delta_s \equiv \sum_{n=-\infty}^{\infty} \delta(n-s). \tag{2.7}$$

Note that if $\beta(s)$ has an extremum, $\dot{\beta} = 0$, at the interaction point (as is usually the case), then, at the interaction point,

$$x = \sqrt{2I}\beta^{1/2}\cos\psi \tag{2.8}$$
$$\dot{x} = -\sqrt{2I}\beta^{-1/2}\sin\psi,$$

which corresponds to equations (1.1), (1.6) and (1.7).

It follows that at the interaction point the action I is related to the normalized amplitude a (1.10) by

$$I = \frac{a^2\sigma^2}{2\beta^*}. \tag{2.9}$$

Fourier transforming (2.6) with respect to ψ, we get

$$H(I,\psi,s) = I\omega_0 + \delta_s\epsilon\sum_{k=0}^{\infty} F_k(I)\cos(k\psi). \tag{2.10}$$

The summation here exhibits explicitly the two roles of the beam-beam interaction. The $k=0$ term creates the nonlinearity while the $k>0$ terms introduce the coupling to the longitudinal motion. Each of these coupling terms represents a specific resonance, the strength of the resonance being proportional to the Fourier amplitude F_k.

We first calculate the "nonlinearity". To do this, we separate the $k=0$ term from the sum to get

$$H(I,\psi,s) = I\omega_0 + \delta_s\epsilon F_0(I) + \delta_s\epsilon\sum_{k=1}^{\infty} F_k(I)\cos(k\psi), \tag{2.11}$$

and define a new nonlinear oscillator by the Hamiltonian

$$H_0(I,s) = I\omega_0 + \delta_s\epsilon F_0(I). \tag{2.12}$$

During one revolution, the phase ψ of this oscillator changes by an amount

$$\omega(I) = \omega_0 + \epsilon\left.\frac{\partial F_0}{\partial I}\right|_I. \tag{2.13}$$

The linear tune shift is then

$$\xi = \frac{\epsilon}{2\pi}\left.\frac{\partial F_0}{\partial I}\right|_{I=0}. \tag{2.14}$$

We may now define a nonlinearity function $\epsilon\lambda(I)$ for this oscillator,

$$\epsilon\lambda(I) = \frac{\partial\omega(I)}{\partial I} = \epsilon\left.\frac{\partial^2 F_0}{\partial I^2}\right|_I. \tag{2.15}$$

To find the resonance widths and the frequencies of small amplitude libration f_L, it is useful to adopt a "reduced" Hamiltonian which is obtained from (2.10) by dropping all of the terms except the one representing the particular resonance to be examined (and, of course, the $k=0$ term). Thus, if we are interested in the properties of the kth resonance, the appropriate Hamiltonian is

$$H(I,\psi,s) = H_0(I) + \delta_0\epsilon F_k(I)\cos(k\psi). \tag{2.16}$$

For computational convenience, we change variables again, this time to "resonance" coordinates[8] p,θ. If the resonance position I_R is defined to be the value of the action at which the phase in the cosine term of (2.14) changes after one revolution by

$$\omega(I_R) = 2\pi\frac{n}{k}, \tag{2.17}$$

then the resonance coordinates are

$$p = \frac{I-I_R}{k} \tag{2.18}$$

$$\theta = k\psi. \tag{2.19}$$

The new Hamiltonian is

$$H(p,\theta,s) = H_o(p,s) + \delta_s \epsilon F_k(p)\cos\theta. \tag{2.20}$$

Because we are interested in trajectories which are close to the resonance, we expand H_o in a power series about I_R

$$H_o(p,s) = H_o(I_R) + \left.\frac{\partial H_o}{\partial I}\right|_{I_R} kp + \frac{1}{2}\left.\frac{\partial^2 H_o}{\partial I^2}\right|_{I_R} k^2p^2 + \cdots \tag{2.21}$$

$$H_o(p,s) \approx \omega(I_R)\, kp + \frac{1}{2}\epsilon\lambda(I_R)\, k^2p^2. \tag{2.22}$$

The full Hamiltonian is then

$$H(p,\theta,s) \approx \omega(I_R)\, kp + \frac{\epsilon}{2}\lambda(I_R)\, k^2p^2 + \delta_s\epsilon F_k\cos\theta. \tag{2.23}$$

From (2.17), the first term on the right hand side produces a change in phase after one revolution of

$$\Delta\theta = 2\pi\, n. \tag{2.24}$$

It is thus invisible to the other terms and may be dropped. For further convenience, we introduce the "effective nonlinearity" Λ_k defined by

$$\Lambda_k = \lambda k^2. \tag{2.25}$$

The final Hamiltonian represents a "kicked" pendulum

$$H(p,\theta,s) = \epsilon\frac{\Lambda_k}{2}p^2 + \delta_s\epsilon F_k\cos\theta. \tag{2.26}$$

This system corresponds exactly to the so-called "standard map".[8] It is characterized by a set of resonances with uniform widths and spacings in p. When $\epsilon^2\Lambda_k F_k \ll 1$ and $H \ll 1$, the delta functions may be averaged out and the system becomes identical to the pendulum,

$$H(p,\theta) = \frac{\epsilon\Lambda_k}{2}p^2 + \epsilon F_k\cos\theta. \tag{2.27}$$

The resonance width corresponds to the maximum separation of the separatrix orbits (fig. 8)

$$\Delta p_w = 4\left(\frac{\epsilon F_k}{\epsilon\Lambda_k}\right)^{1/2} \tag{2.28}$$

or

$$\Delta I_w = 4k\left(\frac{F_k}{\Lambda_k}\right)^{1/2}. \tag{2.29}$$

The resonance spacing for these kth order resonances is (from (2.15), (2.17) and (2.25)),

$$\Delta I_s = \frac{2\pi k}{\epsilon\Lambda_k}, \tag{2.30}$$

and the ratio of resonance width to spacing is

$$\frac{\Delta I_w}{\Delta I_s} = 4k\left(\frac{F_k}{\Lambda_k}\right)^{1/2}\left(\frac{2\pi k}{\epsilon\Lambda_k}\right)^{-1} = \frac{2\epsilon}{\pi}\left(F_k\Lambda_k\right)^{1/2}. \tag{2.31}$$

Apparently, resonances overlap when $\epsilon\left(F_k\Lambda_k\right)^{1/2} > \frac{\pi}{2}$. Several studies[8] have shown, however, that the last stable rotational orbits disappear somewhat sooner, when

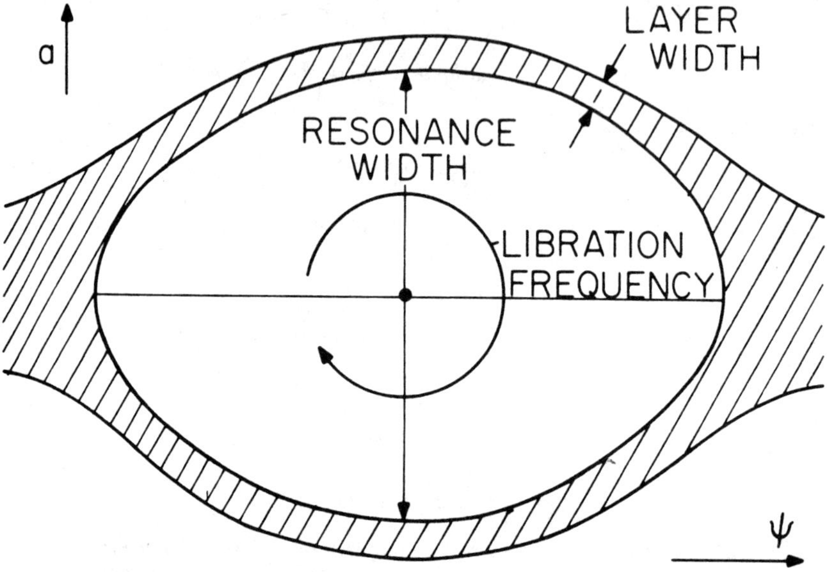

Fig. 8. Three important dimensions of a nonlinear resonance: the resonance width ΔA_w, the frequency of small amplitude libration f_L, and the stochastic layer width.

$$\epsilon^2 F_k \Lambda_k > 1. \tag{2.32}$$

This provides us then, with a rough guide for the onset of strong stochasticity in the system. It must be remembered, however, that the above analysis is for a system of uniform nonlinearity, while our beam-beam system has a very nonuniform nonlinearity. Thus, for example, ϵ may be large enough according to (2.32) to produce overlap of k=2 resonances at $a=\sigma$, and yet still small enough that no k=2 resonance exists in the system at all, much less overlaps a second. Thus, the usefulness of (2.32) is rather limited. As we shall see in the next section, this "resonance overlap criterion" can be more effectively applied to the problem of modulation sideband resonances.

It should be noted that the width (2.29) of the resonance is independent of the strength of the beam-beam interaction ϵ. At first glance, this seems to contradict the clearly established fact that the resonance disappears when $\epsilon = 0$. The paradox is resolved when one notices that the frequency f_L of small amplitude libration in the pendulum (2.26) is given by

$$f_L = \epsilon\sqrt{F_k \Lambda_k}. \tag{2.33}$$

So, as the beam-beam force goes to zero, so does the libration frequency. In this limit, the system is linear but still in resonance; so all trajectories become periodic. Thus, one must wait an infinite amount of time to see the resonance mapped out in the phase plane. Although the resonance widths remain constant as $\xi \rightarrow 0$, the distance between them (2.30), goes to infinity. Consequently, resonances are less of a problem at low tune shifts because they can more easily be avoided.

2.2 Tune Modulation

If a slow parametric modulation is added to the system (2.1), a multiplet of sideband resonances will be generated for each primary resonance. The widths, spacings and libration frequencies of these sidebands can be calculated in a manner similar to that used for the primary resonances. [7,9,10,11]

Modulation effects have been studied in detail by Izrailev et al. at Novosibirsk. [7,10,11] In a real machine, the major source of slow modulation is the synchrotron oscillation. There are three principle mechanisms by which the synchrotron oscillation may couple the beam-beam interaction. Probably the most important is "dispersion coupling" . If the machine has a finite dispersion at the interaction point, the center of the betatron oscillation will oscillate as the particle's energy oscillates. This corresponds, in the particle's frame of reference, to a swaying of the bi-gaussion charge distribution of the opposing beam. It causes both the sizes and positions of the resonances to oscillate in amplitude at twice the synchrotron frequency. There is a significant vertical dispersion in both PEP and PETRA (it accounts for more than half of the beam height in both machines). The second source of synchrotron modulation comes from the finite bunch length. Because the beta funtion $\beta(s)$ varies slightly over a bunch length at the interaction point, particles that happen to be at either end of a bunch will see an attenuated beam-beam force. The ultimate effect is similar to that of dispersion coupling. Finally, synchrotron oscillations may couple to the beam-beam force via chromaticity effects. In this case, an energy variation results in a variation of the tune. The result, again, is an oscillation of the resonances in amplitude (although this modulation has half the frequency of the others). It should be noted that synchrotron modulation most strongly affects those particles with the largest amplitudes of synchrotron oscillation.

If we look, for example, at the case in which the tune of the machine is modulated at some slow frequency $\Omega << 1$, and with some amplitude of modulation $2\pi(\nu_{max}-\nu_{min}) = 2M$, the Hamiltonian becomes

$$H(I,\psi,s) = H_o'(I,s) + \delta_s \epsilon F_k(I)\cos(k\psi), \tag{2.34}$$

where H_o' represents a new nonlinear oscillator with a slowly oscillating frequency

$$H_o' = I[\omega_o + M\cos(\Omega s)] + \delta_o \epsilon F_o(I). \tag{2.35}$$

In one revolution, the phase of this oscillator changes approximately by an amount (using (2.13))

$$\omega(I,s) = \omega(I) + M\cos(\Omega s). \tag{2.36}$$

This may be integrated to give the phase after s revolutions

$$\psi(s) = \omega(I)s + \frac{M}{\Omega}\sin(\Omega s). \tag{2.37}$$

The rate of change of the action, from Hamilton's equations, (2.34), and (2.37) is

$$\dot{I} = \delta_s \epsilon F_k k \sin\left[k(\omega(I)s + \frac{M}{\Omega}\sin(\Omega s))\right] \tag{2.38}$$

or

$$\dot{I} = \delta_s \epsilon F_k k \sum_{l=0}^{\infty} J_l(\frac{Mk}{\Omega}) \sin(k\omega(I)s + l\Omega s), \tag{2.39}$$

where the J_l are Bessel functions. The new resonance condition is

$$k\omega(I_R) + l\Omega = 2\pi n. \tag{2.40}$$

Each of the k, n resonances has been split into a multiplet of small closely spaced resonances. The width of this multiplet in action is determined by the Bessel functions J_l, which fall off quickly when (see fig. 9)

$$l > (\frac{Mk}{\Omega}). \tag{2.41}$$

The spacing between sidebands is (from (2.40))

$$\Delta\omega_s^s = \frac{\Omega}{k} \tag{2.42}$$

$$\Delta I_s^s = \frac{\Omega}{\epsilon k \lambda}. \tag{2.43}$$

From (2.41) and (2.42), it is apparent that the full width of the multiplet is just equal to the modulation depth

$$\Delta\omega_m = 2M. \tag{2.44}$$

If the trajectory is located within the multiplet, the phase of (2.38) will become stationary twice each modulation period. These moments of stationary phase correspond to the crossing of the primary resonance as it oscillates back and forth within the multiplet. Thus, the integral of (2.38) receives a kick with each crossing, the magnitude of which depends on the phase of the sine function at the moment of crossing.

If, after one modulation period, the change of phase $k\psi$ is some multiple of 2π , the system is in a sideband resonance, and successive alternate crossings reinforce one another. This is shown explicitly by the decomposition into Bessel functions (2.39). If the trajectory is close to a sideband resonance, contributions to the integration of (2.39) will come primarily from the resonant term. The Bessel function amplitude of this term represents the combined effect of the two kicks. Sometimes, for some values of l, these two kicks will just cancel each other, a situation which may be identified with the nodes of the Bessel functions

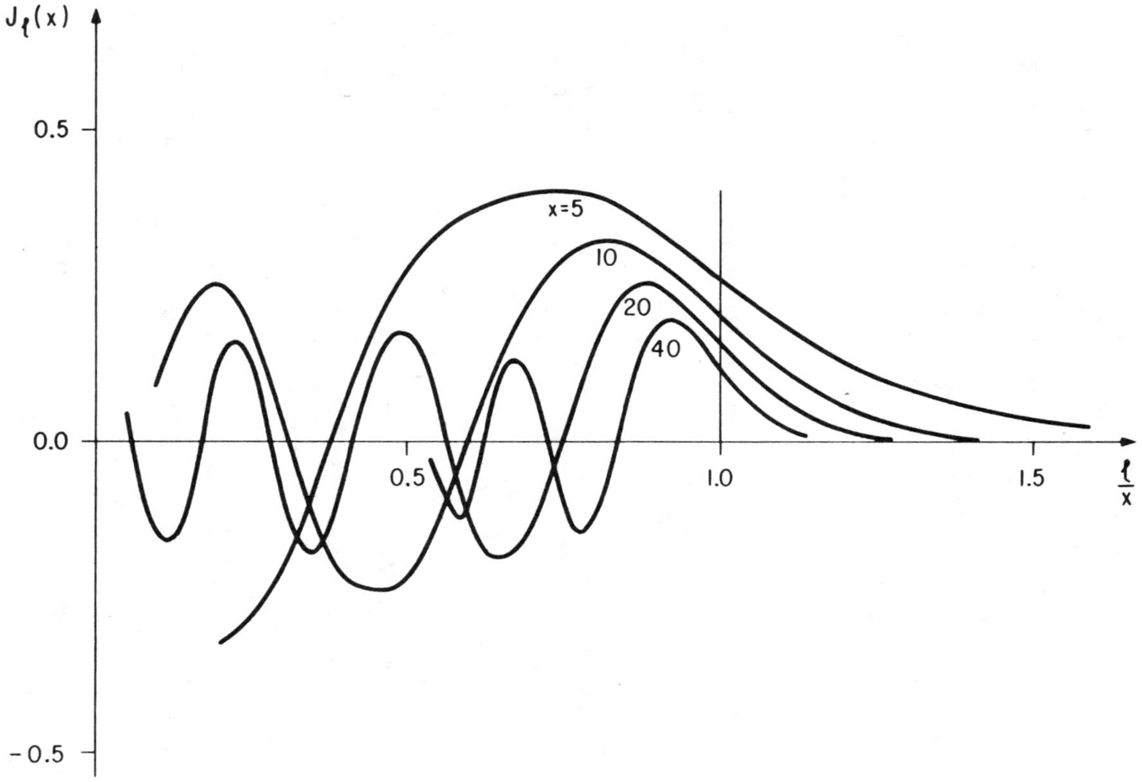

Fig. 9. Bessel functions $J_l(x)$, plotted against $\frac{l}{x}$. The line at $\frac{l}{x} = 1.0$ marks the layer edge (from ref. 4).

in fig. 9.

To find the widths of the sidebands, we take only the resonant term from (2.39), average out the delta function $<\delta_s> = 1$, and change again to resonance coordinates

$$p = \frac{I - I_R}{k} \tag{2.45}$$

$$\theta = k\omega(I)s + l\Omega s. \tag{2.46}$$

The constant I_R is now defined by (2.40). Equation (2.39) then becomes

$$\dot{p} = \epsilon F_k(p)\ J_l(\frac{Mk}{\Omega})\ \sin(\theta). \tag{2.47}$$

Assuming uniform nonlinearity $\epsilon\lambda$ across the sideband, the rate of change of θ is about

$$\dot{\theta} = k\omega(I_R) + \Omega\, l + k^2\lambda\epsilon p \tag{2.48}$$

or, from (2.25) & (2.40)

$$\dot{\theta} = \epsilon\Lambda_k p + 2\pi n. \tag{2.49}$$

Equations (2.47) and (2.49) together, give us again the pendulum,

$$H(p,\theta) = \frac{\epsilon\Lambda_k(I_R)p^2}{2} + \epsilon F_k(I_R) J_l(\frac{Mk}{\Omega}) \cos(\theta). \tag{2.50}$$

The widths of the sidebands are then

$$\Delta I_w^s = 4k\left|\frac{F_k J_l}{\Lambda_k}\right|^{1/2}, \tag{2.51}$$

or just a factor of $\sqrt{J_l}$ smaller than the primary resonance.

When the trajectory is inside the multiplet, the Bessel function has an average value of (see fig. 9)

$$|J_l| = \frac{2}{\pi}\left|\frac{2}{\pi}\frac{\Omega}{Mk}\right|^{1/2}. \tag{2.52}$$

The sideband width is thus, on the average,

$$\Delta I_w^s \approx 4k\left|\frac{F_k}{\Lambda_k}\left(\frac{2}{\pi}\right)^{3/2}\left(\frac{\Omega}{Mk}\right)^{1/2}\right|^{1/2}, \tag{2.53}$$

where the functions $F_k(I)$ and $\Lambda_k(I)$ are evaluated at the resonance $I = I_R$. We are primarily concerned here with the possibility of sideband overlap and the amplitude diffusion that would result. The ratio of sideband width to spacing is

$$\frac{\Delta I_w^s}{\Delta I_s^s} = \frac{4\epsilon}{\Omega}\left(F_k\Lambda_k\left(\frac{2}{\pi}\right)^{3/2}\left(\frac{\Omega}{Mk}\right)^{1/2}\right)^{1/2}. \tag{2.54}$$

Sideband overlap occurs when this is greater than $\frac{2}{\pi}$, or when

$$\frac{4\epsilon}{\Omega}\left|F_k\Lambda_k\left(\frac{\pi}{2}\right)^{1/2}\left(\frac{\Omega}{Mk}\right)^{1/2}\right|^{1/2} > 1. \tag{2.55}$$

If only the modulation frequency is varied (everything else held constant) there is a critical frequency Ω_c below which the sidebands will overlap

$$\Omega_c = (4\epsilon)^{4/3}\,(F_k\Lambda_k)^{2/3}\left(\frac{2}{\pi Mk}\right)^{1/3}. \tag{2.56}$$

An example of sideband overlap in the beam-beam system (produced by a computer iterated map) is shown in fig. 10.

We now calculate the diffusion rate inside a stochastic multiplet caused by overlapping sidebands. There are two distinct diffusion regimes, sometimes called the "fast crossing" regime and the "slow crossing" or "trapping" regime.If the modulation frequency is low enough, some particles will be trapped in the primary resonance as it moves back and forth across the multiplet band. Trapping will occur when the time it takes for the primary resonance to move a distance equal to its width is greater than the libration period $T_L = \frac{2\pi}{f_L}$ of the resonance. The maximum velocity of the resonance in action is, from (2.15)

$$\dot{I}_{\max} = \frac{\dot{\omega}_{\max}}{\epsilon\lambda}. \tag{2.57}$$

The condition for trapping, at the center of the multiplet, is thus (using (2.29), (2.33)&(2.36)),

$$T_L \frac{\dot{I}_{\max}}{\Delta I_w} < 1 \tag{2.58}$$

or

$$\frac{\epsilon^2 \lambda k F_k}{M\Omega} > 1. \tag{2.59}$$

Again, this may be expressed as a condition on Ω. When Ω is less than some value Ω_t, trapping will occur;

$$\Omega_t = \frac{\epsilon^2 \lambda k F_k}{M}. \tag{2.60}$$

The diffusion rate in the fast crossing regime $\Omega_t < \Omega < \Omega_c$ may be derived from (2.47). The change in p during one modulation period is about

$$\Delta p = \frac{2\pi}{\Omega} \epsilon F_k \frac{2}{\pi} \left| \frac{2}{\pi} \frac{\Omega}{Mk} \right|^{1/2} \sin(\theta), \tag{2.61}$$

and most of this comes from the two discrete kicks at the resonance crossings. In the fast crossing regime, the overlapping sidebands create a fast phase mixing, so that the phase θ at each crossing is approximately random. The diffusion of p for $\Omega_t < \Omega < \Omega_c$ (fast crossing) is thus

$$D_p^f = \frac{<\Delta p^2>_\theta}{4\pi/\Omega} = \frac{(2\epsilon F_k)^2}{\pi^2 Mk}, \tag{2.62}$$

or, in action,

$$D_I^f = \frac{(2k\epsilon F_k)^2}{\pi^2 Mk}. \tag{2.63}$$

Note that this is independent of Ω.

In the slow crossing regime, $\Omega < \Omega_t$, the diffusion rate (for $k = 1$) is approximately

$$D_I^s \approx \frac{(Mk)^2 \Omega^3}{\epsilon^2 F_1 \Lambda_1}. \tag{2.64}$$

This was originally derived by Chirikov[12,13] for a simple accelerating wave system. Note that it is somewhat counter-intuitive because it says that the diffusion rate increases as the beam-beam strength ϵ decreases. The potential paradox at $\epsilon \to 0$ is avoided by the fact that at

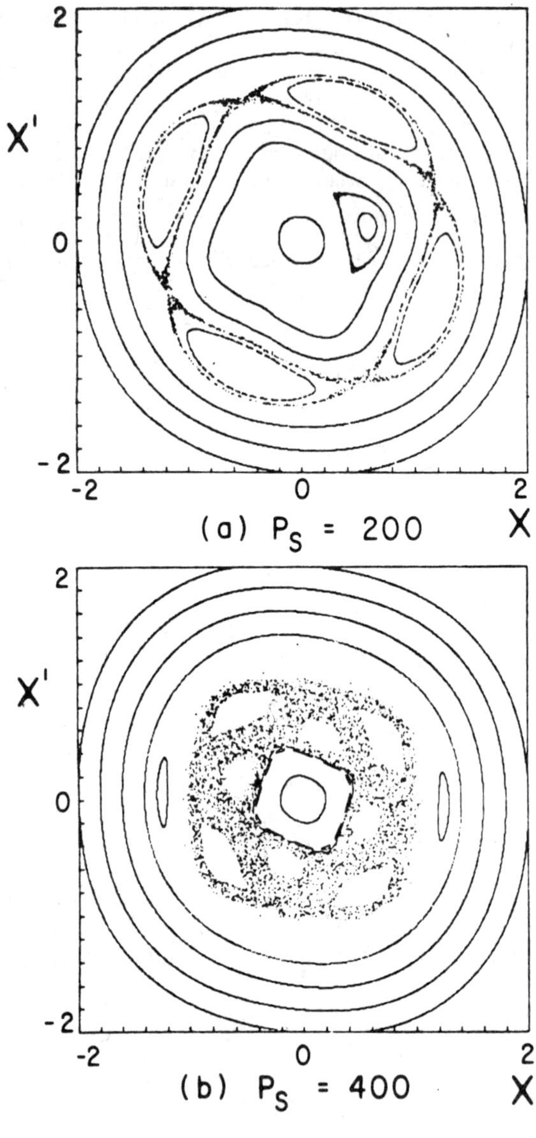

Fig. 10. Sideband resonances in the phase plane. These are sidebands of the 15/4 resonance. The two plots differ only in the modulation period P_s. In a), the modulation period is not large enough to cause the sidebands to overlap. In b), it is.

small enough ϵ, the resonance is no longer in the slow crossing regime and (2.64) is invalid.

In the limit where $\Omega \rightarrow 0$, all untrapped trajectories inside the multiplet band are stochastic, but with a diffusion rate approaching zero according to (2.64).

2.3 Fluctuations

Thus far, we have discussed only those diffusion processes which are induced by dynamical instability, i.e. deterministic stochasticity. Clearly, however, the presence of some small quantity of externally produced randomness will also contribute to the loss of phase correlations. We therefore turn our attention at this time to the problem of extrinsic randomness in the beam-beam interaction.

The sources of extrinsic randomness may be roughly divided into two categories. In the first category are those which directly affect the amplitude of the transverse oscillation (for example, quantum fluctuations in the synchrotron radiation, or intrabeam scattering). The diffusion produced by these sources exists even in the absence of the beam-beam interaction. In the second category are those processes which result only indirectly in amplitude diffusion (for example, tune diffusion due to chromaticity or dispersion). These mechanisms work by decorrelating the evolution of the betatron phase ψ, and thus, of the beam-beam kicks. Their influence is minimal in the absence of the beam-beam interaction. (note that static errors such as magnet imperfections or misallignments are not sources of randomness, but serve only to increase the strength of certain Fourier terms in (2.10)).

It turns out that those fluctuations which affect the amplitude directly (those belonging to the first category) are not significantly altered by the presence of the beam-beam interaction in the one degree of freedom system. We shall therefore reserve discussion of these processes for the next section and concentrate here on the indirect sources of diffusion. These indirect, or "parametric", fluctuations may enter the system (2.10) via either variation of the working tune ν_0 , or of the potential $V(x)$. However, a variation of the potential will usually result in some variation of the tune shift ξ, so that the dominant effect in either case is a variation of the transverse frequency $\omega(I)$. We will examine here the case in which the working tune ν_0 diffuses. It is the simplest case, and an analogous problem (neoclassical transport) has been previously studied in considerable detail by plasma physicists. [14,15,16]

In the discussion that follows, we shall identify three distinct diffusion "regimes" which correspond roughly to slow, intermediate, and fast diffusion rates. These are closely related to the so-called "banana", "plateau", and "Pfirsch-Schluter" regimes investigated in connection with particle loss in Tokamaks. The slow and intermediate regimes here are also roughly equivalent to the slow and fast crossing cases described in the preceding section on tune modulation.

We start with the basic beam-beam system (2.10), and hypothesize the existence of a frequency diffusion rate D_ω.

2.3.1 Slow Diffusion Regime

In the slow diffusion regime, particles may become trapped in the resonances as they diffuse. The condition for trapping, as in the modulation case, is that the time necessary for a resonance to diffuse a distance equal to its width must be greater than the period of small amplitude libration.

$$\frac{(\Delta\omega_w)^2}{2D_\omega} > T_L, \tag{2.65}$$

where the width of the resonance in frequency is

$$\Delta\omega_w = \epsilon\lambda\,\Delta I_w. \tag{2.66}$$

This may be written as a condition on D_ω

$$D_\omega < D_\omega^T, \tag{2.67}$$

where (from (2.25), (2.29) & (2.33)),

$$D_\omega^T \equiv \frac{(\epsilon\lambda\Delta I_w)^2}{2T_L} = \frac{4}{\pi}\epsilon^3 k(F_k\lambda)^{3/2}. \tag{2.68}$$

When $D_\omega < D_\omega^T$, the trapped particles are forced to move with the resonances wherever they go. If a resonance diffuses out of the beam, then the particles trapped inside it are lost. This mechanism was proposed by Chao and Month[17] to explain anomalous particle loss in the ISR at CERN. An average diffusion rate due to particle trapping in kth order resonances may be calculated using the standard map model (2.26). We assume: 1) that trapped particles diffuse at the same rate as the resonances, 2) that untrapped particles don't diffuse at all, and 3) that the time a particle spends inside resonances (in the long run) is proportional to the resonant area in the phase space. Then the average (over initial conditions or time) diffusion rate of the particles is,

$$< D_I^{(3)} > = < D_\omega^{(3)} > \frac{1}{(\epsilon\lambda)^2} = D_\omega \frac{A_t}{(\epsilon\lambda)^2} \tag{2.69}$$

where A_t is the ratio (2.31) of trapped area to untrapped area in the standard map.

$$A_t \approx .7 \frac{2\epsilon}{\pi}(F_k\Lambda_k)^{1/2}. \tag{2.70}$$

The average diffusion rate in the slow diffusion regime is thus

$$< D_I^{(3)} > = D_\omega \frac{1.4}{\pi\epsilon} \left(\frac{F_k}{\lambda^3}\right)^{1/2}. \tag{2.71}$$

The equal-times-equal-areas assumption made above is often invalid, since the tune diffusion is usually restricted to a fairly narrow band in frequency. This can sometimes be corrected for by adjusting the averaging distribution, but more often it necessitates a calculation of the effect of tune diffusion on the nonresonant particles. This is done in ref. 16 for a different application.

2.3.2 Fast Diffusion Regime

In the fast diffusion regime, by definition, the resonance phase θ is randomized after each kick. This will occur if, between kicks, the resonance "diffuses" a distance greater than the resonance spacing $\Delta\omega_s$ (from (2.15)&(2.30)). Thus, the critical value of D_ω, above which the kicks (from the kth Fourier component) are randomized is

$$D_\omega^R = \left(\frac{2\pi}{k}\right)^2. \tag{2.72}$$

From the Hamiltonian (2.26), and (2.18), the kick ΔI received by the particle at the interaction point is

$$\Delta I = k\Delta p = -k\epsilon F_k \sin(\theta). \tag{2.73}$$

Since θ is randomized with each kick, the rate of particle diffusion in the fast (tune) diffusion regime $D_\omega > D_\omega^R$ is about

$$D_I^{(1)} = \frac{(k\epsilon F_k)^2}{4}. \tag{2.74}$$

This does not depend on the position of the resonance. Note that if the kth resonance phase is randomized, the $k' > k$ resonance phases will also be randomized.

2.3.3 Intermediate Regime

Finally, we consider the intermediate regime where the diffusion rate is less than that necessary to randomize the phase, but still too large to allow trapping $D_\omega^T < D_\omega < D_\omega^R$. In this case, a particle close to the resonance receives correlated kicks for a certain period of time T_c less than the libration period of the resonance. Suppose a particle is initially on a resonance (so that the resonance phase (2.19) is stationary). As the resonance diffuses away from the particle, the resonance phase will begin to change. Following ref. 16, the change in the phase after N iterations is about

$$\Delta\theta = \sum_{i=1}^{N} \dot{\theta}_i,$$

where $\dot{\theta}_i$ is the phase velocity at the ith interaction. This may be rewritten

$$\Delta\theta = k\sum_{i=1}^{N}\sum_{k=1}^{i} \Delta\omega_k,$$

where $\Delta\omega_j$ is the change in betatron frequency between interactions due to the tune diffusion. Since $<\Delta\omega_i\Delta\omega_j> = \delta_{ij} D_\omega$, the variance of $\Delta\theta$ after N iterations is

$$<\Delta\theta^2> = k^2 \sum_{i=1}^{N}\sum_{j=1}^{N}\sum_{k=1}^{i}\sum_{l=1}^{j} <\Delta\omega_k \Delta\omega_l>$$

$$<\Delta\theta^2> \approx k^2 N^3 D_\omega. \tag{2.75}$$

The auto-correlation time for the phase is thus, approximately,

$$T_c \approx (k^2 D_\omega)^{-1/3}. \tag{2.76}$$

The highly correlated kicks come within a band which is wider than the resonance itself, but smaller than ΔI_s. The width of this band is given by the average distance that the resonance diffuses in the time T_c,

$$\Delta\omega = \sum_{i=1}^{T_c} \Delta\omega_i. \tag{2.77}$$

The variance is thus

$$<\Delta\omega^2> = \sum_{i=1}^{T_c}\sum_{i=1}^{T_c} <\Delta\omega_j \Delta\omega_i>$$

$$<\Delta\omega^2> = T_c D_\omega = \left(\frac{D_\omega}{k}\right)^{2/3}, \tag{2.78}$$

so the band width is about

$$\Delta\omega_b = (<\Delta\omega^2>)^{1/2}$$

or

$$\Delta\omega_b = \left(\frac{D_\omega}{k}\right)^{1/3}. \tag{2.79}$$

The rate at which the particles diffuse in action when they are inside the band is about

$$D_I^{(2)}(in) = \frac{T_c^2 (k\epsilon F_k)^2}{4T_c} = \frac{(k\epsilon F_k)^2}{4} (k^2 D_\omega)^{-1/3}. \tag{2.80}$$

This rate is higher than the random phase rate (2.74). It is interesting to note, however, that the average (over ω) diffusion rate in this case is independent of D_ω and equal to the random phase rate (2.72). Since the diffusion rate outside the band is approximately zero, this average rate is just

$$<D_I^{(2)}>_\omega = D_I^{(2)}(in)\ \Delta\omega_b\ k \tag{2.81}$$

$$<D_I^{(2)}>_\omega = \frac{(k\epsilon F_k)^2}{4} = D_I^{(1)}. \tag{2.82}$$

A thorough dynamical description of neoclassical transport in the standard map system can be found in ref. 16.

3. THE TWO DEGREES OF FREEDOM SYSTEM

3.1 The Hamiltonian

In a colliding beam system, the two kick functions $\Delta\dot{x}$ and $\Delta\dot{y}$ are, in general, each dependent on both x and y. Thus, the radial betatron motion affects the vertical betatron motion and vice versa. The amplitude and frequency spaces for the two degrees of freedom system are two dimensional and the resonances are curves instead of points. All of the various phenomena discussed previously for the one degree of freedom problem still exist. However, we now have some important additional features. First of all, new resonances called "coupling resonances" appear in the problem. Secondly, the resonant libration may be very large, even for relatively weak resonances. Finally, particles no longer have to move across resonances to leave the beam. They may now travel along both parametric and coupling resonances via one of several different transport mechanisms. In this section, we will first describe the resonances, their positions, widths, libration frequencies, angles, etc. Following this, there will be a discussion of the different transport mechanisms. Two of them, Arnold diffusion and the so-called "modulation" diffusion, will be discussed only briefly (since a detailed treatment has yet to be performed, and would probably be too lengthy anyway). Another, resonance streaming, will be examined with more care.

In analyzing the two degrees of freedom system, it is necessary to first chart the amplitude contours of the beam in frequency space, i.e. we must find the nonlinear frequencies $\nu_x(a_x, a_y)$ and $\nu_y(a_x, a_y)$ as functions of the amplitudes. To do this, we essentially retrace our treatment of the single degree of freedom system. The two dimensional Hamiltonian looks like

$$H(x,\dot{x},y,\dot{y},s) = \frac{\dot{x}^2}{2} + \frac{\dot{y}^2}{2} + K_x(s)\frac{x^2}{2} + K_y(s)\frac{y^2}{2} + \delta_s \epsilon V(x,y). \tag{3.1}$$

In action-angle variables, and with a Fourier transformation, this becomes,

$$H(I_x, I_y, \psi_x, \psi_y, s) = I_x\omega_{ox} + I_y\omega_{oy} + \delta_s\epsilon \sum_{k=0}^{\infty} F_k(I_x, I_y)\cos(\underline{m}_k \cdot \underline{\psi}), \tag{3.2}$$

where $\underline{m}_k \cdot \underline{\psi}$ means $(m_{kx}\psi_x + m_{ky}\psi_y)$ and m_{kx}, m_{ky} are integers. Again, the nonlinearity comes from the k=0 term (where $m_{ox} = m_{oy} = 0$). If we add this ψ independent term to the unperturbed linear system, we get an integrable two dimensional nonlinear oscillator,

$$H_o(I_x, I_y, s) = I_x\omega_{ox} + I_y\omega_{oy} + \delta_s\epsilon F_0(I_x, I_y). \tag{3.3}$$

The changes in phase after one revolution are

$$\omega_x(I_x, I_y) = \frac{\partial H_o}{\partial I_x} = \omega_{ox} + \epsilon \left.\frac{\partial F_o}{\partial I_x}\right|_{I_x, I_y} \tag{3.4}$$

$$\omega_y(I_x, I_y) = \omega_{oy} + \epsilon \left.\frac{\partial F_o}{\partial I_y}\right|_{I_x, I_y}. \tag{3.5}$$

The actions I_x, I_y are related to the amplitudes by (see (2.9)),

$$I_x = \frac{a_x^2\sigma_x^2}{2\beta_x^*} \tag{3.6}$$

$$I_y = \frac{a_y^2\sigma_y^2}{2\beta_y^*}. \tag{3.7}$$

The differentials of action and amplitude are related by

$$dI_x = \frac{a_x\sigma_x^2}{\beta^*} da_x \tag{3.8}$$

$$dI_y = \frac{a_y \sigma_y^2}{\beta_y^*} da_y .$$

For a bi-Gaussian beam,[3]

$$\frac{\xi_y}{\xi_x} = \frac{\sigma_x}{\sigma_y} \frac{\beta_y^*}{\beta_x^*}. \tag{3.9}$$

To find the dependence of frequency on action (and consequently on amplitude) via (3.5), it is necessary to calculate $\frac{\partial F_o}{\partial I_x}$ and $\frac{\partial F_o}{\partial I_y}$. This can be done by computer, providing a suitable analytic expression for the potential $V(x,y)$ is available. In figures 11,12 and 13 the amplitude contours of three beams are plotted in the tune space. The first, fig. 11, represents a flat beam $\frac{\sigma_x}{\sigma_y} = \infty$; the second fig. 12, a round beam, $\frac{\sigma_x}{\sigma_y} = 1$; and the third, fig. 13, an elliptical beam $\frac{\sigma_x}{\sigma_y} = 5$. Unfortunately, a simple analytic model of the elliptical beam potential is difficult to find and the one shown here[18] exhibits an unnatural fold in the contour grid. These grid patterns scale linearly with the tune shifts ξ_x and ξ_y as long as $\xi_x, \xi_y << 1$.

It is also possible to plot the resonances in frequency space. The phases of the cosine terms in (3.2) are stationary when

$$m_{kx}\nu_x + m_{ky}\nu_y = n. \tag{3.10}$$

This "resonance condition" defines the resonance lines in frequency space, fig. 14. If either m_{kx} or m_{ky} is zero, the resonance is simply a parametric resonance of the type described previously. If neither is zero, then the resonance serves to couple all three degrees of freedom, the two transverse and the longitudinal. The thickness of the resonance lines in fig. 14 indicates roughly their strengths.

In fig. 15, the amplitude contour grid for the flat beam fig. 11 is superimposed on the resonance diagram. Given a specific working tune ν_{ox}, ν_{oy} (corresponding to the lower left corner of the contour grid), it is now possible to see which betatron amplitudes will be affected by which resonances. As long as ξ_x and ξ_y are not too big, and the beam doesn't intersect any very low order resonances, the choice of working tune should not affect the shape of the contour grid. The tune chosen here corresponds roughly to that of SPEAR II .

3.2 Symmetries

It is interesting to consider which resonances dissappear when certain symmetries are present. For example, suppose there are N interaction points, but they, and the sectors between them, are identical. In this case, n in (3.10) will be restricted to multiples of N, and the pattern shown in fig. 14 will reappear N times as large (the new pattern will be periodic with period N instead of one). However, the tune shifts, and thus the size of the amplitude contour grid, will also scale up by a factor of N. So the resulting system will be identical to the one interaction case, fig. 15 (the only difference will be the shorter unit of time). Of course, if the N interaction sectors are not identical, those resonances forbidden by the symmetry will not dissappear. If the sectors differ by very much, the density of resonance lines in fig.s 14 & 15 becomes N times as large.

Another symmetry which is important, has to do with the centering of the betatron oscillation on the beam center. If the oscillation is centered, then the beam-beam force will be periodic in ψ_x and ψ_y with period π. In this case, only the resonances (3.10) with even values of m_{kx} and m_{ky} will remain. This will *increase* the periodicity of the resonance pattern fig. 14 (the period becomes .5), while reducing the strength of each line in the pattern.

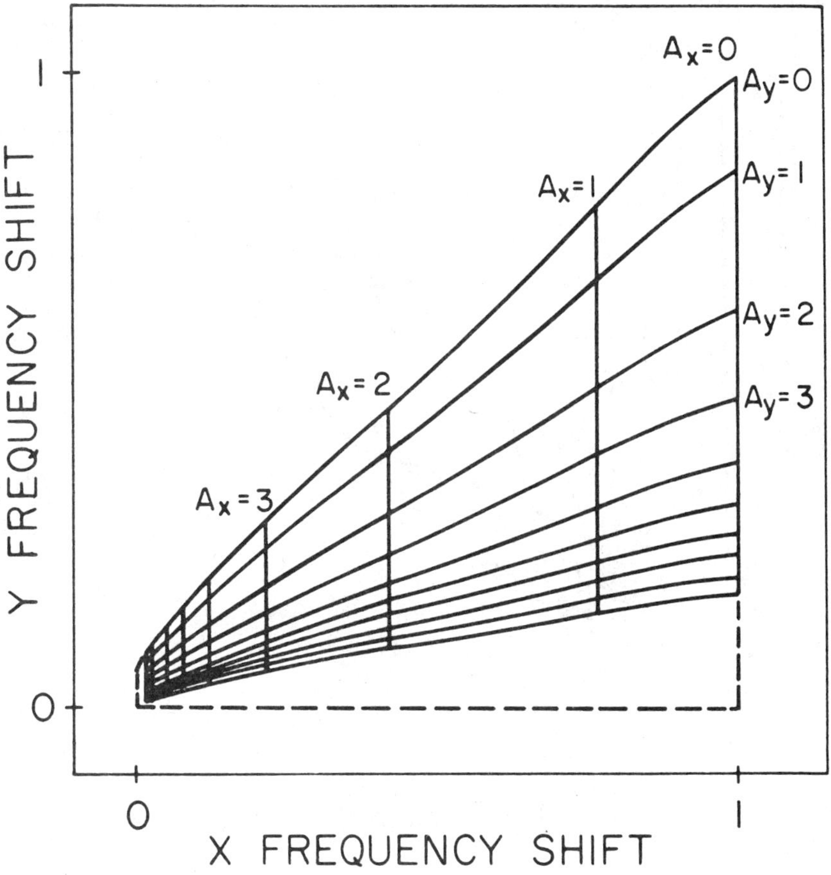

Fig. 11. The amplitude contour grid for a flat beam ($R = \infty$) in tune space. The coordinates give the displacements from the single beam tunes ν_{0x} and ν_{0y}. The grid scales linearly with the coordinates as the tune shifts change.

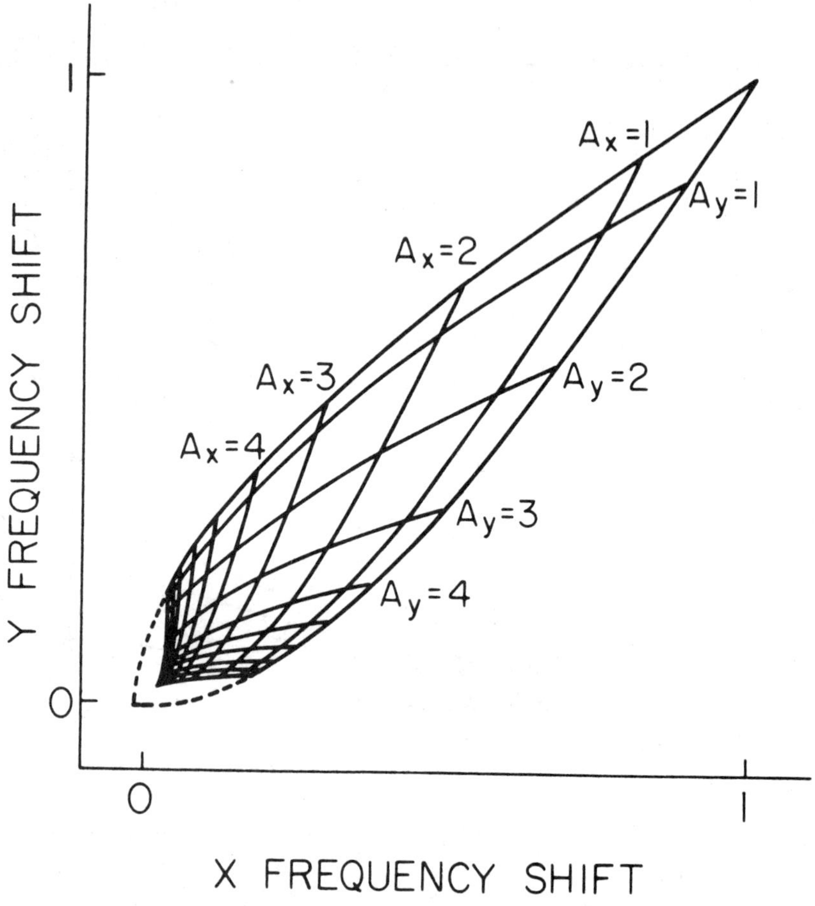

Fig. 12. The amplitude contour grid for a round beam ($R = 1.0$).

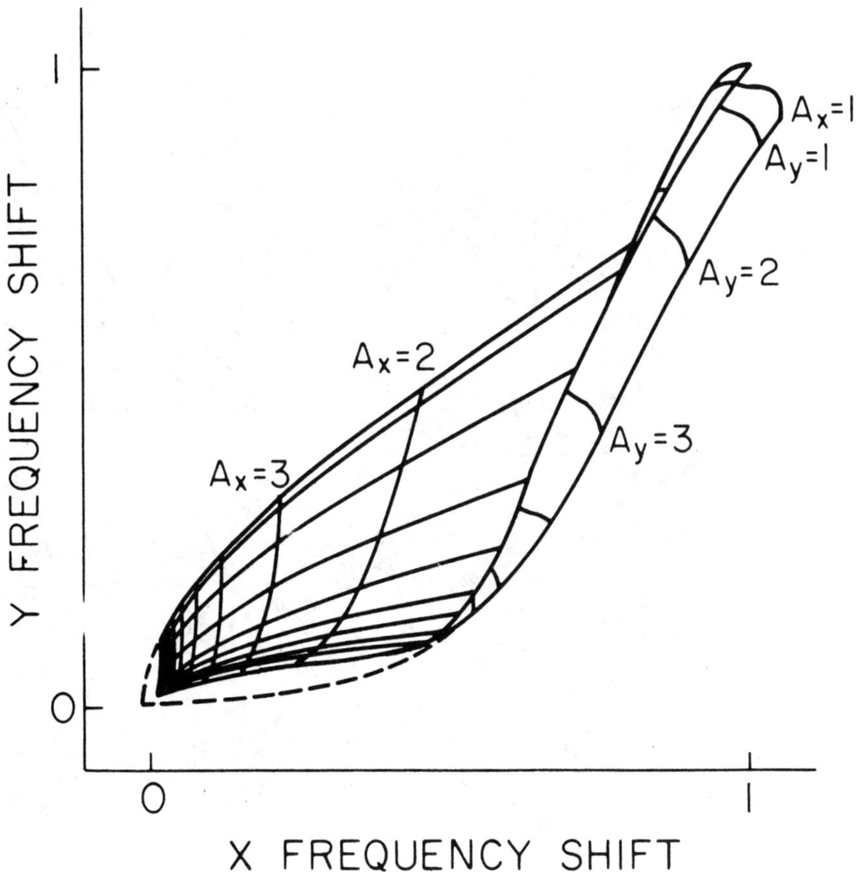

Fig. 13. The amplitude contour grid for an elliptical beam ($R = 5.0$). The fold here is artificial. This grid was calculated for the model presented in [17].

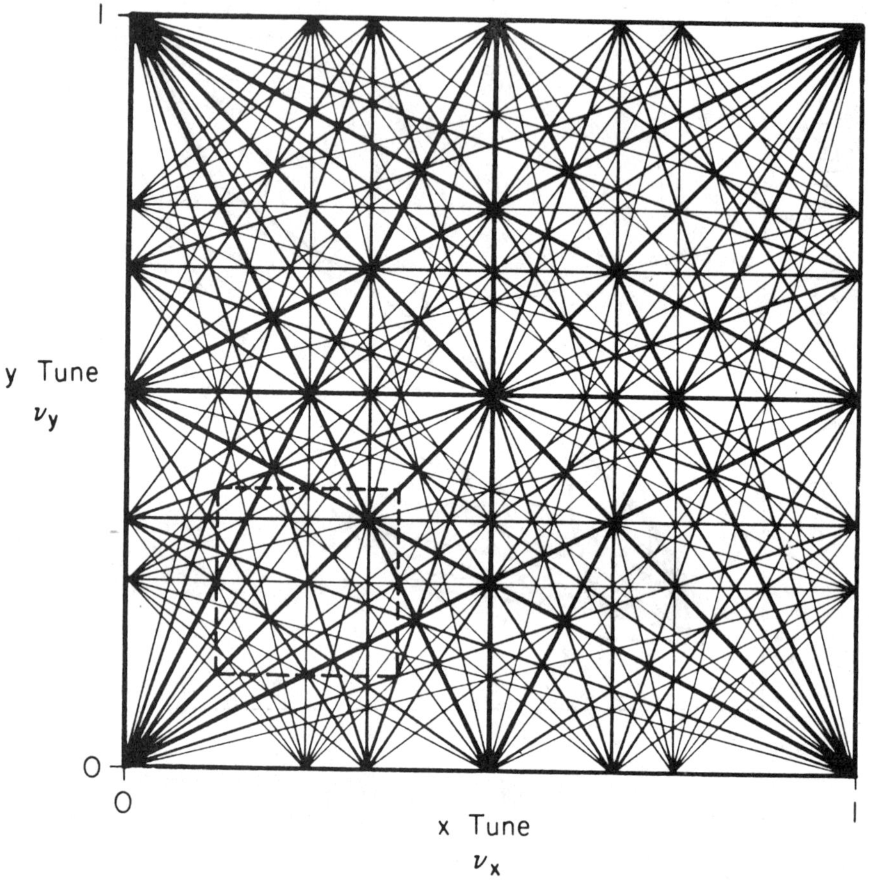

Fig. 14. Coupling and parametric resonances in tune space with $m_i \leq 4$. The dashed box is shown in fig. 16.

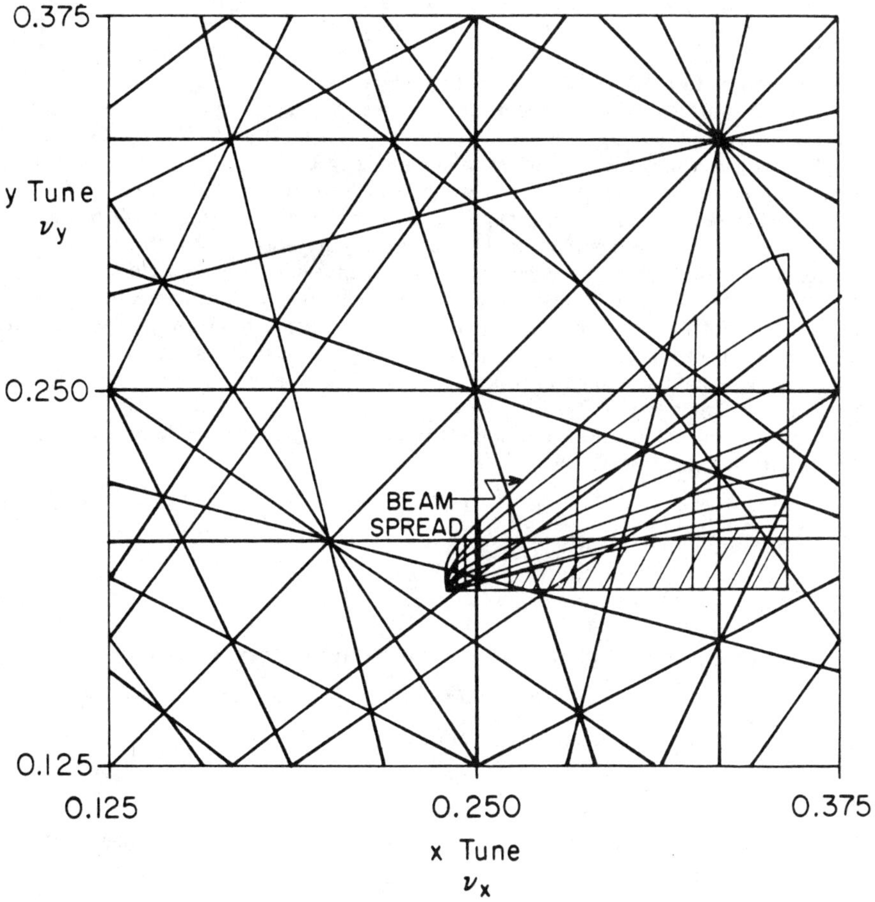

Fig. 15. The superposition of the amplitude contour grid for the flat beam fig. 12 on the resonance lines fig. 15.

In a machine with two identical interaction sectors and zero dispersion at the interaction points, the two scaling effects cancel. However the net tune shift is doubled while the strengths of the individual resonances are reduced.

3.3 Resonances in the Two Dimensional System

Once a working tune has been chosen, the metric of fig. 15 may be adjusted to straighten the amplitude contours into a uniform orthogonal grid. When this is done, the straight resonance lines become curves in amplitude space fig. 16. The coordinates in fig. 16 are normalized to the beam height and width (which for the flat beam are vastly different). The dashed closed curve represents approximately the location of an unperturbed beam with a bi-Gaussian transverse distribution. The resonances are only those with m_{kx}, m_{ky} even, i.e. they reflect centered betatron oscillations and two identical interaction sectors.

To analyze the system further, it is convenient to introduce some small changes in notation. First of all, we drop the subscript k since from now on, we will be considering only a fixed pair of integers m_x, m_y. Also, it will be useful now to use dummy indices and to adopt the convention according to which repeated indices on covariant-contravariant pairs in the same term imply a sum.

The fundamental quantities of interest in the two degrees of freedom system are[8,19]: the Hamiltonian (3.3)

$$H_o(I_x, I_y, s), \tag{3.11}$$

its derivative (the frequency)

$$\omega_i = \frac{\partial H_o}{\partial I_i}, \tag{3.12}$$

the pairing of ω and the "resonance vector" **m** (the resonance frequency)

$$S = \omega_i m^i, \tag{3.13}$$

the derivative of this quantity (the normal covector)

$$n_i = \frac{\partial S}{\partial I_i}, \tag{3.14}$$

and finally, the pairing of **m** and **n** (the effective nonlinearity, in analogy to (2.25))

$$\epsilon\Lambda = n_i m^i. \tag{3.15}$$

Note that in action coordinates, $\epsilon\Lambda$ may be written in the form

$$\epsilon\Lambda = m^i m^j \frac{\partial^2 H_o}{\partial I_i \partial I_j}. \tag{3.16}$$

The resonance condition (3.10) is now just

$$S(I_{Rx}, I_{Ry}) = 2\pi n, \tag{3.17}$$

and the normal covector **n** corresponds to a vector perpendicular to the resonance curve in euclidian action space. The Jacobian matrix for the transformation from action to amplitude coordinates is (from (3.8)),

$$J^i_j = \frac{\partial a_i}{\partial I_j} = \left(\frac{\beta_j^{\bullet}}{a_j \sigma_j^2} \right) \delta^i_j, \tag{3.18}$$

where β_j and a_j are constants and δ^i_j is the Kronecker delta.

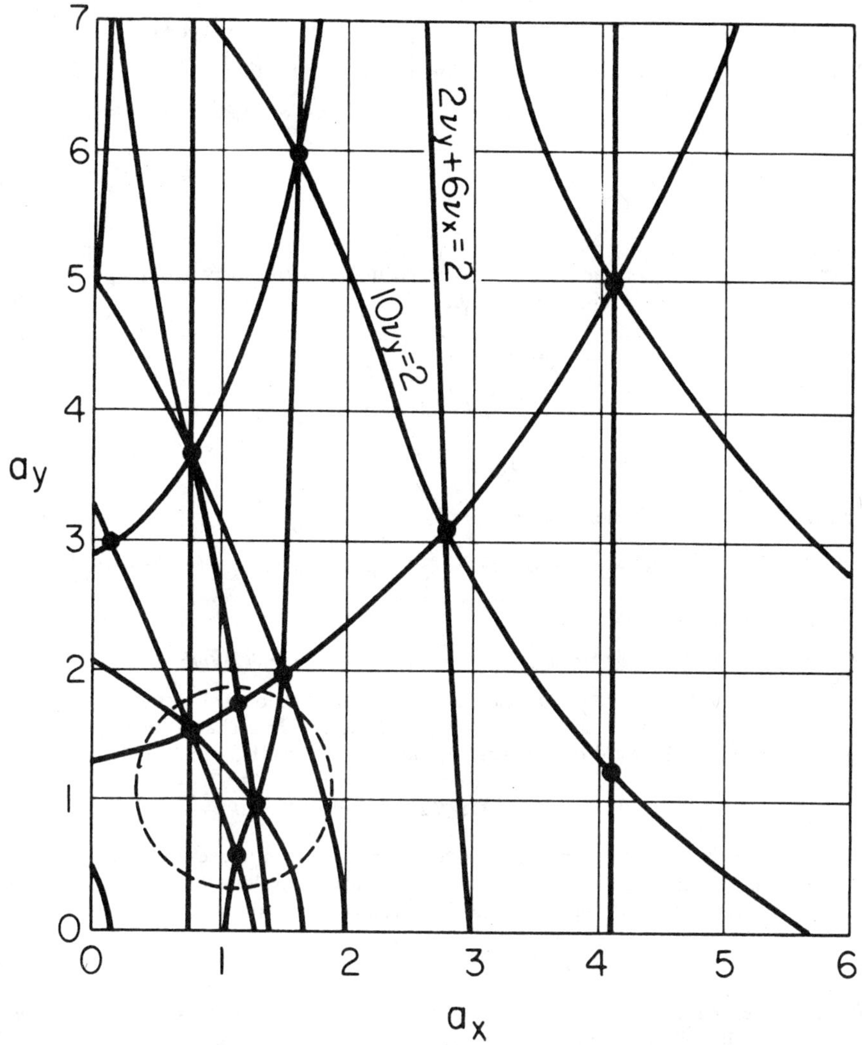

Fig. 16. Resonance lines in amplitude space. The dashed line is the beam core location. This figure is just a stretched version of figure 16.

We are now prepared to calculate the width of the resonant libration. The reduced Hamiltonian from (3.2) is

$$H(I_x, I_y, \psi_x, \psi_y, s) = H_o(I_x, I_y, s) + \delta_s F(I_x, I_y)\cos(\underline{m}\cdot\underline{\psi}). \tag{3.19}$$

The resonance coordinates are defined by the generating function[8]

$$G_2 = (I_R^i - \mu^i_j p^j)\psi_i, \tag{3.20}$$

so

$$\theta_j = \mu^i_j \psi_i \tag{3.21}$$

$$(I^i - I_R^i) = \mu^i_j p^j,$$

where $\mu_1^x = m_x$, $\mu_1^y = m_y$, and μ_2^x, μ_2^y are unspecified. The resonant action I_R is now defined by (3.17). Expanding H_o about I_R, we get

$$H_o = \frac{\partial H_0}{\partial I_i}\Delta I^i + \frac{1}{2}\frac{\partial^2 H_o}{\partial I_i \partial I_j}\Delta I^i \Delta I^j + \cdots \tag{3.22}$$

or

$$H_o = \omega_i \mu^i_j p^j + \frac{\epsilon}{2}\Lambda_{ij} p^j p^i + \cdots \tag{3.23}$$

The Hamiltonian (3.19) then becomes

$$H(p_1, p_2, \theta_1, s) = \omega_i(p{=}0)\mu^i_j p^j + \frac{\epsilon}{2}\Lambda_{ji}(p{=}0)\, p^j p^i + \delta_s \epsilon F(p{=}0)\cos(\theta_1). \tag{3.24}$$

The first sum on the RHS is

$$\omega_i \mu^i_j p^j = (\omega_x m_x + \omega_y m_y)\, p_1 + (\omega_x \mu_2^x + \omega_y \mu_2^y)\, p_2. \tag{3.25}$$

From (3.24), p_2 is a constant of the motion, so the linear-in-p_2 terms may be dropped. From (3.17), the term which is linear in p_1 produces a $2\pi n$ change in θ_1 per revolution, and therefore may also be dropped. The second sum on the RHS of (3.24) is

$$\frac{\epsilon}{2}\Lambda_{ji} p^i p^j = \frac{\epsilon}{2}[\Lambda_{22} p_2 p_2 + 2\Lambda_{12} p_1 p_2 + \Lambda_{11} p_1 p_1]. \tag{3.26}$$

The first term is a constant and the second serves only to produce a constant offset of the resonance in p_1 (this corresponds to a small displacement along the resonance curve). So these terms may also be dropped and the final resonance Hamiltonian is,

$$H(p_1, \theta_1, s) = \frac{\epsilon\Lambda}{2} p_1^2 + \delta_s \epsilon F\cos(\theta_1). \tag{3.27}$$

Thus the resonance produces an oscillation in amplitude space which is approximately confined to a line. The direction of the line is given by the unit vector $\mathbf{p}_1$, which from (3.21) is just the resonance vector $\mathbf{m}$ in resonance coordinates. The maximum libration width is

$$\Delta p_w^1 = 4\left(\frac{F}{\Lambda}\right)^{1/2}. \tag{3.28}$$

In action coordinates, this is (from (3.21))

$$\Delta I_w^i = 4m^i\left(\frac{F}{\Lambda}\right)^{1/2}, \tag{3.29}$$

and in amplitude coordinates

$$\Delta a_w^i = 4r^i\left(\frac{F}{\Lambda}\right)^{1/2}, \tag{3.30}$$

where

$$r^i = m^j J^i_j = m^j \left(\frac{\beta^*_j}{a_j \sigma^2_j} \right) \delta^i_j. \tag{3.31}$$

The libration frequency is, of course, independent of coordinate system,

$$f_L = \epsilon \sqrt{F\Lambda}. \tag{3.32}$$

The direction of resonant libration, given for amplitude coordinates by the resonance vector **r**, is not always perpendicular to the resonance curve, but, in general, makes some angle ϕ with its tangent (see fig. 17). Since the components of the normal covector **n** in amplitude coordinates are

$$c_j = n_i (J^{-1})^i_j = n_j \left(\frac{a_j \sigma^2_j}{\beta^*_j} \right), \tag{3.33}$$

the angle ϕ is given by

$$c_j r^j = |\mathbf{c}||\mathbf{r}| \sin\phi = \epsilon\Lambda, \tag{3.34}$$

or

$$E = \csc(\phi) = \frac{|\mathbf{c}||\mathbf{r}|}{\epsilon\Lambda}. \tag{3.35}$$

With expressions for Δa_w, f_L, and ϕ, we now have a rough picture of the regular motion inside of, and close to, an isolated resonance. This motion can be visualized by imagining a three dimensional space formed by adding a third dimension, the resonance phase ϕ, to the amplitude space fig. 16. In this space, the resonances look like tubes rather than curves (fig. 18). The libration orbits form closed loops inside the tubes, skewed at various angles ϕ relative to the tubes themselves.

Since we were able to reduce the system to one degree of freedom by transforming to the resonance coordinates p,θ, the previously developed treatments of tune modulation and diffusion may be applied almost directly to the two degrees of freedom system.

The existence of coupling resonances, which at each order are approximately as large as their parametric counterparts, increases dramatically the overall density of resonances in the amplitude space. The density of kth order parametric resonances in the 2-dimensional tune space is about

$$d_p = 2k \tag{3.36}$$

lines per $(2\pi)^2$ square radians/revolution. When coupling resonances are included, and defining $K = m_x + m_y$, the density of Kth order resonances is about

$$d_c = K^2 \tag{3.37}$$

lines per $(2\pi)^2$ square radians/revolution. This distinction between the one and two degrees of freedom systems is especially significant when high order resonances are involved.

A second distinction relates to the width of the resonant libration (3.30). The fact that the resonant libration is not perpendicular to the resonance means that the width of the libration may be somewhat greater than the width of the resonance itself. This effect is most pronounced in beams with large aspect ratios

$$R \equiv \frac{\sigma_x}{\sigma_y} >> 1.$$

The ratio of libration width to resonance width is given by the enhancement factor E. To get a feeling for this quantity in a real machine, we notice that the slope of the libration in

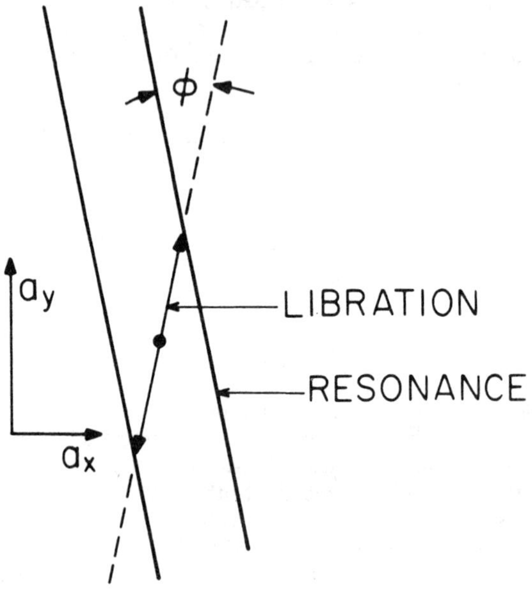

Fig. 17. The angle ϕ between the resonant libration and the resonance line.

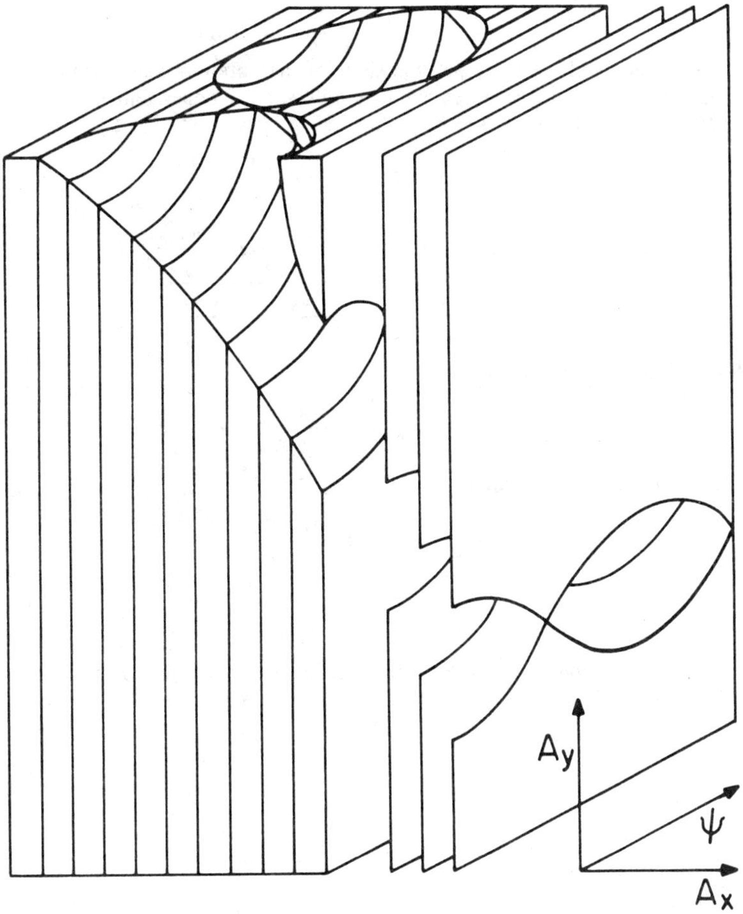

Fig. 18. Resonance tubes in the A_x, A_y, θ space.

amplitude space is just the slope of the resonance vector **r**,

$$s_L = \frac{r_y}{r_x} = \frac{\beta_y^{\bullet} a_x \sigma_x^2 m_y}{\beta_x^{\bullet} a_y \sigma_x^2 m_x}$$

or

$$s_L = \frac{\xi_y}{\xi_x} R \frac{a_x}{a_y} \frac{m_y}{m_x}. \tag{3.38}$$

For fixed tune shifts ξ_y and ξ_x, the slope of the libration is proportional to R. The slope s_R of the resonance, on the other hand, is only slightly dependent on R, and in fact does not depend on R in the limit $R \rightarrow \infty$ (from (3.33)&(3.13)):

$$s_R = -\frac{c_x}{c_y} = -\frac{\partial S}{\partial a_x} \Big/ \frac{\partial S}{\partial a_y}$$

$$s_R = -\frac{m_x \dfrac{\partial \omega_x}{\partial a_x} + m_y \dfrac{\partial \omega_y}{\partial a_x}}{m_x \dfrac{\partial \omega_x}{\partial a_y} + m_y \dfrac{\partial \omega_y}{\partial a_y}}. \tag{3.39}$$

When $R >> 1$ and $a_x \approx a_y \approx 1$,

$$s_R \approx -\frac{m_x \xi_x + m_y \xi_y}{m_y \xi_y}. \tag{3.40}$$

If, in addition, $\xi_x \approx \xi_y$, then

$$s_L \approx R \frac{m_y}{m_x} \tag{3.41}$$

$$s_R \approx -\left(\frac{m_x}{m_y} + 1 \right). \tag{3.42}$$

Assuming the angle between the resonances and the libration is small, the enhancement factor is about

$$E \approx \frac{s_L s_R}{s_L - s_R}$$

or

$$E \approx \frac{R}{R \dfrac{m_x}{m_y} + \dfrac{m_y}{m_x}}. \tag{3.43}$$

This has a maximum when

$$\frac{m_x}{m_y} = R^{-1/2}, \tag{3.44}$$

and the maximum value is

$$E_{\max} = \frac{R^{1/2}}{2}. \tag{3.45}$$

At SPEAR II, $R \approx 30$, so the maximum enhancement at $a_x \approx a_y \approx 1$ is about $E_{\max} = 2.7$.

3.4 Resonance Streaming

The first important feature of the two degrees of freedom system is the possibility of particle transport along the resonances.[19] This transport may be driven, as before, by either intrinsic stochasticity or externally generated fluctutations. In the one degree of freedom system, the "direct" diffusion of amplitude did not play a significant role in the beam-beam interaction. With two degrees of freedom, however, direct diffusion may be amplified inside those resonances where the enhancement factor E is large. The effect is illustrated in fig. 19. If a particle which is initially on a libration orbit is instantaneously displaced to another libration orbit, the displacement **A** of the particle's oscillation center (average position) is, in general, different from the true displacement **B** in both direction and magnitude. The relationship between the magnitudes of the two displacements is,

$$|\mathbf{A}| = |\mathbf{B}| \csc\phi \sin\alpha = |\mathbf{B}| \; E \sin\alpha . \tag{3.46}$$

If the enhancement factor E is large, the oscillation center displacement may be much larger than the actual displacement.

We suppose now that there is an isotropic "classical" diffusion on the amplitude space. This diffusion process is just a collection of small random displacements, each one of which is amplified inside the resonance according to (3.46). Thus, the oscillation center of a trapped particle will diffuse along its resonance at an enhanced rate of

$$D_R = D_C E^2 . \tag{3.47}$$

The situation here is in many ways analagous to the case of "parametric" diffusion in the one degree of freedom system. In particular, the diffusion is characterized by three regimes. If the diffusion is slow enough, a particle may remain inside a resonance for many libration periods, i.e. may be trapped. In this case, it is the diffusion of the particle's oscillation center which is important, not that of the particle itself. Particles outside the resonances will behave classically (according to D_C). Those trapped inside a resonance will diffuse at the enhanced rate D_R. The intermediate regime corresponds to a diffusion rate D_C which is too fast to allow trapping, but still slower than the diffusion induced by resonantly correlated kicks in the vicinity of the resonance. The third, or "classical", regime refers to a diffusion D_C which is faster than that produced by the resonant kicks. In this regime, the resonance is invisible, i.e. has no significant effect on particle motion (for an early treatment of streaming, see ref. 20).

The slow diffusion, or trapping, regime is defined as usual by the requirement that the time necessary to diffuse across the resonance be large compared to the libration period. From (3.30), (3.32), and (3.35), this condition is

$$D_C < \frac{4}{\pi} \frac{(\epsilon^2 F \Lambda)^{3/2}}{|\mathbf{c}|^2} . \tag{3.48}$$

In the fast diffusion, or classical, regime the motion induced by the diffusion is faster than that induced by the resonant libration. Qualitatively, this means that in the period of time necessary for a particle to diffuse across the resonance, the resonant libration moves the particle a distance less than the resonance width. The *rms* speed of the resonant libration is

$$S_{rms} = \frac{\Delta a_w f_L}{\sqrt{8}} , \tag{3.49}$$

so the classical regime is defined by

$$\frac{(\Delta a_w)^2}{2 D_C E^2} \, S_{rms} < \frac{\Delta a_w}{E} \tag{3.50}$$

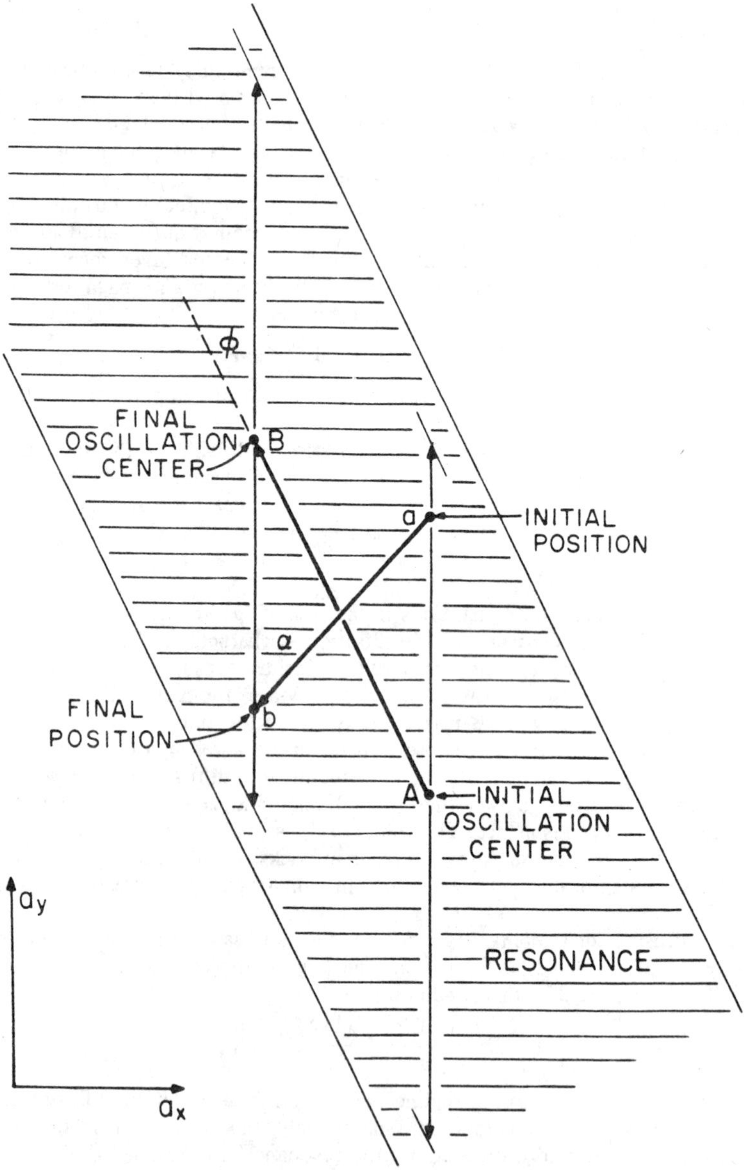

Fig. 19. A "classical" displacement in amplitude and the corresponding jump in oscillation center for a resonant particle.

or, from (3.30), (3.32), and (3.35)

$$D_C > 16\sqrt{2}\,\frac{(\epsilon^2 F\Lambda)^{3/2}}{|\mathbf{c}|^2}\,E. \qquad (3.51)$$

If a resonance is in the classical regime relative to a certain diffusion process, it may be ignored not only with respect to diffusion enhancement, but in every other respect as well. All but a finite number of resonances are in the classical regime for any given set of parameters a_x, a_y, ξ_x, ξ_y, and D_C.

The number of resonances in the enhanced diffusion regimes which have large enhancement factors $E > 2$ is usually small, even for beams with large aspect ratios $R \approx 30$. This is shown in fig. 20 where E is plotted (using (3.43)) as a function of the ratio m_y/m_x for different values of R. Only those resonances within the indicated intervals have enhancement factors larger than two.

In electron-positron colliding beams, diffusion caused by quantum fluctuations is contained by radiation damping. Normally, this damping may be viewed as a simple drift of the particles towards $a_x = a_y = 0$. But for trapped particles, this damping may be partly compensated for (and in some cases actually reversed) by the resonant libration (in a manner similar to the way in which particles are compensated for their loss of longitudinal energy by the electromagnetic "buckets" in the RF cavities). In general, the component of the damping force parallel to the resonant libration is cancelled. The remaining component, which is perpendicular to the libration, serves to pull the particle through the resonance tube. The rate at which the particle is pulled through the tube is E times as fast as the perpendicular component of the drift. Fig. 21 compares the trajectories of resonant and nonresonant particles in a simple two degrees of freedom system with damping.

The interaction of diffusion enhancement and damping "compensation-amplification" in the resonances may critically alter the normal diffusion-damping equilibrium of the resonant particles. The extent to which this local imbalance upsets the overall stability of the beam profile depends on the number of active resonances in or close to the beam, and the total number of particles that they are capable of trapping at any one time. A careful study requires a Fourier analysis of the beam potential to determine the strengths of the individual resonances, and an examination of the compensation-amplification characteristics as they depend on the ratio m_y/m_x and amplitudes a_x, a_y.

An example of resonant transport in a computer simulation of a flat beam ($R = \infty$) system is shown in fig. 22. The tune shifts here are $\nu_x = 5.24$, $\nu_y = 5.18$ and the damping time is $T_D = 50000$ revolutions. The trajectory of an oscillation center is traced in in the amplitude space for 2×10^5 interactions. The particle first becomes trapped in the $(2\nu_y + 6\nu_x = 2)$ resonance and climbs to a large value of a_y. It then becomes untrapped, falls back toward the beam core and is retrapped in the $(10\nu_y = 2)$ resonance. The resonance pattern here corresponds approximately to that of fig. 16.

3.5 Arnold diffusion

The stochastic layers in the two degrees of freedom system form a dense interconnected "web" in the phase space. They appear as thin coatings on the outsides of each of the resonance tubes fig. 18. In theory, a particle whose initial conditions lie within such a layer, can diffuse along this layer (fig. 23) to an intersection where it may turn onto another layer.[21,22] In this manner it is able to, and eventually will, penetrate any finite region of the phase space. In practice, this type of diffusion is rarely seen because the diffusion rates are usually very small and any amount of classical diffusion usually dominates. One of the few physical systems in which Arnold diffusion might be observed is the $p-\bar{p}$ colliding beam machine. Intra-beam scattering in the $p-\bar{p}$ collider can produce a very weak diffusion D_C of the particle actions. If the rate of Arnold diffusion D_A is much larger than D_C, particles will diffuse

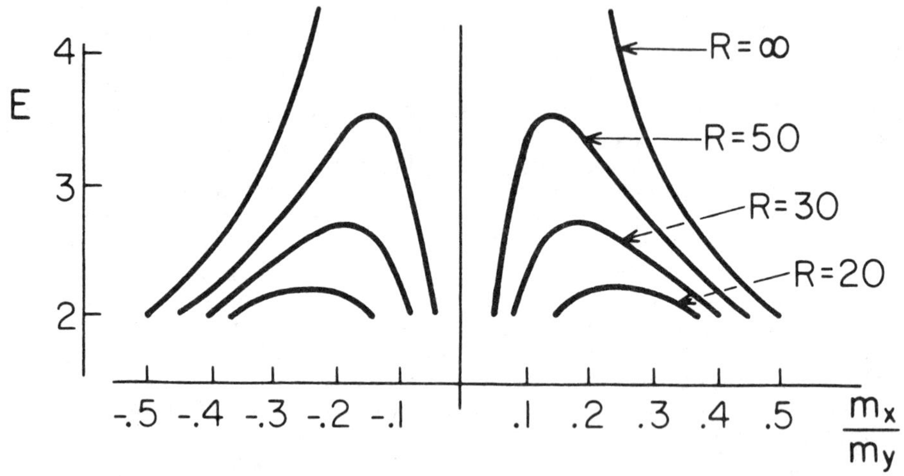

Fig. 20. The range of resonance ratios $\frac{m_x}{m_y}$ with enhancement factors greater than two when $\xi_x = \xi_y$ and $a_x = a_y = 1$. The four curves represent beams with different ellipticities.

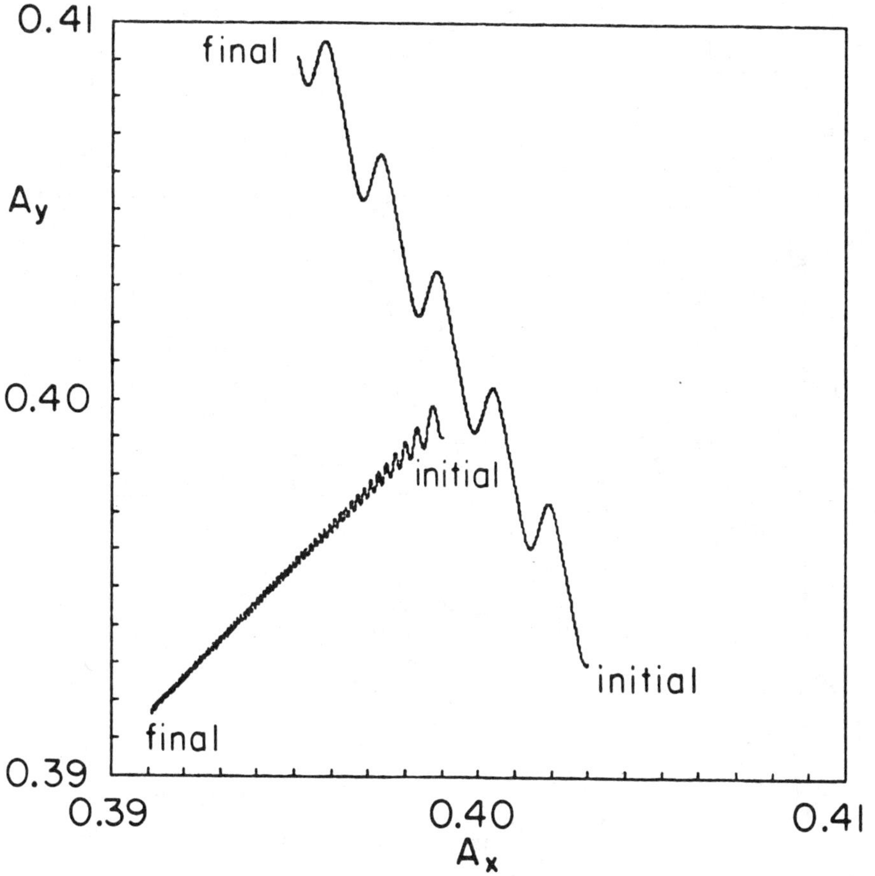

Fig. 21. The compensation-amplification of damping for resonant motion is compared with the usual behavior of a nonresonant particle. Only the initial conditions are different. This was drawn by computer using the mapping

$$A_x' = A_x + K\cos(\psi_x + C\psi_y) - 10^{-4} A_x$$

$$A_y' = A_y + CK\cos(\psi_x + C\psi_y) - 10^{-4} A_y$$

$$\psi_x' = \psi_x + P2\pi A_x'$$

$$\psi_y' = \psi_y + 2\pi A_y'$$

where $K = 10^{-5}$; $P = 10$; $C = 5$. The resonance is defined by $P A_x + C A_y = 5$.

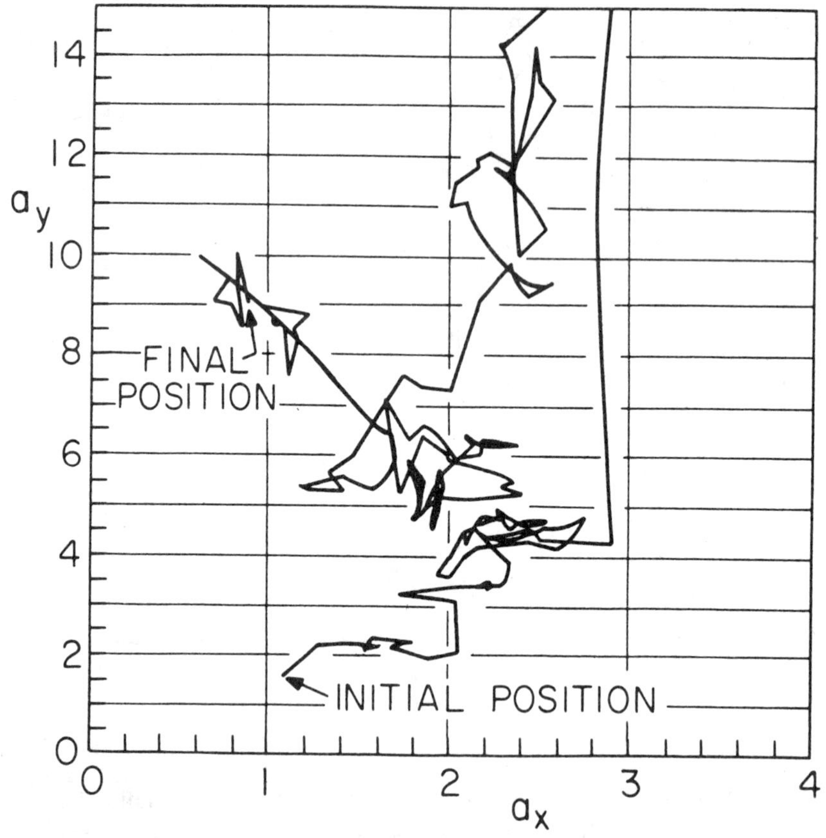

Fig. 22. The trajectory of a single particle in the amplitude space of a flat beam. The length of the run is three damping times. The simulation includes both diffusion (quantum fluctuations) and damping (classical radiation). This particle is exceptional since it becomes trapped in a resonance, leaves the beam, and then returns to get trapped in another.

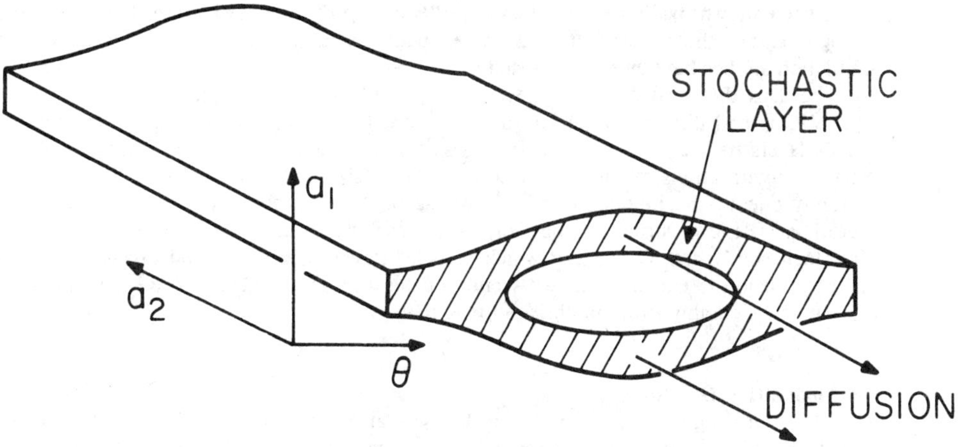

Fig. 23. Arnold diffusion. Particles diffuse along the separatrix stochastic layer.

collisionally until they reach a stochastic layer. They will then experience a greatly accelerated diffusion along the layer and perhaps out of the beam. The extent to which these layers affect the overall diffusion rate depends on the average spacing P between the more significant layers and the average widths Q of these layers in action coordinates. The average diffusion rate D_{av} is

$$D_{av} = \frac{QD_A + PD_C}{(Q+P)}, \tag{3.52}$$

so Arnold diffusion is important if

$$D_C < D_A \frac{Q}{P}. \tag{3.53}$$

It has been shown by Chirikov[8,22] that both D_A and Q drop off exponentially with the beam strength ϵ, but the actual values of these quantities have not yet been calculated. Some simulation studies have been performed.[23]

Perhaps the most difficult aspect of this calculation is the rapid variation of the local layer widths and diffusion rates. A rough picture, fig. 24, of the resonances and their stochastic layers has been constructed from a series of section plots for the flat beam $R = \infty$. Near the intersections of the resonance curves, the size of the stochastic region grows considerably due to the interaction of the intersecting resonances. At each intersection then, there is an island of strong stochasticity, and these islands are connected to one another by thin stochastic layers. The rate at which particles travel from one island to another depends on the diffusion rate at the thinnest section of the stochastic bridge, and these minimal rates can vary considerably from one bridge to another.

3.6 Modulation Diffusion

The third transport mechanism for diffusion along resonances is the so-called "modulation diffusion". If a slow enough modulation of one (or both) tune(s) is added to the system, the resonance lines in fig. 16 will become stochastic bands of overlapping sideband resonances fig. 25. The situation is identical to that of Arnold diffusion except for the fact that in this case the stochastic bridges are much wider. The diffusion rates along the bridges are still very small, and fall off approximately exponentially with the distance to the nearest significant island. This effect, like the Arnold diffusion, has been studied for general models,[24,25] but not for the beam-beam interaction itself.

Acknowledgements

I wish to thank F. M. Izrailev whose criticisms led to a number of changes in the original manuscript, and B. V. Chirikov who patiently explained many of the ideas reviewed here. I am also grateful to A. J. Lichtenberg, M. A. Lieberman, and F. Vivaldi for valuable discussions, as well as to M. Month through whose efforts these studies were made possible. This work was supported by ONR grant N00014-79-C-0674, DOE contract DE-AS05-81ER40003, and by the Academy of Sciences of the Soviet Union.

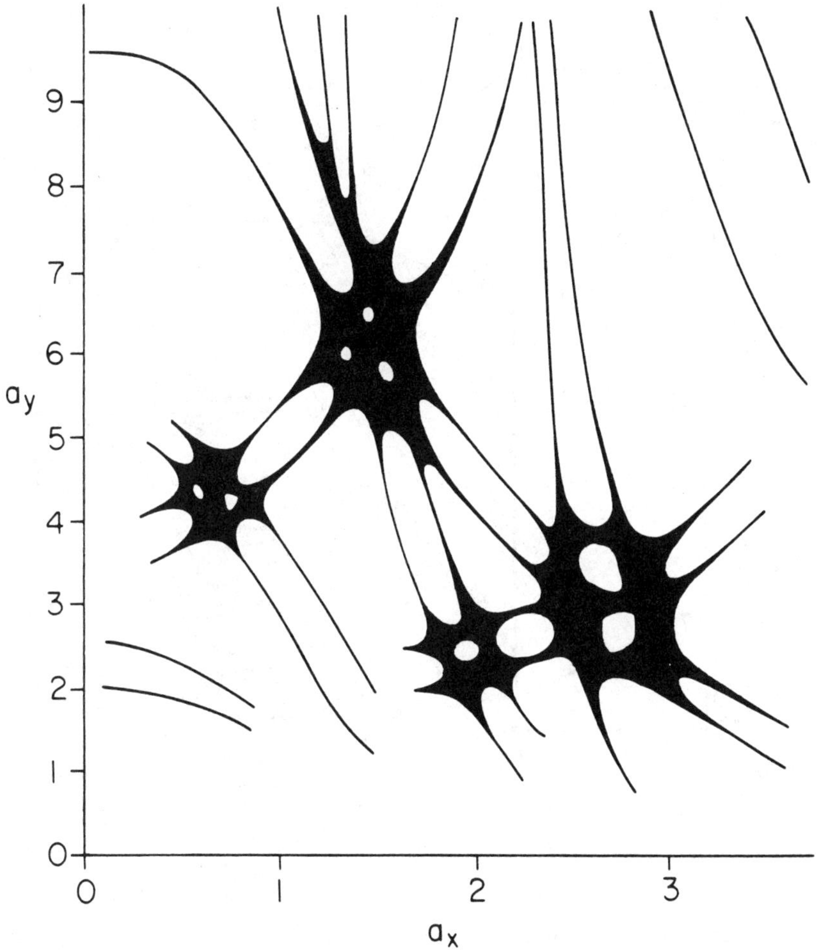

Fig. 24. Resonance bands and their stochastic regions for a typical flat beam model with $\xi_x = \xi_y = 0.06$. Stochastic islands (black areas) appear at the resonance intersections. This figure is somewhat schematic since it is impossible to take a crossection of two different resonances simultaneously. However, the resonance widths and the stochastic layer widths have been observed and are accurate to within about 20%. These resonances correspond to those shown in fig. 17.

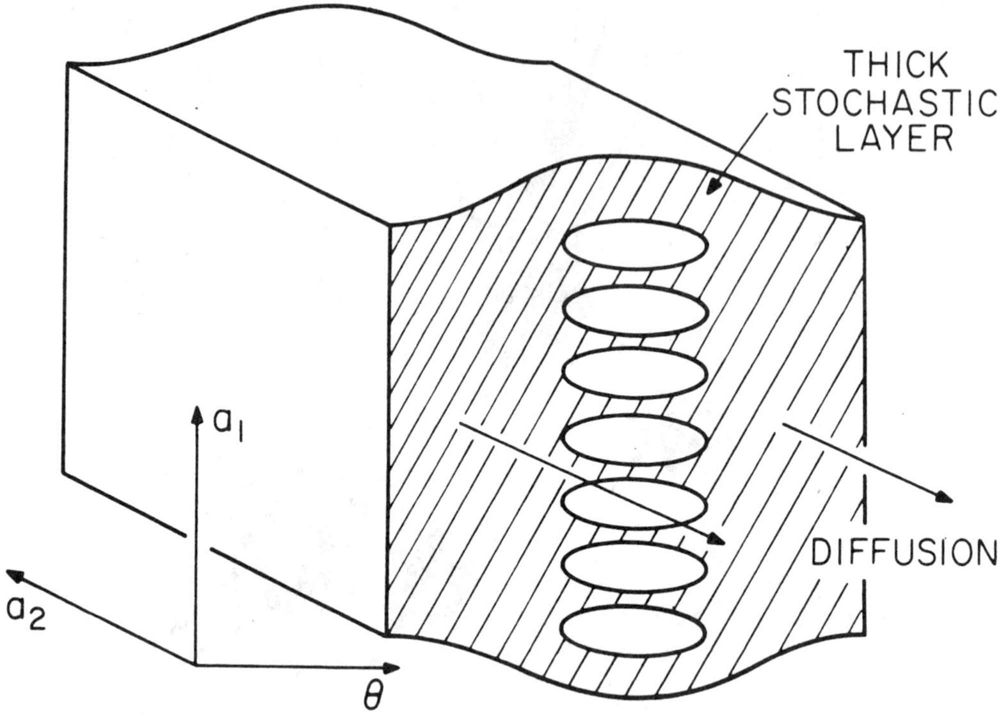

Fig. 25. Modulation diffusion along a thick stochastic layer formed by the overlapping sidebands of a modulation multiplet.

References

1. A. J. Lichtenberg and M. A. Lieberman, **Regular and Stochastic Motion,** (Springer-Verlag, New York 1982).
2. E. D. Courant and H. S. Snyder, "Theory of Alternating Gradient Synchrotrons", Ann. of Physics, **3** 1 (1968).
3. M. Month, "Nature of the Beam-Beam Limit in Storage Rings", IEEE Trans. on Nucl. Scien. NS-**22** 1376 (1976).
4. J. Tennyson, in **Nonlinear Dynamics and the Beam-Beam Interaction,** edited by M. Month and J. C. Herrera, AIP Conf. Proc. **57**, (AIP Press, New York 1979) p. 158.
5. M. Berry, in **Topics in Nonlinear Dynamics,** edited by S. Jorna, AIP Conf. Proc. **46** (AIP Press, New York 1978).
6. J. Moser, **Stable and Unstable Motion in Dynamical Systems,** (Princeton University Press, Princeton 1973).
7. F. M. Izrailev, "Nearly Linear Mappings and Their Applications", Physica 1D **3** 243 (1980).
8. B. V. Chirikov, "Universal Instability of Many Dimensional Oscillator Systems", Physics Reports **52** 263 (1979).
9. B. V. Chirikov, "Stability of the Motion of a Charged Particle in a Magnetic Confinement System", Sov. Jour. Plasma Phys. **4**, May-June, 289 (1978).
10. F. M. Izrailev, S. I. Misnev and G. M. Tumaikin, "Numerical Studies of Stochasticity Limit in Colliding Beams (One Dimensional Model)", preprint No 77-43, Institute of Nuclear Physics, Siberian Branch, Academy of Sciences of the USSR, Novosibirsk (1977).
11. F. M. Izrailev and I. B. Vasserman, "The Influence of Different Types of Modulations on a Decrease of the Stochasticity Limit in Beam-Beam Effects", preprint No 81-60, Institute of Nuclear Physics, Siberian Branch, Academy of Sciences of the USSR, Novosibirsk (1981).
12. B. V. Chirikov, Dokl. Acad. Nauk, 125, 5 (1959).
13. D. V. Chirikov and D. L. Shepalyansky, "Diffusion Due to Repeated Crossings of a Nonlinear Resonance", preprint No 80-211, Institute of Nuclear Physics, Siberian Branch, Academy of Sciences of the USSR, Novosibirsk (1980).
14. A. A. Galeev and R. Z. Sagdeev,"Transport Phenomena in a Collisonless Plasma in a Toroidal Magnetic System", JETP 26 233 (1968).
15. D. D. Ryutov and G. V. Stupakov, "Neoclassical Transport in Ambipolar Confinement Systems", Sov. J. Plasma Phys. 4, 278 (1978).
16. B. V. Chirikov, "Homogeneous Model for Resonant Particle Diffusion in an Open Magnetic Confinement System", Sov. J. Plasma Physics, 5 492 (1979).
17. A. W. Chao and M. Month, "Particle Trapping During Passage Through a High Order Nonlinear Resonance", Nuc. Inst. and Meth. **121** 129 (1974).
18. V. V. Vecheslovov,"A Model for the Potential of Elliptical Colliding Beams", preprint No 80-72, Institute of Nuclear Physics, Siberian Branch, Academy of Sciences of the USSR, Novosibirsk (1980).
19. J. Tennyson,"Resonance Transport in Near Integrable Systems With Many Degrees of Freedom", Physica 3**D**, (1982)
20. H. G. Hereward,"Diffusion in the Presence of Resonances", Internal CERN Report, CERN/ISR-DI/72-26 (1972).
21. J. Tennyson, M. A. Lieberman, A. J. Lichtenberg, "Diffusion in Near Integrable Hamiltonian Systems With Three Degrees of Freedom", in **Nonlinear Dynamics and the Beam-Beam Interaction,** edited by M. Month and J. C. Herrera, AIP Conf. Proc. vol 57, AIP Press, New York (1979).

22. B. V. Chirikov, F. Vivaldi, and J. Ford, "Some Numerical Studies of Arnold Diffusion In a Simple Model," in **Nonlinear Dynamics and the Beam-Beam Interaction**, edited by M. Month and J. C. Herrera, AIP Conf. Proc. vol 57, AIP Press, New York (1979).
23. B. V. Chirikov and E. Keil, unpublished work.
24. B. V. Chirikov, F. M. Izrailev, and D. L. Shepalyansky, "Dynamical Stochasticity in Classical Mechanics," Institute of Nuclear Physics, Siberian Branch, Academy of Sciences of the USSR (1980).
25. B. V. Chirikov, J. Ford, F. M. Izrailev, D. L. Shepalyansky, and F. Vivaldi, "The Modulation Diffusion in Nonlinear Oscillator Systems," preprint No 81-70, Institute of Nuclear Physics, Siberian Branch, Academy of Sciences of the USSR (1981).

POLARIZATION OF A STORED ELECTRON BEAM

A. W. Chao
Stanford Linear Accelerator Center
Stanford University, Stanford, California 94305

TABLE OF CONTENTS

0094-243X/82/870395-55$3.00 Copyright 1982 American Institute of Physics

1. SOME QUANTUM MECHANICAL ASPECTS OF SYNCHROTRON RADIATION

As an electron travels in a circular accelerator, it is accelerated sideways and radiates electromagnetic waves known as synchrotron radiation. The phenomenon of synchrotron radiation is a much studied subject. For instance, one finds in textbooks[1,2] that the instantaneous power radiated by a relativistic electron of energy E is given by classical electrodynamics

$$P_{class} = \frac{2}{3}\,\frac{e^2\gamma^4 c}{\rho^2}, \tag{1}$$

with e the electron charge, $\gamma = E/mc^2$ the Lorentz factor, m the electron mass, c the speed of light and ρ the instantaneous bending radius. The frequency spectrum of the radiation is somewhat complicated. It covers more or less all frequencies up to a critical frequency defined by

$$\omega_c = \frac{3\gamma^3 c}{\rho} \tag{2}$$

which is essentially γ^3 times the revolution frequency of the electron.

The above results assume that the electron follows a prescribed circular trajectory and is unperturbed by its radiation. This is a good approximation if the radiation can be regarded as being continuously emitted rather than being emitted as quantized photons as dictated by quantum mechanics. A more accurate picture is in fact to imagine an electron emitting discrete photons as it circulates along. The photon energies are typically around the value ω_c but their exact values are otherwise unpredictable. As a quantum is emitted, the electron receives a recoil. The effective energy of the electron during the emission process is thus not E, but slightly lower than E by an amount comparable to the energy of the quantum. Assuming $\hbar\omega_c \ll E$, this slight reduction in the effective electron energy means the synchrotron radiation power is slightly reduced from expression (1). A quantum mechanical calculation[3] shows in fact

$$P = P_{class}\left(1 - \frac{55}{16\sqrt{3}}\,\frac{\hbar\omega_c}{E}\right) . \tag{3}$$

The correction term to the classical expression (1) is of the order of $\hbar\omega_c/E$. The fact the correction involves the Planck constant $\hbar$ is a distinct sign of quantum mechanical considerations. Equation (3) has included only the leading term linear in $\hbar$; higher order terms have been ignored.

In practical electron storage rings, $\hbar\omega_c/E$ is very small. Take $E = 5$ GeV and $\rho = 25$ meters, for instance, we find $\hbar\omega_c/E = 4.5 \times 10^{-6}$. This means the quantum correction in Eq. (3) is not easy to detect. However, the discreteness of quantum emissions does have an easily detectable effect in practical accelerators. The equilibrium emittances of a bunch of electrons in an electron storage ring, for example, is determined by the balance between a damping effect (the "radiation damping," which is a purely classical phenomenon) and a diffusion effect (the " quantum diffusion," which is a quantum phenomenon) of the electron trajectories.[4,5] Should all electrons radiate continuously, the electron bunch will eventually shrink into a point bunch of zero dimensions due to radiation damping. The discreteness of photon emissions introduces noise into the electron trajectories and causes the beam dimensions to grow by diffusion, which counteracts and balances the damping at equilibrium. The fact that the beam does have a finite size in a storage ring is therefore a quantum mechanical effect. Indeed, if we take the rms energy spread of the beam for example, it does contain a factor of $\hbar$

$$\left(\frac{\Delta E}{E}\right)^2 = \frac{55}{192\sqrt{3}} \frac{\hbar\omega_c}{E} \tag{4}$$

The energy spread is thus of the order of the square root of $\hbar\omega_c/E$. With $\hbar\omega_c/E = 4.5 \times 10^{-6}$ as in the example mentioned above, we find $\Delta E/E = 0.9 \times 10^{-3}$, which is easy to detect.

In addition to the discreteness of photon emissions, there is another quantum mechanical aspect of synchrotron radiation, namely that associated with the spin of the electron. Since spin is a quantum mechanical quantity with its magnitude in units of $\hbar$, all spin effects of synchrotron radiation involve the Planck constant and are necessarily of quantum mechanical origin. The problem becomes complicated when spin is taken into account; in that case one has to distinquish between two cases whether the electron spin stays in its initial state or flips over after emitting the synchrotron photon.

Let $\hat{n}$ be the spin orientation in the electron's rest frame before photon emission. In case of no spin-flip, the main contribution to the synchrotron radiation power emitted is still given by (1), but the correction term now has a spin-dependent term in addition to that given by Eq. (3)[6]

$$P = P_{\text{class}}\left[1 - \left(\frac{55}{16\sqrt{3}} + \frac{1}{2}\,\hat{n}\cdot\hat{y}\right)\frac{\hbar\omega_c}{E}\right], \tag{5}$$

where $\hat{y}$ is the direction of magnetic field that bends the electron. If we average over all spin orientations $\hat{n}$, we obtain Eq. (3) as expected. Again, this spin correction term is very small in practice and is difficult to observe experimentally. However, the other spin effect that involves spin-flips does have an easily observable effect on the electron beam--that on its spin polarization. One can, of course, calculate the instantantous power radiated with spin-flip and compare with Eq. (5), but the more relevant quantity here is the instantaneous transition rate that involves spin-flip,[7]

$$W = \frac{5\sqrt{3}}{16} \frac{e^2 \gamma^5 \hbar}{m^2 c^2 \gamma^3} \left[1 - \frac{2}{9} (\hat{n} \cdot \hat{z})^2 + \frac{8}{5\sqrt{3}} \hat{n} \cdot \hat{y} \right], \qquad (6)$$

which we note is linear in $\hbar$. In (6), $\hat{z}$ is the unit vector in the direction of motion of the electron. Remembering that the power is equal to the transition rate multiplied by the energy carried by each photon $\hbar\omega$, we note the spin-flip power contains a factor of $\hbar^2$ and is smaller than the classical power by a factor of $(\hbar\omega_c/E)^2$, which typically can be 10^{-11}.

In a storage ring, the guiding magnetic field is in the vertical direction $\hat{y}$. If we specify $\hat{n}$ to be either along the field (the up state) or against the field (the down state), we find that the transition rate from up state to down state, $W_{\uparrow\downarrow}$, is larger than that from a down state to an up state, $W_{\downarrow\uparrow}$

$$W_{\uparrow\downarrow} = \frac{5\sqrt{3}}{16} \frac{e^2 \gamma^5 \hbar}{m^2 c^2 \rho^3} \left(1 + \frac{8}{5\sqrt{3}} \right) \qquad (7)$$

$$W_{\downarrow\uparrow} = \frac{5\sqrt{3}}{16} \frac{e^2 \gamma^5 \hbar}{m^2 c^2 \rho^3} \left(1 - \frac{8}{5\sqrt{3}} \right) .$$

If we inject an unpolarized electron beam into a storage ring, the imbalance between the two transition rates will cause the beam to accumulate a net polarization in the direction against the guiding field. In addition, note that $W_{\uparrow\downarrow}$ and $W_{\downarrow\uparrow}$ are not only different, but also so very different that the net polarization can nearly reach 100%, i.e.,

$$P_0 = \frac{W_{\uparrow\downarrow} - W_{\downarrow\uparrow}}{W_{\uparrow\downarrow} + W_{\downarrow\uparrow}} = \frac{8}{5\sqrt{3}} = 92.38\%. \qquad (8)$$

Furthermore, the time constant that this equilibrium polarization is approached by the initially unpolarized beam, inspite of being proportional to $\hbar^{-1}$, is short enough to be practical. The time constant is

$$\tau_0 = (W_{\uparrow\downarrow} + W_{\downarrow\uparrow}) \qquad (9)$$

$$= \left(\frac{5\sqrt{3}}{8} \frac{e^2 \gamma^5 \hbar}{m^2 c^2 \rho^3} \right)^{-1} .$$

Taking again E = 5 GeV and ρ = 25 m, the time constant is found to be 8 minutes. You can now imagine the excitement when it was realized that the electron beam would polarize itself to a high degree; all we have to do is to inject an unpolarized beam into a storage ring and wait a quarter of an hour or so. For once, we seem to be getting something free from mother nature.

If we draw an anology to how equilibrium emittances are established in a storage ring, saying the beam will polarize to the full value of 92% due to spin-flip radiation is the same as saying the beam will shrink into a dimensionless size due to radiation damping. What we have forgotten here is the fact that the discrete photon emissions have introduced noise into the system, and when taken into account, there is a diffusion effect on both the emittances and the spin orientations of the electrons.

The equilibrium value of beam polarization, just like the emittances, must be determined by a balance between the polarizing effect of spin-flip radiation and the depolarizing effect of quantum diffusion. The analogy is illustrated in Table I. In particular, it is necessary to calculate the quantum diffusion rate of spin orientation. We will find then that the pleasant situation of the beam building up 92% polarization all on its own is subject to a stormy environment in a jungle of what is known as the depolarization resonances, near which the spin diffusion rate becomes large and the beam polarization can be much reduced from 92%.

Table I. The Analogy Between the Mechanisms for the Orbital and Spin Equilibrium

	damping ⟷	diffusion	equilibrium
orbital motion	radiation damping ⟷	quantum diffusion on orbit	emittances
spin motion	radiative polarization ⟷	quantum diffusion on spin	beam polarization

The spin diffusion rate has been treated by several authors[8-11] using different methods. The one we shall adopt utilizes the beam transport matrices discussed in Courant's lecture.[4] The difference here is that those matrices, which describe the orbital motions of electrons, will be generalized to include spin motions as well. The advantage of using matrices is that one can put the orbital and the spin motions of an electron on equal footing. The analogy of Table I is then taken care of more easily. Once quantum diffusion is introduced, this matrix formalism provides the calculations for beam emittances and polarization simultaneously. How to develop those matrices and how to use them are also subjects that we want to cover.

2. SEMI-CLASSICAL DESCRIPTION OF SPIN EFFECTS ON SYNCHROTRON RADIATION

Although spin effects are necessarily quantum mechanical, it is possible to derive most of the results of the previous section semi-classically provided we start with an effective Hamiltonian that includes a term that describes the interaction between electron spin and electromagnetic fields. These derivations will be given in this section. The purpose of doing this is not to replace the more rigorous quantum mechanical calculations[7,8] but to do a calculation that avoids the need of explicitly introducing the Dirac equation or the commutation relations of various operators. The procedure of such a calculation has been discussed in the literature[6,12] and what we will do in this Section 2 is to continue such an effort. Readers who are not interested in detailed derivations can skip Sections 2.2 to 2.7.

2.1 Spin Precession in an Electromagnetic Field

Spin of a particle interacts with an electromagnetic field through the magnetic moment associated with the spin. Let $\vec{S}$ be the spin represented as a 3-dimensional vector, the associated magnetic moment is given by

$$\vec{\mu} = \frac{ge}{2mc} \hbar \vec{S}, \tag{10}$$

where g is the gyromagnetic ratio of the particle. For electrons, g is very close to 2. The deviation of g from 2, attributed to an "anomolous" magnetic moment of an electron, is specified by the parameter

$$a = \frac{g-2}{2} \quad . \tag{11}$$

The value of a is approximately given by the fine structure constant 1/137 divided by 2π. More accurately, it is found both theoretically and experimentally that $a = 0.00115965$.

Consider an electron at rest in a magnetic field $\vec{B}$. The precession equation of motion for the spin is

$$\frac{d\vec{S}}{dt} = \vec{\Omega} \times \vec{S} \tag{12}$$

with the precession angular velocity given by

$$\vec{\Omega} = -\frac{ge}{2mc}\vec{B}. \tag{13}$$

Equations (12) and (13) describe the precession for a stationary electron, but we need an equation for a relativistic electron moving in an electromagnetic field $\vec{E}$ and $\vec{B}$. Let $c\vec{\beta}$ be the instantaneous velocity of the electron, it is obvious that we need to make a Lorentz transformation to the electron's rest frame. When doing so, the form of the precession equation remains to be (12); only $\vec{\Omega}$ needs to be transformed. Note that we are not Lorentz transforming $\vec{S}$, so in the final equation, $\vec{S}$ will be a quantity in the electron's rest frame, while all other quantities t, $\vec{E}$ and $\vec{B}$ refer to the laboratory frame. One may find it necessary to stretch his imagination somewhat here. A covariant description does exist,[1] but for our purpose, it is not necessary.

The magnetic field in the rest frame is given by a Lorentz transformation from the laboratory frame

$$\vec{B}_R = \gamma\vec{B}_\perp + \vec{B}_\parallel - \gamma\beta \times \vec{E} \tag{14}$$

with $\vec{B}_\perp$ and $\vec{B}_\parallel$ the components of $\vec{B}$ perpendicular and parallel to $\vec{\beta}$. The angular velocity $\vec{\Omega}$ in the laboratory frame consists of two terms. The first term is

$$-\frac{1}{\gamma}\frac{ge}{2mc}\vec{B}_R, \tag{15a}$$

where we have included a factor of $1/\gamma$ to take care of the time dilation. The second term is due to Thomas precession* which contributes an additional term to the angular velocity when the electron is accelerated sideways

$$-\frac{(\gamma - 1)}{\beta^2}\vec{\beta} \times \dot{\vec{\beta}} = -\frac{e\gamma}{mc(\gamma + 1)}(\vec{\beta} \times \vec{E} - \beta^2\vec{B}_\perp). \tag{15b}$$

---

*Two successive Lorentz transformations along $\vec{\beta}_1$ and $\vec{\beta}_2$ can be combined into one single Lorentz transformation only if $\vec{\beta}_1 \parallel \vec{\beta}_2$. Otherwise, the two Lorentz transformations can be combined into a Lorentz transformation plus a rotation. The additional rotation needed here is the origin of the Thomas precession.

Adding the two terms and substituting Eq. (14) into the result, we obtain

$$\vec{\Omega} = -\frac{e}{mc}\left[\left(a + \frac{1}{\gamma}\right)\vec{B} - \frac{a\gamma}{\gamma + 1}\vec{\beta}(\vec{\beta}\cdot\vec{B}) - \left(a + \frac{1}{\gamma + 1}\right)\vec{\beta}\times\vec{E}\right] \quad (16)$$

which, when substituted into Eq. (12), is called the BMT equation,[14] where BMT stands for Bargman, Michel, and Telegdi.

To describe the spin motion in a storage ring, it is more convenient to change the time variable t into the distance traveled by the electron s = ct. In a storage ring, we apply several types of electric and magnetic fields to confine the electrons. These fields as seen by a circulating electron are periodic in s with the period equal to the circumference $2\pi R$ of the storage ring. Many of those applied fields, such as those provided by quadrupole and sextupole magnets, have effects on a particle only if its trajectory deviates from the designed circular orbit. An ideal electron traveling along the designed orbit sees only the guiding magnetic field and the accelerating electric field. The accelerating field does not cause spin precession on the ideal electron because the electric field is parallel (or anti-parallel, rather, for negatively charged electrons) to the velocity $\vec{\beta}$ and the precession is, according to (16), proportional to $\vec{\beta}\times\vec{E}$. The guiding field $\vec{B} = B_0(s)\,\hat{y}$, with $B_0\,(s + 2\pi R) = B_0\,(s)$, on the other hand, does give rise to a precession

$$\frac{d\vec{S}}{ds} = -\frac{eB_0(s)}{mc^2}\,(a + \frac{1}{\gamma})\,\hat{y}\times\vec{S}. \quad (17)$$

With the precession axis along $\hat{y}$, the y-component of spin S_y is preserved. If we adopt the coordinate system $(\hat{x},\hat{y},\hat{z})$ that rotates with the ideal electron with $\hat{z}$ along the electron's velocity and $\hat{x}$ the horizontal direction, the other two spin components S_x and S_z rotates with the angular speed $a\gamma eB_0/E$ which is $a\gamma$ times the speed that the coordinate system rotates. As the electron completes one revolution, the coordinate system rotates by 2π and the spin has precessed around $\hat{y}$ by an angle $2\pi\,a\gamma$. In analogy to the definitions of tunes ν_x, ν_y and ν_s for the horizontal and the vertical betatron motions and the longitudinal synchrotron motion, we define

$$\text{spin tune} = a\gamma, \quad (18)$$

which can be rewritten as E/0.44065 GeV.

Consider an electron beam polarized initially in a certain direction. As the beam circulates around, only the polarization projection along $\hat{y}$ is preserved; components perpendicular to $\hat{y}$ precess around $\hat{y}$ and since different particles precess with somewhat different rates, rapidly smear out. As a result, if the beam is polarized at all, the equilibrium polarization can only be in the $\hat{y}$ direction. It is also interesting to note the fact that the spin tune involves not the gyromagnetic ratio g but only the anomalous part of g, i.e., g-2 is a consequence of the Thomas precession.

2.2 The Hamiltonian

For a non-relativistic electron in a magnetic field $\vec{B}$, the Hamiltonian is

$$H = \frac{1}{2m}(\vec{p} - \frac{e}{c}\vec{A})^2 - \vec{\mu} \cdot \vec{B}, \tag{19}$$

where $\vec{\mu}$ is the magnetic moment defined in Eq. (11) and $\vec{A}$ is the vector potential associated with $\vec{B}$. In the semiclassical calculation of electromagnetic radiation, one needs the part of Hamiltonian that describes the interaction between the electron and the field $\vec{B}$

$$H_{int} = -e\vec{\beta} \cdot \vec{A} - \vec{\mu} \cdot \vec{B}. \tag{20}$$

We have dropped from (19) the term $\vec{p}^2/2m$ that describes a free electron and the term $e^2\vec{A}^2/2mc^2$ that describes the negligible two-photon processes.

Equation (20) is the Hamiltonian in the non-relativistic limit. To describe the radiation by a relativistic electron, we need the relativistic generalization of Eq. (20). A rigorous derivation of the relativistic semi-classical Hamiltonian should be obtained by making canonical transformations on the Dirac Hamiltonian,[13] but since this is not a course on relativistic quantum mechanics, we shall content ourselves with something less glorious. The first term in (20) does not require extra work; it remains the same relativistically. To see that, we note that the Hamiltonian $[(\vec{p} - (e\vec{A}/c)]^2/2m$ should be replaced by the relativistic counterpart $[m^2c^4 + c^2(\vec{p} - e\vec{A}/c)^2]^{1/2}$ which, up to the linear order in eA, can be written as a free particle term $[m^2c^4 + c^2\vec{p}^2]^{1/2}$ plus the interaction term $-e\vec{\beta} \cdot \vec{A}$. To generalize the second term of (20), we first rewrite it as

$$\hbar\vec{S} \cdot \vec{\Omega}, \tag{21}$$

where $\vec{\Omega}$ is given by the non-relativistic expression (13) and use has been made of Eq. (11). Generalization is then obtained simply by inserting the relativistic expression (16) to replace (13) into $\vec{\Omega}$. Adding the two terms together, the Hamiltonian reads

$$H_{int} = -e\vec{\beta} \cdot \vec{A} - \frac{e\hbar}{mc}\vec{S}$$

$$\cdot \left[(a + \frac{1}{\gamma})\vec{B} - \frac{a\gamma}{\gamma + 1} \vec{\beta}(\vec{\beta} \cdot \vec{B}) - (a + \frac{1}{\gamma + 1}) \vec{\beta} \times \vec{E} \right] \qquad (22)$$

In the non-relativistic limit, (22) reduces to (20) as it should.

2.3 Power and Transition Rate of Synchrotron Radiation

To describe synchrotron radiation, we let $\vec{A}$, $\vec{E}$ and $\vec{B}$ in the interaction Hamiltonian contain, in addition to an external applied field, the field due to radiation. The interaction Hamiltonian then contains two terms: a time-independent term due to external fields and a time-varying term due to the radiation field. The external-field term is grouped with the free particle term to form an "unperturbed" Hamiltonian (unperturbed by radiation field), H_0. The total Hamiltonian is then written and $H_0 + H_{int}$, where H_{int} is given by (22) with the understanding that $\vec{A}$, $\vec{E}$ and $\vec{B}$ only contain the radiation field

$$\vec{A} = \hat{\varepsilon}(\frac{2\pi\hbar c}{k})^{\frac{1}{2}} e^{-i\vec{k} \cdot \vec{r} + i\omega t}, \qquad (23)$$

where $\hat{\varepsilon}$, ω and $\vec{k}$ are the polarization, the frequency and wave vector of the emitted synchrotron photon, respectively. The normalization constant of $\vec{A}$ is chosen so that there is one such photon per unit volume. The complex conjugate of $\vec{A}$ is not included in (23) since it contributes to a photon absorption process that does not concern us here. From Maxwell's equations, we have

$$\vec{E} = -i\, k\, \vec{A}$$

$$\vec{B} = \hat{k} \times \vec{E}. \qquad (24)$$

To find the synchrotron radiation power and transition rate, we use the standard technique of quantum mechanical time-dependent perturbation theory.[13] Let $|n(t)>$ be the n-th eigenstate of the unperturbed Hamiltonian that evolves in time according to $\exp(-iE_n t/\hbar)$. Let the electron be initially in the state $|i(t)>$. Time-dependent perturbation theory says that the probability amplitude that the electron is found in the state $|f(t)>$ after perturbation, to first order of the perturbation strength, is given by

$$C_{fi} = \frac{1}{i\hbar} \int_{-\infty}^{\infty} dt < f(t)| H_{int}(t)| i(t)> . \qquad (25)$$

H_{int} in Eq. (25) is obtained by inserting (23) and (24) into (22)

$$H_{int} = (-e\hat{\varepsilon} \cdot \vec{\beta} + i \frac{e\hbar k}{mc} \vec{S} \cdot \vec{V}) (\frac{2\pi\hbar c}{k})^{\frac{1}{2}} e^{-i\vec{k} \cdot \vec{r} + i\omega t}, \qquad (26)$$

where we have followed Jackson[12] to define

$$\vec{V} = (a + \frac{1}{\gamma}) \hat{k} \times \hat{\varepsilon} - \frac{a\gamma}{\gamma + 1} \vec{\beta}(\vec{\beta} \cdot \hat{k} \times \hat{\varepsilon}) - (a + \frac{1}{\gamma + 1})\vec{\beta} \times \hat{\varepsilon} . \qquad (27)$$

The first term of the interaction Hamiltonian (26) describes a spinless point charge and is independent of $\hbar$ aside from a normalization constant. The second term involves spin and is linear in $\hbar$.

In expressions (25), (26) and (27), we understand $\vec{\beta}$, $\vec{r}$, $\vec{S}$ and H_{int} are quantum mechanical operators. In our semi-classical calculations, however, they will be substituted by their classical values. Consequently, we avoid most of the troubles in taking expectation values between $|i>$ and $|f>$ and we ignore the commutators of the operators. The only exception will be for the spin $\vec{S}$ when spin-flips are involved which we take care of by using the 2 × 2 Pauli matrices.

The differential probability that a photon of polarization is emitted with wave vector between $\vec{k}$ and $\vec{k} + d\vec{k}$ is

$$dp = | C_{fi}|^2 \frac{d^3\vec{k}}{(2\pi)^3} \qquad (28)$$

where the factor $d^3\vec{k}/(2\pi)^3$ is the number of photon states per unit volume. The power dP is given by the probability dp times the photon energy $\hbar\omega$, times the instantaneous frequency of revolution $c\beta/2\pi\rho$. This gives the instantaneous power radiated per unit solid angle, per unit frequency interval, and summed over the two possibilities of photon polarizations

$$\frac{d^2 P}{d\omega d\Omega} = \frac{\hbar\omega^3}{(2\pi)^4 c^2 \rho} \sum_{\hat{\varepsilon}} | C_{fi}|^2. \qquad (29)$$

Substituting explicitly Eqs. (25) and (26) into the above expression, we find

$$\frac{d^2P}{d\omega d\Omega} = \frac{\omega^2}{(2\pi)^3 \rho} \sum_{\hat{\varepsilon}} | I_1 + I_2|^2, \tag{30}$$

where we have defined a spin-independent integral

$$I_1 = \int_{-\infty}^{\infty} dt \left\langle f(t)\left|-e\hat{\varepsilon} \cdot \vec{\beta}\, e^{-i\vec{k} \cdot \vec{r} + i\omega t}\right|i(t)\right\rangle \tag{31}$$

and a spin dependent integral proportional to $\hbar$

$$I_2 = \int_{-\infty}^{\infty} dt \left\langle f\,(t) \left| \frac{ie\hbar k}{mc} \quad \vec{S} \cdot \vec{V}\, e^{-i\vec{k} \cdot \vec{r} + i\omega t}\right| i(t)\right\rangle. \tag{32}$$

Transition rate is, of course, obtained by taking away a factor of from the power

$$\frac{d^2W}{d\omega d\Omega} = \frac{1}{\hbar\omega} \frac{d^2P}{d\omega d\Omega}. \tag{33}$$

2.4 The Classical Limit

The classical result of synchrotron radiation is obtained by ignoring terms involving $\hbar$'s. The integral I_2 is therefore dropped from Eq. (30), and if we do not care about the electron motion after photon emission, the integral I_1 can be replaced by

$$I_1 = -e\hat{\varepsilon} \cdot \int_{-\infty}^{\infty} dt\, \vec{\beta}(t)\, e^{-i\vec{k} \cdot \vec{r}(t) + i\omega t} \tag{34}$$

with $\vec{\beta}(t)$ and $\vec{r}(t)$ now given by their classical values. Calculation of I_1 using Eq. (34) can be found in textbooks. We first note that there is the identity

$$\sum_{\hat{\varepsilon}} (\hat{\varepsilon} \cdot \vec{Z}_1)(\hat{\varepsilon} \cdot \vec{Z}_2) = (\hat{k} \times \vec{Z}_1) \cdot (\hat{k} \times \vec{Z}_2) \tag{35}$$

for any complex vectors $\vec{Z}_1$ and $\vec{Z}_2$. If we consider both $\vec{Z}_1$ and $\vec{Z}_2$ to be the integral that appears in (34), we realize that the quantity to be evaluated in the classical limit is

$$\hat{k} \times \int_{-\infty}^{\infty} dt\, \vec{\beta}(t)\, e^{-i\vec{k} \cdot \vec{r}(t) + i\omega t}. \tag{36}$$

The coordinate system is shown in Fig. 1. The bending field is along $\hat{y}$; θ and ϕ define the direction of photon emission

$$\hat{k} = \hat{z}\cos\theta + \hat{x}\sin\theta\cos\phi + \hat{y}\sin\theta\sin\phi. \tag{37}$$

In the classical limit, the electron motion is unperturbed by radiation and follows a circular path

$$\vec{\beta}(t) = \beta\left(\hat{z}\cos\frac{\beta ct}{\rho} + \hat{x}\sin\frac{\beta ct}{\rho}\right)$$

$$\vec{r}(t) = \rho\left[\hat{z}\sin\frac{\beta ct}{\rho} + \hat{x}\left(1 - \cos\frac{\beta ct}{\rho}\right)\right]. \tag{38}$$

We recall that synchrotron radiation by a relativistic electron almost always is emitted in the direction of electron motion. The angle between the directions of motion of the electron and the photons is of the order of $1/\gamma$. We therefore expect $\theta \lesssim 1/\gamma$. Also, for a given $\hat{k}$, the time interval during which an electron emits a photon in the $\hat{k}$ direction is very short $|\beta ct/\rho| \lesssim 1/\gamma$. What we do is now straight-forward; substitute Eqs. (37) and (38) into (36), keeping only leading terms in $1/\gamma$. If we define, again following Jackson,[12]

$$t = \theta\gamma\sin\phi$$

$$\eta = \frac{\omega}{\omega_C}\left(1 + t^2\right)^{3/2} \tag{39}$$

where ω_c is given by Eq. (2), then the quantity (36) is found to be (a phase factor has been dropped)

$$\frac{2\rho}{\sqrt{3}\,c\gamma^2}\left(1 + t^2\right)^{\frac{1}{2}}\left[\hat{x}t\,K_{1/3}(\eta) + i\hat{y}\left(1 + t^2\right)^{\frac{1}{2}}K_{2/3}(\eta)\right]. \tag{40}$$

It follows that the classical differential radiation power is

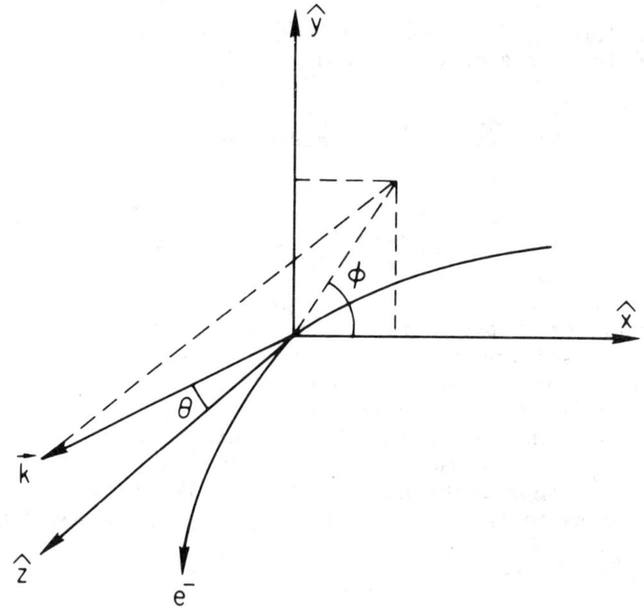

Fig. 1: Relative orientations of the coordinate system, the electron trajectory and the wave vector $\hat{k}$ of an emitted synchrotron photon. The bending magnetic field is in the $\hat{y}$ direction. The polar angle θ is defined relative to $\hat{z}$.

$$\frac{d^2 P_{class}}{d\omega\, d\Omega} = \frac{e^2 \rho\, \omega^2}{6\pi^3 c^2 \gamma^4} (1 + t^2) \left[t^2 K_{1/3}^2 (\eta) + (1 + t^2) K_{2/3}^2 (\eta) \right]. \tag{41}$$

Geometrically, t/γ is the angle between $\hat{k}$ and the orbital plane of the electron. (The modified Bessel functions $K_{1/3}$ and $K_{2/3}$, together with some useful integrals involving them, are given in Table II.) Integrating (40) over ω gives the angular distribution of instantaneous power

$$\frac{dP_{class}}{d\Omega} = \frac{e^2 c \gamma^5}{32\pi\rho^2} \frac{7 + 12\, t^2}{(1 + t^2)^{7/2}}. \tag{42}$$

Table I. Definition and Some Integrals of the Modified Bessel Functions $K_{1/3}$ and $K_{2/3}$

$$\int_{-\infty}^{\infty} du\, e^{i z_0 u + \frac{i}{3} u^3} = 2\left(\frac{z_0}{3}\right)^{\frac{1}{2}} K_{1/3}\left(\frac{2}{3} z_0^{3/2}\right)$$

$$\int_{-\infty}^{\infty} du\, u\, e^{i z_0 u + \frac{i}{3} u^3} = \frac{2i}{\sqrt{3}} z_0 K_{2/3}\left(\frac{2}{3} z_0^{3/2}\right)$$

$$\int_0^{\infty} x^2 K_{1/3}^2 (x)\, dx = \frac{5\pi^2}{144}$$

$$\int_0^{\infty} x^2 K_{2/3}^2 (x)\, dx = \frac{7\pi^2}{144}$$

$$\int_0^{\infty} x^3 K_{1/3}^2 (x)\, dx = \frac{16}{81\sqrt{3}}$$

$$\int_0^{\infty} x^3 K_{1/3}(x)\, K_{2/3}(x)\, dx = \frac{35\pi^2}{864}$$

$$\int_0^{\infty} x^3 K_{2/3}^2 (x)\, dx = \frac{20\pi}{81\sqrt{3}}$$

Making a change of variable

$$\int d\Omega = 2\pi \int_{-\infty}^{\infty} \frac{dt}{\gamma}$$

One can integrate (42) over solid angles. The result is, of course, just Eq. (1).

2.5 Quantum Correction for a Spinless Charge

By quantum correction here we mean correction to the classical results up to first order in $\hbar$. When we wrote down the integrals I_1 and I_2 for the synchrotron radiation power, we were not too careful about the order in which the various operators appeared. Since the commutators of the operators are of the order of $\hbar$, this carelessness is acceptable for I_2, which is already first order in $\hbar$. It is, in fact, also acceptable for I_1 because I_1 is independent of spin and it is only the spin-dependent $\hbar$ correction that we are interested in for later calculations. Nevertheless, one can insist on doing the job right and obtain the quantum correction for a spinless charge (for which I_2 vanishes). This has been done by Schwinger,[3] who showed that the first order correction can be obtained by simply making a replacement

$$\omega \rightarrow \omega \left(1 + \frac{\hbar\omega}{E}\right) \tag{43}$$

in the classical result of (41). One can then integrate the result over frequency to obtain the angular distribution

$$\frac{dP_{class}}{d\Omega} \left[1 - \frac{64}{3\sqrt{3}\,\pi} \frac{\hbar\omega_c}{E} \frac{5 + 9t^2}{(1 + t^2)^{3/2} (7 + 12t^2)} \right] . \tag{44}$$

Integrating again over solid angles gives Eq. (3).

2.6 Radiation Power without Spin-flip

Although the fact that g is not exactly equal to 2 plays an important role in how spin precesses in a storage ring, it is not so essential for the synchrotron radiation of the electron. In the rest of this section, we choose to ignore the difference between g and 2.

Recall that I_1 defined by Eq. (31) can be approximated by a classical integral, Eq. (34). A similar approximation can be made on I_2

$$I_2 = \frac{ie\hbar k}{mc} \int_{-\infty}^{\infty} dt \langle f|\vec{S}(t)|i\rangle \cdot \vec{V}(t) e^{-i\vec{k}\cdot\vec{r}(t) + i\omega t}, \tag{45}$$

where $\vec{V}$, $\vec{r}$ are now classical quantities, $|i\rangle$ and $|f\rangle$ refer to the initial and final spin states of the electron. To find I_2, we need to evaluate $\langle f|\vec{S}(t)|i\rangle$.

Let the electron spin be instantaneously (t = 0) in the direction $\hat{n}$ and define angles θ_0 and ϕ_0 as shown in Fig. 2. Note that θ_0 is defined with respect to $\hat{y}$, while θ of Fig. 1 is defined with respect to $\hat{z}$. We distinguish between two cases according to whether there is a spin flip or not after photon emission. In case of no spin-flip, $\langle f|\vec{S}(t)|i\rangle$ is easy to find. Knowing $\hat{n}$ at time t = 0 and knowing that for g = 2 the spin precesses with the same angular frequency $\omega_0 = |e|B_0/mc\gamma$ as the electron circulates in the field B_0, we find

$$\langle f|\vec{S}(t)|i\rangle = \frac{\hat{x}}{2}\sin\theta_0 \sin(\omega_0 t + \phi_0) + \frac{\hat{y}}{2}\cos\theta_0 + \frac{\hat{z}}{2}\sin\theta_0 \cos(\omega_0 t + \phi_0), \quad (46)$$

where a factor of 1/2 is included on the right hand side because the electron spin is $\hbar/2$.

We insert Eq. (46) and the expression (27) for $\vec{V}$ (remembering we have set g = 2 or equivalently a = 0) into I_2 to obtain

$$I_2 = \frac{ie\hbar k}{2mc}\,\hat{\varepsilon}\cdot\left(\cos\theta_0\,\vec{u}_1 + \frac{1}{2}\sin\theta_0\, e^{i\phi_0}\vec{u}_2 + \frac{1}{2}\sin\theta_0\, e^{-i\phi_0}\vec{u}_3\right), \quad (47)$$

where we have defined three more new symbols

$$\vec{u}_1 = \hat{y}\int_{-\infty}^{\infty} dt\left[\frac{\hat{k}}{\gamma} - \frac{\vec{\beta}(t)}{\gamma+1}\right] e^{-i\vec{k}\cdot\vec{r}(t) + i\omega t}$$

$$\vec{u}_{2,3} = (\hat{z} \mp i\hat{x})\int_{-\infty}^{\infty} dt\left[\frac{\hat{k}}{\gamma} - \frac{\vec{\beta}(t)}{\gamma+1}\right] e^{-i\vec{k}\cdot\vec{r}(t) + i\omega t \pm i\omega_0 t}. \quad (48)$$

In the expression for $u_{2,3}$, the upper signs are for u_2 and the lower signs are for u_3. The reason we factor $\hat{\varepsilon}$ outside of the parentheses in Eq. (47) is so that we can make use of the identity (35).

The $\hbar$ correction to synchrotron radiation power involves, from Eq. (30), the interference between the spin-independent amplitude and the spin-dependent amplitude. Explicitly, it involves the real part of $\sum_{\hat{\varepsilon}} (I_1 I_2^*)$. Making use of Eq. (35), one finds

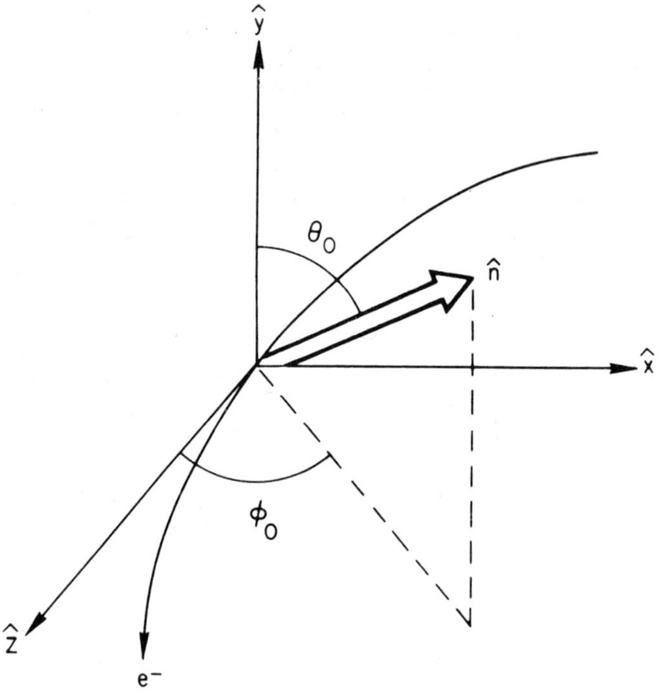

Fig. 2: Relative orientations of the coordinate system, the electron trajectory and the instantaneous direction $\hat{n}$ of electron spin. The bending magnetic field is in the $\hat{y}$ direction. The polar angle θ_0 is defined relative to $\hat{y}$.

$$\sum_{\hat{\varepsilon}} (I_1 I_2^*) = \frac{e^2 \hbar k}{2mc} \left[\hat{k} \times \int_{-\infty}^{\infty} dt\, \vec{\beta}(t)\, e^{-i\vec{k} \cdot \vec{r}(t) + i\omega t} \right]$$
$$\cdot \left[\hat{k} \times \left(\cos\theta_0\, \vec{u}_1^{\,*} + \frac{1}{2} \sin\theta_0\, e^{-i\phi_0} u_2^{\,*} + \frac{1}{2} \sin\theta_0\, e^{i\phi_0}\, \vec{u}_3^{\,*} \right) \right] \quad (49)$$

The quantity in the first pair of square brackets has been evaluated before; it is given by Eq. (40). Similar steps that led to (40) also lead to

$$\hat{k} \times \vec{u}_1^{\,*} = \frac{2}{\sqrt{3}} \frac{\rho \hat{z}}{\gamma^3 c} (1 + t^2)^{\frac{1}{2}} K_{1/3}(\eta)$$

$$\hat{k} \times \vec{u}_{2,3}^{\,*} = -\frac{2}{\sqrt{3}} \frac{\rho}{\gamma^3 c} (1 + t^2)^{\frac{1}{2}}$$
$$\left[(t\hat{y} \mp i\hat{x})\, K_{1/3}(\eta) + i\, \hat{x}\, (1 + t^2)^{\frac{1}{2}} K_{2/3}(\eta) \right] \quad (50)$$

which can be readily substituted into (49) to find the spin-dependent correction to synchrotron radiation power to first order in $\hbar$. If we put this result together with our previous spin-independent results, we obtain

$$\frac{d^2 P}{d\omega\, d\Omega} = \left(\frac{d^2 P_{class}}{d\omega\, d\Omega} \right)_{\omega \to \omega + \frac{\hbar\omega}{E}}$$
$$+ \frac{e^2 \rho^3 \hbar \omega^3}{6\pi^3 mc^4 \gamma^5} (1 + t^2)^{3/2} (-\hat{n} \cdot \hat{y} + 2t\, \hat{n} \cdot \hat{z}) K_{1/3}(\eta) K_{2/3}(\eta). \quad (51)$$

From (51), we have

$$\frac{dP}{d\Omega} = \frac{dP_{class}}{d\Omega} \left[1 - \frac{64}{3\sqrt{3}\,\pi} \frac{\hbar\omega_c}{E} \frac{5 + 9t^2}{(1 + t^2)^{3/2} (7 + 12t^2)} \right.$$
$$\left. + \frac{35}{6} \frac{\hbar\omega_c}{E} \frac{-\hat{n} \cdot \hat{y} + 2t\, \hat{n} \cdot \hat{z}}{(1 + t^2)(7 + 12t^2)} \right] \quad (52)$$

in which the spin-independent terms are those that appeared in Eq. (44). For a longitudinally polarized electron, the term proportional to $\hat{n} \cdot \hat{z}$ gives rise to an up-down asymmetry of synchrotron radiation. With positive helicity ($\hat{n} = \hat{z}$), there is more radiation in the upper plane, while with negative helicity ($\hat{n} = -\hat{z}$), more radiation is found in the lower plane. Integrating over solid angles averages out the up-down asymmetry and we get Eq. (5). We now see an asymmetry with respect to whether the spin is up ($\hat{n} = \hat{y}$) or down ($\hat{n} = -\hat{y}$); more energy is radiated if the electron spin points against the bending magnetic field. As we will see in section 3.5, the $\hat{n} \cdot \hat{y}$ term in Eq. (5) plays a role in determining the beam polarization in an electron storage ring.

2.7 Transition Rate with Spinflip

The spin-independent integral I_1 does not contribute to spin-flip radiation. To evaluate I_2 using Eq. (45), we need first to find the spin transition amplitude $\langle f|\vec{S}(t)|i\rangle$. Unlike the case without spin-flip, this cannot be done in two lines.

Let us choose the spin operator at time t = 0 to be $\vec{S}(0) = \vec{\sigma}/2$ with $\vec{\sigma}$ the Pauli matrices

$$\sigma_x = \begin{bmatrix} 0 & -i \\ i & 0 \end{bmatrix} \qquad \sigma_y = \begin{bmatrix} 1 & 0 \\ 0 & -1 \end{bmatrix} \qquad \sigma_z = \begin{bmatrix} 0 & 1 \\ 1 & 0 \end{bmatrix} \tag{53}$$

(We have made a cyclic permutation upon the more familiar definition of Pauli matrices. This choice is more convenient because our magnetic field $\vec{B}_0$ is in the $\vec{y}$ direction.) The spin operator at other times can be obtained from $\vec{S}(0)$ by considering the precession

$$\begin{aligned} S_x(t) &= \frac{1}{2}(\sigma_x \cos\omega_0 t + \sigma_z \sin\omega_0 t) \\ S_y(t) &= \frac{1}{2}\sigma_y \\ S_z(t) &= \frac{1}{2}(-\sigma_x \sin\omega_0 t + \sigma_z \cos\omega_0 t). \end{aligned} \tag{54}$$

Again let $\hat{n}$ be the direction of the electron spin before radiation. The initial and final states $|i\rangle$ and $|f\rangle$ in the matrix representation are

$$|i\rangle = \begin{bmatrix} \cos\frac{\theta_0}{2}\, e^{-i\phi_0/2} \\ \sin\frac{\theta_0}{2}\, e^{i\phi_0/2} \end{bmatrix} \qquad |f\rangle = \begin{bmatrix} -\sin\frac{\theta_0}{2}\, e^{-i\phi_0/2} \\ \cos\frac{\theta_0}{2}\, e^{i\phi_0/2} \end{bmatrix} \tag{55}$$

which are eigenstates of the operator $\hat{n} \cdot \vec{\sigma}$ with eigenvalues +1 and -1, respectively. Angles θ_0 and ϕ_0 are the same as before (see Fig.

2). The matrix representation of $|f\rangle$ can be obtained from that of $|i\rangle$ by replacing $(\theta_0, \phi_0) \to (\pi - \theta_0, \pi + \phi_0)$. Having obtained (54) and (55), it is straightforward by matrix multiplication to find

$$\langle f|\vec{S}(t)|i\rangle = -\frac{\hat{y}}{2}\sin\theta_0 + \cos^2\frac{\theta_0}{2}\left(\frac{\hat{z}}{2} + i\frac{\hat{x}}{2}\right)e^{-i\phi_0 - i\omega_0 t}$$

$$- \sin^2\frac{\theta_0}{2}\left(\frac{\hat{z}}{2} - i\frac{\hat{x}}{2}\right)e^{i\phi_0 + i\omega_0 t}. \tag{56}$$

We are now in the position to calculate I_2. Equation (45) gives

$$I_2 = \frac{ie\hbar k}{2mc}\hat{\varepsilon}\cdot\left(-\sin\theta_0\vec{u}_1 - \sin^2\frac{\theta_0}{2}e^{i\phi_0}\vec{u}_2 + \cos^2\frac{\theta_0}{2}e^{-i\phi_0}\vec{u}_3\right) \tag{57}$$

using $\vec{u}_{1,2,3}$ already defined in Eq. (48). The next step, by now familiar, is to sum over the photon polarizations $\hat{\varepsilon}$ using the identity (35). After doing so, we get an expression that contains $k \times \vec{u}_{1,2,3}$. Substituting from (50) then yields

$$\frac{d^2W}{d\omega d\Omega} = \frac{e^2\hbar\omega^3\rho}{24\pi^3 m^2 c^6 \gamma^6}(1 + t^2)\left\{\sin^2\theta_0 K^2_{1/3}\right.$$

$$+ 1 + t^2\left(\frac{1 + \cos^2\theta_0}{2}\right)\cdot\left(K^2_{1/3} + K^2_{2/3}\right)$$

$$+ 2\cos\theta_0 (1 + t^2)^{\frac{1}{2}} K_{1/3} K_{2/3} + t\cos\phi_0 \sin 2\theta_0 K^2_{1/3} \tag{58}$$

$$\left. - \frac{1}{2}\cos 2\phi_0 \sin^2\theta_0\left[\left(1 + t^2\right)K^2_{2/3} - \left(1 - t^2\right)K^2_{1/3}\right]\right\}.$$

We have given the transition rate rather than the power. The reason has been explained when we discussed Eq. (7). Integrating over frequency gives

$$\frac{dW}{d\Omega} = \frac{2}{3\sqrt{3}\pi^2}\frac{e^2\hbar\gamma^6}{m^2c^2\rho^3}\frac{1}{(1 + t^2)^5}$$

$$\left[\sin^2\theta_0 + \frac{9}{8}(1 + t^2)(1 + \cos^2\theta_0)\right. \tag{59}$$

$$+ \frac{105\sqrt{3}\pi}{256}\cos\theta_0 (1 + t^2)^{\frac{1}{2}} + t\cos\phi_0 \sin 2\theta_0$$

$$\left. - \frac{1}{8}\cos 2\phi_0 \sin^2\theta_0 (1 + 9t^2)\right]$$

We have kept the five terms in the curly brackets in the same order as we had them in Eq. (58). The fourth term, being proportional to t, gives an up-down asymmetry to spin-flip radiation. This asymmetry disappears if the spin direction is in the xz-plane or in the xy-plane. For example, one does not observe up-down asymmetry if $\hat{n}$ is along $\hat{x}$, or $\hat{y}$ or $\hat{z}$. The total spin-flip transition rate is obtained by integrating (59) over solid angles. Using the fact that $\cos\theta_0 = \hat{n} \cdot \hat{y}$ and $\sin\theta_0 \cos\phi_0 = \hat{n} \cdot \hat{z}$, we discover Eq. (6).

2.8 Radiative Polarization

We briefly mentioned the mechanism for the beam to polarize itself naturally in a storage ring when we discussed equations (7), (8) and (9). We will now do it more systematically.

In the above semi-classical treatments, we have been considering the radiation by a single electron. Polarization, of course, is the net spin of a group of many electrons. Let $\vec{\zeta}$ be the polarization vector. Its direction is along the direction of the net spin and its magnitude ζ ($\leq$ 100%) is the beam polarization. The equation of motion of $\vec{\zeta}$ contains, of course, precession described by the BMT terms, Eq. (16). In addition, it must also take into account the polarizing effect of spin-flip synchrotron radiation. In fact, it even has to include the various depolarization effects so far not yet described. Here, let us consider an idealistic situation in which the electrons form a point bunch of zero emittances and no energy spread; all electrons follow the circular designed trajectory and see only a guiding magnetic field in the vertical direction $\hat{y}$. The only relevant terms are then the BMT precession and the spin-flip transition rate, Eq. (6),

$$\frac{d\vec{\zeta}}{dt} = \frac{(a\gamma + 1)c}{\rho} \quad \hat{y} \times \vec{\zeta} - \frac{1}{\tau_0} \left[\vec{\zeta} - \frac{2}{9} \hat{z} \, (\vec{\zeta} \cdot \hat{z}) + \frac{8}{5\sqrt{3}} \hat{y} \right], \quad (60)$$

where the factor $a\gamma$, we recognize, is the spin tune; c/ρ is the revolution frequency of the electron and τ_0 has been defined in Eq. (9). An additional factor of 2 appears in the transition rate term because in one spin-flip event, polarization changes by 2 units of electron spin.

Admittedly Eq. (60) is somewhat awkward since in the first precession term, we have included, and indeed we must include, the fact that $g \neq 2$. In the second spin-flip term, however, we have insisted to set $g = 2$. The justification is that taking into account of $g \neq 2$ in the second term, in the first place, does not change our final result much (since after all g is very close to 2), and second, the mathematics becomes much more complicated. Those who are interested in the general case for arbitrary g should refer to the literature.[6,12] See also Eq. (65) below.

Let us rewrite (60) in terms of the three components ζ_x, ζ_y, and ζ_z of the polarization in a coordinate system that rotates with the beam.

$$\dot{\zeta}_x = a\gamma \frac{c}{\rho} \zeta_z - \frac{1}{\tau_0} \zeta_x$$
$$\dot{\zeta}_y = -\frac{1}{\tau_0} \left(\zeta_y + \frac{8}{5\sqrt{3}} \right) \tag{61}$$
$$\dot{\zeta}_z = -a\gamma \frac{c}{\rho} \zeta_x - \frac{7}{9\tau_0} \zeta_z$$

Note that ζ_x and ζ_z are coupled while ζ_y is independent. From Eq. (61), we observe that at equilibrium when $\dot{\zeta}_x = \dot{\zeta}_y = \dot{\zeta}_z = 0$, we must have $\zeta_x = \zeta_z = 0$ and $\zeta_y = -8/5\sqrt{3} = -92.38\%$.

In order to get a feeling about how this polarization is reached in time, let us simplify the problem by considering a uniform magnetic field; ρ and τ_0 are then constants. We readily solve ζ_y

$$\zeta_y(t) = \left[\zeta_y(0) + \frac{8}{5\sqrt{3}} \right] e^{-t/\tau_0} - \frac{8}{5\sqrt{3}} . \tag{62}$$

The vertical component of polarization thus approaches its equilibrium with time constant τ_0. To find the time evolution for ζ_x and ζ_z we first note that if we ignore spin precession, ζ_x will approach 0 with a rate τ_0^{-1} while ζ_z will take a slightly lower rate, $7\tau_0^{-1}/9$, to reach its 0. Both rates are very slow compared with the rate $a\gamma c/\rho$ at which ζ_x and ζ_z rotate and mix into each other. It is therefore a good approximation if we replace $1/\tau_0$ in the $\dot{\zeta}_x$ equation and $7/9\tau_0$ in the $\dot{\zeta}_z$ equation by their average value $8/9\tau_0$. After doing so, we can solve ζ_x and ζ_z[8]

$$\zeta_x(t) = \left[\zeta_x(0) \cos \frac{a\gamma ct}{\rho} + \zeta_z(0) \sin \frac{a\gamma ct}{\rho} \right] e^{-8t/9\tau_0}$$
$$\zeta_z(t) = \left[-\zeta_x(0) \sin \frac{a\gamma ct}{\rho} + \zeta_z(0) \cos \frac{a\gamma ct}{\rho} \right] e^{-8t/9\tau_0} . \tag{63}$$

Equations (62) and (63) describe the time evolution of polarization if we inject into a storage ring a beam with initial polarization $\vec{\zeta}(0)$. In particular, if the injected beam is unpolarized, the spin-flip synchrotron radiation will cause the beam to build up its polarization against the field

$$\vec{\zeta}(t) = -\frac{8}{5\sqrt{3}}\hat{y}\,(1 - e^{-t/\tau_0}). \tag{64}$$

Up to now, we have been considering electrons. For positrons, the equilibrium polarization will be parallel to the magnetic field. One may try to draw a more intuitive picture of the effect of polarization build-up. For that, one imagines a magnetic moment $\vec{\mu}$ in a magnetic field $\vec{B}$. Two energy states are generated, one with $\vec{\mu}$ parallel to $\vec{B}$, another with $\vec{\mu}$ anti-parallel to $\vec{B}$. Particles, of course, prefer to stay in the lower energy state, namely the one with $\vec{\mu}$ parallel to $\vec{B}$. One concludes then that electrons must polarize against $\vec{B}$ while positrons are polarized along $\vec{B}$. The difficulty with such a picture has been discussed by Jackson. The two states cannot be regarded as isolated states; orbital motion of the electron or the positron must be considered together with the spin as one coupled system. During the time interval it takes an electron to complete the process of emitting a photon, the electron trajectory has rotated by an angle $\sim 1/\gamma$. In the mean time, the electron spin has precessed by an angle $a\gamma$ times as much, i.e., it has precessed by an angle $\sim a$. In order for the two energy states to be regarded as being isolated, the spin must complete at least one turn of precession during the photon emission process. This is true only if $|a| \gtrsim 2\pi$, or equivalently, $|g| \gtrsim 4\pi$. For electrons and positrons, this is far from being valid. The above intuitive picture remains not too much more than a quick way to memorize the direction of polarization correctly for both electrons and positrons. In fact, even for this limited purpose, the fact that it does work is only accidental. According to this picture, electron polarization will be in the $-\hat{y}$ direction if $g > 0$ and $+\hat{y}$ direction if $g > 0$. The general calculation, which is not only valid for $g = 2$ as we have done, but also valid for arbitrary values of g, shows differently: the electron polarization switches direction between $-\hat{y}$ and $+\hat{y}$ not at $g = 0$ but at $g = 1.198$. More explicitly, let us copy the result for the case of arbitrary g[6,12]

$$\frac{\tau_0(a)}{\tau_0} = \Bigg[(1 + \frac{41}{45}a - \frac{23}{18}a^2 - \frac{8}{15}a^3 + \frac{14}{15}a^4)\,e^{-\sqrt{12}|a|}$$

$$- \frac{8}{5\sqrt{3}}\frac{|a|}{a}(1 + \frac{11}{12}a - \frac{17}{12}a^2 - \frac{13}{24}a^3 + a^4)\,e^{-\sqrt{12}|a|} \tag{65}$$

$$+ \frac{8}{5\sqrt{3}}\frac{a}{|a|}(1 + \frac{14}{3}a + 8a^2 + \frac{23}{3}a^3 + \frac{10}{3}a^4 + \frac{2}{3}a^5)\Bigg]^{-1}$$

$$P_0(a) = -\frac{8}{5\sqrt{3}}\frac{\tau_0(a)}{\tau_0}(1 + \frac{14}{3}a + 8a^2 + \frac{23}{3}a^3 + \frac{10}{3}a^4 + \frac{2}{3}a^5).$$

Plotted in Figs. 3(a) and 3(b) are the values of $\tau_0(a)/\tau_0$ and $P_0(a)$ versus a. For large $|a|$, the magnetic moment is large; the polarization time constant becomes short and the level of polarization approaches 100% as we would expect. Our results, however, correspond only to the values at a = 0. If we insist on using the right value of a = 0.00116, the equilibrium polarization would have been 92.44%, somewhat higher than the value 92.38% we have been talking about; and the polarization time constant would have been shorter by about half a percent.

Since protons have a = 1.793, a proton beam will be fully radiative polarized. The problem is its polarization time; with $\tau_0(1.793)/\tau_0 = 1/116.5$ and τ_0 from Eq. (9), the polarization time is about 3×10^{16} minutes for a 500-GeV ring with 1 Km radius.

3. BEAM POLARIZATION

When we discussed spin precession in the previous section, we mentioned that if an electron follows the designed trajectory exactly, its spin will precess around the vertical direction $\hat{y}$; and if all electrons do so, the net beam polarization direction $\hat{n}$ will have to be along $\hat{y}$. We defined a spin tune as the rate of spin precession and found it is equal to $a\gamma$. Then in section 2.8, we concluded that under this same condition the radiative beam polarization will be 92%. In other words, we showed

$$\hat{n} = \hat{y} \tag{66a}$$

$$\text{spin tune} = a\gamma \tag{66b}$$

and

$$P_0 = 92\%. \tag{66c}$$

We know the assumption that all electrons follow the designed trajectory is never fulfilled because the beam distribution has a finite size. Even if we build a storage ring for which all electric and magnetic devices are constructed and installed perfectly, the designed trajectory is followed only by the center of beam distribution and not by all electrons. One urgent question to be answered is what happens to the polarization properties (66) if we take into account the finite size of the beam.*

3.1 Polarization for a Perfect Storage Ring

Finite beam sizes in an electron storage ring come from the recoil perturbations that electrons receive as they radiate synchrotron photons. Let us define the orbital state of an electron by a vector

*For this discussion, we assume that the perfect storage ring does not have skew quadrupole and sextupole fields. Those fields will be later discussed as error fields.

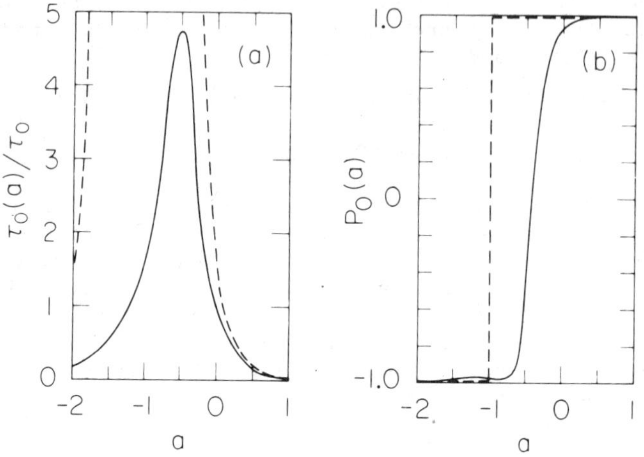

Fig. 3: (a) The characteristic time $\tau_0(a)$ for radiative polarization build-up, normalized by τ_0 of Eq. (9), vs. the magnetic anomaly parameter a. (b) The equilibrium beam polarization $P_0(a)$ vs. the parameter a. Our results of τ_0 and P_0 correspond to the values at a = 0. The dotted curves indicate what one would expect from an intuitive picture that is valid for large $|a|$.

$$X = \begin{bmatrix} x \\ x' \\ y \\ y' \\ z \\ \Delta E/E \end{bmatrix} \quad (67)$$

where x, y and z are the displacements of the electron relative to the center of particle distribution; x' and y' are the corresponding conjugate momenta defined to be the slopes of the electron's trajectory; $\Delta E/E$ is the relative energy deviation. Immediately after radiation, only the $\Delta E/E$-coordinate of the electron is perturbed. As the electron keeps on circulating, this perturbation in $\Delta E/E$ in general propagates into the other five orbital coordinates, giving the beam finite sizes in all six dimensions. In a perfect storage ring, however, the perturbation on $\Delta E/E$ propagates only into the x-, x'- and z-coordinates, leaving y- and y'-coordinates free from being perturbed. As a result, the beam distribution is an infinitely thin ribbon with finite width and length but zero height.

A particle in such a ribbon distribution sees, in addition to the bending magnetic field and the rf accelerating electric field seen along the designed trajectory, the perturbing magnetic fields in the quadrupoles. However, since y = 0, these quadrupole fields are all along $\hat{y}$. If we look at the BMT Eq. (16), we find that the spin precession angular velocity $\vec{\Omega}$ is also along $\hat{y}$. [The second term in Eq. (16) vanishes, the other two terms are along y.] This establishes (66a) since any polarization components perpendicular to y will disappear rapidly due to the different precession phases and rates of different particles. We also find from Eq. (16) that the contributions from quadrupoles and rf cavities oscillate between positive and negative values as x and x' executes betatron and synchrotron oscillations.* As a result, the average rate of spin precession is determined by the bending magnets alone. This establishes (66b). (It is in fact a general result that the spin tune is always determined from the EM field seen along the closed orbit.) But we shall keep in mind that the actual spin precession angle per turn deviates slightly from the average value $2\pi a\gamma$ by an amount that depends on the betatron and synchrotron motions of the electron. This effect of "frequency modulation," as we will see later, is in fact one of the mechanisms that depolarize the beam.

The level of radiative polarization, Eq. (66c), is also unaffected by the finite size of the beam. This is because the rate of spin-flip transition, which is the mechanism responsible for polarization build-up, is proportional to the magnetic field to the cubic power [see Eq. (60)]. The magnetic field in quadrupoles is too weak to have an appreciable effect on radiative polarization build-up. We thus conclude that in a perfectly constructed storage ring, beam polarization satisfies the nice properties listed in (66).

---

*In general, x contains a betatron part and a synchrotron part.

3.2 The Case of a Ribbon Beam

A real storage ring is never perfect. So we want to know what happens to beam polarization if the storage ring contains error fields. Let us first consider two types of error fields: those due to sextupole magnets and those accidental dipole fields that cause a horizontal closed orbit distortion, for example a horizontally misaligned quadrupole magnet. These error fields are special because they do not cause particles to execute vertical motions and the beam keeps its ribbon distribution. The perturbing magnetic fields seen by particles always point in the $\pm\hat{y}$ directions. Most of the previous discussions for a perfect machine still apply. In particular, (66a) follows from the fact that $\vec{\Omega}$ is always along $\pm\hat{y}$ directions. Remembering that the beam distribution center always follows the closed orbit which rotates 2π radians per turn--no matter how distorted the closed orbit may be--and that spin precesses $a\gamma$ times faster than the coordinate does, we find the spin tune is always equal to a , i.e., (66b) is assured. Equation (66c) again follows provided the perturbing fields are weaker than the main bending fields, which is satisfied for almost all practical cases.* We thus conclude that as long as particles do not execute y-motions, the beam (a ribbon beam!) will happily polarize itself according to (66).

3.3 Integer Resonances

Problems occur as soon as we include error fields that cause y excursions in particle motion. One might think these fields are weak and question why should they do any harm. For example, if an electron passes through a quadrupole of strength $G\ell$ = 50 kilogauss off-centered by 1 mm, the spin precesses by an angle of 0.3 mrad, which looks harmless. The answer to this question lies in two facts: (1) the electron passes through this quadrupole not just once but again and again as it circulates around. The innocent-looking 0.3 mrad may add up every time the electron passes through the quadrupole. The conditions for those small spin rotations to add up are referred to as the depolarization resonance conditions. (2) Even more importantly, the strengths of some of those depolarization resonances are greatly enhanced due to the presence of a noise source--the synchrotron radiation. The enhancement factor involved is typically as large as 10^6.

Before we go on to discuss depolarization effects, let me make a comment here. In storage rings typical error fields coming from, for example, magnet misalignments are proportional to beam energy E. (This is because strengths of all magnets scale with E.) This means as a particle passes through the error field, its angular deflection θ

---

*That is all except one. If we insert a wiggler device--a series of bending magnets with alternating positive and negative polarities in the storage ring, the beam remains ribbon-shaped but the associated "error" fields are strong enough to have an appreciable effect on polarization level. In fact, it always makes the polarization lower. See problem 6.

is independent of E. The same thing does not happen for spin; it is perturbed by an angle $a\gamma\theta$ which is proportional to E. In other words, the higher the beam energy is, the more sensitive are the particle spins to magnet misalignments and therefore the more vulnerable is the beam polarization to the depolarization resonances.

One type of depolarization resonances occurs when the spin tune $a\gamma$ is close to an integer. To see that, let us start with Eq. (66a), i.e., the polarization direction $\hat{n}$ is along the vertical direction $\hat{y}$. The reason Eq. (66a) is important is that the radiative polarization built up painstakingly by the spin-flip synchrotron radiation is along $\hat{y}$. If $\hat{n} \neq \hat{y}$, the beam will keep only the polarization component along $\hat{n}$ and the net beam polarization will be reduced by a cosine factor $\hat{n} \cdot \hat{y}$. Clearly, one loses polarization if $\hat{n}$ deviates appreciably from $\hat{y}$.*

Consider a particle at the center of beam distribution. It follows the closed orbit and sees an external EM field that is periodic in s with period $2\pi R$. Its spin therefore precesses with a periodic angular velocity. Starting with s, we can integrate this angular velocity through one turn to obtain a net rotation on spin. If the spin is represented as a 3-dimensional vector

$$\vec{S} = \begin{bmatrix} S_x \\ S_y \\ S_z \end{bmatrix}, \tag{68}$$

the net rotation can be written as a 3×3 matrix R(s). The beam polarization direction $\hat{n}(s)$ is then given by the rotational axis of R(s) (use right-hand rule)

$$R(s)\,\hat{n}(s) = \hat{n}(s). \tag{69}$$

If the particles do not execute y-motions (the ribbon beam case), R(s) is simply a rotation about $\hat{y}$

$$R(s) = \begin{bmatrix} \cos 2\pi a\gamma & 0 & \sin 2\pi a\gamma \\ 0 & 1 & 0 \\ -\sin 2\pi a\gamma & 0 & \cos 2\pi a\gamma \end{bmatrix}. \tag{70}$$

The rotational axis of (70) is $\hat{n}(s) = \hat{y}$, which, of course, is just

* Strictly speaking, such a loss of polarization is not a "depolarization" mechanism. It is rather a "lack of polarization."

(66a). The calculation of R(s) and $\hat{n}(s)$ for the general case will be described in detail later. Here let us consider a somewhat idealized case in which the spin precession from s = 0 to s = $2\pi R$ is given by Eq. (70) but at s = 0 there is a perturbing magnetic field B_x along the horizontal direction. Such a field may come from a vertical closed orbit distortion at a quadrupole. The spin rotation across the perturbing field is described by

$$\begin{bmatrix} 1 & 0 & 0 \\ 0 & \cos\theta & \sin\theta \\ 0 & -\sin\theta & \cos\theta \end{bmatrix} \qquad (71)$$

with $\theta = (1 + a\gamma)\, B_x \ell / B\rho$. The total rotation matrix for one revolution is given by the product of (70) and (71). It can easily be shown that the corresponding rotational axis $\hat{n}(s)$ everywhere outside the perturbing field region has the cosine factor $\hat{n} \cdot \hat{y} = 1/(1 + \tan^2\theta/2 \csc^2\pi a\gamma)^{\frac{1}{2}}$. It follows that the beam polarization vanishes ($\hat{n} \cdot \hat{y} = 0$) on an "integer resonance," i.e., when the spin tune $a\gamma$ is equal to an integer k. One can also calculate the width in $a\gamma - k$ within which beam polarization is significantly reduced. The width is found to be of the order of $\theta/2\pi$. Taking again θ = 0.3 mrad as a typical value from our numerical example mentioned before, the resonance width is about 0.5×10^{-4}, which is much narrower than the spacing between the integer resonances. This means integer depolarization resonances are easy to avoid. Furthermore, the integer resonances depolarize the beam through the cosine factor $\hat{n} \cdot \hat{y}$. Unlike other depolarization resonances to be mentioned later, they are not enhanced by the noise due to synchrotron radiation. We thus expect that integer resonances are not a serious problem in storage rings.

3.4 Sideband Resonances

In the previous idealized example, we have followed a particle at the beam center to obtain the polarization direction $\hat{n}$. Consider now instead a particle that executes a horizontal betatron oscillation x_β. The part of precession described by Eq. (70) needs to be modified; the angle $2\pi a\gamma$ now contains an additional term that is proportional to the betatron amplitude x_β and is oscillatory with the betatron tune ν_x. One might say that the spin precession motion is "frequency modulated" by the x_β-motion. A result of such a frequency modulation is the occurrence of frequency sidebands. In other words, to first order in x_β, the system now contains, in addition to the natural frequency $a\gamma$, two more frequencies $a\gamma \pm \nu_x$. If we now introduce the perturbation (71), the spin motion will be seriously influenced if $a\gamma \pm \nu_x$ is equal to an integer k.

A similar thing happens if the electron executes a synchrotron oscillation. The spin precession motion described by Eq. (70), is frequency modulated by the synchrotron motion at the synchrotron tune ν_s. Two sidebands at $a\gamma \pm \nu_s$ occur and the spin motion is seriously influenced by the perturbation (71) if $a\gamma \pm \nu_s = k$.

The spin motion is also strongly perturbed at the vertical betatron sidebands $a\gamma \pm \nu_y = k$. The mechanism, however, is different from the frequency modulation mechanism for the previous cases. In the idealized example, the source of the problem is now not Eq. (70) but Eq. (71). As the particle executes a y_β-oscillation, the magnetic field it experiences at the quadrupole contains two terms: the static B_x that causes the spin to precess according to Eq. (71) and an additional B_x that oscillates with y_β. One might now say that the simple harmonic spin precession is "driven" by an oscillatory driving force every time the electron passes through the quadrupole. If the frequency ν_y of the driving force and the natural simple harmonic frequency $a\gamma$ satisfy the resonance condition $a\gamma \pm \nu_y = k$, we expect a strong response of spin to the driving.

Once we deviate from our idealized case, the situation rapidly becomes complicated. For example, if there is a skew quadrupole field somewhere, it produces a perturbation (71) when the electron has an x-displacement. The resonance driving mechanism now also applies to the $a\gamma \pm \nu_x = k$ and the $a\gamma \pm \nu_x = k$ sidebands. One can also imagine that the simple harmonic precession will be frequency modulated by y_β-motion if there are vertical bending dipoles in the storage ring. It is clear that studying these effects case by case is cumbersome, if not impossible. What is needed is a general, more formal description, which we will offer in section 3.7.

I have not yet explained the role of synchrotron radiation in enhancing the depolarization resonances. Imagine an electron following the closed orbit with its spin $\vec{S}$ happily polarized along $\hat{n}$. Now suddenly it emits a photon of energy δE at time $t = 0$. After the emission the electron starts to execute orbital oscillations around the closed orbit. The oscillations can be decomposed into three modes, which we somewhat loosely refer to as the horizontal and vertical betatron modes and the synchrotron mode. We know that these excited orbital oscillations are damped by radiation damping. The damping times τ_{rad} for the three modes are somewhat different but they are all comparable, typically about several milliseconds.

A few τ_{rad} after the radiation, the electron damps to the closed orbit and quiets down again. Meanwhile, $\vec{S}$ starts to precess away from n due to the perturbing EM fields seen away from the closed orbit. Similar to the orbital motion, this excited spin motion will also quiet down. The time constant, however, is not τ_{rad} but the polarization time constant τ_0 given by Eq. (9), which is typically at least several minutes. We have illustrated in Figs. 4(a)-(d) the spin motion during this whole process.

The perturbing EM field that acts on the spin from t = 0 to t = a few τ_{rad} is oscillatory with frequencies ν_x, ν_y and ν_s. This field perturbs the spin through both the frequency modulation and the driving mechanisms mentioned before. In case the spin tune $a\gamma$ is such that one of the sideband conditions is fulfilled or nearly fulfilled, this photon emission event will destroy the polarization of this electron. One can imagine doing a calculation of the widths of the sideband resonances just like we did for the integer resonances. Within the widths, the angle θ of Fig. 4 is of the order of 1 radian. One then probably finds that the widths are very narrow and concludes that sideband resonances are not a serious problem for beam polarization. What happens, however, is that photons are constantly being emitted. Staying outside of such a resonance width does not necessarily guarantee a good polarization. For example, if each photon emission causes the spin to deviate from $\hat{n}$ by an angle θ of, say, 10^{-6} rad, then the spin will random-walk away from $\hat{n}$ in about 10^{12} emissions. For a 5-GeV storage ring of 25m radius, this means depolarization time of about 10 minutes. To guarantee good polarization, the depolarization time must be longer than the polarization time. This means one must stay away from the sideband resonance far enough so that θ is less than something like 10^{-6} rad rather than 1 rad. Synchrotron radiation thus greatly enhances the sideband depolarization resonances.

Since τ_{rad} is so much shorter than τ_0, one can ignore the time period $0 < \tau \lesssim \tau_{rad}$ [Fig. 4(b)] as far as spin polarization is concerned. For t < 0, we have $\vec{S} - \hat{n} = 0$. For t > 0, the deviation of $\vec{S}$ from $\hat{n}$ is proportional to the perturbation $\delta E/E$. If we extrapolate the spin precession motion of Fig. 4(c) backwards in time to the moment of emission, t = 0, we can write

$$\vec{S} - \hat{n} = \frac{\delta E}{E} \cdot \gamma \frac{\partial \hat{n}}{\partial \gamma} \quad \text{at } t = 0, \tag{72}$$

where we have defined a proportionality vector $\gamma \partial\hat{n}/\partial\gamma$.

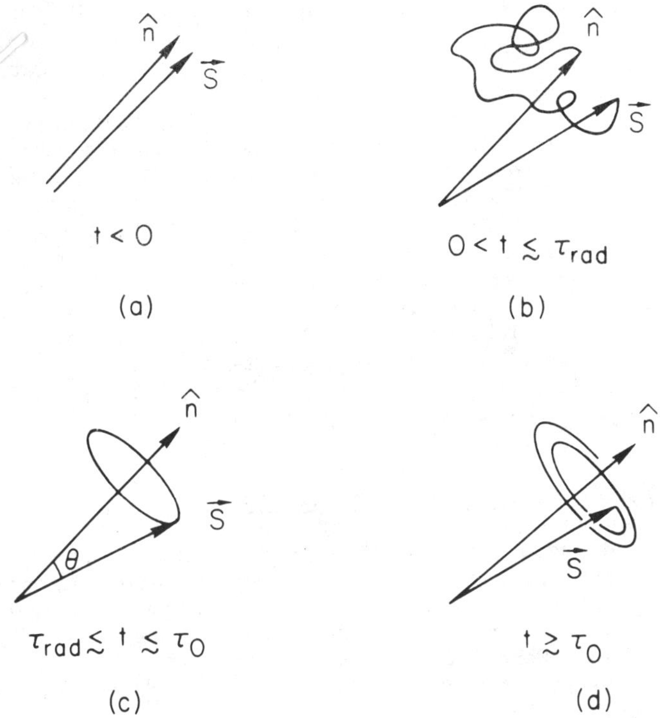

Fig. 4: The motion of an electron spin $\vec{S}$ following the sudden emission of a synchrotron photon of energy δE. (a) Before emission ($t < 0$), the electron is polarized with $\vec{S}$ along the direction $\hat{n}$ of the net beam polarization. (b) Photon emission excites the orbital motions of the electron, which cause the electron to see some perturbing EM fields. After the emission ($t > 0$) and before the orbital motions are radiation damped ($t \lesssim \tau_{rad}$), $\vec{S}$ precesses according to the perturbing fields in some complicated manner. The radiation damping time τ_{rad} is typically several milliseconds. (c) After the orbital motions are damped ($t \gtrsim \tau_{rad}$), $\vec{S}$ sees no perturbing fields and starts to execute a simple precession motion around $\hat{n}$. The angle θ is an important parameter that determines the strength of depolarization due to synchrotron radiation. If $\theta \gtrsim 10^{-6}$, one expects loss of polarization. (d) The precessing $\vec{S}$ slowly spirals in towards $\hat{n}$ due to the polarizing effect of synchrotron radiation. Significant spiralling occurs after a time τ_0 given by the polarization time, typically at least several minutes. A few τ_0 later, $\vec{S}$ damps to $\hat{n}$. The excitation-damping process (a) to (d) is repeated every time a synchrotron photon is emitted.

The vector $\gamma\partial\hat{n}/\partial\gamma$ is a crucial quantity in determining the radiative polarization of the beam. It is a 2-dimensional vector perpendicular to $\hat{n}$. For a ribbon beam, perturbations due to synchrotron radiation are decoupled from the spin motion and we have $\gamma\partial\hat{n}/\partial\gamma = 0$. In general, it is a vector completely determined by the storage ring lattice depending only on the location s where the photon is emitted, independently of synchrotron radiation and spin. Following Buon,[15] we shall call $\gamma\partial\hat{n}/\partial\gamma$ the "spin chromaticity." The notation used here follows that of Derbenev, Kondrateno and Skrinsky.[9] It should be mentioned that although this notation suggests more or less its physical meaning, it is not to be taken too literally to mean the partial derivative of $\hat{n}$ relative to γ. Note also that the angle θ shown in Fig. 4 is equal to $|(\delta E/E)\gamma\partial\,\hat{n}/\partial\gamma|$. It specifies the random walk step-size of quantum diffusion on spin motion.

3.5 Determining the Beam Polarization

We assume that the storage ring fields, including the error fields, and the associated closed-orbit distortion are known. From this information, one can obtain the polarization direction n(s) and the spin chromaticity $\gamma\partial\hat{n}/\partial\gamma(s)$ around the storage ring (see Fig. 5). A matrix formulation will be described in sections 3.6 and 3.7 for this purpose. Here let us assume $\hat{n}$ and $\gamma\partial\hat{n}/\partial\gamma$ are already known and we will look for an expression of beam polarization in terms of these quantities.

Consider an unpolarized beam stored at time $t = 0$. Due to synchrotron radiation, with all the polarizing as well as depolarizing effects, the beam slowly acquires a polarization $\zeta(t)\,\hat{n}$ along $\hat{n}$. We expect $\zeta(t)$ to approach an equilibrium value P with a time constant τ. The ideal case has been worked out in section 2.8. Here we want to find the general expressions for P and τ.

We start with Eq. (60). The first term in (60) describes the precession motion. For a polarization along $\hat{n}$, it can be dropped since $\hat{n}$ is the precession axis. The second term comes from spin-flip radiation. It, of course, must be kept and we have

$$\dot{\zeta}(t) = -\frac{1}{\tau_0}\left[\zeta(t) - \frac{2}{9}\zeta(t)\ \ (\hat{n}\cdot\hat{z})^2 + \frac{8}{5\sqrt{3}}\,\hat{n}\cdot\hat{y}\right], \qquad (73)$$

where $\hat{z}$ is along the beam motion, $\hat{y}$ is along the bending magnetic field.

Since we expect the polarization to be very slowly changing, it is a good approximation to average the right-hand side of Eq. (73) over the circumference of the ring. Inserting τ_0 from Eq. (9), this gives

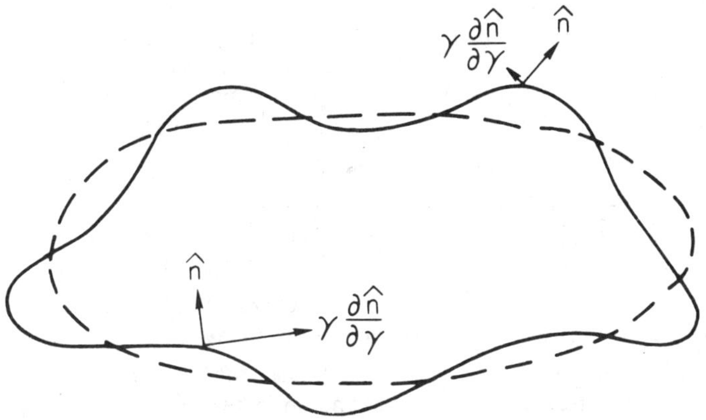

Fig. 5: A schematic drawing of the direction of polarization $\hat{n}$ and the spin chromaticity $\gamma\partial\hat{n}/\partial\gamma$. The dotted line indicates the designed trajectory. The solid line is the distorted closed orbit. Note that $\hat{n}(s)$ is a unit vector but the magnitude of $\gamma\partial\hat{n}/\partial\gamma(s)$ varies with s.

$$\dot{\zeta}(t) = -\frac{5\sqrt{3}}{8}\frac{e^2\gamma^5\hbar}{m^2c^2}\frac{1}{2\pi R}\left\{\zeta(t)\oint ds\,\frac{\left[1 - \frac{2}{9}(\hat{n}\cdot\hat{z})^2\right]}{|\rho|^3} + \frac{8}{5\sqrt{3}}\oint ds\,\frac{\hat{n}\cdot\hat{y}}{|\rho|^3}\right\} \tag{74}$$

Eq. (74) is incomplete; it contains only the polarizing effect. We will have to include more terms coming from the spin chromatic effects due to a nonzero $\gamma\partial n/\partial\gamma$. Let us, however, ignore $\gamma\partial n/\partial\gamma$ for a short moment. The equilibrium level of polarization would then be given by

$$-\frac{8}{5\sqrt{3}}\frac{\oint ds\,\hat{n}\cdot\hat{y}/|\rho|^3}{\oint ds\left[1 - \frac{2}{9}(\hat{n}\cdot\hat{z})^2\right]/|\rho|^3} \tag{75}$$

The factor $\hat{n}\cdot\hat{y}$ is the cosine reduction factor mentioned when we discussed integer depolarization resonances. The (less important) factor $\left[1 - 2/9\,(\hat{n}\cdot\hat{z})^2\right]$ in the denominator comes from the slight dependence of spin-flip radiation on the z-component of electron spin.

The effects of spin chromaticity $\gamma\partial\hat{n}/\partial\gamma$ are associated with synchrotron radiation without spin-flips. Consider an electron polarized along $\hat{n}$. As a photon of energy δE is emitted, its spin starts to precess around $\hat{n}$ with a small rotating deviation $\vec{d}$. After emission, the electron has lost a polarization

$$\frac{\Delta P}{P} = -\frac{1}{2}\,|\vec{d}|^2 = -\frac{1}{2}\left|\frac{\delta E}{E}\,\gamma\,\frac{\partial\hat{n}}{\partial\gamma}\right|^2 . \tag{76}$$

Let $\dot{N}$ be the number of photon emissions per unit time, we obtain the quantum diffusion rate on polarization

$$\dot{\zeta}(t) = \frac{\zeta(t)}{2\pi R}\oint ds\,\left(-\frac{1}{2}\right)\left\langle \dot{N}\frac{\delta E^2}{E^2}\right\rangle_s \left|\gamma\frac{\partial\hat{n}}{\partial\gamma}\right|_s^2 , \tag{77}$$

where we have averaged over s and $\left\langle \dot{N}\delta E^2/E^2\right\rangle$ is given by integrating $\hbar\omega d^2P/d\omega d\Omega$ [see Eq. (40)] over ω and Ω

$$\left\langle \dot{N}\frac{\delta E^2}{E^2}\right\rangle_s = \frac{55}{24\sqrt{3}}\,\frac{\hbar e^2\gamma^5}{m^2c^2|\rho^3(s)|} . \tag{78}$$

There is another effect on polarization due to the spin chromaticity. Consider now an electron that is not perfectly polarized before radiation. Let its spin be $\hat{n} + \vec{D}$ with $\vec{D}$ a small rotating vector orthogonal to $\hat{n}$. Now the electron emits a photon of energy δE. After emission, the spin acquires another rotating deviation $\vec{d}$. Let

$\vec{D}_0$ and $\vec{d}_0$ be the values of $\vec{D}$ and $\vec{d}$ extrapolated to the moment of emission. If δE does not depend on $\vec{D}_0$, i.e., if the synchrotron radiation does not depend on the instantaneous spin, $\vec{d}_0$ is uncorrelated with $\vec{D}_0$ and we simply have observed a random walk in spin motion. The story is quite different if δE does depend on $\vec{D}_0$. Then $\vec{d}_0$ correlates with $\vec{D}_0$ and the original amount of polarization will decrease or increase according to how $\vec{d}_0$ and $\vec{D}_0$ are correlated.

More quantitiavely, the polarization of the electron before and after the radiation are equal to $1 - 1/2|\vec{D}_0|^2$ and $1 - 1/2|\vec{D}_0 + \vec{d}_0|^2$, respectively. Change of polarization due to the radiation is therefore

$$\Delta P = -\vec{D}_0 \cdot \vec{d}_0 - \frac{1}{2}|\vec{d}_0|^2. \tag{79}$$

The second term in (79) is the random walk term already discussed. Summing up on photon emission events, the contribution of the first term in (79) to the polarization process is found to be

$$\begin{aligned} \dot{\zeta}(t) &= \langle -\dot{N}\,\vec{D}_0 \cdot \vec{d}_0 \rangle \\ &= -\frac{\langle \dot{N}\delta E \rangle}{E}\,\vec{D}_0 \cdot \gamma\frac{\partial \hat{n}}{\partial \gamma}, \end{aligned} \tag{80}$$

where, as before, $\langle\ \rangle$ means averaging over the radiation spectrum. In addition, an averaging over s is understood. Expression for $\langle \dot{N}\delta E \rangle$ has been obtained before; it is given by Eq. (5). The spin-independent terms in Eq. (5) do not concern us here. Keeping only the spin-dependent term gives

$$\langle \dot{N}\delta E \rangle = -\frac{1}{2} P_{class}\, \vec{S} \cdot \hat{y}\, \frac{\hbar\omega_c}{E}, \tag{81}$$

where the spin direction in Eq. (5) has been replaced by the instantaneous value $\vec{S} = \hat{n} + \vec{D}_0$. Also since $\hat{n}$ is perpendicular to $\gamma\partial\hat{n}/\partial\gamma$, the vector $\vec{D}_0$ in Eq. (80) can be replaced by $\vec{S}$. Substituting Eq. (81) into Eq. (80) yields

$$\dot{\zeta}(t) = \frac{\hbar\omega_c}{2E^2}\, P_{class}\, (\vec{S} \cdot \hat{y})\, (\vec{S} \cdot \gamma\,\frac{\partial n}{\partial \gamma}). \tag{82}$$

Since $\vec{S}$ can in principle point in any arbitrary direction, the next step is to average over its solid angles, keeping its magnitude constant. When this is done, the factor $(\vec{S} \cdot \hat{y})(\vec{S} \cdot \gamma\partial\hat{n}/\partial\gamma)$ becomes $1/3\ (\hat{y} \cdot \gamma\partial\hat{n}/\partial\gamma)\ |\vec{S}|^2$. Now the question is what to use for $|\vec{S}|^2$. One may argue that since $\vec{S}$ is the unit spin direction, it obviously has $|\vec{S}|^2 = 1$. The correct answer, however, is $|\vec{S}|^2 = 3$, which comes from the fact that the magnitude of the electron spin must be determined from the quantum mechanical relation $|1/2\hbar\vec{S}|^2 = 1/2\ (1/2 + 1)\hbar^2$. Thus, Eq. (82) becomes, after averaging over $\vec{s}$

$$\dot{\zeta}(t) = \frac{1}{2\pi R} \oint ds \ \frac{\hbar\omega_c}{2E^2}\ P_{class}\ \hat{y} \cdot \gamma \frac{\partial\hat{n}}{\partial\gamma} \tag{83}$$

We have now obtained three separate contributions to $\zeta(t)$; they are given by Eqs. (74), (77) and (83). Adding them up gives the final expression obtained by Derbenev and Kondratenko

$$\dot{\zeta}(t) = -\frac{5\sqrt{3}}{8} \frac{e^2\gamma^5\hbar}{m^2c^2} \left[\alpha_+ \ \zeta(t) - \frac{8}{5\sqrt{3}} \alpha_- \right], \tag{84}$$

where

$$\alpha_+ = \frac{1}{2\pi R}\oint \frac{ds}{|\rho(s)|^3} \left[1 - \frac{2}{9}(\hat{n} \cdot \hat{v}) + \frac{11}{18}\left|\gamma \frac{\partial\hat{n}}{\partial\gamma}\right|^2 \right]_s$$

$$\alpha_- = \frac{1}{2\pi R}\oint \frac{ds}{|\rho(s)|^3} \left[\frac{\dot{\hat{v}} \times \hat{v}}{|\dot{\hat{v}}|} \cdot (\hat{n} - \gamma\frac{\partial\hat{n}}{\partial\gamma}) \right]_s \ . \tag{85}$$

In Eq. (84), the symbol for the instantaneous direction of beam motion has been changed (hopefully for clarity) from $\hat{z}$ to $\hat{v}$; the symbol for the magnetic field direction has been changed from $\hat{y}$ to $-\dot{\hat{v}} \times \hat{v}/|\dot{\hat{v}}|$ with $\dot{\hat{v}}$ along the direction of acceleration. The later change of symbol has the advantage that it also takes care of positrons. From Eq. (84), it follows that the equilibrium beam polarization is equal to

$$P = \frac{8}{5\sqrt{3}} \ \frac{\alpha_-}{\alpha_+} \ , \tag{86}$$

and the time constant to reach the equilibrium is

$$\tau = \left(\frac{5\sqrt{3}}{8} \frac{e^2\gamma^5\hbar}{m^2\ c^2} \ \alpha_+ \right)^{-1} . \tag{87}$$

In the case of a perfect planar storage ring, Eqs. (86) and (87) reduce to the results of section 2.8. An indication of why the symbols α_+ and α_- were chosen can be found by comparing the expressions (86) and (87) with (8) and (9).

We have thus obtained an expression for the equilibrium level of beam polarization. The integer resonances $a\gamma = k$ show their effect in causing $\hat{n}$ to deviate from $\hat{y}$. They are not enhanced by synchrotron radiation noise. The sideband resonances $a\gamma \pm \nu_{x,y,s} = k$, on the other hand, cause the spin chromaticity to become large. They are enhanced by synchrotron radiation noise and are responsible for most of the loss of beam polarization in electron storage rings.

An inspection of Eqs. (85) and (86) shows that the spin chromaticity appears as a quadratic term in the denominator of P and only linearly in the numerator. Loss of polarization occurs if $|\gamma\partial\hat{n}/\partial\gamma| \gtrsim 1$; in this case, the angle θ of Fig. 4(c) will be bigger than $\delta E/E$, which is of the order of $\hbar\omega_c/E$. The spin chromaticity term $\hat{y} \cdot \gamma\partial\hat{n}/\partial\gamma$ in α_- is small for most practical cases. This follows from the fact that $\hat{n}$ is nearly equal to $\hat{y}$ and that $\gamma\partial\hat{n}/\partial\gamma$ is perpendicular to $\hat{n}$. Finally, skeptical readers who wonder if P, as given by Eq. (86), could be larger than unity (then something is obviously wrong!) should work out problem 9.

3.6 The Polarization Direction $\hat{n}(s)$

We assume that the 6-dimensional closed-orbit vector $X_0 = (x_0, x_0', y_0, y_0', z_0, \delta_0)$ in the presence of various error fields has been obtained around the storage ring. From the electric and magnetic field along the closed orbit, one obtains from Eq. (16) the angular velocity $\vec{\Omega}(X_0)$. Adopting the thin-lens approximation, we let $\vec{\Omega}(X_0)$ to be uniform in a given lattice element. The matrix which transforms the spin components (68) as the particle travels through a distance s in a uniform EM field is given by

$$\begin{bmatrix} \alpha^2(1 - C) + C & \alpha\beta(1 - C) - \gamma S & \alpha\gamma(1 - C) + \beta S \\ \alpha\beta(1 - C) + \gamma S & \beta^2(1 - C) + C & \beta\gamma(1 - C) - \alpha S \\ \alpha\gamma(1 - C) - \beta S & \beta\gamma(1 - C) + \alpha S & \gamma^2(1 - C) + C \end{bmatrix}, \quad (88)$$

where α, β and γ are the direction cosines $\hat{\Omega} \cdot \hat{x}$, $\hat{\Omega} \cdot \hat{y}$ and $\Omega \cdot \hat{z}$; and $C = \cos(\Omega s)$, $S = \sin(\Omega s)$. Knowing $\vec{\Omega}(X_0)$, one can find the 3×3 matrix which transforms the spin components through a given lattice element.

One then multiplies all 3×3 matrices successively to obtain the total spin precession transformation R_{tot} for one revolution around $s = 0$. A right-handed orthonormal base $(\hat{n}, \hat{m}, \hat{\ell})$ with $\hat{n}$ rotation axis of R_{tot} is then chosen. Successive transformations bring this base to other positions. In one revolution, $\hat{n}$ comes back to its starting value; but $\hat{m}$ and $\hat{\ell}$ have rotated around $\hat{n}$ by an angle $2\pi\nu$, where $\exp(\pm i2\pi\nu)$ are the two nontrivial eigenvalues of R_{tot}. The quantity ν gives the spin precession tune and $\hat{n}$ gives the direction of beam polarization. For a storage ring with planar geometry and without error fields, ν is equal to $a\gamma$ and $\hat{n}$ is along $\hat{y}$. For rings with error fields, $\nu \simeq a\gamma$ and $\hat{n} \simeq \hat{y}$ to a high degree of accuracy provided $a\gamma$ is a distance $\gtrsim 10^{-3}$ away from integers.

3.7 Spin Chromaticity $\gamma\partial\hat{n}/\partial\gamma$

We assume that the closed orbit X_0 and the spin base vectors $(\hat{n}, \hat{m}, \hat{\ell})$ are now obtained. The spin of a nearly polarized electron is written as

$$\vec{S} \simeq \hat{n} + \alpha\hat{m} + \beta\hat{\ell}, \quad |\alpha,\beta| << 1. \tag{89}$$

The quantities α and β thus describe the spin to a linear approximation and $1/2\ (\alpha^2 + \beta^2)$ specifies the degree of depolarization of this electron. The assumption $|\alpha, \beta| << 1$ is acceptable since, as explained before, we are interested in cases down to the $|\alpha, \beta| \lesssim 10^{-6}$ level.

For an electron that deviates from the closed orbit by the state vector X, given by Eq. (67), the angular velocity is given by $\vec{\Omega}(X_0 + X)$. In a linear approximation, $\vec{\Omega}$ can be decomposed into $\vec{\Omega}(X_0) + \vec{\omega}(X)$, where the perturbation ω is small compared with Ω.

We need now to know how the orbital coordinates X and the spin coordinates α,β evolve in time. We know that[4] the orbital motion of a particle in an accelerator is most conveniently described by the transport matrices. In the absence of coupling, transport matrices of a small dimension (2×2 for y-motion, 3×3 for x-motion, etc.) will be sufficient. With x-y coupling, one uses 4×4 matrices and in case x-, y- and z-motions are all coupled together, one must deal with 6×6 matrices. It does not require too much imagination to realize that the next step is to construct an 8-dimensional state vector

$$\begin{bmatrix} x \\ x' \\ y \\ y' \\ z \\ \Delta E/E \\ \alpha \\ \beta \end{bmatrix} \tag{90}$$

The corresponding transport matrices are then 8×8.

To appreciate the need of dealing with such a generality, we remember that the spin motion of an electron depends on the electric and magnetic fields it experiences according to the BMT equation; and that those fields, in turn, depend on its orbital coordinates. This means coupling effects between spin and orbital coordinates play an important role as far as spin motion is concerned.

Concerning the spin-orbit coupling, we mentioned that the spin motion is influenced by the orbital motion. In fact, orbital motion of an electron is also influenced by its spin. The influence is expected to be extremely weak (of the order of $\hbar$) and will be ignored. To see how small these effects are, let us consider a vertically polarized electron with magnetic moment $\vec{\mu} = \mu\hat{y}$. A quadrupole magnet, which is a focusing element for an electric charge, acts on $\vec{\mu}$ as a bending element. The bending is done in the horizontal plane and the bending angle is $\Delta x' = G\ell/E$ with G the field gradient and ℓ the magnet length. Taking a typical quadrupole magnet in an electron storage ring, we might have $G\ell$ = 50 kilogauss, the kicking angle $\Delta x'$ is found to be about 10^{-13} rad. Similarly, a sextupole magnet which produces a nonlinear field for an electric charge acts on $\vec{\mu}$ as a linear focusing element. The focal length is given by $f^{-1} = \mu G'\ell/E$ with $G' = \partial^2 B_y/\partial x^2$ the sextupole strength. Again taking a typical sextupole strength $G'\ell$ = 500 kilogauss/m, we find the focal length is about 10^{12} meters. Both the bending by quadrupoles and focusing by sextupoles are indeed exceedingly weak.

Noting that $\hat{n}$, $\hat{m}$ and $\hat{\ell}$ satisfy

$$\frac{d}{ds}\vec{S} = \vec{\Omega}(X_0) \times \vec{S}$$

one obtains by substituting Eq. (89) into the precession equation that

$$\frac{d}{ds}\alpha \simeq \vec{\omega}(X) \cdot \hat{\ell}$$
$$\frac{d}{ds}\beta \simeq -\vec{\omega}(X) \cdot \hat{m} \qquad (91)$$

The 8 × 8 transport matrix looks like

$$\left[\begin{array}{c|cc} T_{6\times 6} & & 0 \\ \hline D & 1 & 0 \\ & 0 & 1 \end{array}\right] \qquad (92)$$

where $T_{6\times 6}$ means the 6 × 6 transport matrices describing the transformation among the orbital coordinates; the upper right corner is a 6 × 2 matrix filled by 0's because we are ignoring the effects of spin on orbital motions; the spin-orbit coupling coefficients in the 2 × 6 matrix D are obtained from Eq. (91). Explicit expressions of the 8 × 8 matrices are given in the Appendix.

One must not forget that, due to the discontinuous transition in the definition of the base vectors as the electron travels across $s = 2\pi R$, an extra transformation for the spin components is required

$$\begin{bmatrix} \alpha \\ \\ \beta \end{bmatrix}_{s = 2\pi R^+} = \begin{bmatrix} \cos 2\pi\nu & -\sin 2\pi\nu \\ \\ \sin 2\pi\nu & \cos 2\pi\nu \end{bmatrix} \times \begin{bmatrix} \alpha \\ \\ \beta \end{bmatrix}_{s = 2\pi R^-} \qquad (93)$$

where ν is the spin tune found in section 3.6. Starting from s we multiply matrices to obtain a transformation matrix T(s) for one revolution. Let the eignvalues and eigenvectors of T(s) be λ_k and $E_k(s)$, respectively, with

$$\begin{aligned} &T(s)\, E_k\, (s) = \lambda_k\, E_k\, (s) \\ &\lambda_k^* = \lambda_{-k} \\ &E_k^* = E_{-k},\ k = \pm I,\ \pm II,\ \pm III,\ \pm IV. \end{aligned} \qquad (94)$$

Eigenvectors at other positions, $E_k(s')$, are obtained from $E_k(s)$ by successive transformations from s to s'. The first three sets of eigenmodes are the orbital modes. The three eigenvalues give the orbital tunes. The corresponding eigenvectors in general carry nonzero spin components. The fourth pair of eigenvectors, $E_{\pm IV}$, on the other hand, contains only spin components and no orbital components. The corresponding eigenvalues are given by $\exp(\pm i2\pi\nu)$ with ν the spin tune.

Consider an electron that follows the closed orbit with spin along $\hat{n}$. A photon of energy δE is emitted at s_0. Immediately after the emission, the electron is left in a state

$$X(S_0) = \begin{bmatrix} 0 \\ 0 \\ 0 \\ 0 \\ 0 \\ -\delta E/E \\ 0 \\ 0 \end{bmatrix} \qquad (95)$$

Decomposed into eigenstates, it can be written as

$$X(s_0) = \sum_k A_k E_k(s_0)$$

$$= \sum_{k = \pm I, \pm II, \pm III} A_k E_k(s_0) + \begin{bmatrix} 0 \\ 0 \\ 0 \\ 0 \\ 0 \\ 0 \\ \bar{\alpha} \\ \bar{\beta} \end{bmatrix}_{s_0} \tag{96}$$

where we have used the fact that $E_{\pm IV}$ contain no orbital coordinates.

Equation (96) contains 8 unknowns $A_{\pm I, \pm II, \pm III}$ and $\bar{\alpha}$, $\bar{\beta}$, and 8 equations to determine them. The first six equations obtained by equating the orbital components of Eq. (96) can be written as

$$\begin{bmatrix} 0 \\ 0 \\ 0 \\ 0 \\ 0 \\ -\delta E/E \end{bmatrix} = \sum_k A_k e_k(s_0) \tag{97}$$

where e_k is a 6-dimensional vector whose components are the orbital coordinates of E_k. To find A_k, some orthogonality condition on the eigenvectors e_k is needed. This condition is provided by the symplecticity property of the transport matrices $T_{6 \times 6}(s)$, i.e.,

$$\tilde{T}_{6 \times 6} S \, T_{6 \times 6} = S \quad , \tag{98}$$

where a tilde means taking the transpose of a matrix and

$$S = \begin{bmatrix} 0 & -1 & 0 & 0 & 0 & 0 \\ 1 & 0 & 0 & 0 & 0 & 0 \\ 0 & 0 & 0 & -1 & 0 & 0 \\ 0 & 0 & 1 & 0 & 0 & 0 \\ 0 & 0 & 0 & 0 & 0 & -1 \\ 0 & 0 & 0 & 0 & 1 & 0 \end{bmatrix} \tag{99}$$

From Eq. (98), one can find the orthogonality condition

$$\tilde{e}_j \, S \, e_i = 0 \quad \text{unless} \quad j = -i. \tag{100}$$

When j = -i, we choose the normalization so that

$$\tilde{e}_{-k}\, S\, e_k = i, \quad k = I, II, III. \tag{101}$$

The reason we normalize it to i rather than 1 is, as one can easily show, this quantity must be purely imaginary. For k = -I, -II and -III, they are normalized to -i. Conditions (100) and (101) are preserved as a function of s due to the symplecticity of T(s). Using equations (100) and (101), Eq. (97) yields

$$A_k = -i\,\frac{\delta E}{E}\; E^*_{k5}\,(s_0), \tag{102}$$

where E_{ki} means the ith component of the vector E_k.

Having found A_k, one then solves the remaining 2 equations corresponding to the spin part of Eq. (96)

$$\begin{bmatrix} \bar{\alpha} \\ \\ \bar{\beta} \end{bmatrix}_{s_0} = -2\,\frac{\delta E}{E} \sum_{k=I,\,II,\,III} \begin{bmatrix} \mathrm{Im}(E^*_{k5}\,E_{k7}) \\ \\ \mathrm{Im}(E^*_{k5}\,E_{k8}) \end{bmatrix}_{s_0} \tag{103}$$

After the photon emission, the orbital components of Eq. (96) are rapidly damped by the radiation damping, leaving the spin to precess around $\hat{n}$ as if it started at s_0 with an initial deviation $\vec{S}-\hat{n} = \bar{\alpha}\hat{m} + \bar{\beta}\hat{\ell}$. Associating with the definition of the spin chromaticity, we find

$$\left(\gamma\,\frac{\partial \hat{n}}{\partial \gamma}\right)_s = -2 \sum_{k=I,\,II,\,III} \left[\mathrm{Im}(E^*_{k5}\,E_{k7})\,\hat{m} + \mathrm{Im}\,(E^*_{k5}\,E_{k8})\hat{\ell}\,\right]_s \tag{104}$$

in which we have dropped the subscript on s_0 since spin chromaticity is defined for all s.

3.8 Numerical Examples

To calculate the polarization numerically, we have to know the arrangement of all lattice elements around the storage ring, including elements coming from imperfections. These computational steps are then followed:

(1) the closed orbit X_0

(2) polarization direction $\hat{n}$

(3) 8×8 transport matrices

(4) spin chromaticity $\gamma \partial \hat{n} / \partial \gamma$

(5) τ and P

A computer code has been developed following these steps and has been applied to estimate the beam polarizations for the storage rings SPEAR and PEP. The lattice elements for the ideal rings include horizontal bending magnets, quadrupole magnets, sextupole magnets, rf cavities and drift spaces. Without field imperfections, the ideal lattice produces an equilibrium polarization of 92%. To simulate field imperfections, we introduce a random distribution of vertical orbit kickers. The resulting vertical closed orbit distortion makes sextupoles behave like skew quadrupoles and quadrupoles behave like additional vertical kickers. In the presence of these field imperfections, the degree of polarization P can be plotted as a function of the beam energy E. Results will, of course, be different for different simulations of field imperfections; however, typical results for SPEAR and PEP are shown in Figs. 6 and 7, respectively.

The SPEAR lattice used is specified by the lattice parameters: $\nu_x = 5.28$, $\nu_y = 5.18$, $\nu_s = 0.022$, $\beta_x^* = 1.2$ m, $\beta_y^* = 0.10$ m and $\eta_x^* = 0$, where β_x^*, β_y^* and η_x^* are[4] the horizontal beta-function, vertical beta-function and the energy dispersion function at the points where positron and electron beams collide. The strengths of the vertical kickers are normalized such that the rms closed orbit distortion after orbit correction is $\Delta y_{rms} = 1.2$ mm, which is typical for SPEAR operation. Locations of the depolarization resonances are indicated by arrows at the top of Fig. 6. Each integer resonance is surrounded by six sideband resonances. The integer resonances and the two nearby synchrotron side-band resonances overlap and are shown as single depolarization dips.

We have expanded the energy scale of Fig. 6 and plotted the result again in Fig. 8. Superimposed are polarization measurements performed by the SPEAR polarization team.[16] The agreement between calculation and measurements is acceptable except that the calculation has missed the depolarization resonance located at $a\gamma - \nu_x + \nu_s = 3$. In general, there are depolarization resonances located at

$$a\gamma + n_x \nu_x + n_y \nu_y = n_s \nu_s = k \qquad (105)$$

for all integers $n_{x,y,s}$ and k. The matrix formalism we developed takes care of only the integer and the linear sideband resonances.

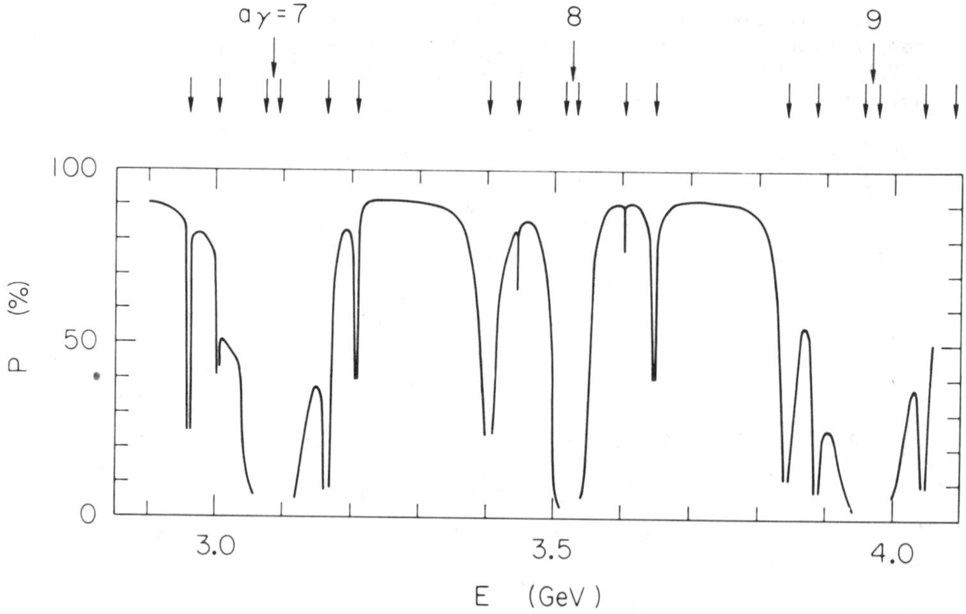

Fig. 6: Expected beam polarization P vs. beam energy E around 3-4 GeV for a typical lattice configuration of the SPEAR storage ring. The simulated field imperfections contribute to an rms vertical closed orbit distortion of 1.2 mm. Locations of depolarizaiton resonances are indicated by arrows.

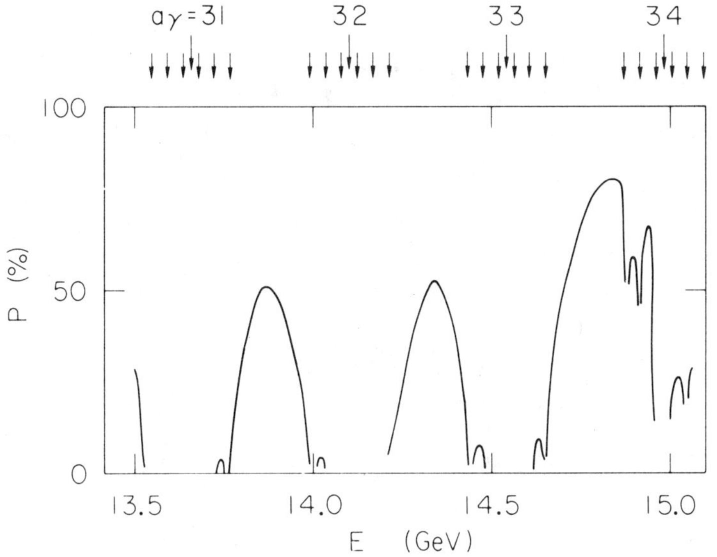

Fig. 7: Expected beam polarization P vs. beam energy E around 13.5-15 GeV for a typical lattice configuration of the PEP storage ring. The simulated field imperfections contribute to an rms vertical closed orbit distortion of 0.6 mm.

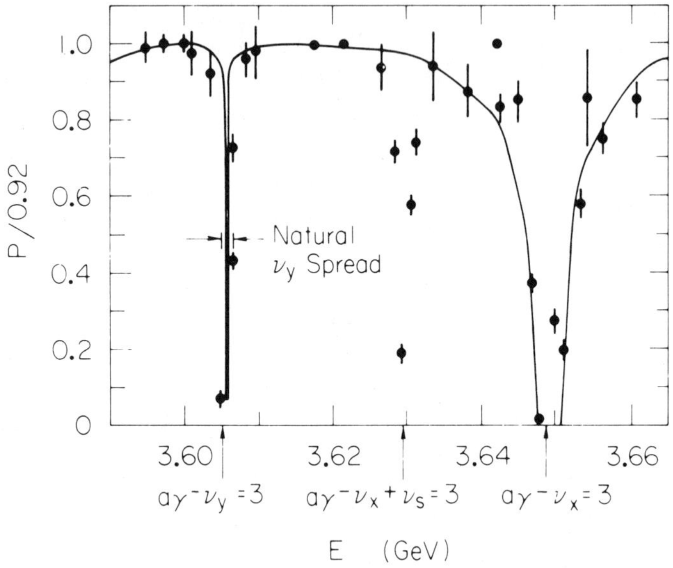

Fig. 8: Comparison of calculation and measurements for SPEAR. Agreement near the two linear sideband resonances $a\gamma-\nu_y = 3$ and $a\gamma - \nu_x = 3$ is acceptable. The third resonance $a\gamma - \nu_x + \nu_s = 3$, however, is missed by the calculation.

The PEP configuration used has $\nu_x = 21.15$, $\nu_y = 18.75$ and $\nu_s = 0.05$. The corresponding rms closed orbit distortion is set to be 0.6 mm, which is half of what we use for SPEAR because PEP has a more sophisticated orbit correction scheme. Nevertheless, the expected PEP polarization is lower than for SPEAR due to its higher beam energies.

APPENDIX
GENERALIZED TRANSPORT MATRICES

The generalized transport matrices for the state vector (x, x', y, y', z, δ, α, β) are listed below for various lattice elements. Thin-lens approximation has been used. For the rf cavity, ϕ_s is the synchronous phase and $\hat{V}$ is the peak voltage.

<u>Drift Space</u>

$$\begin{bmatrix} 1 & \ell & 0 & 0 & 0 & 0 & 0 & 0 \\ 0 & 1 & 0 & 0 & 0 & 0 & 0 & 0 \\ 0 & 0 & 1 & \ell & 0 & 0 & 0 & 0 \\ 0 & 0 & 0 & 1 & 0 & 0 & 0 & 0 \\ 0 & 0 & 0 & 0 & 1 & 0 & 0 & 0 \\ 0 & 0 & 0 & 0 & 0 & 1 & 0 & 0 \\ 0 & 0 & 0 & 0 & 0 & 0 & 1 & 0 \\ 0 & 0 & 0 & 0 & 0 & 0 & 0 & 1 \end{bmatrix}$$

<u>Horizontal Bend Magnet or Kicker:</u> $q = B_y \ell / B\rho$

$$\begin{bmatrix} 1 & 0 & 0 & 0 & 0 & 0 & 0 & 0 \\ 0 & 1 & 0 & 0 & 0 & q & 0 & 0 \\ 0 & 0 & 1 & 0 & 0 & 0 & 0 & 0 \\ 0 & 0 & 0 & 1 & 0 & 0 & 0 & 0 \\ -q & 0 & 0 & 0 & 1 & 0 & 0 & 0 \\ 0 & 0 & 0 & 0 & 0 & 1 & 0 & 0 \\ 0 & 0 & 0 & a\gamma q \ell_z & 0 & q\ell_y & 1 & 0 \\ 0 & 0 & 0 & -a\gamma q m_z & 0 & -q m_y & 0 & 1 \end{bmatrix}$$

<u>Vertical Bend Magnet or Kicker:</u> $q = B_x \ell / \beta\rho$

$$\begin{bmatrix} 1 & 0 & 0 & 0 & 0 & 0 & 0 & 0 \\ 0 & 1 & 0 & 0 & 0 & 0 & 0 & 0 \\ 0 & 0 & 1 & 0 & 0 & 0 & 0 & 0 \\ 0 & 0 & 0 & 1 & 0 & -q & 0 & 0 \\ 0 & 0 & q & 0 & 1 & 0 & 0 & 0 \\ 0 & 0 & 0 & 0 & 0 & 1 & 0 & 0 \\ 0 & a\gamma q \ell_z & 0 & 0 & 0 & q\ell_x & 1 & 0 \\ 0 & -a\gamma q m_z & 0 & 0 & 0 & -q m_x & 0 & 1 \end{bmatrix}$$

Quadrupole: $q = \frac{\ell}{\beta\rho} \frac{\partial}{\partial x} B_y$

$$\begin{bmatrix} 1 & 0 & 0 & 0 & 0 & 0 & 0 & 0 \\ -q & 1 & 0 & 0 & 0 & qx_o & 0 & 0 \\ 0 & 0 & 1 & 0 & 0 & 0 & 0 & 0 \\ 0 & 0 & q & 1 & 0 & -qy_o & 0 & 0 \\ -qx_o & 0 & qy_o & 0 & 1 & 0 & 0 & 0 \\ 0 & 0 & 0 & 0 & 0 & 1 & 0 & 0 \\ -(1+a\gamma)q\ell_y & 0 & -(1+a\gamma)q\ell_x & 0 & 0 & 0 & 1 & 0 \\ (1+a\gamma)qm_y & 0 & (1+a\gamma)qm_x & 0 & 0 & 0 & 0 & 1 \end{bmatrix}$$

Skew Quadrupole: $q = \frac{\ell}{B\rho} \frac{\partial}{\partial y} B_y$

$$\begin{bmatrix} 1 & 0 & 0 & 0 & 0 & 0 & 0 & 0 \\ 0 & 1 & -q & 0 & 0 & qy_o & 0 & 0 \\ 0 & 0 & 1 & 0 & 0 & 0 & 0 & 0 \\ -q & 0 & 0 & 1 & 0 & qx_o & 0 & 0 \\ -qy_o & 0 & -qx_o & 0 & 1 & 0 & 0 & 0 \\ 0 & 0 & 0 & 0 & 0 & 1 & 0 & 0 \\ (1+a\gamma)q\ell_x & 0 & -(1+a\gamma)q\ell_y & 0 & 0 & 0 & 1 & 0 \\ -(1+a\gamma)qm_x & 0 & (1+a\gamma)qm_y & 0 & 0 & 0 & 0 & 1 \end{bmatrix}$$

RF Cavity: $q = e\,\hat{V}\cos\phi_s/RE$; $r = (1+a\gamma)\,e\,\hat{V}\sin\phi_s/E$

$$\begin{bmatrix} 1 & 0 & 0 & 0 & 0 & 0 & 0 & 0 \\ 0 & 1 & 0 & 0 & 0 & 0 & 0 & 0 \\ 0 & 0 & 1 & 0 & 0 & 0 & 0 & 0 \\ 0 & 0 & 0 & 1 & 0 & 0 & 0 & 0 \\ 0 & 0 & 0 & 0 & 1 & 0 & 0 & 0 \\ 0 & 0 & 0 & 0 & q & 1 & 0 & 0 \\ 0 & -r\ell_y & 0 & r\ell_x & 0 & 0 & 1 & 0 \\ 0 & rm_y & 0 & -rm_x & 0 & 0 & 0 & 1 \end{bmatrix}$$

Sextupole: $q = \frac{\ell}{B\rho} \frac{\partial^2}{\partial x^2} B_y \quad ; \quad r = (1 + a\gamma)q$

$$\begin{bmatrix}
1 & 0 & 0 & 0 & 0 & 0 & 0 & 0 \\
-qx_o & 1 & qy_o & 0 & 0 & \frac{q}{2}(x_o^2 - y_o^2) & 0 & 0 \\
0 & 0 & 1 & 0 & 0 & 0 & 0 & 0 \\
qy_o & 0 & qx_o & 1 & 0 & -qx_o y_o & 0 & 0 \\
-\frac{q}{2}(x_o^2 - y_o^2) & 0 & qx_o y_o & 0 & 1 & 0 & 0 & 0 \\
0 & 0 & 0 & 0 & 0 & 1 & 0 & 0 \\
-r(y_o \ell_x + x_o \ell_y) & 0 & -r(x_o \ell_x - y_o \ell_y) & 0 & 0 & 0 & 1 & 0 \\
r(y_o m_x + x_o m_y) & 0 & r(x_o m_x - y_o m_y) & 0 & 0 & 0 & 0 & 1
\end{bmatrix}$$

PROBLEMS

1. Spin precesses ahead of the coordinate system by a factor of $a\gamma$ if we bend the trajectory by a magnetic field. What if we bend by an electric field? What will be the spin tune? Along which direction is the beam polarized?
2. Show that the Hamiltonian (21), together with $\dot{\vec{S}} = i/\hbar\, [H,\vec{S}]$ and $[S_i, S_j] = i \sum_k \varepsilon_{ijk} S_k$ gives the precession equation (13). This is true whether $\vec{\Omega}$ is nonrelativistic [Eq. (14)] or has been generalized [Eq. (16)].
3. Show Eq. (35). Make use of the fact that the two polarization vectors $\hat{\varepsilon}_1$ and $\hat{\varepsilon}_2$ are orthogonal to $\hat{k}$.
4. Follow the calculation of $\langle f|\vec{S}(t)|i\rangle$ of section 2.7 to evaluate the case without spin flip. You should get Eq. (46).
5. If the gyromagnetic ratio $|g| \gg 1$, Eq. (65) becomes

$$\tau_0^{-1} = \frac{|g|^5\, e^2 \hbar \gamma^5}{48\, m^2\, c^2\, \rho^3} \text{ and } P_0 = -\frac{g}{|g|} .$$

Show that these results agree with the intuitive picture discussed before Eq. (65). Refer to Ref. 12.
6. If there are wiggler devices (see the footnote of section 3.2) in the storage ring, show that the equilibrium beam polarization is still along y but with a reduced magnitude

$$P_0 = -\frac{8}{5\sqrt{3}} \frac{\oint ds/\rho^3(s)}{\oint ds/|\rho^3(s)|} ,$$

where $\rho < 0$ for a reversed bending. Refer to section 2.8. Remember that the "up" and the "down" states switch roles in a reversed bending magnet. See also Eq. (75).
7. Let $\vec{S}_n$ be the spin of an electron as it circulates the n-th revolution. Transform the spin according to

$$\vec{S}_{n+1} = M_1(a\gamma + \varepsilon \cos n\nu) M_2(\theta) \vec{S}_n ,$$

where $M_1(a\gamma)$ is the precession matrix (70); $M_2(\theta)$ is the perturbation matrix (71); $|\varepsilon| \ll 1$ is a small parameter; ν is the frequency at which the spin precession frequency $a\gamma$ is modulated. Starting with $\vec{S}_0$ along the rotational axis of the transformation $M_1(a\gamma)M_2(\theta)$, show that $\vec{S}_n$ deviates from $\hat{y}$ significantly if $a\gamma \pm \nu$

is close to an integer and then estimate the resonance width. Do the analysis to first order in ε. Repeat the problem for the case when the spin is driven by an oscillating perturbation

$$\vec{S}_{n+1} = M_1(a\gamma)\, M_2(\varepsilon \cos n\nu)\vec{S}_n .$$

8. To obtain Eq. (83) from Eq. (82), we have used a semi-classical argument to replace the quantity $(\vec{S} \cdot \vec{A})\, (\vec{S} \cdot \vec{B})$ by $\vec{A} \cdot \vec{B}$ by averaging over the solid angles of spin $\vec{S}$, where $\vec{A}$ and $\vec{B}$ are arbitrary vectors. Prove the quantum mechanical counterpart of this argument

$$\{(\vec{\sigma} \cdot \vec{A}) \;,\; (\vec{\sigma} \cdot \vec{B})\} = \vec{A} \cdot \vec{B} \;,$$

where $\vec{\sigma}$ is a vector whose components are the Pauli matrices, $\{\ ,\}$ is the symmetric anti-commutator of two operators.

9. If one can design a storage ring with arbitrary $\hat{n}$ and $\gamma\partial\hat{n}/\partial\gamma$, how should he choose these quantities so that the beam polarization given by equations.(85) and (86) is maximized? The answer is given in Ref. 6

$$\hat{n} = \sqrt{\frac{7}{11}}\,\frac{\hat{v}\times\dot{\hat{v}}}{|\dot{\hat{v}}|} \pm \sqrt{\frac{4}{11}}\,\hat{v}$$

and

$$\gamma\,\frac{\partial\hat{n}}{\partial\gamma} = \frac{2\sqrt{7}}{11}\left(-\sqrt{\frac{4}{11}}\,\frac{\hat{v}\times\dot{\hat{v}}}{|\dot{\hat{v}}|} \pm \sqrt{\frac{7}{11}}\,\hat{v}\right).$$

The maximum value is $P_{max} = 72/5\sqrt{231} = 94.7\%$.

REFERENCES

[1]J. D. Jackson, Classical Electrodynamics (Wiley, New York, 1975).

[2]L. D. Landau and E. M. Lifshitz, Classical Theory of Fields (Addison-Wesley, Mass., 1971).

[3]J. S. Schwinger, Proc. Natl. Acad. Sci. U.S.A. 40, 132 (1954).

[4]E. D. Courant, lecture for this summer school. See also E. D. Courant and H. S. Snyder, Ann. Phys. 3, 1 (1958).

[5]M. Sands, "The Physics of Electron Storage Rings," SLAC-121 (1970).

[6]Ya. S. Derbenev and A. M. Kondratenko, Sov. Phys. JETP 37, 968 (1973).

[7]A. A. Sokolov and I. M. Ternov, Sov. Phys. Dokl. 8, 1203 (1964). A. A. Sokolov and I. M. Ternov, Synchrotron Radiation (Pergamon Press, 1968).

[8]V. N. Baier, Sov. Phys. Ups. 14, 695 (1972).

[9]Ya. S. Derbenev, A. M. Kondratenko, and A. N. Skrinsky, Particle Accel. 9, 247(1979).

[10]R.F. Schwitters, Nucl. Instrum. Methods 118, 331 (1974).

[11]A. W. Chao, Nucl. Instrum. Methods 180, 29 (1981).

[12]J. D. Jackson, Rev. Mod. Phys. 48, 417 (1976).

[13]See Textbooks on quantum mechanics: L. I. Schiff, Quantum Mechanics (McGraw-Hill, New York, 1968).

[14]V. Bargmann, L. Michel and V. L. Telegdi, Phys. Rev. Lett. 2, 435 (1959). Our derivation follows W. K. H. Panofsky, SPEAR Note 142 (1972) unpublished.

[15]J. Buon, Proc. Intern. Symp. on High Energy Physics with Polarized Beams and Polarized Targets, Lausanne (1980), p. 1.

[16]SPEAR Polarization team, J. Johnson, R. Schwitters, et al. (1980).

HIGH ENERGY ELECTRON LINACS; APPLICATION TO STORAGE RING RF SYSTEMS AND LINEAR COLLIDERS*

Text: Perry B. Wilson
Stanford Linear Accelerator Center, Stanford, California 94305

Appendix: James E. Griffin
Fermi National Accelerator Laboratory, Batavia, Illinois 60510

TABLE OF CONTENTS

*Work supported by the Department of Energy, contract DE-AC03-76SF00515.

0094-243X/82/870450-133$3.00 Copyright 1982 American Institute of Physics

HIGH ENERGY ELECTRON LINACS: APPLICATION TO STORAGE RING RF SYSTEMS AND LINEAR COLLIDERS

Perry B. Wilson
Stanford Linear Accelerator Center
Stanford University, Stanford, California 95305

1. INTRODUCTION

The theory of electron linacs will be developed with two particular applications in mind: the use of standing-wave linacs as rf accelerating structures in high energy electron-positron storage rings, and the (potential) future application of traveling-wave linacs to the acceleration of intense single bunches in very high gradient linear colliders. These topics are of special interest for the future of high energy particle accelerators, and in addition they are not covered in a coherent manner in the existing literature. Excellent and complete references to the theory of conventional high energy traveling-wave linacs, such as the SLAC two-mile accelerator, do of course exist. In Refs. 1 and 2, for example, topics such as structure design, particle dynamics and beam break-up in traveling-wave electron linacs are given extensive treatment. Although we cannot hope to duplicate the completeness of the coverage in these references, some of the main features of the theory will be summarized here. Hopefully these lecture notes will complement this previous work, especially in the area of beam loading by single bunches of charge.

Because of the broad scope of the material being covered, the treatment of some topics (for example, standing-wave rf structure design) must remain superficial. However, an attempt will be made to present a thorough and comprehensive treatment of the general problem of beam-structure interactions; that is, the problem of beam loading in all of its many manifestations. The interaction of intense single bunches with longitudinal and transverse modes in the rf structure sets fundamental limits on the performance of both linear colliders and electron storage rings, and will be given particular attention.

It is in principle possible to solve many beam-structure interaction problems by simply setting up an appropriate differential equation and turning the mathematical crank. In these notes we prefer to take a more visual approach, using the principle of superposition and the geometry of phasor diagrams to analyze rather complex multiple-bunch beam loading problems under transient conditions. In the case of single-bunch beam loading, we are able to bypass some messy details in the direct solution of Maxwell's equations by a careful application of basic principles such as superposition, conservation of energy and causality. These concepts are useful not only as aids in developing physical thinking; they also provide techniques for solving important real-world problems in a relatively simple way.

The initial charge for this particular set of lectures was to cover both linacs and rf power sources. We have already chosen to limit the discussion of linacs to high energy electron linacs, and in fact to only a portion of this subject area. The theory and design of rf power sources, usually klystrons, for conventional cw and pulsed electron linacs is a separable subject that will not be treated here. The reader is referred to Refs. 3 and 4 for an introduction to klystron theory and design. We will, however, consider briefly some possible high peak power rf sources for very high gradient linear colliders. The emphasis will be on non-conventional sources; for example, the use of energy storage cavities with fast switching of the stored energy into the accelerating structure.

2. BASIC CONCEPTS

2.1 Phasors

Fields and voltages in standing-wave rf structures are taken to be complex (phasor) quantities, written with a tilde. For example,

$$\tilde{V} = V\, e^{j\omega t} \tag{2.1}$$

where $V = |\tilde{V}|$ is the absolute value of $\tilde{V}$. Here $\tilde{V}$ might represent the voltage gain for a particle crossing a cavity driven at rf angular frequency ω. In this case eV is the maximum energy that can be gained by a non-perturbing charge traversing the cavity; that is, the charge is assumed to be sufficiently small so that the beam-induced voltage is negligible compared to the rf driving voltage. The trajectory of a particle or bunch of particles in usually taken to be the axis of symmetry of a cavity or structure, except when considering dipole (deflecting) modes. In that case the trajectory is assumed to be displaced from, but parallel to, the axis. The real part of $\tilde{V}$, $\mathrm{Re}\,\tilde{V} = V\cos\theta$ where $\theta = \tan^{-1}[\mathrm{Im}\,\tilde{V}/\mathrm{Re}\,\tilde{V}]$, gives the energy gain for a charge crossing the cavity or structure at an arbitrary phase with respect to the cavity field. The position of a charge at time t can be written $z = z_0 + ct$. The position z_0 at time $t = 0$ for a point charge which receives the maximum possible energy gain defines a reference position or plane inside the cavity. It is often useful to take this reference plane as the origin for the axial coordinate z.

For problems concerning resonant cavities driven by an external generator, it is useful to view the phasor in a frame of reference rotating at the driving frequency ω. Thus if the phase of the rf voltage is varying with time as $\theta = \omega t + \theta_0$, the phasor is written in this reference frame as

$$\tilde{V} = Ve^{j\theta_0} \quad . \tag{2.2}$$

The importance of choosing a reference frame determined by the external generator will become apparent in the discussion of the longitudinal stability of the beam in a storage ring against phase oscillations.

Phasors are mainipulated using the usual rules of complex algebra. In particular, it is useful to recall that multiplying a phasor by $e^{j\psi}$ rotates the phasor through angle ψ without changing its magnitude.

2.2 The "Big Four" Basic Principles

Four basic principles that will often be of use in the development to follow are: superposition, conservation of energy, orthogonality of modes and causality. Superposition will be called upon most frequently. As a typical example, consider a standing-wave resonant cavity driven by an external rf generator and loaded by a beam current. The total cavity voltage $\tilde{V}_c$ can be considered to be the superposition of a voltage component $\tilde{V}_g$ produced by the rf generator acting alone (beam current off), and a component $\tilde{V}_b$ due to the beam (generator off):

$$\tilde{V}_c = \tilde{V}_g + \tilde{V}_b \quad . \tag{2.3}$$

Conservation of energy will be called upon to establish some basic theorems concerning beam loading. Conservation of total energy is straightforward. If a charge q with energy U_i enters a cavity with no initial stored energy, and if after the charge leaves the cavity the stored energy is W_c, then clearly

$$W_c = U_i - U_f \quad , \tag{2.4}$$

where U_f is the final energy of the charge. Conservation of energy can also be applied to differential energy exchanges. Suppose, for example, that a charge q at position z = z' moves a distance dz' along a trajectory (taken to be the z coordinate axis) such that the electric field for a given mode has a z component $E_z(z')$. The change in the energy stored in the mode is then

$$dW = -qE_z(z')\ dz' \quad . \tag{2.5}$$

The field at position $\vec{r}$ for the mode in question is related to the energy W stored in the mode by $E^2(\vec{r}) = f(\vec{r})W$, where the function $f(\vec{r})$ depends on the cavity geometry. Thus we have on the cavity axis

$$f(z)\ dW = 2\ E_z(z)\ dE_z \quad . \tag{2.6}$$

From these two expressions, together with the fact that time is related to the position of the charge through ct' = z', an expression is obtained for dE_z as a function of z at time t'. Treated as a phasor, the field element dE_z at some later time t will be described by

$$d\tilde{E}_z(z,t) = d\tilde{E}_z(z,t')e^{j\omega_o(t-t')} \tag{2.7}$$

where ω_o is the resonant frequency of the mode. Using the concept of differential superposition, the total beam-induced cavity field at any position z and time t can now be obtained by adding up all the differential field elements induced at previous times through an integration which takes proper account of the phase relationships between elements. But as will be seen later, one more ingredient — causality — must be added to complete the picture.

Implicit in the foregoing analysis is the concept of normal modes. It is assumed that each mode in the cavity can be treated independently in computing the fields induced by a charge crossing the cavity. The total stored energy is taken as the sum of the energies in the separate modes. The total field is the vector (phasor) sum of all the individual mode fields at any instant.

Causality is a somewhat more subtle principle that must also be taken into account in computing the field induced by a charge passing through an rf cavity or structure. By causality we mean simply that there can be no disturbance ahead of a charge moving at the velocity of light. Thus, in a mode analysis of the growth of the beam-induced field, the field must vanish ahead of the moving charge for each mode. As we will see in more detail in Sec. 9.3, this is accomplished if the charge also induces imaginary differential field components in addition to the real field components as obtained from the energy interchange described by Eqs. (2.5) and (2.6). These imaginary components, which lie at $\pm 90^\circ$ with respect to the real component at time $t = t'$, must have an amplitude distribution as a function of frequency such that they add up, when integrated over frequency, to cancel the real induced components ahead of the charge ($t < t'$) and to double the real components behind the charge ($t > t'$).

Real high energy electrons and positrons move at velocities which are close to, but not exactly equal to, the velocity of light. Subtle questions arise as to how close is close enough so that the $v \approx c$ approximation is sufficiently accurate in any given situation. There will not be space here to go into this problem in detail; in fact, some aspects of the causality problem are still controversial and have not yet been adequately resolved to everyone's satisfaction. For our purposes here, we will assume that causality is absolute for point charges moving through rf cavities and traveling-wave structures.

2.3 Differential Superposition

Because of the importance of the concept of differential superposition, let us use it here to compute the answer to a practical question: what is the voltage induced in a cavity by a Gaussian charge distribution with total charge q, if the voltage induced by a point charge q is V_o? A charge element dq will induce a voltage $dV = V_o(dq/q)$. Assume that the charge element dq crosses the cavity reference plane at time t_o. At some other time t the voltage induced by this charge element will be

$$d\tilde{V} = \frac{V_o}{q} e^{j\omega_o(t - t_o)} dq(t_o) \quad . \tag{2.8}$$

For a Gaussian charge distribution

$$dq(t_o) = I(t_o)dt_o = \frac{q}{\sqrt{2\pi}\sigma} e^{-t_o^2/2\sigma^2} dt_o \quad , \tag{2.9}$$

and therefore

$$d\tilde{V} = \frac{V_o}{\sqrt{2\pi}\sigma} e^{j\omega_o t} e^{-t_o^2/2\sigma^2} (\cos\omega_o t_o - j \sin\omega_o t_o) dt_o \quad . \qquad (2.10)$$

We invoke differential superposition and integrate over all arrival times t_o, noting that the integral of the second term in the preceding expression vanishes by symmetry. The result is

$$\tilde{V} = V_o e^{j\omega_o t} e^{-\omega_o^2\sigma^2/2} \quad , \qquad (2.11)$$

where $V_o e^{j\omega_o t}$ is just the voltage induced by a point charge. Thus for a Gaussian charge distribution,

$$V = V_o e^{-\omega_o^2\sigma^2/2} \quad . \qquad (2.12)$$

Since bunch distributions in storage rings and linacs are usually Gaussian, or nearly so, the result given by Eq. (2.12) is of broad applicability.

3. STANDING WAVE LINACS

3.1 Shunt Impedance

The shunt impedance R_a for an rf cavity is a figure of merit which relates the accelerating voltage V to the power P dissipated in the cavity walls through the expression $V = (R_a P)^{1/2}$. For a mode with stored energy W, both the power dissipation $P = \omega W/Q$ and the longitudinal electric field on the cavity axis $E_z(z) = [f(z)W]^{1/2}$ are specified in terms of the geometry-dependent factors Q and f(z). Assuming that these factors are known, it remains to compute V in terms of $E_z(z)$.

Assume that the path of an electron (positron) lies along the z coordinate in an arbitrary standing wave structure driven by an external generator at frequency ω. The z component of the electric field along the axis is then

$$E_z(z,t) = E(z) e^{j\omega t} \quad . \qquad (3.1)$$

Assume that a positive E_z produces an accelerating force on the particle in question, and that the particle velocity is $v_e \approx c$. The particle position at time t is

$$z_e = c(t - t_o) \qquad (3.2)$$

where $z_e = 0$ at $t = t_o$. The accelerating field seen in a reference frame moving with the particle (the co-moving frame) is then

$$E_z(\text{cmf}) = E(z)\, e^{j\omega(t_o + z/c)} = E(z)\, e^{j\omega t_o}\, e^{jkz} \tag{3.3}$$

where $k \equiv \omega/c$. The voltage gained by the particle in moving from $z = z_1$ to $z = z_2$ (z_1 and z_2 would normally be at the cavity entrance and exit) is

$$\tilde{V} = \int_{z_1}^{z_2} E_z(\text{cmf})dz = e^{j\omega t_o} \int_{z_1}^{z_2} E(z)\, e^{jkz} dz = e^{j\omega t_o} [C + jS] \quad . \tag{3.4}$$

Here C and S are the cosine and sine integrals

$$C = \int_{z_1}^{z_2} E(z) \cos kz \, dz \tag{3.5a}$$

$$S = \int_{z_1}^{z_2} E(z) \sin kz \, dz \quad . \tag{3.5b}$$

We then have

$$\tilde{V} = V\, e^{j(\omega t_o + \theta)} \quad , \tag{3.6}$$

where

$$V = |\tilde{V}| = \left| \int_{z_1}^{z_2} E(z)\, e^{jkz} dz \right| = (C^2 + S^2)^{\frac{1}{2}} \tag{3.7a}$$

$$\theta = \tan^{-1}(S/C) \quad . \tag{3.7b}$$

If E(z) is symmetric about a point on the z axis, the S integral in Eq. (3.5b) will vanish if the symmetry point is chosen to be the origin $z = 0$. Even if the structure is not symmetric, we can make the transformations

$$\omega t_o' = \omega t_o + \theta, \qquad kz' = kz - \theta \quad , \tag{3.8}$$

where $z' = c(t - t_0')$ is the position of the charge with respect to the new coordinate origin. Then $\tilde{V} = V e^{j\omega t_0'}$ and the point $z' = 0$ defines the reference plane for the cavity. The shunt impedance is now defined as

$$R_a = \frac{V^2}{P} \quad , \tag{3.9}$$

where

$$P = \frac{1}{2} R_s \int_{\substack{\text{cavity} \\ \text{surface}}} H^2 dA \quad . \tag{3.10}$$

Here

$$R_s = (\omega\mu/2\sigma)^{\frac{1}{2}} = \pi Z_o (\delta/\lambda) \tag{3.11}$$

is the surface resistance, μ the permeability, σ the dc conductivity, δ the skin depth, Z_o the impedance of free space and $\lambda = 2\pi c/\omega$.

The above definition of shunt impedance, R_a, is the so-called accelerator definition, which is used in most of the modern literature on linac structure design. The shunt impedance is, however, occasionally defined with a factor of 2 in the denominator as $R = V^2/2P$. The reader should be aware of this potential source of confusion.

3.2 Transit-Time Factor

An "uncorrected" shunt impedance R_u is sometimes defined in terms of a voltage V_u, the integral of the electric field along the cavity axis:

$$R_u = \frac{V_u^2}{P} \tag{3.12a}$$

$$V_u = \int E(z)\, dz \quad . \tag{3.12b}$$

The shunt impedance defined in this way does not take into account the variation in the field during the time it takes a particle to cross an accelerating gap or pass through an rf cavity; that is, the effect of transit time is ignored. To obtain the true shunt impedance, a transit-time factor T is applied to the uncorrected shunt impedance so that*

$$R_a = R_u T^2 \quad , \tag{3.13a}$$

*In the literature Eq. (3.13) is often written $Z_{sh} = ZT^2$. In these notes we reserve Z for the rf impedance.

where

$$T = \frac{V}{V_u} = \frac{\left|\int E(z)\, e^{jkz} dz\right|}{\int E(z)\, dz} \quad . \tag{3.13b}$$

Problem 3.1: Show that the transit-time factor for a gap of length L with a uniform field E_z along the particle trajectory is

$$T = \frac{\sin(\theta/2)}{\theta/2} \quad , \tag{3.14}$$

where $\theta = kL = 2\pi L/\lambda$ is the transit angle. Use the definition in Eq. (3.13b) and compute the transit-time factor in two ways: with the origin $z = 0$ at the center of the gap, and with the origin such that the gap extends from $z = 0$ to $z = L$.

The transit-time factor is introduced here for historical reasons, and because it is often found in the literature. Since the voltage V as given by Eq. (3.7a) has to be computed in any case, the attentive reader might wonder why the shunt impedance is not computed directly using Eq. (3.9), rather than through the circular process of Eqs. (3.12a), (3.13a) and (3.13b). Indeed, the transit-time factor does not need to be calculated to obtain the shunt impedance, and it is sometimes even misleading. Consider, for example, a cavity of length L operating in a mode such that the axial field is

$$E(z,t) = E_o \cos kz \cos \omega t \quad . \tag{3.15}$$

If the cavity is exactly one-half wavelength long, then $kL = \pi$ and

$$V_u = E_o \int_0^L \cos kz\, dz = 0$$
$$R_u = 0 \quad . \tag{3.16}$$

The axial field for such a cavity is shown by the solid curves in Fig. 3.1 at time $t = 0$ and $t = L/c$. (Can this cavity be a cylindrical "pillbox" cavity of finite radius? Why not?) On the other hand, the field in a co-moving frame ($kz = \omega t$) for a particle which enters the cavity at $t = 0$ varies as

$$E(cmf) = E_o \cos^2 kz \quad . \tag{3.17}$$

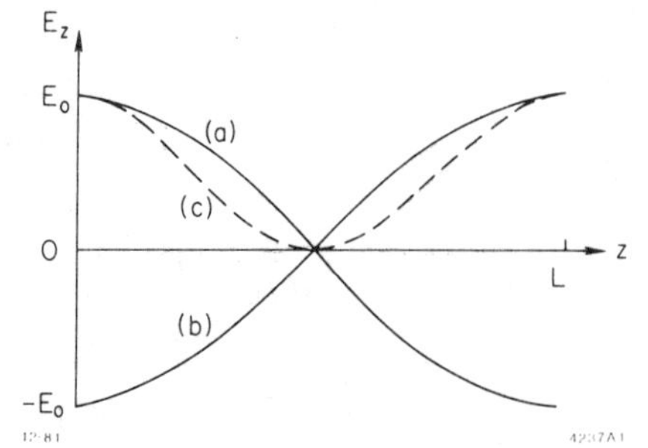

Fig. 3.1. Axial electric field in a TM_{011}-mode cavity one-half free-space wavelength long; (a) at t = 0, (b) at t = L/c and (c) in a co-moving frame for a particle with $v \approx c$.

The field seen in such a frame moving with a relativistic particle is shown by the dashed line in Fig. 3.1. The voltage gained by the particle is

$$V = E_o \int_0^L \cos^2 kz \, dz = E_o L/2 \quad , \tag{3.18}$$

and

$$T = V/V_u = \infty$$
$$R_a = R_u T^2 = (0)(\infty) \quad . \tag{3.19}$$

In this case, it is meaningless to define the shunt impedance through Eq. (3.13a). Although the concept of a transit-time factor breaks down in this case, it is sometimes helpful in giving a better physical feeling for the process of optimizing the shunt impedance of accelerating cavities (see Sec. 3.4).

3.3 Bunch Form Factor

Real bunches in real accelerators and storage rings are not point bunches, but extend over some finite length. It is clear for this case that not all particles in the bunch can achieve the maximum energy gain, but that some particles must cross the cavity reference plane earlier or later than the time for peak gain. Suppose the current in the bunch flowing past a fixed point is I(t), and that the total charge in the bunch is

$$\int_{-\infty}^{\infty} I(t)dt = q \quad . \tag{3.20}$$

Suppose also that a reference plane is again chosen such that the maximum voltage V_0 is gained by a particle which crosses the plane at t = 0. Then the average voltage gained by all the charge elements dq = I(t)dt is

$$\tilde{V}_a = \frac{\int_{-\infty}^{\infty} V_o \, e^{j\omega t} \cdot I(t)dt}{\int_{-\infty}^{\infty} I(t)dt} = V_o(C' + jS') \tag{3.21}$$

where C' and S' are the cosine and sine integrals

$$C' = \frac{1}{q}\int_{-\infty}^{\infty} I(t)\cos\omega t\,dt \qquad (3.22a)$$

$$S' = \frac{1}{q}\int_{-\infty}^{\infty} I(t)\sin\omega t\,dt \quad . \qquad (3.22b)$$

The magnitude of the average voltage gain is

$$V_a = |\tilde{V}_a| = FV_o \quad , \qquad (3.23)$$

where F is the bunch form factor,

$$F = (C'^2 + S'^2)^{1/2} \qquad (3.24)$$

For a Gaussian bunch with rms bunch length σ_t and for a uniform bunch of time length t_b we have

$$F(\text{Gaussian}) = e^{-\frac{1}{2}\omega^2\sigma_t^2} \qquad (3.25a)$$

$$F(\text{rectangular}) = \frac{\sin\left(\frac{1}{2}\,\omega t_b\right)}{\frac{1}{2}\,\omega t_b} \quad . \qquad (3.25b)$$

For F = 0.9, we have $\sigma_z/\lambda = 0.073$ and $\ell_b/\lambda = 0.25$, where $\sigma_z = c\sigma_t$ and $\ell_b = ct_b$. Note that the form factor for the case of a Gaussian bunch is the same as obtained previously in Sec. 2.3, where the voltage induced in a cavity by such a bunch was calculated using the principle of differential superposition.

3.4 Standing-Wave Structures

The longitudinal and transverse modes in a chain of cylindrical "pillbox" cavities provide an approximate yet often surprisingly accurate model for the accelerating and deflecting fields in more realistic accelerating structures. The properties of a single cylindrical resonator are simple to treat analytically, and will serve as a starting point for a discussion of standing-wave accelerating structures.

Consider a pillbox cavity with radius b and axial length L. The axial electric and azimuthal magnetic field components for the lowest-order accelerating mode (TM_{010} mode) are

$$E_z = E_o J_o(kr)\cos\omega t$$

$$H_\phi = -\frac{E_o}{Z_o} J_1(kr)\sin\omega t \quad , \qquad (3.26)$$

where $Z_o = 377$ ohms, $k = 2\pi/\lambda = p_{01}/b$ and $p_{01} = 2.405$ is the first root of J_o. The stored energy and power dissipation are computed to be

$$W = \frac{\varepsilon_o}{2} \int_V E_z^2 \, dV = \frac{\pi}{2} \varepsilon_o b^2 L \, E_o^2 J_1^2 \, (p_{01})$$

$$P = \frac{R_s}{2} \int_A H_\phi^2 \, dA = \frac{\pi \, b \, R_s E_o^2}{Z_o^2} \, (b + L) \; J_1^2(p_{01}) \quad . \tag{3.27}$$

The accelerator parameters of interest are

$$Q = \frac{\omega W}{P} = \frac{G_1}{R_s} \sim \omega^{-\frac{1}{2}} \tag{3.28a}$$

$$\frac{r}{Q} = \frac{V^2}{\omega WL} = \frac{G_2 T^2}{\lambda} \sim \omega \tag{3.28b}$$

$$r = \frac{V^2}{PL} = \frac{G_1 \, G_2 \, T^2}{\lambda R_s} \sim \omega^{\frac{1}{2}} \quad . \tag{3.28c}$$

Here $r \equiv R_a/L$ is the shunt impedance per unit length, $R_s = (\omega\mu_o/2\sigma)^{\frac{1}{2}}$ is the surface resistance, T is the transit angle factor and G_1 and G_2 are two constants, independent of frequency and cavity material, given by

$$G_1 = \frac{p_{01}}{2} \left(\frac{L}{b + L}\right) Z_o = 453 \left(\frac{L}{b + L}\right) \text{ohms} \tag{3.29a}$$

$$G_2 = \frac{4 \, Z_o}{p_{01}^2 \; J_1^2(p_{01})} = 967 \text{ ohms} \tag{3.29b}$$

$$T = \frac{\sin \, (\pi L/\lambda)}{(\pi L/\lambda)} \quad . \tag{3.29c}$$

Problem 3.2: Show that the shunt impedance per unit length r for a pillbox cavity is maximum at $L/b = 0.75$, and that the total shunt impedance rL is maximum at $L/b = 1.15$. What are the corresponding values of r, rL and Q in these two cases for a room-temperature copper cavity at 500 MHz?

Designers of accelerating structures have been working for many years to increase the shunt impedance as much as possible beyond that which can be obtained from a chain of simple pillbox cavities. Initially this was accomplished by a combination of intuition and laborious rf measurements in the laboratory. In more recent years, powerful computer programs have greatly facilitated the process of optimizing the design of standing-wave accelerating structures. The first of these codes, LALA,[5] was developed at the Los Alamos Scientific Laboratory to aid in the design of structures for high-energy proton linacs. A more recent and more powerful code, SUPERFISH,[6] is now available at many accelerator laboratories. SUPERFISH can calculate higher-order cavity modes as well as the lowest frequency accelerating mode, although both programs are limited to axially-symmetric modes in axially-symmetric structures. However, a new program, ULTRAFISH,[7] is now under development which can compute the frequencies and fields of modes which vary as cos $m\phi$ (where ϕ is the aximuthal angle and $m > 0$) in axially-symmetric structures. Modes with $m > 0$ can cause deflection and defocusing of bunches and trains of bunches in an accelerating structure, leading to emittance growth and to beam breakup.

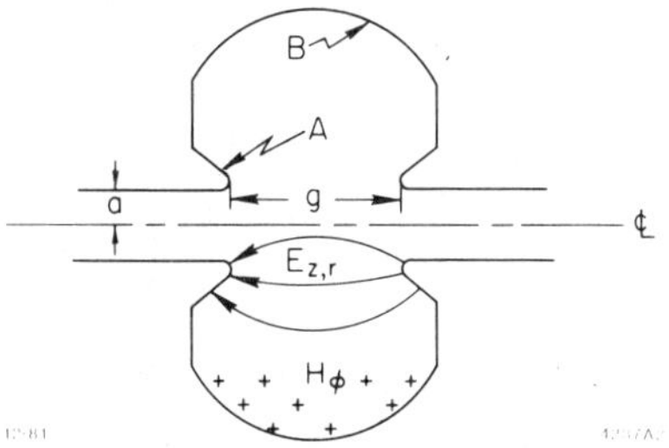

Fig. 3.2. Single Cell of a π-mode accelerating structure.

Figure 3.2 illustrates some of the factors entering into the design of a single cell of a standing-wave accelerating structure. The most characteristic features are the so-called "nose cones," as shown at A. For a given stored energy, the nose cones help to concentrate the electric field in the region of the beam, thus increasing the factor $R_a/Q = V^2/\omega W$. The gap length g between nose cones is adjusted for maximum R_a/Q. As g is decreased, the transit time factor T increases, but the integral of the axial field, Eq. (3.12b), decreases for a given stored energy. After the R_a/Q factor has been optimized by shaping the nose cones and adjusting the gap length, the shunt impedance can be increased further by maximizing the Q. The Q is controlled largely by losses at the outer surfaces of the cavity, shown at B in Fig. 3.2, where the magnetic field is greatest. The highest Q is obtained if this part of the cavity surface can be made approximately spherical in shape. This, however, increases the complexity in manufacturing the cavity. It is often a reasonable trade-off to keep a cylindrical outer boundary with a consequent 10% or so reduction in shunt impedance.

It is usually awkward to feed each cavity separately with rf in a long linac strcture. Thus a number of cavities, or cells, are usually coupled together to form a coupled-cavity strcture with a single rf feed point. Such a structure is shown schematically in Fig. 3.3. A structure consisting of N coupled cells (resonators) will have N normal modes, as shown in the dispersion diagram of Fig. 3.4. The frequencies of the normal modes can be obtained by solving an equivalent circuit[8,9] consisting of a chain of coupled LRC resonators as shown in Fig. 3.5. For a structure with weak magnetic cell-to-cell coupling and vanishingly small losses, the normal mode frequencies are given by

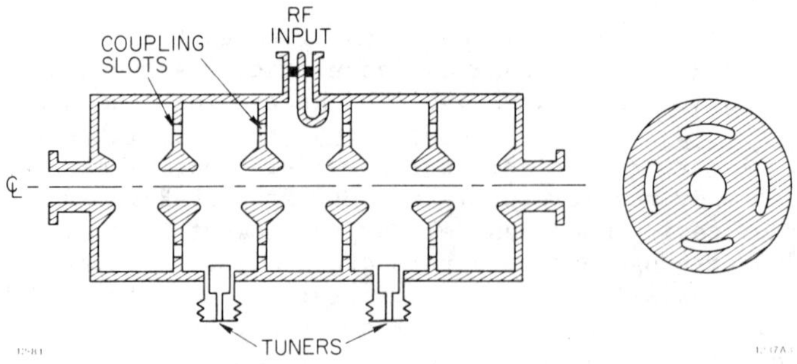

Fig. 3.3. Diagram showing the important features of a five-cell π-mode structure with magnetic field coupling.

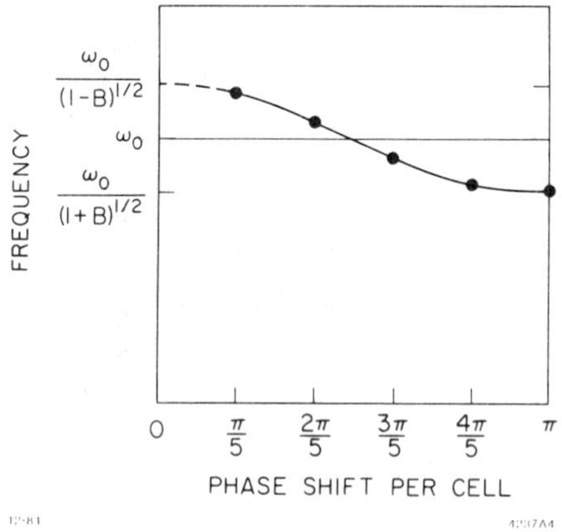

Fig. 3.4. Dispersion diagram for a five-cell structure with "flat" π-mode.

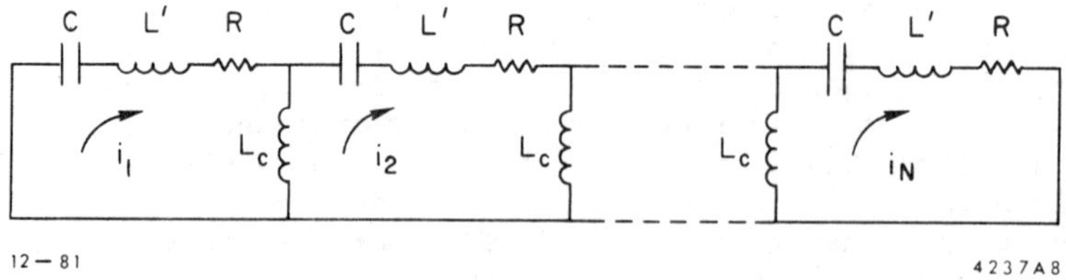

Fig. 3.5. Equivalent circuit representation for a chain of coupled resonant cavities. For a flat π-mode, $L' = L + 2L_c$ and there is then no mode with zero phase shift per cell.

$$\omega(m) = \frac{\omega_o}{(1 - B\cos m\pi/N)^{\frac{1}{2}}} \approx \omega_o\left(1 + \frac{B}{2}\cos m\pi/N\right) \qquad (3.30)$$

where B is the bandwidth of the structure, m is the mode number (m = 1, 2, -- N for a structure with full-length end cells) and $m\pi/N$ is the phase shift per period.

RF structures for storage rings usually operate in the π mode (m = N). In order to obtain a "flat" π-mode (field amplitude equal in all N cells) in a structure with full-length end cells, the two end cells must be tuned lower in frequency (for magnetic-field coupling) by an amount $\delta\omega/\omega \approx B/2$. The field amplitude in the nth cell for the mth normal mode is for this case

$$E_n = A_m \sin\left[m\pi(2n-1)/2N\right] \quad , \qquad (3.31)$$

where n = 1, 2, -- N.

Problem 3.3: Draw the equivalent curcuit for a chain of N coupled resonators with half-length end cells (metal boundaries at the planes of symmetry in each end cell). Show that the normal-mode frequencies are given by $\omega(m) = \omega_o[1 - B\cos m\pi/(N-1)]^{-\frac{1}{2}}$ and the corresponding field amplitudes by $E_n = A_m \cos[m\pi\,(n-1)/(N-1)]$, where n = 1, 2, -- N and m = 0, 1, -- (N - 1). The $\pi/2$ mode is obtained for m = (N - 1)/2. Compare the field amplitudes E_n for this case with the $\pi/2$-mode fields given by Eq. (3.31) with m = N/2.

It is important to know the sensitivity of the field amplitudes in the individual cells to errors in tuning, due either to unwanted perturbations or to the presence of tuners. It is usually not practical to put a remotely-controlled tuner in each cell of a multicell structure. If, for example, we attempt to adjust the frequency of a multicell structure with a single tuner in one cell, an error in field flatness will be introduced. A mathematically elegant approach to this problem is given by the application of perturbation theory to the equivalent circuit representation.[10] The problem can also be treated in certain simple cases by considering the multiple reflections of a wave traveling on a finite-length chain of coupled resonators.[11] Suppose we have a chain of N π-mode cells with both the rf feed point and a single tuner located in the center cell in a structure with an odd number of cells. If f(n) is the flatness function, defined as the ratio of the perturbed field amplitude to the unperturbed field along the structure, then the maximum deviation from flatness is given by

$$\delta f \approx \frac{(N-1)^2}{2B} \cdot \left(\frac{\delta\omega}{\omega}\right) \quad , \qquad (3.32)$$

where $\delta\omega$ is the change in structure resonant frequency produced by the tuner, and N = 3, 5, 7 etc. We see that for the π mode the sensitivity of the field flatness to tuning varies quadratically with the number of cells, and is inversely proportional to the bandwidth. A similar analysis for the $\pi/2$ mode shows that the field flatness is less sensitive to tuning errors. The deviation from flatness varies as

$$\delta f \approx \frac{(N-1)^2}{2B^2} \cdot \left(\frac{\delta\omega}{\omega}\right)^2 \tag{3.33}$$

where N = 5, 9, 13 etc. As shown in Fig. 3.6a, every other cell in an unperturbed $\pi/2$ mode is unexcited for a lossless structure.* The main effect of a detuning error is to introduce a field in the nominally unexcited cells. The maximum value of this field is in the two cells adjacent to the center cell with tuner and is given by

$$\delta f \approx \frac{(N-3)}{B} \cdot \left(\frac{\delta\omega}{\omega}\right) \quad . \tag{3.34}$$

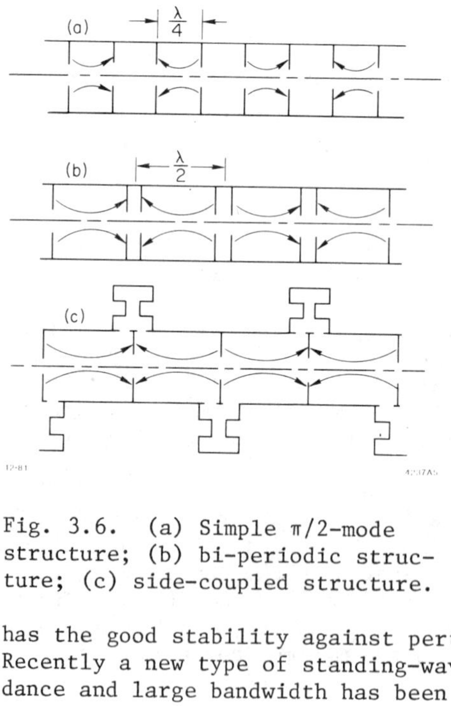

Fig. 3.6. (a) Simple $\pi/2$-mode structure; (b) bi-periodic structure; (c) side-coupled structure.

A comparison of Eqs. (3.33) and (3.34) with Eq. (3.32) shows the superiority of the $\pi/2$ mode against tuning perturbations. However, from Fig. 3.6(a) it is apparent that the shunt impedance of the $\pi/2$ mode will be poor, since every other cavity is unexcited and will not contribute to the acceleration of particles. One solution is to shrink down the length of the unexcited cavities, as shown in Fig. 3.6(b), resulting in a so-called bi-periodic structure. A more elegant solution is to remove the unexcited cavities from the beam line entirely, as shown in Fig. 3.6(c). This results in the side-coupled structure, exploited extensively at Los Alamos.[9] The field on the axis looks like that for a π-mode, but the structure has the good stability against perturbations of the $\pi/2$ mode. Recently a new type of standing-wave structure with good shunt impedance and large bandwidth has been under development, particularly at

---

*This is true for a structure terminated in half-length end cells (see problem 3.3). Full-length end cells can also be used if they are properly detuned.

Los Alamos.[12,13] This is the disk and washer (DAW) structure, shown schematically in Fig. 3.7. The r/Q of this structure is less than that of a chain π-mode cells with nose cones, but the Q is significantly higher. The reason for this is that the structure has evolved from a chain of pillbox cavities operating in the next higher-order radial mode. The stored energy is therefore higher for a given field on the axis, leading to a lower r/Q. However, the current tends to flow as a lossless displacement current between the disks and the washers, rather than as a physical current in the surface at the outer boundary. This leads to a much greater value for Q. Like the side-coupled structure, the DAW structure works in a π-like mode, but with resonant coupling in the region of the disks. The coupling is very heavy, giving the structure a large bandwidth and great stability against perturbations.

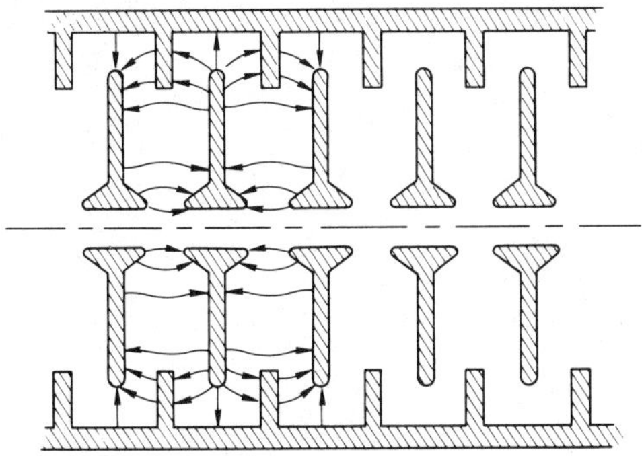

Fig. 3.7. Disk and Washer (DAW) structure with sketch of electric field lines.

Properties of several structures operating at 357 MHz are compared in the table below: the DAW structure just described, a π-mode structure proposed for the rf system for the LEP storage ring at CERN,[14] and for comparison a chain of pillbox cavities λ/2 in length. The DAW and LEP structures have beam aperture radii of 5 cm, while the pillbox cavity, of course, has no beam opening. This brings up a very important point: structures should always be compared at the same value of beam hole radius, since the shunt impedance is a strong function of the size of the beam aperture. Figure 3.8 shows the variation in shunt impedance per unit length as a function of beam-hole radius a for a simple disk-loaded structure and for a shaped π-mode cell with nose cones. Note that the shunt impedance for these structures is reduced by a factor of two at a ≈ 0.15 λ.

Table 3.1 A Comparison of Several Copper Structures at 350 MHz

	r/Q (Ω/m)	Q (cm)	r (MΩ/m)	B
LEP[14]	635	49,000*	31*	≈0.01
DAW[13]	325	130,000*	42*	≈0.5
Pillbox	465	52,000	24	--

*These Q and r values should be reduced by about 15% for a practical structure to take into account losses due to washer supports (DAW), coupling slots (LEP), and imperfect surfaces.

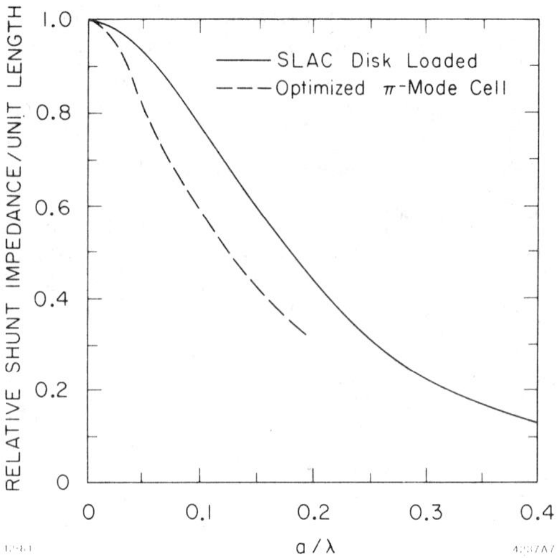

Fig. 3.8. Variation in shunt impedance per unit length as a function of beam-hole radius for two typical structures.

3.5 Equivalent Circuit for a Cavity with Beam Loading

Figure 3.9(a) shows the equivalent circuit for an rf source (usually a klystron) connected to a linac or storage ring cavity by a transmission line. Since such equivalent circuits are basic to the analysis of rf system design and performance, several comments are in order. First, note that the rf cavity and the kystron output cavity are represented by resonant LRC circuits. While this circuit representation may be intuitively obvious, a rigorous justification of the use of lumped-element circuits to model resonant modes in metal cavities is given in Ref. 15. Second, note that the beam in the rf cavity is represented by a current generator. This is an excellent representation for a relativistic beam, since the velocity of the particles passing through the cavity is independent of the cavity voltage. The situation is different for the case of the klystron output cavity. The velocities of the electrons as they pass through the gap of the output cavity can change in response to the cavity fields, and as a consequence a current-dependent beam loading admittance, Y_{bk}, is needed in the equivalent circuit (see, for example, Ref. 3). Third, note that the transmission line connecting cavity and klystron has both forward and backward traveling waves. These waves must satisfy the boundary condition $V_k^+ + V_k^- = V_k/n_k$ at the klystron, and a similar condition at the cavity. Since there may be a number of transmission line elements between A and B, each with reflection, phase shift and possibly loss, the solution of the general problem can be quite complex.[16] For our purposes here, we can simplify the problem considerably by assuming that there is an isolator or circulator just before the cavity. Thus, any power which is reflected from the cavity and which travels back toward the klystron will be

(a)

KLYSTRON OUTPUT CAVITY   TRANSMISSION LINE   R F CAVITY

(b)

$$\beta = \frac{1}{n_c{}^2}\left(\frac{G_o}{G_c}\right)$$

$$P_g = \frac{1}{8}\frac{i_g^2}{\beta G_c}$$

12—81   4237A9

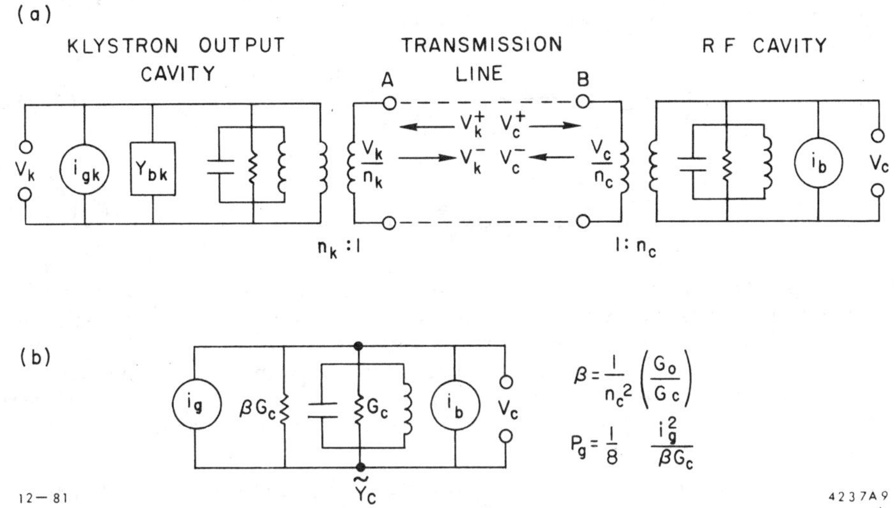

Fig. 3.9. (a) Equivalent circuit for a beam-loaded cavity coupled to a klystron; (b) simplified circuit assuming a matched RF source.

absorbed. The simplified equivalent circuit in Fig. 3.9(b), in which the transmission line impedance G_0 and the current generator representing the rf source are transformed to the cavity side of the transformer representing the transmission-line-to-cavity coupling network, can now be used. Here β is termed the cavity coupling coefficient. If the source generator is off and the cavity is excited internally by the beam, β is then seen to be the ratio of the power radiated out of the cavity through the coupling loop or aperture to the power dissipated in the cavity walls.

In using the simplified equivalent circuit, the available power from the generator, P_g, is to be identified with the incident klystron power. Also watch out for factors of two. In terms of the accelerator definition of shunt impedance introduced previously, and the dc current I_0, we have

$$G_c = \frac{2}{R_a}$$

$$P_c = \frac{1}{2}G_c V_c^2 = \frac{V_c^2}{R_a} \tag{3.34'}$$

$$i_b = 2\,I_0\,e^{-\omega_o^2\sigma_t^2/2} \approx 2\,I_0 \quad .$$

Assuming short bunches ($\omega_o\sigma_t << 1$), we have from Fig. 3.9(b) that the voltages at resonance produced by the beam and the rf source, if each one acts independently on the circuit, are

$$V_{gr} = \frac{i_g}{G_c(1+\beta)} = \frac{2\sqrt{\beta}}{1+\beta} \cdot \sqrt{R_a P_g} \tag{3.35a}$$

$$V_{br} = \frac{i_b}{G_c(1+\beta)} = \frac{I_o R_a}{1+\beta} \quad . \tag{3.35b}$$

It is instructive to consider the accelerating voltage V_a, the power dissipated the cavity walls P_c, the efficiency η for the conversion of generator power into beam power, and the reflected power P_r for the case of a linac operating on resonance and in phase (bunches receive maximum acceleration such that $V_a = V_c$). In terms of a beam-loading parameter $K = (I_o/2)(R_a/P_g)^{\frac{1}{2}}$, these quantities are:

$$V_a = \sqrt{R_a P_g} \left\{ \frac{2\sqrt{\beta}}{1+\beta} \left(1 - \frac{K}{\sqrt{\beta}}\right) \right\} = \sqrt{R_a P_c} \tag{3.36a}$$

$$\eta = \frac{I_o V_a}{P_g} = \frac{2\sqrt{\beta}}{1+\beta} \left[2\,K\left(1 - \frac{K}{\sqrt{\beta}}\right) \right] \tag{3.36b}$$

$$\frac{P_r}{P_g} = \frac{\left[(\beta-1) - 2K\sqrt{\beta}\right]^2}{(\beta+1)^2} \quad . \tag{3.36c}$$

Problem 3.4: Show that Eq. (3.36c) follows from conservation of energy: $P_r = P_g - \eta P_g - P_c$.

The important feature of Eq. (3.36a) is that the accelerating voltage decreases linearly with increasing current. These "load lines" are shown in Fig. 3.10 for various values of β. For a given beam current, the maximum accelerating voltage is determined by the condition $\partial V_a/\partial\beta = 0$ at $K_m = (\beta-1)/(2\sqrt{\beta})$. The conversion efficiency, shown in Fig. 3.11 with β as a parameter, is seen to vary parabolically as a function of beam current, reaching a maximum at $K_m = \sqrt{\beta}/2$. The beam voltage is then one-half of the voltage at zero current. From Eq. (3.36c), note that the condition for zero reflected power is given by $K = (\beta-1)/(2\sqrt{\beta})$, but that this is not the condition for optimum efficiency as a function of beam current.

Problem 3.5: What is the condition for optimum efficiency at a fixed current as a function of β? Why is this different than the condition for maximum efficiency at fixed β as a function of beam current? What would a contour plot showing lines of constant efficiency in the K - β plane look like? (See Ref. 16a.) Show also that for $P_r = 0$ the coupling coefficient is $\beta = 1 + P_b/P_c$, and $P_g = \beta P_c$.

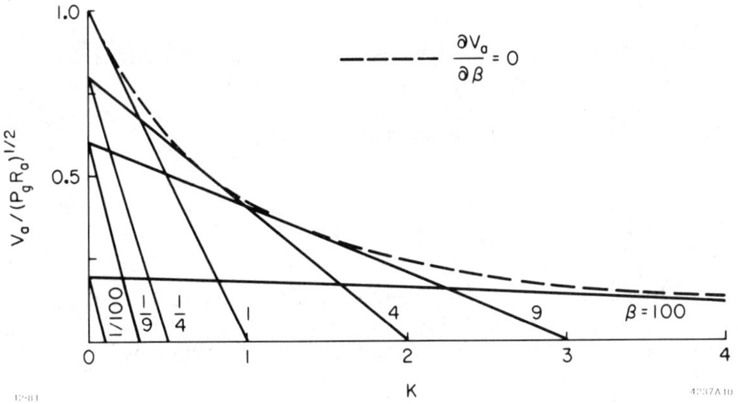

Fig. 3.10. Normalized energy gain as a function of the beam-loading parameter for various values of the coupling coefficient.

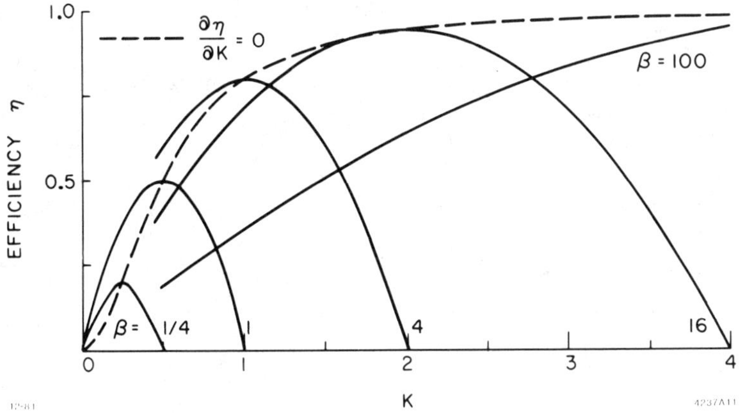

Fig. 3.11. Conversion efficiency as a function of the beam-loading parameter for various values of the coupling coefficient.

We next consider beam loading in a standing-wave structure which is tuned to be off resonance. The admittance of the parallel resonant circuit representing the cavity without coupling ($\tilde{Y}_c$ in Fig. 3.9(b)) is

$$\tilde{Y}_c = G_c \left[1 + jQ_o \left(\frac{\omega}{\omega_o} - \frac{\omega_o}{\omega}\right)\right] \quad , \tag{3.37}$$

where $\omega_o = 1/\sqrt{LC}$ is the resonant frequency, $W = 1/2\ CV_c^2$ is the stored energy, and $Q_o \equiv \omega_o W/P_c = \omega_o C/G_c$. We limit the following discussion to the case of a high Q cavity such that $\delta \equiv (\omega - \omega_o)/\omega_o << 1$. Introducing δ, Eq. (3.37) becomes

$$\tilde{Y}_c \approx G_c(1+j\ 2\ Q_o\delta) \quad . \tag{3.38}$$

The total admittance seen by the beam must include the loading by the coupled admittance of the input transmission line. This external admittance is taken into account by adding βG_c to the preceding expression to obtain the loaded cavity impedance

$$\tilde{Z}_L = \frac{1}{\tilde{Y}_L} = \frac{R_o}{1+j\ 2\ Q_L\delta} \quad , \tag{3.39}$$

where $R_o = [G_c(1+\beta)]^{-1}$ is the loaded impedance at resonance and $Q_L = Q_o/(1+\beta)$ is the loaded Q.

We now define a quantity ψ, termed the tuning angle for reasons that will be clear shortly, by

$$\tan\psi \equiv -2\ Q_L\delta \quad . \tag{3.40}$$

A simple manipulation of Eq. (3.39) gives

$$\tilde{Z}_L = R_o(\cos^2\psi)(1+j\ \tan\psi) = R_o\ \cos\psi\ e^{j\psi} \quad . \tag{3.41}$$

In terms of the beam-loading voltage and the generator voltage at resonance, given by Eqs. (3.35), we have

$$\tilde{V}_g = i_g\ \tilde{Z}_L = V_{gr}\ \cos\psi\ e^{j\psi} \tag{3.42a}$$

$$\tilde{V}_b = i_b\ \tilde{Z}_L = V_{br}\ \cos\psi\ e^{j\psi} \quad . \tag{3.42b}$$

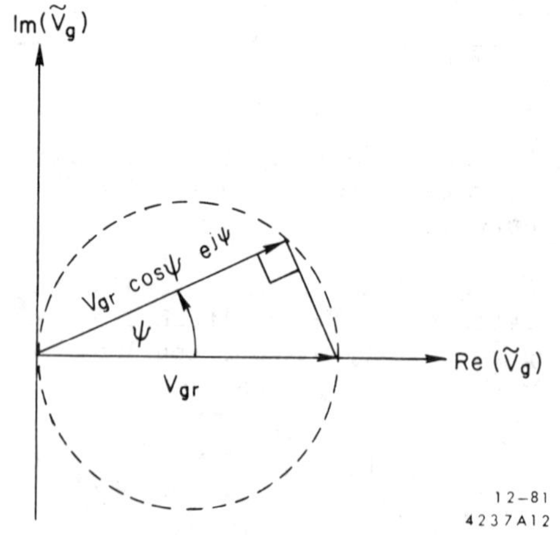

Fig. 3.12. Diagram showing how both generator and beam-loading voltages vary in the complex plane as a function of the tuning angle.

Thus, as the tuning angle increases from zero, the magnitudes of both $\tilde{V}_g$ and $\tilde{V}_b$ decrease as $\cos\psi$, and the phases rotate through angle ψ. This is illustrated in Fig. 3.12. Note especially that the tip of the phasor $\tilde{V}_g$ or $\tilde{V}_b$ traces out a circle in the complex plane as the tuning angle ψ is varied.

We are now ready to consider the superposition of the generator and beam-loading voltages to obtain the net cavity voltage. For convenience, the reference phase (positive real axis) is taken in the direction of $-\tilde{i}_b$. The accelerating voltage V_a is then simply the real component of the net cavity voltage. The superposition $\tilde{V}_c = \tilde{V}_g + \tilde{V}_b$ in this reference frame is shown in Fig. 3.13. Note that two additional important angles have been defined: the phase angle ϕ between $\tilde{V}_c$ and $-\tilde{i}_b$, and the angle θ between $\tilde{i}_g$ and $-\tilde{i}_b$. In storage ring applications, ϕ is termed the synchronous phase angle. In a linac ϕ is the angle between the current bunches and the crest of the rf voltage wave. The angle θ is under external control in an rf linac; it can be adjusted by means of a phase shifter in the input drive to a klystron feeding a cavity or group of cavities. In a storage ring θ is determined if the beam-current (or V_{br}), the cavity voltage V_c, the voltage gain per turn V_a and the tuning angle ψ are specified.

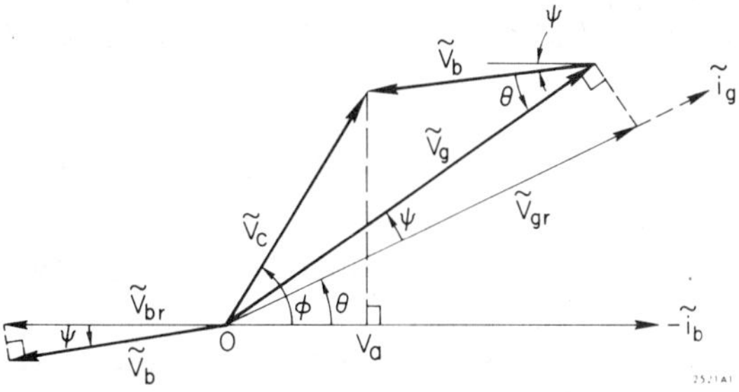

Fig. 3.13. Diagram showing the vector addition of generator and beam-loading voltages in an RF cavity.

A thorough understanding of the vector diagram in Fig. 3.13 is the key to steady-state beam loading calculations. The diagram will be exploited in the following section to compute, as an example, the optimum tuning and coupling for a storage-ring rf cavity. Before proceeding, however, we should recall what is meant by "steady-state." First of all, it is assumed that both the beam current and the rf generator have been turned on for a time which is long compared to the cavity filling time, which is given by

$$T_f = \frac{2\,Q_L}{\omega_o} = \frac{2\,Q_o}{\omega_o(1+\beta)} \quad . \tag{3.43}$$

If the beam current is turned on at t = 0, then for t less than several filling times the cavity fields are in a transient state.

However, another type of transient behavior is also possible. Suppose the bunches in either a linac or storage ring are spaced apart by time T_b. In this section we have implicitly assumed that all bunches contain equal charge, and that $T_b \ll T_f$. The case where T_b is comparable to T_f will be dealt with in a later section.

4. APPLICATION TO STORAGE RING RF SYSTEMS

4.1 Beam Loading in Storage Ring RF Systems

In high energy electron linacs, bunches are accelerated at the peak of the rf voltage wave in order to achieve the maximum possible energy gain. On the other hand, in an electron-positron storage ring it is necessary to operate off the crest of the accelerating voltage waveform in order to insure stability against phase oscillations, and to contain the energy fluctuations due to the quantum nature of synchrotron radiation. The rf cavities must as a consequence be detuned off resonance in order to minimize the reflected power and the required generator power.

Let us compute first the generator power required if the cavity shunt impedance R_a, the coupling coefficient β, the beam current I_o, the cavity tuning angle ψ, the accelerating voltage $V_a = V_c \cos\phi$, and the desired synchronous phase angle ϕ are specified. From Fig. 3.13,

$$V_a = V_c \cos\phi = V_{gr} \cos\psi \cos(\theta + \psi) - V_{br} \cos^2\psi \tag{4.1a}$$

$$V_c \sin\phi = V_{gr} \cos\psi \sin(\theta + \psi) - V_{br} \cos\psi \sin\psi \quad . \tag{4.1b}$$

Eliminate $(\theta + \psi)$ from these two equations, and rewrite the result using Eqs. (3.35) to obtain

$$P_g = \frac{V_c^2}{R_a} \cdot \frac{(1+\beta)^2}{4\beta} \cdot \frac{1}{\cos^2\psi} \left\{ \left[\cos\phi + \frac{I_o R_a}{V_c(1+\beta)} \cos^2\psi \right]^2 + \left[\sin\phi + \frac{I_o R_a}{V_c(1+\beta)} \cos\psi \sin\psi \right]^2 \right\} \quad . \tag{4.2}$$

By choosing the tuning angle ψ correctly, we can make the cavity voltage look "real"; that is, just as is the case at resonance with no beam current, the net cavity voltage $\tilde{V}_c$ must have the same phase as $\tilde{i}_g$. From Fig. 3.13 this implies that

$$\theta = \phi \quad . \tag{4.3}$$

Using the law of sines on the vector triangle in Fig. 3.13, we have

$$\frac{V_{br}\cos\psi}{V_c} = \frac{\sin(\phi - \theta - \psi)}{\sin\theta} = -\frac{\sin\psi}{\sin\phi}$$

$$\tan\psi = \frac{I_o R_a}{V_c(1+\beta)} \sin\phi \quad . \tag{4.4}$$

Problem 4.1: Show that the condition in Eq. (4.4) is also obtained by minimizing the generator power with respect to the tuning angle; that is, take $\partial P_g/\partial\psi = 0$ using Eq. (4.2).

Using Eq. (4.4) in Eq. (4.2), the generator power at optimum tuning is

$$P_g = \frac{(1+\beta)^2}{4\beta} \cdot \frac{(V_c + V_{br}\cos\phi)^2}{R_a} \quad . \tag{4.5}$$

By differentiating this expression with respect to β (don't forget that V_{br} is also a function of β), the minimum generator power at $\beta = \beta_o$ is found to be

$$\beta_o = 1 + \frac{I_o R_a \cos\phi}{V_c} = 1 + \frac{P_b}{P_c} \tag{4.6a}$$

$$P_{go} = \frac{V_c^2 \beta_o}{R_a} = P_c\beta_o = P_c + P_b \quad . \tag{4.6b}$$

Here $P_b = I_o V_a = I_o V_c \cos\phi$ is the power transferred to the beam, and $P_c = V_c^2/R_a$ is the power dissipated in the cavity walls. By conservation of energy, the reflected power is $P_r = P_g - P_c - P_b$. From the above expression for P_{go}, we see that the reflected power is zero when both ψ and β are set to their optimum values. At optimum coupling, Eq. (4.4) becomes

$$\tan\psi_o = \frac{\beta_o - 1}{\beta_o + 1} \tan\phi \quad . \tag{4.7}$$

4.2 Phase Stability and Robinson Damping

As shown in E. Courant's lecture (see also Ref. 17, Ch.3), there is an effective restoring force in a storage ring for deviations in the energy or phase of a particle away from the synchronous energy or phase. A non-synchronous particle undergoes harmonic oscillations at the synchrotron frequency given by (for small amplitude oscillations).

$$\omega_s = \left[\frac{\alpha |dV_a/dt|}{V_o T_o}\right]^{1/2} \quad . \tag{4.8}$$

Here α is the momentum compaction factor, $U_0 = eV_0$ is the particle energy, T_0 is the revolution time and dV_a is the change in accelerating voltage per turn for a particle which is delayed by time dt per turn with respect to a synchronous particle. Above transition (always the case for high-energy electron storage rings), a particle with too much energy will follow a longer path compared to a synchronous particle, and will therefore return to a given point in the ring at a later time after one revolution. For stability, such a particle must gain less energy than a synchronous particle, or $dV_a/dt < 0$. In the absence of beam loading, this condition leads to

$$\frac{dV_a}{dt} = -\omega V_c \sin\phi < 0 \quad , \tag{4.9}$$

or $\phi > 0$ for stability. That is, the synchronous phase is on the time-falling side of the rf cavity voltage. However, at high currents where the beam-induced voltage component is large, the situation is more complicated. As the arrival time varies due to phase oscillations, the beam-induced voltage component moves with the bunch and hence cannot contribute to phase stability; only the generator voltage component can provide an effective restoring force against phase perturbations. From Fig. 3.13, recalling that the phasors rotate counterclockwise with angular velocity ω, the condition $dV_g/dt < 0$ implies

$$0 < (\theta + \psi) < \pi \quad . \tag{4.10}$$

An equivalent way to obtain this same condition is to compute $dV_a/d\theta$ directly from Eq. (4.1a), recognizing that t must be measured by an external clock which is independent of phase oscillations, and that the phase θ of the external rf generator $\tilde{i}_g$ provides such a clock.

Problem 4.2: Draw a phasor diagram, similar to that in Fig. 3.13, with a large beam voltage component, with $\phi > 0$ and with $(\theta + \psi) < 0$. Show from the geometry of the figure that a positive Δt in arrival time results in a positive ΔV_a.

From Eq. (4.1b), using the condition in Eq. (4.10) that $\sin(\theta + \psi)$ is positive, we obtain

$$2V_c \sin\phi + V_{br} \sin 2\psi > 0 \quad . \tag{4.11}$$

This is the condition for the high-current limit on phase stability first derived by Robinson.[18] Robinson's derivation involves setting up a set of linear equations in terms of perturbations to the variables of the system. He then applies Routh's criterion to the determinant of the coefficients to test for solutions which grow exponentially. However, it is well to remember that the result is completely equivalent to the simple condition in Eq. (4.10), which is almost immediately obvious from a carefully constructed phasor diagram.

If the cavity tuning is adjusted to make the beam-cavity impedance look "real" according to Eq. (4.4), then the condition for phase stability reduces to

$$V_{br} \cos\phi < V_c \quad . \tag{4.12}$$

Problem 4.3: If the cavity coupling is also optimized according to Eq. (4.6a), show that the condition in Eq. (4.12) is met for any value of beam current.

We next want to compute the damping time for phase oscillations (sometimes termed Robinson damping). A derivation in the frequency domain of the damping time is given in Ref. 19. Some interesting physics, however, is highlighted in a time-domain analysis.[20] Assume a beam current with phase modulation of the form $\tilde{i}_b = \tilde{i}_0\,(1 + jA\cos\omega_s t)$, where $A \ll 1$. The response of a parallel resonant circuit to this driving current is

$$\tilde{V}_b(t) = R_0\tilde{i}_0\left\{\frac{1}{1+j\xi} + \frac{jA}{2}\left[\frac{e^{j\omega_s t}}{1+j(\xi+\eta)} + \frac{e^{-j\omega_s t}}{1+j(\xi-\eta)}\right]\right\} \tag{4.13}$$

where $\xi = -\tan\psi = (\omega - \omega_0)T_f$ and $\eta = \omega_s T_f$. The terms in $e^{\pm j\omega_s t}$ represent two counter-rotating vectors with origins at the tip of the steady-state beam loading vector $\tilde{V}_0 = R_0\tilde{i}_0\cos\psi\, e^{j\psi}$ where $R_0 i_0 = V_{br}$.

Problem 4.4: Show that the resultant of the two vectors is a vector whose tip moves on an ellipse in the complex plane with semi-major axes

$$a = (A/2)\left\{\left[1 + (\xi+\eta)^2\right]^{-\frac{1}{2}} + \left[1 + (\xi-\eta)^2\right]^{-\frac{1}{2}}\right\} V_{br}$$

$$b = (A/2)\left\{\left[1 + (\xi+\eta)^2\right]^{-\frac{1}{2}} - \left[1 + (\xi-\eta)^2\right]^{-\frac{1}{2}}\right\} V_{br} \quad . \tag{4.14}$$

Show further that the angle γ in Fig. 4.1 is given by $\gamma = \pi/2 + (\psi_+ + \psi_-)/2$, where $\tan\psi_+ = -(\xi+\eta)$ and $\tan\psi_- = -(\xi-\eta)$.

The phasor diagram in Fig. 4.1 illustrates the response $V_b(t)$ to a driving current resulting from a phase oscillation of the bunch center of charge. Note from the result of Problem 4.4 that as ω_s approaches zero the ellipse collapses to a line perpendicular to $\tilde{V}_0$, while for $\omega_s T_f \gg 1$ it collapses to a point at the tip of $\tilde{V}_0$.

The ellipse in Fig. 4.1 is quite suggestive. In analogy with similar diagrams in the force-displacement plane, or the pressure-volume plane in thermodynamics, we conjecture that the area of the ellipse is proportional to the power transfer to or from the oscillation. The conjugate coordinates in the present case are voltage and charge, given by $\delta V = (dV/d\phi)\delta\phi = (V_c \sin\phi)(A\cos\omega_s t)$ and

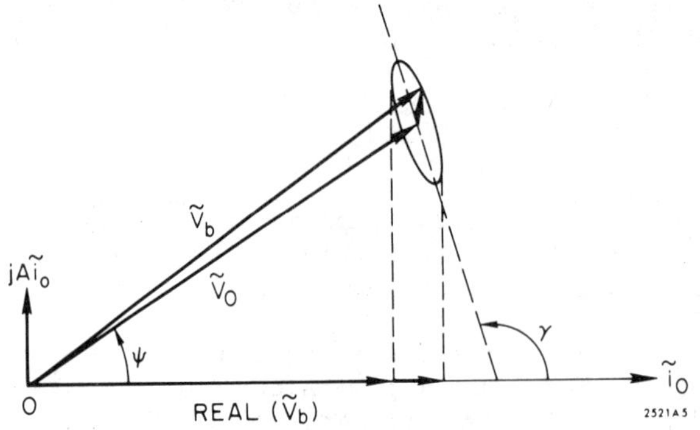

Fig. 4.1. Response of a resonant circuit to a phase-modulated driving current.

$\delta q = (\pi A i_0/\omega_s)\cos\omega_s t$ (the relation between charge and current is given in Sec. 9.1). Assume small damping and integrate $\delta V dq = \delta V[d/dt(\delta q)]dt$ to obtain the energy in the oscillation,

$$\delta W = \frac{\pi A^2 i_0 V_c \sin\phi}{2\omega_s} \quad .$$

Likewise, the average power transfer to the oscillation can be shown to be $\bar{P} = (i_0/V_{br})$ times the area πab of the ellipse, where a and b are given by Eqs. (4.14).

Problem 4.5: Show that $\bar{P} = \pi i_0 ab/V_{br}$ using the following procedure. First, take the real part of Eq. (4.13) to find $\delta V_b(t)$. Then $P(t) = \delta q[d/dt(\delta V_b)]$ is the instantaneous power transfer during the oscillation. Average over one cycle of $\omega_s t$ to find $\bar{P}$.

The damping time is now obtained from

$$\frac{1}{\tau_d} = \frac{1}{2}\frac{\bar{P}}{\delta W} = \frac{V_{br}\,\omega_s}{V_c \sin\phi} \cdot \frac{-\xi\eta}{[1+(\xi+\eta)^2][1+(\xi-\eta)^2]} \quad . \qquad (4.15)$$

Here a negative τ_d implies damping, and a positive τ_d growth of the phase oscillation. Whether there is growth or damping of the oscillation depends on the direction that the ellipse in Fig. 4.1 is followed with time, and in turn this depends on the sign of ξ. Positive ξ (or negative tuning angle) gives damping. The origin of the damping can be traced to the inertia of the stored energy in the rf cavities. Because of the finite filling time, the beam-induced voltage cannot follow changes in beam current instantaneously. A phase difference between the induced voltage and driving current appears, which in turn leads to an energy interchange between the oscillation and the cavity fields.

A somewhat different derivation of Eq. (4.15) is given in Ref. 20. It is also shown there that the synchrotron oscillation frequency is shifted as the beam loading increases. The limit of zero frequency is just the stability limit in Eq. (4.11). The condition $\xi > 0$ is the dynamic stability condition, also derived by Robinson.[18] It is worth noting that the dynamic condition $\xi > 0$ and the "static" condition in Eq. (4.11) have analogs in any high frequency resonant system in which the stored energy is modulated by a low frequency parametric variation. For example, Ceperly[21] has analyzed the electromechanical oscillations which result from the modulation of the resonant frequency of a cavity by mechanical vibrations. In this case, the mechanical oscillation is coupled to the rf stored energy through the force exerted by the rf fields on the cavity walls. Ceperly concludes that in this case the oscillations are antidamped for $\omega > \omega_0$, and that for $\omega < \omega_0$ a static instability occurs as the cavity fields increase and the modulation frequency goes to zero, corresponding to the limit in Eq. (4.11).

As a final comment, we note that Robinson damping operates only on the center of charge of the bunch as a whole. Radiation damping, on the other hand, acts on the incoherent synchrotron oscillations of the individual particles within the bunch.

5. TRAVELING-WAVE LINACS

5.1 Basic Principles

Consider a traveling wave for a given mode of propagation in a structure of arbitrary cross section with periodic length p along the z axis. By Floquet's theorem,[22] at a given frequency the fields at one cross section differ from those one period away only by a complex constant. Thus

$$\vec{E}(r,\phi,z,t) = \vec{E}_p(r,\phi,z)\, e^{-\gamma z}\, e^{j\omega t} \quad , \tag{5.1}$$

where $\gamma = j\beta + \alpha$ is the propagation constant and $\vec{E}_p(r,\phi,z)$ is periodic in z with period p. Expanding $\vec{E}_p(r,\phi,z)$ in a Fourier series,

$$\vec{E}(r,\phi,z,t) = \sum_{n=-\infty}^{\infty} \vec{E}_n(r,\phi)\, e^{j(\omega t - \beta_n z)}\, e^{-\alpha z} \quad , \tag{5.2}$$

where

$$\beta_n = \beta_o + \frac{2\pi n}{p} \tag{5.3}$$

and

$$\vec{E}_n(r,\phi) = \frac{1}{p} \int_z^{z+p} \vec{E}_p(r,\phi,z)\, e^{j(2\pi n/p)z} dz \quad . \tag{5.4}$$

Thus the total traveling-wave field has been expanded in a series of space harmonics, each with its own propatation constant β_n and phase velocity $v_{pn} = \omega/\beta_n$, but with all space harmonic having the same group velocity $v_g = d\omega/d\beta$. These relationships are illustrated by the dispersion curve (also called a Brillouin diagram or $\omega - \beta$ diagram) in Fig. 5.1.

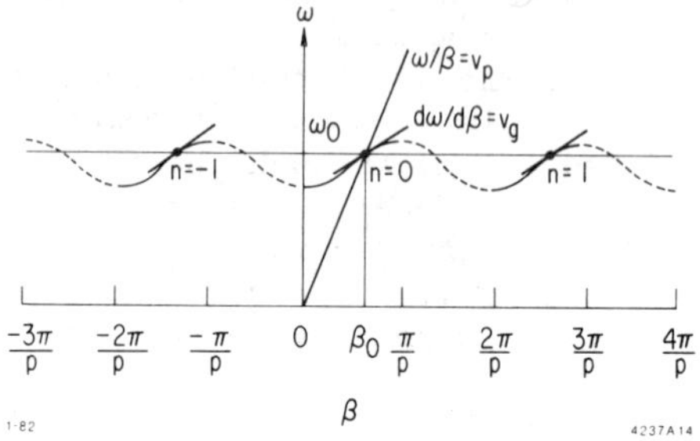

Fig. 5.1. Dispersion diagram for a periodic accelerating structure.

Consider the specific case of a cylindrically symmetric structure. In the neighborhood of the axis, the accelerating field component for a given propagating mode in a lossless structure has the form (see for example Ref. 23),

$$E_z(r,\phi,z,t) = \cos m\phi \sum_{n=-\infty}^{\infty} A_n J_m(X_n r)\, e^{j(\omega t - \beta_n z)} \qquad (5.5)$$

where

$$X_n^2 = (\omega/c)^2 - \beta_n^2 \quad . \qquad (5.6)$$

For a space harmonic component which is synchronous with a velocity of light particle, $\beta_n = \omega/c$ and $E_z \sim r^m$ in the region of the beam aperture.[23] Thus for a synchronous wave in the accelerating mode ($m = 0$), the accelerating field is independent of transverse position within the beam aperture. The structure design problem now consists of several parts. First, at the operating frequency the transverse dimensions of the structure are adjusted to obtain synchronism with the fundamental space harmonic component ($n = 0$). Second, the geometry of the structure is chosen, in so far as possible, to reduce the amplitudes of the non-synchronous space harmonic components. These components carry energy which can play no part in the acceleration of particles. Third, the geometry is adjusted to reduce the stored energy per unit length for a given synchronous accelerating field. Finally, the Q of the structure is maximized by choosing a structure material, usually copper, with good conductivity. As in the case of standing wave cavities, the Q of the structure does not depend very strongly on the shape of the individual periodic cells. The Q does, however, increase if there are fewer periods per unit length (for example, fewer disks per wavelength in a disk-loaded structure). But then the amplitudes of the non-synchronous space harmonic compenents tend also to increase for a given synchronous component amplitude. These trade-offs are explored in detail in Ref. 2, Ch. B.1.1, for the case of the SLAC-type disk-loaded structure.

5.2 Structure Parameters

If $E_a = E_{zn}$ is the accelerating field for the synchronous traveling-wave space harmonic component and w the total stored energy per unit length in the propagating wave with power flow P, then the shunt impedance per unit length and the structure Q are defined by

$$r \equiv \frac{E_a^2}{|dP/dz|} \tag{5.7a}$$

$$Q \equiv \frac{\omega w}{|dP/dz|} \tag{5.7b}$$

$$\frac{r}{Q} = \frac{E_a^2}{\omega w} \quad . \tag{5.7c}$$

We can define an energy flow velocity by $v_E = P/w$. In Ref. 22, Sec. 1.5, it is proven that $v_E = v_g$, where v_g was defined as $d\omega/d\beta$. Thus, from the expression for Q,

$$\frac{dP}{dz} = -\frac{\omega P}{v_g Q} = -2\alpha P \tag{5.8a}$$

$$\frac{dE_a}{dz} = -\alpha E_a \tag{5.8b}$$

$$\alpha = \frac{\omega}{2v_g Q} \tag{5.8c}$$

where α is the attenuation parameter per unit length. The relation between power flow and accelerating field is now obtained as

$$E_a^2 = r|dP/dz| = 2\alpha r P \tag{5.9a}$$

$$\alpha r = \frac{\omega r}{2v_g Q} \quad . \tag{5.9b}$$

A structure which has uniform parameters along its length is called a constant impedance structure.* For such a structure, Eqs. (5.8) can be integrated to yield

$$E_a = E_o \, e^{-\alpha z} \tag{5.10a}$$

$$P = P_o \, e^{-2\alpha z} \tag{5.10b}$$

where E_o and P_o are the accelerating field and power flow at the imput to the structure. The field and power flow at the end of a structure of length L are then $E_L = E_o \, e^{-\tau}$ and $P_L = P_o \, e^{-2\tau}$, where

$$\tau = \alpha L = \frac{\omega L}{2 v_g Q} \tag{5.11}$$

is the attenuation parameter for the structure.

Consider now an accelerating mode (no variation with azimuthal angle) propagating in a disk-loaded structure with disk hole radius a. In the disk hole region, both H and E_r are proportional to r near the axis. Thus, if the disk opening is not too large, the power flow per unit area for a given stored energy per unit length is proportional to r^2. Integrating from r = a, the total power flow, and thus the group velocity, will be proportional to a^4. From Eq. (5.8c), it is therefore possible to change α over a wide range by varying the disk aperture over a relatively small range. Of course, the shunt impedance per unit length will also vary as the disk opening is changed, but its dependence on the disk hole radius is much weaker. From Eq. (5.9a) the possibility now exists, as the power flow along the structure decreases due to dissipation in the structure walls, to keep E_a constant by increasing $\alpha \sim 1/P$. This is the basis for the constant gradient structure.

Let us ignore the weak variation in r along the length of such a structure. Then from Eq. (5.7a) dP/dz = constant, or

$$P = P_o - (P_o - P_L)(z/L) \quad . \tag{5.12}$$

If the attenuation parameter τ is again defined from the expression $P_L = P_o \, e^{-2\tau}$, the above relation gives

$$\frac{P}{P_o} = 1 - (z/L)(1 - e^{-2\tau}) \tag{5.13a}$$

---

*Note that $\alpha r = E_a^2/2P$ has dimensions of ohms/m^2. A closely related quantity used in microwave circuit theory, $E_n^2/(2\beta_n^2 P)$ is called the coupling impedance, or sometimes the interaction impedance for the nth space harmonic component.

$$\frac{dP}{dz} = - \frac{P_o - P_L}{L} = - \frac{P_o}{L}(i - e^{-2\tau}) \quad . \tag{5.13b}$$

From Eqs. (5.8a), (5.8c), (5.13a) and (5.13b) the variation in group velocity with length required to produce a constant gradient is seen to be

$$v_g(z) = \frac{\omega L}{Q} \frac{[1 - (z/L)(1 - e^{-2\tau})]}{1 - e^{-2\tau}} \quad . \tag{5.14}$$

Problem 5.1: The filling time for a constant impedance (and hence constant group velocity) structure is simply $T_f = L/v_g$. By integrating $dt = dz/v_g$ from $z = 0$ to $z = L$ using Eq. (5.14), show that for a constant gradient structure

$$T_f = \tau\left(\frac{2Q}{\omega}\right) \quad . \tag{5.15}$$

From Eq. (5.11), note that this is exactly the same as the filling time for a constant impedance structure.

5.3 Energy Gain and Beam Loading

By integrating Eq. (5.10a) from $z = 0$ to $z = L$, and substituting for $E_o = (2\alpha r P_o)^{\frac{1}{2}}$ according to Eq. (5.9a), the unloaded energy gain of a constant impedance (CZ) accelerating section is calculated to be

$$\text{CZ:} \quad V_o = (rLP_o)^{\frac{1}{2}} \left[(2/\tau)^{\frac{1}{2}}(1 - e^{-\tau})\right] \quad . \tag{5.16}$$

The unloaded energy gain of a constant gradient (CG) section is, using Eq. (5.9a),

$$V_o = E_o L = (rLP_o)^{\frac{1}{2}} (2\alpha_o L)^{\frac{1}{2}} \quad . \tag{5.17}$$

Using Eqs. (5.8a) and (5.13b), the above expression becomes

$$\text{CG:} \quad V_o = (rLP_o)^{\frac{1}{2}} (1 - e^{-2\tau})^{\frac{1}{2}} \quad . \tag{5.18}$$

As a function of τ, Eq. (5.16) has a broad maximum at $\tau = 1.26$ where $V_o/(rLP_o)^{\frac{1}{2}} = 0.90$. For the case of a constant gradient structure, V_o approaches $(rLP_o)^{\frac{1}{2}}$ for large τ.

We next compute the beam induced field in a traveling-wave structure, assuming that there is no input power from the rf generator. If there is a generator-produced field component, the net accelerating voltage is readily obtained using superposition. From conservation of energy, at any point in the structure

$$\frac{dP}{dz} = I_o E_b - 2\alpha P \quad , \tag{5.19}$$

where E_b is the peak beam-induced field which opposes the motion of a beam of short bunches with dc current I_o. Using $E_b^2 = 2\alpha r P$, this becomes

$$\frac{dE_b}{dz} = I_o \alpha r - \alpha E_b \quad . \tag{5.20}$$

Now assume a constant impedance structure (α independent of z) and integrate to obtain

$$E_b(z) = I_o r(1 - e^{-\alpha z}) \quad . \tag{5.21}$$

Integrate again to find the energy,

$$V_b = I_o rL \, [1 - (1 - e^{-\tau})/\tau] \quad . \tag{5.22}$$

Problem 5.2: Start from Eq. (5.20) and compute the beam-loading voltage for the case of a constant gradient structure. Use $\alpha(z) = \omega/2Qv_g(z)$, where $v_g(z)$ is given by Eq. (5.14). Check your result in Eq. (5.23b).

The results of Eq. (5.22) and Problem 5.2 can be used, together with superposition, to express the net voltage gain in a beam-loaded structure as

$$V = V_o \cos\theta - mI_o$$

$$\text{CZ:} \quad m = rL \left[1 - \frac{1 - e^{-\tau}}{\tau} \right] \tag{5.23a}$$

$$\text{CG:} \quad m = rL \left[\frac{1}{2} - \frac{e^{-2\tau}}{1 - e^{-2}} \right] \tag{5.23b}$$

where V_o is the unloaded energy gain given by Eqs. (5.16) and (5.18), and θ is phase of the current bunches with respect to the crest of the generator-produced wave. If the bunches are not short compared to the rf wave length, the energy gain is reduced by the same bunch form factor computed in Sec. 3.3.

In Fig. 5.2 the energy gain for constant gradient and constant impedance structures is plotted as a function of current for several values of τ. Note the linear load lines, similar to those in Fig. 3.10 for the case of a standing wave structure. Note, in addition, that τ and $1/\beta$ play similar roles in the two types of structures. This can also be seen from the expressions for the filling time,

$$T_f(SW) = \frac{2Q_o}{\omega_o(1+\beta)} \tag{5.24a}$$

$$T_f(TW) = \frac{2Q\tau}{\omega} \quad . \tag{5.24b}$$

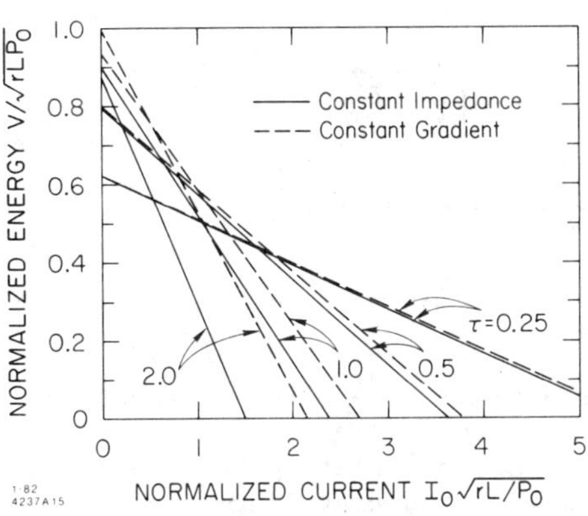

Fig. 5.2. Beam-loaded energy as a function of beam current for constant impedance and constant gradient structures for several values of the attenuation parameter τ.

Multiplying Eq. (5.23) by I_o, the power transferred to the beam, and hence the conversion efficiency, is seen to be quadratic as a function of beam current. Recall that this was also the case for standing wave structures (see Fig. 3.11). The maximum efficiency is reached when the beam energy is reduced to one-half of its unloaded value at $I_o = V_o/2m$. The maximum conversion efficiency is then

$$\eta_{max} = \frac{V_o^2}{4\,mP_o} \quad . \tag{5.25}$$

As an example, η_{max} at $\tau = 1$ is 63% for a constant gradient structure and 54% for a constant impedance structure. These efficiencies increase to 76% and 73% respectively at $\tau = 0.5$, and both increase toward 100% as τ approaches zero according to $\eta_{max} \approx (1 - 2\tau/3)$.

The power flowing into the output termination of a beam-loaded traveling-wave section can be computed by first finding the net field at the load. For example, using Eqs. (5.21), (5.10a) and (5.9a) in the case of a constant impedance structure,

$$\begin{aligned} E_L &= E_o e^{-\tau} - I_o r(1-e^{-\tau}) \\ P_L &= E_L^2\, L/2\tau r \quad . \end{aligned} \tag{5.26}$$

The power dissipated in the structure is then obtained as $P_s = P_o - P_L - I_o V$.

5.4 Non-Synchronous Operation

If a traveling-wave structure is operated at a frequency different than the synchronous frequency, the bunches will slip in phase with respect to the traveling wave. The total phase slip in length L for an electron with velocity $v_e \approx c$ is described by the parameter

$$\delta = \omega \left(\frac{L}{v_p} - \frac{L}{v_e} \right) \approx \beta_o L \left(1 - \frac{v_p}{c} \right) \quad . \tag{5.27}$$

In a constant gradient structure (or in a constant impedance structure for small τ) without beam loading we expect

$$V = V_o \left[\frac{\sin (\delta/2)}{(\delta/2)} \right] \approx V_o \left(1 - \frac{\delta^2}{24} \right) \tag{5.28}$$

for $\delta << 1$. Recall that for a standing-wave cavity,

$$V = V_o \cos\psi \approx V_o \left(1 - \frac{\psi^2}{2} \right) \quad . \tag{5.29}$$

Thus the phase-slip parameter plays a similar role for a traveling-wave structure as the tuning angle does for a standing-wave cavity. This correspondence is evident also from the relation between ψ and δ and the filling times for standing-wave and traveling-wave structures. For a frequency derivation $\Delta\omega = \omega - \omega_o$ and using $\Delta\beta \approx \Delta\omega / v_g$,

$$\begin{aligned} \psi &= \tan^{-1} \left(\frac{2Q_L}{\omega_o} \Delta\omega \right) \approx T_f(SW) \cdot \Delta\omega \\ \delta &= L(\beta_o - \beta) = L\, \Delta\beta \approx T_f(TW) \cdot \Delta\omega \quad . \end{aligned} \tag{5.30}$$

In both cases, the sensitivity to tuning errors is seen to be proportional to the filling time.

A detailed discussion of non-synchronous beam loading in constant impedance structures is given in Ref. 24.

6. SINGLE-BUNCH BEAM LOADING

6.1 The Fundamental Theorem of Beam Loading

Consider a point charge crossing a cavity initially empty of energy. After the charge has passed out of the cavity, a beam-induced voltage V_{bn} remains in each mode. What fraction of V_{bn} does the charge itself see? Since the induced voltage for mode n starts at zero as the charge enters the cavity, and ends up at V_{bn} as the charge exits from the cavity, the most naive assumption is to take the average, or $1/2\ V_{bn}$, as the effective fraction of the induced voltage acting on the charge. In this section we prove that this factor of one-half is indeed exact for any cavity. The fact that a charge "sees" exactly one-half of its own beam-induced voltage we will call the fundamental theorem of beam loading. The theorem provides the key which relates the energy loss by a charge crossing a cavity or passing through a structure to the electromagnetic properties of modes in the cavity or structure computed in the absence of any charge. By superposition, the beam-induced voltage in a cavity is the same whether or not a generator voltage component is present. Thus the theorem is also basic to the computation of the effective voltage acting on a bunch when both a generator voltage and a beam-induced voltage are present. Following is one of several possible proofs of the theorem.

Let a charge pass through a cavity in which the stored energy is related to the cavity voltage in a given mode by

$$W = \alpha V^2 \quad . \tag{6.1}$$

Assume that a fraction f of the beam-induced voltage V_b acts on the particle, or $V_e = fV_b$ where V_e is the effective voltage seen by the charge. Assume further that the beam-induced voltage is not necessarily at such a phase as to maximally oppose the motion of the charge; that is, assume it might lie at an angle ε with respect to V_e. Now let the charge be bent back around in a lossless manner, for example by magnetic fields, such that it passes through the cavity a second time. Let the time for the charge to traverse the external path be any multiple n of the rf period, plus a residual time θ/ω_0 where θ is an arbitrary angle and ω_0 is the resonant frequency of the mode. When the particle crosses the cavity reference plane a second time, we have the phasor addition of voltages shown in Fig. 6.1.

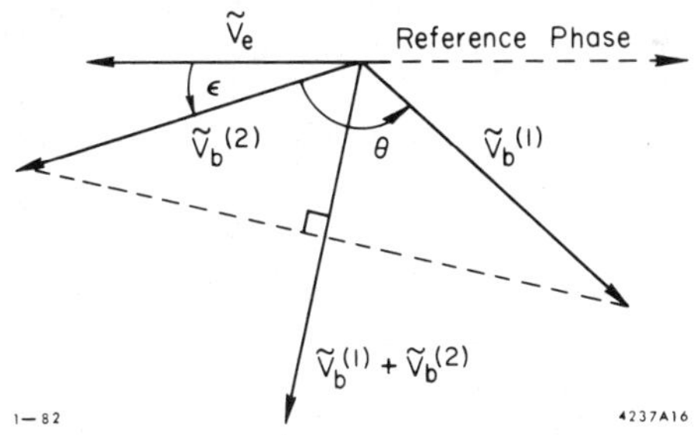

Fig. 6.1. Diagram showing addition of beam-induced voltages for two passes by the same charge through a cavity.

Here $\tilde{V}_b(2)$ is voltage induced on the second pass by the charge, while the voltage induced on the first pass, $\tilde{V}_b(1)$, has rotated with respect to $\tilde{V}_b(2)$ by an angle $2\pi n + \theta$. We can assume the cavity losses are very small so that $V_b(2) = V_b(1)$. Thus the net energy stored in the cavity is

$$\begin{aligned} W_c &= \alpha \left(2\, V_b \cos \frac{\theta}{2}\right)^2 \\ &= 2\alpha V_b^2 (1 + \cos\theta) \quad . \end{aligned} \tag{6.2}$$

On the other hand, the energy lost by the particle on the two passes is

$$\Delta U = 2\, q\, V_e + q\, V_b \cos(\varepsilon + \theta) \quad . \tag{6.3}$$

That is, on the first pass the charge experiences a retarding voltage V_e, while on the second pass it sees the sum of V_e plus the component of $V_b(1)$ which lies along the negative real axis in the phasor diagram. By conservation of energy W_c and ΔU must be equal. Letting $V_e = fV_b$ and equating Eqs. (6.2) and (6.3) we have

$$2(qf - \alpha V_b) + (q \cos\varepsilon - 2\alpha V_b)\cos\theta - (q \sin\varepsilon)\sin\theta = 0 \ .$$

The left-hand side can only vanish for arbitrary θ if

$$\sin\varepsilon = 0, \qquad \varepsilon = 0 \tag{6.4a}$$

$$V_b = q/2\alpha \tag{6.4b}$$

$$f = \alpha V_b/q = 1/2 \quad . \tag{6.4c}$$

Eq. (6.4a) expresses the fact that the beam-induced voltage must have a phase such as to maximally oppose the motion of the inducing charge. (Is $\varepsilon = \pi$ a valid solution to Eq. (6.4a)?) Equation (6.4c) tells us that the charge sees exactly one-half of its own beam induced field. Combining Eqs. (6.1) and (6.4b), we obtain

$$W = \alpha V_b^2 = \frac{q^2}{4\alpha} \equiv kq^2 \tag{6.5}$$

for the energy left behind in a cavity by a charge q. The quantity k is called the loss parameter, and, of course, each resonant mode has its own value of k. From Eqs. (6.5) and (6.1) we have

$$\alpha = \frac{1}{4k} \tag{6.6a}$$

$$k = \frac{V^2}{4W} \quad . \tag{6.6b}$$

Further, from Eqs. (6.4b) and (6.6a),

$$V_b = 2kq \tag{6.7a}$$

$$V_e = \frac{V_b}{2} = kq \quad . \tag{6.7b}$$

Thus the loss parameter k relates the beam-induced voltage to the charge, by Eq. (6.7a), and the energy loss by a charge passing through a cavity initially empty of energy, by Eq. (6.5). It is important to note that superposition applies and Eqs. (6.7) are valid even if a voltage is already present in the cavity before the charge arrives. We can therefore construct the basic phasor diagram in Fig. 6.2 for single-bunch beam loading for the accelerating mode ($k = k_0$, $V_b = V_{bo}$), or for any mode with an externally applied generator voltage. Here $\widetilde{V}_e = -k_0 q$ is the effective beam loading voltage seen by the charge. The reference phase is taken in the direction $-\widetilde{V}_e$. Thus the net accelerating voltage acting on the charge is

$$V_a = V_c \cos\phi = V_g \cos\theta_g - k_0 q \quad , \tag{6.8}$$

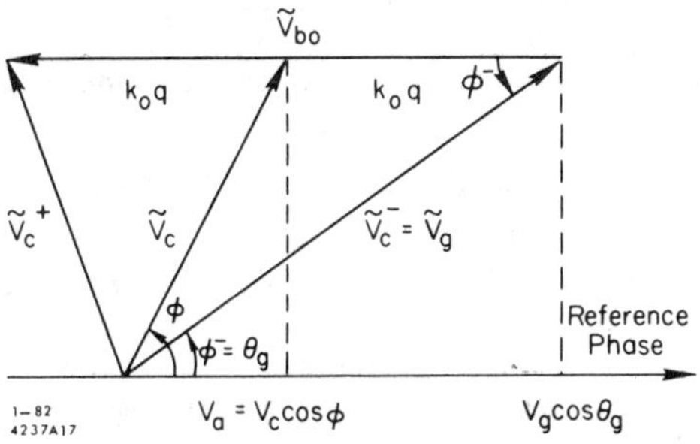

Fig. 6.2. Phasor diagram showing the net single-bunch energy gain for a cavity driven by an external rf source.

where $\phi^- = \theta_g$ is the phase of the generator voltage component just before the charge crosses the cavity reference plane.

Problem 6.1: Prove that relations (6.4), (6.6) and (6.7) are also valid when a generator voltage component is present. Using Fig. 6.2, compute the decrease in cavity stored energy, $\Delta W = \alpha[(V_c^-)^2 - (V_c^+)^2]$. Using conservation of energy, equate this to the energy gained by charge q. Write the result in the form $V_g = f_1(V_b)/f_2(V_b)$ where both f_1 and f_2 must vanish, since V_g cannot depend on V_b.

As a final comment, note that the parameter k_n describes the single-bunch beam loading properties of the nth cavity mode, and that it can be computed in terms of the charge-free properties of the cavity from Eq. (6.6b). As described in Sec. 3.4, the programs LALA[5] and SUPERFISH[6] compute the quantity $R_a/Q = V_a^2/\omega W$. Then from Eq. (6.6b)

$$k = \frac{\omega}{4} (R_a/Q) \quad . \tag{6.9}$$

6.2 Higher-Order Cavity Modes and the Loss Impedance

Consider the energy lost by a charge to all modes in an rf accelerating cavity, assuming the cavity is initially empty of stored energy before the arrival of the charge. Let ΔU_o be the energy lost to the fundamental (accelerating) mode, and

$$\Delta U_t = B\Delta U_o \tag{6.10}$$

be the total energy lost to all modes, where B is called the beam loading enhancement factor. The energy lost to higher-order cavity modes only is

$$\Delta U_{hm} = (B-1)\ \Delta U_o \quad . \tag{6.11}$$

After the charge has exited from the cavity, a beam-induced voltage $\tilde{V}_{bo}$ and corresponding stored energy $\Delta U_o = \alpha_o V_{bo}^2$ remain in the fundamental mode. Then From Eqs. (6.11), (6.6a) and (6.7a),

$$\Delta U_{hm} = \alpha_o (B-1) V_{bo}^2 = (B-1)\ k_o q^2 \quad . \tag{6.12}$$

As discussed earlier, $V_{bo} = k_o q$ is a voltage which, by superposition, is the same whether or not there is energy stored in the fundamental mode before the arrival of the charge. Equation (6.12) therefore is valid also when the fundamental mode is driven by an external generator.

Consider now a linac or storage ring with equal bunches of charge q spaced apart in time by T_b. If the fields in each cavity mode decay away completely between bunches ($T_b \gg T_{fn}$ for all modes), and using also $q = I_o T_b$ where I_o is the average current, Eq. (6.12) gives

$$P_{hm} = \frac{1}{T_b} \Delta U_{hm} = I_o^2 Z_{hm} = I_o V_{hm} \tag{6.13a}$$

$$V_{hm} = I_o Z_{hm} \tag{6.13b}$$

$$Z_{hm} \equiv (B-1)\ k_o T_b = T_b \sum_{n>o} k_n \quad . \tag{6.13c}$$

In a storage ring the presence of higher-order cavity modes mean that, in addition to the synchrotron radiation loss per turn V_s, the rf system must supply an accelerating voltage V_{hm}. There are also losses to other vacuum chamber components outside the rf system. If the sum of all the loss parameters for these components is k_{vc}, and if it is again assumed that the induced fields decay away between bunches, then

$$\begin{aligned} P_{vc} &= I_o^2 Z_{vc} \\ V_{vc} &= I_o Z_{vc} \\ Z_{vc} &= k_{vc} T_b \quad . \end{aligned} \tag{6.14}$$

Thus the total accelerating voltage that must be supplied by the rf system to each beam in a storage ring is

$$V_a = V_s + V_{hm} + V_{vc} \quad . \tag{6.15}$$

If the beam induced fields do not decay away between bunches for a particular mode, the situation is more complicated. The resonance function, described in Sec. 6.5, is then needed to compute the voltage lost to that mode.

For simplicity, the expressions in this and the preceding section have been written assuming a point bunch. For a bunch of non-zero length, the bunch form factor must be taken into account. For a Gaussian bunch, the loss parameter for each mode must be multiplied by $e^{-\omega_n^2 \sigma_t^2}$ (see Sec. 9.4).

6.3 Efficiency for Energy Extraction from a Cavity

In a linac or storage ring rf system, the beam takes energy from the driven fundamental mode, but dumps some of it back into the higher cavity modes. It is of interest to compute the net energy extracted from the cavity. If we apply the law of cosines to the vector triangle ($\tilde{V}_c^+$, $\tilde{V}_c^-$, $\tilde{V}_{bo}$) in Fig. 6.2,

$$(V_c^+)^2 = (V_c^-)^2 + V_{bo}^2 - 2V_c^- V_{bo} \cos\phi^- \quad . \tag{6.16}$$

By conservation of energy, the energy extracted from the accelerating mode is $\Delta U_o = \alpha_o[(V_c^-)^2 - (V_c^+)^2]$. Using Eq. (6.16),

$$\Delta U_o = \alpha_o\left(2V_c^- V_{bo} \cos\phi^- - V_{bo}^2\right) \quad . \tag{6.17}$$

To obtain the net energy extracted from the cavity, we subtract off the energy put back into higher-order modes, as given by Eq. (6.12), to obtain

$$\Delta U_{net} = \Delta U_o - \Delta U_{hm} = \alpha_o\left(2V_c^- V_{bo} \cos\phi^- - BV_{bo}^2\right) \quad . \tag{6.18}$$

The efficiency for net energy extraction is now

$$\eta = \frac{\Delta U_{net}}{\alpha_o (V_c^-)^2} = 2\left(\frac{V_{bo}}{V_c^-}\right)\cos\phi^- - B\left(\frac{V_{bo}}{V_c^-}\right)^2 \quad . \tag{6.19}$$

The maximum efficiency as a function of V_{bo} for a given initial stored energy is readily obtained to be

$$\eta_{max} = \frac{\cos\phi^-}{B} \tag{6.20}$$

at a beam-induced voltage

$$V_{bo} = \frac{V_c^- \cos\phi^-}{B} \quad . \tag{6.21}$$

Note that angle ϕ^- in Eq. (6.14) is <u>not</u> the synchronous phase angle for a storage ring. It is the phase angle of the cavity voltage just before the arrival of the bunch. From Fig. 6.2 it is related to the synchronous phase angle by

$$\tan\phi^- = \frac{\tan\phi}{1 + k_o q/V_a} \quad . \tag{6.22}$$

Problem 6.2: A storage ring is often operated with two counter-circulating beams of opposite charge and equal intensity. The rf cavities are located so that the fields induced in the fundamental mode by the q^+ and q^- charges are coherent; that is, the cavities are located at distances from the interaction point which are integral multiples of a half-wavelength at the accelerating mode frequency. On the other hand, it is reasonable to assume that the fields induced in the higher cavity modes are, on the average, incoherent for the two beams (see discussion in Sec. 6.5). Show that for this case the maximum efficiency for energy extraction is

$$\eta_{max} = \frac{\cos^2\phi^-}{(B+1)/2} \quad . \tag{6.23}$$

6.4 Beam Loading by a Bunch Train with $T_b \sim T_f$

We next calculate the build-up of the beam-induced voltage when bunches pass repetitively through a cavity, as in the rf system of a storage ring or for a train of equally-spaced bunches in a linac. A cavity filling time is assumed which is not necessarily short compared to the bunch spacing. The situation is illustrated graphically in Fig. 6.3. Here $\tilde{V}_{bo}$ is the single-pass beam-induced voltage, $e^{-\tau}$ gives the decay of the cavity fields during one turn, δ is the net phase shift per turn (subtracting off multiples of 2π) and $\tilde{V}_b^-$ and $\tilde{V}_b^+$ are the cavity of voltages for $t \to \infty$ just before and just after the passage of a bunch. The decay parameter τ and phase angle δ can be written

$$\tau = \frac{T_b}{T_f} \tag{6.24a}$$

$$\delta = T_b\omega_o - 2\pi h_b \tag{6.24b}$$

$$= T_b(\omega_o - \omega) \quad .$$

Here ω_o is the resonant frequency of the cavity and h_b, an integer, is the harmonic number for a single-bunch machine, or the number of rf wavelengths between bunches for a linac or for a ring with more than one bunch.

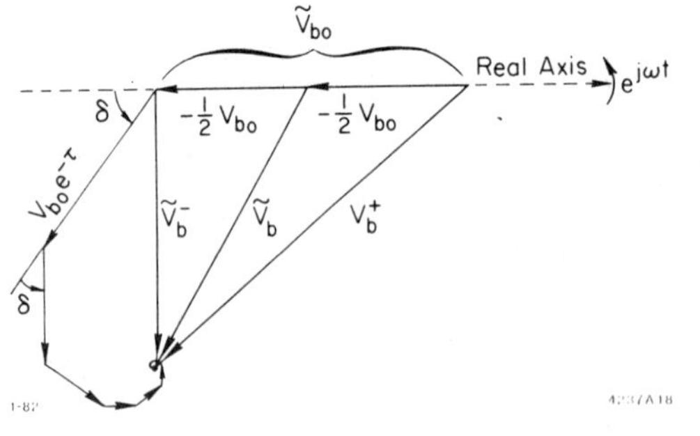

Fig. 6.3. Phasor diagram showing the buildup of the beam-induced voltage by a train of bunches of equal charge.

In constructing Fig. 6.3, we again consider a reference frame which is rotating at the angular frequency ω of the external rf generator. It is natural to use the external generator as the basic clock for describing field variations in the cavity, since the spacing of bunches in a storage ring is determined by the driving frequency of the generator and not by the cavity resonant frequency.

The final ($t \to \infty$) voltage just after a bunch passage is now readily obtained as the sum of the geometric series

$$V_b^+ = \tilde{V}_{bo}(1 + e^{-\tau} e^{j\delta} + e^{-2\tau} e^{j2\delta} + \ldots.)$$

$$\frac{V_b^+}{\tilde{V}_{bo}} = \frac{1}{1 - e^{-\tau} e^{j\delta}} \quad . \tag{6.25a}$$

To obtain the effective beam-loading field $\tilde{V}_b$ in the limit $t \to \infty$, we take the field $\tilde{V}_b^-$ induced by all the previous bunch passages at a time just before the arrival of a bunch at the cavity reference plane, and add to it a phasor $\tilde{V}_e = -1/2\ V_{bo}$ to account for the effective self-field seen by the bunch in question to obtain

$$\tilde{V}_b = \tilde{V}_b^- - \frac{1}{2} V_{bo} = \tilde{V}_b^+ + \frac{1}{2} V_{bo} \quad . \tag{6.25b}$$

Using this expression together with Eq. (6.25a) and the fact that $\tilde{V}_{bo} = -V_{bo}$,

$$\frac{\tilde{V}_b}{(-V_{bo})} = \frac{1}{1 - e^{-\tau} e^{j\delta}} - \frac{1}{2} = F_R(\tau,\delta) + jF_I(\tau,\delta) \tag{6.26a}$$

$$F_R(\tau,\delta) = \frac{1 - e^{-2\tau}}{2(1 - 2e^{-\tau} \cos\delta + e^{-2\tau})} \tag{6.26b}$$

$$F_I(\tau,\delta) = \frac{2e^{-\tau} \sin\delta}{2(1 - 2e^{-\tau} \cos\delta + e^{-2\tau})} \quad . \tag{6.26c}$$

These expressions give the real and imaginary parts of the enhancement of the single-bunch beam loading voltage due to resonant build-up.

The quantities τ, δ and V_{bo} in Eqs. (6.26) can be expressed in terms of more usual cavity parameters. The voltage decay parameter per turn is

$$\tau = \tau_o(1 + \beta), \qquad \tau_o = T_b/T_{fo} \tag{6.27}$$

where $T_{fo} = 2Q_o/\omega_o$. From the definition of the tuning angle, $\psi = 2Q_L (\omega_o - \omega)/\omega_o = T_f(\omega_o - \omega)$ and Eq. (6.24b), we have

$$\delta = \tau \tan\psi \quad . \tag{6.28}$$

The single-bunch beam loading voltage can also be written as

$$V_{bo} = 2k_o q = \frac{\omega_o}{2}\left(\frac{R_a}{Q}\right) q = \frac{I_o R_a}{1+\beta}\tau = I_o R_a \tau_o \quad , \tag{6.29}$$

where I_o is dc current (assuming short bunches) or the total circulating current for both beams in a storage ring. Equation (6.26a) can now be written in the form

$$\tilde{V}_b = -I_o R_a \tau_o \, [F_R(\tau_o,\beta,\psi) + jF_I(\tau_o,\beta,\psi)] \quad . \tag{6.30}$$

In a storage ring the desired net cavity voltage, including the effect of beam loading, is usually specified; that is, a certain accelerating voltage $V_c \cos\phi$ and synchronous phase angle ϕ are required. If the beam current and cavity parameters are specified, then the generator voltage can be obtained from the phasor relation

$$\tilde{V}_g = \tilde{V}_c - \tilde{V}_b \quad . \tag{6.31}$$

This is illustrated in Fig. 6.4, in which a constant generator voltage has been added to the beam-induced voltages shown in Fig. 6.3.

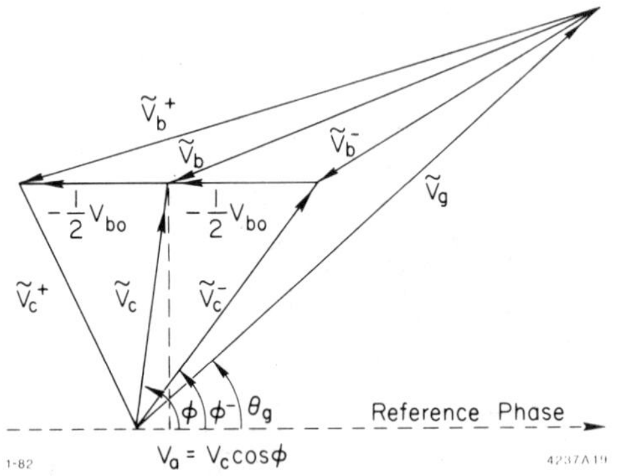

Fig. 6.4. Vector sum of voltages in a beam-loaded cavity driven by an external generator.

Let us now compute the required generator power for a linac or storage ring rf system with beam loading under the condition $T_b \sim T_f$. Taking the real and imaginary components of the preceding phasor relation and using also the notation in Fig. 6.4, together with Eq. (6.26a), we obtain

$$V_g \cos\theta_g = V_c \cos\phi + V_{bo} F_R(\tau,\delta) \tag{6.32a}$$

$$V_g \sin\theta_g = V_c \sin\phi + V_{bo} F_I(\tau,\delta) \quad . \tag{6.32b}$$

Squaring and adding these two expressions to eliminate θ_g, then using Eqs. (3.35a) and (3.42a) to express V_g^2 in terms of P_g, we have

$$P_g = \frac{V_c^2}{R_a \cos^2\psi} \cdot \frac{(1+\beta)^2}{4\beta} \cdot \left\{ \left[\cos\phi + \frac{I_o R_a \tau_o}{V_c} F_R(\tau_o,\beta,\psi) \right]^2 + \left[\sin\phi + \frac{I_o R_a \tau_o}{V_c} F_I(\tau_o,\beta,\psi) \right]^2 \right\} \quad . \tag{6.33}$$

Problem 6.3: Show that, in the limit $\tau_o \to 0$, the result in Eq. (6.33) approaches that in Eq. (4.2).

The phase angle of the generator voltage is obtained by dividing Eq. (6.32b) by Eq. (5.32a),

$$\tan\theta_g = \frac{V_c \sin\theta + V_{bo} F_I(\tau_o,\beta,\psi)}{V_c \cos\phi + V_{bo} F_R(\tau_o,\beta,\psi)} \quad . \tag{6.34}$$

For a given τ_o, the generator power in Eq. (6.33) can be minimized by varying β and ψ, although it is not possible to obtain simple analytic expressions as was the case for the minimization of Eq. (4.2) for $\tau << 1$. However, the minimum value of Eq. (6.33), and the corresponding values of β and ψ, are easily found numerically. It is found that the transient nature of the beam loading between bunches increases the minimum generator power by a few percent for typical cavity parameters for τ_o up to about 0.5. For $\tau_o > 1$ the generator power increases rapidly, and for large τ_o, where the time between bunches becomes large compared to the cavity filling time, it is clear that some sort of pulsed rf system is desirable. In such a system, power is applied to the cavities for about a filling time preceding the arrival of the bunch. For most of the period between bunches there is no stored energy in the rf cavities and hence no power dissipation. A discussion of pulsed rf systems for large storage rings is given in Ref. 25.

6.5 The Resonance Function

From Fig. 6.4 and Eq. (6.32a), the net accelerating voltage acting on a charge passing through an rf cavity is

$$V_a = V_c \cos\phi = V_g \cos\theta_g - k_o q[2F_R(\tau,\delta)] \quad . \tag{6.35}$$

Since $k_0 q$ is the effective beam loading voltage seen by a charge making a single passage through a cavity initially empty of energy, the factor $2F_R(\tau,\delta)$ takes into account the resonant build-up of fields due to successive bunch passages, either for a storage ring or for a train of bunches is a linac. For large τ, $2F_R$ is seen to approach unity, as expected. For small τ, Eq. (6.35) can be rewritten in a form which is more natural for a nearly continuous beam,

$$V_a = V_g \cos\theta_g - \frac{i_0 R_a}{1+\beta}\left[\tau F_R(\tau,\delta)\right] \quad .$$

In Problem 6.3 it was shown that τF_R approaches $\cos^2\psi$ in the limit $\tau \to 0$. This can be compared with the result of Eq. (4.1a), noting that $\theta_g = \theta + \psi$ (see Fig. 3.13).

In Fig. 6.5, the resonance function

$$2F_R(\tau,\delta) = \frac{1 - e^{-2\tau}}{1 - 2e^{-\tau}\cos\delta + e^{-2\tau}}$$

is plotted as a function of δ for two values of τ. Note that the maximum amplitude at resonance ($\delta = 0$) is given by

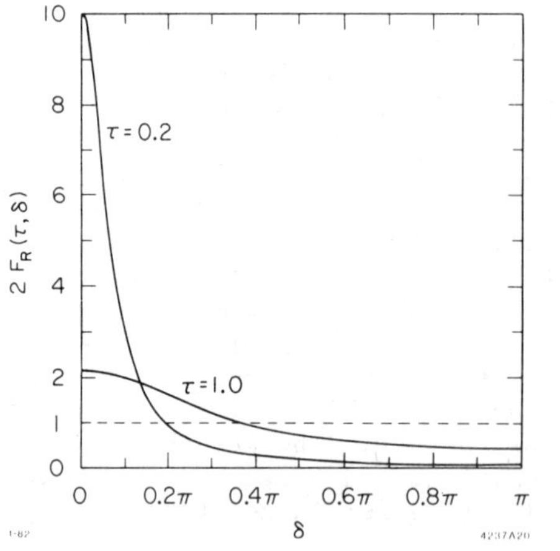

Fig. 6.5. The resonance function $2F_R(\tau,\delta)$ as δ for two values of the decay parameter τ.

$$2F_R(\tau,0) = \frac{1 + e^{-\tau}}{1 - e^{-\tau}}$$

$$2F_R(\tau,0) \approx \frac{2}{\tau} \qquad \tau \ll 1 \quad . \tag{6.37}$$

At anti-resonance ($\delta = \pi$),

$$2F_R(\tau,\pi) = \frac{1 - e^{-\tau}}{1 + e^{-\tau}}$$

$$2F_R(\tau,\pi) \approx \frac{\tau}{2} \qquad \tau \ll 1 \quad . \tag{6.38}$$

The phase angle which divides resonance and anti-resonance, that is, the value of δ at which $2F_R(\tau,\delta_1) = 1$, is seen to be

$$\cos\delta_1 = e^{-\tau} \tag{6.39}$$

$$\delta_1 \approx (2\tau)^{1/2} \qquad \tau \ll 1 \quad .$$

An important property of the resonance function has been pointed out by Sands.[26] The average value of the resonance function is computed to be

$$\langle 2F_R \rangle = \frac{1}{\pi} \int_0^{\pi} 2F_R(\tau,\delta)\, d\delta = 1 \quad . \tag{6.40}$$

Thus, if the phase shift δ is chosen at random, the expectation value of the resonance function is unity. For small τ, the maximum value of the resonance function is indeed very large, but the chance of finding $\delta < \delta_1$ is very small. In a storage ring, therefore, since the exact frequencies of the higher modes and hence the values of δ_n are never precisely known, it is reasonable to compute the higher-mode losses in the single-pass limit as $\Delta U_n = k_n q^2$, even though the factors T_b/T_{fn} might be small compared to unity. On the average, the single-pass limit ($2F_R = 1$) will be correct, although in any particular machine there is always the possibility of hitting a high-Q resonance with a consequent large enhancement of the beam loading voltage for that particular mode.

The condition (6.40) also has an important implication for the higher-mode losses in the rf cavities in a storage ring with two counter-rotating beams. The cavities are placed an appropriate distance from the interaction points so that the q^+ and q^- bunches pass through the cavities (in opposite directions) with a time difference that is an integral multiple of the rf period for the accelerating mode. The higher-mode frequencies, however, are in general not rational multiples of the fundamental mode frequency. Thus the angle δ_n for the passage time between the counter-circulating bunches is effectively random for any particular higher-order cavity mode. In other words, the voltages induced in the higher-order modes do not add coherently. The induced voltage and power loss for the higher-order modes can therefore be computed for each beam separately, ignoring the presence of the other beam. Thus the total power lost to both higher-order cavity modes and to parasitic modes in the vacuum chamber components if there are two beams with circulating currents I_o^+ and I_o^- is

$$P_{hm} + P_{vc} = \left[\left(I_o^+\right)^2 + \left(I_o^-\right)^2\right](Z_{hm} + Z_{vc}) \quad , \tag{6.41}$$

where Z_{hm} and Z_{vc} are defined by Eqs. (6.13c) and (6.14) and we assume that the k_n's contain the factor $e^{-\omega_n^2\sigma_t^2}$. However, in computing the required generator power for the fundamental accelerating mode using Eq. (6.33), I_o must be replaced by $(I_o^+ + I_o^-)$.

7. TRANSIENT BEAM LOADING

7.1 Transient Response of a Resonant Cavity

We want first to compute the response of a resonant cavity to a step change in driving voltage. This result will be used to find the transient variation in the voltage and reflected power between bunches for a cavity loaded by a periodic bunch train. The response of a resonant circuit to a step change in driving voltage can, of course, be obtained by applying standard mathematical techniques to solve an appropriate differential equation. Here, however, let us use our phasor approach to find the answer in a very simple way.

Consider first an undriven cavity with resonant frequency ω_0 and damping time T_f. Suppose the cavity is initially charged to voltage $V_d(0)$, and that this voltage then decays as e^{-t/T_f} for $t > 0$ while viewed in a reference frame rotating at angular frequency ω (the rf driving frequency). The time variation of the cavity voltage is

$$\tilde{V}_d(t) = \tilde{V}_d(0)\, e^{-t/T_f}\, e^{jt\Delta\omega} \quad , \tag{7.1}$$

where $\Delta\omega = \omega_0 - \omega$. The time variation of $V_d(t)$ [the reason for the subscript will become clear shortly] is illustrated in Fig. 7.1.

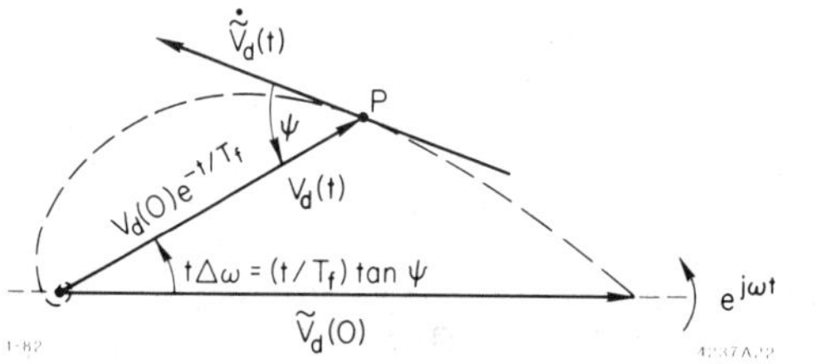

Fig. 7.1. Discharge of a cavity resonant at frequency ω_0 viewed in a coordinate frame rotating at frequency ω.

The relevance of this seemingly simple physical picture may not be obvious at first glance. In a storage ring or linac we are dealing with driven rf cavities, and the bunch repetition frequency is also a sub-harmonic of the driving frequency ω. Thus all steady-state driven voltages are phasors viewed in a coordinate system rotating at the driving frequency ω. Transient variations can, however, be viewed as the superposition of a final steady-state voltage plus an undriven discharge toward this voltage, which occurs at the natural cavity resonant frequency ω_0. Thus, by adding a final steady-state vector $\tilde{V}(\infty)$ to the diagram in Fig. 7.1, we obtain the general transient variation of the cavity voltage $\tilde{V}(t)$, as shown in Fig. 7.2. Equation (7.1) now gives the time variation of the "difference vector," $V_d(t)$, where

$$\tilde{V}_d(t) = \tilde{V}(t) - \tilde{V}(\infty) \tag{7.2a}$$

$$\tilde{V}_d(0) = \tilde{V}(0) - \tilde{V}(\infty) \quad . \tag{7.2b}$$

Using the definition of the tuning angle, $\tan\psi = T_f\Delta\omega$, Eq. (7.1) becomes

$$\tilde{V}_d(t) = \tilde{V}_d(0)\, e^{-(t/T_f)(1 - j\tan\psi)} \quad . \tag{7.3}$$

Substituting for $\tilde{V}_d(t)$ and $\tilde{V}_d(0)$ in this expression using Eqs. (7.2), we obtain

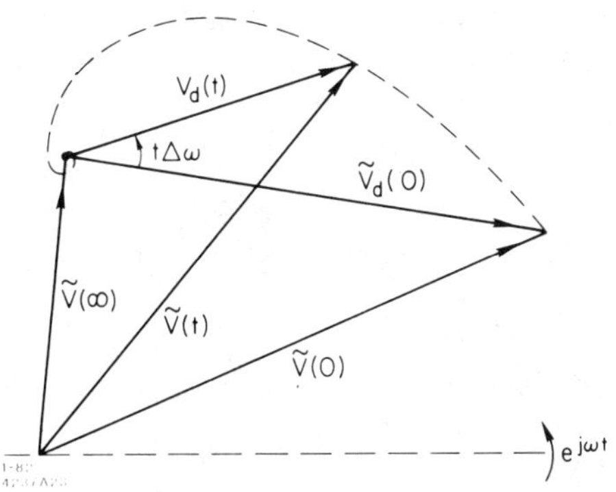

Fig. 7.2. Transient response of a resonant cavity to a step change in driving voltage $\Delta\tilde{V} = -\tilde{V}_d(0)$ applied at $t = 0$.

$$\tilde{V}(t) = \tilde{V}(\infty) + [\tilde{V}(0) - \tilde{V}(\infty)]\, e^{-(t/T_f)(1 - j\tan\psi)} \quad . \tag{7.4}$$

This expression can also be considered as giving the transient response of a resonant cavity to a step change in driving voltage $\Delta\tilde{V} = \tilde{V}(\infty) - \tilde{V}(0) = -\tilde{V}_d(0)$, applied at time $t = 0$.

It is interesting to show that Eq. (7.3) represents an equi-angular spiral; that is, the tangent to the curve at any point P in Fig. 7.1 makes a constant angle with respect to the difference vector joining point P to the origin. The derivative $\dot{\tilde{V}} = d\tilde{V}/dt$ is tangent to the curve $\tilde{V}(t)$. From Eq. (7.3),

$$\dot{\tilde{V}}_d(t) = -\tilde{V}_d(t)(1 - j\tan\psi)/T_f \quad .$$

Since

$$e^{-j\psi} = (1 - j\tan\psi)\cos\psi \quad ,$$

we have

$$\dot{\tilde{V}}_d(t) = -\tilde{V}_d(t)\,\frac{e^{-j\psi}}{T_f\cos\psi} \tag{7.5}$$

Thus if $\dot{\tilde{V}}_d(t)$ is rotated by angle $+\psi$, it will lie along the direction of $-\tilde{V}_d(t)$ as shown in Fig. 7.1.

7.2 Transient Variation of Cavity Voltage and Reflected Power Between Bunches

Let us now apply Eq. (7.4) to find the transient variation of the cavity voltage between bunches for the case of a bunch train in which the time between bunches is not necessarily small compared to the cavity filling time. We start with the vector diagram in Fig. 6.4, showing the cavity and beam loading voltages just before and just after the passage of a single bunch through a cavity driven by a generator voltage $\tilde{V}_g$. These voltages are redrawn in Fig. 7.3.

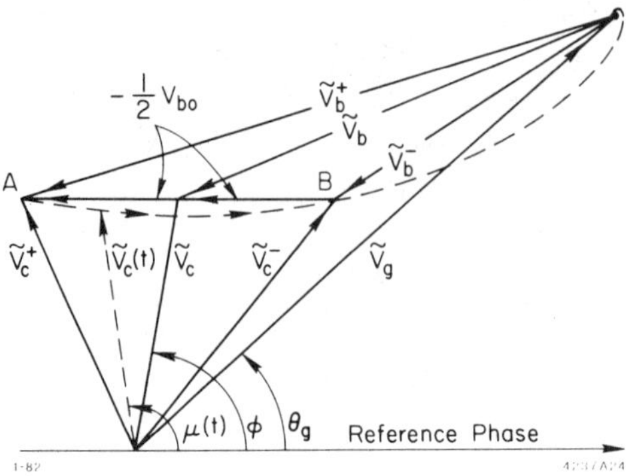

Fig. 7.3. Transient response of a driven cavity to a train of equal bunches.

When the bunch crosses the cavity reference plane, the cavity voltage changes instantaneously (in our model) from $\tilde{V}_c^-$ to $\tilde{V}_c^+$. The magnitude of the change is $-V_{bo}$. The voltage then begins to charge toward $\tilde{V}_g$ along the spiral path shown. At the precise moment the voltage once again reaches $\tilde{V}_c^-$, another bunch comes by to repeat the cycle. We can now make the following correspondences between the voltages in Eq. (7.4) and those in Fig. 7.3:

$$\begin{aligned} \tilde{V}(t) &\sim \tilde{V}_c(t) \\ \tilde{V}(0) &\sim \tilde{V}_c^+ \\ \tilde{V}(\infty) &\sim \tilde{V}_g \quad . \end{aligned} \tag{7.6}$$

We have therefore

$$\tilde{V}_c(t) = \tilde{V}_g + \left(\tilde{V}_c^+ - \tilde{V}_g\right) e^{-(t/T_f)(1 - j\tan\psi)} \quad . \tag{7.7}$$

But from the diagram in Fig. 7.3,

$$\begin{aligned} \tilde{V}_c^+ - \tilde{V}_g &= \tilde{V}_b^+ \\ \tilde{V}_g &= \tilde{V}_c - \tilde{V}_b^+ - \frac{1}{2}V_{bo} \quad . \end{aligned} \tag{7.8}$$

Therefore

$$\tilde{V}_c(t) = \tilde{V}_c + \tilde{V}_b^+ \left[e^{-(t/T_f)(1 - j\tan\psi)} - 1\right] - \frac{1}{2}V_{bo} \quad . \tag{7.9}$$

To simplify the notation, we introduce a normalized time $x = t/T_b$, such that $x = 1$ when t is equal to the arrival time of the next bunch. Recall also that $\tan\psi = \delta/\tau$. Substituting for $\tilde{V}_b^+$ from Eq. (6.25), again taking into account that $\tilde{V}_{bo} = -V_{bo}$, we find

$$\tilde{V}_c(x) = \tilde{V}_c - \frac{V_{bo}\left[e^{-x\tau}\, e^{jx\delta} - 1\right]}{1 - e^{-\tau}\, e^{j\delta}} - \frac{V_{bo}}{2} \quad . \tag{7.10}$$

Separating this expression into real and imaginary components with the aid of Fig. 7.3,

$$V_c(x)\cos\mu = V_c\cos\phi + V_{bo}F_A(x) \tag{7.11a}$$

$$V_c(x)\sin\mu = V_c\sin\phi + V_{bo}F_B(x) \tag{7.11b}$$

where

$$F_A(x) = \left[1 - e^{-2\tau} - 2e^{-x\tau}\cos x\delta + 2e^{-(1+x)\tau}\cos\delta(1-x)\right]/2D \tag{7.12a}$$

$$F_B(x) = \left[e^{-\tau}\sin\delta - e^{-x\tau}\sin x\delta - e^{-(1+x)\tau}\sin\delta(1-x)\right]/D \tag{7.12b}$$

$$D = 1 - 2e^{-\tau}\cos\delta + e^{-2\tau} \quad .$$

Squaring and adding Eqs. (7.11a) and (7.11b), using also $V_{bo} = i_o R\tau_o$,

$$\frac{V_c^2(x)}{V_c^2} = \left[\cos\phi + \frac{i_o R\tau_o}{V_c}F_A(x)\right]^2 + \left[\sin\phi + \frac{i_o R\tau_o}{V_c}F_B(x)\right]^2 \quad . \tag{7.13}$$

For a fixed τ_o, the optimum values of β and ψ can be obtained by minimizing the generator power as given by Eq. (6.33). Equation (7.13), together with the definitions of F_A and F_B given by Eqs. (7.12), then determines the transient variation between bunches in the amplitude of the cavity voltage. The transient variation in the phase of the cavity voltage is obtained by taking the ratio of Eqs. (7.11b) and (7.11a),

$$\tan\mu(x) = \frac{V_c\sin\phi + V_{bo}F_B(x)}{V_c\cos\phi + V_{bo}F_A(x)} \quad . \tag{7.14}$$

The reflected power P_r can now be computed using conservation of energy:

$$P_r = P_g - P_c - dW/dt \quad , \tag{7.15}$$

where P_g is the incident generator power, $P_c = V_c^2(t)/R_a$ is the instantaneous cavity dissipated power and W is the stored energy given by

$$W(t) = \frac{V_c^2(t)}{\omega_o(R_a/Q)} = \frac{1}{2}T_{fo}P_c(t) \quad .$$

Here $T_{fo} = 2Q_0/\omega_0$ is again the unloaded filling time. Equation (7.15) now becomes

$$P_r(t) = P_g - P_c(t) - \frac{1}{2} T_{fo} \frac{d}{dt} [P_c(t)] \quad . \tag{7.16}$$

If a normalized cavity voltage $v(t) = V_c(t)/V_c$ is introduced, the above expression can be written in normalized form, again using $x = t/T_b$ and $\tau_o = T_b/T_{fo}$, as

$$P_r(x) = P_g - \frac{V_c^2}{R_a} \left\{ v^2(x) + \frac{1}{2\tau_o} \frac{d}{dx} \left[v^2(x)\right] \right\} \quad . \tag{7.17}$$

The function $v^2(x)$ is just that given by Eq. (7.13).

The above derivation does not give the phase of the reflected voltage wave in the input transmission line to the cavity, which may sometimes be of interest. An alternative derivation, which solves for both the magnitude and the phase of the reflected wave, is given in Ref. 27.

7.3 Transient Beam Loading in Traveling-Wave Linacs

The concepts introduced in Ch. 6 to deal with single-bunch beam loading in standing-wave structures can also serve as the starting point for an analysis of transient beam loading in traveling-wave structures. Assume an element of charge dq passes through a traveling-wave structure at a velocity $v_q \approx c$. Assume also that the group velocity is low, $v_g << c$, so that the induced wave of amplitude dE_b travels a negligible distance during the time $\Delta t = L/c$ it takes for dq to transit through the structure. By analogy to Eq. (6.7a), the induced wave will have amplitude

$$dE_b = 2\, k_1\, dq \quad , \tag{7.18}$$

where k_1 is the traveling-wave loss parameter per unit length given by

$$k_1 = (\omega/4)(r/Q) = \frac{1}{2}\, \alpha r v_g \quad . \tag{7.19}$$

Here r is the shunt impedance per unit length for a synchronous wave as defined by Eq. (5.7a), α is the attenuation parameter per unit length, and Eq. (5.8c) has been used to eliminate ω/Q.

Assume now a constant impedance structure in which v_g does not vary with length. The analysis for the case of a constant gradient structure would diverge at this point. For a constant impedance structure of length L, the voltage induced by dq is, using Eq. (7.18)

$$dV_b = L(dE_b) = 2\, k_1 L\, dq \quad . \tag{7.20}$$

Using Eq. (7.19) for k_1 in this expression, together with $dq = I_0 dt$, $T_f = L/v_g$, $\tau \equiv \alpha L$ and $x \equiv t/T_f$, we obtain

$$dV_b = I_0 r\tau v_g (dt) = I_0 r L \tau (dx) \quad . \tag{7.21}$$

As time proceeds, this induced field element propagates downstream through the structure and slips out of the downstream end into the terminating load. If dE_b is induced at $t = 0$, then at time $t = xT_f$ the above voltage element dV_b is reduced by a factor $(1 - x)$ due to this downstream propagation. In addition, the voltage element will also decay by a factor $\exp(-\omega t/2Q) = \exp(-\tau x)$ because of wall losses. The preceding expression for dV_b thus becomes at time x,

$$dV_b(x) = I_0 r L \tau (1 - x) \, e^{-\tau x} \, dx \quad . \tag{7.22}$$

Integrating to add up all the induced voltage elements from $t = 0$ to $t = x$, we obtain

$$V_b(x) = I_0 r L \left[\left(1 - \frac{1}{\tau}\right)\left(1 - e^{-\tau x}\right) + x \, e^{-\tau x} \right] \quad . \tag{7.23}$$

For $\tau << 1$, this reduces to

$$V_b(x) \approx I_0 r L \tau (x - x^2/2) \quad . \tag{7.24}$$

In this limit the beam-loading voltage increases parabolically with time. In general, the beam loading voltage starts off linearly for $x << 1$ with slope $dV_b/dt \approx I_0 r v_g \tau$ and approaches the steady-state limit

$$V_b = I_0 r \; L \; [1 - (1 - e^{-\tau})/\tau] \tag{7.25}$$

with slope $I_0 r v_g \tau^2$ at $t = T_f$.

The transient energy gain from the generator voltage component can be obtained by integrating Eq. (5.10a) from $z = 0$ to $z - xL$. The result is, assuming V_g is turned on at time $t' = 0$,

$$V_g(x') = (r L P_0)^{1/2} \, [(2/\tau)^{1/2} \, (1 - e^{-x'\tau})] \quad . \tag{7.26}$$

When a generator-produced wave and beam-induced wave are both present, the net energy gain as a function of time can be obtained by a superposition of Eqs. (7.23) and (7.26). The two voltage components can, of course, be turned on at different times. There may be a phase difference also, which can be taken into account by multiplying $V_g(x')$ by $\cos\theta$ where again θ is the phase angle of $\tilde{V}_g$ with respect to a reference phase taken in the direction $-\tilde{V}_b$.

As a final comment, note that by setting $dq = I_0 dt$ we have implicitly assumed a train of bunches, each of which is short compared to an rf wavelength, and which are spaced closely compared to the filling time. If this is not the case, then from Eq. (7.22) the net voltage at time t can be obtained by summing the beam induced voltages

$$V_{bn}(t) = r\, v_g\, \tau(1 - x_n)\, e^{-\tau x_n}\, q_n \tag{7.27}$$

due to charges q_n passing through the structure at times t_n, where $x_n = (t - t_n)/T_f$ and $V_{bn} = 0$ for $x_n > 1$.

8. BEAM BREAKUP

Both the theoretical and experimental aspects of beam breakup in electron linacs are discussed in detail in Refs. 1 and 2, and we will not attempt to duplicate this coverage here. However, since we have set ourselves the task in these notes of reviewing the main features of the beam-structure interaction problem, a brief summary follows giving a few of the important analytic results of beam breakup theory.

8.1 Regenerative Beam Breakup

Regenerative beam breakup is an oscillation within a single accelerating section due to the interaction of the beam with a dipole (deflecting) mode. In these modes the E_z field component varies linearly with distance r from the axis, and as cosϕ in the azimuthal direction (see Sec. 5.1). Regions of transverse magnetic deflecting fields lie displaced from the region of maximum E_z by $\pm\lambda/4$ in a synchronous wave moving at velocity c. The field pattern for such a "TM_{11}-like," or HEM hybrid mode, is sketched in Fig. 8.1. In the usual disk-loaded structure, these deflecting modes are often of the backward wave type; that is, the phase velocity and the group velocity are in opposite directions. The interaction between a synchronous particle and a deflection mode can be characterized by the transverse shunt impedance per unit length, defined by

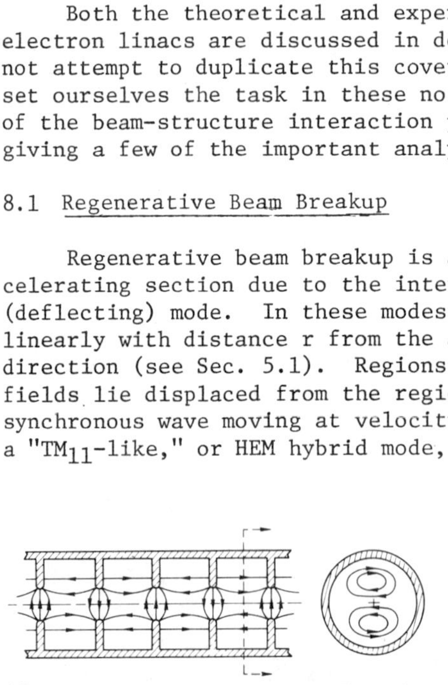

Fig. 8.1. Approximate electric and magnetic field lines for the TM_{11}-like deflection mode in a disk-loaded structure with π-phase shift per cavity. Maximum H-field occurs a quarter-cycle after maximum E-field at the cross-section shown.

$$r_\perp = \frac{(1/k^2)\left(\partial E_z^+/\partial r\right)^2}{dP_s/dz} \; . \tag{8.1}$$

Here $k = \omega/c$ and E_z^+ is the synchronous forward-wave field component. For a standing-wave mode, dP_s/dz is the average power dissipated in the structure walls per unit length. For a typical disk-loaded structure, the relation

$$\frac{r_\perp}{Q} \approx \frac{100 \text{ ohms}}{\lambda} \tag{8.2}$$

can be used to get a rough estimate of the transverse shunt impedance of the lowest-order deflection mode. In a typical structure the frequency of this mode is 40-50% higher than the accelerating mode frequency. Therefore, the Q can be expected to be somewhat lower (70-80%) compared to the Q for the accelerating mode.

Consider now a traveling-wave structure with fields proportional to an amplitude factor A_1, and assume a continuous electron beam entering the structure on axis. The particles in the region of transverse magnetic field will experience a deflecting force, and the transverse displacement of these particles will tend to increase as the square of the distance along the structure. Since the sign of the deflecting field alternates every half wavelength, the beam viewed from the side will look like a wave of growing amplitude, something like the wiggles in a stream of water from a hose nozzle which is shaken sideways. The mechanism for energy interchange depends upon the beam velocity being slightly non-synchronous with respect to the wave. If the electrons in the regions of maximum displacement (maximum deflecting H field) begin to slip ahead of the wave, they enter a region of the wave having a deaccelerating E_z electric field component. We would, consequently, expect maximum energy to be extracted from the beam if the electrons slip ahead by about a half a wavelength in the length of the structure. A detailed calculation shows that the phase slip parameter defined in Eq. (5.27) is $\delta = 2.65$ for maximum energy extraction. The power extracted from the beam is proportional to the beam current. This power propagates toward the upstream end of the section, since we are dealing with a backward wave, where it produces a field with an amplitude factor A_2. The condition for an oscillation is that A_2 be equal to the assumed initial field amplitude A_1. Detailed calculations[28] give a starting current

$$I_s(TW) = \frac{V_o \lambda^2 (v_g/c)}{8\, g_2(\delta)\, L^3 (r_\perp/Q)} \quad . \tag{8.3a}$$

Here eV_o is the energy of the beam in electron volts, and $g_2(\delta)$ is a function of the phase slip parameter. This function has a maximum value of 1.04 at $\delta = 2.65$, giving the minimum starting current. This expression was derived assuming a constant electron energy in the section. However, the first section in a linac is most likely to oscillate since the energy is lowest, and the energy is far from constant over the length of this section. If V_i and V_f are the input and output energies for such a section, and if $V_f >> V_i$, then it can be shown that the starting current is reduced by a factor of 3 below that in Eq. (8.3) when $V_f = V_o$, giving

$$I_s(TW) = \frac{V_f \lambda^2 (v_g/c)}{25\, L^3 (r_\perp/Q)} \qquad V_f >> V_i \quad . \tag{8.3b}$$

Attenuation in the section was not taken into account in deriving these expressions. Thus, measured threshold currents tend to exceed the computed thresholds by perhaps 50%. It is also assumed that the phase slip condition is accurately maintained over the entire length of the section, which may not be the case if the group velocity is very small.

The above relations were derived for a traveling-wave section in which it was assumed that the backward-wave deflecting mode is not reflected at the input coupler (upstream end) of the structure. If the structure is short with bad reflections, or if we are considering deflecting mode fields trapped within a short region of a constant gradient structure, then a standing-wave analysis is more appropriate. In such an analysis, the condition for oscillation is that the power extracted from the beam be equal to the power dissipated in the structure walls. This leads to a starting current

$$I_s(SW) = \frac{\pi^2 V_o \lambda}{4\, g_2\, r_\perp L^2} \quad . \tag{8.4}$$

Again, $g_2(\max) = 1.04$, and if the energy gain in the section is large compared to the input energy the starting current is expected to be lower. By using $r_\perp/Q$ from Eq. (8.2) in Eq. (8.4), and setting g_2 equal to $g_2(\max)$, we have

$$I_s(SW) \approx .025 \frac{V_o \lambda^2}{Q L^2} \quad . \tag{8.5}$$

Note that $I_s(TW)$ varies as $(\lambda/L)^3$ since $r_\perp/Q \sim 1/\lambda$, and that $I_s(SW)$ varies as $(\lambda/L)^2$ for a given Q.

The above starting currents were derived assuming a continuous beam. For a beam pulse of finite length t_p, the starting current is increased by the ratio[28]

$$\frac{I_s(t_p)}{I_s(\infty)} = 1 + \frac{F_e T_f}{t_p} \quad . \tag{8.6}$$

Here $T_f = 2Q/\omega$ is the filling time and e^{F_e} is the amplification factor from noise required to produce breakup. Experimental data indicate that F_e is in the range 10-20.

8.2 Cumulative Beam Breakup

The mechanism for cumulative beam breakup is quite different. In a multi-section accelerator, each section acts like an amplifier which provides a small increase in the amplitude of the transverse displacement wave. Even though the "gain" per stage is close to unity, $(1+\varepsilon)$ say, the total gain in an accelerator such as SLAC with many sections can be very large. Thus for the SLAC accelerator $(1+\varepsilon)^N = \exp(F_e)$, where $N = 960$. F_e (the e-folding factor) ≈ 20 and $(1+\varepsilon) \approx 1.02$ at the threshold for breakup. Assume that the deflecting mode occupies a length ℓ in a structure of total length L. The total transverse shunt impedance per section is then $R_\perp = \ell r_\perp$, where, for the particular case of the SLAC constant gradient structure $\ell \approx 25$ cm, L = 3 m and $R_\perp/Q$ has been measured to be 400 ohms.

Details of the beam-cavity interaction are relatively easy to calculate from first principles in the steady-state limit (cw beam). At each amplifying cavity (regions in the structure) there is a transverse displacement modulation and a transverse momentum modulation on the beam. The transverse displacement modulation excites the cavity through the interaction with the off-axis E_z field component, and the resulting H_ϕ field component provides an additional momentum kick to the beam. In the drift space between cavities, the transverse momentum is converted into additional displacement. For maximum gain, it can be shown[29] that the momentum "wave" lags the displacement wave by 30°. Furthermore, the frequency of the modulation for maximum gain is such as to drive the cavities off resonance with a tuning angle $\psi = 30°$. For an accelerator with a uniform accelerating gradient $V' = dV/dz$, the e-folding factor in the asymptotic limit ($F_e >> 1$) can be shown to be[28,29]

$$F_e(CW) = (3)^{3/4} \cdot \left[\frac{\pi I_o z R_\perp}{2V' \lambda L}\right]^{1/2} . \tag{8.7}$$

For the transient case (pulse length t_p less than or comparable to the filling time), the analysis is more complex. For times which are not too long ($t_p < F_e T_f$), the e-folding factor can be written[28]

$$F_e(t) = \frac{(3)^{3/2}}{2} \cdot \left[\frac{\pi^2 I_o z\ ct(R_\perp/Q)}{V' \lambda^2 L}\right]^{1/3} . \tag{8.8}$$

The preceding expressions were all derived assuming no focusing. If the focusing is not too strong, the e-folding factor can be modified[28] to take focusing into account. For the case of an accelerating gradient and a focusing strength which are constant along the accelerator,

$$F_e' = F_e \left[1 - C k_\beta^2 z^2/F_e^2\right] . \tag{8.9}$$

Here k_β is the betatron wave number of the focusing system and the constant C has the values for the steady-state and transient cases

$$C_{ss} = 1/2$$

$$C_t = 3/4 .$$

In Ref. 30 an analysis is given for cumulative beam breakup in the presence of solenoidal focusing. An asymptotic expression (z must be sufficiently large) is developed which is valid for strong focusing and arbitrary pulse length compared to the cavity filling time.

9. IMPEDANCES AND WAKES

9.1 Longitudinal Impedance Function and Wake Potential

If a sinusoidal current at frequency ω having a peak value $I(\omega)$ induces a voltage with peak value $V(\omega)$ in a component or chain of components, then the impedance is defined as

$$Z(\omega) = V(\omega)/I(\omega) \quad .$$

The impedance is complex, since $V(\omega)$ can be out of phase with $I(\omega)$. The chain of components can be, in particular, the components in one complete turn for a storage ring. Similarly, if a unit point charge passes through a component or chain of components, the wake potential $w(\tau)$ is defined as the potential experienced by a test particle following a distanct $c\tau$ behind the unit charge. In the following discussion we assume high-energy electrons or positrons traveling close to the speed of light, such that space charge forces between particles can be neglected. Both the impedance function and wake potential are therefore identically zero for a beam of particles in free space. As we will see, either $Z(\omega)$ or $w(\tau)$ are sufficient to completely characterize the longitudinal effects produced by the beam environment.

The concepts of an impedance function and a wake potential apply both to particles and currents passing through vacuum chamber components and to currents and charges in lumped equivalent circuits. In the case of an equivalent circuit, the wake potential is the voltage across the circuit as a function of time following the application of a unit current impulse $I(t) = \delta(t)$. The response of a component to a unit current step is also useful in certain calculations. If $s(\tau)$ is the response to a unit current step applied at $\tau = 0$, then the relation between the step and impulse response functions is

$$s(\tau) = \int_0^{\tau} w(\tau')d\tau' \quad .$$

Note that $w(\tau)$ is in units of volts per coulomb or ohms per second, while $s(\tau)$ is in ohms. The forms for $w(\tau)$ and $s(\tau)$ for several common circuit elements are shown in Fig. 9.1. A resistive (deaccelerating) wake is taken to be positive.

9.2 Transform Relations

Problems in accelerator theory can often be viewed within the conceptual framework provided by either the frequency or the time domains. Sometimes the framework provided by one domain or the other is more useful for viewing or solving a particular problem. In the past, there may have been some preference for the frequency domain as being the more fundamental. In these notes, however, problems have been approached wherever possible in a time-domain framework, with phasors providing a graphic aid in describing the physics of such processes as, for example, beam loading by bunch trains. It is

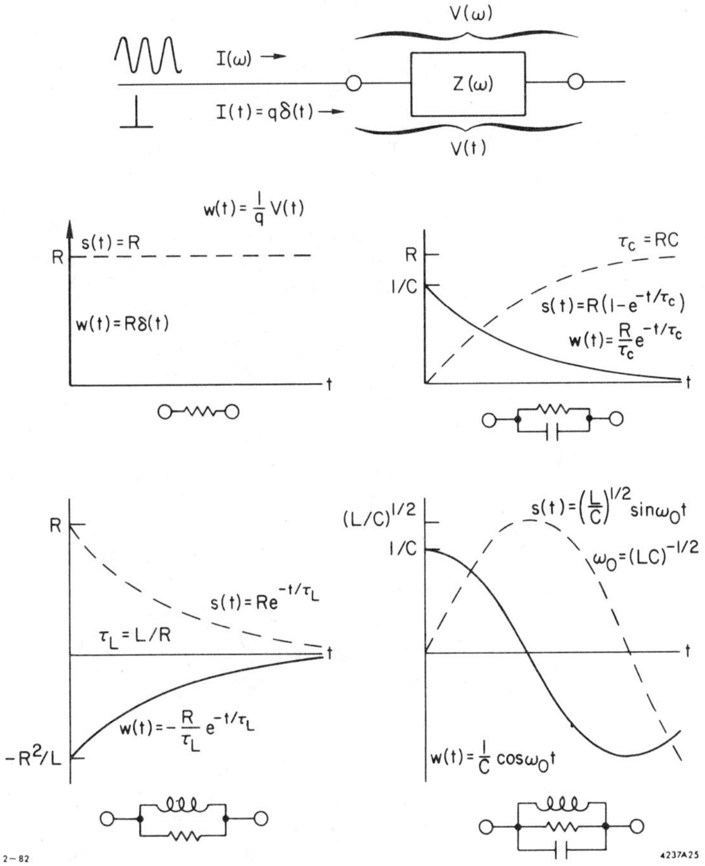

Fig. 9.1. Impulse wake and step response function for four circuit elements.

clearly desirable to be able to view a problem in either domain, and to transform physical quantities back and forth between these two worlds.

Consider the Fourier expansion for a periodic time-domain function, such as the current I(t) for a bunch train with period Δt:

$$I(t) = \frac{1}{2}\sum_{-\infty}^{\infty} \alpha_n \, e^{jn\omega_0 t} = I_0 + \sum_{1}^{\infty} I_n(\omega_n t) \tag{9.1}$$

where

$$\alpha_n = \frac{\omega_0}{\pi} \int_{-\Delta t/2}^{\Delta t/2} I(t) \, e^{-jn\omega_0 t} \, dt$$

$$I_n = a_n \cos\omega_n t + b_n \sin\omega_n t; \quad I_o = \frac{\alpha_o}{2}$$

$$\omega_n = n\omega_o = 2\pi n/\Delta t$$

$$a_n = (\alpha_n + \alpha_{-n})/2, \qquad b_n = j(\alpha_n - \alpha_{-n})/2 \quad .$$

Consider the limit $\Delta t \to \infty$, $\omega_o \to 0$. Set $n\omega_o = \omega$ and $\omega_o = d\omega$, and let $I(\omega) = \alpha_n/\omega_o$ be the density at frequency ω of Fourier components in the expansion of $I(t)$ as ω_o approaches zero. We then obtain the Fourier transform relations for the general function $f(t)$:

$$F(t) = \frac{1}{2} \int_{-\infty}^{\infty} F(\omega)\, e^{j\omega t}\, d\omega \equiv \overset{\frown}{F(\omega)} \tag{9.2a}$$

$$F(\omega) = \frac{1}{\pi} \int_{-\infty}^{\infty} f(t)\, e^{-j\omega t}\, dt \equiv \overset{\smile}{f(t)} \quad . \tag{9.2b}$$

For the particular case of a Gaussian bunch with charge q and bunch length σ_t, these relation become

$$I(t) = \frac{q}{\sqrt{2\pi}\,\sigma_t} \exp\left(-t^2/2\sigma_t^2\right) \tag{9.3a}$$

$$I(\omega) = \frac{q}{\pi} \exp\left(-\omega^2\sigma_t^2/2\right) \quad . \tag{9.3b}$$

The particular choice of normalization in Eqs. (9.2) is convenient because it allows us to interpret $I(\omega)$ as the spectral density of lines in the Fourier expansion for a train of symmetric bunches as $\omega_o = (2\pi)(I_o/q)$ approaches zero, without having to take into account negative frequencies. The amplitude of each line is then the density times the separation per line. For the case of Gaussian bunches,

$$a_n = \omega_o\, I(\omega) = 2\, I_o \exp\left(-n^2\omega_o^2\sigma_t^2/2\right) \tag{9.4}$$

Note that, with the normalization chosen, n is a positive integer.

Let us now apply these transform relations to a bunch with current distribution $I(t)$ passing through a structure with a wake function $w(\tau)$. The potential at time t due to the interaction of the bunch with its environment then follows directly from the definition of the wake potential,

$$V_b(t) = \int_{-\infty}^{t} w(t-\tau)\, I(\tau)\, d\tau = \int_{0}^{\infty} w(\tau)\, I(t-\tau)\, d\tau \quad . \tag{9.5}$$

The total energy loss by charge q, in terms of the loss parameter k introduced in Eq. (6.5), is $\Delta U = kq^2$. Thus

$$k = \frac{1}{q^2} \int_{-\infty}^{\infty} V_b(t)\ I(t)dt \quad . \tag{9.6}$$

Now take the transform of Eq. (9.5) to obtain

$$\widetilde{V_b(t)} = \frac{1}{\pi} \int_{-\infty}^{\infty} e^{-j\omega t}\ dt \int_{0}^{\infty} w(\tau)\ I(t-\tau)d\tau \quad . \tag{9.7}$$

Reverse the order of integration, let $t = \tau + t'$ and find

$$V(\omega) \equiv \widetilde{V_b(t)} = \pi\ I(\omega)\ \widetilde{w(\tau)} \quad . \tag{9.8}$$

Since $Z(\omega) \equiv V(\omega)/I(\omega)$, we have

$$Z(\omega) = \pi\ \widetilde{w(\tau)} \tag{9.9a}$$

$$V(\omega) = \widetilde{V_b(t)} \tag{9.9b}$$

$$I(\omega) = \widetilde{I(t)} \quad . \tag{9.9c}$$

9.3 Properties of the Impedance Function

We define the wake function to be a real function of time. This then imposes a condition on the impedance function, $Z(\omega) = Z_R(\omega) + j\ Z_I(\omega)$. Thus

$$w(t) = \frac{1}{\pi}\widetilde{Z(\omega)} = \frac{1}{2\pi}\int_{-\infty}^{\infty} Z(\omega)\ e^{j\omega t}\ d\omega$$

$$= \frac{1}{2\pi}\int_{-\infty}^{\infty} \left[Z_R(\omega)\ \cos\omega t - Z_I(\omega)\ \sin\omega t\right] d\omega \tag{9.10}$$

$$+ \frac{j}{2\pi}\int_{-\infty}^{\infty} \left[Z_R(\omega)\ \sin\omega t + Z_I(\omega)\ \cos\omega t\right] d\omega \quad .$$

If the imaginary part is to vanish for arbitrary $Z(\omega)$, it is necessary for $Z_R(\omega)$ to be an even function of frequency and for $Z_I(\omega)$ to be an odd function of frequency. We can confine ourselves to positive frequencies only, to obtain

$$w(t) = \frac{1}{\pi} \int_0^\infty [Z_R(\omega) \cos\omega t - Z_I(\omega) \sin\omega t] \, d\omega \quad . \qquad (9.11)$$

In addition, the wake potential must be causal; that is, $w(t) \equiv 0$ for $t < 0$. Therefore

$$w(-t) = \frac{1}{\pi} \int_0^\infty [Z_R(\omega) \cos\omega t + Z_I(\omega) \sin\omega t] \, d\omega \equiv 0 \quad , \qquad (9.12)$$

leading to

$$\int_0^\infty Z_R(\omega) \cos\omega t \, d\omega \equiv - \int_0^\infty Z_I(\omega) \sin\omega t \, d\omega \quad . \qquad (9.13)$$

Substituting Eq. (9.13) in Eq. (9.11),

$$w(\tau) = \frac{2}{\pi} \int_0^\infty Z_R(\omega) \cos\omega t \, d\omega \quad . \qquad (9.14)$$

Problem 9.1: Show that Eq. (9.13) is equivalent to the Hilbert transform,

$$Z_I(\omega) = \frac{1}{\pi} \int_{-\infty}^\infty \frac{Z_R(\omega')}{\omega' - \omega} \, d\omega' \quad . \qquad (9.15)$$

Hint: Rewrite Eq. (9.13) with limits of integration between $-\infty$ and ∞. Substitute for $Z_I(\omega)$ using Eq. (9.15), then reverse the order of integration. $Z_R(\omega)$ is obtained from $Z_I(\omega)$ by the inverse transform (above transform multiplied by -1).

Thus, if either the real or the imaginary component of the impedance function is specified, the other component is also determined as a consequence of the causality condition.

The preceding relations can be visualized using phasor concepts. Consider a unit point charge interacting with an impedance $Z(\omega)$. By Eq. (9.3b), the spectral density of the current in the frequency domain is $I(\omega) = 1/\pi$ at all frequencies. At time $t = 0$, due to the interaction of the charge with the real component of the impedance, beam-induced voltage elements

$$d\tilde{V}_R^+(0) = Z_R(\omega) \, I(\omega) d\omega = \frac{1}{\pi} \, Z_R^+(\omega) d\omega$$

$$d\tilde{V}_R^-(0) = \frac{1}{\pi} \, Z_R^-(-\omega) d\omega = \frac{1}{\pi} \, Z_R^+(\omega) d\omega \qquad (9.16)$$

are produced in the frequency interval $d\omega$ at $\pm\omega$. A positive real impedance component indicates that the induced voltage elements oppose the motion of the charge and extract energy from it. At $t = 0$, imaginary beam-induced voltage components

$$
\begin{aligned}
d\tilde{V}_I^+(0) &= \frac{j}{\pi} Z_I^+(\omega)d\omega = -j\left[-\frac{1}{\pi} Z_I^+(\omega)d\omega\right] \\
d\tilde{V}_I^-(0) &= \frac{j}{\pi} Z_I^-(-\omega)d\omega = +j\left[-\frac{1}{\pi} Z_I^+(\omega)d\omega\right]
\end{aligned}
\qquad (9.17)
$$

are also produced which are at right angles to the real components and hence play no part in the energy interchange with the charge. These real and imaginary voltage elements are shown schematically by the solid phasors in Fig. 9.2. The total self-voltage acting on the charge is obtained by integrating the real components over all frequency at $t = 0$,

$$
w_s = \frac{1}{2}\int_{-\infty}^{\infty} d\omega\left[dV_R^+(0) + dV_R^-(0)\right] = \frac{1}{\pi}\int_0^{\infty} Z_R^+(\omega)d\omega \quad . \qquad (9.18)
$$

At some later time t, the phasor voltage elements will have rotated to the positions shown by the dashed phasors in Fig. 9.2. The total wake voltage, including the contribution from the imaginary (at $t = 0$) components, is obtained by integrating over frequency,

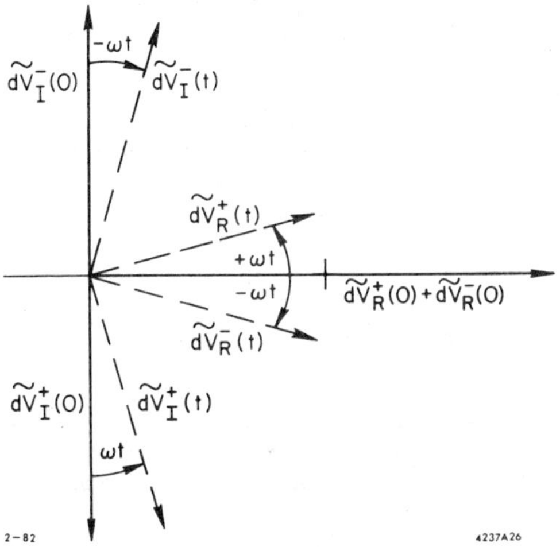

Fig. 9.2. Diagram showing how imaginary voltage elements induced by a point charge at $t = 0$ in the frequency interval $d\omega$ rotate to produce a causal wake.

$$w(t>0) = \int_0^\infty \left[dV_R^+ \cos\omega t + dV_I^+ \sin\omega t\right] d\omega$$
$$= \frac{1}{\pi}\int_0^\infty \left[Z_R^+(\omega)\cos\omega t - Z_I^+(\omega)\sin\omega t\right] d\omega \quad . \tag{9.19}$$

For $t < 0$, the four phasors rotate in the opposite sense, and

$$w(t<0) = \frac{1}{\pi}\int_0^\infty \left[Z_R^+(\omega)\cos\omega t + Z_I^+(\omega)\sin\omega t\right] d\omega \equiv 0 \quad . \tag{9.20}$$

This last relation is equivalent to the causality condition (9.13), and together with the preceding relation leads again to Eq. (9.14).

Thus, to satisfy causality, the imaginary voltage components induced at $t = 0$ in a frequency interval $d\omega$ rotate and always add up for $t < 0$ so as to cancel the wake produced by the sum of the real components induced at $t = 0$ when integrated over all frequencies. For $t > 0$, the imaginary components rotate so as to produce a total real wake which is exactly double that due to the sum of all real components. Note that by extrapolating Eq. (9.14) to $t = 0$, we obtain

$$w(0) = 2w_s \quad . \tag{9.21}$$

The factor of two in this expression is essentially the same as that in Eq. (6.7b), which was obtained by applying conservation of energy and superposition to resonant modes in a cavity. To see this more clearly, let us calculate the total loss parameter k, using Eq. (9.6):

$$k = \frac{1}{q^2}\int_{-\infty}^\infty V_b(t)I(t) = \frac{1}{2q^2}\int_{-\infty}^\infty V_b(t)dt \int_{-\infty}^\infty I^*(\omega)\, e^{-j\omega t}\, d\omega \quad . \tag{9.22}$$

Reversing the order of integration,

$$k = \frac{\pi}{2q^2}\int_{-\infty}^\infty I^*(\omega)V(\omega)d\omega = \frac{\pi}{2q^2}\int_{-\infty}^\infty I^2(\omega)Z(\omega)d\omega \quad . \tag{9.23}$$

Since $I(t)$ is a real function of time, we have by the same argument that lead to Eq. (9.11) that the real and imaginary part of $I(\omega)$ are symmetric and anti-symmetric respectively. Thus $I^2(\omega)$ is symmetric, and

$$k = \frac{\pi}{q^2}\int_0^\infty I^2(\omega)Z_R(\omega)d\omega \quad . \tag{9.24}$$

For a Gaussian bunch, using Eq. (9.3b),

$$k = \frac{1}{\pi} \int_0^\infty Z_R(\omega)\, e^{-\omega^2 \sigma_t^2}\, d\omega \quad . \tag{9.25}$$

For a point charge this agrees with the self-wake derived in Eq. (9.18).

9.4 Application to Resonant Modes

Let us now compute the wake potential and loss parameter for a single resonant mode. The impedance for such a mode is given by Eq. (3.39),

$$Z(\omega) = Z_R(\omega) + jZ_I(\omega) = \frac{R_o}{1 + j\xi}$$

$$Z_R(\omega) = \frac{R_o}{1 + \xi^2} \qquad Z_I(\omega) = \frac{-R_o \xi}{1 + \xi^2} \quad , \tag{9.26}$$

where, as before, $\xi = (\omega - \omega_o)T_f$ and $T_f = 2Q_L/\omega_o$.

> Problem 9.2: Show that $Z_R(\omega)$ and $Z_I(\omega)$ above obey the causality condition in Eq. (9.13), assuming $\omega_o T_f = 2Q_L >> 1$.

Applying Eq. (9.14), assuming a high Q mode so that $\cos\omega t \approx \cos\omega_o t$ over the range of ω where $Z_R(\omega)$ is appreciable, we obtain

$$w(\tau) \approx \cos\omega_o\tau \cdot \frac{2R_o}{\pi T_f} \int_{-\infty}^{\infty} \frac{d\xi}{1 + \xi^2} = \frac{\omega_o R_o}{Q_L} \cos\omega_o\tau \quad . \tag{9.27}$$

Using $R_o = [G_c(1+\beta)]^{-1}$, $Q_o = (1+\beta)Q_L$ and the fact that the accelerator shunt impedance is $R_a = 2/G_c$, we have

$$w(\tau) = \frac{\omega_o}{2}\left(\frac{R_a}{Q}\right) \cos\omega_o\tau \tag{9.28}$$

for the wake function for a mode resonant at frequency ω_o. The total loss parameter for a Gaussian bunch is obtained in a similar fashion using Eq. (9.25),

$$k(\sigma_t) = \frac{\omega_o}{4}\left(\frac{R_a}{Q}\right) e^{-\omega_o \sigma_t^2} \quad . \tag{9.29}$$

Note that for a point bunch $w(0) = 2k_o$, where $k_o = (\omega_o/4)(R_a/Q)$. This gives

$$w(\tau) = 2k_o \cos\omega_o\tau \quad , \tag{9.30a}$$

$$k(\sigma_t) = k_o e^{-\omega_o \sigma_t^2} \quad . \tag{9.30b}$$

These results are readily extended to find the total wake function and total loss parameter by summing over all modes (assumed non-overlapping) in a resonant cavity or traveling-wave structure:

$$w(\tau) = 2 \sum_n k_n \cos\omega_n \tau \tag{9.31a}$$

$$k(\sigma_t) = \sum_n k_n e^{-\omega_n^2 \sigma^2} \quad . \tag{9.31b}$$

For the case of a resonant cavity, using Eq. (6.6b),

$$k_n = \frac{\omega_n}{4}\left(\frac{R_a}{Q}\right)_n = \frac{V_n^2}{4W_n} \tag{9.32}$$

where V_n is the maximum voltage gain for the nth mode for a velocity of light test particle when the stored energy in the mode is W_n. Similar concepts apply to traveling-wave modes in a periodic structure. Using Eq. (5.7c),

$$k_n = \frac{\omega_n}{4}\left(\frac{r}{Q}\right)_n = \frac{E_n^2}{4w_n} \quad , \tag{9.33}$$

where E_n is the amplitude of the synchronous space harmonic component of the axial electric field for the nth mode, and w_n is the stored energy per unit length summed over all space harmonic components for that mode.

In order to compute the wake potential using Eq. (9.31a), values for ω_n and k_n are needed for as many modes as possible, either resonant modes in the case of a cavity or traveling-wave modes for a periodic accelerating structure. Values of ω_n and k_n are obtained by solving the boundary value problem for a charge-free cavity or structure. Two computer programs are generally available at the present time which accomplish this purpose. The program KN7C[31] solves for traveling-wave modes in a round pipe loaded by disks with flat, parallel faces. This structure is described by four parameters: the radius of the beam aperture in the disk, the inside radius of the pipe, the length of a period, and the length of the pipe between disk faces. The program SUPERFISH[6] solves for resonant modes in an axially-symmetric cavity having an arbitrary boundary as a function of the axial coordinate z; that is, on the boundary $r(\phi)$ is constant but $r(z)$ is an arbitrary function.

As frequency increases, the number of modes per unit frequency interval also increases. Since there is a limit on the total number of modes that can be calculated with reasonable computer time, there is a corresponding maximum frequency for the sums in Eqs. (9.31). If this frequency is ω_m, details in the wake will not be accurate for time intervals $\Delta\tau \lesssim \omega_m^{-1}$, and the loss parameter will not be accurate for bunch lengths $\sigma_t \lesssim \omega_m^{-1}$. For high frequencies where the mode density is large, it is only the statistical properties of the modes that are important.

Problem 9.3: For a pillbox cavity of radius b and length L, show that the density of modes approaches $dn/d\omega = \omega bL/2\pi c^2$.

In the case of a disk-loaded structure, loss by a point charge into high frequency traveling-wave modes can be considered as a diffraction loss by an equivalent plane wave having the same power spectrum and Poynting vector at the disk radius as the actual field due to the charge. This is the so-called optical resonator model[32] for the energy loss by a point charge passing through a periodic sequence of thin plates with circular holes. In the limit of high-energy ($\gamma \gg \omega a/c$ where a is the hole radius), the loss parameter per unit frequency interval predicted by this model is[32]

$$\frac{dk}{d\omega} = \frac{1}{\pi} Z_R(\omega) = \frac{A_o}{\omega^{3/2}} \quad . \tag{9.34}$$

The wake potential due to this "analytic extension" for loss at all frequencies $\omega > \omega_m$ is then

$$\begin{aligned} w_a(\tau) &= 2A_o \int_{\omega_m}^{\infty} \frac{\cos\omega\tau}{\omega^{3/2}}\, d\omega \\ &= \frac{4A_o}{\omega_m^{3/2}} \left\{ \cos x - \sqrt{\frac{\pi x}{2}} \left[1 - 2S\left(\sqrt{\frac{2x}{\pi}}\right)\right]\right\}_{x=\omega_m\tau} \end{aligned} \tag{9.35}$$

where S is the Fresnel integral. The constant A_o can, in principle, be specified analytically, at least for a structure with thin disks. In practice, it is better to obtain A_o for a particular structure by making a fit of Eq. (9.34) to a log-log plot of computed modes for $\omega < \omega_m$.

The preceding concepts have been applied to compute the wake for the SLAC disk-loaded structure having a periodic length $\lambda/3 = 3.50$ cm. The disk thickness is 0.58 cm, the radius of the outer wall is 4.13 cm and the disk hole radius is 1.16 cm for an average cell near the center of each constant gradient structure of 3 m length. The wake for the first 10 ps is shown in Fig. 9.3. The dashed curve gives the wake due to 416 computed modes, using Eq. (9.31a). The total wake is obtained by adding an analytic extension given by Eq. (9.35). Note that, because of the analytic extension, the total wake has a vertical tangent at $\tau = 0$ but a finite value of $w(0) = 8$ V/pC/period. The wake due to the excitation of the fundamental accelerating mode

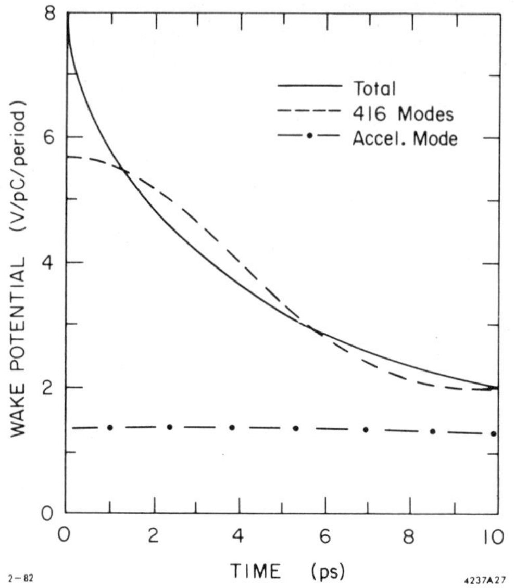

Fig. 9.3. Longitudinal wake per cell for the SLAC disk-loaded structure (0-10 ps). Cell length = 3.5 cm; beam aperture radius = 1.163 cm.

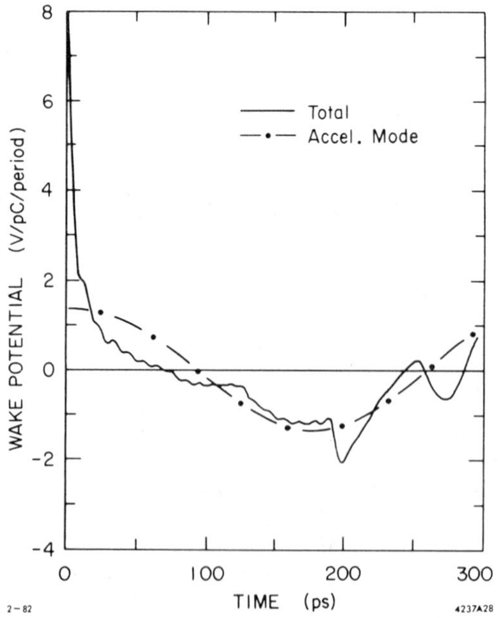

Fig. 9.4. Longitudinal wake per cell for the SLAC disk-loaded structure (0-300 ps).

only, is shown for comparison. Note that on this scale it is almost constant (the period is 350 ps), with an amplitude of about one-sixth of w(0).

Figure 9.4 shows the longitudinal wake for the SLAC structure out to 300 ps. The large negative (accelerating) spike at 200 ps is the first reflection from the outer wall arriving at the structure axis. Shown again is the wake due to the fundamental mode, which undergoes almost a full period of oscillation. After 5 to 10 periods, it will be the dominant term in the wake, the higher modes having almost entirely decohered. On the time scale of interest, damping of the modes has also been ignored. Damping could be taken into account by multiplying each term in Eq. (9.31a) by $\exp(-\alpha_n\tau)$, where α_n is the damping constant for the nth mode.

For a number of years there was a controversy as to whether the modal analysis leading to Eq. (9.31a) was giving the complete wake. It was thought that this modal approach might be neglecting terms in the wake due to the scalar potential of the charge. Bane[33] has recently shown analytically that the modal analysis gives a wake which agrees with that derived from the vector and scalar potentials for a charge with v = c. Weiland and Zotter[34] have shown that the modal wake is in agreement with that obtained by a direct integration of Maxwell's equations, using the program BCI, for a bunch moving through a cavity.

9.5 The Transverse Wake

In this section we work out the transverse wake for the specific case of an axisymmetric disk-loaded structure. For such a structure, the synchronous space harmonic component of the nth traveling-wave mode has an axial electric field variation described by[23]

$$E_{zn} = E_{on}\left(\frac{r}{a}\right)^m \cos m\phi \; \cos\omega_n(t - z/c) \quad , \tag{9.36}$$

where E_{on} is the field strength at the radius of the disk opening. For each mode a loss parameter

$$k_n \equiv \frac{E_{on}^2}{4w_n} \tag{9.37}$$

can again be defined in terms of the "cold" (no charge present) electromagnetic properties of the structure. Using the same formalism developed for the case of longitudinal modes, k_n will also describe the interaction of a point charge with the mode in question. Specifically, the beam-induced energy deposited per unit length in the nth mode by a charge q traveling parallel to the axis at radius $r = r_q$ is

$$w_n = k_n\left(\frac{r_q}{a}\right)^{2m} q^2 \quad . \tag{9.38}$$

Eliminating w_n using the preceding two expressions,

$$E_{on} = -2\left(\frac{r_q}{a}\right)^m k_n q \quad . \tag{9.39}$$

The minus sign indicates that the induced field is such as to oppose the motion of the charge. Substituting Eq. (9.39) in Eq. (9.36), the induced field at position r, azimuth ϕ (assuming the driving charge q is at $\phi = 0$) and position $\Delta z = c\tau$ behind a charge q at radius r_q is

$$E_{zn} = -2k_n q\left(\frac{r}{a}\right)^m \left(\frac{r_q}{a}\right)^m \cos m\phi \; \cos\omega_n\tau \quad . \tag{9.40}$$

For m = 0, we see that the longitudinal wake potential per unit length is recovered.

Now define the transverse (deflecting) wake per unit length of structure by

$$\vec{w}_t(\tau) = (c/e)\; d\vec{p}_t/dz = (\vec{E}_t + c\,\vec{B}_t)^{(\mathrm{cmf})} \tag{9.41}$$

where $d\vec{p}_t/dz$ is the transverse momentum kick experienced per unit length of structure by a particle following at distance $c\tau$ behind a unit driving charge. The superscript indicates as before that the transverse fields are to be evaluated in a reference frame which is co-moving with the particle. In a theorem due to Panofsky and Wenzel,[35] it is shown that the momentum kick in such a co-moving frame can be expressed in terms of the E_z field component only:

$$(\vec{E}_t + c\vec{B}_t)^{(cmf)} = j(c/\omega)\vec{\nabla}_t E_z^{(cmf)} \quad . \tag{9.42}$$

Problem 9.4: Prove the Panofsky-Wenzel theorem.
Hint: Express $\vec{E}_t$ and $(\vec{c} \times \vec{B})_t$ in terms of the vector potential $\vec{A}$. Expand $\vec{c} \times \vec{\nabla} \times \vec{A}$, find the total derivative $d\vec{A}/dt$, and set this equal to zero for a synchronous wave.

For a synchronous wave, putting Eq. (9.40) in the form $E_{zn} = -|E_{zn}|e^{j\omega_n\tau}$ and using the preceding theorem, the transverse wake becomes

$$\vec{w}_{tn} = (c/\omega)\ \sin\omega\tau \left(\hat{r}\,\frac{\partial|E_{zn}|}{\partial r} + \hat{\phi}\,\frac{1}{r}\,\frac{\partial|E_{zn}|}{\partial\phi}\right) \quad , \tag{9.43}$$

where $\hat{r}$ and $\hat{\phi}$ are unit vectors. Evaluating these two components,

$$\hat{r}:\quad w_{tn}(r,\phi,\tau) = 2m\left(\frac{k_n c}{\omega_n a}\right)\left(\frac{r}{a}\right)^{m-1}\left(\frac{r_q}{a}\right)^m \quad \cos m\phi\ \sin\omega_n\tau \tag{9.44a}$$

$$\hat{\phi}:\quad w_{tn}(r,\phi,\tau) = -2m\left(\frac{k_n c}{\omega_n a}\right)\left(\frac{r}{a}\right)^{m-1}\left(\frac{r_q}{a}\right)^m \quad \sin m\phi\ \sin\omega_n\tau \quad . \tag{9.44b}$$

The $\hat{r}$ component of the dipole $(m = 1)$ wake at $\phi = 0$ is of most interest:

$$w_{dn}(\tau) = 2\left(\frac{k_n c}{\omega_n a}\right)\left(\frac{r_q}{a}\right)\ \sin\omega_n\tau \quad . \tag{9.45}$$

Note that the amplitude of the dipole wake depends on the transverse coordinate r_q of the exciting unit charge, but that behind r_q the wake itself is independent of r. Again, for a sum of modes

$$w_{dn}(\tau) = 2\left(\frac{r_q}{a}\right)\sum_n \frac{k_n c}{\omega_n a}\sin\omega_n\tau \quad . \tag{9.46}$$

Assuming $dk_n/d\omega = A_1/\omega^{3/2}$, we can compute an analytic extension to the above sum over modes following the same procedure as that which led to Eq. (9.35). The result to be added to a sum over modes up to a maximum frequency ω_m is

$$w_{da}(\tau) = \left(\frac{r_q}{a}\right)\left(\frac{4\ A_1 c}{\omega_m^{3/2} a}\right)\left\{ \frac{x}{3}\left[2\cos x + \frac{\sin x}{x} - \sqrt{2\pi x}\left(1 - 2S\left(\sqrt{\frac{2x}{\pi}}\right)\right)\right]\right\}_{x = \omega_m \tau} \quad (9.47)$$

Again, S is the Fresnel integral and the constant A_1 is obtained by fit to modal results for $\omega < \omega_m$.

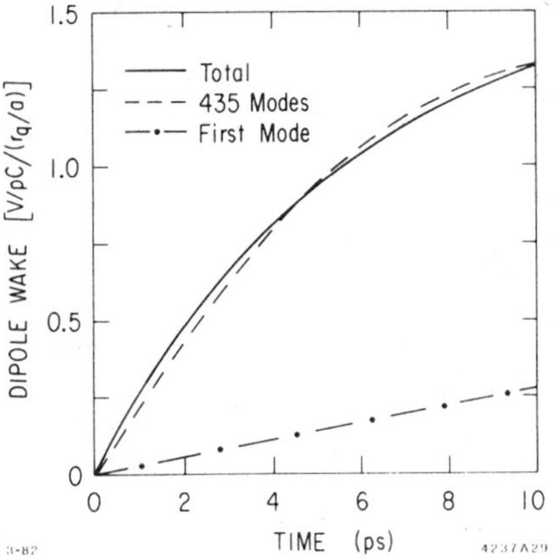

Fig. 9.5. Dipole wake per cell for the SLAC disk-loaded structure (0-10 ps).

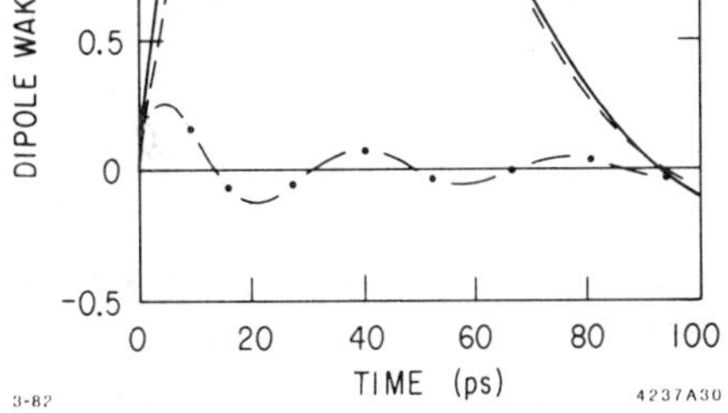

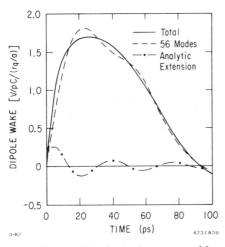

Fig. 9.6. Dipole wake per cell for the SLAC disk-loaded structure (0-100 ps).

Values for ω_n and k_{dn} can be computed for an axisymmetric disk-loaded structure using the program TRANSVERS.[36] The resulting transverse wake per period for the SLAC structure is shown in Figs. 9.5 to 9.7. Note in Fig. 9.5 that for very short times the total wake increases almost linearly at the rate of 0.25 V/pC/ps per period. This is about 10 times the slope due to the lowest frequency mode, which is responsible for beam-breakup in SLAC. In Fig. 9.6, note how the analytic extension combines with the modal contribution to produce a smooth total wake. If more modes are used, together with a contribution from the analytic extension which is consequently smaller, essentially the same total wake is obtained. The long-range wake is shown in Fig. 9.7. The high frequency modes, all of which add coherently at $\tau = 0$, have nearly decohered on this time scale. The main contribution to the wake is the lowest frequency mode, which has a period of 235 ps and an amplitude of 1.0 V/pC.

It is sometimes useful to define a dipole transverse impedance per unit length of structure by

$$Z_d(\omega) = -j\,\frac{(E_t + cB_t)}{I_o\ \Delta r} \quad (9.48)$$

where E_t and B_t are the transverse deflecting field components produced by a current filament of strength I_o having a sinusoidal

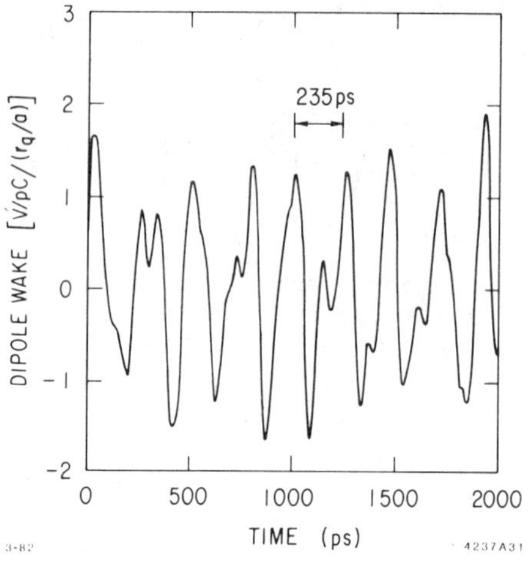

Fig. 9.7. Dipole wake per cell for the SLAC disk-loaded structure (0-2000 ps).

transverse modulation of amplitude Δr at frequency ω. For a component or for a storage ring of circumference L, the transverse impedance is $Z_d(\omega) = V_t(\omega)/I_d(\omega)$ where $V_t(\omega) = \int_0^L (\vec{E} + \vec{c} \times \vec{B})_t dz$ and $I_d(\omega) = q\Delta r/\pi$.

Problem 9.5: Show that, for a resonant mode,

$$\frac{Z_d}{Q} = \frac{2c\ k_d}{\omega^2 a^2} \tag{9.49}$$

Hint: Multiply Eq. (9.45) by $e^{-\omega t/2Q}$, then take the transform using Eq. (9.26) to find $V_t(\omega)$, noting also that $I_d(\omega) = q\ r_q/\pi$ for a point charge.

For typical dipole and longitudinal modes having the same E_z at the disk radius in a disk-loaded structure, we expect $k_d \approx 4\ k_\ell$. However, the density of transverse modes per unit frequency interval is twice as large, since both TE- and TM-like modes can be excited.[37] Using also the fact that $k_\ell = (\omega/2)(Z_\ell/Q)$ for a longitudinal mode (setting $R_a = 2Z_\ell$ in Eq. (6.9)), we obtain

$$Z_d = \frac{2c}{\omega a^2} Z_\ell \quad . \tag{9.50}$$

This expression is often used to estimate the broadband dipole impedance if the longitudinal impedance is known.

9.6 The Quadrupole Wake

Evaluating the expression for the $\hat{r}$ component of $w_{tn}(r,\phi,\tau)$ in Eq. (9.44a) for m = 2, we obtain the quadrupole wake potential

$$w_{qn}(\tau) = 4\left(\frac{r_q}{a}\right)^2\left(\frac{r}{a}\right)\sum_n \frac{k_n c}{\omega_n a}\sin\omega_n\tau \quad . \tag{9.51}$$

The wake varies with azimuthal angle as $\cos 2\phi$, where again it is assumed that $\phi = 0$ is the aximuth of the exciting charge at radius r_q. The expression for the analytic extension is that given by Eq. (9.25) multiplied by $2(r_q/a)(r/a)$. The quadrupole wake computed for the SLAC disk-loaded structure is shown in Figs. 9.8 and 9.9. The wake is again normalized to the disk hole radius a = 1.163 cm and to the periodic length p = 3.50 cm. To convert to a wake per unit length of structure, $w_q(\tau)/r_q^2 r$ in units of $V/C/m^4$, the ordinate must be multiplied by $10^{12}/a^3 p = 1.82 \times 10^{19}$. (To obtain the dipole wake $w_d(\tau)/r_q$ in units of $V/C/m^2$, the corresponding factor is $10^{12}/ap = 2.45 \times 10^{15}$.)

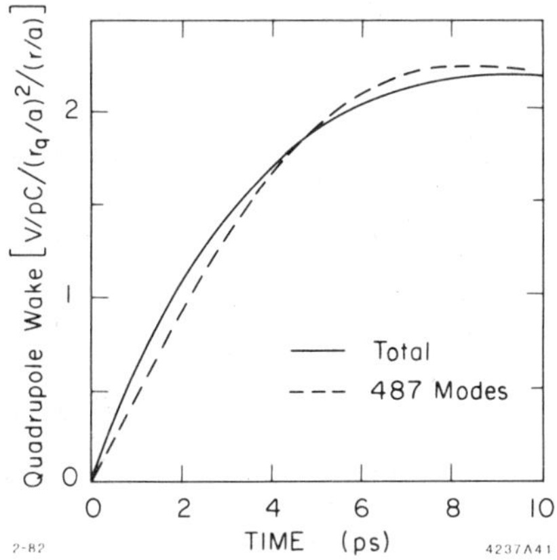

Fig. 9.8. Quadrupole wake per cell for the SLAC structure (0-10 ps).

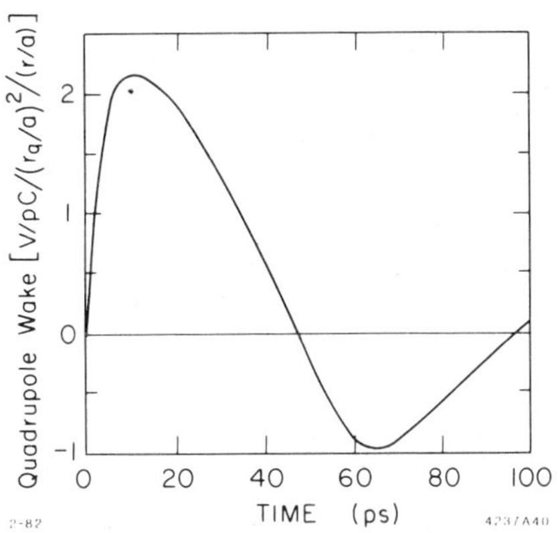

Fig. 9.9. Quadrupole wake per cell for the SLAC structure (0-100 ps).

It can be argued that effects due to the quadrupole wake will be small, since its strength is smaller than the strength of the dipole wake by roughly a factor $r_q r/a^2$. Note, however, that the quadrupole wake sets a fundamental limit on emittance growth in a linac with alternating gradient focusing. In such a machine the beam cannot always be round. Even if the beam is exactly on axis and there are no misalignments, there will still be an emittance growth due to the quadrupole wake because the beam will necessarily have a quadrupole moment. In a recent calculation, Chao and Cooper[38] have found such a situation in which effects due to the quadrupole wake can be significant. In the first sector of the SLAC accelerator, there is a non-negligible emittance growth for an injected bunch of 5×10^{10} particles with $\sigma_{x,y,z} \approx 1$ mm.

9.7 Scaling of the Wake With Frequency and Structure Parameters

The scaling with frequency of the amplitude of the wake potential for a resonant mode, or the magnitude of some characteristic feature such as the intercept at $\tau = 0$ for the longitudinal wake or the value of the first maximum for the deflection wake, is per unit of structure,

$$\begin{aligned} w(\text{longitudinal}) &\sim \omega^2 \\ w(\text{dipole}) &\sim \omega^3 \\ w(\text{quadrupole}) &\sim \omega^5 \quad . \end{aligned} \tag{9.52}$$

The time at which some characteristic feature occurs, such as the first zero crossing of the longitudinal wake or the first maximum of the deflection wake, scales of course as ω^{-1}. The magnitude of the impedance for a resonant mode, again per unit length of structure, then scales as

$$\begin{aligned} \frac{Z_n}{Q}\,(\text{longitudinal}) &\sim \omega \\ \frac{Z_{dn}}{Q}\,(\text{dipole}) &\sim \omega^2 \\ \frac{Z_{qn}}{Q}\,(\text{quadrupole}) &\sim \omega^4 \quad . \end{aligned} \tag{9.53}$$

The amplitude of the so-called broad-band impedance function scales with the same frequency dependence. The impedance or the wake for a specific vacuum chamber component (as opposed to the impedance or wake per unit length of structure) scales as one power of frequency less than given above.

The dependence of both the longitudinal and transverse wakes on beam aperture has been investigated by K. Bane[39] for the SLAC disk-loaded structure. The intercept at $\tau = 0$ of the longitudinal wake was found to vary as

$$w_\ell(0) \sim a^{-1.68} \tag{9.54a}$$

over the range in aperture radii for the SLAC structure. The time at which the longitudinal wake falls to one-half its value at $\tau = 0$ is

$$\tau_{1/2} \approx 0.09\ a/c \quad . \tag{9.54b}$$

The amplitude of the first maximum of the dipole wake was computed to vary as

$$w_d(\tau_m) \sim a^{-2.25} \quad . \tag{9.55a}$$

However, the time at which the wake reaches its maximum value also varies with the beam aperture radius as

$$\tau_m \approx 0.64\ a/c \quad . \tag{9.55b}$$

Thus the initial slope of the wake was found to vary more strongly with a than the value of the first maximum:

$$\frac{dw_d}{d\tau} \sim a^{-3.48} \qquad \tau \to 0 \quad . \tag{9.55c}$$

These scaling relations would not be expected to hold when extrapolating to beam apertures significantly different from a = 1.163 cm, which is the aperture radius for an average cavity in the SLAC constant gradient structure. If the scaling law is written in the form a^{-n}, then the value n is larger than given above when scaling to larger values of a, and smaller when scaling to smaller a.

A structure filling factor can be defined by $f \equiv (p-t)/p$, where p is the periodic length and t is the disk thickness. For the SLAC structure, p = 3.5 cm and f = 0.83. When scaling to a structure with a different filling factor, computations indicate that the amplitude of the wake scales roughly in direct proportion to f.

10. SINGLE-BUNCH ACCELERATION AND BEAM LOADING

10.1 Structure Parameters

In this section we consider the acceleration of single bunches in traveling-wave linac structures. As will be discussed later, traveling-wave structures are to be preferred over standing-wave structures for single-bunch acceleration because the stored energy per unit length required to produce a given accelerating gradient is in general lower.

Using the notation in Ch. 5, the average accelerating gradient $\bar{E}_a$ for a structure of length L with input power P and unloaded energy gain V_o can be written in the form

$$\bar{E}_a^2 = \left(\frac{V_o}{L}\right)^2 = \frac{P_o r}{L} f(\tau) \quad . \tag{10.1}$$

For constant impedance (CZ) and constant gradient (CG) structures,

$$\text{CZ:} \quad f(\tau) = (2/\tau)(1-e^{-\tau})^2 \approx 2\tau(1-\tau+\tau^2/2) \tag{10.2a}$$

$$\text{CG:} \quad f(\tau) = (1-e^{-2\tau}) \approx 2\tau(1-\tau+2\tau^2/3) \tag{10.2b}$$

For single-bunch acceleration it is also of interest to relate the energy stored per unit length of structure, w_s, to the energy gradient. The loss parameter k introduced previously and defined in Eq. (9.33) provides the desired relation:

$$k_1 \equiv \frac{E_a^2}{4w_s} = \frac{\omega}{4}\left(\frac{r}{Q}\right) \sim \omega^2 \quad ,$$

where the subscript emphasizes that k_1 is per unit length of structure. The average gradient for a structure of length L can now be written as

$$\bar{E}_a^2 = 4\, k_1 w_o\, \eta_s \quad , \qquad (10.3)$$

where $w_o = P_o T_f / L$ is the imput energy per unit length and η_s is a structure efficiency, given by

$$\text{CZ:} \quad \eta_s = (1 - e^{-\tau})^2/\tau^2 \qquad (10.4a)$$

$$\text{CG:} \quad \eta_s = (1 - e^{-2\tau})/2\tau \quad . \qquad (10.4b)$$

Problem 10.1: Derive the expressions for η_s in Eqs. (10.4). Hint: Recall that the filling time for both constant impedance and constant gradient structures is given by $T_f = \tau(2Q/\omega)$.

The structure efficiency η_s and normalized power $P_n = P_o/(\bar{E}_a^2 L/r) = 1/f(\tau)$ are plotted in Fig. 10.1 as a function of the attenuation parameter τ for a constant impedance structure. Note that a high structure efficiency and the lowest peak power requirement are mutually exclusive. The best compromise for both high η_s and low P_n is reached for a τ on the order of 0.3-0.4.

The structure parameter k_1 is a strong function of the beam aperture radius a. The dependence of k_1 on beam aperture is shown in Fig. 10.2 for the SLAC disk-loaded structure at 2856 MHz. The solid curve can be approximated analytically by the expression

$$k_1 \approx \frac{27 \text{ V/pC/m}}{1 + 30.5(a/\lambda)^2} \quad . \qquad (10.5)$$

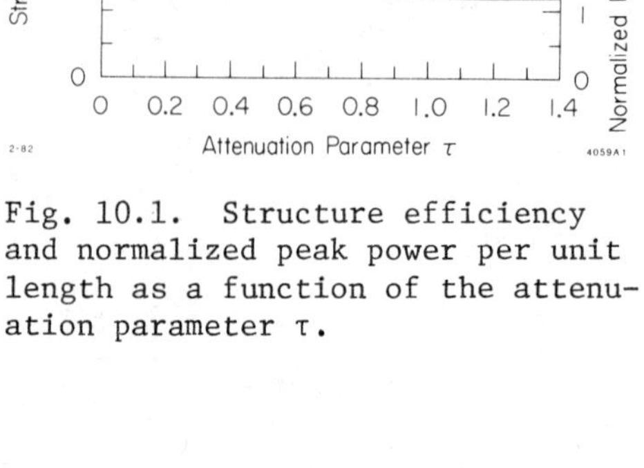

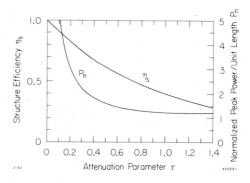

Fig. 10.1. Structure efficiency and normalized peak power per unit length as a function of the attenuation parameter τ.

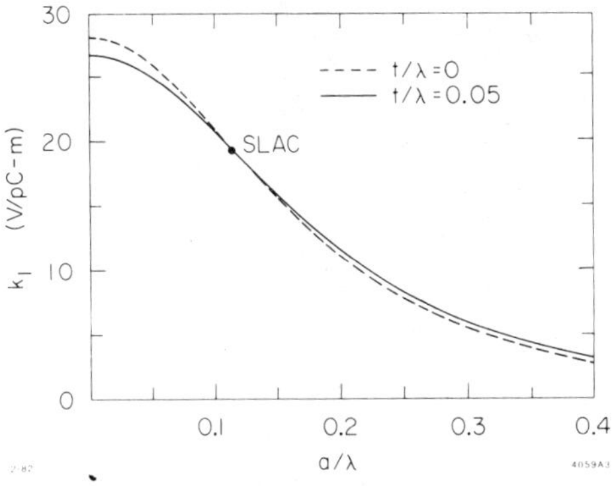

Fig. 10.2. Structure parameter k_1 as a function of beam aperture radius for the SLAC disk-loaded structure for two values of disk thickness t ($t/\lambda = 0.056$, $\lambda = 10.5$ cm for the SLAC structure).

Other structures might be expected to have a similar dependence of k_1 on a/λ.

The familiar disk-loaded structure does not necessarily have the highest value of k_1 at a given frequency and beam aperture. An alternative structure is the jungle gym structure, shown in Fig. 10.3. The group velocity of the accelerating mode in the jungle gym structure tends to be considerably higher than is the case for the disk-loaded structure. In addition, the jungle gym is a backward wave structure (phase velocity and group velocity have opposite signs for the accelerating mode). Typically, $v_g/c \approx 0.20$ for the $\pi/2$ mode ($\pi/2$ phase shift between adjacent bar), and $v_g/c \approx 0.10$ for the $\pi/3$ mode. Table 10.1 compares the jungle gym and disk-loaded structures at three frequencies that might be of interest for a high-energy linear collider. Values of r, k_1, Q and v_g/c for the $\pi/2$-mode jungle gym are scaled from values measured[40] at 714 MHz for a structure used for several years as an rf cavity in the Cornell University electron synchrotron. Values for the $\pi/3$-mode jungle gym are estimated from some old measurements[41] made at the Stanford University Microwave Laboratory. The k_1 value for the disk-loaded structure with a wider beam aperture, a = 1.50, is obtained from Fig. 10.2; v_g/c is scaled as a^4. From Eq. (9.55c), the slope of transverse wake for this structure for $t \to 0$ should be lower by a factor of about 2.5 compared to a structure with a = 1.16 cm. Finally, values for the standing-wave disk-and-washer structure, described in Sec. 3.4, are given for comparison. Note that although the Q and shunt impedance for this structure are very high, the value of k_1 is low compared to both the jungle gym and disk-loaded structures.

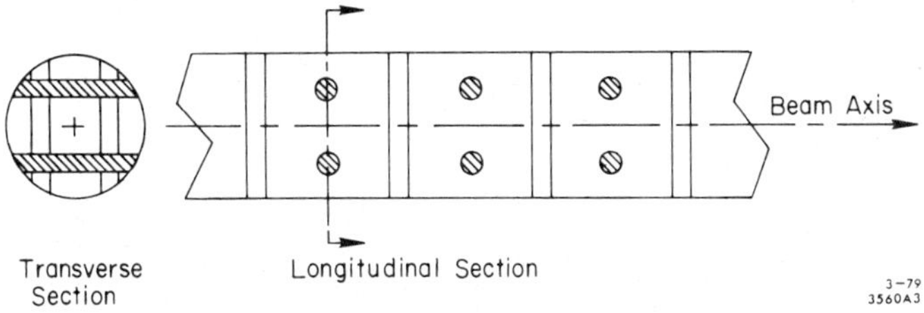

Fig. 10.3. The jungle gym bar-loaded accelerating structure.

Table 10.1 Comparison of Accelerating Structures

	r (MΩ/m)	k_1 (V/pC/m)	Q	v_g/c	L (m)	T_f (μs)	τ
2856 MHz							
Disk-Loaded (a = 1.6 cm)	56	19	13,300	.012	3	.83	.57
Disk-Loaded (a = 1.50 cm)	46	16	13,000	.035	6	.57	.40
Disk and Washer (a = 1.16 cm)	85	10	40,000	--	--	--	--
Jungle Gym (π/2) (1/2 diag. = 0.84 cm)	51	25	9,000	.22	6	.09	.09
Jungle Gym (π/3) (1/2 diag. = 0.90 cm)	60	30	9,000	.10	6	.20	.20
4040 MHz							
Jungle Gym (π/2)	61	50	7,500	.22	6	.09	.15
Jungle Gym (π/3)	71	60	7,500	.10	6	.20	.35
5712 MHz							
Jungle Gym (π/2)	72	100	6,500	.22	6	.09	.26
Jungle Gym (π/3)	85	120	6,500	.10	6	.20	.57

10.2 Beam Loading for a Gaussian Bunch

Equation (9.5) gives the beam loading potential within a bunch in terms of the wake potential $w(\tau)$. As a specific illustration, consider the disk-loaded structure for the SLAC linac. The computation of the wake potential for this structure was described in the last chapter; the resulting wake is shown again in Fig. 10.4 for the range 0-20 ps. For this time range, the wake is described quite closely by the expression

$$w(\tau) = A \exp[-(\tau/B)^n] \quad , \tag{10.6}$$

where

$$A = 226 \text{ V/pC/m} \sim \omega^2$$

$$B = 6.13 \text{ ps} \sim \omega^{-1}$$

$$n = 0.605 \quad .$$

For a Gaussian bunch, Eq. (9.5) can be written in the form

$$E_b(t) = \frac{e\,c\,A\,N_b}{\sqrt{2\pi}\,\sigma_z} \int_{-\infty}^{t} \exp\left[-\left(\frac{t-t'}{B}\right)^n\right] \exp\left[-t'^2 c^2/2\sigma_z^2\right] dt' \quad . \tag{10.7}$$

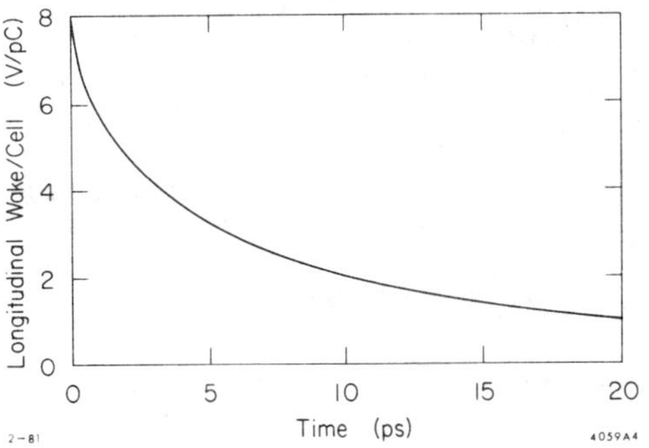

Fig. 10.4. Longitudinal wake potential per period, p = 3.50 cm, for the SLAC structure.

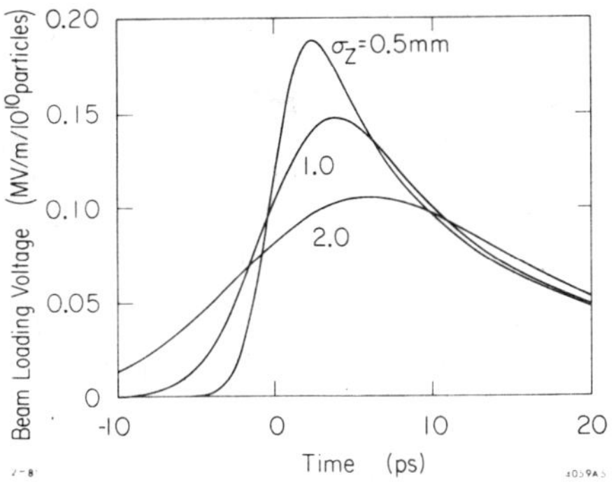

Fig. 10.5. Beam loading voltage within a Gaussian bunch for the SLAC structure for three values of bunch length.

Results for the SLAC structure at 2856 MHz for several values of σ_z are shown in Fig. 10.5.

The total energy gain per unit length by a particle at time t in the bunch can be obtained by adding the external accelerating voltage to the beam loading voltage,

$$E(t) = \bar{E}_a \cos(\omega t - \theta) - E_b(t) \quad . \tag{10.8}$$

Here θ is the phase angle by which the bunch center leads the crest of the accelerating wave produced by the external rf source. By adjusting this phase angle, the rising slope of the accelerating voltage waveform can be made to cancel, at least in part, the negative-going beam loading waveform, resulting in a reduction in the energy spread of the particles in the bunch below the energy spread for the case $\theta = 0$. This is shown schematically in Fig. 10.6. Note, however, that the decrease in energy spread is achieved at the expense of a reduction in the average energy gain per unit length per particle, given by

$$\bar{E} = \frac{1}{q} \int_{-\infty}^{\infty} E(t)\, I(t)\, dt \quad . \tag{10.9}$$

As an example, consider the case of the SLAC structure operating at $\bar{E}_a$ = 17 MV/m (V_0 = 50 GeV total energy) to accelerate a single bunch of particles with $N_b = 5 \times 10^{10}$ and σ_z = 1.0 mm. The energy spread $\Delta V/V_0$ which contains 90% of the particles, and the average particle

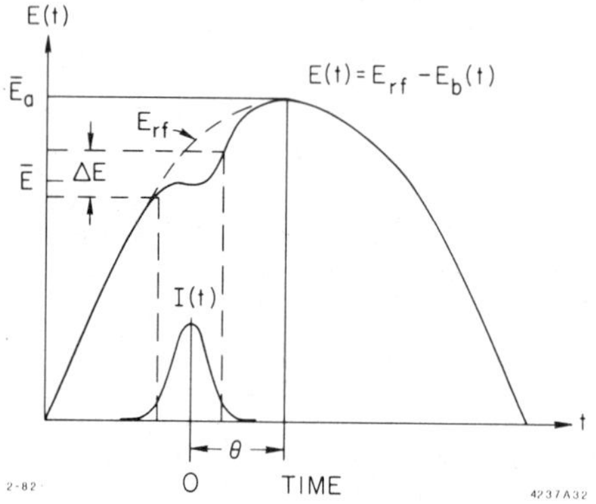

Fig. 10.6. Diagram showing how the single-bunch beam loading gradient $E_b(t)$ subtracts from the rf accelerating wave $E_{rf} = \bar{E}_a \cos(\omega t - \theta)$ to give the net gradient E(t).

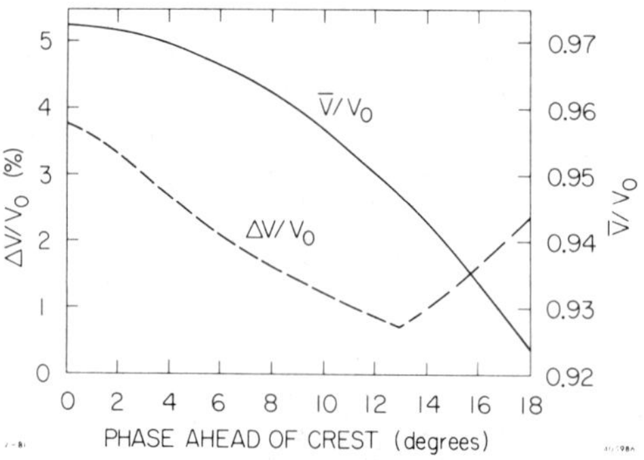

Fig. 10.7. Energy spread and average gain per particle for the SLAC structure with $\bar{E}_a$ = 17 MV/m, $N_b = 5 \times 10^{10}$ and σ_z = 1.0 mm.

energy $\bar{V}/V_0$, are plotted as a function of the phase offset θ in Fig. 10.7. In this example, the energy spread at θ = 13° is reduced by a factor of four below the spread at θ = 0, but at the expense of an additional 2-1/2% loss in average energy per particle.

The detailed energy distribution function for the particles within the bunch may sometimes be of interest. The charge dq in the energy range dV is given by

$$\frac{dq}{dV} = \sum_n \frac{I(t_n)}{(dV/dt)_{t = t_n}} \qquad (10.10)$$

The sum is necessary because, as can be seen from Fig. 10.6, there may be up to four values of time $t = t_n$ giving the same energy $V(t_n)$. The energy spectrum actually observed in a linac is modified further because of the finite energy range accepted by the energy defining slits. This effect can be taken into account by convolving the above distribution function with an appropriate slit function. This removes the infinite spikes at energies where the derivative dV/dt vanishes, resulting in a smoothed distribution function. Examples of energy spectra for beam loading by single bunches in the SLAC linac are given in Ref. 42 for various values of the phase offset parameter θ. Agreement between the measured and computed distribution functions is very good, indicating that the functional form of the wake as shown in Fig. 10.4 has a basis in physical reality. The measurements indicate, however, that the amplitude of the computed wake may be about 30% low.[42]

The efficiency for transfer of stored energy in the structure into beam energy is also of interest:

$$\eta_b \equiv q\, \bar{E}/\bar{w}_s \qquad (10.11)$$

where $\bar{w}_s = \bar{E}_a^2/4k_1$ is defined as the effective stored energy per unit length in the structure. Thus

$$\eta_b = \frac{4\, e\, N_b\, k_1\, \bar{E}}{\bar{E}_a^2} \sim \frac{\omega^2\, N_b}{\bar{E}_a} \left(\frac{\bar{E}}{\bar{E}_a}\right) \quad . \tag{10.12}$$

Note that, from Eqs. (10.3) and (10.11),

$$q\,\bar{E} = \eta_b\, \eta_s\, w_o \quad . \tag{10.13}$$

Thus, the product of the beam efficiency and the strcture efficiency gives the net efficiency for the conversion of the applied input energy per unit length from the rf sources into beam energy.

It is informative to introduce the beam loading enhancement factor B, defined in Sec. 6.2 for a point bunch, into the expressions for average energy gain per particle and beam efficiency. By definition,

$$\bar{E} = \bar{E}_a F_1 \cos\theta - \Delta E_b \equiv \bar{E}_a - \Delta E \tag{10.14a}$$

$$\Delta E_b = k_1 q\, B(\sigma) \quad . \tag{10.14b}$$

In Eq. (10.14b) we note explicitly that B is a function of bunch length. The bunch form factor for the accelerating mode, F_1, is also introduced in Eq. (10.14a), although usually it will be quite close to unity (see Sec. 3.3). If a given relative energy reduction per particle $\Delta E/\bar{E}_a$ is specified, the number of particles that can be accelerated is, from Eqs. (10.14),

$$N_b = \frac{\bar{E}_a}{e\, B\, k_1} \left[\frac{\Delta E}{\bar{E}_a} - (1 - F_1 \cos\theta)\right] \quad . \tag{10.15}$$

For $F_1 \cos\theta \approx 1$, note that N_b scales as $\bar{E}_a/B\omega^2$. The beam efficiency, Eq. (10.12), can also be written in terms of B, using Eqs. (10.14), as

$$\eta_b = \frac{4}{B} \left(1 - \frac{\Delta E}{\bar{E}_a}\right) \left[\frac{\Delta E}{\bar{E}_a} - (1 - F_1 \cos\theta)\right] \quad . \tag{10.16}$$

The maximum possible efficiency as a function of $\Delta E/\bar{E}_a$ is

$$\eta_b(\max) = \frac{F_1^2 \cos^2\theta}{B} \tag{10.17}$$

at $\Delta E/\bar{E}_a = 1 - (F_1 \cos\theta)/2$.

The enhancement factor can be computed using

$$B(\sigma) = \frac{1}{F_1^2 k_1 q^2} \int_{-\infty}^{\infty} E_b(t)\, I(t)\, dt \tag{10.18}$$

where F_1 is the bunch form factor for the accelerating mode (normally $F_1 \approx 1$). The enhancement factor is shown as a function of bunch length in Fig. 10.8 for the SLAC disk-loaded structure (λ = 10.5 cm, k_1 = 19 V/pC/m). As an example, consider a 1 mm bunch with B = 3.1. For $\Delta E/\bar{E}_a = 0.1$, $F_1 \cos\ = 1$ and $\bar{E}_a$ = 100 MV/m, the number of particles that can be accelerated is, from Eq. (10.15), $N_b = 1.1 \times 10^{12}$. Using Eq. (10.16), the beam efficiency is 12%.

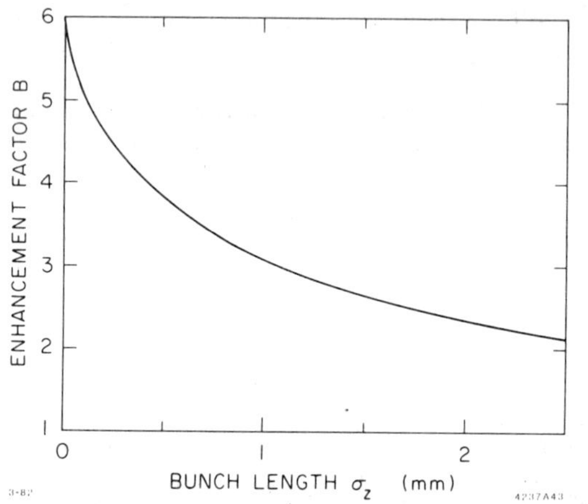

Fig. 10.8. Beam loading enhancement factor as a function of bunch length for the SLAC structure.

For a Gaussian bunch, B(σ) can be written directly in terms of the values k_n and ω_n for the structure modes as

$$B(\sigma) = \left(k_1 e^{-\omega_1^2\sigma_t^2}\right)^{-1} \sum_{\text{all } n} k_n e^{-\omega_n^2\sigma_t^2} \quad . \tag{10.19}$$

11. TRANSVERSE BEAM DYNAMICS

Transverse beam dynamics effects which do not depend on current are treated in detail in Ref. 43, and this coverage will not be repeated here. Beam transport in the presence of an external focusing field, the dependence of transverse emittance on energy, and the effect of transverse rf fields (due, for example, to fringe fields at the entrance and exit of an accelerator section and to asymmetry in the imput and output rf couplers) are among the topics discussed in Ref. 43 which are of importance in the design of high energy electron linacs. In this chapter we limit the discussion of transverse beam dynamics to a simple model which predicts the approximate emittance growth due to the transverse wake in a linac structure or storage ring rf system.

11.1 Two-Particle Model for Transverse Emittance Growth

In a continuous distribution of charge, each particle is affected by the transverse wakes from all other particles in the bunch which are ahead of the particle in question. To solve the transverse equation of motion for the case of an arbitrary charge distribution, also taking into account acceleration and external fucusing, is a formidable task. Chao *et al.*,[44] give a solution for the case of a linear wake and a rectangular charge distribution, but the result is too complex to be cast in a form permitting simple scaling. The simplest model that still contains the essentials of the physical situation is a two-particle model for the bunch. Consider a model for a Gaussian bunch of total charge q in which a head charge q/2 is located at $z' = +\sigma_z$ and a tail charge q/2 is located at $z' = -\sigma_z$, where z' is the coordinate relative to the bunch center. Assume a uniform external focusing field of strength k_β, such that the head particle moves on an orbit described by $x_0 \cos k_\beta z$ as a function of distance z along the accelerator. The transverse force acting on the tail particle due to the dipole wake is then

$$F_1 = \frac{1}{2}\, e\, q\, x_0\, w_d(2\sigma_z) \cos k_\beta z \quad ,$$

where $w_d(2\sigma_z)$ is the dipole wake evaluated at $2\sigma_z$. The displacement x_1 for the tail particle obeys the transverse equation of motion

$$\frac{d}{dz}\left[V(z)\,\frac{dx_1}{dz}\right] + V(z)\, k_\beta^2 x_1 = F_1(z)/e \quad , \tag{11.1}$$

where eV(z) is the energy. For the case of constant energy eV_0, Eq. (11.1) simplifies to

$$x_1'' + k_\beta^2\, x_1 = C \cos k_\beta z \quad , \tag{11.2}$$

where $x_1' \equiv dx_1/dz$, $x_1'' \equiv d^2x_1/dz^2$ and

$$C = \frac{q\, x_0\, w_d(2\sigma_z)}{2\, V_0} \quad . \tag{11.3}$$

Equation (11.2) is the equation for the amplitude of a lossless harmonic oscillator driven at resonance. Assuming $x_1 = x_1' = 0$ at $z = 0$, the solution is

$$x_1 = \frac{C}{2k_\beta}\,(z \sin k_\beta z) \tag{11.4a}$$

$$x_1' = \frac{C}{2k_\beta}\,(\sin k_\beta z + k_\beta z \cos k_\beta z) \quad . \tag{11.4b}$$

Two limits are of special interest. For $k_\beta = 0$ we obtain

$$x_1 = Cz^2/2 \tag{11.5a}$$

$$x_1' = Cz \tag{11.5b}$$

$$\varepsilon = \pi\, x_1 x_1' = \pi \left(\frac{C^2}{2}\right) z^3 \quad . \tag{11.5c}$$

In the limit of strong focusing, $k_\beta z >> 1$,

$$x_1 = \left(\frac{Cz}{2k_\beta}\right) \sin k_\beta z \tag{11.6a}$$

$$x_1' = \left(\frac{Cz}{2}\right) \cos k_\beta z \tag{11.6b}$$

$$\varepsilon \equiv \pi |x_1||x_1'| = \pi \left(\frac{C^2}{4k_\beta}\right) z^2 \quad . \tag{11.6c}$$

Thus, the ratio of the maximum displacement with strong focusing to the displacement with no focusing is

$$\frac{|x_1|\,(\text{focusing})}{x_1\ (\text{no focusing})} = \frac{1}{k_\beta z} = \frac{1}{2\pi\, N_\beta} \quad , \tag{11.7}$$

where $N_\beta = z/\lambda_\beta$ is the number of betatron wavelengths. The corresponding ratio of emittances is $1/(4\pi\, N_\beta)$.

Note that the motion of the tail particle is 90° out of phase with respect to the motion of the head particle. In phasor notation, Eq. (11.4a) gives

$$\tilde{x}_1(z) = \tilde{x}_1(0) - j\,A\,\tilde{x}_o \tag{11.8}$$

where

$$A = \frac{Cz}{2k_\beta x_o} = \frac{q\, z\, w_t(2\sigma_z)}{4\, V_o k_\beta} \quad . \tag{11.9}$$

As an example, consider the SLAC linac ($z = 3 \times 10^3$ m), with λ_β = 100 m and a bunch with 5×10^{10} particles ($q = 8 \times 10^{-9}$ C). Assume a bunch length σ_z = 1 mm, or σ_t = 3.3 ps (these are parameters for the proposed SLAC Linear Collider). From Fig. 9.5 the transverse wake at 6.6 ps is 1.0×10^{14} V/C-m/period, recalling that the dimension <u>a</u> of the disk opening is 1.163×10^{-2} m. Since the length of a cell is 3.5×10^{-2} m, the wake $w_t(2\sigma_z) = 2.8 \times 10^{15}$ V/C-m^2. Let us approximate acceleration to 50 GeV (again, the energy for the SLAC Linear Collider) by a constant energy of 25 GeV. Putting these numbers in Eq. (11.9), we obtain A = 10, or $|x_1| = 10|x_o|$, for the amplitude of the oscillation of the tail charge as driven by the head charge. The solution obtained by Chao <u>et al</u>.,[44] for these same parameters, but assuming a continuous rectangular charge distribution and uniform acceleration from 1.2 to 50 GeV, is A = 6. Thus, Eq. (11.9) provides a simple but reasonably accurate expression for estimating the single-bunch emittance growth due to the transverse wake in a linac with focusing.

Problem 11.1: Write the equation of transverse motion, Eq. (11.1), for the case of a linac with uniform acceleration, $V(z) = V_o + V'z$. Let the strength of the focusing force scale with beam energy so that k_β remains constant, independent of z. Can you find an asymptotic expression analogous to Eq. (11.9) for the growth ratio A?

If there is an energy spread for the particles within the bunch, there will also be spread in betatron frequencies, since $k_\beta^2 \sim 1/\gamma$ for a given focusing strength. This corresponds, in the two-particle model, to a head particle with frequency (wave number) $k_{\beta o}$ driving a tail particle with frequency $k_{\beta 1} = k_{\beta o} + \delta k$. The tail particle is now a harmonic oscillator being driven off-resonance, and we might expect a reduction in the growth of the amplitude of the oscillation compared to that given in Eq. (11.6a). This is, in essence, Landau damping.

The equation of motion for this case and its solutions, assuming $x_1 = x_1' = 0$ at $z = 0$, is

$$x_1'' + k_{\beta 1}^2 x_1 = C \cos k_{\beta o} z \qquad (11.10a)$$

$$x_1 = \frac{C}{k_{\beta 1}^2 - k_{\beta o}^2} (\cos k_{\beta o} z - \cos k_{\beta 1} z) \qquad (11.10b)$$

$$= \frac{2C}{k_{\beta 1}^2 - k_{\beta o}^2} \left\{ \left[\sin \frac{1}{2} (k_{\beta 1} + k_{\beta o}) z \right] \left[\sin \frac{1}{2} (k_{\beta 1} - k_{\beta o}) z \right] \right\}$$

$$x' = \frac{C}{k_{\beta 1}^2 - k_{\beta o}^2} (k_{\beta 1} \sin k_{\beta 1} z - k_{\beta o} \sin k_{\beta o} z) \quad . \qquad (11.10c)$$

The maximum amplitudes of x and x', assuming $\delta k / k_\beta$ is small, are

$$|x_1| = \frac{C}{k_\beta \, \delta k} \qquad (11.11a)$$

$$|x_1'| = \frac{C}{\delta k} \qquad (11.11b)$$

$$\varepsilon = \pi \left[\frac{C^2}{k_\beta (\delta k)^2} \right] \quad . \qquad (11.11c)$$

By comparing Eq. (10.11a) with Eq. (11.6a), the reduction in $|x_1|$ due to the head-tail frequency difference is

$$\frac{|x_1|\,(\text{with } \delta k)}{|x_1|\,(\delta k = 0)} = \frac{2}{z\ \delta k} \quad . \tag{11.12}$$

The emittance is reduced by the square of this factor.

11.2 Strong Head-Tail Instability in a Storage Ring

In a storage ring, the head and the tail of the bunch change places periodically due to synchrotron oscillations. The head first drives the tail for half a synchrotron period, and on the next half-period the tail moves forward to drive the former head, which has now become the new tail. Thus a feedback mechanism exists which can lead to a possible instability. Let us assume that the transverse deflection wake is confined to the rf structure only, of length L_{rf}. There will certainly be transverse wakes associated with other vacuum chamber components in the ring, but the rf structure is often the major impedance source. In one-half synchrotron period the bunch will pass through the rf structure $f_r/2f_s$ times, where f_r is the revolution frequency and f_s is the synchrotron frequency. Since the current per bunch is $I_b = q\ f_r$, the growth factor A in Eq. (11.9) is, after one-half synchrotron period,

$$A = \frac{I_b\ L_{rf}\ \bar{w}\ \beta_{rf}}{8\ V_o\ f_s} \quad . \tag{11.13}$$

Here we have introduced the beta function, $\beta_{rf} = 1/k_\beta$, which is normal in storage ring nomenclature. The wake function in Eq. (11.13) must also be averaged from $\tau = 0$ to $\tau_m = 2\sigma_z$, where $\tau = \tau_m \sin\omega_s t$.

Problem 11.2: Show that the average wake seen by the tail particle during one-half period of synchrotron oscillation is

$$\bar{w} = \frac{2}{\pi} \int_o^{\tau_m} \frac{w(\tau)\ d\tau}{\left(\tau_m^2 - \tau^2\right)^{1/2}} \quad . \tag{11.14}$$

The phasor diagram in Fig. 11.1 illustrates how the phasors representing the betatron oscillations of the head and the tail particles change during each half synchrotron period. Let $\tilde{x}_o^{(1)}$ and $\tilde{x}_1^{(1)}$ be the head and tail particles during the first half-period. If $\tilde{x}_1^{(1)} = \tilde{x}_o^{(1)}\ e^{j\alpha}$ at the beginning of the half-period, then it will be driven to $\tilde{x}_1^{(1)}\ e^{-j\alpha}$ at the end of the half-period, where (see Fig. 11.1)

$$\sin\alpha = \frac{A}{2} \quad . \tag{11.15}$$

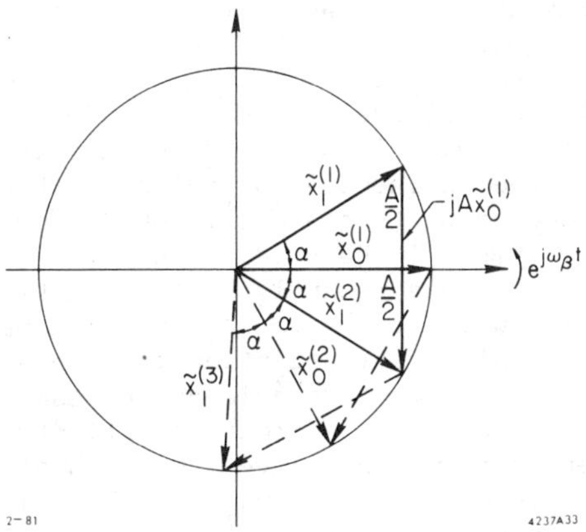

Fig. 11.1. Diagram showing the eigenvectors on successive half periods of synchrotron oscillation for the motion of two particles driven by the transverse wake in a storage ring.

Thus $\tilde{x}_0(1)$ and $\tilde{x}_1(1)$, when chosen in this manner, are eigenvectors which differ from one turn to the next by a real phase shift only. During the next half period the roles of the particles are reversed, and particle $\tilde{x}_0$ is changed according to $\tilde{x}_0^{(2)} = \tilde{x}_1^{(2)} e^{-j\alpha} = \tilde{x}_0^{(1)} e^{-j2\alpha}$.

Equation (11.15) has two important consequences. First, if $(A/2) \leq 1$, it corresponds to a real shift in betatron frequency given by

$$\Delta\omega = -\left(\frac{\alpha}{\pi}\right)\omega_s = -\left(\frac{1}{\pi}\sin^{-1}\frac{A}{2}\right)\omega_s \tag{11.16}$$

Secondly, for $A > 2$ the frequency shift becomes imaginary, indicating unstable growth. From the geometry of Fig. 11.1 it is also seen that $A = 2$, $\alpha = \pi/2$ is a limit for stable amplitudes of oscillation for the two particles. This limit on stability using the two-particle model was first derived by Kohaupt.[45] The threshold current for the instability can be found by setting $A = 2$ in Eq. (11.13),

$$I_b(\text{threshold}) = \frac{16\, V_o\, f_s}{\bar{w}\, L_{rf}\, \beta_{rf}} \quad . \tag{11.17}$$

Measurements of threshold currents for the single-bunch transverse instability observed in PEP, PETRA and SPEAR are in reasonable agreement with Eq. (11.17).

12. HIGH-GRADIENT LINEAR COLLIDER DESIGN

The energy realistically attainable by an electron-positron storage ring is limited to an energy on the order of 100 GeV by the RF voltage and power requirements imposed by synchrotron radiation. The proposed LEP storage ring, for example, would have a circumference of 27 km and would attain an energy of 85 GeV with an rf power of 96 MW. The energy could be extended to 125 GeV by replacing the room-temperature rf cavities by superconducting cavities. To reach still higher energies in a storage ring, the circumference, rf power requirement and cost must be scaled roughly as the square of the energy. At 300 GeV the circumference of a storage ring with normal conducting cavities would be on the order of 300 km. Therefore, to reach energies on the order of 300 × 300 GeV or higher in an electron colliding beam machine of reasonable dimensions, it is clear that one must look to a high gradient linear collider — two linear accelerators firing intense bunches toward each other, operating at a gradient of at least 100 MV/m.

A summary of the history of linear colliders is given in Ref. 46. The SLAC Linear Collider, a variant of this class of accelerator in which e^+ and e^- bunches are accelerated in the SLAC 3 km linac, then separated at the end of the linac and brought into collision through separate beam transport arcs, is described in Ref. 47. Finally, a general review of linear collider design has recently been given by Wiedemann.[48] In this chapter we will not attempt to review all aspects of linear collider design, but will confine our attention to high gradient acceleration, the rf power requirement, single-bunch beam loading and transverse wake effects. However, in order to estimate beam loading and transverse wake effects, we must have some idea of the bunch length and the charge per bunch. These parameters are, in turn, set by the desired luminosity and the allowable energy spread arising from synchrotron radiation during the beam-beam interaction (beamstrahlung). Some relationships for estimating the most important bunch parameters are given in the following section.

12.1 Bunch Parameters

The luminosity for a linear collider is

$$\mathscr{L}_o = \frac{N^2 f_r}{4\pi\ \sigma_x^* \sigma_y^*} \tag{12.1a}$$

where N is the number of particles per bunch, f_r is the repetition rate and σ_x^* and σ_y^* are the transverse beam dimensions at the collision point. This can be written in normalized form, assuming a round beam,

$$\mathscr{L}_o = 5.3 \times 10^{31}\ \mathrm{cm}^{-2}\mathrm{s}^{-1} \left[\frac{(N/10^{11})^2 (f_r/1\ \mathrm{kHz})(E_o/100\ \mathrm{GeV})}{(\beta^*/1\ \mathrm{cm})\left(\varepsilon_n/3 \times 10^{-3}\ \mathrm{cm\text{-}rad}\right)} \right] . \tag{12.1b}$$

Here E_o is the beam energy, ε_n the invariant emittance $\varepsilon_{nx,y} = \gamma\, \sigma_{x,y}\, \sigma_{x',y'}$, and $\sigma^*_{x,y} = (\beta^*_{x,y}\varepsilon_n/\gamma)^{1/2}$. The normalization constants $\beta^* = 1$ cm and $\varepsilon_n = 3 \times 10^{-3}$ cm-rad are the design values for the SLAC Linear Collider,[47] and are believed to be reasonably attainable.

Another important bunch parameter is the beamstrahlung parameter δ, defined as the fractional energy loss due to synchrotron radiation from the particles in one bunch passing through the opposing bunch. It is given by[48]

$$\delta = \frac{2\, r_e^3}{3} \cdot \frac{N^2 \gamma}{\sigma_x^* \sigma_y^* \sigma_z} \cdot F(R) \quad , \tag{12.2a}$$

where $r_e = 2.82 \times 10^{-13}$ cm is the classical electron radius and $R \equiv \sigma_x^*/\sigma_y^*$. For a round Gaussian beam $F(R) = 0.325$, and $F(R) \approx 1.3/R$ for $R >> 1$. The above relation becomes, in normalized form for a round beam,

$$\delta = 0.66 \times 10^{-2} \left[\frac{(N/10^{11})^2 (E_0/100\ \text{GeV})^2}{(\sigma_z/1\ \text{mm})(\beta^*/1\ \text{cm})\left(\varepsilon_n/3 \times 10^{-3}\ \text{cm-rad}\right)} \right] . \quad (12.2b)$$

A third important bunch parameter is the disruption parameter D. It is defined as σ_z/F, where F is the focal length of the lens produced by the electromagnetic fields in one bunch acting on a particle near the axis in the opposing beam. Thus, for bunches consisting of particles of the opposite sign, there is a pinch effect due to this focusing which reduces the beam cross section and enhances the luminosity. For D = 1, the luminosity is increased by a factor[49] of 3.5. For D = 2.5 - 15, the enhancement is nearly constant at a factor of six. This enhancement in luminosity is a desirable effect. However, if D is too large, the particles emerging from the collision may have an angular spread large enough to interfere with the detection of small-angle events. In terms of the bunch parameters,

$$D_{x,y} = \frac{2\, r_e\, N\, \sigma_z}{\gamma\, \sigma^*_{x,y}\left(\sigma^*_y + \sigma^*_x\right)} . \quad (12.3a)$$

In normalized form for a round beam,

$$D = 0.94 \left[\frac{(N/10^{11})(\sigma_z/1\ \text{mm})}{(\beta^*/1\ \text{cm})\left(\varepsilon_n/3 \times 10^{-3}\ \text{cm-rad}\right)} \right] . \quad (12.3b)$$

Finally, by eliminating N^2 between Eqs. (12.1a) and 12.2a), we obtain a useful relation for the repetition rate,

$$f_r = \frac{8\pi}{3} r_e^3 \cdot \frac{\gamma\, \mathscr{L}}{\sigma_z\, \delta} \cdot \frac{F(R)}{f(D)} . \quad (12.4a)$$

Here we have introduced the enhancement factor $f(D) \equiv \mathscr{L}/\mathscr{L}_0$ due to the pinch effect described above. The repetition rate is important because it determines the average rf power requirement for a linear collider. In normalized form,

$$f_r(\text{Hz}) = 3750 \left[\frac{(E_0/100\ \text{GeV})(\mathscr{L}/10^{32}\ \text{cm}^{-2}\text{s}^{-1})}{(\sigma_z/1\ \text{mm})(\delta/.01)} \right] \cdot \frac{F(R)}{f(D)} . \quad (12.4b)$$

For a given energy and luminoisity, it is seen that a long bunch length and a large beamstrahlung parameter δ give the lowest repetition rate and hence the lowest average power consumption. As will be discussed later, the maximum tolerable bunch length will also depend on the operating wavelength which is chosen.

Using the preceding relations, we can work out a consistent set of parameters for a collider with 300 GeV per beam and a luminosity of 10^{32} $cm^{-2}s^{-1}$. Let us also choose a gradient of 100 MV/m and a length per linac of 3 km. As will be discussed later, a gradient of 100 MV/m is realistically attainable in a properly designed accelerating structure for short RF pulses (<500 ns) at S-band frequencies (3000 MHz) or higher. These collider parameters are summarized in Table 12.1 below. For simplicity we have assumed a round beam; R = 1 and F(R) = 0.325. Weidemann[48] considers the case for a beam with a large aspect ratio, which results in some reduction in required repetition rate. A luminosity enhancement factor f(D) = 3.5 at D = 0.95 has also been used.

Table 12.1 Parameters for a 300 × 300 GeV Collider

Accelerating Gradient	100 MV/m
Length (both linacs)	6 km
Luminosity	10^{32} $cm^{-2}s^{-1}$
Particles per Bunch N	10^{11}
Repetition Rate f_r	210 Hz
Invariant Emittance $\gamma\,\sigma_x\sigma_{x'} = \gamma\,\sigma_y\sigma_{y'}$	3×10^{-3} cm-rad
Beta Function at Collision Point $\beta_x^* = \beta_y^*$	1 cm
Beam Dimensions at Collision Point $\sigma_x^* = \sigma_y^*$	0.7 μm
Bunch Length σ_z	1 mm
Disruption Parameter D	0.95
Beamstrahlung Parameter δ	0.05
Average Current eNf_r	3.4 μA
Average Beam Power (each linac)	1.0 MW

12.2 RF Power Requirement

The average rf power will be determined by the choice of structure, the total energy, the accelerating gradient and the repetition rate. The total average power for two linacs, each of length L_1, is

$$\bar{P}_{rf} = 2\,w_o L_1 f_r = 2\,w_o V_1 f_r/\bar{E}_a \tag{12.5}$$

where w_o is the average power per unit length delivered by the rf sources and V_1 is the unloaded energy of each linac. Combining this relation with Eq. (10.4),

$$\bar{P}_{rf} = \frac{V_1 f_r \bar{E}_a}{2 k_1 \eta_s} \quad . \qquad (12.6)$$

For a given energy and repetition rate (which was set by the luminosity requirement), it is clear that a low gradient and a high value for the structure parameter k_1 are desirable. Since $k_1 \sim \omega^2$, this implies that a high operating frequency is also desirable. However, a low gradient uses up real estate, and there will be a limit on frequency set by the choice of $\sigma_z = 1$ mm. Also, there must be sufficient stored energy per unit length so that the energy spread due to single-bunch beam loading, and the energy reduction factor $\Delta E/\bar{E}_a$ discussed in Sec. 10.2, are not too large.

The maximum gradient will in any case be limited by breakdown initiated by electric fields at the surface of the structure. This breakdown field may be a function of frequency, pulse length, and vacuum conditions. It may also depend on the detailed physical and chemical properties of the metal surface. If $\hat{E}_s$ is the peak electric field anywhere on the surface of a structure, and if $\bar{E}_a$ is the average gradient in the structure, then the ratio $\hat{E}_s/\bar{E}_a$ will be important in determining the maximum gradient that can be achieved. For the SLAC disk-loaded structure, $\hat{E}_s/\bar{E}_a = 2.1$. The SLAC structure has been tested in a traveling-wave resonant ring to $E_a = 40$ MV/m without breakdown. The maximum surface field was therefore about 85 MV/m at an effective pulse length on the order of 1 μs at 2856 MHz. The value of $\hat{E}_s/\bar{E}_a$ can be lowered to about 1.6 in a disk-loaded structure by proper disk shaping, and still lower values of this ratio have been reported.[50,51] It is also generally expected that breakdown fields should increase with increasing frequency and decreasing pulse length. A gradient of 100 MV/m therefore does not seem unreasonable for very short pulse lengths, perhaps at a frequency somewhat higher than 3000 MHz, assuming also that careful attention is paid to the details of structure geometry and surface conditions.

Assuming that a gradient of 100 MV/m is feasible, Eq. (12.6) can now be applied, using the structure parameters given in Table 10.1, to compute the average RF power requirement of a 300 × 300 GeV collider with a length of 3 × 3 km and a repetition rate of 210 Hz. The result is given in Table 12.2. Also listed is the peak power $\hat{P}_0$ and energy per pulse $W_0 = w_0 T_f$ required from each rf source, assuming there are 1000 sources spaced 6 m apart. The expressions in Sec. 10.1 are used to compute η_s, w_0, and $\hat{P}_0$ from the structure parameters, and the pulse length T_p is taken to be equal to the filling time, T_f.

12.3 Beam Loading and Efficiency

The expressions in Sec. 10.2 have been used to compute single-bunch beam loading effects in the SLAC disk-loaded structure at a gradient of 100 MV/m. The disk-loaded structure has been chosen for analysis rather than the jungle gym because the non-axisymmetric jungle gym structure is not at present amenable to an analytic calculation of the wake potentials. Results are shown in Figs. 12.1 through 12.4.

Table 12.2 RF Power Requirements for a 3 × 3 km Collider with E_a = 1000 MV/m

	T_p (μs)	W_o (J)	η_s	$\hat{P}_o$ (MW)	$\bar{P}_{rf}$ (MW)
2856 MHz					
Disk-Loaded (a = 1.16 cm)	.83	1320	.58	1600	275
Disk-Loaded (a = 1.50 cm)	.57	1340	.68	2350	280
Jungle Gym (π/2)	.09	650	.92	7000	135
Jungle Gym (π/3)	.20	620	.82	3100	130
4040 MHz					
Jungle Gym (π/2)	.09	350	.86	3600	75
Jungle Gym (π/3)	.20	350	.71	1750	75
5712 MHz					
Jungle Gym (π/2)	.09	190	.78	2000	40
Jungle Gym (π/3)	.20	210	.58	1050	45

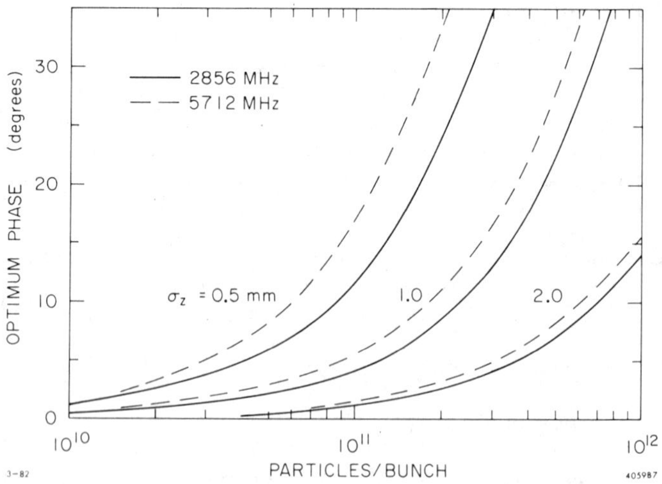

Fig. 12.1. Phase angle ahead of crest which minimizes the single-bunch beam loading energy spread as a function of number of particles per bunch for the SLAC disk-loaded structure at a gradient of 100 MV/m.

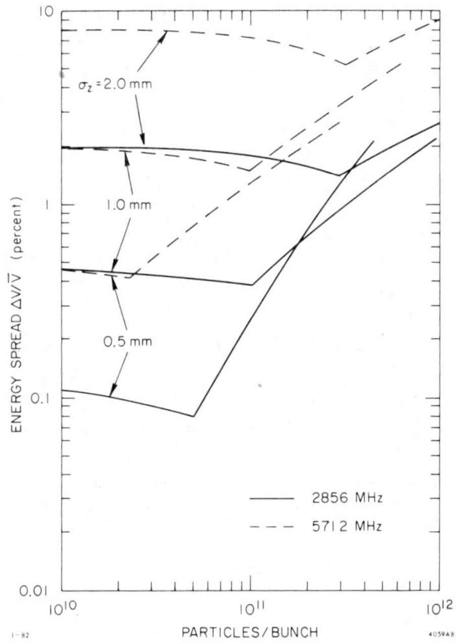

Fig. 12.2. Minimum energy spread as a function of number of particles per bunch.

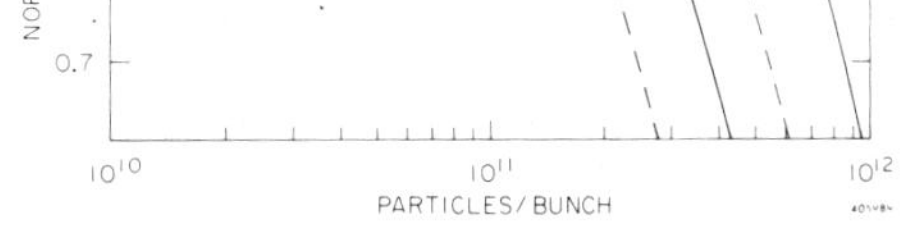

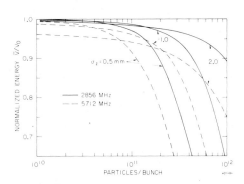

Fig. 12.3. Average energy per particle at minimum energy spread.

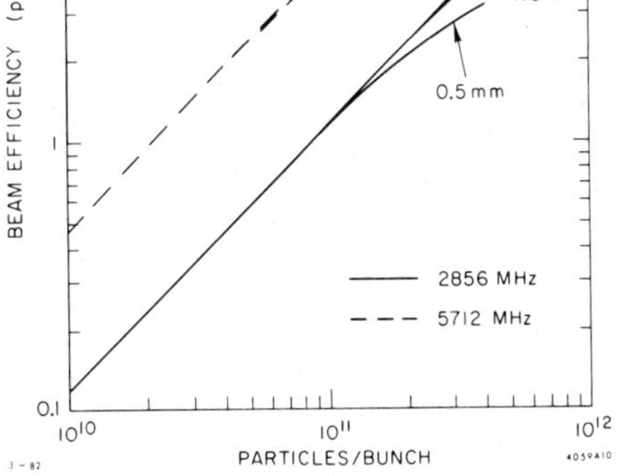

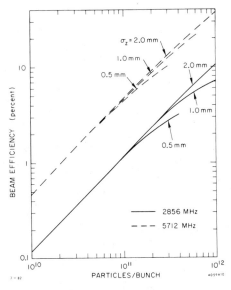

Fig. 12.4. Beam efficiency at minimum energy spread.

In Fig. 12.1 the phase angle which minimizes the energy spread is shown as a function of the number of particles per bunch for three bunch lengths and two rf frequencies. The energy spread at the optimum phase angle is shown in Fig. 12.2. The energy spread is defined such that 90% of the particles are contained within it. The normalized average energy gain per particle at the phase for minimum energy spread is shown in Fig. 12.3. It is, of course, possible to trade off a higher average energy for a larger energy spread by choosing a phase closer to the wave crest. Finally, Fig. 12.4 shows the beam efficiency, as defined by Eq. (10.11).

The preceding results can be applied to estimate the energy spread, normalized energy and beam efficiency for the collider parameters given in Table 12.1: $\bar{E}_a = 100$ MV/m, $N = 10^{11}$ and $\sigma_z = 1$ mm. Assume also a $\pi/3$-mode jungle gym structure. We can extrapolate from the computed results for a disk-loaded structure, at least roughly, by assuming that the wake function scales in proportion to the structure parameter k_1 for the fundamental mode: $k_1(JG)/k_1(DL) = 1.6$. Since the amplitude of the wake also scales as N, Figs. 12.2, 12.3 and 12.4 can be used by taking the effective number of particles to be 1.6×10^{11}. Results at three frequencies are given in Table 12.3.

Table 12.3 Collider Performance at Three Frequencies

Frequency (MHz)	w_o (J/m)	η_s	$\bar{w}_s$ (J/m)	η_b (%)	$\bar{E}/E_a$	$\Delta E/E_a$ (%)
2856	103	.82	84	2.2	.98	0.5
4040	58	.71	41	4.1	.93	1.0
5712	35	.58	20	7.5	.88	1.9

12.4 Bunch Trains

It is possible to improve the effective luminosity and beam efficiency of a collider by accelerating trains of bunches spaced apart by about ten wavelengths or more. At this bunch spacing, higher modes have effectively decohered, and for each following bunch only the superposition of the fundamental beam-loading voltages from those bunches which have already passed through the structure need by considered. Successive bunches can be directed, using fast kickers, to different interaction regions. This allows a number of experiments to be run in parallel, although at successively lower energies for each successive bunch.

The long-range fundamental-mode wake per bunch is $\Delta E_1 = 2k_1q = 2eNk_1$. The average energy of the n-th bunch is therefore

$$\bar{E}(n) = \bar{E}_1 - (n-1)\Delta E_1 = \bar{E}_a\left[\frac{\bar{E}_1}{\bar{E}_a} - (n-1)\frac{\Delta E_1}{\bar{E}_a}\right] . \quad (12.7)$$

The efficiency for m bunches is, using $q/\bar{w}_s = 2\Delta E_1/\bar{E}_a^2$,

$$\eta_b(m) = \frac{q}{\bar{w}_s} \sum_1^m \bar{E}(n) = 2m \frac{\Delta E_1}{\bar{E}_a} \left[\frac{\bar{E}_1}{\bar{E}_a} - \frac{(m-1)}{2} \frac{\Delta E_1}{\bar{E}_a} \right] . \quad (12.8)$$

Results for a train of four bunches with $N = 10^{11}$, again assuming a $\pi/3$-mode jungle gym structure at $E_a = 100$ MV/m, are given in Table 12.4. Note that the beam efficiency is quite reasonable at the two higher frequencies, while the energy of the fourth bunch is still an appreciable fraction of the unloaded energy. The peak power per rf source and the total average power are repeated from Table 12.2.

Table 12.4 Collider Parameters for a Train of Four Bunches

Frequency (MHz)	$\eta_b(4)$ (%)	$\bar{E}(4)/E_a$	$\bar{P}_{rf}$ (MW)	$\hat{P}_o$ (MW)
2856	7.5	.95	130	3100
4040	14	.87	75	1750
5712	25	.76	45	1050

12.5 Transverse Emittance Growth

The computation of the head-tail growth factor due to the transverse wake, based on the two-particle model of the bunch, can be scaled to the present collider parameters using the calculation following Eq. (11.9). Assume that the focusing strength remains the same, that the energy is up by a factor of 6, and that the charge per bunch is up by a factor of two with respect to the parameters used in the previous computation. The growth factor A therefore is reduced by a factor of 3, to A = 3.5 at 2856 MHz. However, the growth factor at 5712 is a different story. To find the wake for a 1 mm bunch at 5712 MHz, first find the wake for a 2 mm bunch ($2\sigma_z$ = 13.3 ps) from Fig. 9.6. The result is 1.5×10^{14} V/C-m/period, up by a factor of 1.5 from the wake at $2\sigma_z$ = 6.7 ps. Thus at 5712 MHz the transverse wake is larger by a factor $1.5 \times 2^3 = 12$, giving a value for the growth factor of A = 40. To bring this down to an acceptable level, the focusing strength would have to be increased.

Another method has been suggested[52] which would reduce the transverse emittance growth due to the dipole wake. If there is an energy spread over the length of the bunch, there will be a spread in the betatron frequencies (Landau damping), and a consequent reduction in the growth of amplitude of the oscillation of the tail particles. The required energy spread can be estimated using the two-particle model of Sec. 11.1. If we want to reduce the growth factor to $A' = fA$, then according to Eq. (11.12) an energy spread

$$\frac{\delta\gamma}{\gamma} \approx 2\,\frac{\delta k_\beta}{k_\beta} \approx \frac{4}{f z\, k_\beta} \tag{12.9}$$

is needed. To reduce A from 40 to 10 in the previous example, again using $k_\beta = 2\pi/100$ m and $z = 3000$ m, an energy spread $\delta\gamma/\gamma \approx 0.08$ is required.

13. HIGH PEAK POWER RF SOURCES

As shown in Table 12.2, a peak power on the order of 1 - 3 GW is required to attain a gradient of 100 MV/m in an accelerating structure operating in the frequency range 2.8 - 5.7 GHz. No sources exist at present which can reach this power level, but several promising possibilities are discussed in the following sections.

13.1 Conventional Sources

A klystron, scaled from existing designs to a higher beam voltage and current, is a potential candidate for a high peak power source. There are, however, some potential problems. Space charge effects in electron beam tubes scale with a parameter called the beam perveance, defined in terms of the beam voltage V_o and current I_o as

$$K \equiv I_o/V_o^{3/2} \quad . \tag{13.1}$$

As the perveance increases, space charge effects become more severe, making it more difficult to achieve sharp mono-energetic beam bunches at the gap of the output cavity. The efficiency of a klystron, therefore, tends to fall off with increasing perveance. If the efficiency is to be greater than 60%, the perveance should be[53] less than about 2×10^{-6} A/$V_o^{3/2}$. From Eq. (13.1), this implies an rf output power given by

$$P_{rf} = \eta K V_o^{5/2} \quad . \tag{13.2}$$

For $P_{rf} = 1.2 \times 10^9$W, $K = 2 \times 10^{-6}$ and $\eta = 60\%$, a beam voltage V_o = 1.0 MV is required. However, this high a beam voltage leads to other design difficulties. At high beam voltages, more rf input power or a longer drift length, or both, are required to overcome the increased stiffness of the beam against the required longitudinal modulation in the electron velocities. Perhaps more severe limitations are imposed by possible dc beam interception, by rf breakdown in the output cavity, and by the problem of designing an adequate window for the output waveguide. In spite of these difficulties, there are no fundamental limitations standing in the way of a 1 GW peak power klystron. In fact, a klystron with a very short pulse length (≈15 ns), operating at 3.35 GHz, has already been built[54] with this design power level (design parameters: V_o = 1 MV, I_o = 2kA, η = 50%). However, the tube failed before it could be tested at full output power, and the development has not been persued further. At SLAC, a klystron capable of delivering 150 MW at 2856 MHz with a 1 μs pulse length is being designed.[55] Parameters for this tube are given in the first row of Table 13.1. By placing seven cathodes and beams of this design within the same vacuum envelope, with six beams arranged in a hexagonal manner around a central beam, the power output could be increased to the 1 GW level. The cost and focusing power requirements for such a tube would be considerably less than seven times that of a single unit. The beam parameters for this device are listed in Table 13.1.

Crossed-field devices are also potentially capable of delivering very high peak output powers. A power of 500 MW at 3.2 GHz, with a pulse length of 30 ns, has already been obtained from a coaxial magnetron.[56] The design parameters for this tube are shown in Table 13.1. Note that a crossed field device can achieve a high output power at a modest beam voltage because the perveance is much greater than for a klystron. The cathode is cylindrical, coaxial with the anode, and can have a large area. Conversion of dc to rf power is accomplished by the growth of a density modulation in a rotating space charge cloud in the relatively narrow cathode-anode gap. Although this device as presently constructed is an oscillator, it might be possible to produce an amplifier counterpart.

Table 13.1 High Power RF Sources (2.8 - 5.7 GHz)

	Micro-Perveance $\left(\frac{10^{-6}A}{V^{3/2}}\right)$	Beam Voltage (kV)	Beam Current (kA)	Beam Impedance (Ω)	Est. Efficiency (%)	RF Power (MW)
Conventional Klystron	2	450	0.6	750	55	150
Multiple-Beam Klystron	7 × 2	450	4.2	110	55	1050
Crossed-Field Amplifier	13	600	6.0	100	50	1800
Gyracon	0.5	1500	1.0	1500	80	1200
Photocathode Device	1.5	1200	2.0	600	75	1800

A third conventional high peak power source (conventional in the sense that operating versions presently exist) is the gyracon. In this device, a dc electron beam is deflected transversely by a circularly polarized field in an rf input cavity, such that the beam follows a conical trajectory. At an appropriate distance, the beam enters a slit in the top of a doughnut-shaped output waveguide. The phase velocity of the traveling-wave field in this waveguide ring is synchronized to move with the rotating electron beam. If the electric field has the correct amplitude, each electron in the beam can be brought nearly to rest as it transits across the guide, since the beam is monoenergetic (unlike the modulated beam in a klystron). The efficiency of this device can, therefore, in principle approach 100%. In addition, since the modulation is transverse, the required modulation voltage increases only in proportion to γ.

Problem 13.1: Show that the rf voltage V_1 required to produce a longitudinal peak velocity variation $\Delta v/v_o$, where v_o is the dc beam velocity, varies as $[\gamma(\gamma+1)/2]V_o$.

The gyracon has been extensively exploited at Novosibirsk.[57] It is planned that gyracons will be used to power a very high energy linear collider (VLEPP project[52]) now under design by this group. The beam parameters[58] for a prototype of this tube are given in Table 13.1.

13.2 New Ideas for Electron Beam RF Sources

The concept of using parallel beams to achieve a high effective perveance was introduced in connection with the seven-beam klystron proposed in the previous section. This idea has been carried still further in the form of the Meqatron,[59] an rf amplifier which would employ an electrostatically-focused array of small-diameter beamlets. The success of this device depends on the remarkable fact that the total current carried by a single beam in a channel with quadrupole focusing is independent of the channel radius if all dimensions are reduced by the same scale factor.[60] Although no device of this kind has yet been designed at the required power level, the basic concept seems promising.

A Free Electron Laser (FEL) has also been proposed[61] as a means for generating very high peak rf power levels at centimeter wavelengths. In the FEL, an electron beam is "wiggled" transversely by a transverse magnetic field with alternating polarity. With proper synchronization between the velocity of the beam and the phase velocity of a traveling wave, the transverse component of the electron velocity can interact with the transverse electric field in the wave to produce a transfer of energy from beam to wave. An FEL driven by an induction linac with a beam current of 1 kA and an energy gain of 1 MV/m could transfer a peak power of 10^9 W/m to an accelerating structure.

Another potential rf source which could produce a high peak output power is illustrated in Fig. 13.1. In this device a laser which is intensity modulated at the desired rf frequency illuminates a photocathode. Bunches of electrons emitted during the peaks of the laser illumination are accelerated by a high voltage, and then compressed in lateral dimensions to a sufficiently small fraction of an rf wavelength (transverse dimensions less than about $\lambda/5$) for good coupling to the fields in the rf output cavity. Because the electrons are emitted in bunches from the cathode, the long drift length required by a klystron for longitudinal bunching at relativistic velocities is eliminated. By emitting the bunches at a low density from a cathode with a relatively large area, and then compressing the beam laterally after the electrons have attained a high velocity, the effects of longitudinal and transverse space-charge defocusing are reduced. Preliminary calculations taking into account space-charge effects indicate that the parameters given in the last row of Table 13.1 could be attained.

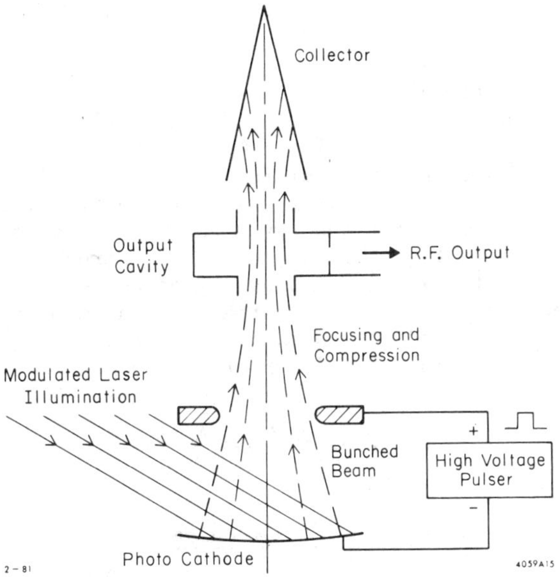

Fig. 13.1. Schematic of a photocathode microwave power source.

13.3 SLED

SLED (SLAC Energy Development) is essentially a pulse compression scheme which has been applied at SLAC to enhance peak rf power at the expense of pulse width. High-Q resonant cavities (two identical cavities are used) store energy during a large fraction of each klystron pulse, then discharge this energy rapidly into the accelerator during the final part of the pulse (equal to the accelerator filling time). The energy compression is accomplished with an efficiency of about 65% for the present SLAC system. The reader is referred to Ref. 62 for a detailed description of the SLED concept, and for a derivation of the energy gain enhancement ratio R for the case of a constant gradient structure. The enhancement ratio is defined as the energy gain with SLED, divided by the energy gain that would be obtained with the klystron power flowing directly into the accelerating structure.

To illustrate the enhancement ratios that can be achieved with SLED, we give without detailed derivation the expression for R for the case of a constant impedance structure:

$$R(\text{SLED}) = \frac{1-\beta}{1+\beta} + \frac{2\beta}{1+\beta}\,(2-e^{-\mu'})\left[\frac{e^{-\nu'}-e^{-\tau}}{\tau-\nu'}\cdot\frac{\tau}{1-e^{-\tau}}\right]$$

$$\mu' \equiv \frac{T_p - T_s}{T_o}\,(1+\beta) \qquad (13.3)$$

$$\nu' \equiv \frac{T_s}{T_o}\,(1+\beta) \quad .$$

Here τ is the attenuation parameter for the accelerating structure, β is the coupling coefficient for the storage cavities, $T_o = 2Q_o/\omega$ is the unloaded filling time for the cavities, T_p is the total klystron pulse length and T_s is the filling time of the structure. If τ is not too large, Eq. (13.3) will also be a good approximation for a constant gradient structure. We see that the maximum enhancement ratio that can be attained with SLED is R = 3 for $T_o \gg \beta T_s$, $T_p \gg T_o$, and $\beta \gg 1$. The efficiency of the compression is given by

$$\eta = \frac{R^2 T_s}{T_p} = R^2\left(\frac{\nu'}{\mu'+\nu'}\right) \quad . \qquad (13.4)$$

13.4 Energy Storage Cavities with Switching

Another form of pulse compression with the potential capability of yielding very high peak pulsed rf power is energy storage with active switching. As in SLED, energy from a klystron pulse is stored over a relatively long time in a high-Q resonator, and then switched out rapidly in about a filling time of the structure. In the case of SLED, this switching is accomplished passively by making a 180° phase reversal in the input drive to the klystron. In the case of switched energy storage (SES), the switching is accomplished by making a large change in the coupling coefficient of the output port of the storage cavity by means of an active switching element. One possible form for an SES device is shown in Fig. 13.2. The length of the waveguide stub on the left is chosen to be about $\lambda_g/2$ so that, before the switch is fired, there is a voltage null at the side branch leading to the accelerator. Thus, in this state there is very little coupling between the cavity and the accelerator. When the switch is fired, the shorting plane at the end of the stub is effectively moved by $\lambda_g/4$, which now produces a heavy coupling between the cavity and the accelerating structure.

The idea of energy storage and switching was introduced by Birx, Little, Mercereau and Scalapino.[63] The critical component in this technique is the switch. Birx and Scalapino[64] describe an electron beam switch and show that an electron density of about $10^{13}/\text{cm}^3$ is needed to produce an adequate short. More recently, a low pressure plasma switch, which may provide a simpler way to achieve the necessary electron density, has been investigated[65] at Lawrence Livermore Laboratory. In a recent experiment,[65] 160 MW has been switched using

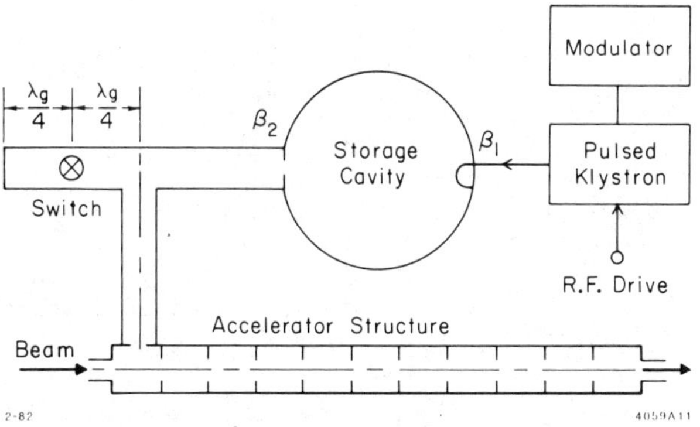

Fig. 13.2. RF pulse compression using an energy storage cavity with switching.

a high-pressure gas switch and a shortened length of S-band waveguide as a storage cavity.

It is useful to consider the efficiency of SES pulse compression in two steps: first, the efficiency η_c for the transfer of energy from the klystrong pulse to energy stored in the cavity, and second, the efficiency η_s for the transfer of energy from the cavity to effective energy stored in the accelerating structure. In terms of the energy stored in the cavity W_c, the energy in the klystron pulse W_k, the effective structure stored energy $\bar{W}_s$, the structure parameter k_1, and the average accelerating gradient $\bar{E}_a$ in an accelerating structure of length L, these efficiencies are

$$\eta_c = \frac{W_c}{W_k} = \frac{W_c}{T_p P_o} \qquad (13.5a)$$

$$\eta_c = \frac{W_s}{W_c} = \frac{E_a^2 L}{4 k_1 W_c} \quad . \qquad (13.5b)$$

The energy gain from an accelerating section is therefore

$$V_a = \bar{E}_a L = (\eta_c \eta_s \cdot 4k_1 L\, T_p P_o)^{1/2} \quad . \qquad (13.6)$$

In Ref. 66 these efficiencies are calculated to be

$$\eta_c = \frac{2(1 - e^{-\mu})^2}{\mu} \cdot \frac{\beta_1}{1 + \beta_1 + \beta_{20}} \qquad (13.7a)$$

$$\eta_s = \frac{2\nu(e^{-\tau} - e^{-\nu})^2}{(\nu - \tau)^2} \cdot \frac{\beta_2}{1 + \beta_1 + \beta_2} \qquad (13.7b)$$

where

$$\mu = \mu_o(1 + \beta_1 + \beta_{20}) \qquad (13.8a)$$

$$\nu = \nu_o(1 + \beta_1 + \beta_2) \qquad (13.8b)$$

$$\mu_o = T_p/T_o \tag{13.8c}$$

$$\nu_o = T_s/T_o \quad . \tag{13.8d}$$

Here β_1 is the input coupling coefficient for the storage cavity, β_{20} is the output coupling coefficient with the switch off, β_2 is the output coupling coefficient when the switch is fired, and as before $T_o = 2Q_o/\omega$ is the unloaded cavity filling time.

By adjusting β_2 in Eq. (13.7b), the transfer efficiency η_s can be optimized for fixed values of ν_o and τ. This optimized efficiency is shown in Fig. 13.3 as a function of ν_o assuming $\beta_2 \gg \beta_1$.

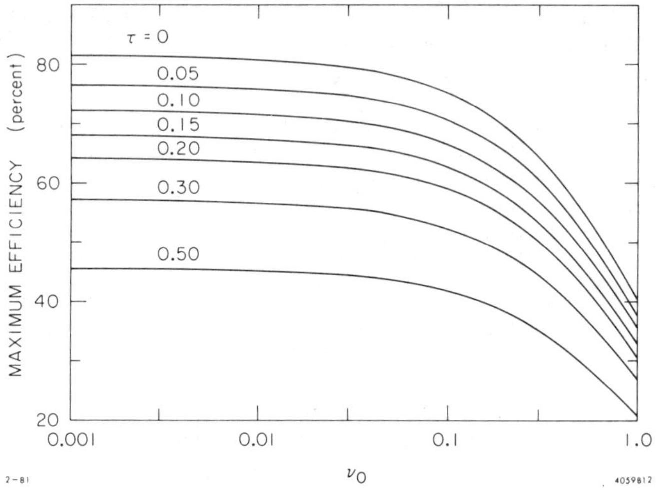

Fig. 13.3. Maximum energy transfer efficiency η_s as a function of ν_o.

The value of β_2 which maximizes η_s is shown in Fig. 13.4. The efficiency η_c for the transfer of energy from the klystron pulse to stored energy in the storage cavity can be optimized in a similar fashion by adjusting the input coupling β_1. Since Eq. (13.7b) reduces to the same form as Eq. (13.7a) in the limit $\tau = 0$ (assuming $\beta_{20} \ll 1$ and $\beta_1 \ll \beta_2$), the top curve in Fig. 13.3 can also be used to find the maximum value of η_c as a function of μ_o. Likewise, the optimum value of β_1 is given by the curve for $\tau = 0$ in Fig. 13.4.

Problem 13.2: Show that for fixed β_1 the maximum efficiency η_c as a function of klystron pulse length is

$$\eta_c = .814 \left(\frac{\beta_1}{1 + \beta_1 + \beta_{20}} \right) \tag{13.9}$$

at pulse length $T_p = 1.26\ T_f$, where T_f is the loaded filling time given by $T_f = T_o/(1 + \beta_1 + \beta_{20})$.

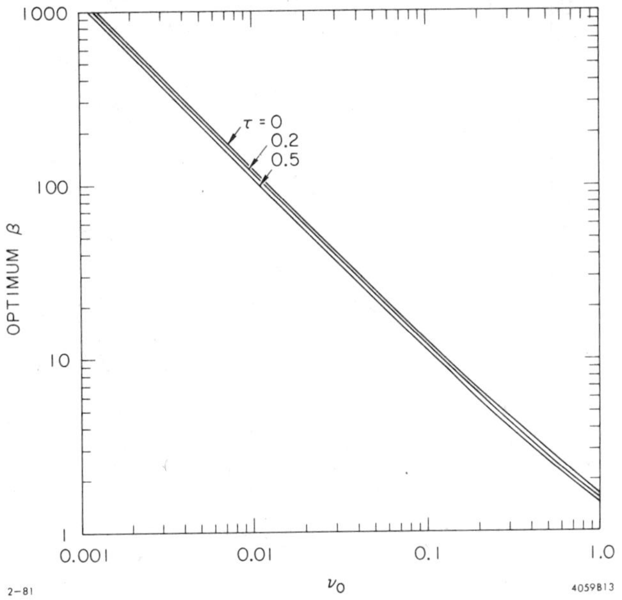

Fig. 13.4. Cavity coupling coefficient β_2 which maximizes the transfer efficiency η_s.

The energy gain enhancement ratio for an SES system is also of interest. For power P_0 flowing directly into a constant impedance structure, the energy gain V_0 is given by Eq. (5.16). From the definition $R \equiv V_a/V_0$, where V_a is given by Eq. (13.6), we obtain

$$R(SES) = \frac{2(\beta_1\beta_2)^{1/2}}{1+\beta_1+\beta_{20}} \cdot (1-e^{-\mu}) \left[\frac{e^{-\nu} - e^{-\tau}}{\tau - \nu} \cdot \frac{\tau}{1 - e^{-\tau}} \right] . \qquad (13.10)$$

This can be compared with R(SLED) given by Eq. (13.3). While R(SLED) cannot exceed three, R(SES) can, in principle, be made arbitrarily large.

As an example of an SES system, consider an accelerator powered by klystrons with a peak power output of 80 MW and pulse length of 4 μs, operating at 5712 MHz. A copper energy storage cavity with a radius of 30 cm would have an unloaded TE-mode Q of 3.4×10^5 at this frequency (the unloaded Q of a spherical resonator in a TE mode is just the radius divided by the skin depth). Using $T_p = 4$ μs and $T_0 = 2Q_0/\omega = 19$ μs in Eq. (13.8c), $\mu_0 = 0.21$. The top curve in Fig. 13.3 shows that energy can be transferred to the storage cavity with an efficiency of 67%. Assuming two π/3-mode jungle gym structures 3 m in length ($\tau = 0.28$, $T_s = 0.1$ μs, $\nu_0 = .005$) we have from Fig. 13.3 that $\eta_s = 58\%$. Thus the effective stored energy per unit length is 21J, and using $k_1 = 1.2 \times 10^{14}$ V^2/J-m the unloaded gradient is 100 MV/m. Figure 13.5 shows the energy that could be obtained from a 3-km accelerator of this design for other pulse lengths. An ideal switch is assumed, and breakdown limitations are ignored.

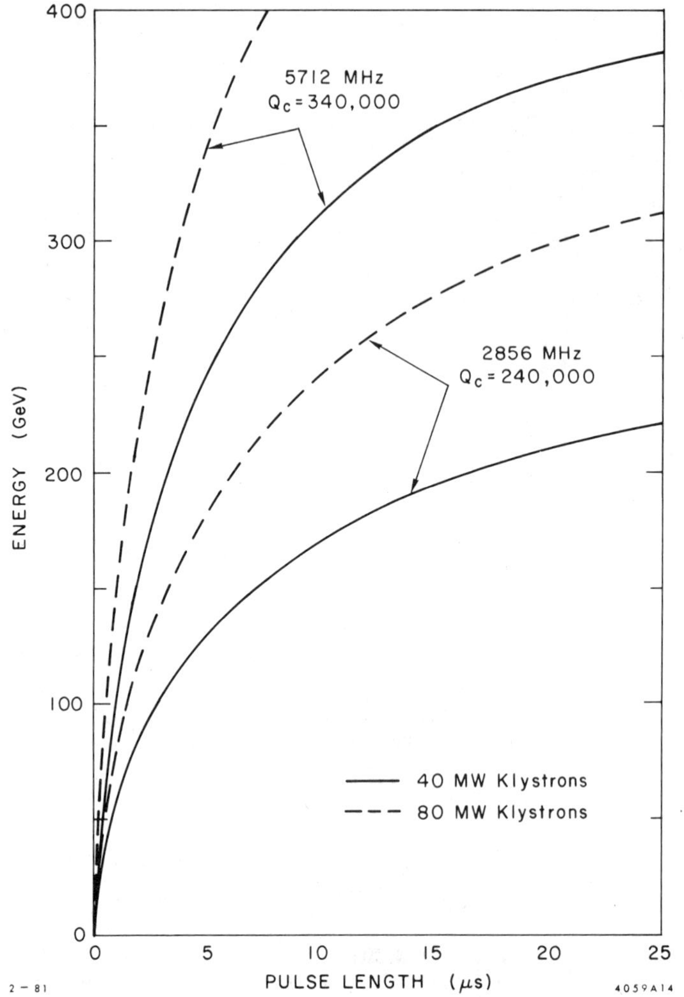

Fig. 13.5. Energy as a function of pulse length for a 3-km accelerator using energy storage and switching.

Problem 13.3: Find an appropriate expression for the minimum time in which energy can be switched out of a storage cavity of volume V having output ports with total area A. Hint: the power emitted from an aperture of area A in a cavity full of photons is $P = 1/4\ c\,\bar{u}\,A$, where $\bar{u}$ is the mean energy density in the cavity. Is this likely to pose a problem in achieving a decay time of 0.1 μs from a spherical storage cavity of 30 cm radius at a frequency of 5712 MHZ?

ACKNOWLEDGEMENTS

In developing the material presented in these lectures, I have profited through interactions with many colleagues over a period of years. While it is not possible to acknowledge the many relevant discussions in detail, I would like to single out a few contributions which have been especially helpful. The proof of the fundamental theorem of beam loading outlined in Problem 6.1 was suggested by Klaus Halbach. Karl Bane has been involved in all aspects of the theory and computation of the wake potentials, and in particular was responsible for computing the wake potentials shown in Figs. 9.3 through 9.9. Discussions with Alex Chao and Phil Morton have been helpful in clarifying various aspects of the impedance-wake potential formalism presented in Ch. 9. The two-particle model for transverse emittance growth in a linac, as given in Sec. 11.1, is largely based on an analysis suggested by Alex Chao and Phil Morton. The analysis in Sec. 11.2 of the strong head-tail instability in storage rings is the result of a collaboration with Alex Chao, Phil Morton, John Rees and Matthew Sands. A paper with a more formal derivation of the theory, together with a comparison with experimental results in PEP, will be published. Finally, I would like to thank Ronald Ruth for his efforts in editing the manuscript for these lectures, and for his numerous helpful comments and suggestions.

REFERENCES

1. R. B. Neal, ed. The Stanford Two-Mile Accelerator, (W. A. Benjamin, New York, 1968).
2. P. Lapostolle and A. Septier, eds. Linear Accelerators (North Holland, Amsterdam, 1970). See Chapters B.1.1, Accelerating Structures; B.1.2, Particle Dynamics; B.1.3, Beam Loading and Transient Behavior; B.1.4 Beam Breakup.
3. Marvin Chodorow and Charles Susskind, Fundamental of Microwave Electronics (McGraw-Hill, New York, 1964).
4. A. Staprans, E. W. McCune and J. A. Ruetz, "High Power Linear Beam Tubes," Proc. IEEE 61, 299 (1973).
5. H. C. Hoyt, D. D. Simmonds and W. F. Rich, Rev. Sci. Instrum. 37, 755 (1966).
6. K. Halbach and R. Holsinger, Particle Accel. 7, 213 (1976).
7. R. L. Gluckstern, K. Halbach, R. F. Holsinger and G. N. Minerbo, 1981 Linear Accelerator Conference (Santa Fe, New Mexico, October 19-23, 1981). To be published.
8. D. E. Nagle, E. A. Knapp and B. C. Knapp, Rev. Sci. Instrum. 38, 1583 (1967).
9. E. A. Knapp, Ch. C.1.1c in Linear Accelerators, P. Lapostolle and A. Septier, eds. (North Holland, Amsterdam, 1970).
10. J. R. Rees, "A Perturbation Approach to Calculating the Behavior of Multi-cell Radiofrequency Accelerating Structures" PEP-255, Stanford Linear Accelerator Center (1976).
11. P. B. Wilson, IEEE Trans. Nucl. Sci. NS-16, No. 3, 1092 (1969).

12. J. J. Manca, E. A. Knapp and D. A. Swenson, IEEE Trans. Nucl. Sci. NS-24, No. 3, 1087 (1977).
13. S. O. Schriber, Proc. 1979 Linear Accelerator Conference (BNL-51134, Brookhaven National Laboratory, 1979), p. 164.
14. H. Henke, "The LEP Accelerating Cavity," LEP Note 143, CERN (1979).
15. C. G. Montgomery, R. H. Dicke and E. M. Purcell, eds. Principles of Microwave Circuits. Radiation Laboratory Series Vol. 8 (McGraw-Hill, New York, 1948), Ch. 7.
16. M. Lee and L. Smith, PEP-Note-222, Stanford Linear Accelerator Center (1977).
16a. P. B. Wilson, Ref. 2, Ch. E.2.
17. M. Sands, "The Physics of Electron Storage Rings: An Introduction," in Physics with Intersecting Storage Rings, B. Touschek, ed. (Academic Press, New York, 1971); p. 257f. Also available as SLAC-121, Stanford Linear Accelerator Center (November 1970).
18. K. W. Robinson, "Stability of Beam in Radiofrequency System." Cambridge Electron Accelerator Report CEAL-1010 (February 1964).
19. A. Hofmann in Theoretical Aspects of the Behavior of Beams in Accelerators and Storage Rings. (CERN 77-13, July 1977), pp. 163-165.
20. P. B. Wilson, Proc. 9th Int. Conf. on High Energy Accelerators (Stanford Linear Accelerator Center, 1974), p. 60.
21. P. H. Ceperly, IEEE Trans. Nucl. Sci. NS-19, No. 2, 217 (1972).
22. D. A. Watkins, Topics in Electromagnetic Theory (John Wiley, New York, 1958), Ch. 1.
23. B. Zotter and K. Bane, "Transverse Resonances of Periodically Widened Cylindrical Tubes with Circular Cross Section." PEP-Note 308, Stanford Linear Accelerator Center (September 1979).
24. G. A. Loew, "Non-Synchronous Beam Loading in Linear Electron Accelerators," Microwave Laboratory Report No. 740, Hansen Laboratories, Stanford University (August 1960).
25. P. B. Wilson, IEEE Trans. Nucl. Sci. NS-26, No. 3, 3255 (1979).
26. M. Sands, PEP-Note-90, Stanford Linear Accelerator Center (July 1974).
27. P. B. Wilson, "Transient Beam Loading in Electron-Positron Storage Rings," PEP-Note-276, Stanford Linear Accelerator Center (December 1978).
28. R. H. Helm and G. A. Loew, Ref. 2, Ch. B.1.4.
29. P. B. Wilson, "A Simple Analysis of Cumulative Beam Breakup for the Steady State Case." HEPL-TN-67-8, High Energy Physics Laboratory, Stanford University (September 1967).
30. V. K. Neal, L. S. Hall and R. K. Cooper, Part. Accel. 9, 213 (1979).
31. E. Keil, Nucl. Instrum. Methods 100, 419 (1972).
32. See, for example, Sec. 4.3 in Ref. 31.
33. K. Bane, "Constructing the Wake Potentials from the Empty Cavity Solutions of Maxwell's Equations." CERN/ISR-TH/80-47 (November 1980).
34. T. Weiland and B. Zotter, "Wakefield of a Relativistic Current in a Cavity." CERN/ISR-TH/80-36 (July 1980).
35. W.K.H. Panofsky and W. A. Wenzel, Rev. Sci. Instrum. 27, 967 (1956).

36. K. Bane and B. Zotter, 11th Int. Conf. on High Energy Accelerators (Birkhäuser Verlag, Basel, Switzerland, 1980), p. 581.
37. K. Bane, "The Transverse Cavity Impedance in LEP," CERN/ISR-TH/80-48 (November 1980).
38. A. W. Chao and R. K. Cooper, "Beam Breakup Due to Quadrupole Wake Field for Sector 1." Internal Note CN-142, Stanford Linear Accelerator Center (January 1982).
39 K. Bane, private communication.
40. M. Tigner, IEEE Trans. Nucl. Sci. NS-18, No. 3, 249 (1971).
41. See, for example, Microwave Laboratory Reports ML-416, ML-432, ML-520, ML-557, and ML-581, Stanford University (April 1957-February 1959).
42. R. F. Koontz, G. A. Loew, R. H. Miller and P. B. Wilson, IEEE Trans. Nucl. Sci. NS-24, No. 3, 1493 (1977).
43. R. H. Helm and R. Miller, Ref. 2, Ch. B.1.2.
44. A. W. Chao, B. Richter and C. Y. Yao, 11th Int. Conf. on High Energy Accelerators (Birkhäuser Verlag, Basel, Switzerland, 1980), p. 597.
45. R. D. Kohaupt, "Simplified Presentation of Head-Tail Turbulence." DESY Internal Report M-80/19 (October 1980).
46. U. Amaldi, Proceedings of the 9th International Symposium on Lepton Photon Interactions at High Energies (Fermilab, August 1979), p. 314. Also CERN-EP/79-136 (November 1979).
47. B. Richter, 11th Int. Conf. on High Energy Accelerators (Birkhäuser Verlag, Basel, Switzerland, 1980), p. 168.
48. H. Wiedemann, 1981 Summer School on Particle Physics, SLAC-PUB-2849, Stanford Linear Accelerator Center (1981).
49. R. Hollebeek, Nucl. Instrum. Methods 184, 333 (1981).
50. V. A. Vaguine, IEEE Trans. Nucl. Sci. NS-24, No. 3, 1048 (1977).
51. V. E. Balakhin et al., "Accelerating Structure of a Colliding Linear Electron-Positron Beam (VLEPP)." Proc. 6th All Union Conf. on Charged Particle Accelerators, Dubna, USSR, 1978. See also SLAC-TRANS-187, Stanford Linear Accelerator Center (1978).
52. V. E. Balakhin and A. N. Skrinsky, " VLEPP Project (Status Report)." Preprint 81-129, Institute for Nuclear Physics, Novosibirsk, USSR (1981).
53. P. J. Tallerico, IEEE Trans. Nucl. Sci. NS-26, No. 3, 3877 (1979).
54. Rome Air Development Center, Report No. RADC-TR-70-101 (July 1970).
55. G. Konrad, private communication.
56. W. M. Black et al., International Electron Devices Meeting, Washington D.C., December 1979 (IEEE Electron Devices Society, IEEE, New York), p. 175.
57. G. I. Budker et al., Particle Accel. 10, 41 (1979).
58. A. N. Skrinsky, private communication.
59. A. W. Maschke, "The Meqatron: A High-Power High Frequency Source." Report BNL-51094, Brookhaven National Laboaratory, (September 1979).
60. M. Reiser, J. Appl. Phys. 52, 555 (1981).
61. A. M. Sessler, Proc. of the Workshop on the Laser Acceleration of Particles, Los Alamos, NW, February 18-23, 1982. To be published.
62. Z. D. Farkas et al., Proc. of the 9th Int. Conf. on High Energy Accelerators (Stanford Linear Accelerator Center, 1974), p. 576.

63. D. L. Birx et al., Appl. Phys. Lett. 32, 68 (1978).
64. D. L. Birx and D. J. Scalapino, J. Appl. Phys. 51, 3629 (1980).
65. R. A. Alvarez et al., Particle Accel. 11, 125 (1981).
66. P. B. Wilson, Internal Note AATF/80/20, Stanford Linear Accelerator Center (April 1980).

LIST OF SYMBOLS

Symbols are listed by chapter, in order of first use. The equation number in which, or immediately following which, the symbol is first defined or used is given in parentheses after each definition. Symbols are omitted when they have a clear, conventional meaning (e.g., t, q, γ), or when they are used a single time only without possibility of confusion. Note that several symbols have different meanings in different chapters.

Chapter 2

ω	Angular frequency of rf generator (2.1)
$\tilde{V}$, $\tilde{V}_c$	Complex (phasor) cavity voltage (2.1)(2.3)
$\tilde{V}_g$	Generator voltage component in a cavity (2.3)
$\tilde{V}_b$	Beam loading voltage component in a cavity (2.3)
W_c, W	Stored energy in a cavity or in a given cavity mode (2.4)(2.5)
f(z)	Function relating cavity field to stored energy (2.6)
ω_0	Resonant frequency for a particular cavity mode (2.7)
V_0	Voltage induced in a cavity by a point charge (2.8)
σ, σ_t	RMS bunch length in time (2.9)

Chapter 3

E_z(cmf)	Axial field in a co-moving frame (3.3)
k	Free-space propagation constant $k \equiv \omega/c = 2\pi/\lambda$ (3.3)
C, S	Cosine and sine integrals for cavity voltage (3.5)
R_a	Cavity shunt impedance (accelerator definition)(3.9)
P	Cavity power dissipation (3.10)
R_s	Surface impedance (3.11)
Z_0	Impedance of free space (3.11)
R_u	Uncorrected shunt impedance (3.12a)
V_u	Uncorrected cavity voltage (3.12b)
T	Transit-time factor (3.13a)
L	Length of cavity or gap (3.14)
θ	Transit angle, $\theta = kL$ (3.14)
$\tilde{V}_a$	Average voltage gain per particle; accelerating voltage (3.21)
C', S'	Cosine and sine integrals for bunch form factor (3.22)
F	Bunch form factor (3.24)
t_b	Total length in time of a rectangular bunch (3.25b)
b	Radius of a pillbox cavity (3.27)
Q, Q_0	Unloaded cavity Q (3.28a)
r	Shunt impedance per unit length (3.28c)

G_1, G_2	Geometry constants for a pillbox cavity (3.29a,b)
N	Number of cells, coupled resonator model (3.30)
m	Mode number, coupled resonator model (3.30)
B	Bandwidth, coupled resonator model (3.30)
ω_o	Center frequency of passband, coupled resonator model (3.30)
n	Cell number, coupled resonator model (3.31)
f(n)	Field flatness function, coupled resonator model (3.32)
β	Cavity coupling coefficient (3.34')
G_c	Cavity shunt conductance, equivalent circuit model (3.34')
P_c	Cavity power dissipation, equivalent circuit model (3.34')
i_b	Peak value of rf current (3.34')
I_o	DC current (3.34')
i_g	RF Generator current, equivalent circuit model (3.35a)
P_g	Generator power, equivalent circuit model (3.35a)
V_{gr}	Generator voltage component at resonance (3.35a)
V_{br}	Beam loading voltage component at resonance (3.35b)
K	Beam loading parameter, $K \equiv (I_o/2)(R_a/P_g)^{1/2}$ (3.36a)
η	Beam conversion efficiency (3.36b)
$\underset{\sim}{P}_r$	Reflected power
$\tilde{Y}_c$	Unloaded cavity admittance for a resonant mode (3.37)
δ	Tuning parameter, $\delta \equiv (\omega - \omega_o)/\omega_o$ (3.38)
$\tilde{Z}_L$, $\tilde{Y}_L$	Loaded cavity impedance, admittance (3.39)
R_o	Loaded impedance at resonance, $R_o \equiv [G_c(1+\beta)]^{-1}$ (3.39)
Q_L	Loaded Q, $Q_L \equiv Q_o/(1+\beta)$ (3.39)
ψ	Tuning angle, $\psi \equiv \tan^{-1}[-2Q_L\delta]$ (3.40)
T_f	Loaded filling time, $T_f \equiv 2Q_L/\omega_o$ (3.43)

Chapter 4

θ	Phase of V_{gr} with respect to $-\tilde{i}_b$ (see Fig. 3.13) (4.1a)
ϕ	Phase of V_c with respect to $-\tilde{i}_b$ (synchronous phase angle) (4.1a)
P_b	Power transferred to beam, $P_b = I_oV_a$ (4.6a)
β_o	Cavity coupling for zero reflected power (4.6a)
P_{go}	Generator power at optimum tuning and coupling (4.6b)
ψ_o	Tuning angle at optimum tuning and coupling (4.7)
ω_s	Synchrotron frequency (4.8)
ξ	Tuning parameter, $\xi \equiv (\omega - \omega_o)T_f$ (4.13)
$\underset{\sim}{\eta}$	Filling-time parameter, $\eta \equiv \omega_s T_f$ (4.13)
$\tilde{V}_o$, $\tilde{\underline{i}}_o$	Equilibrium beam loading voltage and current (4.13)
δW, $\overline{P}$	Stored energy and average power transfer for a phase oscillation (4.14)

Chapter 5

ϕ — Azimuthal angle, cylindrical coordinates (5.1)
γ — Complex propagation constant per unit length (5.1)
β — Propagation constant per unit length (5.2)
α — Attenuation parameter per unit length (5.2)
p — Periodic length (5.3)
β_0 — Propagation constant for the fundamental space harmonic (5.3)
β_n — Propagation constant for the nth space harmonic (5.3)
v_p — Phase velocity, $v_p \equiv \omega/\beta$ (5.4)
E_a — Accelerating field in a traveling-wave structure (5.7a)
P — Power flow in a traveling-wave structure (5.7a)
r — Traveling-wave shunt impedance per unit length (5.7a)
w — Stored energy per unit length (5.7b)
v_g — Group velocity, $v_g \equiv d\omega/d\beta$ (5.8a)
E_0, E_L — Field at $z = 0$ and $z = L$ in a traveling-wave structure (5.10a)
P_0, P_L — Power flow at $z = 0$ and $z = L$ (5.10b)
τ — Attenuation parameter, $\tau \equiv \alpha L$, for structure of length L (5.11)
T_f — Filling time for a traveling-wave structure (5.15)
V_0 — Unloaded energy gain for a structure of length L (5.16)
E_b — Beam-induced field in a traveling-wave structure (5.20)
V_b — Beam-induced voltage in a structure of length L (5.22)
m — Beam-loading coefficients for a traveling-wave structure (5.23)
δ — Phase slip parameter (5.27)

Chapter 6

α — Parameter relating stored energy and voltage, $\alpha \equiv W/V^2$ (6.1)
θ, ε — Angles in the proof of the fundamental theorem of beam loading (see Fig. 6.1) (6.2)(6.3)
V_b — Single-pass beam induced voltage (6.2)
V_e — Effective voltage seen by a point charge (5.3)
f — Self-voltage factor, $f \equiv V_e/V_b$ (6.4c)
k — Energy loss parameter, $k \equiv w/q^2$ (6.5)
θ_g — Phase of the generator voltage component (see Fig. 6.2) (6.8)
k_0 — Loss parameter for the accelerating mode (6.8)
B — Beam-loading enhancement factor (6.10)
ΔU_0, ΔU_{hm} — Energy loss to the fundamental and to higher-order modes (6.11)
α_0 — Parameter α for the fundamental (accelerating) mode (6.12)
V_{bo} — Single-pass beam induced voltage for the fundamental mode (6.12)
Z_{hm}, V_{hm} — Higher-order mode loss impedance and voltage (6.13a)
k_n — Loss parameter for the nth mode (6.13c)

T_b	Time between bunches (6.13c)
V_s	Synchrotron radiation loss (in volts) per turn (6.15)
$\tilde{V}_c^-$, $\tilde{V}_c^+$	Cavity voltage just before and just after arrival of bunch (see Fig. 6.2) (6.16)
ϕ^-	Phase of cavity voltage just before arrival of bunch (6.16)
τ	Decay parameter, $\tau \equiv T_b/T_f$ (6.24a)
δ	Phase shift between bunches, $\delta \equiv T_b(\omega_o - \omega)$ (6.24b)
$\tilde{V}_b^-$, $\tilde{V}_b^+$	Beam-induced voltage component just before and just after arrival of bunch (see Fig. 6.3) (6.25a)
$\tilde{V}_b$	Effective beam-induced voltage (see Fig. 6.3) (6.25b)
F_R, F_I	Real and imaginary components of $\tilde{V}_b$ (6.26)
T_{fo}	Unloaded filling time, $T_{fo} \equiv 2Q_o/\omega_o$ (6.27)
τ_o	Unloaded decay parameter, $\tau_o \equiv T_b/T_{fo}$ (6.27)

Chapter 7

$\tilde{V}_d(t)$	Transient difference vector (see Fig. 7.1) (7.1)
$\tilde{V}(t)$, $\tilde{V}_c(t)$	Transient cavity voltage (see Fig. 7.2) (7.2a) (7.6)
x	Normalized time, $x \equiv t/T_b$ (Sec. 7.2) (7.10)
τ	Decay parameter, $\tau \equiv T_b/T_f$ (Sec. 7.2) (7.10)
$\mu(t)$	Transient phase angle of cavity voltage (7.11)
F_A, F_B	Real and imaginary components of the transient part of $V_c(x)$ (7.11)
k_1	Loss parameter per unit length for the accelerating mode in a traveling-wave structure (7.19)
x	Normalized time, $x \equiv t/T_f$ (Sec. 7.3) (7.21)
τ	Traveling-wave attenuation parameter (Sec. 7.3) (7.21)

Chapter 8

$r_\perp$	Transverse shunt impedance per unit length (8.1)
eV_o	Beam energy in electron volts (8.3)
F_e	e-folding factor (8.6)
t_p	Beam pulse length (8.6)
V'	Energy gradient, dV/dz (8.7)
k_β	Focusing strength (betatron wave-number) (8.9)

Chapter 9

$Z(\omega)$	Complex impedance function (longitudinal) (Sec. 9.1)
$w(\tau)$	Time domain wake potential (Sec. 9.1)
$s(\tau)$	Step response function (Sec. 9.1)
τ	Time following a unit point charge (Sec. 9.1)
$V_b(t)$	Beam-induced voltage within a bunch (9.5)
w_s	Self-wake seen by a point charge (9.18)
k_n, ω_n	Loss parameter and frequency for the nth mode (9.31)
$S(\sqrt{2x/\pi})$	Fresnel integral, with $x = \omega_m\tau$ (9.35)
ω_m	Maximum frequency for sum over modes (9.35)
a	Disk-hole radius (9.36)
m	Azimuthal mode index (9.36)

k_n	Transverse loss parameter (Sec. 9.5) (9.37)
E_{on}	Field at disk-hole radius (9.37)
r_q	Radius at location of wake-producing charge (9.38)
$w_t(\tau)$	Transverse wake (any m > 1) (9.41)
$w_d(\tau)$	Dipole wake potential (m = 1) (9.45)
$Z_d(\omega)$	Transverse impedance (m = 1) (9.48)
k_d	Dipole loss parameter (9.49)
$w_q(\tau)$	Quadrupole wake potential (m = 2) (9.51)

Chapter 10

$\bar{E}_a$	Unloaded accelerating gradient averaged over structure length (10.1)
η_s	Structure efficiency (10.3)
N_b	Particles per bunch (10.7)
σ_z	RMS bunch length, $\sigma_z = c\sigma_t$ (10.7)
θ	Phase angle ahead of crest (10.8)
$\bar{E}$	Average energy gain per particle (10.9)
$\bar{w}_s$	Effective stored energy per unit length (10.11)
η_b	Beam efficiency (10.11)
w_o	Energy from rf source per unit length (10.13)

Chapter 11

k_β	Wave number for focusing field, $k_\beta = 2\pi/\lambda_\beta$ (11.1)
A	Amplitude growth factor, $A = x_1(z)/x_o$ (11.9)
f_s	Synchrotron frequency, $f_s = \omega_s/2\pi$ (11.13)
f_r	Revolution frequency (11.13)
I_b	Current per bunch, $I_b = q\,f_r$ (11.13)
β_{rf}	Beta-function in rf system, $\beta = 1/k_\beta$ (11.13)
2α	Phase shift per half synchrotron period (11.15)

Chapter 12

f_r	Repetition rate (12.1a)
σ_x^*, σ_y^*	Bunch dimensions at collision point (12.1a)
ε_n	Invariant emittance (12.1b)
eE_o	Beam energy (12.1b)
N	Particles per bunch (12.1b)
β^*	Beta-function at collision point (12.1b)
δ	Beamstrahlung parameter (12.2a)
R	The ratio σ_x^*/σ_y^* (12.2a)
D	Disruption parameter (12.3b)
$\bar{P}_{rf}$	Total average rf power, two linacs (12.5)
V_1	Unloaded energy, each linac (12.5)
L_1	Total length, each linac (12.5)
$\hat{P}_o$	Peak power, each rf source (Table 12.2)
W_o	Total energy, each rf source (Table 12.2)

Chapter 13

V_o, I_o	DC beam voltage and current (13.1)
K	Beam perveance, $K \equiv I_o/V_o^{3/2}$ (13.1)
R	Energy gain enhancement ratio (13.3)
β	Cavity coupling coefficient (SLED) (13.3)
T_p	Klystron pulse length (13.3)
T_o	Unloaded cavity filling time, $T_\sigma \equiv 2Q_o/\omega$ (13.3)
T_s	Structure filling time (13.3)
P_o	Peak rf power (13.5a)
η_c	Storage cavity filling efficiency (13.5a)
β_1	Storage cavity input coupling coefficient (13.7a)
β_{20}	Output coupling coefficient, switch off (13.7a)
β_2	Output coupling coefficient, switch on (13.7b)

APPENDIX. A NUMERICAL EXAMPLE OF AN RF ACCELERATING SYSTEM

James E. Griffin
Fermi National Accelerator Laboratory, Batavia, Illinois 60510

INTRODUCTION

In this note we develop a simple numerical example for an accelerator rf system based on the concepts presented by E. Courant[1] and P. Wilson.[2] The accelerator to be considered is a large proton machine similar in many respects to the Fermilab Main Ring. RF energy is used here to provide the required longitudinal phase space area (bucket area) and to increase the proton energy at some specified rate. Energy loss due to synchrotron radiation will be completely negligible and the system will be required to change frequency slightly during acceleration to compensate for changing β of the protons. This situation would be reversed if we were to consider instead an electron (or e^+e^-) storage ring, where β is essentially constant and the rf energy is required to compensate for synchrotron radiation and higher mode excitation of the surrounding structures by the electrons. The rf hardware required for the two tasks would be quite different.

Parameters of the machine to be considered are:

Injection energy (kinetic)	8 GeV
Transition gamma, γ_t.	20
Acceleration rate, $\dot{c}p$	100 GeV/sec
Maximum Energy	500 GeV
Required frequency shift $df/f = d\beta/\beta$	5.55×10^{-3}
Average Radius	954.27 meters
Rotation Period (for β=1), F_∞^{-1}	20×10^{-6} sec
Injected Longitudinal Emittance (total)	75 eV-sec

RF VOLTAGE - HARMONIC NUMBER

A proton moving at speed c would have a rotation frequency $c/2\pi R$ at the average machine radius. In this example the rotation frequency F_∞ = 50 kHz. The rf system is to be operated at a frequency $h\beta F_\infty$ where the integer h is the "harmonic number" and β represents the proton speed with respect to c at any energy. We assume the proton beam consists of h equal bunches during acceleration without regard to how the bunches were previously generated. Each bunch consists of an ensemble of protons near the "synchronous" momentum where the off-momentum protons execute momentum (and conjugate phase) oscillations about the instantaneous synchronous momentum. Thus each bunch constitutes the population of a longitudinal phase space bucket, the area of which is the total injected longitudinal emittance divided by h. Each bunch is accelerated at the required rate by crossing the rf accelerating gap or gaps at an appropriate voltage phase angle ϕ_s. One voltage-phase angle relationship is established by writing the force on a synchronous proton resulting from the average electric field along the proton orbit established by the rf voltage.

$$\dot{cp} = c[\text{Force}] = ceE = \frac{ceV\sin\phi_s}{2\pi R} = eVF_\infty \sin\phi_s \tag{A.1}$$

[We will use the convention $\Gamma \equiv \sin\phi_s$.]

$$V\Gamma = \frac{c\dot{p}[\text{eV/sec}]}{e\,F_\infty} = \frac{10^{11}}{(5)(10^4)} = 2 \times 10^6 \ \frac{\text{volts}}{\text{turn}} \ . \tag{A.2}$$

Another relationship between the voltage and the phase angle ϕ_s is established by the requirement that the rf system provide the required phase space area or 'bucket area' during acceleration.

$$\begin{bmatrix}\text{Single Stationary} \\ \text{Bucket Area}\end{bmatrix} = A_1(0) = \frac{8R}{c}\left[\frac{2E_sV}{\pi h^3 \eta}\right]^{1/2} , \tag{A.3}$$

where h = harmonic number and

$$\eta = \left| \frac{1}{\gamma_t^2} - \frac{1}{\gamma^2} \right| . \tag{A.4}$$

$$\begin{bmatrix}\text{Total Area in} \\ \text{h stationary} \\ \text{buckets}\end{bmatrix} = A_t(0) = hA_1(0) = \frac{8R}{c}\left[\frac{2E_sV}{\pi h \eta}\right]^{1/2} .$$

The area of a "moving" bucket, i.e., $\phi_s \neq 0$, is smaller than that of a stationary bucket by the "moving bucket factor" $\alpha(\Gamma)$.

$$A_t(\Gamma) = \alpha(\Gamma)A_t(0). \tag{A.5}$$

The function $\alpha(\Gamma)$ decreases monotonically from unity at $\phi_s = 0$ to zero at $\phi_s = \pi/2$. These values are tabulated as functions of ϕ_s and of Γ in CERN/MPS-SI/Int, DL/70/4.[3]

We have the boundary condition that the total longitudinal emittance to be accelerated is 75 eV-sec. In order to minimize loss of charge from the buckets we propose to provide a minimum bucket area of 100 eV-sec at any time during acceleration. Then Eq. (A.5) becomes

$$\alpha(\Gamma)\frac{8R}{c}\left[\frac{2E_sV}{\pi h \eta}\right]^{1/2} = 100 \text{ eV sec} \tag{A.6}$$

This expression can be rewritten

$$\frac{\alpha^2(\Gamma)}{\Gamma} = \frac{10^4\ \pi h\eta}{(8R/c)^2 2E_s V\Gamma} . \tag{A.7}$$

This expression relates the synchronous phase angle ϕ_s (through Γ) to the harmonic number at any energy and, since $V\Gamma$ is known from the required acceleration rate, the required rf voltage is established also.

Problem. Show that, for constant rf voltage, the bucket area is minimum at $E_s = (3)^{1/2}E_t$. (E_t is the lattice transition energy.)

The maximum rf voltage requirement will occur at 32.5 GeV, so at that energy Eq. (A.7) becomes

$$\frac{\alpha^2(\Gamma)}{\Gamma} = 6.3 \times 10^{-7}\ h. \tag{A.8}$$

The rf requirements at different harmonic numbers can now be tabulated.

h	f_{rf}	Γ	$\alpha(\Gamma)$	ϕ_s	V
1	50 kHz	0.995	7.9×10^{-4}	84.5°	2.01×10^6
10	500 kHz	0.989	2.5×10^{-3}	81.5°	2.02×10^6
10^2	5 MHz	0.976	7.8×10^{-3}	77.5°	2.05×10^6
10^3	50 MHz	0.929	0.024	68.3°	2.15×10^6
10^4	500 MHz	0.819	0.072	55°	2.44×10^6

We see that, even though the voltage required to make stationary buckets of fixed total area varies linearly with harmonic number, the additional requirements imposed by moving buckets and constant acceleration rate result in a relatively small variation of voltage with harmonic number. How, then, is one to choose an appropriate harmonic number? The physical space to be occupied by the accelerating cavities, the initial cost, operating power requirements and feasibility of assembly and operation are all valid considerations. At rf frequencies below a few tens of MHz the rf accelerating cavities almost certainly must be ferrite loaded in order to fit the accelerating system into the available lattice. The volume, cost, and power requirements of ferrite cavities capable of developing megavolts may preclude their use.

We consider here the parameters of a simple ferrite accelerating system operating at h = 20, or 10 MHz. Ni-Zn "soft" ferrite with relative permeability around 100 works well at this frequency.

Large volumes of ferrite are difficult to cool uniformly so the average power density to be dissipated in the ferrite is chosen not to exceed 50 mW/cm^3. At 10 MHz this power is deposited in Ni-Zn ferrite by a spatially averaged peak rf flux density of about 40 G

(4 mT). The accelerating structure would be assembled from ferrite rings encircling the beam orbit. Since the rf flux falls off linearly with increasing radius little is gained by using very large diameter cores. We choose here rings of 24 cm inner diameter and 64 cm outer diameter (and perhaps 2.5 cm in thickness).

The rf voltage is equal to the time derivative of the total flux linking the orbit, so assuming sinusoidal flux

$$V_{peak} = A\ B_{peak}\ \omega_{rf}$$

$$A = \frac{2\times10^6}{(4\times10^{-3})(2\pi)(10^7)} = 7.95\ m^2, \tag{A.9}$$

where A is the total area of ferrite linking the orbit. Since the cores are 0.2 m in radial extent the total length of ferrite required along the orbit is 39.75 meters. Because of supports, cooling requirements, accelerating gaps, rf power inputs, etc., the total structure length would probably be around 75 meters. The frontal area of each ferrite ring is 0.276 m^2 so the total volume of ferrite is close to 11 cubic meters and the total weight will be about 51000 kg. (At this writing ferrite ring finished cost is about \$75.00 per kg.) At the assumed power density the ferrite will dissipate 550 kW. If the rf power amplifier is 50 percent efficient, then about 1 MW of input power (exclusive of cooling power) will be required just to deliver the specified rf voltage, before any energy is delivered to the proton beam.

The quality factor, or Q, of the ferrite accelerating system is dominated by the Q of the ferrite itself, nominally around 300. The bandwidth resulting from such a Q, broad as it is, is not adequate to allow for excitation over the required acceleration frequency range without somehow changing the resonant frequency of the structure. This can be accomplished rather easily by arranging current carrying busses which encircle the ferrite but encircle no net rf flux. By adjusting the current in the busses the incremental permeability of the ferrite is changed so that the system can be tuned always to some required frequency near resonance.

A similar exercise, using Mn-Zn ferrite operating at 800 G at h = 1, 50 kHz, yields almost exactly the same volume of ferrite and power level.

At higher harmonic numbers the physical size of the rf structure (nominally one-half wavelength) becomes small enough so that a resonant structure containing little or no ferrite may be considered. Such structures can be built with significantly lower power loss, higher Q, and narrower resonant bandwidth. However, at frequencies above a few hundred MHz, no convenient way of tuning the accelerating structure over the required range presents itself. So one is drawn toward h = 1000, 50 MHz, λ = 6 m. At this frequency the main portion of the accelerating structure can be made of copper or aluminum, adequate and efficient rf power sources are available, and a small amount of Ni-Zn ferrite can still be used for tuning.

ACCELERATING STRUCTURE FOR h = 1000, 50 MHz

P. Wilson[2] has described accelerating structures useful for electron machines and operated at a few hundred MHz. The idea of a cylindrical or spherical structure with re-entrant nose cones can be extended to lower frequencies by making the cavity very long with long re-entrant nose cones, as shown in Figs. 1, a, b, c. The long thin structure becomes a transmission line operating in a transverse electric and magnetic field (TEM) standing wave mode. Each half of the structure has the characteristics of a quarter-wave TEM resonator slightly foreshortened by a capacitance at the high voltage end. Calculation of the salient properties of such an accelerating structure is straightforward. We will work through a representative example here. We are primarily interested in the Q, the shunt impedance at the accelerating gap (as seen by the beam) and the physical size of the structure.

Consider one half of the structure including a gap capacitance which is twice the capacitance between the re-entrant pipes. (This capacitance may result from the presence of corona rolls at the accelerating gap, a ceramic vacuum seal at that point, electric field concentration at the ends of the pipe, etc.) Because the two mirror images will operate in push-pull mode, the voltage across the capacitor will be equal to that at the open end of the line and it may be considered to be in parallel with the line.

The Q of the structure depends upon its physical dimensions, the material used, and the operating frequency. A structure with a beam aperture of 10 cm might reasonably have inner and outer radii $r_1 = 6.25$ cm, $r_2 = 17$ cm. We assume the structure is built from copper with conductivity $\sigma = 8.8 \times 10^7$ (ohm-meters).$^{-1}$ The surface resistivity of the copper at 50 MHz is

$$R_s = \left[\frac{\mu_0 \pi f}{\sigma}\right]^{1/2} = 1.8 \times 10^{-3} \text{ ohms} \tag{A.10}$$

The line resistance per unit length is

$$R = \frac{R_s}{2\pi}\left[\frac{1}{r_1} + \frac{1}{r_2}\right] = 3.5\ R_s = 6.5 \times 10^{-3}\ \frac{\text{ohms}}{\text{meter}}\ . \tag{A.11}$$

We have also inductance and capacitance per unit length.

$$L = \frac{\mu_0}{2\pi} \ln \frac{r_2}{r_1} = 2 \times 10^{-7} \text{ H/m}$$

$$C = \frac{2\pi\varepsilon_0}{\ln r_2/r_1} = 5.5 \times 10^{-11} \text{ F/m}\ . \tag{A.12}$$

Neglecting attenuation due to the small line resistance we have a real characteristic impedance for the line

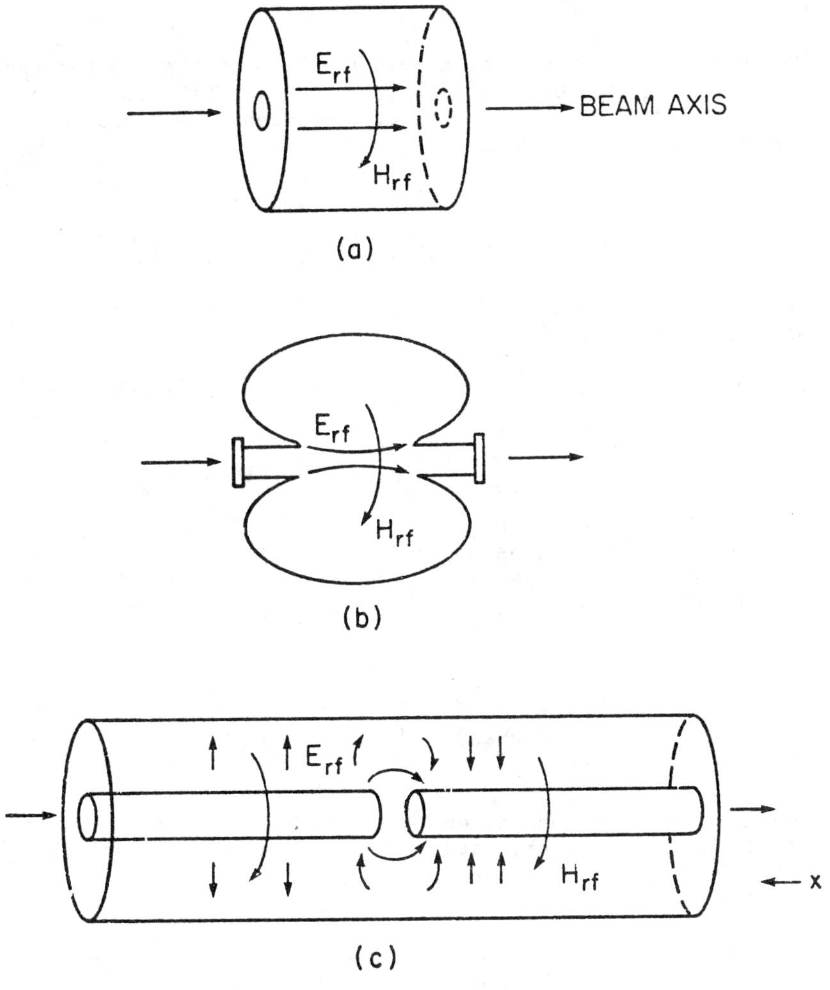

Fig. 1 (a) Simple pill box resonator TM_{010} mode.
(b) Spherical resonator with nose-cones.
(c) Lower frequency resonator, TEM mode. Suitable for MHz operation.

$$Z_c = R_c = \left[\frac{L}{C}\right]^{1/2} = \frac{1}{2\pi}\left[\frac{\mu_o}{\varepsilon_o}\right]^{1/2} \ln \frac{r_2}{r_1} = 60 \text{ ohms.} \qquad (A.13)$$

We assume that the voltage across the line at the shorted end is zero and that some current $i(0,t)$ exists there. The voltage and current in the standing wave on the line, expressed as a function of distance ℓ from the shorted end, are

$$i(\ell,t) = i(t) \cos \beta\ell$$
$$v(\ell,t) = j\, i(0,t)\, R_c \sin \beta\ell, \qquad (A.14)$$

where $\beta = 2\pi/\lambda = \omega\,[LC]^{1/2} = \omega/c = 1.05 \text{ meters}^{-1}$.

The impedance presented by the shorted line of length ℓ is

$$Z_{in} = \frac{v(\ell,t)}{i(\ell,t)} = j\, R_c \tan\beta\ell, \qquad (A.15)$$

so for $\beta\ell < \pi/2$ the line appears inductive. The line is adjusted in length so that the system is resonated by the gap capacitance at the required frequency.

$$X_{gap} + X_{line} = 0$$
$$\text{whence} \quad \frac{1}{\omega R_c C_{gap}} = \tan\beta\ell_r, \qquad (A.16)$$

where ℓ_r is the resonant length of the half cavity. If the total capacitance is 10 pF, then the half capacitance attributed to the half cavity under consideration is 20 pF and

$$\tan\beta\ell_r = \frac{1}{(2\pi)(5\times10^7)(60)(20\times10^{-12})} = 2.65$$
$$\beta\ell_r = 1.21 \qquad (A.17)$$
$$\ell_r = 1.15 \text{ meters.}$$

The cavity Q is defined

$$Q = \frac{(2\pi)(\text{Stored Energy})}{\text{Energy Dissipated per Cycle}} = \frac{2\pi f\, \overline{W}_{st}}{\text{Power loss}} . \qquad (A.18)$$

The time averaged stored magnetic energy is

$$d\overline{W}_m = \frac{1}{4}ii^*L d\ell = \frac{1}{4}i^2(0,0)L \cos^2 \beta\ell d\ell$$

$$\overline{W}_m = \frac{i^2(0,0)L}{4} \int_0^{\ell r} \cos^2 \beta\ell \, d\ell \quad . \tag{A.19}$$

At resonance the time averaged electric stored energy must be equal to its magnetic counterpart so the total stored energy is twice the above calculated value.

The power dissipated in the line is

$$dP = \frac{1}{2} Ri^2(\ell,0)d\ell$$

$$P = \frac{1}{2} Ri^2 \int_0^{\ell r} \cos^2 \beta\ell \, d\ell \ . \tag{A.20}$$

The power dissipated in the shorted end-wall is neglected at this point for a reason which will appear later.

The Q of this structure is simply

$$Q = \frac{\omega L}{R} = \frac{2(\pi\mu_o\sigma f)^{1/2}}{1 + r_2/r_1} r_2 \ln\frac{r_2}{r_1} \ . \tag{A.21}$$

For the case at hand $Q \simeq 10^4$. It is evident that the Q is not affected by the (relatively lossless) gap capacitance or the amount of foreshortening introduced by it.

Of more practical importance is the shunt resistance of the structure at resonance. This is the resistance presented by the structure to the beam current in the accelerator (at resonance). The shunt resistance can be thought of as that resistance which dissipates all of the power lost in the structure when the gap voltage appears across it.

$$P = \frac{(VV^*)_{gap}}{2 R_{sh}} \text{ or } R_{sh} = \frac{(VV^*)_{gap}}{2P} \tag{A.22}$$

For convenience we define a foreshortening factor $q \equiv \omega R_c C_{gap}$ so $\beta\ell_r \equiv \tan^{-1}(1/q)$. From Eq. (A.14) the (half) gap voltage, normalized to the endwall current, is

$$\frac{V_{gap}}{i(0)} = j\,R_c \sin\beta\ell_r = \frac{jR_c}{(1+q^2)^{1/2}} \quad . \tag{A.23}$$

Similarly, from Eq. (A.20) the total power dissipated is

$$\frac{P}{i^2(0)} = \frac{R}{2\beta}\int_0^{\beta\ell_r} \cos^2\beta\ell\, d\beta\ell = \frac{R}{4\beta}\,[\beta\ell_r + \sin\beta\ell_r \cos\beta\ell_r]$$

$$= \frac{R}{4\beta}\left[\tan^{-1}\frac{1}{q} + \frac{q}{1+q^2}\right] \tag{A.24}$$

Then

$$R_{sh} = \frac{2\beta R_c^{\,2}}{R[(1+q^2)\tan^{-1}1/q + q]} = \frac{4}{\pi}R_cQ\left\{\frac{\pi/2}{(1+q^2)\cot^{-1}q + q}\right\} . \tag{A.25}$$

The expression in brackets represents the factor by which R_s is reduced by foreshortening the structure by gap capacitance loading. If $C_{gap} = 0$ then the half shunt resistance is

$$R_{sh} = \frac{4}{\pi}R_cQ \simeq 764 \text{ kOhm} \tag{A.26}$$

In the case at hand, q = 0.377 and the reduction factor is 0.893 so the half shunt resistance is reduced to 682 kOhm.

Note that if q = 1 (26.5 pF total gap capacitance) the accelerating structure length is reduced by half and the shunt impedance reduction factor is 0.61. If the gap voltage is reduced by half, yielding the same accelerating gradient per unit length, the power per cavity is reduced by a factor of 0.41 so the power required per unit accelerating gradient is reduced to 82 percent of that required by an unloaded structure. The total shunt resistance per unit gradient is, however, increased by about 20 percent and this may be an undesirable consequence with regard to beam loading stability.

The two face to face shorted transmission lines which constitute our accelerating structure may now be modified in such a way that the total length is reduced by almost half at the expense of transverse dimensions by folding the two shorted ends around the outside of the outer conductor to meet each other. Since the two sides are operating out of phase, the currents at the low impedance end are exactly opposite at any instant so the need for a real metallic short is eliminated. This configuration is shown in Fig. 2(a). Each cavity may now be made less than 2 m in length so the entire system,

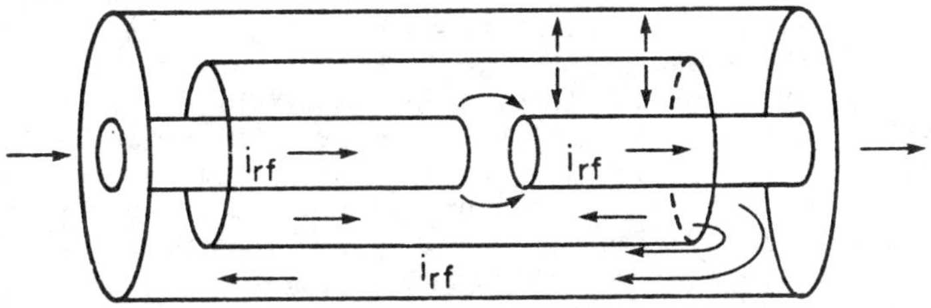

(a)

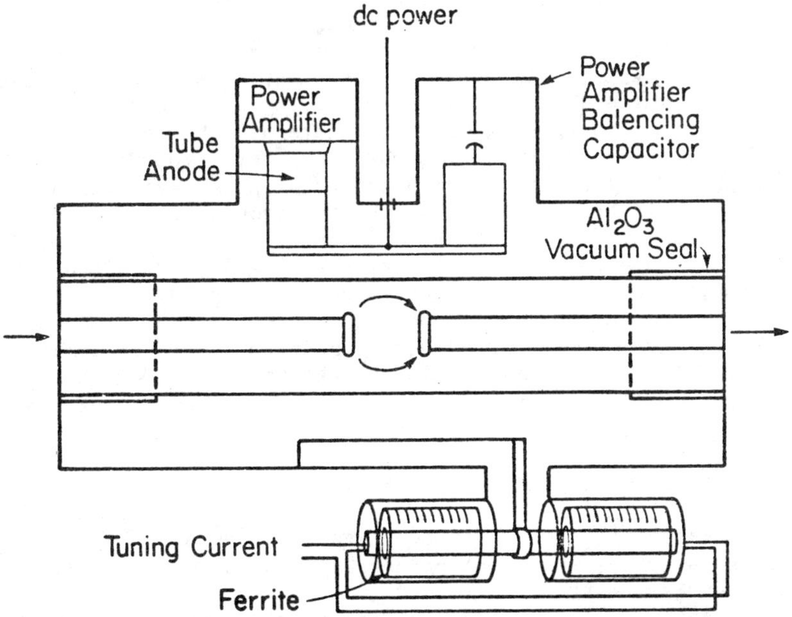

(b)

Fig. 2 (a) Folded face to face TEM resonator.
(b) Method of coupling rf power in and tuning.

consisting of perhaps ten cavities, may easily be installed in a straight section of about 25 m. This geometric configuration has several practical advantages, shown in Fig. 2(b). The "intermediate cylinder," shown suspended in space in Fig. 2(a), can be supported at each end by cylindrical insulators (probably high purity Alumina, Al_2O_3) which serve also as vacuum seals. This results in a rigid mechanical design in which the very high voltage regions are evacuated with seals which need not be located at points of maximum electric field stress. The outer region is a convenient place to locate input power coupling loops and tuning loops. In this way the power amplifier can be operated at a conveniently low rf voltage ($\sim$ 15 kV) with a voltage step-up from input loop to gap of perhaps 12:1. Tuning is accomplished by coupling a small fraction of the cavity stored energy into externally located inductors containing ferrite. The ferrite geometry is such that the ferrite rings can be linked by wires carrying dc or low frequency signals which change the ferrite incremental permeability, the inductor stored energy, and the structure resonant frequency.

Energy dissipated in the lower Q tuners and in the input power and tuning coupling systems will reduce the Q and the shunt impedance to about 8000 and 600 kOhm respectively. We assume that the accelerating system will consist of ten cavities, each one developing a total gap voltage of 200 kV. Each cavity will then dissipate 33 kW so the total power required to develop the voltage will now be 330 kW. Each cavity is driven by a power amplifier capable of delivering 100 kW of rf power so the total power available for proton acceleration is 666 kW. At 100 GeV per second, each proton requires (10^{11}) (1.6×10^{-19}) $= 1.6 \times 10^{-8}$ watts. Consequently the designed system is capable of accelerating 4.2×10^{13} protons on each acceleration cycle. This corresponds to a dc beam current of 0.336 A. This current may exceed some space charge limit and may not be achievable for reasons not related to the rf system. However, there might exist rf limitations related to the rf component of the beam current interacting with the total shunt impedance of the accelerating system.

BEAM LOADING-STABILITY-COMPENSATION

The "moving bucket factor" tables[3] reveal that, for a synchronous phase angle of 68 degrees, the total bucket length will be 66 degrees. Bunches contained within these buckets will have bunch length (full width at half maximum) of about 5 percent of the rf period, or about 1 nsec in the 20 nsec, 50 MHz rf system. Such bunches will have a Fourier component of rf current almost two times the dc current. If we assume here 0.5 A. rf current in ten rf cavities, each with 600 kOhm shunt impedance, we find a total beam induced voltage of 3 MV which is larger than the design rf voltage and obviously at the wrong phase angle, since the beam induced voltage must remove energy from the beam. This is the steady-state beam loading problem described by P. Wilson.[2]

At frequencies very near parallel resonance (sometimes called anti-resonance), a distributed system is well represented by a lumped

RLC circuit. Each of the cavities described in this note is driven by a tetrode power amplifier tube and such tubes are remarkably good current sources. For the purpose of making a simple example of the fundamental beam loading problem, we represent the entire accelerating system by an RLC circuit driven by two current sources, the rf power source and the beam image current. This representation is shown in Fig. 3(a). The total shunt resistance here is 8 M ohm. The currents are represented on a phasor diagram in Fig. 3(b). It is convenient to locate the beam current on the positive real axis, and the beam image current, which actually excites the rf accelerating system, is shown equal and opposite the beam current. The rf generator current, $\tilde{I}_g$, is shown leading the beam current by angle θ and we assume that the phase angle of the rf generator current is the correct angle for development of rf voltage for acceleration. We assume in this example that the accelerator is above transition so that the beam bunches (and the phase of the corresponding Fourier current) lag the proper accelerating rf voltage by an angle ϕ_s greater than $\pi/2$ as shown on the phasor diagram. (The synchronous phase angle ϕ_s mentioned earlier, 68°, would obtain below transition.) The impedance of the RLC circuit is expressed

$$Z_{rf} = R_{sh} \cos\psi e^{-j\psi} \tag{A.27}$$

where $\psi = \tan^{-1} (2Q\ \Delta\omega/\omega)$; "detuning angle."

If the accelerating system were tuned to resonance (purely resistive load) the generator current $i_g e^{j\theta}$ would develop a voltage at angle θ, the required angle for acceleration. A representation of this voltage is shown dashed on the phasor diagram and labelled v_g. (For clarity the voltage is normalized to twice the magnitude of the exciting current.) The accelerating voltage is the projection of the cavity voltage on the beam current (real axis), so here it is $V_{acc} = v_g \sin\phi_s$ as it should be. However, the beam image current also develops a voltage in the (real) cavity impedance, again shown dashed, and the resulting gap voltage is no longer at the required angle, nor does it have the required magnitude. In fact, since the beam current may be larger than the generator current required to develop the required voltage magnitude, the resultant voltage can easily be a decelerating voltage as shown.

One obvious solution to the beam loading problem is to add to the generator current a component exactly opposite to the beam image current so that the beam cavity excitation is cancelled. This requires the rf current generator to deliver unnecessarily large current at a phase angle other than that of the rf voltage. This results in a degradation of the efficiency of the rf excitation system and an increase in cost. Another, more standard, solution is to de-tune the accelerating structure, in this case below the excitation frequency, so that it becomes a partially capacitive load. The voltage developed by each of the exciting currents lags that current in time. The detuning angle and the generator current are now adjusted so that the resultant voltage is of the correct

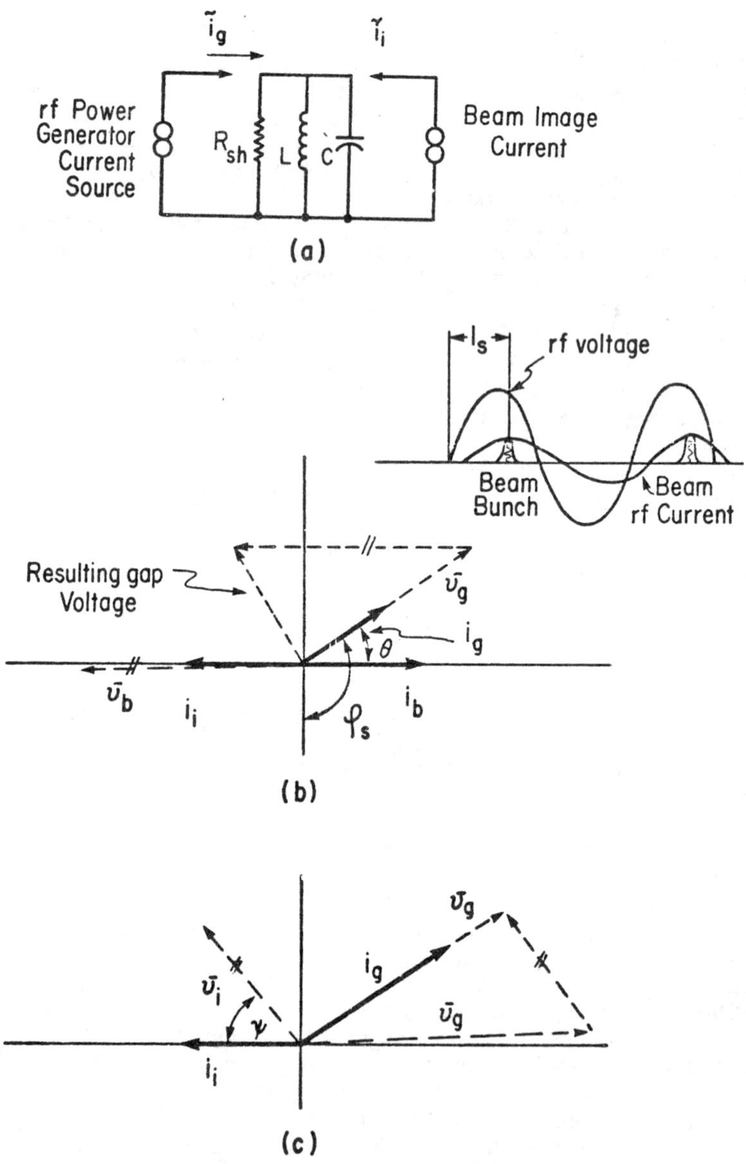

Fig. 3 (a) Lumped RLC Circuit representing accelerating system being driven by two current sources, generator and beam.

(b) Phasor diagram of generator and beam image current with resulting gap voltage if resonator tuned to resonance.

(c) Phasor diagram of beam image and generator current delivered to detuned resonator such that resulting voltage is the required accelerating voltage.

magnitude and at the correct angle. When this is done the rf voltage appearing opposite the generator is in phase with the generator current and the load appears real to the generator. This situation is shown in Fig. 3(c). For analysis, let

$\tilde{I}_o = i_o e^{j\theta} \equiv$ generator current necessary to develop required voltage in real R_{sh} in absence of beam,

$\tilde{I}_g = i_g e^{j\theta} \equiv$ required generator current with beam

$\tilde{I}_i = -\tilde{I}_b = i_i e^{j\pi} = -i_i \equiv$ beam image current.

$\psi = 2Q\ \Delta\omega/\omega =$ detuning angle,

$\tilde{V}_g = v_g e^{i\theta} \equiv$ required rf voltage.

Equate the required voltage with that developed by the separate currents in the detuned accelerating structure.

$$\tilde{V}_g = i_o R_{sh} e^{j\theta} = (i_g e^{j\theta} - i_i)\ R_{sh} \cos\psi e^{-j\psi}\ . \tag{A.28}$$

By equating the real and imaginary parts separately we get explicit expressions for the required detuning angle and generator current in terms of the beam current,

$$\begin{aligned} i_g &= i_o + i_i \cos\theta && \text{(a)} \\ \tan\psi &= \frac{i_i}{i_o} \sin\theta && \text{(b)} \end{aligned} \tag{A.29}$$

By making these adjustments the required steady state conditions are established. There remains a question of stability. This question has been examined by K. Robinson[4] in 1964 and by many others since.[5-7] A simplified sketch of the analysis is given here. Consider a small perturbation of the system in which all of the bunches are uniformly displaced by a small angle η so that the beam current phasor is similarly displaced as shown in Fig. 4. These bunches are now "non-synchronous" and in the normal course of things they would execute synchrotron phase oscillations about the synchronous phase. If the rf accelerating voltage were unaffected by this displacement, the instantaneous accelerating voltage would be

$$\begin{aligned} v_{acc} &= v_g \cos(\theta-\eta) \simeq v_g \cos\theta + \eta v_g \sin\theta \\ &= v_g \sin\phi_s + \eta v_g \cos\phi_s\ . \end{aligned} \tag{A.30}$$

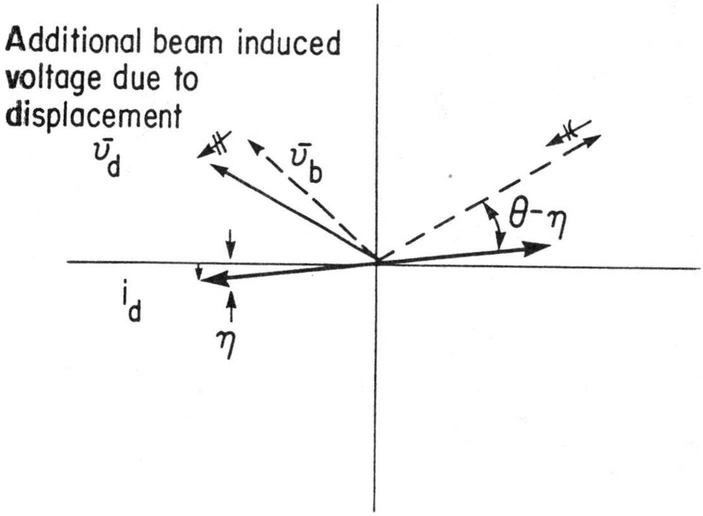

Fig. 4 (a) Beam and image current displaced from equilibrium position by small angle η resulting in degradation of accelerating voltage.

The first term is the synchronous energy gain term while the second term enters the synchrotron phase motion equation, which generates the phase oscillation frequency, the Hamiltonian for the phase motion, and the bucket area. Unfortunately the gap voltage does not remain constant during the perturbation. The rf accelerating system time constant is $T = 2Q\omega^{-1} = 51\ \mu sec$ so the system responds to excitation changes in a few turns, certainly a time short with respect to a phase oscillation period. The displacement of the beam phasor results in an additional small excitation current $\tilde{i}_d$ for the rf system which generates an additional voltage $\tilde{v}_d$.

$$\tilde{i}_d = j\eta\tilde{i}_i = -j\eta i_i \qquad \text{(a)}$$

$$\tilde{v}_d = -j\eta i_i \tilde{Z}_{sh} = \eta v_g \left(\frac{i_i}{i_o}\right) \cos \psi \; e^{-j(\psi + \pi/2)} \qquad \text{(b)} \qquad \text{(A.31)}$$

where $i_o R_{sh}$ has been replaced by v_g. This term contributes an additional voltage to the instantaneous acceleration voltage

$$v_{acc} = \eta v_g \left(\frac{i_i}{i_o}\right) \cos\psi \cos\left(\theta + \psi + \frac{\pi}{2}\right)$$

$$= -v_g \eta \left(\frac{i_i}{i_o}\right) \sin\psi \cos\psi \quad . \qquad \text{(A.32)}$$

Now the term to be entered into the phase motion equation becomes

$$\eta v_g \cos\phi_s \left(1 - \frac{i_i}{i_o} \frac{\sin\psi \cos\psi}{\cos\phi_s}\right) \quad . \qquad \text{(A.33)}$$

Eq. (A.33) (exclusive of the variable η) enters the expression for a moving bucket, where it appears raised to the 1/2 power. If the term in parenthesis vanishes as the beam current is increased, the bucket area also vanishes and the beam is lost. This is not properly a "threshold" for instability, but rather a self-adjusting system. Excessive beam current reduces the bucket area and beam is lost until an equilibrium obtains. If we let $i_i R_{sh} = V_B$, the beam induced voltage, and $i_o R_{sh} = V_{rf}$, then the requirement generated by Eq. (A.33) may be written

$$\frac{2V}{V_b} \cos \phi_s > \sin 2\psi > 0 \qquad \text{(A.34)}$$

The inequality $\sin 2\psi > 0$ represents the initial condition of detuning which was established, but it is also an additional condition for damping the phase motion. Eq. (A.34) is precisely the result obtained by Robinson.[4]

If Eq. (A.29a) is multiplied through by $v_g/2$, an expression for power results.

$$\frac{1}{2} v_g i_g = \frac{1}{2} v_g i_o + \frac{1}{2} v_g i_i \cos\theta \tag{A.35}$$

$$\begin{bmatrix}\text{Total input}\\ \text{rf power}\end{bmatrix} = \begin{bmatrix}\text{Power dissipated}\\ \text{in rf cavities}\end{bmatrix} + \begin{bmatrix}\text{Power delivered}\\ \text{to protons}\end{bmatrix}$$

Let P be the ratio of power delivered to the beam to power delivered to the rf system.

$$P = \frac{i_i}{i_o} \cos\theta = \frac{i_i}{i_o} \sin\phi_s \tag{A.36a}$$

$$i_g = (1 + P) i_o \tag{A.36b}$$

In terms of this power ratio the parenthetic expression (33) may be rewritten

$$1 - \frac{i_i}{i_o} \frac{\sin\psi \cos\psi}{\cos\phi_s} = \frac{(1 - P^2)}{1 + P^2 \cot^2\phi_s} > 0. \tag{A.37}$$

This expression states that the bucket area vanishes as the power delivered to the beam approaches that delivered to the rf system. But in earlier development we proposed to violate this condition by a factor of two.

This difficulty may be alleviated by placing two active feedback systems on the accelerating cavities. One loop examines the phase relationship between the generator current and the gap voltage and adjusts the cavity detuning angle such that the two phases are always equal (or differ by π radians, depending upon details of circuit design). The other feedback loop examines the amplitude of the rf gap voltage and adjusts the generator current so that the gap voltage amplitude remains unchanged under changes in beam loading. The performance of the accelerator is extremely sensitive to the response functions of these two feedback loops and their design requires careful analysis.

Suppose, for instance, that the phase loop is very effective at frequencies exceeding the phase oscillation frequency but that the rf amplitude control loop does not respond with effectiveness at the phase oscillation frequency. This means that the cavity system is always tuned such that the gap voltage is in phase with the excitation current but that its amplitude cannot be maintained at the current level. The modified gap voltage v_g' is expressed

$$v'_g e^{j\theta} = \left(i_0(1 + P)e^{j\theta} - (1 + j\eta)i_i\right) \tilde{Z}' , \qquad (A.38)$$

where $\tilde{Z}'$ is the adjusted cavity impedance. This results in an instantaneous accelerating voltage

$$v'_{acc} = v_g \sin \phi_s + v_g \eta \cos \phi_s (1 - P). \qquad (A.39)$$

In this case, the second term, entering the bucket area equation, carries the bucket area to zero under beam loading conditions even more quickly than the uncompensated case, Eq. (A.37).

In the opposite extreme case, where the phase control loop is ineffective at the synchrotron frequency but the amplitude control loop is able to maintain the magnitude of the gap voltage constant, even though it may be at an inappropriate instantaneous phase, we find

$$v_g e^{j(\theta+\delta)} = \left[\left(i_0(1+P) + i'_g\right)e^{j\theta} - (1+j\eta)i_i\right] \tilde{Z} , \qquad (A.40)$$

where i'_g is additional generator current required. This gives an instantaneous accelerating voltage

$$v''_{acc} = v_g \sin\phi_s + \eta v_g \cos\phi_s (1+P). \qquad (A.41)$$

Here the bucket area increases with beam loading (because the constant amplitude gap voltage phasor is rotated toward the beam current phasor). This may, of course, result in growing amplitude phase oscillation and longitudinal emittance growth.

These results, coupled with the fact that the phase oscillation changes during acceleration, re-enforce the idea that careful design of these local feedback loops and of the more global low level loop, which sets the phase of the generator current, is imperative (and sometimes difficult) if the accelerator is to reach its maximum design beam intensity with good beam quality.

ACKNOWLEDGEMENTS

The ideas and concepts expressed in this note are mostly quite old and state of the art. It would be impossible to mention specifically all who have contributed to my approach to them. Certainly a few, Quentin Kerns, Wolfgang Schnell, Lee Teng, Matt Allen, and Martin Plotkin must be mentioned. To them I express special gratitude as well as to the very many unmentioned contributors.

REFERENCES

1. E. Courant - Accelerator Theory - these proceedings.

2. P. Wilson - RF Power and Linacs - these proceedings.

3. C. Bovet, R. Gouiran, I. Gumowski, and K.H. Reich, CERN/MPS-SI/Int. DL/70/4 (1970).

4. K. Robinson, CEAL-1010 (1964).

5. J. Griffin, IEEE Trans. Nucl. Sci. NS-22, 1910 (1975).

6. F. Pederson, IEEE Trans. Nucl. Sci. NS-22, 1906 (1975).

7. M. Allen and R. McConnell, IEEE Trans. Nucl. Sci. NS-20, 373 (1973).

STOCHASTIC COOLING

Joseph Bisognano and Christoph Leemann
Lawrence Berkeley Laboratory, University of California
Berkeley, California 94720

TABLE OF CONTENTS

0094-243X/82/870583-72$3.00 Copyright 1982 American Institute of Physics

STOCHASTIC COOLING

Joseph Bisognano and Christoph Leemann
Lawrence Berkeley Laboratory
University of California
Berkeley, CA 94720

1. INTRODUCTION

Stochastic cooling is the damping of betatron oscillations and momentum spread of a particle beam by a feedback system. In its simplest form, a pickup electrode detects the transverse positions or momenta of particles in a storage ring, and the signal produced is amplified and applied downstream to a kicker. The time delay of the cable and electronics is designed to match the transit time of particles along the arc of the storage ring between the pickup and kicker so that an individual particle receives the amplified version of the signal it produced at the pick-up. If there were only a single particle in the ring, it is obvious that betatron oscillations and momentum offset could be damped. However, in addition to its own signal, a particle receives signals from other beam particles. In the limit of an infinite number of particles, no damping could be achieved; we have Liouville's theorem with constant density of the phase space fluid. For a finite, albeit large number of particles, there remains a residue of the single particle damping which is of practical use in accumulating low phase space density beams of particles such as antiprotons. It was the realization of this fact that led to the invention of stochastic cooling by S. van der Meer in 1968.[38]

Since its conception, stochastic cooling has been the subject of much theoretical and experimental work. The reader is directed to the references for a thorough review of its development. The earliest experiments were performed at the ISR in 1974, with the subsequent ICE studies firmly establishing the stochastic cooling technique. This work directly led to the design and construction of the Antiproton Accumulator at CERN and the beginnings of $p\bar{p}$ colliding beam physics at the SPS. Experiments in stochastic cooling have been performed at Fermilab in collaboration with LBL, and a design is currently under development for a $\bar{p}$ accumulator for the Tevatron.

2. A QUALITATIVE DESCRIPTION OF STOCHASTIC COOLING

To clarify the basic issues of stochastic cooling, we will first give a qualitative description of the betatron oscillation cooling system sketched in Fig. 1. A pickup detects the transverse position y_i of each of N particles in a storage ring. Ideally, the pickup response will be proportional to the transverse position of the particles. The signal is amplified and applied to a kicker structure one quarter of a betatron wavelength downstream. If the electronics has a bandwidth W, the effective duration of a single particle pulse at the kicker is approximately $1/2W$. (This

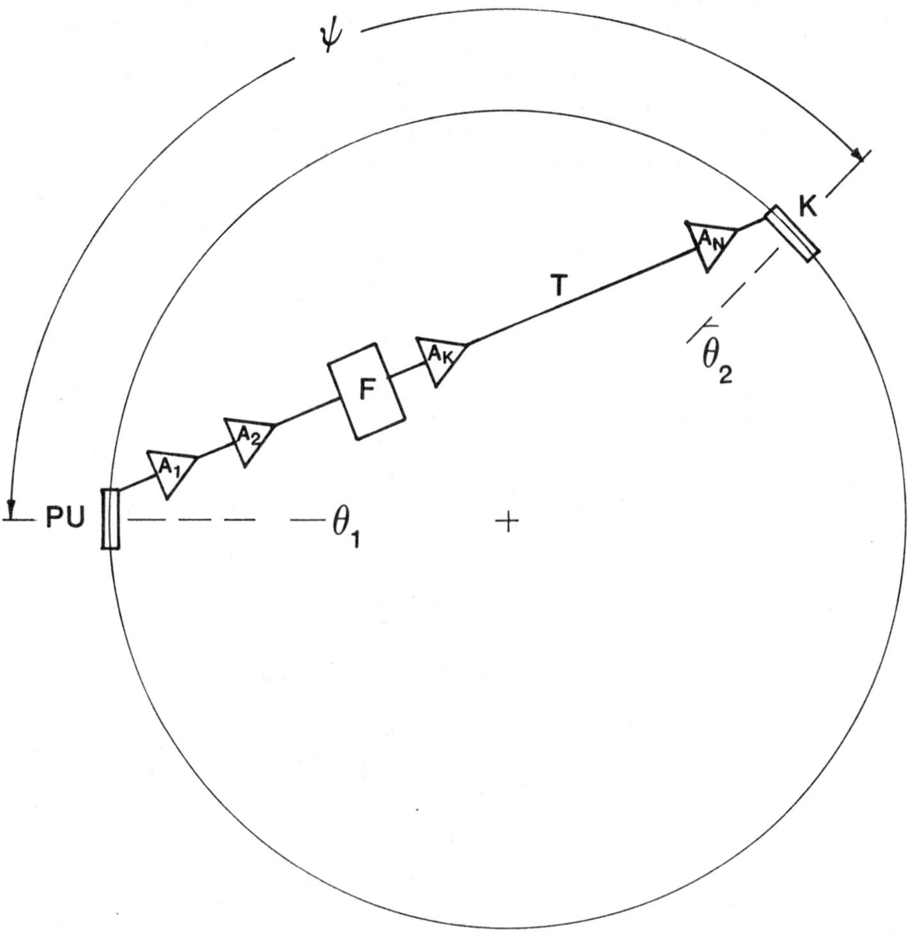

XBL 823-8197

Fig. 1. Schematic of Stochastic Cooling System

$A_1, A_2 \ldots A_N$:	Amplifiers
PU:	Pick-Up Electrode Array
K:	Kicker Electrode Array
F:	Filter(s) (for momentum cooling)
$\psi(\theta_2, \theta_1)$:	Betatron phase advance from θ_1 to θ_2 $\left(= \frac{2n+1}{2}\pi \text{ for transverse cooling}\right)$

is just a manifestation of the Fourier uncertainty principle between time t and its conjugate variable frequency f: $\Delta t \Delta f \geqslant$ constant.) Thus, when the ith particle passes through the kicker, it receives an impulse from $n = N/2WT$ particles, where T is its revolution period. This impulse changes the betatron angle y_i' which, because of the quarter betatron wavelength separation of the pickup and kicker, effectively corrects the betatron oscillation at the pickup by

$$y_i \rightarrow y_i - gy_i - g \sum_{\substack{j=1 \\ j \neq i}}^{n} y_j \tag{1}$$

where g models the gain of an idealized pickup-amplifier-kicker system. The change in y_i^2 is then

$$\Delta y_i^2 = - 2gy_i^2 - 2g \sum_{\substack{j=1 \\ j \neq i}}^{n} y_i\, y_j + g^2 \sum_{j=1}^{n} y_j^2 + g^2 \sum_{k} \sum_{\substack{j \\ k \neq j}} y_j\, y_k \tag{2}$$

Amplifier noise will introduce an additional positive term in (2) proportional to the noise power of the system; we ignore this effect for the moment. We can now attempt to average this expression over the particles in the ring. For the first pass through the system, the betatron positions are independent and uncorrelated, that is $\langle y_i y_j \rangle = 0$ for $i \neq j$. Then equation (2) reduces to

$$\langle \Delta y^2 \rangle = - 2g \langle y^2 \rangle + g^2 n \langle y^2 \rangle \tag{3}$$

The largest reduction in $\langle y^2 \rangle$ occurs when $g = 1/n = 2WT/N$. For this value of g, the rms betatron amplitude damps at a rate

$$\frac{1}{\tau} = \frac{W}{2N} = \frac{1}{4nT} \tag{4}$$

where a factor of 2 is lost in averaging over the phase of the betatron oscillation and another factor of 2 in changing variables from mean squared values $\langle y^2 \rangle$ to rms amplitude $\langle y^2 \rangle^{1/2}$. Equation (4) is the basic cooling relation between rate, bandwidth, and number of particles. Physically, n represents the number of particles which interact during a correction step, and $1/n$ is the optimal single particle damping per step. Greater bandwidth reduces the size of this "sample" and allows faster damping. In the limit of an infinite number of particles (where the beam becomes a continuous fluid), there can be no damping. Typical systems will have bandwidths of hundreds of megahertz to several gigahertz, revolution frequencies of the order of a megahertz, particle number between 10^6 to 10^{11}, and cooling times in seconds to hours. A similar analysis can be applied to momentum cooling with the y parameter now interpreted as the energy offset from some desired central energy.

The above results were derived under the condition that $\langle y_i y_j \rangle = 0$ and are strictly applicable only on the first pass of the beam through the feedback system. After the first correction, we have

$$y_i\, y_j \to \left(y_i - gy_i - g \sum_{\ell \neq i} y_\ell\right)\left(y_j - gy_j - g \sum_{k \neq j} y_k\right) \tag{5}$$

and even if $y_i y_j$ is initially zero, there now appear negative cross terms; e.g., $- gy_j^2 - gy_i^2$. In other words, the feedback system has produced correlations among the beam particle positions. It is apparent from equation (2) that such negative correlations can act to degrade the cooling. In fact, if there is no revolution frequency spread of the particles, these correlations will grow and ultimately stop the cooling process. Recall that only a fraction n/N of the particles in the ring effectively interact during a single correction. If there is a spread in revolution frequency, the members of this sample of n particles will change, and the growth and long term effects of the correlations will be limited. As we shall see, if there is sufficient frequency spread, the correlation becomes negligible and the optimal rate of equation (4) can be obtained.

The development of correlation is a coherent effect which is closely related to beam instabilities. In fact, if the sign of the gain of the feedback system were reversed, instabilities could be induced. In this context the notion of Landau damping describes how frequency spread "breaks up" incipient instabilities, and we will find in the detailed mathematical analysis of cooling dispersion integrals of the same form as encountered in instability theory. An estimate of the rate of growth of these correlations can be obtained by summing equation (1) over the n particles which interact. We have

$$\frac{1}{n} \sum_{i=1}^{n} y_i \equiv \bar{y} \to \bar{y} - ng\bar{y} \tag{6}$$

If there were no frequency spread, the average transverse position of the beam would damp at a rate

$$\left(\frac{1}{\tau}\right)_{\text{coherent}} = - ng \tag{7}$$

which is n times faster than the single particle cooling. The signal would damp away and the cooling would stop. If the sign of the gain were reversed, there would be a coherent instability at this rate.

The theoretical analysis of stochastic cooling can be approached from a variety of perspectives with a fundamental dichotomy between frequency domain and time domain analyses. At the heart of the matter is the existence of two distinct time scales. One corresponds to the single particle cooling rate of equation (4), with characteristic times of the order of seconds or longer. The other scale is that of the coherent effects of Eq. (7) -- instabilities, signal suppression, and Landau damping -- with times typically of the order of milliseconds. The coherent phenomena are rapid compared to the gross cooling rate and are best treated in the frequency domain with the revolution frequencies and betatron amplitudes treated as constant. The frequency domain notions of power spectrum, bandwidth, filters, and phase provide the basis for analysis. The particle dynamics, on the other hand, which describe the slow damping of phase space are clearest in the time domain with the Fokker-Planck equation as the main tool. This "two-timing" approach is basic to understanding stochastic cooling, and is implicit in most of the literature on the subject. It also provides constraints on computer simulations, where care must be taken with the use of relatively few pseudo particles in a simulation that these two time scales do not become comparable. Otherwise, extrapolation to actual cooling scenarios, which do have two disparate time scales, becomes highly questionable.

3. SCHOTTKY SIGNALS AND SAMPLING

As a particle is cooled, it experiences a succession of kicks from other beam particles. Since the relative positions of the particles change slowly (the frequency spread is often a fraction of 1% of the revolution frequency), there is a coherence among the signals produced by a given particle over many passes through the pick-up as experienced by another particle over many passes through the kicker. This correlation over time is translated in the frequency domain into a non-constant power spectrum of the particle signals; that is, the beam Schottky spectrum.

Consider a single particle circulating in a ring at angular revolution frequency $\omega = 2\pi/T$. The current at a pick-up can be described as a series of delta functions of the form

$$I(t) = e \sum_{n=-\infty}^{\infty} \delta\left(t-t_0-nT\right) \tag{8}$$

where t_0 is some arbitrary time at which the particle is in the pick-up. Since this current is periodic in t, it may be expanded in a simple Fourier series

$$I(t) = \frac{e\omega}{2\pi} \sum_{n=-\infty}^{+\infty} e^{in\omega(t-t_0)} \tag{9}$$

and we have the basic result that a circulating particle produces a signal at each harmonic of its revolution frequency. This is the single particle longitudinal Schottky signal. Now consider a collection of N particles in the ring. From (9) we immediately have that the total current is

$$I(t) = \sum_{n=-\infty}^{+\infty} \sum_{j=1}^{N} \frac{e\omega_j}{2\pi} e^{in\omega_j(t-t_{oj})}$$

$$= \sum_{j=1}^{N} \frac{e\omega_j}{2\pi} \left(1 + \sum_{n=1}^{\infty} 2\cos n\omega_j(t-t_{oj})\right) \tag{10}$$

The signal at the pickup is composed of bands in frequency corresponding to each harmonic n, with a width $n\Delta\omega$, where $\Delta\omega$ is the full width of the revolution frequency distribution $f(\omega)$ (normalized to unity). For sufficiently large n, these Schottky bands will touch; there is band overlap. A current sensitive pickup will produce a signal with this frequency structure over its bandwidth.

Suppose that the current is analyzed at frequency $n\omega_0$, where ω_0 is inside the particle frequency distribution, and assume there is no band overlap. Then the squared current in Schottky band n is simply

$$I_n^2(t) = 4\left(\sum_j \frac{e\omega_j}{2\pi} \cos n\omega_j(t-t_{oj})\right)\left(\sum_i \frac{e\omega_i}{2\pi} \cos n\omega_i(t-t_{oi})\right) \tag{11}$$

If there is no correlation in the t_{oi} from a feedback system, the cross terms average to zero and

$$\langle I_n^2(t)\rangle = 4\sum_j \left(\frac{e\omega_j}{2\pi}\right)^2 \langle\cos^2 n\omega_j(t-t_{oj})\rangle \tag{12}$$

and

$$I_n^{rms} = \left(\frac{e\bar{\omega}}{\pi}\right)\sqrt{2N} \tag{13}$$

where $\bar{\omega}$ is the average revolution frequency. If this current were analyzed by a spectrum analyzer with resolution $\Delta\Omega$, about frequency $\Omega = n\omega_0$, then only particles such that $|n(\omega_0-\omega_j)| < \Delta\Omega/2$ would enter into the sum and we have

$$I_n^{rms}(\Omega) = \frac{e\omega_0}{\pi}\sqrt{2N}\sqrt{f\left(\frac{\Omega}{n}\right)\frac{\Delta\Omega}{n}} \tag{14}$$

Thus, for uncorrelated beam parameters, the power (proportional to $(I^{rms}(\Omega))^2$) in the nth Schottky band mirrors the frequency distribution, with its width and height proportional to n and $1/n$, respectively. However, if there is correlation among the t_{oj}, the cross terms can act to deform the Schottky spectrum. As we have seen in the previous section, a stochastic cooling feedback system can create correlations, and, in fact, rapid "Schottky signal suppression" is commonly observed when cooling systems are turned on. Other particle interactions through space charge and wall impedances can also induce correlations which modify the Schottky signals; this will be discussed quantitatively in Section 4. This effect is analogous to the polarization and Debye shielding of plasma physics, where a "dielectric function" is used to describe the details. We will derive a similar "suppression factor."

For the remainder of this section, we will use our knowledge of the Schottky spectrum to analyze longitudinal stochastic cooling without shielding in order to introduce some important physical issues. Betatron cooling can be analyzed similarly, with averaging over rapid betatron oscillations and slowly varying amplitudes. In practice, cooling systems often operate in this regime where the details of feedback shielding are relatively unimportant.

The signal produced at the pick-up by I_T can be amplified and applied to a kicker for longitudinal (momentum) cooling. The correction signal at the kicker at time t will have the spectral form of the current I_T, modified by the gain of the system. The kicker signal $k(t)$ will generally be of the form

$$k(t) = \frac{2\pi}{\omega_0} \sum_j \sum_n G(n\omega_j, x_j)\, e^{in\omega_j(t-t_{oj})} \tag{15}$$

where $G(\Omega, x_j)$ represents the electronic transfer character of the system at electronic frequency (that is, the Fourier conjugate of time) Ω and particle energy x_j. With this functional form we are allowing both a direct energy variation of the gain (e.g., by transverse position variations due to momentum dispersion) and electronic variation (e.g., through a filter structure). A similar expression can be derived for a transverse dipole signal, with multiplication by the betatron position yielding sideband structure at $\pm Q\omega_j$. In general, G will roll-off for high frequencies and the sum in (12) will converge absolutely.

Let us now focus attention on a single particle, say the ith. It does not experience the signal at the kicker continuously, but rather samples it once every revolution. If the particle is at the kicker at time τ_i plus multiples of the revolution period, the correction signal it receives is actually

$$k_i(t) = \sum_m \delta(t-\tau_i-mT_i)\, k(t) \tag{16}$$

That is, it samples the kicker signal only when it is in the kicker (assumed very short). Again, we may expand the series of delta functions in a Fourier series and obtain

$$k_i(t) = \sum_j \sum_n \sum_m G(n\omega_j, x_j)\, e^{i(n\omega_j - m\omega_i)t}\, e^{i(n\omega_j t_{oj} - m\omega_i \tau_i)} \tag{17}$$

Note that k_i is the sum of k's translated in frequency by all multiples of ω_i. Expression (17) is a special case of the "sampling theorem" of signal processing theory.

The energy x_i of particle i will be changed according to

$$\frac{dx_i}{dt} = k_i(t) \tag{18}$$

Since cooling is a slow process relative to the revolution frequency, the effect of the rapidly oscillating part of k_i is negligible unless there is a rapidly growing instability. To simplify matters further, let us assume that there is no band overlap, so the only slowly varying terms occur when $n = m$ in the sum. (This assumption will be dropped in Section 5.) With this approximation, we have

$$\frac{dx_i}{dt} = \sum_j \sum_n G(n\omega_j, x_j)\, e^{in(\omega_j - \omega_i)t}\, e^{in(\omega_j t_{oj} - \omega_i \tau_i)} \tag{19}$$

First consider the self-interaction term; i.e., when $i = j$. We have

$$\left.\frac{dx_i}{dt}\right|_{self} = \sum_{n=-\infty}^{+\infty} G(n\omega_i, x_i)\, e^{in\omega_i (t_{oj} - \tau_i)} \tag{20}$$

The phase factor represents the particle transit time between the pick-up and kicker. The gain function G contains within it a phase factor corresponding to the signal transit time through the amplifier chain and the cable. This phase factor is of the form $\exp(-in\omega_i(\tau))$, where $\tau = \theta/\omega_0$, θ is the azimuthal distance between pickup and kicker, and ω_0 is the frequency of some particle centered in the distribution, with energy x_0. For this particle the phase factors are made to cancel. If the residual energy variation of the gain is of the form $G(n\omega, x) = -g(x - x_0)$, the self-term moves all particles toward the central energy x_0. More complicated gain variations are found in stack cooling systems and will be discussed later. The cancellation of the transit time and electronic delay time factor is only approximate for all but particles with frequency ω_0, and care must be taken to avoid residual phases exceeding 90°, which would heat the beam. In the literature this effect is described as mixing between pick-up and kicker.

The rms effect of other beam particles requires averaging of x_i^2. As a first approximation, let us average over times short compared to the cooling time, so that the parameters describing k may be considered constant. Then the average change of x^2 per unit time due to other beam particles in time 2S is simply

$$\frac{\Delta x_i^2}{\Delta t} = \left(\int_{-S}^{+S} k_i(t')dt' \right)^2 \frac{1}{2S} \tag{21}$$

If we make the further approximation that the t_{0j} are statistically independent, we have

$$\frac{\Delta x_i^2}{\Delta t} = \sum_{\ell} \sum_{j} |G(\ell\omega_j, x_j)|^2 \frac{\sin^2 \ell(\omega_i - \omega_j)S}{\ell^2(\omega_i - \omega_j)^2 S^2} (2S) \tag{22}$$

For short times S, all particles contribute equally to increasing x_i^2. However, for times long compared to a revolution frequency but still short compared to a cooling time, particles within $\pi/S\ell$ of ω_i in frequency dominate. Thus, we have for small perturbations, that a particle reacts principally to a signal in a small neighborhood of harmonics of its revolution frequency. This enhancement of beam noise is a first example of correlation effects in beam cooling.

Equations (20) and (22) can be combined to yield an average correction per unit time of x_i^2 of

$$\frac{\Delta x_i^2}{\Delta t} = \sum_{\ell} 2G(\ell\omega, x_i)\, x_i e^{i\ell\omega_i (t_{0i} - \tau)}$$
$$+ (2S) \sum_{\ell} \sum_{j} |G(\ell\omega_j, x_j)|^2 \frac{\sin^2 \ell(\omega_i - \omega_j)S}{\left(\ell(\omega_i - \omega_j)S\right)^2} \tag{23}$$

Equation (23) is similar to Eq. (3), but contains some correlation effects due to the long time coherence of the sampled Schottky signal. A few simplifying approximations will make the similarities between these two equations more apparent. Let G be constant over a finite bandwidth $(2\pi)W$ in angular frequency. There will be approximately m = Integer part$(4\pi W/\omega_0)$ terms in the sums in (22). Let $mG = -g(x - x_0)$ and let the mixing between pick-up and kicker be negligible. With the sum over j approximated by an integral over the frequency distribution $f(\omega)$ (normalized to unity) and assuming that near neighbors in frequency dominate, we have

$$\frac{\Delta(x_i-x_o)^2}{\Delta t} = \left\{-2g + \frac{g^2 N}{m^2}\sum_{\ell}\int d\omega'\ f(\omega')\ \frac{\sin^2(\omega_i-\omega')S}{\ell^2(\omega_i-\omega')^2 S^2}\cdot 2S\right\}\cdot(x_i-x_o)^2 \tag{24}$$

For times S short compared to the damping time (so that the revolution frequencies remain approximately constant) but long compared to the revolution period, the $\sin^2 x/\pi x^2$ function approximates a delta function, and we have

$$\frac{1}{\tau}\bigg|_{\Delta(x_i-x_o)^2} = \left\{-2g + g^2\left(\frac{N}{m}\right)2\pi\ f(\omega_i)\ \left[\frac{1}{m}\sum_{\ell}\frac{1}{\ell}\right]\right\} \tag{25}$$

The factor N/m is just the sample size n of Eq. (3). For harmonics ℓ where the Schottky bands touch, $f(\omega)\ 1/m \sum_{\ell} 1/\ell \approx 1/\omega_o$, and (3) and (25) agree, when g of (25) is replaced by g/T and $1/T$ is factored out to obtain the same units. On the other hand, if the Schottky bands are narrow, the effect of the Schottky signal is enhanced over the result in (3), and the cooling is degraded. The rate $1/\tau$ depends on $f(\omega)$. Since $f(\omega)$ will be changing as the longitudinal cooling progresses, an equation for the time development of $f(\omega)$ is necessary for a full description. As we shall see in the next section this time evolution is described by a Fokker-Planck equation. For betatron cooling without simultaneous longitudinal cooling, $f(\omega)$ is constant, and moment equations similar in form to (24) and (25) are often useful.

The analysis so far has singled out one particle (the "cooled" particle), and we have studied in detail the effect of the unperturbed signals of the other particles of the beam, including their long time coherence. In addition, these particles are kicked by the signal from the cooled particle, and thus become carriers for information about it; that is, there is feedback through the beam. The beam becomes polarized and shields the cooled particle from its self interaction. Of course, this assymmetry between cooled particle and other beam particles is not physical, but is an artifact of the description. In the next section an approach is developed which includes both Schottky signal noise enhancement and feedback.

4. KINETIC EQUATIONS FOR LONGITUDINAL COOLING

In this section we will derive a Fokker-Planck equation for longitudinal stochastic cooling which includes both the time coherence of Schottky signals and the shielding induced by the feedback system. Both of these effects are manifestations of correlations. In the usual derivation of the Vlasov equation which describes instabilities in particle accelerators correlation effects are assumed negligible. In addition, there is no dissipative self-interaction; the equations of motion are Hamiltonian. In stochastic cooling systems, it is such a self-interaction term which increases the phase space density. In what follows, equations are derived for the one particle and two-particle (correlation) density functions. The one particle function is just the usual distribution of the Vlasov equation. The two-particle distribution will describe the effects of correlation. The argument follows closely that used in a plasma physics context to derive the Vlasov equation from "first principles", but allows the system to be non-Hamiltonian.

A. Kinetic Equations for a Non-Liouvillian System

The stage for our analysis is a 2N-dimensional ensemble space whose elements are vectors $(q_1, p_1, \ldots, q_n, p_n)$. Each vector represents one whole system of N particles with positions q_i and momenta p_i. Consider the ensemble distribution $D(q_1, p_1, \ldots, q_n, p_n)$ describing a collection of these systems of N particles, normalized so that

$$\int dq_1 \ldots dp_n \; D(q_1, \ldots, p_n) = 1 \tag{26}$$

Conservation of the number of ensemble systems is expressed by

$$\frac{\partial D}{\partial t} + \vec{\nabla} \cdot (\vec{u} \; D) = 0 \tag{27}$$

where $\vec{u} = (\dot{q}_1, \dot{p}_1, \ldots, \dot{q}_n, \dot{p}_n)$. If the system dynamics are described by Hamiltonian equations, then (27) reduces to the condition of incompressible fluid flow[24],

$$\frac{dD}{dt} = \frac{\partial D}{\partial t} + \vec{u} \cdot \vec{\nabla} D = 0 \tag{28}$$

that is, Liouville's Theorem. The Vlasov equation may be derived by integrating (28) over 2(N-1) variables with D assumed symmetric in the particle variables.

A Hamiltonian description is not applicable to longitudinal cooling. Instead, the dynamics are of the form

$$\dot{p}_i = \sum_j G(q_i, q_j, p_j) \qquad \dot{q}_i = Q(p_i) \tag{29}$$

Equation (27) may still be integrated over 2(N-1) or 2(N-2) variables to yield equations for one and two particle distribution functions, but, as we shall see, Liouville's theorem for the one particle distribution will be lost.

Define the one particle distribution by

$$f_1(q_1,p_1,t) = \int dq_2 \ldots dp_n \; D(q_1,p_1,\ldots,q_n,p_n,t) \tag{30}$$

and similarly

$$f_2(q_1,p_1,q_2,p_2,t) = \int dq_3 \ldots dp_n \; D(q_1,p_1,\ldots,q_n,p_n,t) \tag{31}$$

From Eq. (27) we have

$$\frac{\partial f_1}{\partial t} = - \int dq_2 \ldots dp_n \left(\frac{\partial}{\partial q_1} (\dot{q}_1 D) + \frac{\partial}{\partial p_1} (\dot{p}_1 D) \right) \tag{32}$$

With dynamics of the form (29), this reduces under the usual (particle index) symmetry assumption for D to

$$\frac{\partial f_1}{\partial t} + \dot{q}_1 \frac{\partial f_1}{\partial q_1} + (N-1) \frac{\partial}{\partial p_1} \int dq_2 \, dp_2 \; G(q_1,q_2,p_2) \; f_2(q_1,p_1,q_2,p_2,t) + $$

$$+ \left[\frac{\partial}{\partial p_1} \left[G(q_1,q_1,p_1) \; f_1(q_1,p_1,t) \right] \right] = 0 \tag{33}$$

The bracketed term is the addition to the usual kinetic equation without self-interaction and expresses the violation of Liouville's Theorem. It is of the form of the coherent part of a Fokker-Planck equation and induces compression of phase space. The integral in (33) describes interaction with other beam particles and includes the "usual" Vlasov average field effects and also correlation effects that tend to increase the phase space volume and suppress cooling through beam feedback. Both terms together may be interpreted as the divergence of a particle flux. The corresponding equation for f_2 is

$$\frac{\partial f_2}{\partial t} + \dot{q}_1 \frac{\partial f_2}{\partial q_1} + \dot{q}_2 \frac{\partial f_2}{\partial q_2} + (N-2) \frac{\partial}{\partial p_1} \int dq_3 dp_3\, G(q_1,q_3,p_3)\, f_3$$

$$+ (N-2) \frac{\partial}{\partial p_2} \int dq_3 dp_3\, G(q_2,q_3,p_3)\, f_3$$

$$+ \frac{\partial}{\partial p_1} G(q_1,q_2,p_2)\, f_2 + \frac{\partial}{\partial p_2} G(q_2,q_1,p_1)\, f_2$$

$$+\left[\frac{\partial}{\partial p_1} G(q_1,q_1,p_1) f_2 + \frac{\partial}{\partial p_2} G(q_2,q_2,p_2) f_2\right] = 0 \qquad (34)$$

where f_3 is the three particle distribution function.

As is the case with the usual kinetic equation of plasma physics (BBGKY theory),[24] an infinite hierarchy of relations among the f_n is developing, and some approximation is needed to terminate the sequence. We write

$$f_2 = f_1(q_1,p_1,t)\, f_1(q_2,p_2,t) + g(q_1,p_1,q_2,p_2,t) \qquad (35)$$

$$f_3 = f_1(q_1,p_1,t)\, f_1(q_2,p_2,t)\, f_1(q_3,p_3,t)$$

$$+ f_1(q_1,p_1,t)\, g(q_2,p_2,q_3,p_3,t) + f_1(q_3,p_3,t)\, g(q_1,p_1,q_2,p_2,t)$$

$$+ f_1(q_2,p_2,t)\, g(q_3,p_3,q_1,p_1,t) + h(q_1 \ldots p_3,t) \qquad (36)$$

and now assume that correlation effects are small but not negligible; in particular $h \approx 0$. With this assumption (33) and (34) yield[2,3] (with the distinction N, $N-1$, $N-2$ dropped)

$$\frac{\partial f_1}{\partial t} + \dot{q}_1 \frac{\partial f_1}{\partial q_1} + N \frac{\partial f_1}{\partial p_1} \int dq_2\, dp_2\, G(q_1,q_2,p_2)\, f_1(q_2,p_2,t) =$$

$$- N \frac{\partial}{\partial p_1} \int dq_2\, dp_2\, G(q_1,q_2,p_2)\, g(q_1,p_1,q_2,p_2,t)$$

$$- \left[\frac{\partial}{\partial p_1} \left[G(q_1,q_1,p_1)\, f_1(q_1,p_1,t)\right]\right] \qquad (37)$$

and

$$\frac{\partial g}{\partial t} + \dot{q}_1 \frac{\partial g}{\partial q_1} + q_2 \frac{\partial g}{\partial q_2} + N \frac{\partial g}{\partial p_1} \int dq_3\, dp_3\, G(q_1,q_3,p_3)\, f_1(q_3,p_3,t)$$

$$+ N \frac{\partial g}{\partial p_2} \int dq_3\, dp_3\, G(q_2,q_3,p_3)\, f_1(q_3,p_3,t) =$$

$$- N \frac{\partial f_1}{\partial p_1} \int dq_3\, dp_3\, G(q_1,q_3,p_3)\, g(q_2,p_2,q_3,p_3,t)$$

$$- N \frac{\partial f_2}{\partial p_2} \int dq_3\, dp_3\, G(q_2,q_3,p_3)\, g(q_3,p_3,q_1,p_1,t)$$

$$\left.\begin{array}{l} - \dfrac{\partial}{\partial p_1} G(q_1,q_2,p_2)\, f_1(q_1,p_1,t)\, f_1(q_2,p_2,t) \\ - \dfrac{\partial}{\partial p_2} G(q_2,q_1,p_1)\, f_1(q_1,p_1,t)\, f_1(q_2,p_2,t) \end{array}\right\} \text{①}$$

$$-\left\{\frac{\partial}{\partial p_1} G(q_1,q_2,p_2)\, g(q_1,\ldots,p_2,t) + \frac{\partial}{\partial p_2} G(q_2,q_1,p_1)\, g(q_1,\ldots,p_2,t)\right\} \text{②}$$

$$-\left[\frac{\partial}{\partial p_1} G(q_1,q_1,p_1)\, g(q_1,\ldots,p_2,t) + \frac{\partial}{\partial p_2} G(q_2,q_2,p_2)\, g(q_1,\ldots,p_2,t)\right] \text{③} \tag{38}$$

The terms labeled (3) are an addition from the violation of Liouville's Theorem and are of the same order as terms (2). At this level of approximation, terms (2) are normally dropped as second order relative to (1)[21] (they are of the same order as h), and we likewise drop (3) from the analysis. The integrals in this equation are multiplied by the particle number N and in general cannot be considered negligible. With these approximations (38) is formally identical to the usual kinetic equations for two point correlations. The cooling of phase space appears only in the last term of the one-particle distribution Eq. (37). Equation (38), except for the explicit form of interaction, is identical with that of the Lenard-Balescu analysis of plasma physics[21].

B. Momentum Cooling

We take as our variables the azimuthal angle θ, and $x = (E-E_0)$, where we assume small energy changes relative to E_0. For simplicity, consider the situation of nonoverlapping Schottky bands. Then the p equation is of the form

$$\dot{x}_i = \sum_j \sum_n G_n(x_j)\, e^{in(\theta_i-\theta_j)} \tag{39}$$

(see Eq. (19), Section 3) where $G_n \equiv G(n\omega_j, x_j)$ is the system transfer function and $G_0(x) = 0$. We also assume that f_1 is independent of θ and g is a function of $\theta_1-\theta_2$ only, that is

$$g(\theta_1,\theta_2,x_1,x_2,t) = \sum_\ell g_\ell(x_1,x_2,t)\, e^{i\ell(\theta_1-\theta_2)} \tag{40}$$

With these simplifying assumptions, (37) reduces to

$$\frac{\partial f_1}{\partial t}(x,t) = -\frac{\partial}{\partial x}\left(\sum_n G_n(x)\, f_1(x,t)\right)$$

$$- N\frac{\partial}{\partial x}\int dx' \sum_n G_n(x')\, g_{-n}(x,x',t) \tag{41}$$

and (35) reduces to

$$\frac{\partial g_\ell}{\partial t} + i\,\ell(\omega_1-\omega_2)\, g_\ell(x_1,x_2,t) =$$

$$-\frac{\partial}{\partial x_1}\left[G_\ell(x_2)\, f_1(x_1,t)\, f_1(x_2,t)\right] - \frac{\partial}{\partial x_2}\left[G_{-\ell}(x_1)\, f_1(x_1,t)\, f_1(x_2,t)\right]$$

$$- N\frac{\partial f_1}{\partial x_1}\int dx_3\, G_\ell(x_3)\, g_{-\ell}(x_2,x_3) - N\frac{\partial f_1}{\partial x_2}\int dx_3\, G_{-\ell}(x_3)\, g_\ell(x_1,x_3) \tag{42}$$

The first term on the RHS of Eq. (41) describes the cooling of phase space by self interaction. The second term describes Schottky signal and feedback effects. In Eq. (42), the first two terms on the right hand side describe the direct effects of beam particles perturbing each other; the last two terms describe how existing coherence limits the correlation growth. The $(\omega_1-\omega_2)$ terms on the LHS effect mixing through frequency spread and enhance the interaction of particles neighboring in frequency.

Equation (42) may be solved exactly under the assumption that the relaxation time of the correlation g is fast on the scale of variation of f. In terms of our previous analysis, the time scale of coherent effects is rapid compared to the cooling time. We now Laplace transform with respect to time (s being the transform variable), treating f as constant and examine solutions in the dc limit; i.e., taking $s = 0$ poles of the Laplace transform of the "constant" f. On inverting the transform, we have

$$g_\ell(x_1,x_2) = \frac{1}{s + i\ell(\omega_1-\omega_2)} \left\{ -\frac{\partial f_1}{\partial x_1} G_\ell(x_2)\, f(x_2) - \frac{\partial f_1}{\partial x_2} G_{-\ell}(x_1)\, f(x_1) \right.$$

$$\left. - N \frac{\partial f_1}{\partial x_1} \int dx_3\, G_\ell(x_3)\, g_{-\ell}(x_2,x_3) - N \frac{\partial f}{\partial x_2} \int dx_3\, G_{-\ell}(x_3)\, g_\ell(x_1,x_3) \right\} \tag{43}$$

in the limit $s \to 0+$. Define

$$\varepsilon_{\pm|\ell|}(x_1) = 1 + \frac{N}{|\ell|} \int_{\eta \to 0_+} dx_2 \frac{\frac{\partial f}{\partial x_2} \quad G_{\mp|\ell|}(x_2)}{\eta \pm i(\omega_1-\omega_2)} \; ; \qquad \omega \equiv \omega(x) \tag{44}$$

and

$$H_\ell(x_1) = N \int dx_2\, G_{-\ell}(x_2)\, g_\ell(x,x_2) + f(x_1)\, G_{-\ell}(x_1) \tag{45}$$

Then after multiplying (43) by $G_{\mp|\ell|}(x_2)$ and integrating we have

$$\varepsilon_{\pm|\ell|}(x_1)\, H_{\pm|\ell|}(x_1) = G_{\mp|\ell|}(x_1)\, f(x_1)$$

$$- \frac{N}{|\ell|} \frac{\partial f_1}{\partial x_1} \int_{\eta \to 0_+} dx_2 \frac{H_{\mp|\ell|}(x_2)}{\eta \pm i(\omega_1-\omega_2)} G_{\mp|\ell|}(x_2) \tag{46}$$

An iterative solution for H with the second term on the RHS assumed small yields

$$H_{\pm|\ell|}(x_1) = \frac{G_{\mp|\ell|} f(x_1)}{\varepsilon_{\pm|\ell|}}$$

$$- \frac{N}{|\ell|} \frac{\partial f_1}{\partial x_1} \frac{1}{\varepsilon_{\pm|\ell|}} \int_{\eta \to 0_+} \frac{dx_2 |G_{\pm\ell}(x')|^2 f(x_2)}{[\eta \pm i(\omega_1 - \omega_2)]\ \varepsilon_{\mp|\ell|}(x_2)} \tag{47}$$

as the second approximation.

In the limit $\eta \to 0+$, the Cauchy integral yields π times the integrand plus a principal value integral; i.e.,

$$H_{\pm|\ell|}(x_1) = \frac{G_{\mp|\ell|}\ f(x_1)}{\varepsilon_{\pm|\ell|}} - \frac{N}{|\ell|} \frac{\partial f_1}{\partial x_1} \frac{|G_{\pm\ell}(x_1)|^2}{|\varepsilon_{\pm\ell}(x_1)|^2} f_1(x_1) \tag{48}$$

$$+ \text{(principal value integral)}$$

For this iterative solution there is approximate cancellation of the principal value integral between $\pm|\ell|$, when the ℓ sum of (41) is performed. For the exact solution (which requires complex plane gymnastics of Wiener–Hopf),[46] the cancellation is complete. The exact result is

$$N \int dx_2\ G_{\pm|\ell|}(x_2)\ g_{\mp|\ell|}(x_1, x_2) =$$

$$\left[\frac{G_{\pm|\ell|}(x_1)\ f_1(x_1)}{\varepsilon_{\mp|\ell|}(\omega_1)} - G_{\pm|\ell|}(x_1)\ f_1(x) \right]$$

$$- \frac{N}{|\ell|} \int_{\eta \to 0_+} d\omega' \left|\frac{dx'}{d\omega'}\right| \left|\frac{G_{\pm|\ell|}(x')}{\varepsilon_{\mp|\ell|}(\omega')}\right|^2 \frac{\partial f_1}{\partial x'} f_1(x') \frac{1}{\eta \pm i(\omega_1 - \omega')} \tag{49}$$

If this relation is inserted into (41), after cancellations between $\pm\ell$, we have

$$\frac{\partial f_1}{\partial t}(x,t) = -\sum_\ell \left\{ \frac{\partial}{\partial x} \left[\frac{G_\ell(x)\ f_1(x,t)}{\varepsilon_{-\ell}(\omega)} \right] - \frac{\partial}{\partial x} \left[\frac{N\pi}{|\ell|} \left|\frac{\partial x}{\partial \omega}\right| \left|\frac{G_\ell(x)}{\varepsilon_{-\ell}(\omega)}\right|^2 \frac{\partial f_1(x)}{\partial x} f_1(x) \right] \right\} \tag{50}$$

This result is of the form of a nonlinear Fokker-Planck equation. (See Appendix A.) The first term on the right hand side is the coherent cooling of a particle's energy error; the ε_ℓ factor describes the feedback of the coherent signal through the beam. The second term contains the diffusion effects of the beam signal, again including feedback. Note that only the value of $f/|\ell|$ (the Schottky noise density for harmonic ℓ) at energy x enters throughout. This is the full statement of Eq. (11) that particles sample the noise at harmonics of their revolution frequencies. The form of interaction (39) is directly applicable to the Palmer[31)] method of momentum cooling, where the weighting function $G_n(x)$ derives from a transverse position sensitive pickup and the electronic gain is essentially constant over a Schottky band. For the filter method, in which energy information is obtained through variation of the electronic gain with frequency, the ε_ℓ factors are modified, with the corresponding $G_n(x)$ outside the integration in (41). If amplifier noise is included there will be an additional term on the right hand side:

$$\sum_\ell \frac{\partial}{\partial x}\left[\pi\left(\frac{e\omega}{2\pi}\right)^2 \frac{P(\ell\omega)}{|\varepsilon_{-\ell}(\omega)|^2}\frac{\partial f_1}{\partial x}\right] \tag{51}$$

where P is the noise voltage power density. The amplifier noise acts like additional beam particles, but does not enter into the ε feedback factors. The condition $\varepsilon_\ell = 0$ corresponds to the onset of coherent motion and is the analog of the dispersion relations for space charge and wall impedance instabilities. In this kinetic equation approach, these additional forces can be included in Eqs. (41) and (42) and will lead to modification of the ε_ℓ factors and cooling rates.

C. Schottky Signal Suppression

The ℓth Schottky signal at frequency $\ell\omega(x)$ is proportional to

$$f(x) + \mathrm{Re}\int_{\text{resolution}} dx'\; g_\ell(x,x') \tag{52}$$

where the integral is over the resolution of the analyzing device. Since the correlation function is highly peaked near $x = x'$, this integral is well approximated by

$$\mathrm{Re}\int dx'\; g_\ell(x,x') \approx \left(\frac{1}{|\varepsilon_\ell|^2} - 1\right) f(x) \tag{53}$$

(see Eq. (49)). The resulting Schottky signal is modified to

$$\frac{f(x)}{|\varepsilon_\ell|^2} \tag{54}$$

The factor $|\varepsilon_\ell|^{-2}$ describes Schottky signal suppression and is a direct consequence of Eq. (42). This equation remains valid for space charge and wall effects (since it is independent of the cooling self-interaction term) with the appropriate impedance substituted for G. The solution of (39), or equivalently (44) will determine associated ε_ℓ which describe correlations due to the impedance.

5. FEEDBACK WITH SCHOTTKY BAND OVERLAP

From the analysis of the previous section, we have seen that the longitudinal cooling process is described by the Fokker-Planck equation (50). The single derivative term describes the self-interaction of particles which increases the phase space density. The second derivative term describes the effect of other beam particles and produces diffusion or heating. Also, note that it is the Schottky signal at harmonics of the "cooled" particle revolution frequency that enters into the equation. The ε-factors describe feedback through the beam and affect both the self-correction term and the Schottky noise term.

The form of the ε-factor is closely related to the dispersion integrals of coherent instabilities. The condition $\varepsilon_\ell = 0$ is just the usual dispersion relation for longitudinal instability from a Vlasov equation analysis. In fact, the signal suppression was first derived by Sacherer[35] from a Vlasov equation with the Schottky random fluctuation as the initial value. In the model of Section 4 no Schottky band overlap was assumed. In addition, the discrete nature of the correction was lost in dropping rapidly oscillating terms. Here we will derive an expression for the signal suppression ε including these effects. The approach will be through a Vlasov equation.

Consider a longitudinal feedback system consisting of a pick-up at θ_p and a kicker downstream at θ_k. Let both the pick-up and kicker be short (i.e., approximated by δ-functions). Let $f(\theta,x,t)$ be the distribution function for longitudinal coherent motion. This distribution satisfies the linearized Vlasov equation

$$\frac{\partial f}{\partial t} + \omega \frac{\partial f}{\partial \theta} + \frac{F(\theta,t)}{\omega}\frac{\partial}{\partial x}(\omega f_0) = 0 \tag{55}$$

where F is the longitudinal kick of the feedback system and f_0 is the unperturbed energy distribution. The kick F is typically of the form

$$F(\theta,t) = -\,2\pi\delta(\theta-\theta_k)N\int dx'\omega' G(x',t-t')\, f(x',\theta_p,t) \tag{56}$$

Note the δ-function character of the kick and that only values of f at the pick-up location enter into the signal. G models the electrical character of the amplifier. A Fourier transform with respect to θ and a Laplace transform with respect to t yields

$$i\Omega\tilde{f}_\ell - i\ell\omega\tilde{f}_\ell + \frac{e^{i\ell\theta_k}}{\omega}\frac{\partial}{\partial x}(\omega f_0)N \cdot \int dx'\omega'\tilde{G}(x',\Omega)\left(\sum_m e^{-im\theta_p} f_m(x')\right) + I_\ell(x,\Omega) = 0 \qquad (57)$$

where $I_\ell(x,\Omega)$ represents some initial perturbation. For cooling it is the random fluctuation of a single particle. We immediately have

$$f_\ell = \frac{-e^{i\ell\theta_k}N}{\omega(i\Omega - i\ell\omega)}\frac{\partial}{\partial x}(\omega f_0)\int dx'\omega'\,\tilde{G}(x',\Omega)\sum_m e^{-im\theta_p} f_m(x') + \hat{I}_\ell \qquad (58)$$

Multiplying by $e^{-i\ell\theta_p}\omega\, G(x,\Omega)$, summing over ℓ, and integrating over x yields

$$H(\theta_p,\Omega) = I(\theta_p,\Omega) - \left[\sum_\ell \frac{e^{i\ell(\theta_k-\theta_p)}N}{(i\Omega - i\ell\omega)}\frac{\partial}{\partial x}(\omega f_0)\,dx\right] H(\theta_p,\Omega) \qquad (59)$$

with

$$H(\theta_p,\Omega) = \int dx'\omega'\tilde{G}(x',\Omega)\sum_m e^{-im\theta_p} f_m(x') \qquad (60)$$

and finally

$$H(\theta_p,\Omega) = \frac{I(\theta_p,\Omega)}{\left(1 + \sum_\ell \int dx \frac{NG(x,\Omega)\frac{\partial}{\partial x}(\omega f)}{i\Omega - i\ell\omega} e^{i\ell(\theta_k-\theta_p)}\right)} \qquad (61)$$

The denominator is the generalization of the ε factor, and we write

$$\varepsilon(\Omega) = 1 + \sum_\ell \int dx\, NG(x,\Omega)\frac{\partial}{\partial x}(\omega f)\frac{e^{i\ell(\theta_k-\theta_p)}}{i\Omega - i\ell\omega} \qquad (62)$$

The correspondence to our previous analysis is $\epsilon_m(\omega_0) \rightarrow \epsilon(m\omega_0)$. Now note that there are poles in this integral when $m\omega_0 = \ell\omega$. If there is Schottky band overlap, this condition is satisfied for some $\ell \neq m$. The sum over ℓ may be carried out by standard techniques[45] and yields

$$\sum_\ell \frac{e^{i\ell(\theta_k - \theta_p)}}{i\Omega - i\ell\omega} = -\frac{2\pi}{\omega} \frac{\exp i(\theta_k - \theta_p)\frac{\Omega}{\omega}}{1 - \exp 2\pi i\left(\frac{\Omega}{\omega}\right)} \tag{63}$$

The final form of the signal suppression factor is[44,4]

$$\epsilon(\Omega) = 1 + \int dx\, NG(x,\Omega)\, \frac{2\pi}{\omega} \frac{\exp i(\theta_k - \theta_p)\frac{\Omega}{\omega}}{1 - \exp 2\pi i\frac{\Omega}{\omega}} \frac{\partial}{\partial x}(\omega f_0)$$

$$= 1 + \int dx\, NG(x,\Omega)\, \frac{\pi}{\omega}\left[1 + i\,\text{ctn}\,\pi\frac{\Omega}{\omega}\right]\left(\exp i(\theta_k - \theta_p)\frac{\Omega}{\omega}\right)\frac{\partial}{\partial x}(\omega f_0) \tag{64}$$

The Fokker-Planck equation (50) is modified by replacing the $\epsilon_\ell(\omega)$ factors by $\epsilon(\ell\omega)$. In addition, if there is band overlap, additional diffusion (second derivative) terms appear corresponding to values of x' satisfying $\ell\omega(x') = m\omega(x)$ for some ℓ and m. Physically, if there is band overlap, particles at different revolution frequencies produce signals that have the same frequency and these particles can interfere with each other's cooling.

The analysis of betatron cooling proceeds from a similar analysis with the complication of averaging over rapid betatron oscillations and slowly varying amplitudes. However, the basic notions of correlation, Schottky noise, and feedback remain (with associated ϵ-factors). The practical details will be discussed in subsequent sections.

6. BASIC SYSTEMS

A. Introduction

Pick-up electrodes (PU's), signal combiners, amplifiers, transmission lines, frequency filters, power splitters and kicker electrodes (K's) are the basic hardware components of a cooling system. We postpone discussion of most component properties to later sections and concentrate on the derivation of expressions for the gain functions $G_\ell(\omega)$, $G_{\ell\pm Q}(\omega)$ for longitudinal and transverse cooling systems, respectively, in terms of component transfer functions. For simplicity we will mostly concentrate on simple cooling systems acting on a fixed number of particles deferring discussion of the more complex stochastic accumulation process, and most of the complications arising from coupling between different phase planes to later sections. We seek expressions of the form:

$$G = F_{PU} \cdot F_A \cdot F_K \, e f_o^2$$

where F_{PU}, F_A, F_K are transfer functions of PU, Amplifier (including filters and transmission lines) and K.

B. PU-Description.

We want to know the signal voltage spectrum generated by a single particle circulating in a ring. We will show later that the most general spectrum consists of frequencies $f_{q,r_x,r_z} = (q+r_xQ_x+r_zQ_z)f_o$, where q,r_x,r_z are integers, $-\infty < q,r_x,r_z < +\infty$, and Q_x,Q_z are the horizontal and vertical betatron tunes, respectively. For the cooling process only $f_{q,0,0}$, $f_{q,\pm1,0}$ and $f_{q,0,\pm1}$ are useful, and we restrict our discussion to these. PU response is most conveniently defined as an impedance. For a short PU there exists a well defined equilibrium position x_0,z_0 (usually $z_0 \equiv 0$) at the PU location and for a particle with betatron amplitudes x_β,z_β, revolution frequency f_0 and arrival time t_{PU} (modulo T, $T=1/f_0$), the signal voltage is (to first order in x_β,z_β):

$$U_s(t) = U_{s,\parallel}(t) + U_{s,x}(t) + U_{s,z}(t) \tag{65a}$$

with

$$U_{s,\parallel}(t) = e f_o \sum_{q=-\infty}^{+\infty} Z_{PU}(x_0,q\omega)\, e^{iq\omega(t-t_{PU})} \tag{65b}$$

$$U_{s,x}(t) = e f_o \frac{x_\beta}{2} \sum_{q=-\infty}^{+\infty} \sum_{\pm Q_x} \frac{\partial}{\partial x} Z_{PU}(x_0,q\omega\pm Q_x\omega)\, e^{-iq\omega t_{PU}}\, e^{i(q\pm Q_x)\omega t} \tag{65c}$$

$$U_{s,z}(t) = e f_o \frac{z_\beta}{2} \sum_{q=-\infty}^{+\infty} \sum_{\pm Q_z} \frac{\partial}{\partial z} Z_{pu}(x_0,q\omega\pm Q_z\omega)\, e^{-iq\omega t_{PU}}\, e^{i(q\pm Q_z)\omega t} \tag{65d}$$

It is possible to build PU structures delivering signals (65b)-(65d) simultaneously but often specialization to only one of them is a more advantageous choice. In practice quite long PU's or arrays of many (hundreds) of PU's are used. Still, the long PU can be reduced to the form given above and for arrays of many PU's the same holds. For ideal signal combiners we may write:

$$U_{total} = \sum_{n=1}^{N} U_n/\sqrt{N} \tag{66a}$$

If all U_n are equal; $U_n = U$

$$U_{total} = \sqrt{N}\, U \tag{66b}$$

The signal voltage as used in Eqs. (65) is the voltage of an outgoing wave in a properly terminated wave guiding structure, typically a 50Ω coaxial line feeding the preamplifier. As defined above the PU impedance relates particle current and signal voltage. Sometimes a power relation is desired in the form $P(\Omega) = (ef_0)^2\, Z_p(\Omega)$. $Z_p(\Omega)$ is easily obtained as $Z_p = |Z_{PU}(\Omega)|^2/Z_c$, where Z_c stands for the characteristic impedance of the coaxial line (= amp input impedance).

C. Amplifiers, Filters, Transmission Lines

The amplifier chain is unambiguously characterized by its complex gain $g_A(\Omega)$ (amplitude and phase) as a function of frequency. We include the transfer functions of filters and cables (with one exception) into the net electronic gain, including effects of impedance changes etc. If several K's are used the gain is defined from preamp input to a terminal plane at the entrance of the splitting network. It can be verified by the reader that the case of several power amplifiers driving different kickers can also be reduced to this description. Clearly the installed total amplifier gain is different in general from the net electronic gain used here. The only factor in the total electronic transfer function which we do not absorb in the net overall gain is the one associated with delays:

$$T(\Omega) = \exp\left(-i\,\frac{\Omega}{c}\,L\right) \tag{67}$$

where L represents the total electrical length of the system. The total electronic transfer function then is:

$$g_A(\Omega)\,\exp\left(-i\,\frac{\Omega}{c}\,L\right), \quad \text{i.e.} \tag{68a}$$

$$U_{K,in}(\Omega) = g_A(\Omega)\,\exp\left(-i\,\frac{\Omega}{c}\,L\right)\,U_{PU,out}(\Omega) \tag{68b}$$

For the previously introduced single particle signals:

$$U_{K,\parallel}(t) = ef_o \sum_q Z_{PU}(x_o, q\omega)\; g_A(q\omega)\; \exp\, i\left(q\omega t - iq\omega t_{PU} - iq\omega\, \frac{L}{c}\right) \quad (69a)$$

and

$$U_{K,\perp}(t) = ef_o\, \frac{A}{2} \sum_q \sum_{\pm Q} Z'_{PU}(x_o, q\omega \pm Q\omega)\; g_A\left((q\pm Q)\omega\right)$$

$$\exp(-iq\omega t_{PU})\; \exp\, i\left((q\pm Q)\omega t - (q\pm Q)\omega\, \frac{L}{c}\right) \quad (69b)$$

with $A = x_\beta,\ z_\beta$; $' \equiv \frac{\partial}{\partial x}, \frac{\partial}{\partial z}$, respectively.

D. Kicker Transfer Function

We define this in terms of the voltage gain $\Delta U(\Omega)$ or transverse kick $\Delta x'(\Omega)$ experienced by a particle in a single pass through the structure driven by $U_K = U_0 e^{i\Omega t}$.

$$\Delta U(\Omega) = \int_0^L e^{i\Omega \frac{s}{v}}\, E_s(\Omega)\, ds = U_0 \int_0^L \frac{E_s(\Omega)}{U_0}\, e^{i\Omega \frac{s}{v}}\, ds = U_0\, K_\parallel(\Omega) \quad (70a)$$

and

$$\begin{Bmatrix} \Delta x' \\ \Delta z' \end{Bmatrix} = \frac{\Delta p_{x,z}}{p} = \frac{1}{vp} \int_0^L e^{i\Omega \frac{s}{v}}\, (E_{x,z} \pm vB_{z,x})\, ds$$

$$= \frac{U_0}{\beta^2 E} \int_0^L e^{i\Omega \frac{s}{v}}\, \frac{E_{x,z} \pm vB_{z,x}}{U_0}\, ds = \frac{U_0}{\beta^2 E}\, K_{x,z}(\Omega) \quad (70b)$$

where $\beta^2 E$ is in [eV], a Cartesian coordinate system $\{x,z,s\}$ is used, $\vec{E},\vec{B}$ are the fields generated by U_K and the particle trajectory is assumed parallel to the kicker axis. As in the case of PU's, refinements will be treated later, and similar remarks as in the PU case apply to long structures and/or arrays of kickers. In particular for N identical kickers:

$$K_{total} = \sum_{n=1}^{N} U_n / U_0 \cdot K\,. \quad (71a)$$

For ideal power splitters $U_n = \text{const.} = U_0/\sqrt{N}$ and:

$$K_{total} = \sqrt{N}\, K \quad (71b)$$

We will see later that for any structure $K_\parallel, K_\perp$ are not independent but at present we will take them as separate quantities. Other ways to describe kicker efficiency are possible and in use. We may, e.g., relate the voltage gain to the driving current of the kicker, this leads to a kicker impedance Z_K, $\Delta U = Z_K I_K$ with $Z = K\ Z_c$, where Z_c is the characteristic impedance of the line feeding the kicker. Furthermore a kicker shunt impeadance Z_{Sh}, may be used relating ΔU and power, $(\Delta U)^2 = Z_{Sh}\,P$, with $Z_{Sh} = |K|^2 Z_c$.

E. Overall Gain

The single particle gain is now obtained. In the longitudinal case:

$$\dot{x} = e f_0^{\,2} \sum_{q,p} Z_{PU}(x_0, q\omega)\ g_A(q\omega)\ K_\parallel(p\omega)\ e^{iq\omega\left(t-t_{PU}-\frac{L}{c}\right)}\ e^{ip\omega(t-t_K)} \tag{72}$$

obtained from the discrete (delta function like) action of the kicker. In the long time average:

$$p = -\,q, \qquad \text{yielding:}$$

$$\dot{x} = e f_0^{\,2} \sum_q Z_{PU}(x_0, q\omega)\ K_\parallel(x_0, q\omega)\ g_A(q\omega)\ \ e^{iq\omega(t_K-t_{PU})}\ e^{-iq\omega\frac{L}{c}} \tag{73}$$

or with $\tau = t_K - t_{PU}$, $\tau_0 = L/c$

$$\dot{x} = e f_0^{\,2} \sum_q Z_{PU}(x_0, q\omega)\ K_\parallel(x_0, q\omega)\ g_A(q\omega)\ e^{iq\omega(\tau-\tau_0)} = \sum_q G_q(\omega) \tag{74}$$

In view of applying this expression in the evaluation of signal suppression factors we might carefully rewrite $G_q(\omega)$ as $G(\omega, \Omega)$ in the sense of the theoretical section of this paper:

$$G_q(\omega) = G(\omega,\Omega) = e f_0^{\,2}\ Z_{PU}\big(x_0(\omega),\Omega\big)\ K_\parallel\big(x_0(\omega),\Omega\big)\ g_A(\Omega)\ e^{i\Omega\left(\frac{\theta}{\omega}-\tau_0\right)} \tag{75a}$$

or

$$G_q(\omega) = e f_0^{\,2}\ Z_{PU}\big(x_0(\omega),\ q\omega\big)\ K_\parallel\big(x_0(\omega),\ q\omega\big)\ g_A(q\omega) e^{iq\theta - iq\omega\tau_0} \tag{75b}$$

where θ is the azimuthal separation (in radians) between PU and K.

The dimension of G is Vs^{-1}, if we express particle energies in eV.

For the transverse case we evaluate the rate of change of a single particle's emittance $\varepsilon_\perp$, defined by $x = (\varepsilon_\perp \beta_x)^{1/2} \cos(\psi)$, where β_x is the transverse betatron function, ψ the betatron phase and x is the particle transverse position.

For a single kick to first order in $\Delta x'$:

$$\Delta\varepsilon_\perp = -\,2\varepsilon_\perp^{1/2}\,\beta_K^{1/2}\,\sin\psi\Delta x' \tag{76}$$

Taking $x_{PU} = \varepsilon_\perp^{1/2} \cdot \beta_{PU}^{1/2} \cos\psi$ and going through the same steps as for the longitudinal case:

$$\dot{\varepsilon}_\perp = \varepsilon_\perp \frac{(\beta_K \beta_{PU})^{1/2}}{2} \frac{e f_0^{\,2}}{\beta^2 E} \sum_{q,Q} (\mp i)\, Z'_{PU}\Big(x_0, (q\pm Q)\omega\Big)\, g_A\Big((q\pm Q)\omega\Big)$$
$$K_\perp\Big(x_0, (q\pm Q)\omega\Big) \cdot \exp\, iq\omega(\tau-\tau_0)\, \exp\Big(\mp iQ\omega\tau_0\Big) \tag{77}$$

or

$$G\Big((q\pm Q)\omega\Big) = (\mp i)\, \frac{(\beta_K\beta_{PU})^{1/2}}{2} \frac{e f_0^{\,2}}{\beta^2 E}\, Z'_{PU}\Big(x_0, (q\pm Q)\omega\Big)\, g_A\Big((q\pm Q)\omega\Big)$$
$$K_\perp\Big(x_0, (q\pm Q)\omega\Big)\, \exp\, iq\omega(\tau-\tau_0)\, \exp\,(\mp iQ\omega\tau_0) \tag{78}$$

The required betatron phase advance from PU to K, ψ_{PK} follows directly from this:

$$\text{set } \phi = \arg(Z\cdot g\cdot K), \quad \text{and} \quad \psi_{PK} = Q\omega\tau_0 .$$

Then

$$\mathrm{Re}\Big\{G\Big((q\pm Q)\omega\Big)\Big\} \propto \pm\, \sin(\phi \mp \psi_{PK}) \tag{79}$$

The optimum solutions are:

$$\psi_{PK} = \frac{2n+1}{2}\,\pi, \quad n = 0,2,4 \quad \text{for} \quad \phi = 0$$

$$\psi_{PK} = \frac{2n-1}{2}\,\pi, \quad n = 2,4, \quad \text{for} \quad \phi = \pi$$

It is not possible to compensate a substantially "wrong" phase advance ψ_{PK} by adjusting the electrical phase of the system. It is left as an exercise to the reader to show that the proper phase advance is a multiple of π if the transverse kicker is replaced by a longitudinal (uniform) kicker placed at a point in the lattice with $\alpha_p \neq 0$. Also note, by inspecting (77) and (78), that if $\psi \neq ((2n\pm1)/2)\pi$, the phase of the gain at the $\pm$ Q side-bands is different. This is clearly visible experimentally in beam transfer function measurements.

Rewriting (78) again to emphasize the distinction between ω and Ω, with $\Omega_\pm = (q\pm Q)\omega$, and assuming ψ_{PK} to be an odd multiple of $\pi/2$:

$$G(\omega,\Omega_\pm) = -\frac{(\beta_K\beta_{PU})^{1/2}}{2}\,\frac{ef_0^{\,2}}{\beta^2 E}\,Z'_{PU}\left(x_0(\omega),\Omega_\pm\right)$$

$$g_A(\Omega_\pm)\;K_\perp\left(x_0(\omega),\Omega_\pm\right)\,e^{i\Omega_\pm\left(\frac{\theta}{\omega}-\tau_0\right)} \qquad (80)$$

We proceed to apply these expressions to some simple design considerations, further simplifying the expressions in the process.

F. Transverse Cooling, Simplified Design Procedures

We consider a system with PU and K in "$\alpha_p = 0$ straight sections," where α_p is the horizontal dispersion function, to reduce maximally the energy dependence of the cooling process. (This is not always experimentally desirable but simplifies the example calculation.) Before entering into calculations we shall briefly and without proof establish the connection between the moment equation, which we will use here, and the more general framework involving the Fokker-Planck equation, previously derived for longitudinal cooling. The results follow after transforming from (q,p) space to the action-angle variables I,ψ and integrating over ψ. The relevant distribution is denoted by $F(\omega,I)$, where $\int_0^\infty F(\omega,I)dI = f(\omega)$, $\int f(\omega)d\omega = 1$. It can then be shown that for a linear gain, as introduced above, and non-overlapping Schottky bands the following equation holds:

$$\frac{\partial F}{\partial t} = -\frac{\partial}{\partial I}\left\{\sum_{q,Q}\left[\frac{G((q\pm Q)\omega)}{\varepsilon_{-q\mp Q}(\omega)}\,IF + \frac{\pi N}{|q\pm Q|}\,\frac{|G((q\pm Q)\omega)|^2}{|\varepsilon_{q\pm Q}(\omega)|^2}\int_0^\infty FI'dI'\left(I\,\frac{\partial F}{\partial I}\right)\right]\right\}$$

$$= -\frac{\partial}{\partial I}\left\{\sum_{q,Q}\left[\frac{G((q\pm Q)\omega)}{\varepsilon_{-q\mp Q}(\omega)}\,IF(\omega,I) + \frac{\pi Nf(\omega)}{|q\pm Q|}\,\langle I_\omega\rangle\,\frac{|G((q\pm Q)\omega)|^2}{|\varepsilon_{q\pm Q}(\omega)|^2}\,I\,\frac{\partial F(\omega,I)}{\partial I}\right]\right\} \qquad (81)$$

The interesting difference between (81) and the corresponding equation for momentum cooling is that a particle with ω, I' experiences heating from all other particles with the same ω and any I: therefore the appearance of $\langle I \rangle$ in (81). Simplifying (81) we can write:

$$\frac{\partial F(\omega,I)}{\partial t} = -\frac{\partial}{\partial I}\left\{ gIF(\omega,I) + D(\omega)\langle I_\omega \rangle I \frac{\partial F(\omega,I)}{\partial I} \right\} \tag{82}$$

From this the following equation for $\langle I_\omega \rangle$ is derived:

$$\frac{\partial \langle I_\omega \rangle}{\partial t} = g\langle I_\omega \rangle + D(\omega)\langle I_\omega \rangle \tag{83}$$

Obviously this equation holds also if we replace $\langle I \rangle$ by any quantity proportional to $\langle I \rangle$, such as $\varepsilon_\perp$ (rms or full beam) or $\langle A^2 \rangle$. If, in contrast to the assumptions leading to Eqs. (80) and (81), the rate of change of $\varepsilon_\perp$ for a single particle is not linear in $\varepsilon_\perp$, a more general gain function, $G(\omega,\varepsilon)$ with dimensions $[\mathrm{ms}^{-1}]$ must be introduced. A moment equation is then no longer very useful and a correspondingly modified form of Eq. (81) must remain the basis for calculation. For the present purposes however we assume $\dot{\varepsilon}_\perp \propto g\varepsilon_\perp$.

The cooling rate equation in the absence of noise is then:

$$S_\omega = \frac{1}{\langle A^2 \rangle_\omega} \frac{d\langle A^2_\omega \rangle}{dt} = \frac{1}{\varepsilon_{\perp,\omega}} \frac{d\varepsilon_{\perp,\omega}}{dt}$$

$$= \sum_{q,Q} \left\{ \frac{G\,(q\pm Q)\omega}{\varepsilon_{-q\mp Q}(\omega)} + \frac{\pi N f(\omega)}{|q\pm Q|} \frac{|G((q\pm Q)\omega)|^2}{|\varepsilon_{q\pm Q}(\omega)|^2} \right\} . \tag{84}$$

It can be shown that this can be rewritten as:

$$S_\omega = \sum_{q,Q} \frac{G((q\pm Q)\omega)}{|\varepsilon_{q\pm Q}(\omega)|^2} \tag{85}$$

for the case where G can be pulled out of the dispersion integral used in the evaluation of $\varepsilon_{q\pm Q}(\omega)$. With substantial pick-up - kicker delay this is not truly the case but for reasonable phase shifts across the distribution the error remains small.

We evaluate the cooling rate at the center frequency ω of a symmetric distribution where only the pole term enters into the evaluation of ε. The expression for $\varepsilon_{q\pm Q}(\omega)$ is obtained from that for the longitudinal case by substituting $|q\pm Q|$ for $|q|$, and $f(x)$ for $\partial f/\partial x$ respectively.

We then obtain:

$$S_\omega = \sum_{q,Q} \frac{G\big((q\pm Q)\omega\big)}{\left|1 + \frac{\pi N f(\omega)}{|q \pm Q|} G\big((q\pm Q)\omega\big)\right|^2} \tag{86}$$

For simple gain models this expression can be evaluated in closed form. Let

$$G\big((q\pm Q)\omega\big) = G, \text{ real}, \quad n_1 \leqslant |q| \leqslant n_2$$

$$G\big((q\pm Q)\omega\big) = 0, \text{ otherwise}$$

and

$$n = q \cong q \pm Q$$

Then

$$s_\omega \xi \cong -4 \int_{n_1}^{n_2} dn \frac{G}{\left|1 + \frac{\pi N f(\omega)}{n} G\right|^2} = -n_T R(G\xi) \tag{87}$$

where

$$n_T = n_2 - n_1 = q_{max} - q_{min}$$

and

$$\xi = \frac{\pi N f(\omega)}{n_T}$$

where R stands for the somewhat lengthy but elementary expression for the integral (84). R is plotted in Fig. 2 for one and two octave bandwidth systems. The cooling rate is not very sensitive to the frequency dependence of G: if the plotted curves are used, e.g., for a system with $G_k = g_0 k$, i.e. linearly rising, only a few precent error occurs, provided $g_0(n_2+n_1)/2$ is substituted for G.

So far the effect of thermal noise has been neglected. The presence of an external (independent of particle amplitude) heating term modifies the equation for the evolution of the emittance:

$$\frac{d\varepsilon_\perp}{dt} = S\varepsilon_\perp + a \tag{88a}$$

with the solution

$$\varepsilon_\perp(t) = e^{s_\omega t}\Big(\varepsilon_\perp(0) - \varepsilon_\perp(\infty)\Big) + \varepsilon_\perp(\infty), \quad \text{with} \quad \varepsilon_\perp(\infty) = \frac{a}{s_\omega} \tag{88b}$$

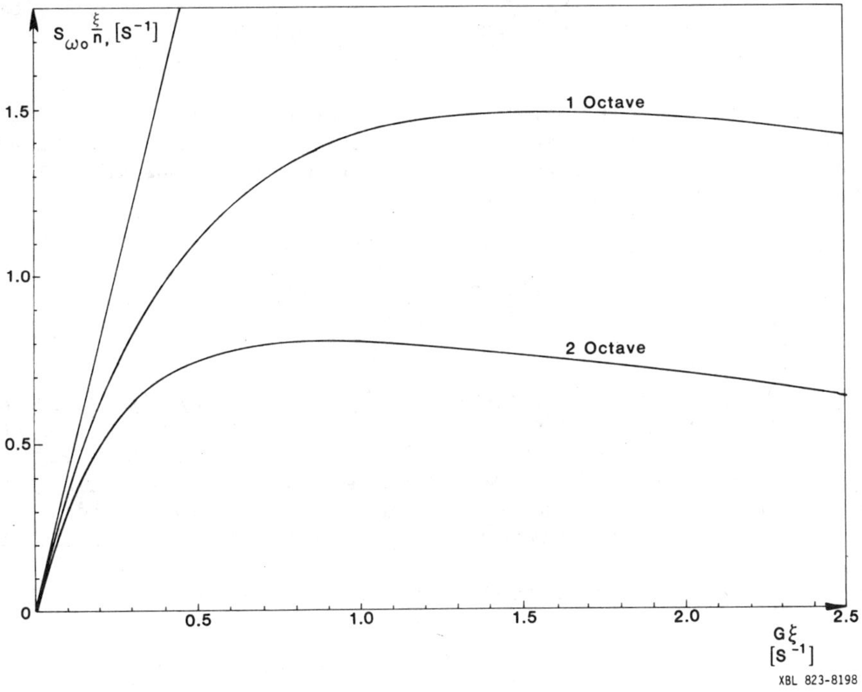

Fig. 2. Transverse cooling rate s_ω at center of symmetric distribution $f(\omega)$; $n = q_{max} - q_{min}$, $\xi = \frac{\pi N f(\omega)}{n}$, G = system gain [see Eq. (87)].

where a represents the thermal noise heating term. An expression for a, and therefore $\varepsilon_\perp(\infty)$, may be obtained by explicitly writing the Schottky heating term in Eq. (84) and substituting the thermal power density at the preamp input for the Schottky power density. The result is:

$$\varepsilon_\perp(\infty) = \frac{\beta_k f_o^2}{S_\omega} \frac{R_{in}}{\left(\beta^2 E\right)^2} \sum_{q,\pm Q} \frac{|K_\perp(q\pm Q)\omega|^2 \frac{dP}{df}\left(-(n\pm Q)f_o\right)}{|\varepsilon_{q\pm Q}(\omega)|^2} |g_A|^2 \quad (89)$$

As written, Eq. (89), gives the asymptotic emittance for the full beam, i.e. $\sigma(\infty) = 1/2\ \sqrt{\beta\varepsilon_\perp(\infty)}$. In many applications total power becomes an important issue and we need simple equations to estimate the total system power. Generally,

$$P = \int df\ |g_A|^2 \frac{dP_{in}}{df}(f) . \quad (90)$$

For an amplifier at temperature T_0 (typically $T_0 \sim 300°K$) whose input is properly back terminated with resistors at temperature T the thermal input power density is given by $dP/df = 1/2(T/T_0 + (10^{NF/10}-1))\ kT_0$, where NF is the amplifier noise figure in dB and k is Boltzmann's constant. For $T = T_0$ and NF = 3dB this reduces to the familiar approximate expression $dP/df \cong kT$.[11)]

The Schottky input power density for transverse cooling is given by

$$dP/df \cong (ef_o Z'_{pu})^2 \langle A^2\rangle (dN/df_o)/(4 R_{in}|q|)$$

$$\simeq (ef_o Z'_{PU})^2 \beta_{PU}\, \varepsilon_\perp (dN/df_o)/(16 R_{in}|q|)$$

where $\varepsilon_\perp$ is the full beam emittance.

Note that the power densities as given are appropriate for our use of both positive and negative frequencies, otherwise they must be multiplied by 2. These equations allow a quick evaluation or preliminary design of transverse cooling systems. One possible way might follow these lines:

1) Determine an appropriate frequency range. This follows from N, η, $f(\omega)$, i.e. $\Delta p/p$, and the desired cooling rate [Fig. 2]. It is desirable to achieve the design rate with less than the optimum gain, since this results quickly in substantial power savings. Fractional bandwidths will typically be limited to one octave or less at frequencies in excess of 1 GHZ. Apart from practical

considerations the upper frequency limit may be determined by the phase error occurring due to the PU-K azimuthal separation. This phase error is given by $\Delta\psi = \pm 1/2\ \alpha\theta$, where θ is the PU-K separation, and αf_0 is the frequency width of the distribution at the top harmonic.

2) Determine g_A, and therefore, $\varepsilon_\perp(\infty)$, P_N, P_{Sch} as a function of n_{PU}, n_K for a given type of PU's and K's. $\varepsilon_\perp(\infty)$ should be $\lesssim 1/2$ the desired final emittance. Both $\varepsilon_\perp(\infty)$ and P_N may be lowered by increasing n_{PU}. P_{Sch} and, again, P_N will decrease with increased n_K.

It is suggested as an exercise to consider the following case: $N = 10^8$, $\gamma = 9.5$, $m_p = .938$ GeV, $|n| = 0.005$, $f_0 = 6\cdot 10^5$ Hz, $\Delta p/p = \pm 1.5\ 10^{-3}$, $\varepsilon_\perp(0) = 20\ \mu m$. Discuss the relative merits of systems working in the 1-2, 2-4 and 4-8 GHz range respectively. Determine n_{PU}, n_K, g_A, for a 2-4 GHz system which will reduce the initial emittance by a factor 4 in 2s. Keep the total power below 500 W. Assume
NF = 3 dB and refer to section 8E for electrode properties.

G. Longitudinal (Momentum) Cooling

The relevant equations are:

$$\frac{\partial\psi(x)}{\partial t} = -\frac{\partial\phi(x)}{\partial x} \tag{91}$$

$$\phi(x) = \psi(x)\sum_\ell \frac{G_\ell(x)}{\varepsilon_{-\ell}(x)} - \psi(x)\frac{\partial\psi}{\partial x}\frac{1}{2}\frac{\beta^2 E}{|n| f_0}\sum_\ell \frac{|G_\ell(x)|^2}{|\ell|\,|\varepsilon_{-\ell}(x)|^2}$$

$$-\frac{\partial\psi}{\partial x}\sum_\ell \frac{|H_\ell(x)|^2}{|\varepsilon_\ell(x)|^2} \tag{92}$$

$|H_\ell(x)|^2$ is given by:

$$|H_\ell(x)|^2 \doteq \frac{1}{2} R_{in} f_0^2 \frac{dP_N}{df}(\ell\omega)\,|g_A(\ell\omega)|^2\,|K_\ell(x)|^2 . \tag{93}$$

Using the previously given expressions for $G_\ell(x)$ Eq. (75a,b), these equations completely determine the evolution of the beam density function under the action of a given cooling system. In contrast to transverse cooling these equations are less amenable to reasonably accurate analytical approximations, and in general, numerical evaluation is needed to determine $\psi(x,t)$. Nevertheless a number of approximate estimates are often possible.

i) A lower limit for the required gain may be obtained by keeping only the coherent term and equating it to the required rate of change of energy, i.e.

$$\left|\sum_{\ell} G_{\ell}(x)\right| \geqslant \left|\frac{dx}{dt}\right| \simeq \frac{x_{max}}{\tau_{cool}}$$

ii) an average instantaneous cooling rate for particles with $x_1 \leqslant x \leqslant x_2$ may be estimated:

$$\frac{\frac{\partial}{\partial t}\int_{x_1}^{x_2} \psi dx}{\int_{x_1}^{x_2} \psi dx} = \frac{[\phi(x_1) - \phi(x_2)]}{\int_{x_1}^{x_2} \psi dx} \tag{94}$$

whose evaluation of course requires that we postulate a certain $\psi(x)$ to exist at time t.

iii) Assuming a certain initial $\psi(x,t)$ we can, neglecting signal suppression and using a simple gain model, also derive an equation for the rate of change of the second moment $\langle x^2 \rangle$.

While all these approaches have a certain usefulness for initial estimates they must remain "crutches", the real answer will come from numerical evaluation tracking the evolution of $\psi(x,t)$.

Two essentially different methods to shape the gain function have been developed, the Thorndahl or filter method,[9] and the Palmer[31] method in various modifications. The filter method makes use of a Σ-PU and periodic frequency filters. Some more detailed comments on filters will be made in a later section. At present it suffices to state that electronic networks may be realized with transfer functions of the form:

$$T(\Omega) = 1 - Ae^{-i\Omega \frac{L}{c}} = 1 - Ae^{-i\Omega\tau_0} = 1 - Ae^{-i\Omega/f_0} \tag{95}$$

The quantities A and c are not in reality completely frequency independent over a wide band, and one of the outstanding problems of filter design is compensation for that dependence. Writing $\Omega = \ell\omega = 2\pi\ell f$

$$T(\ell\omega) = 1 - Ae^{-2\pi i \ell f/f_0} = 1 - A\cos 2\pi\ell f/f_0 + i\, A\sin 2\pi\ell f/f_0 \tag{96}$$

and

$$|T(\ell\omega)| = \left[1 + A^2 - 2A\cos 2\pi\ell f/f_0\right]^{1/2},$$

$$\tag{97}$$

$$\arg\left\{T(\ell\omega)\right\} = \mathrm{arctg}\,\frac{A\sin 2\pi\ell f/f_0}{1 - A\cos 2\pi\ell f/f_0}$$

This results in a transfer function periodic in f, with minimum transmission and a rapid phase jump ($-90°$ to $+90°$ for $A = 1$) whenever $\ell f = f_0$. By introducing an overall $\pi/2$ phase off-set in the electronics we obtain

$$\mathrm{Re}\left\{G_\ell\right\} \propto \sin\left(2\pi\ell f/f_0\right) = \sin\left(2\pi\ell\,\frac{x\eta}{\beta^2 E}\right) \tag{98}$$

In the Palmer method,[31)] by contrast, a position sensitive PU is placed at a location with $\alpha_p \neq 0$, resulting in a signal voltage, and therefore overall gain, proportional to $Z_{PU}(\alpha_p x/\beta^2 E)$. In the simplest case a difference PU is used and $Z_{pu}(\alpha_p x/\beta^2 E) \cong (\alpha_p x/\beta^2 E)Z'_{pu}$. Such a pick-up, of course, also detects horizontal betatron motion, a fact which, depending on band overlap, location and nature of kicker, can be used to achieve simultaneous betatron cooling, or may become a nuisance in that betatron signals from particles at one energy longitudinally heat particles at another energy. Both methods may be combined of course (although the role of filters will then be different from the one just outlined). Finally it may be pointed out that one might of course combine a sum PU with a K in a dispersed location and a transfer function $K_\parallel(x) = -K_\parallel(-x)$. Such a kicker will, however, always introduce transverse kicks and will not be pursued further here, since some of the associated problems in computing the appropriate signal suppression factors have not yet been solved to our satisfaction.

Apart from the differences in apparatus there are some more fundamental differences between the two methods, one having to do with thermal noise, the other with signal suppression. The filter method is more effective in reducing the effects of thermal noise, since the overall electronic gain, and therefore, the rms noise voltage seen by a particle is reduced proportional to $[1 + A^2 - 2A \cos 2\pi\ell f/f_0]$ tending towards $1 + A^2 - 2A$ as f approaches f_0, i.e., as the particles converge towards the desired equilibrium energy $x = 0$.

With regard to signal suppression we note that the following expressions apply:

a) Filter Method:

$$\varepsilon_{\mp|\ell|}(x) = 1 + \frac{\pi N}{|\ell|}\frac{\beta^2 E}{|n|\omega} G_{\mp|\ell|}(\omega) \int_{\sigma\to 0_+} \frac{dx' \frac{\partial\psi}{\partial x'}}{\sigma \pm i(x-x')} \qquad (99a)$$

b) Palmer Method:

$$\varepsilon_{\mp|\ell|}(x) = 1 + \frac{\pi N}{|\ell|}\frac{\beta^2 E}{|n|\omega} \int_{\sigma\to 0_+} \frac{dx' \frac{\partial\psi}{\partial x'}}{\sigma \pm i(x-x')} G_{\mp|\ell|}(x') \qquad (99b)$$

For illustration we use a simple, well behaved function, such as

$$\psi(x) = \frac{15}{16}\frac{1}{x_0}\left(1 - (x/x_0)^2\right)^2, \qquad |x| \leqslant x_0 \qquad (100)$$

$$\psi(x) = 0, \qquad |x| > x_0$$

The evaluation will be made at a single harmonic with the following gain functions:

a) Filter: $$G = -g\sqrt{2}\left(\sin\frac{\pi}{4}\frac{x}{x_0} + i\left(\frac{1}{A} - \cos\frac{\pi}{4}\frac{x}{x_0}\right)\right)x_0 \qquad (101b)$$

with $A = 0.9, \quad g > 0, \quad \text{real}, \quad [g] = s^{-1}$

b) Palmer: $$G = -gx \qquad (101b)$$

This choice corresponds to a distribution filling one fourth of a Schottky band at the selected harmonic ℓ, a filter with ~ -25 dB notch depth and the same single particle cooling rate at the edges of the distribution for the two cases.

With the distribution function (100) the PV-parts can be evaluated in closed form:

a) $$PV\int \propto \left\{\frac{2}{3} - \left(1 - \left(\frac{x}{x_0}\right)^2\right)\left(2 + \left(\frac{x}{x_0}\right)\right) \ln\left|\frac{1-x/x_0}{1+x/x_0}\right|\right\} \qquad (102a)$$

$$\text{b)} \quad PV\int \propto \left(\frac{x}{x_0}\right) \left\{\frac{1}{3} - \left(1 - \left(\frac{x}{x_0}\right)^2\right)\left(2 + \left(\frac{x}{x_0}\right)\right) \ell n \left|\frac{1-x/x_0}{1+x/x_0}\right|\right\} \quad (102b)$$

The ε factor can then be rewritten as

$$\varepsilon(x) = 1 + \frac{Ng}{\Delta f}(\alpha + i\beta) = 1 + \frac{4Ng}{f_0}(\alpha + i\beta) \quad (103)$$

where Δf is the full frequency spread of the distribution at the ℓth harmonic and where we recall that in the example we assumed one fourth of the Schottky band to be occupied. Figure 3 shows the quantity $(\alpha + i\beta)$ for both cases. It can be seen from the figure that not only signal suppression but also enhancement is possible, and that with sufficiently high gain, instability will occur. This is an immediate consequence of the substantial imaginary part of the gain with the filter method, which is particularly pronounced near the zero of $\mathrm{Re}\{G\}$. In practice phase shifts will occur with any method and it should be remembered that we neglected to include the unavoidable phase error across the distribution due to different PU-K transit times of particles with different momenta.

H. Stochastic Stacking

The capability to increase the phase-space density of a circulating beam, or to maintain it in the presence of intrabeam and gas scattering, weak resonances or beam-beam effects, by means of stochastic cooling[43)] is of great interest but the most important application to date is the stochastic accumulation of antiprotons. In machines such as the CERN-AA ring or the proposed FNAL accumulator a density increase of $\gtrsim 10^4$ in momentum space and 10 to 100 in each transverse phase plane are required.

We will only consider longitudinal cooling in this section. This presents the most demanding design task and requires the bulk of the cooling hardware. The process is schematically shown in Fig. 4 and is analyzed using the Fokker-Planck equation. A detailed study requires numerical solution, integrating with pulse by pulse injection. This proves to be time consuming, largely due to the need for frequent evaluation of the dispersion integrals needed for the ε-factor. Fortunately it is observed that over a substantial portion of the stack a nearly static situation with time independent ϕ, $\partial\psi/\partial x$ and ψ is established. This allows solution of the equation with boundary conditions on ψ and $\partial\psi/\partial x$, without a precise calculation of pulse injection. At various times through the accumulation of a stack a few single pulses might be calculated to verify that they are properly accumulated.

To gain a semi-quantitative insight we follow van der Meer,[41)] observing that over most of the stack width we want an increase in ψ while maintaining a constant (energy and time independent) flux ϕ_0. (However, near injection and near the core this will not be

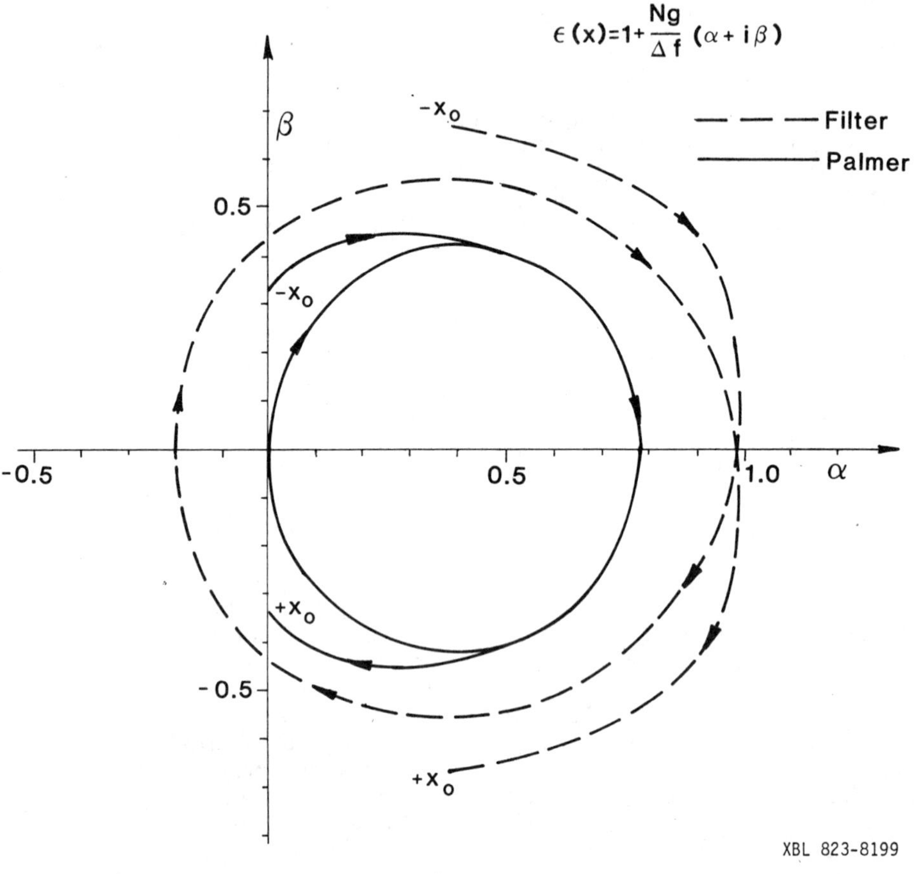

Fig. 3. Example for signal suppression factor ε(x): see text, Eqs. (99) through (103) for system parameters.

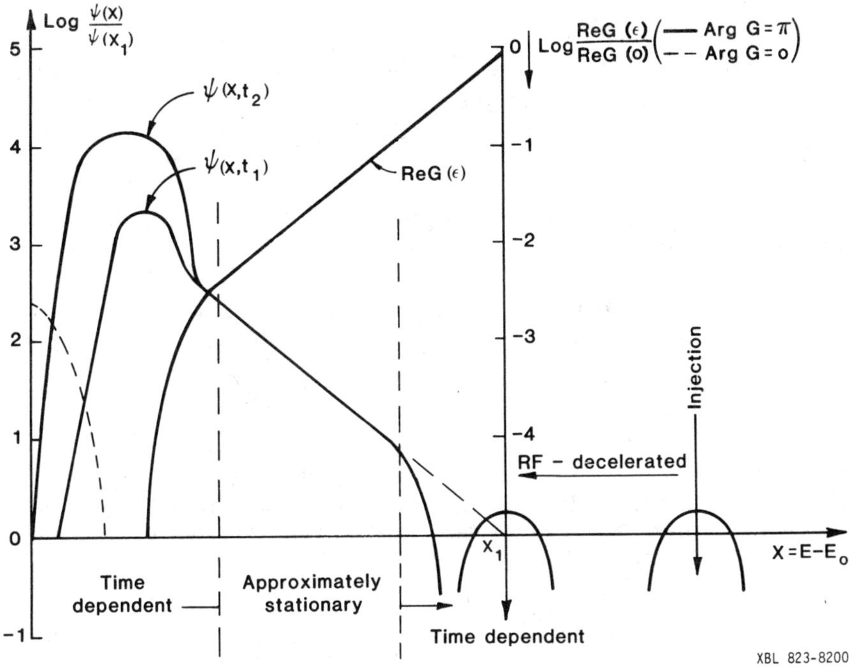

Fig. 4. Schematic description of stochastic stacking process: A pulse is injected on an injection orbit, then decelerated (or accelerated) to x_1 by an RF-system. From x_1 further acceleration and compression of longitudinal phase space density is achieved by a stochastic cooling system. Shown is the density function $\psi(x,t)$ at two different times t_1, t_2 and the system gain.

correct.) Neglecting signal suppression and thermal noise we obtain

$$\phi_0 = \psi(x) \sum_{\ell} G_{\ell}(x) - \frac{1}{2} \frac{\beta^2 E}{|n| f_0} \psi \frac{\partial \psi}{\partial x} \sum_{\ell} \frac{|G_{\ell}|^2}{|\ell|} \tag{104}$$

or

$$\frac{\partial \psi}{\partial x} = \frac{\psi(x) \sum_{\ell} G_{\ell}(x) - \phi_0}{\frac{1}{2} \frac{\beta^2 E}{|n| f_0} \psi \sum_{\ell} \frac{|G_{\ell}|^2}{|\ell|}} \tag{105}$$

Assuming $G_{\ell} = G_0$, a constant independent of ℓ over the working band, we obtain:

$$\frac{\partial \psi}{\partial x} = \frac{\psi(x)\ 2(\ell_{max} - \ell_{min})\ \rho(x)\ |G_0| - \phi_0}{\frac{\beta^2 E}{|n| f_0} \psi |G_0|^2\ \ell n(\ell_{max}/\ell_{min})} \tag{106}$$

where the factor $\rho(x)$ takes into account that $G_{\ell}(x)$ may not be purely real. Equation (106) cannot be solved analytically for arbitrary $G_0(x)$ but we can find a special choice of $G_0(x)$, yielding an "implicit" solution. Let

$$\rho(x)\ G_0(x) = \frac{\alpha(x)\ \phi_0}{\psi(x)\ 2(1-r)\ \ell_{max}} \tag{107}$$

where $r = \ell_{min}/\ell_{max}$. Then

$$\frac{\partial \psi}{\partial x} = \frac{\alpha(x) - 1}{\alpha^2(x)} \rho^2(x) \frac{4(1-r)^2}{\ell n \left(\frac{1}{r}\right)} \frac{\ell_{max}^2 |n| f_0}{\beta^2 E\ \phi_0} \psi = \lambda(x) \psi \tag{108}$$

with the obvious solution

$$\psi(x_2) = \psi(x_1) \exp \int_{x_1}^{x_2} \lambda(x')\, dx' \tag{109}$$

The steepest density increase is obtained with $\rho(x) = 1$ (real gain), $\alpha(x) = 2$ yielding

$$\lambda = E_D^{-1} = \frac{(1-r)^2}{\ell n\left(\frac{1}{r}\right)} \frac{\ell_{max}^2 |\eta| f_0}{\beta^2 E \, \phi_0} \tag{110}$$

There will be some restrictions imposed on ℓ_{max}, η, and the stack width x_s, either from the use of frequency filters or the need to control PU-K mixing. It can be put in the form

$$\frac{\ell_{max} |\eta| f_0}{\beta^2 E} = \frac{k f_0}{x_s}$$

where k is some number dependent on the details of the system, typically $k < 1$. This leads to:

$$\frac{x_s}{E_D} = \frac{(1-r)^2}{\ell n\left(\frac{1}{r}\right)} k \frac{f_{max}}{\phi_0} = \frac{(1-r)}{\ell n\left(\frac{1}{r}\right)} k \frac{W}{\phi_0} \tag{111}$$

showing that the number of e-foldings depends on the ratio of system bandwidth to particle flux and the well known fact that a given bandwidth is more effective if centered at higher frequency (higher r). In practice several systems are used, with the highest frequency operating in the core where also a sign reversal not included above is introduced (this leads to a nonstationary configuration and some additional "peaking" at the core).

For gain shaping pick-up electrodes in straight sections with high dispersion α_p and filters are used. In principle kickers in sections with high dispersion could be used too, possibly eliminating the need for filters but their use, complicated by the transverse kicks they impart, has not been fully analyzed. The role of the filters is predominantly to protect the high density core from the thermal noise generated by the high power tail system. Stability must be carefully considered in splitting the gain roll-off between the electrodes and filters. High demands are also put on the lattice design: η is typically rather strictly defined, and long straight sections (~15 m in the FNAL design) with minimized β-functions in both planes and $\alpha_p \simeq 10$ m are desired.

7. ELECTRONIC COMPONENTS

The essential components are signal combiners, transmission lines, amplifiers, power splitters and filters. All of these except the filters are usually commercially available.

Signal combiners may be of the "hybrid" type, i.e. 4 port junctions with the property that any port connects to only two of the other three and is isolated from the remaining one. They are used to create sum and difference signals from two incoming signals. They have the property that they always provide an impedance matched input independent of the relative magnitude and phase of the incoming signals. The simple addition of equal signals, or arbitrary signals in a situation where reflections are not critical, may be achieved by leading N incoming lines of impedance Z together into a line of impedance Z/N. Tapers or multistep transformers are used to make the impedance of input and output lines equal, i.e. typically 50 Ω. The number of input lines need not be a binary number. All these devices come very close to "ideal performance;" they provide an output voltage $U_{out} = 1/\sqrt{N}\ \Sigma\ U_{n,in}$. Power splitters are essentially the same networks, the main practical difference is the particular power level.

Preamplifiers, and power amplifiers below 1 GHz, are usually solid state, with noise figures for preamps between 2.5 and 3 dB in the 1 to 2 and 2 to 4 GHz range, and lower ($\lesssim$ 2 dB) in the few hundred MHz range. Travelling wave tube amplifiers (TWTA) will be required in the GHz range for power levels exceeding a fraction of 1 W. Flat phase and gain characteristics are required, in particularly substantial phase deviations should not occur in the high gain region. This criterion is quite well satisfied by TWTA's, while many solid state power amplifiers often show a very sharp gain cut-off with substantial phase errors (~ 2π) within the working band.

Periodic filters are usually "custom made". Most of the pioneering work in that area has been performed by L. Thorndahl et al. at CERN.[9] We will not discuss all the intricacies involved here but merely illustrate the principles. Three basic configurations are common, as depicted in Figs. 5. Consider first the configuration shown in Fig. 5a. Applying standard "S-matrix" formalism[11] to the junction shown yields:

$$S_{21} = \frac{2}{2 + \bar{y}_L} \tag{112}$$

where $\bar{y}_L$ is the normalized admittance of the shorted, lossy line.

$$\bar{y}_L = \bar{y}\ \mathrm{ctgh}(i\beta L + \alpha L), \tag{113}$$

where β, α are the (frequency dependent in real cables) propagation and loss constants of the line. Therefore

$$S_{21} = \frac{2(1 - e^{-2(i\beta L + \alpha L)})}{2 + \bar{y} - (2 - \bar{y})\ e^{-2(i\beta L + \alpha L)}} \tag{114}$$

or if we use, for simplicity, $\bar{y} = 2$

$$S_{21} = \frac{1}{2}\left(1 - e^{-2\alpha L} \cdot e^{-2i\beta L}\right) \tag{115}$$

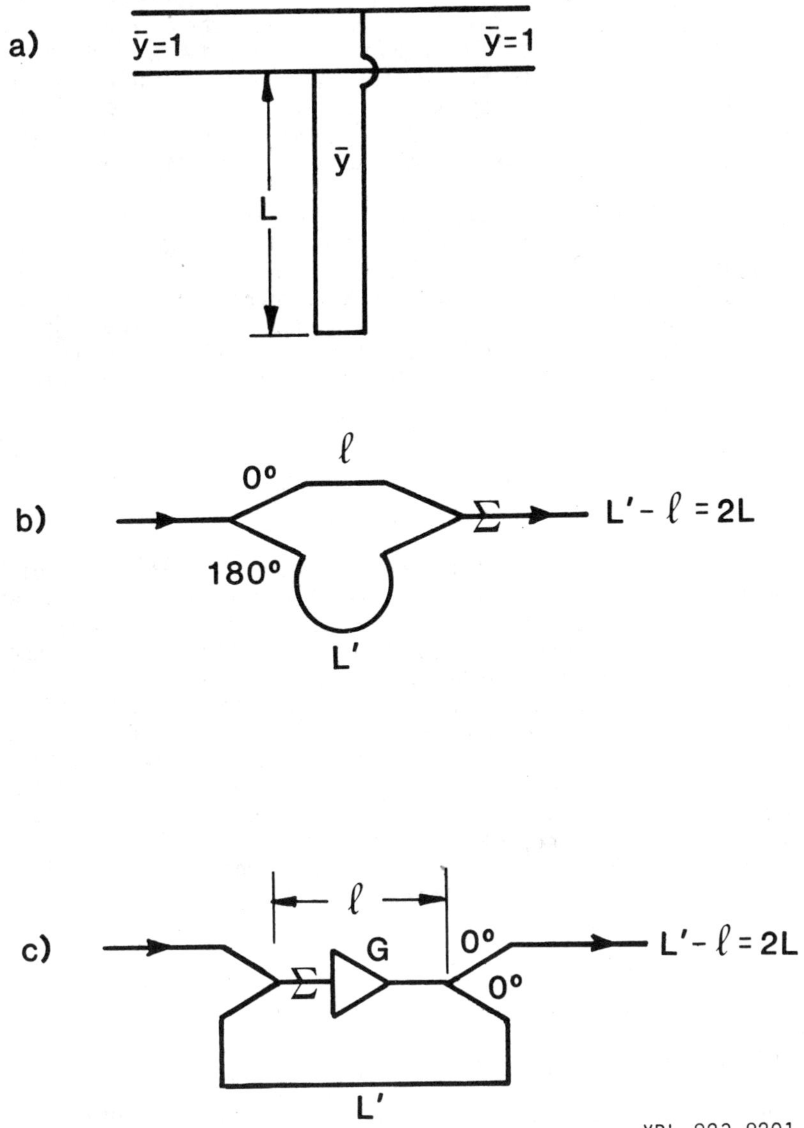

XBL 823-8201

Fig. 5. Schematic filter configurations:

5a: shorted line, minimum transmission at $f_n = nc/2L$

5b: splitter-combiner network, minimum transmission at $f_n = nc/2L$

5c: "active" filter, maximum transmission at $f_n = nc/2L$ (ℓ,L,L' are "electrical lengths" at f_n)

which is of the same form previously used in the context of momentum cooling. Inspection of Eq. (114) shows that the ratio of maximum to minimum transmission for a given attenuation α increases as $\bar{y} \to \infty$. This is not possible in practice, also minimum α for a given outer diameter is reached at $Z \cong 77\,\Omega$. In practice, at frequencies of a few 100 MHz, series resistors are inserted in the input and output lines, which has essentially the same effect as increasing $\bar{y}$ in Eq. (114).

A simple analysis of the next configuration yields exactly the same result for the transfer function. Using an open line, or equivalently an adding instead of a subtracting network in the second case yields filters with the same characteristics but translated by $f_0/2$. These filters produce, within the limits of their cable losses, sharp minima. The third illustrated configuration generates a sharp maximum. Analyzing the network we obtain for the transmission:

$$T(\omega) = \frac{G}{1 - Ge^{-(i\beta+\alpha)L}} \; . \tag{116}$$

Care must be exercised to avoid instability in the choice of the gain G. Typical phase/amplitude characteristics are shown in Fig. 6. Common to all is a falling phase characteristic at maximum transmission which may be a problem if the filter is to be used at maximum transmission such as in stochastic stacking.

The most fundamental problems in filter construction are posed by cable losses and velocity dispersion. High uniformity, large diameters (controlled by the cut-off frequencies of modes other than the TEM-mode), and possibly cryogenic temperatures are required. In addition compensating networks may be used. In these schemes typically a signal proportional to the cable loss, generated by taking the difference signal between a high and a low loss line, is subtracted from the signal transmitted through the filter after suitable attenuation.

8. PICK-UP AND KICKER ELECTRODES, GENERAL SINGLE PARTICLE SCHOTTKY SPECTRUM

A. Introduction and Overview

We will discuss pick-up and kicker electrodes in greater depth and more detail than all other system components. Their performance affects important quantities such as signal to noise ratio, total amplifier power requirements, and amplitude and phase characteristics of the overall system gain. Unlike most components, they are not commercially available but must be tailored to the specific application needs. Furthermore, their requirements for space and specific locations in a cooling ring lattice interact strongly with the overall machine design. Reliable calculations and measurements often pose a nontrivial task.

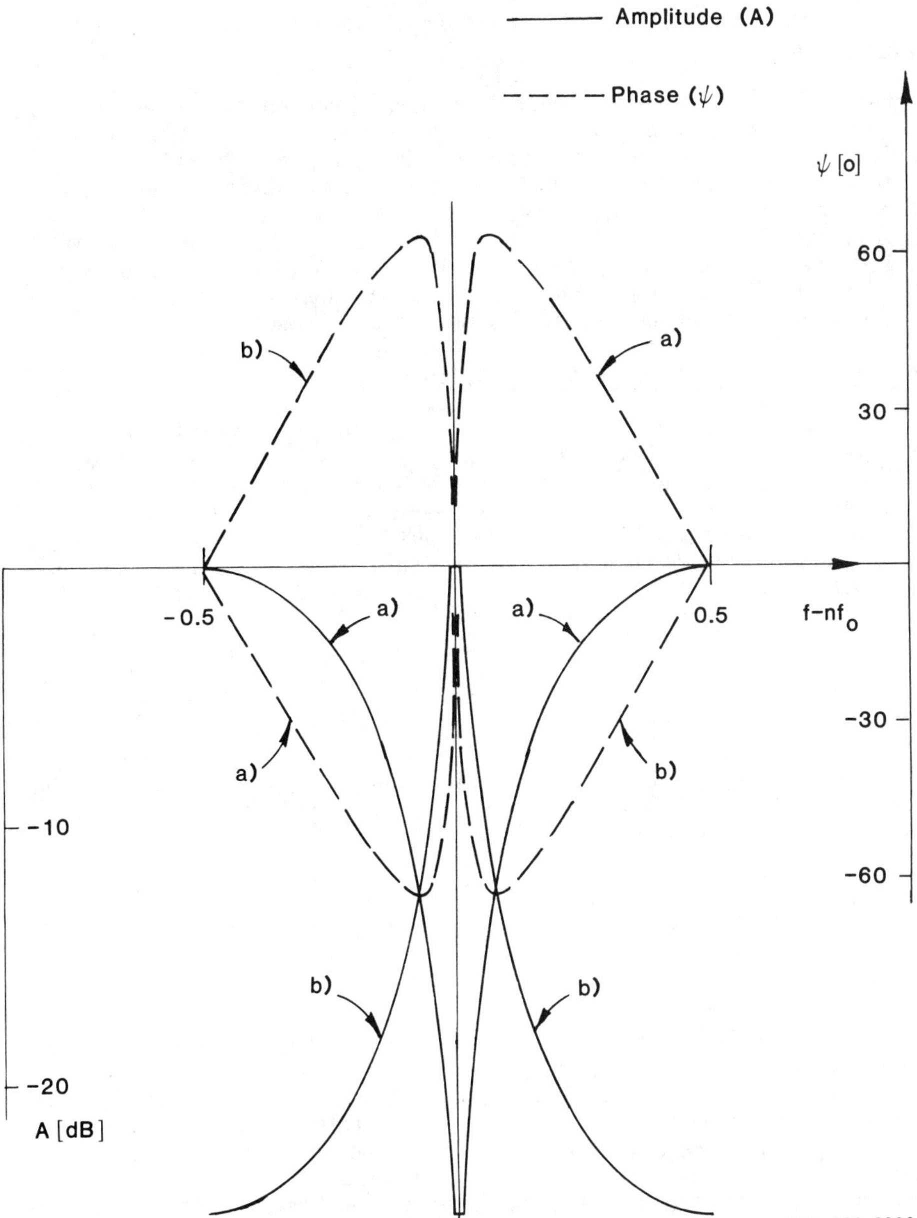

XBL 823-8203

Fig. 6. Filter transfer functions, shown for $(n-1/2)f_0 \leqslant f \leqslant (n+1/2)f_0$; amplitude is normalized to maximum transmission.

a) type (a) with $\bar{y}_L = 2$ or type (b)
b) type (c)

In all cases $e^{-2\alpha L} \cong 0.9$.

In this chapter we will proceed along the following lines of development:

First, we will describe general expressions for the PU-response. We will base this discussion on the Lorentz reciprocity theorem which, in our view, provides some conceptual simplifications and is related closely to the second objective, the discussion of kicker transfer functions. These discussions will provide us with general expressions for the pick-up and kicker transfer functions. We will use these then to obtain the signal spectrum of a single particle circulating in a ring. Finally we apply the results obtained to estimate the characteristics of some example structures.

B. PU-Characteristics, Reciprocity Theorem

Figure 7 schematically shows a pick-up (or kicker) structure: a discontinuity of some sort in the beam chamber connected to the "outside world" by one or more wave guiding structures, in practice transmission lines. Viewed as pick-up, we want to calculate the complex voltage amplitudes in the output port(s) generated by a particle travelling through the structure. The relevant equations are, of course

$$\nabla^2 \vec{A} - \frac{1}{c^2}\frac{\partial^2 \vec{A}}{\partial t^2} = -\mu_0 \vec{j} \quad \text{and} \quad \Delta^2 \phi - \frac{1}{c^2}\frac{\partial^2 \phi}{\partial t^2} = -\rho/\varepsilon_0 \qquad (117)$$

where $\vec{j},\rho$ are the current and charge density corresponding to the particle. These, to be exact, must be complemented by the relativistic equations of motion, i.e.:

$$\frac{d\vec{p}}{dt} = e(\vec{E} + \vec{v} \times \vec{B}) \quad \text{and} \quad \frac{dE}{dt} = e(\vec{v} \cdot \vec{E}) \qquad (118)$$

where $\vec{E},\vec{B}$ are the fields corresponding to $\vec{A},\phi$ of Eq. (117). For PU response calculations we may, in contrast to stability calculations or electron beam tube evaluations, e.g., neglect (118), i.e, we need not take into account the effects on the particles trajectories by the fields generated by the particle. (This is borne out by the numerical results and, for stochastic cooling, is true in the case of kicker electrodes also.) Instead of solving Eq. (117) directly we use the Lorentz reciprocity theorem, which of course must lead to the same result since it is based solely on Maxwell's equations, but seems to provide some practical simplification and provides us with a unified view of PU and kicker response. To derive the theorem[11]) we make use of Maxwell's curl equations for fields depending on time as $e^{i\Omega t}$, i.e.

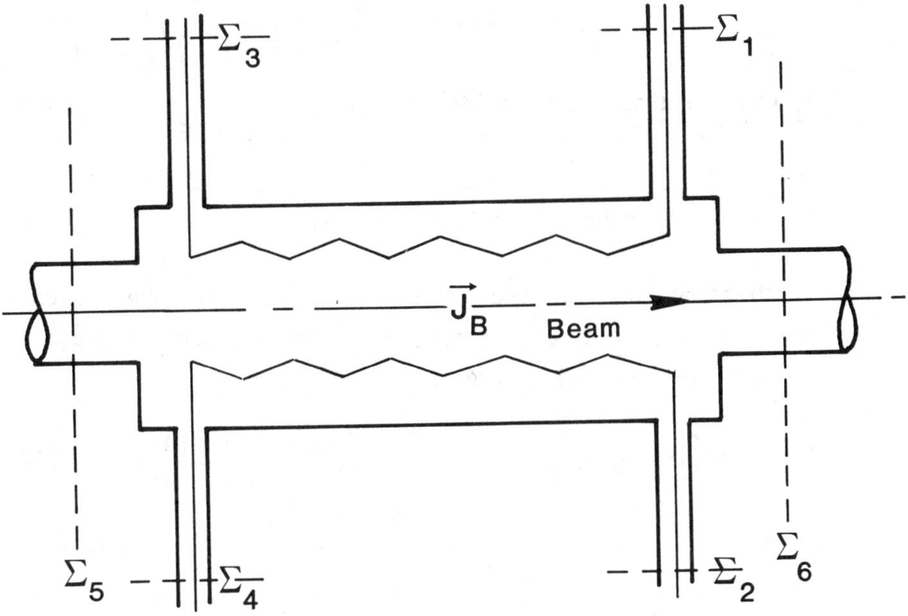

XBL 823-8202

Fig. 7. Schematic representation of a PU or K; figure relevant to discussion of reciprocity theorem.

$$\mathrm{curl}\ \vec{E} = -\,i\Omega\vec{B} \quad \text{and} \quad \mathrm{curl}\ \vec{B} = \mu_0\left(\vec{J} + \mathrm{curl}\ \vec{M} + i\Omega\vec{D}\right) \tag{119a}$$

and the vector identity:

$$\mathrm{div}\left(\vec{E}_1 \times \vec{B}_2 - \vec{E}_2 \times \vec{B}_1\right) = \left(\mathrm{curl}\ \vec{E}_1\right)\cdot\vec{B}_2 - \left(\mathrm{curl}\ \vec{B}_2\right)\cdot\vec{E}_1 - \left(\mathrm{curl}\ \vec{E}_2\right)\vec{B}_1 + \left(\mathrm{curl}\ \vec{B}_1\right)\vec{E}_2 \tag{119b}$$

For isotropic and constant (independent of $\vec{B},\vec{E}$) μ and ε we combine (119a) and (119b) to obtain with $\vec{B} = \mu\vec{H}$, $\vec{D} = \varepsilon\vec{E}$:

$$\mathrm{div}\left(\vec{E}_1 \times \vec{H}_2 - \vec{E}_2 \times \vec{H}_1\right) = \vec{E}_2 \cdot \vec{J}_1 - \vec{E}_1 \cdot \vec{J}_2 \tag{120}$$

Note furthermore that if the current densities $\vec{J}_k$ are expressed $\vec{J}_k = \vec{J}_k^{(0)} + \sigma\vec{E}_k$, where $\sigma\vec{E}_k$ represent conduction currents in a resistive conductor Eq. (120) becomes:

$$\mathrm{div}\left(\vec{E}_1 \times \vec{H}_2 - \vec{E}_2 \times \vec{H}_1\right) = \vec{E}_2 \cdot \vec{J}_1^{(0)} - \vec{E}_1 \cdot \vec{J}_2^{(0)} \tag{121}$$

Integrating over a volume V bounded by a surface Σ we obtain:

$$\oint_\Sigma d\vec{\Sigma}\left(\vec{E}_1 \times \vec{H}_2 - \vec{E}_2 \times \vec{H}_1\right) = \int_V dV\left(\vec{E}_2 \cdot \vec{J}_1^{(0)} - \vec{E}_1 \cdot \vec{J}_2^{(0)}\right) \tag{122}$$

where the surface integral vanishes if:

a) Σ is at infinity, or more generally includes all the sources $\vec{J}_1,\vec{J}_2$

b) Σ is a perfectly conducting surface

c) Σ is characterized by a surface impedance Z_m, $\vec{E}_{tang} = -\,Z_m\,\vec{n} \times \vec{H}$

For our application we make the following choices: $\{\vec{E}_1,\vec{H}_1,\vec{J}_1\} \equiv \{\vec{E}_B,\vec{H}_B,\vec{J}_B\}$ are the electromagnetic fields at frequency Ω generated by the particle represented by the current density $\vec{J}_B$. At the output ports this field must correspond to an outgoing

TEM wave if we select the location of the terminal planes Σ_1, Σ_2, Σ_{N-2} properly and operate at frequencies Ω below the cut-off for other modes, therefore:

$$\vec{E}_B(\Sigma_k) = V_B \text{ grad } \phi \quad \text{and} \quad \vec{H}_B(\Sigma_k) = \frac{V_B}{Z} (\vec{n} \times \text{grad } \phi) \tag{123}$$

where $Z = (1/\varepsilon')^{1/2} Z_0$, $Z_0 = 377\,\Omega$ and ϕ is the two-dimensional potential function appropriate for the particular transmission line geometry.

For the fields $\{\vec{E}_2,\vec{H}_2\} = \vec{E}_T,\vec{H}_T$ we select a TEM wave with an incoming voltage amplitude V_T in the kth guide. This creates the fields $\vec{E}_T,\vec{H}_T$ in the interior of V, its sources are obviously outside the volume V, $\vec{J}_T = 0$ therefore and at the terminal planes we have:

$$\vec{E}_T(\Sigma_k) = V_T(1 + S_{kk}) \text{ grad } \phi$$

$$\vec{H}_T(\Sigma_k) = -\frac{V_T}{Z} (1 - S_{kk}) (\vec{n} \times \text{grad } \phi) \tag{124a}$$

$$\vec{E}_T(\Sigma_n) = V_T S_{nk} \text{ grad } \phi \quad n \neq k, N-1, N-2$$

$$\vec{H}_T(\Sigma_n) = \frac{V_T}{Z} S_{nk} (\vec{n} \times \text{grad } \phi) \tag{124b}$$

where S_{nk} are the elements of the S-matrix of the considered N-port junction. To complete the description of our configuration we have to define $\vec{E}_T,\vec{H}_T$ at Σ_{N-1} and Σ_N. We require $\{\vec{E}_T,\vec{H}_T\} \cong \{\vec{0},\vec{0}\}$ at these locations. This implies that the frequency is below cut-off of the beam pipe or that we insert attenuating materials and structures for any propagating modes. Performing the surface integral we then obtain immediately from b) or c) above that only Σ_ℓ, $\ell = 1,2,...N$ can contribute. The contributions for Σ_{N-1}, Σ_N vanish with the requirements just made and on further inspection, it is seen that a non zero result is obtained only at those ports where both incoming and outgoing TEM waves are present, i.e. only at S_k. Therefore

$$\oint_{\Sigma} \left(\vec{E}_B \times \vec{H}_T - \vec{E}_T \times \vec{H}_B\right) d\vec{\Sigma} = \int_{\Sigma_k} \left(\vec{E}_B \times \vec{H}_T - \vec{E}_T \times \vec{H}_B\right) d\vec{\Sigma}$$

$$= - 2 \frac{V_T V_B}{Z} \int_{\Sigma_k} \text{grad } \phi \times (\vec{n} \times \text{grad } \phi) \, d\vec{\Sigma}$$

$$= - 2 \frac{V_T V_B}{Z} \int_{\Sigma_k} |\text{grad } \phi|^2 \, d\Sigma \tag{125}$$

$$= - 2 \frac{V_T V_B}{Z_C}$$

where Z_C is the characteristic impedance of the output line.

Comparing this finally with Eq. (122) we obtain the result

$$V_B = - \frac{Z_c}{2V_T} \int dV \, \vec{E}_T \cdot \vec{J}_B^{(0)} \tag{126}$$

In words: the output voltage amplitude at a terminal plane Σ_k in the kth output port, generated by a current density $\vec{J}_B$ (at frequency Ω) in the interior of the structure, is obtained by performing the volume integral of Eq. (126)) where $\vec{E}_T$ is the field excited by an incoming wave with amplitude V_T at the terminal plane Σ_k. Once the current density $\vec{J}_B$ is known the only non-trivial task is to calculate the fields inside the structure for a given excitation. This task must be performed anyhow in order to establish the kicker properties and, as we shall see, the integral in Eq. (126) is closely related to the energy gain of a particle in a kicker of identical electrical properties.

We conclude this paragraph by calculating the signal voltage due to a single passage of a particle with velocity v and charge e. We make the usual paraxial ray approximation, assume the PU to be located in a drift space, and as mentioned above, neglect any effects on the particle trajectory due to the interaction with the PU, i.e.:

$$\frac{ds}{dt} = v\left(x'^2 + z'^2 + 1\right)^{-1/2} \cong v, \quad x' \equiv \frac{dx}{ds} = \text{const.} \quad z' \equiv \frac{dz}{ds} = \text{const.}$$

Then

$$\vec{j}(x,z,s,t) = e\vec{v}\ \delta\big(x-x(s)\big)\ \delta\big(z-z(s)\big)\ \delta(s-vt) \tag{127}$$

assuming $s(t=0) = 0$. By Fourier transform:

$$\vec{j} = e\vec{v}\ \delta\big(x-x(s)\big)\ \delta\big(z-z(s)\big)\ \frac{1}{\sqrt{2\pi}} \int_{-\infty}^{+\infty} dk\ e^{ik(s-vt)}$$

or for the quantity $\vec{J}(\Omega,x,z,s)$ which we need for use in Eq. (126):

$$\vec{J}(\Omega,x,z,s) = \frac{e\vec{v}}{v}\ \delta\big(x-x(s)\big)\ \delta\big(z-z(s)\big)\ e^{-i\frac{\Omega}{v}s} \tag{128}$$

Insertion in (126) finally yields:

$$V_B(\Omega) = -\frac{Z_c}{2}\frac{e}{V_T(\Omega)}\int ds$$

$$\Big\{x'E_x\big(x(s),z(s),s\big) + z'E_z\big(x(s),z(s),s\big) + E_s\big(x(s),z(s),s\big)\Big\}\ e^{-i\frac{\Omega}{v}s} \tag{129}$$

or equivalently:

$$V_B(\Omega) = -\frac{Z_c}{2}\frac{e}{V_1(\Omega)}\ \left\{x'\ \tilde{E}_x\left(\frac{\Omega}{v}\right) + z'\ \tilde{E}_x\left(\frac{\Omega}{v}\right) + \tilde{E}_s\left(\frac{\Omega}{v}\right)\right\}\sqrt{2\pi}\ ^{*)}$$

where the tilde indicates the Fourier transform with respect to the (longitudinal) spatial coordinate s.

C. Expressions for Kickers

The evaluation of the kicker transfer functions is relatively straightforward if we know the fields inside the structure. The starting point is provided by Eq. (118).

The energy change experienced by a particle in a single pass with harmonic excitation of the kicker $(\vec{E} \propto e^{i\Omega t})$ is easily evaluated.

*$[V_B(\Omega)] = V/Hz$ in Eq. (129). Later, for discrete spectra, we will use $[V_B(\Omega)] = V$. Equation (126) remains valid with either convention provided all quantities are treated consistently.

$$\Delta E = e \int_{s_1}^{s_2} ds \left\{ x' E_x\big(x(s),z(s),s\big) + z' E_z\big(x(s),z(s),s\big) + E_s\big(x(s),z(s),s\big) \right\} e^{i\Omega \frac{s}{v}} \tag{130}$$

where we made the substitution $s = vt$. Or in form of the Fourier transforms of the fields:

$$\Delta E = e \left\{ x' \tilde{E}_x\left(-\frac{\Omega}{v}\right) + z' \tilde{E}_z\left(-\frac{\Omega}{v}\right) + \tilde{E}_s\left(-\frac{\Omega}{v}\right) \right\} \sqrt{2\pi} \tag{131}$$

where the relation to the PU sensitivity expression (129) is evident.

We also need to know the transverse effects, $\Delta p_x, \Delta p_z$, from a transverse kicker for betatron cooling or in order to calculate the transverse kicks due to a longitudinal kicker. Expressions for $\Delta p_{x,z}(\Omega)$ may be described in analogy to the longitudinal case, evaluating $(\vec{E}+\vec{v}\times\vec{B})$ along the particle's trajectory. Instead of this explicit approach, we will derive an equation relating the transverse to the longitudinal kicks.

Assuming again an $e^{i\Omega t}$ time dependence of the fields we rewrite $d\vec{p}/dt$ by substituting i/Ω curl $\vec{E}$ for $\vec{B}$ in the expression for $d\vec{p}/dt$. We obtain:

$$\frac{dp_x}{dt} = e\left(E_x + \dot{y}\,\frac{i}{\Omega}\left(\frac{\partial E_z}{\partial x} - \frac{\partial E_x}{\partial z}\right) - \dot{z}\,\frac{i}{\Omega}\left(\frac{\partial E_x}{\partial z} - \frac{\partial E_s}{\partial x}\right)\right) \tag{132}$$

where dp_z/dt, dp_s/dt are obtained by cyclic permutation of the variables (x,z,s). With $dE_x/dt \equiv d/dt\ E_x\big((x(t),z(t),s(t),t\big)$ where x,z,s represent the particles trajectory through the kicker we arrive at:

$$\frac{dp_x}{dt} = e\,\frac{i}{\Omega}\left(-\frac{d}{dt}E_x + \vec{v}\,\frac{\partial}{\partial x}\vec{E}\right) \tag{133}$$

After substituting $s = vt$, we obtain:

$$\frac{dp_x}{ds} = e\,\frac{i}{\Omega}\big(x',z',1\big)\cdot\frac{\partial}{\partial x}\vec{E} - e\,\frac{i}{\Omega}\,\frac{1}{v}\,\frac{dE_x}{ds} \tag{134}$$

or

$$\Delta p_x = e\,\frac{i}{\Omega}\int_{s_1}^{s_2} ds\left\{x'\,\frac{\partial E_x}{\partial x} + z'\,\frac{\partial E_z}{\partial x} + \frac{\partial}{\partial x}E_s\right\} \tag{135}$$

since $E_x(s_1) = E_x(s_2) = 0$ by definition. This expression reveals the close relation between transverse kick strength and the transverse variation of the longitudinal kick strength. If the distances $x'L$, $z'L$, $L = s_2 - s_1$ are short compared to the scale of transverse changes of the fields in the structure we may simplify further:

$$\Delta p_x = \frac{i}{\Omega}\frac{\partial}{\partial x}\Delta E(x), \qquad \Delta p_z = \frac{i}{\Omega}\frac{\partial}{\partial z}\Delta E(z) \tag{136}$$

Often the terms $x'E_x$, $z'E_z$ may be safely neglected compared with E_s. In that case we are left with a single expression to be evaluated:

$$\Delta E = e\int_{s_1}^{s_2} E_s(x,z,s)\,e^{i\frac{\Omega}{v}s}\,ds \tag{137}$$

from which all pick-up and kicker properties of a given structure may be derived.

Δp_s and ΔE are of course closely related and one may be obtained from the other without use of the equations described above. It is left as an exercise to the reader to show that application of Eqs. (135,137) yields the correct answer.

D. General Single Particle Spectrum

We derive the signal spectrum of a particle circulating with frequency $\omega/2\pi$, energy deviation x, $\Delta p/p = x/\beta^2E$, and betatron amplitudes $x_\beta = \sqrt{\epsilon_x\beta_x}$, $z_\beta = \sqrt{\epsilon_z\beta_z}$.

The signal frequencies are determined by the current density which is the driving term, while the respective amplitudes must depend on the PU structure.

The current density is given by:

$$\vec{j}(x,z,s,t) = e\Big(x'(s,t),z'(s,t),1\Big)\;\delta\Big(x-x(s,t)\delta(z-z(s,t))\Big)$$

$$\sum_m \delta\Big(t - \frac{s}{v} - mT\Big) \tag{138}$$

where s is the distance along the trajectory within the PU, $-L/2 \leqslant s \leqslant L/2$, where L is the PU length. We Fourier transform the δ-functions to obtain:

$$\vec{j}(x,z,s,t) = e\left(x'(s,t), z'(s,t), 1\right) \frac{1}{2\pi}$$

$$\iint_{-\infty}^{+\infty} dk_x \, dk_z \, e^{ik_x\left(x-x(s,t)\right)} e^{ik_z\left(z-z(s,t)\right)} f_0 \sum_q e^{iq\omega\left(t-\frac{s}{v}\right)} \tag{139}$$

In order to shorten the expressions we concentrate on j_s:

$$j_s = \frac{ef_0}{2\pi} \iint_{-\infty}^{+\infty} dk_x \, dk_z \, e^{ik_x\left(x-x(s,t)\right)} e^{ik_z\left(z-z(s,t)\right)} \sum_q e^{iq\omega\left(t-\frac{s}{v}\right)} \tag{140}$$

For $x(s,t)$ we write:

$$x(s,t) = x_p(s) + x_\beta(s) \cos\left(Q_x \omega t + \delta_x(s)\right), \tag{141a}$$

$$x_p(s) = \frac{\alpha_p(s)x}{\beta^2 E}, \tag{141b}$$

$$\delta_x(s) = \psi_x(s) - Q_x \, \omega \, \frac{s}{v} \tag{141c}$$

where $\alpha_p(s)$ is the dispersion function and where $\psi_x(s)$ is the betatron phase advance from the reference point $s = 0$ in the centre of the PU.

Similarly:

$$z(s,t) = z_\beta(s) \cos\left(Q_z \, \omega t + \delta_z(s)\right)$$

We make use of the expansion:

$$e^{ia \cos \theta} = \sum_r e^{ir\frac{\pi}{2}} e^{ir\theta} J_r(a) \tag{142}$$

where $J_r(a)$ is a Bessel function to obtain:

$$j_s(x,z,s,t) = \frac{ef_0}{2\pi} \sum_{q,r_x,r_z} e^{ir_x \frac{\pi}{2}} e^{ir_z \frac{\pi}{2}}$$

$$\iint_{-\infty}^{+\infty} dk_x\, dk_z\, e^{ik_x x}\, e^{ik_z z}\, e^{-iq\omega \frac{s}{v}}\, e^{-ik_x x_p(s)}\, e^{ir_x Q_x \omega t}\, e^{ir_x \delta_x(s)}$$

$$e^{ir_z Q_z \omega t}\, e^{ir_z \delta_z(s)}\, J_{r_x}\left(-k_x x_\beta(s)\right) J_{r_z}\left(-k_z z_\beta(s)\right) e^{iq\omega t} \tag{143}$$

At any location s in the PU therefore:

$$j_s = \sum_{q,r_x,r_z} j_{q,r_x,r_z}(s)\, e^{i(q+r_x Q_x + r_z Q_z)\omega t} \tag{144}$$

with

$$j_{q,r_x,r_z}(s) = \frac{ef_0}{2\pi} e^{i(r_x + r_z)\frac{\pi}{2} + ir_x \delta_x(s) + ir_z \delta_z(s) - iq\omega \frac{s}{v}}$$

$$\iint_{-\infty}^{+\infty} dk_x\, dk_z\, e^{ik_x x}\, e^{ik_z z}\, e^{-ik_x x_p(s)}\, J_{r_x}\left(-k_x x_\beta(s)\right)\, J_{r_z}\left(-k_z z_\beta(s)\right) \tag{145}$$

In general therefore a spectrum $\Omega_{q,r,s} = (q + r_x Q + r_z Q)\omega$ is possible, with $-\infty < q, r_x, r_z < +\infty$. It is easy to see that the com- components j_x, j_z will not introduce new frequencies since all they imply is essentially another multiplier $\exp(\pm i Q_x \omega t)$, $\exp(\pm i Q_z \omega t)$. How strong the different sideband signals are depends, of course, on the PU structure.

To demonstrate the principle we evaluate the $J_s \cdot E_s$ contribution to Eq. (126) only. With $\Omega = (q + r_x Q_x + r_z Q_z)\omega$:

$$V_B(\Omega) = -\frac{Z_c}{2V_T(\Omega)} \int dxdzds\ e^{i(r_x+r_z)\frac{\pi}{2}} e^{-i\frac{\Omega}{v}s} \frac{ef_o}{2\pi}$$

$$\iint dk_x\ dk_z\ e^{ik_x(x-x_p(s))}\ e^{ik_z z}\ e^{ir_x\psi_x(s)}\ e^{ir_z\psi_z(s)}\ J_{r_x}(-k_x x_\beta)$$

$$J_{r_z}(-k_z z_\beta)\ E_s(x,z,s) \qquad (146)$$

Upon introducing the two-dimensional Fourier transform of $E_S(x,z,s)$:

$$\tilde{\tilde{E}}_s(m,\ell,s) = \frac{1}{2\pi} \iint dxdz\ e^{imx}\ e^{i\ell z}\ E_s(x,z,s)$$

we obtain:

$$V_B(\Omega) = -\frac{Z_c}{2V_T(\Omega)} \frac{ef_o}{2\pi} \iiint dsdmd\ell\ e^{i(r_x+r_z)\frac{\pi}{2}}\ e^{-i\frac{\Omega}{v}s}\ e^{ir_x\psi_x}\ e^{ir_z\psi_z}$$

$$e^{imx_p(s)}\ J_{r_x}(mx_\beta)\ J_{r_z}(\ell z_\beta)\ \tilde{\tilde{E}}_s(m,\ell,s) \qquad (147)$$

Cumbersome as it appears, Eq. (147) represents one possible general formulation for the signal spectrum generated by a circulating particle in an arbitrary PU (or PU array). We examine a few of its aspects to establish the connection with previously introduced concepts.

i) For $x_\beta = z_\beta = 0$ all Bessel functions except J_0, ($J_0(0) = 1$), vanish and therefore also all sidebands with $r_x, r_z \neq 0$. Recalling that we always assume the PU to be located in a drift space, we obtain:

$$V_B(q\omega) = -\frac{Z_c}{2V_T(q\omega)}\, ef_o \int_{-L/2}^{+L/2} ds\ e^{-i\frac{q\omega s}{v}}\ E_s\left(x_p(0)+x_p's, 0, s\right) \qquad (148a)$$

$$= ef_o\ Z_{pu}(x_o, q\omega) \qquad (148b)$$

in agreement with the PU-impedance introduced, somewhat arbitrarily, in Eq. (65b). The suppression of the argument x_p' in (148b) needs no further justification for a short PU; for a long PU or an array of many PU's it implies a statement about the lattice functions at the location of the electrode.

ii) For the betatron sidebands, $(q \pm Q_x)\omega$, $(q \pm Q_z)\omega$, the terms proportional to J_1 must be inspected. For sufficiently small arguments mx_β, ℓz_β, we write:

$$V_B(q\omega \pm Q_x\omega) = -\frac{Z_c}{2V_T((q\pm Q_x)\omega)} \, ef_o \, \frac{(\pm i)}{2}$$

$$\int_{-L/2}^{+L/2} ds \; e^{-i(q\pm Q_x)\frac{\omega}{v}s} \; e^{i\psi_x(s)} \; x_\beta(s) \frac{\partial}{\partial x} E_s\left(x_p(s),0,s\right) \tag{149a}$$

and

$$V_B(q\omega \pm Q_z\omega) = -\frac{Z_c}{2V_T((q\pm Q_z)\omega)} \, ef_o \, \frac{(\pm i)}{2}$$

$$\int_{-L/2}^{+L/2} ds \; e^{-i(q\pm Q_z)\frac{\omega}{v}s} \; e^{i\psi_z(s)} \; z_\beta(s) \frac{\partial}{\partial z} E_s\left(x_p(s),0,s\right) \tag{149b}$$

Again the expressions reduce immediately to Eqs. (65b,c) for a short PU, and provide the basis for the correct extension of Eqs. (65b,c) to a long PU structure.

iii) For a short PU we can write approximately

$$V_B(\Omega) \cong ef_o \, Z_{pu}(x_p,\Omega) \tag{150}$$

for any frequency Ω and obtain a simplified version of (147) in the form

$$V_B(\Omega) = \frac{ef_o}{2\pi} \iint dm d\ell \; e^{i(r_x+r_z)\frac{\pi}{2}}$$

$$J_{r_x}(mx_\beta)\, J_{r_z}(\ell z_\beta)\, e^{ikx_p}\, \tilde{\tilde{Z}}_{pu}(m,\ell,\Omega) \tag{151}$$

It should be pointed out that there are configurations where these expressions are of more than academic interest. In stochastic accumulation, e.g., variation of Z_{pu} with x_p is essential to shape the longitudinal system gain. With finite x_β, z_β not only higher order sidebands appear but also the response at $\Omega = q\omega$ is modified and becomes dependent on betatron amplitudes. Lattice functions must be chosen to minimize this effect. Their magnitude is strikingly evident in Figs. 8 and 9, calculated from Eq. (151).

It is left as an exercise for the reader to verify that the same frequencies $\Omega = (q + r_x Q_x + r_z Q_z)\omega$ exert an influence in a long time average on a particle traversing a kicker of arbitrary field configuration.

E. A Few Specific Electrode Models

We will now apply these rather theoretical arguments to the approximate evaluation of some practical structures. As our first example we select the simple device depicted in Fig. 10, which is sometimes referred to as a wall current monitor and which might typically serve as a Σ-PU, i.e., a device with minimized spatial (transverse) variation of response which simply detects the passage of a particle.

Let us first consider the voltage gain of a particle travelling parallel to the axis at radius r. From Eq. (131)

$$\Delta U = \sqrt{2\pi}\, \tilde{E}\left(r, -\frac{\Omega}{v}\right) \tag{152}$$

The field must satisfy the vector Helmholtz equation: $\nabla^2 \vec{E} + k_0^2 \vec{E} = 0$. In cylindrical coordinates the longitudinal component is:

$$\frac{1}{r}\frac{\partial}{\partial r}\left(r \frac{\partial E_s}{\partial r}\right) + \left(k_0^{\,2} - k^2\right) E_s = 0 \tag{153}$$

and with $k^2 = \Omega^2/v^2$, $k_0^2 = \Omega^2/c^2$:

$$\frac{1}{r}\frac{\partial}{\partial r}\left(r \frac{\partial E_s}{\partial r}\right) - \frac{k_0^{\,2}}{\beta^2\gamma^2} E_s = 0 \tag{154}$$

The solution is $I_0(k_0 r/\beta\gamma)$, where I_0 is the modified Bessel function. Therefore:

$$\Delta U(r_1)/\Delta U(r_2) = I_0\left(\frac{k_0 r_1}{\beta\gamma}\right) / I_0\left(\frac{k_0 r_2}{\beta\gamma}\right) \tag{155}$$

In particular:

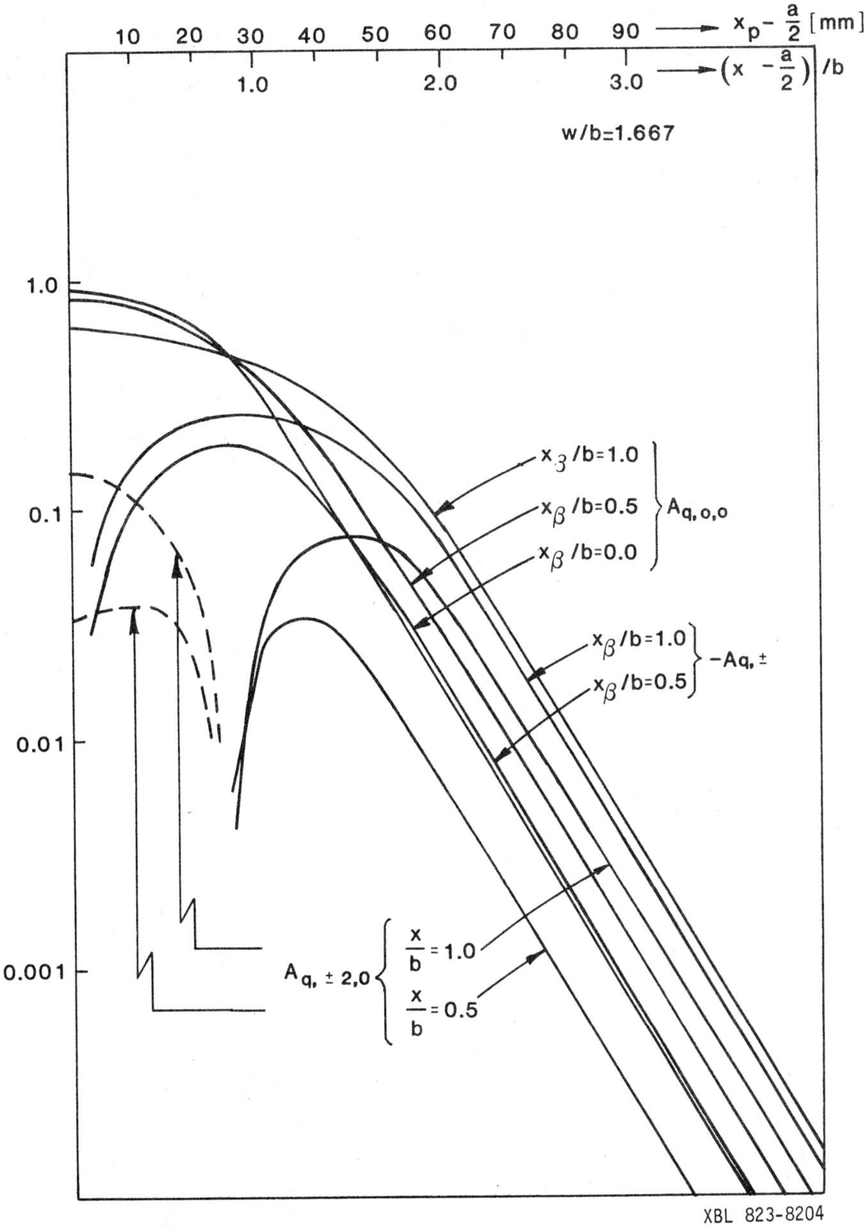

Fig. 8. Relative signal amplitudes $A_{q,r_x,0}$ at frequencies $(q \pm r_x Q_x)\omega$ vs. equilibrium position x_p for common mode operation of the PU structure shown in Fig. 11. Shown are the nominal response $A_{q,0,0}$ and the sidebands $r_x = \pm 1, \pm 2$ for $z_\beta = 0$, $x_\beta = 0, 0.5b$ and b.

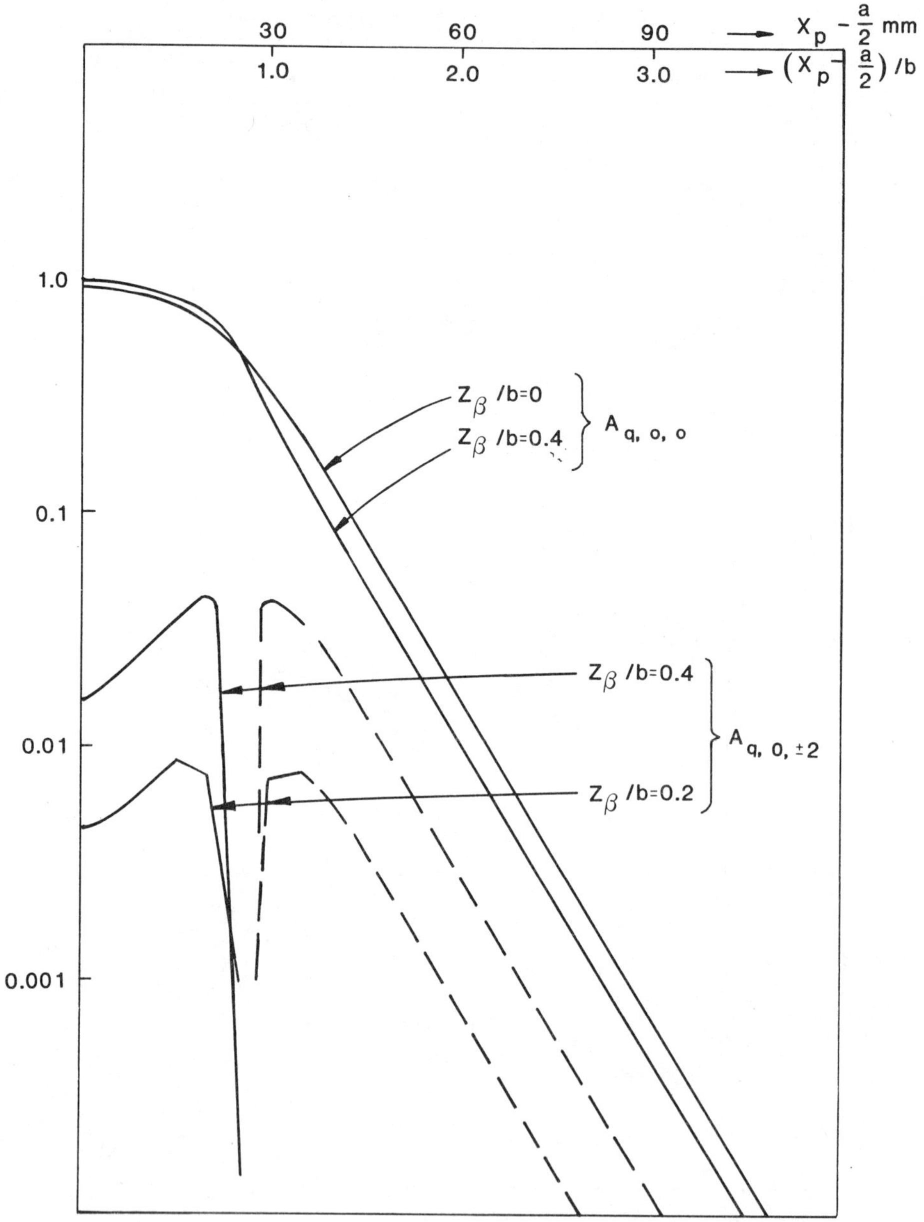

XBL 823-8205

Fig. 9. Relative signal amplitudes A_{q,o,r_z} for same PU as Fig. 8. $x_\beta = 0$, $z_\beta = 0, 0.2b, 0.4b$. Note disappearance of $r_z = \pm 1$ sidebands in common mode operation of PU.

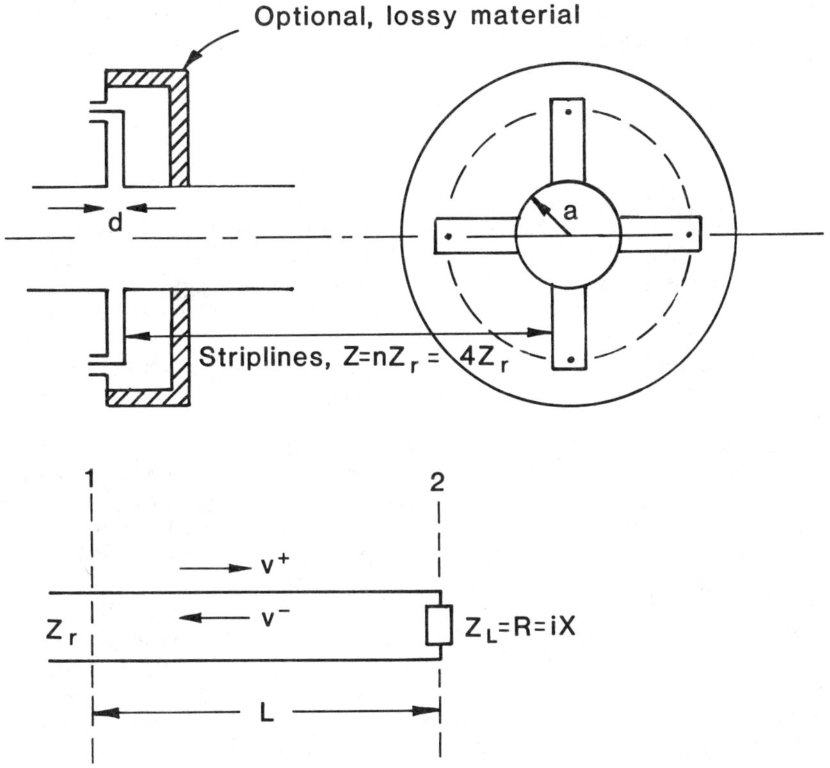

$$V_{gap} = V^{+}(2)\left(1 + S_{11}(2)\right) = V_1(1)\left(e^{-i\beta L} + e^{i\beta L} S_{11}(1)\right),\quad S_{11}(2) = \frac{Z_L - Z}{Z_L + Z}$$

XBL 823-8206

Fig. 10. Schematic description of "wall current monitor," Σ-PU.

$$\Delta U(0) = \Delta U(a)/I_0\left(\frac{k_0 a}{\beta\gamma}\right) \tag{156}$$

To calculate $\Delta U(a)$ exactly we must know $E_s(a,s)$. An approximate result is obtained by noting that $\Delta U(a) \cong U$, the voltage across the gap, provided $\Omega << \beta c/d$. Then

$$\Delta U(0) = U_{gap}/I_0\left(\frac{k_0 a}{\beta\gamma}\right) \tag{157}$$

We observe that the frequency response of the device is limited a) by the gap length d, and b) the appearance of the Bessel function $I_0(k_0 a/\beta\gamma)$. The physical origin of the latter becomes evident if we look at the device as a PU which, by virtue of Eqs. (129) and (131), must exhibit the same behavior: it is just the expression, appropriate for the boundary conditions of this configuration, of the fact that the fields of a particle passing an observer at a distance a from the trajectory fall off rapidly for frequencies in excess of $\Omega = v\gamma/a$.

In this derivation it is implicitly assumed that we are below cut-off for any propagating waveguide modes (at least TM modes with $E_z \neq 0$). Above cut-off we could still follow this analysis but obviously the electric fields at the wall $(r=a)$ would then not only be given by those across the gap but also those along attenuating surfaces necessary to damp out these modes.

The second part of the problem at hand is the calculation of the gap voltage in terms of some externally applied voltage. This calculation can rarely be performed exactly for a pick-up or kicker electrode. We use a simplified model also in this case. With regard to Fig. 10 we consider the effect of the entire configuration as a load impedance concentrated at the ends of the transmission (strip) lines.

To this impedance there corresponds a reflection coefficient S_{11}. The voltage across the gap is then $(e^{-i\beta L} + e^{i\beta L}S_{11})V_0$, where V_0, S_{11} refer to the reference terminal plane location and L is the electrical length from the reference plane to the gap. S_{11} is an experimentally measurable quantity and we obtain:

$$\Delta U = \left(e^{-i\beta L} + e^{i\beta L}S_{11}\right)V_0/I_0\left(\frac{k_0 a}{\beta\gamma}\right) \tag{158}$$

and

$$Z_{pu} = \frac{-Z_c}{2}\left(e^{-i\beta L} + e^{i\beta L}S_{11}\right)/I_0\left(\frac{k_0 a}{\beta\gamma}\right) \tag{159}$$

for a particle trajectory on and parallel to the cylinder axis. The weakness of the model is the assumption that the reflection is completely due to a localized impedance at the end of the strip lines.

Again, however, a measurement of the phase of S_{11} can help to determine how good this assumption is.

Another pick-up or kicker device of practical importance is the loop coupler shown in Fig. 11. Driven in common mode it provides a Σ-PU or longitudinal kicker, in push-pull we obtain a vertical Δ-PU or vertical transverse kicker. As derived previously all quantities of interest will follow from $\int(x'E_x+z'E_z+E_s)\exp(i\Omega s/v)ds$. The most important contribution to the integral comes from E_s, even in the push-pull configuration and we drop therefore the terms $x'E_x, z'E_z$ for our approximate analysis.

For a particle with $x = a/2$, $z = b$, $x' = z' = 0$ we obtain

$$\Delta U(a/2,b) = \sqrt{2}\, i \left(\frac{Z_L}{Z_R}\right)^{1/2} U \sin\left[\left(k_0 + \frac{\Omega}{v}\right)\frac{\ell}{2}\right] \tag{160}$$

where U is the voltage applied at the input to the splitting network. In deriving (160) it was assumed that the signal propagates through the loop as a TEM-wave (with propagation constant k_0) without reflections and that the longitudinal field may be represented as a δ-function at the two gaps. In general we expect:

$$\Delta U_c(x,z) = \sqrt{2}\, i \left(\frac{Z_L}{Z_R}\right)^{1/2} U\, \alpha_c(x,z,\Omega) \sin\left[\left(k_0 + \frac{\Omega}{v}\right)\frac{\ell}{2}\right] \tag{161a}$$

and

$$\Delta U_{pp}(x,z) = \sqrt{2}\, i \left(\frac{Z_L}{Z_R}\right)^{1/2} U\, \alpha_{pp}(x,z,\Omega) \sin\left[\left(k_0 + \frac{\Omega}{v}\right)\frac{\ell}{2}\right] \tag{161b}$$

where the subscripts c,pp refer to common mode and push-pull excitation, respectively. Equations (161) are completely general and can be applied to an analysis of nonlinearities and higher order sidebands in the sense of section 8D. The simpler expressions for "nominal" PU impedances and kicker transfer functions follow:

$$K_{\parallel}(x,\Omega) = \sqrt{2}\, i (Z_L/Z_R)^{1/2} \alpha_c(x,0,\Omega) \sin\left[\left(k_0 + \frac{\Omega}{v}\right)\frac{\ell}{2}\right] \tag{162a}$$

$$Z_{pu,\parallel}(x,\Omega) = \frac{i}{\sqrt{2}} (Z_L Z_R)^{1/2} \alpha_c(x,0,\Omega) \sin\left[\left(k_0 + \frac{\Omega}{v}\right)\frac{\ell}{2}\right] \tag{162b}$$

$$K_x(x,\Omega) = -\sqrt{2}\,\frac{c}{\Omega\beta^2 E} (Z_L/Z_R)^{1/2} \frac{\partial}{\partial x} \alpha_c(x,0,\Omega) \sin\left[\left(k_0 + \frac{\Omega}{v}\right)\frac{\ell}{2}\right] \tag{163a}$$

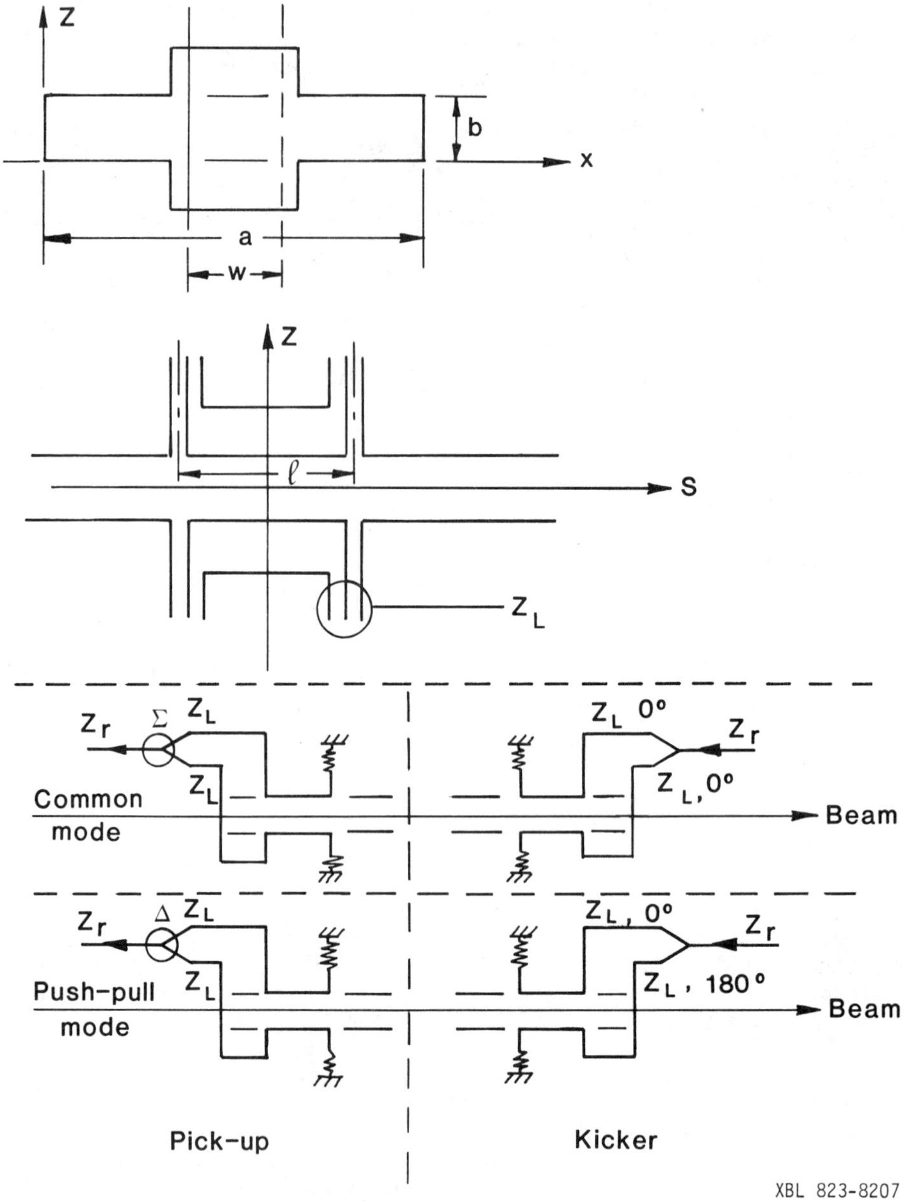

Fig. 11. "Loop coupler" type PU or K. Shown are the geometric configuration (top) and the electrical configuration for PU and K operation in either common or push-pull mode.

$$K_z(x,\Omega) = -\sqrt{2}\,\frac{c}{\Omega\beta^2 E}\,(Z_L/Z_R)^{1/2}\,\frac{\partial}{\partial z}\,\alpha_{pp}(x,0,\Omega)\,\sin\left[\left(k_0 + \frac{\Omega}{v}\right)\frac{\ell}{2}\right] \tag{163b}$$

$$\frac{\partial Z}{\partial x}(x,\Omega) = \frac{i}{\sqrt{2}}\,(Z_L Z_R)^{1/2}\,\frac{\partial}{\partial x}\,\alpha_c(x,0,\Omega)\,\sin\left[\left(k_0 + \frac{\Omega}{v}\right)\frac{\ell}{2}\right] \tag{164a}$$

$$\frac{\partial Z}{\partial z}(x,\Omega) = \frac{i}{\sqrt{2}}\,(Z_L Z_R)^{1/2}\,\frac{\partial}{\partial z}\,\alpha_{pp}(x,0,\Omega)\,\sin\left[\left(k_0 + \frac{\Omega}{v}\right)\frac{\ell}{2}\right] \tag{164b}$$

For the evaluation of $\alpha(x,z,\Omega)$ we make use of standard equations for the excitation of the eigenmodes of the rectangular wave guide by fields on coupling apertures. Since we restrict our attention to E_s we need only concern ourselves with the TM modes which we write as follows:[37)]

$$\vec{E} = a_{m,p}(s,t)\ \text{grad}\ \phi_{m,p} + c_{m,p}(s,t)\ \phi_{m,p}\cdot\vec{u}_s \tag{165}$$

$$\vec{H} = \alpha_{m,p}(s,t)\ \vec{u}_s \times \text{grad}\ \phi_{m,p}$$

with

$$\phi_{m,p}(x,z) = \sin\frac{m\pi x}{a}\,\sin\frac{p\pi z}{b}$$

where the coefficients $\alpha_{m,p}$, $a_{m,p}$, $c_{m,p}$ satisfy the following equations if the driving terms arise from fields on coupling holes:[37)]

$$\frac{d^2\alpha_{m,p}}{dz^2} + \left(k_0^2 - k_{m,p}^2\right)\alpha_{m,p} = i\,\frac{k_0}{Z_0}\,I_s(s),$$

$$a_{m,p} = i\,\frac{Z_0}{k_0}\,\frac{d\alpha_{m,p}}{dz},$$

$$c_{m,p} = i\,\frac{Z_0}{k_0}\,k_{mp}^2\,\alpha_{o\,m,p} \tag{166}$$

where

$$k_{mp}^2 = \left(\frac{m\pi}{a}\right)^2 + \left(\frac{p\pi}{b}\right)^2$$

and

$$I_s(s) = \frac{1}{N_{mp}^2}\frac{p\pi}{b}\int_0^a dx\left\{e_s(x,0,s) + (-1)^{p-1}e_s(x,b,s)\right\}\sin\frac{m\pi x}{a} \quad (167)$$

where e_s is the longitudinal component of the electric field on the coupling aperture and $N_{mp}^2 = k_{mp}^2\ ab/4$.

Analogous expressions could be derived for the TE modes, including both e_s and e_x, if we were to take the $x'E_x$, $z'E_z$ terms into account as well. The solution of (166) is immediately obtained for the Fourier transform $\tilde{c}_{m,p}(\Gamma)$ of $c_{m,p}(s)$:

$$\tilde{c}_{m,p}(\Gamma) = \frac{k_{m,p}^2}{\Gamma^2 + k_{m,p}^2 - k_o^2}\tilde{I}_s(\Gamma) \quad (168)$$

The expression for the longitudinal field is:

$$\tilde{E}_s(\Gamma) = \sum_{m,p}\tilde{c}_{m,p}(\Gamma)\ \phi_{m,p}(x,z) = \sum_{m,p}\tilde{c}_{m,p}(\Gamma)\sin\frac{m\pi x}{a}\sin\frac{p\pi z}{b} \quad (169)$$

With

$$\Gamma = -\frac{\Omega}{v}, \quad K_m^2 = \frac{k_o^2}{(\beta\gamma)^2} + \left(\frac{m\pi}{a}\right)^2 \quad (170)$$

the voltage gain follows as:

$$\Delta U(x,z) = \sqrt{2\pi}\sum_{m,p}\frac{1}{a}\frac{4\pi p}{b^2}\sin\frac{m\pi x}{a}\sin\frac{p\pi z}{b}\left(K_m^2 + \left(\frac{p\pi}{b}\right)^2\right)^{-1}$$

$$\int_0^a dx'\sin\frac{m\pi x'}{a}\left\{\tilde{e}_x\left(x',0,-\frac{\Omega}{v}\right) + (-1)^{p-1}\tilde{e}_s\left(x',b,-\frac{\Omega}{v}\right)\right\} \quad (171)$$

From (171) we see that symmetric (common mode) excitation leads to odd p only while anti-symmetric (push-pull) excitation yields even p only. The sum (169) over the index p is not uniformly convergent since the expansion includes functions which all go to 0 at the boundary while (169) must reproduce the non-zero tangential field on the coupling apertures. Therefore we seek a closed form solution for the summation over p. The Fourier series for the functions $f(z) = -\cosh(K(z+b/2))$, $-b < z \leqslant 0$, $f(z) = \cosh(K(x-b/2))$, $0 \leqslant z \leqslant +b$, and $f(z) = \sinh(K(z+b/2))$, $-b \leqslant z < 0$, $f(z) = \sinh(K(z-b/2))$, $0 < z \leqslant b$, provide us with the solutions:

$$\Delta U_c(x,z) = \sqrt{2\pi}\sum_m \frac{2}{a}\frac{\cosh(K_m(z-b/2))}{\cosh(K_m b/2)} \sin\frac{m\pi x}{a}$$

$$\int_0^a dx' \sin\frac{m\pi x'}{a}\,\tilde{e}_s\left(x',b,-\frac{\Omega}{v}\right) \tag{172a}$$

$$\Delta U_{pp}(x,z) = \sqrt{2\pi}\sum_m \frac{2}{a}\frac{\sinh(K_m(z-b/2))}{\sinh(K_m b/2)} \sin\frac{m\pi x}{a}$$

$$\int_0^a dx' \sin\frac{m\pi x'}{a}\,\tilde{e}_s\left(x',b,-\frac{\Omega}{v}\right) \tag{172b}$$

As can be seen this solution has the required properties

$$\Delta U_c(x,b) = \Delta U_c(x,0) = \sqrt{2\pi}\,\tilde{e}_s\left(x',b,-\frac{\Omega}{v}\right)$$

and

$$\Delta U_{pp}(x,b) = -\Delta U_{pp}(x,0) = \sqrt{2\pi}\,\tilde{e}_s\left(x',b,-\frac{\Omega}{v}\right).$$

We notice furthermore that the γ-dependence required for a true 3-dimensional and relativistic solution is contained in the quantity K_m (Eq. (170)).

Convergence of the sum over m is quite good, except for $z \to b$, $z \to 0$. Connection with Eqs. (161) is established by setting

$$\tilde{e}_s\left(x',b,-\frac{\Omega}{v}\right) = (2\pi)^{-1/2}\, i g(x') \sin\left(k_0 + \frac{\Omega}{v}\right)\frac{\ell}{z}$$

leading to

$$\alpha_c(x,z,\Omega) = \sum_m \frac{2}{a}\sin\frac{m\pi x}{a}\frac{\cosh(K_m(z-b/2))}{\cosh(K_m b/2)}\int_0^a g(x')\sin\frac{m\pi x'}{a}\,dx' \tag{173a}$$

$$\alpha_{pp}(x,z,\Omega) = \sum_m \frac{2}{a} \sin \frac{m\pi x}{a} \frac{\sinh(K_m(z-b/2))}{\sinh(K_m b/2)} \int_0^a g(x') \sin \frac{m\pi x'}{a} dx' \tag{173b}$$

Curves for α, for a given simple model of $g(x)$, are shown in Fig. 12.

It should be pointed out that this is an exact "solution" in terms of the assumed fields on the coupling apertures. We have, however, strictly speaking, no a priori knowledge about these fields, they constitute part of the problem. A complete solution would require a simultaneous, self-consistent solution of the field problem both inside the beam chamber and the strip lines of the loop which is a formidable task for this particular geometry.

It is seen that the performance of a single wideband electrode is limited to $Z_{PU} \leq 50\,\Omega$, $K \leq 2$. Applications such as fast cooling (precooling) or tail cooling in stochastic stacking require approximately one order of magnitude higher values. By summing the outputs from N individual electrodes $Z_{PU} \cong \sqrt{N}\, Z_{PU,1}$ $K \cong \sqrt{N}\, K_1$ is obtained, i.e. the impedance of the array increases with $\sqrt{L}$, where L is its length.

Several attempts have been made to design and build travelling wave electrodes, based on helices or periodically loaded transmission lines. The voltage gain in such a structure is

$$\Delta U \cong E_s(\Omega) L \cdot \frac{\sin x}{x}, \qquad x = \left[\frac{\Omega}{v} - \Gamma(\Omega)\right] \frac{L}{2}$$

where $\Gamma(\Omega)$ is the propagation constant of the structure. The design problem is to match the phase velocity $(v_{ph} = \Omega/\Gamma(\Omega))$ to the particle velocity v over a sufficiently wide frequency range and to maximize E for a given total power flow. The advantages seem obvious: a voltage gain $\propto L$, rather than $\sqrt{L}$, and no need for extensive signal combining (or power splitting) networks. The drawback is that (at least for broadband operation and structures based on loaded transmission lines and helices) the ratio of E_s/V_{in} seems to decrease approximately as $1/\gamma^2$, eventually overcoming the advantages. The first such electrode used for stochastic cooling is the "slot-box" designed by L. Faltin[16], successfully operated at the ISR and the CERN AA-ring. Also, very high impedances were obtained with the helical and ladder line electrodes built by LBL for use at the FNAL 200 MeV cooling ring.[22,23]

Finally, it should be observed that in special applications where narrow band operation is possible (slow cooling of a small number of particles) very high impedances are achieved by using PU's based on resonant cavities.[10]

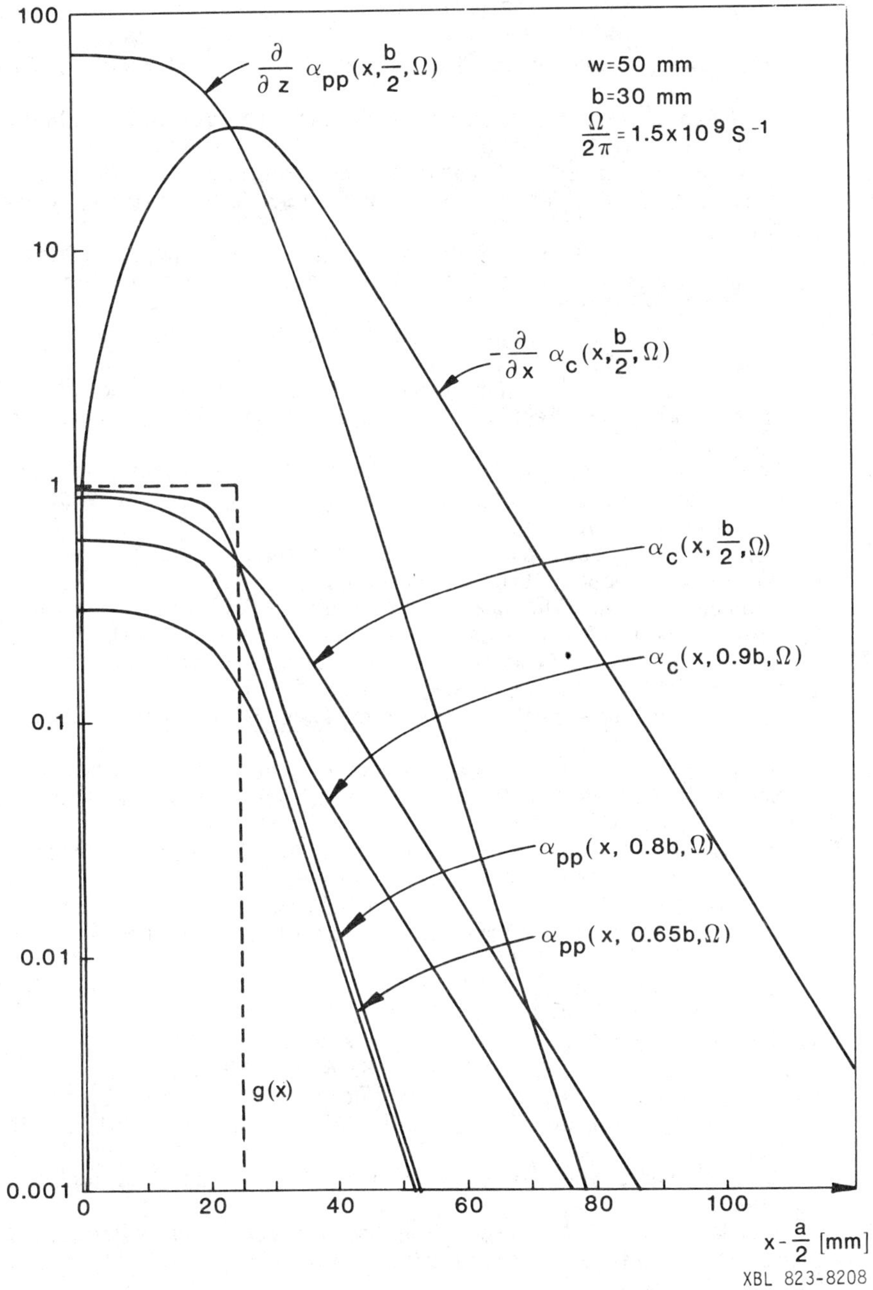

Fig. 12. $\alpha_c(x,z,\Omega)$, $\alpha_{pp}(x,z,\Omega)$ and their derivatives vs. x (see text) for PU of Fig. 11.

BIBLIOGRAPHY

1. J. Bisognano, "Transverse Stochastic Cooling," BECON-9, (LBL) (1979).
2. J. Bisognano, "Vertical Transverse Stochastic Cooling," BECON-10, LBID-119, (1979).
3. J. Bisognano, "Kinetic Equations for Longitudinal Stochastic Cooling," XIth Int. Conf. on High Energy Accelerators, CERN, Geneva, (LBL-10753) (1980).
4. J. Bisognano, "On Signal Suppression with Overlapping Schottky Bands and Large Gain Instability of Feedback Systems for Stochastic Cooling" (1981).
5. J. Bisognano and S. Chattopadhyay, Bunched Beam Stochastic Cooling, BECON-18, (1981).
6. J. Bisognano and S. Chattopadhyay, "Stochastic Cooling of Bunched Beams," 1981 Particle Accelerator Conference, Washington D.C., 2462 (1981).
7. M. Borer et al., "Non-destructive Diagnostics of Coasting Beams with Schottky Noise," IXth Int. Conf. on High Energy Accelerators, SLAC (1974).
8. D. Boussard, "Effects of RF Noise, Workshop on p̄p in the SPS," March 1980, SPS-pp-I, CERN, 87 (1980).
9. G. Carron and L. Thorndahl, "Stochastic Cooling of Momentum Spread by Filter Techniques," CERN-ISR-RF/78-12, (1978).
10. G. Carron et al., "Experiments on Stochastic Cooling in ICE," IEEE Transactions in Nucl. Sci., NS-26 3, 3456 (1979).
11. R. E. Collin, Foundations for Microwave Engineering, McGraw-Hill (1964).
12. E. N. Dement'ev et al., "Measurement of the Thermal Noise of a Proton Beam in the NAP-M Storage Ring," Sov. Phys. Tech. Phys. 25(8), 1001 (1980).
13. Ya S. Derbenev and S. A. Kheifhets, "On Stochastic Cooling," Particle Accelerators, Vol. 9, 237 (1979).
14. Ya S. Derbenev and S. A. Kheifhets, "Damping of Incoherent Motion by Dissipative Elements in a Storage Ring," Soviet Phys. Tech. Phys. 24(2), 209 (1979).
15. Ya S. Derbenev and S. A. Kheifhets, "Stochastic Cooling," Sov. Phys. Tech. Phys. 24(2), 209 (1979).
16. L. Faltin, "Slot-type Pick-up and Kicker for Stochastic Beam Cooling," Nucl. Inst. Methods, 148, 449 (1978).
17. T. Hardek et al., "ANL Stochastic Cooling Experiments Using the FNAL 200-MeV Cooling Ring," 1981 PAC, Washington D.C., 2455 (1981).
18. H. G. Hereward, "The Elementary Theory of Landau Damping," CERN 65 (1965).
19. H. G. Hereward, "Statistical Phenomena-Theory," in: Proc. Int. School of Particle Accelerators, Erice 1976, CERN 77-13, 284 (1977).
20. H. Herr and D. Mohl, "Bunched Beam Stochastic Cooling," CERN-EP-NOTE/79-34, (1979).

21. S. Ichimaru, Basic Principles of Plasma Physics. W. A. Benjamin (1973).
22. G. Lambertson et al., "Stochastic Cooling of 200 MeV Protons," Proceedings of the XIth International Conf. on High Energy Accelerators, CERN, Geneva, (1980).
23. G. Lambertson et al., "Experiments on Stochastic Cooling of 200 MeV Protons," 1981 PAC, Washington D.C., 2471 (1981).
24. R. Liboff, Introduction to the Theory of Kinetic Equations. John Wiley and Sons, (1969).
25. L. Jackson Laslett, "Evolution of the Amplitude Distribution Function for a Beam Subjected to Stochastic Cooling," LBL-6459, (1977).
26. T. Linnecar and W. Scandale, "A Transverse Schottky Noise Detector for Bunched Proton Beams," 1981 Particle Accelerator Conference, Washington D.C. (1981).
27. F. Mills, High Energy Beam Cooling, Workshop on $\bar{p}p$ in the SPS, March 1980, SPS-pp-I, CERN, 157 (1980).
28. F. Mills, Fermilab Summer School (1981).
29. F. Mills and F. Cole, Review Article in Annals of Nucl. Sci., Annual Review of Nuclear and Particle Science, 31, 295, (1981).
30. D. Mohl et al., "Physics and Techniques of Stochastic Cooling," Physics Reports 58, 2, 73 (1980).
31. R. B. Palmer (unpublished).
32. V. V. Parkhomchuk and D. V. Pestrikov, "Thermal Noise in an Intense Beam in a Storage Ring," Sov. Phys. Tech. Phys. 25(7), 818 (1980).
33. A. G. Ruggiero, "Are We Beating Liouville's Theorem," Proceedings of the Workshop on Producing High Luminosity High Energy Proton-Antiproton Collisions, Berkeley, CA, 123 (1978).
34. A. G. Ruggiero, "Stochastic Cooling with Noise and Good Mixing," Proceedings of the Workshop on Producing High Luminosity High Energy Proton-Antiproton Collisions, Berkeley, CA, 150 (1978).
35. F. Sacherer, "Stochastic Cooling Theory," CERN-ISR-TH/78-11, (1978).
36. W. Schnell, "About the Feasibility of Stochastic Damping in the ISR," CERN/ISR/RF/72-46 (1972).
37. J. Van Bladel, Electromagnetic Fields, McGraw-Hill (1964).
38. S. Van der Meer, Stochastic Damping of Betatron Oscillations in the ISR, CERN/ISR-PO/72-31, (1972).
39. S. Van der Meer, Influence of Bad Mixing on Stochastic Acceleration, CERN/ SPS/DI/PP/Int. NOTE/77-8, (1977).
40. S. Van der Meer, "Stochastic Cooling Theory and Devices," Proceedings of the Workshop on Producing High Luminosity High Energy Proton-Antiproton Collisions, Berkeley, CA, 73, (1978).
41. S. Van der Meer, "Stochastic Stacking in the Antiproton Accumulator," CERN/PS/AA/78-2, (1978).
42. S. Van der Meer, "Precooling for the Antiproton Accumulator," CERN/PS/AA/78-26 (1978).
43. S. Van der Meer, "Debunched $\bar{p}$-p operation of the SPS," CERN/PS/ AA/9-42, (1979).
44. S. Van der Meer, "A Different Formulation of Longitudinal and Transverse Beam Response," CERN/PS/AA/80-4 (1980).

45. J. M. Wang and C. Pellegrini, Proceedings of the 1979 Workshop on Beam Current Limitations on Storage Rings, BNL-51236 (1979).
46. Klimontovich Yu, Statistical Theory of Non-equilibrium Processes in a Plasma. MIT Press (1967).

APPENDIX A

Comments on the Fokker-Planck Equation

The solution of Eq. (47) is a nontrivial task requiring a computer. The basic equation is nonlinear through the $f \cdot \partial f/\partial x$ term, the gain functions are complex numbers, and the ϵ factors contain Cauchy singular integrals. However, some insight can be generated from an examination of a simplified version of this Fokker-Planck equation.

Consider the equation

$$\frac{\partial f}{\partial t} = + \frac{\partial}{\partial x} (g\ x\ f) + A \frac{\partial^2 f}{\partial x^2} \qquad (A.1)$$

It would describe cooling with a perfectly linear gain with Schottky noise and signal suppression negligible compared to amplifier noise.

If we drop the second derivative term, (A.1) reduces to

$$\frac{\partial f}{\partial t} - gx \frac{\partial f}{\partial x} = gf \qquad (A.2)$$

The left hand side is of the form of a total time derivative

$$\frac{df(x_0)}{dt} = gf(x_0) \qquad (A.3)$$

on the trajectory defined by

$$\frac{dx}{dt} = - gx, \qquad x(0) = x_0 \qquad (A.4)$$

We have immediately for an initial distribution f_0

$$f(x,t) = e^{gt}\ f_0(x_0) = e^{gt}\ f_0(e^{gt}x) \qquad (A.5)$$

This "method of characteristics" is also of use for more general gain functions. We have that the first derivative term increases the density and narrows the distribution; i.e., it cools.

If we keep the second derivative term of (A.1) alone, we have

$$\frac{\partial f}{\partial t} = A \frac{\partial^2 f}{\partial x} \tag{A.6}$$

which has as one solution

$$f(x,t) = \frac{1}{(4\pi At)^{1/2}} e^{-(x-x_0)^2/4At} \tag{A.7}$$

The standard deviation of this distribution is simply

$$<x^2> = 2At \tag{A.8}$$

In the limit $t \to 0$, $f(x,t) \to \delta(x-x_0)$ and (A.7) is actually the Green's function for (A.6). We have that the second derivative term causes diffusion, increasing $<x^2>$ with time.

ACKNOWLEDGEMENT

This work was supported by the Director, Office of Energy Research, Office of High Energy and Nuclear Physics, High Energy Physics Division, U. S. Dept. of Energy, under Contract No. DE-AC03-76SF00098.

ELECTRON COOLING

W. Kells
Fermi National Accelerator Laboratory, Batavia, Illinois 60510

TABLE OF CONTENTS

0094-243X/82/870656-43$3.00 Copyright 1982 American Institute of Physics

INTRODUCTION

The electron cooling of charged particle beams was conceived of (1966) and demonstrated (1974-75) by G.I. Budker and his collaborators at the INP at Novosibirsk.[1] Under the impetus of having $\bar{p}$-p colliding beams at the large accelerator centers, further electron cooling studies were begun at CERN (on the "ICE" machine) in 1979 and at Fermilab (on the "electron cooling ring") in 1980.[2,3]

The basic electron cooling theory is here outlined in an intuitive way with emphasis on standard terminology of the field: beam "temperature"; "disc" distribution; "magnetization" of the electron beam; "friction" drag force, electrostatic analogy, adiabatic collisions, etc.[4] Several complications to this basic cooling theory are pointed out, but only qualitatively discussed.

I attempt to describe electron cooling as an accelerator system technique. This is necessary since the drag force of electron beams on ion beams is far to small for cooling to occur in a single pass. In all cases the electron beam is incorporated as a segment of an ion storage ring. The cooling is adiabatic with respect to the revolution period. Thus, there is an intimate interplay between the storage ring properties and electron beam properties. The resulting design complexity makes any exact optimization of parameters for a specific electron cooling application impossible. I will simply point out various cases to give a flavor of the possibilities and difficulties.

The Fermilab research on electron cooling concentrates on use for $\bar{p}$ accumulation.[5] This discussion will be regretably biased in this direction. There are now several active plans for employing electron cooling at experimental energy (<1 GeV) to allow high precision/luminosity nuclear investigations.

To simplify expressions the following notation conventions are used. Velocities will only be used in the mean beam rest frame: v (lower case) will be for electrons, and V (upper case) for $\bar{p}$ (or any other ion being cooled). u will be used for relative electron ion velocity (i.e. V-v). Laboratory kinematics will be expressed in terms of $\beta,\gamma,\varepsilon_x,\varepsilon_z$ and p. To resolve possible ambiguities "starred" (*) quantities indicate mean rest frame. Finally, m denotes the electron mass and M the mass of the ion being cooled.

I. TEMPERATURE OF A BEAM

Temperature is often used as a measure of a beam's quality rather than energy spread and emittance. At best there will only be a formal analogy to the usual definition of temperature for a Maxwell-Boltzman distribution. Often the beams being referred to can neither be described by a single T (for e^- beams, one can have $T_{\parallel} \ll T_{\perp}$) nor is the velocity distribution within the beam position independent. However, as evidenced by the term "cooling" in this talk's title, some effective temperature is a useful parameter for describing relative beam quality. By convention I always use T as a mean rest frame quantity (natural definition!).

Storage ring beams (p or $\bar{p}$) are conveniently pictured as having Gaussian profiles

$$dn \propto \exp\,(-X^2/2\sigma_x^2)\; dX \tag{1}$$

and as having Gaussian momentum distribution

$$dn \propto \exp\,(-(p-p_o)^2/2\sigma_p^2)\; dp \tag{2}$$

First consider the transverse distribution and neglect dispersion. Then using the Courant-Snyder invariant plus the usual assumption of random betatron phases(β functions will be distinguished from $\beta = p/\gamma c$ by subscript).

$$X^2/\beta_x + \beta_x X'^2 \equiv \epsilon_x \tag{3}$$

we derive for the X distribution,

$$dn \propto \exp\,(-X'^2\beta_x^2/2\sigma_x^2) \tag{4}$$

The Maxwell-Boltzmann distribution is,

$$dn \propto \exp\,(-1/2\; MV_i^2/KT_i) \tag{5}$$

for the iTh degree of freedom, so comparing with (4) gives

$$\sigma_x^2 = \frac{\sigma_x^2}{\beta_x^2}(\beta c)^2 = \frac{KT_x}{M}$$

combining the two transverse degrees

$$\sigma_x^2 + \sigma_y^2 \equiv \sigma_{V_\perp}^2 = \frac{2KT_\perp}{M}$$

and, similarly for the longitudinal distribution,

$$\sigma^2_{V_{\parallel}} = \frac{KT_{\parallel}}{M}$$

EXAMPLE 1 - The Fermilab Electron Cooling Ring runs at p = 645 MeV/c.[4] When filled with $\bar{p}$ it contains a phase space of $\Delta p/p = 2\%$, $\varepsilon_x = 20\pi$ μm, $\varepsilon_y = 40\pi$ μm. Such a beam has temperatures $T_{\parallel} \sim 700$ev, $T_{\perp} \sim 1000$ev.

EXAMPLE 2 - For actual $\bar{p}$ production, p_o is in the vicinity of 5 GeV/c. Here $d\sigma/dp_{\perp} \propto \exp-(1.6\ P_{\perp})^{o}$ so that a $\bar{p}$ beam formed from a target of radius ~0.5 mm would have a $T_{\perp}$ > 100 MeV ((0.5 mm/(0.5 mm β_x (100/5000)$^{1/2}$)^2 ~200 KeV, for a machine capturing substantially all the produced $\bar{p}$'s and with β_x ~ 10 m.

EXAMPLE 3 - A thermionically generated electron beam starts at a ~1000°C cathode so that it is born with $T_{\perp} \sim T_{\parallel} \sim 0.1$ev. If such electrons are electrostatically accelerated to high energy without any optical (lens) or space charge perturbation, then the velocity distribution becomes flattened. The Fermilab electron beam is V_o ~ 120 kV ($\beta = 0.6$) so if $1/2m\ (V_o + \delta V)^2 - 1/2m\ V_o^2 \sim 0.1$ ev, then $1/2m\ \delta V^2 = 1/2\ (0.1\text{ev}/120\text{kV})\ 0.1\text{ev} \sim 10^{-6}$eV. Of course, $T^*_{\perp} \sim 0.1$eV is unchanged. If such an unisotropic beam were to continue long enough (say circulate in a ring) $T_{\perp}$ and $T_{\parallel}$ would eventually come back into equilibrium due to intra beam scattering (IBS).

II. ELECTRON COOLING

1. One way to "cool" a $\bar{p}$ beam is to shoot it into a lead block. The $\bar{p}$'s are all ranged out. They come to rest and $\Delta p/p = 0$. The idea of electron cooling is to have the block move at a desired lab velocity and to eliminate nuclear absorption (yet retain dE/dx) by removing the nuclei. That is, one substitutes an electron beam for the metal.

One may wonder why some arrangement of (thin) foils properly arranged in a $\bar{p}$ storage ring would not "cool" the circulating beam. After all the dE/dx friction is non-linear ($\sim 1/V^2$), so that the velocity distribution would shrink in time. This is a famous problem. It is left to the reader to convince himself that nuclear interaction always wins out. Indeed, in the absence of nuclear interaction, e.g. for muons, such a cooling method is feasible.

2. In fact, the analogy to dE/dx loss is exact. Consider dE/dx in a frame where the electrons (density n*) are at rest

---

*Starting with this section $T_{\perp}$, $T_{\parallel}$, $\sigma_{\perp}$, $\sigma_{\parallel}$, etc., refer to electrons. Ion parameters will be identified by subscript in ambiguous situations.

$$dE/dx = - \frac{4\pi Z^2 e^4 n^*}{m(V)^2} \ln \left(\frac{b_{max}}{b_{min}}\right) \tag{6}$$

We can define a friction drag force on the ion by

$$\underset{\sim}{F}^* \cdot \underset{\sim}{V} \equiv \dot{E} = \underset{\sim}{\nabla} (E) \cdot \underset{\sim}{V} \tag{7}$$

Clearly, dE/dx is just one component of $\underset{\sim}{\nabla}$ (E) along the direction of the ion motion so that,

$$\underset{\sim}{F}^* = \underset{\sim}{\nabla}(E) = - \frac{4\pi e^4 n^*}{m} \ln\left(\frac{b_x}{b_n}\right) \frac{\underset{\sim}{V}}{(V)^3} .$$

Of course distinguishing the ion direction is superfluous when all the electrons are at rest. We assume now some distribution f (v) of electron velocities, thus departing from the usual dE/dx picture. Consider a subgroup of this electron distribution, f(v) d^3v. They contribute friction

$$dF^* = - \frac{4\pi e^4}{M} \ln\left(\frac{b_x}{b_n}\right) \frac{V - v}{(V-v)^3} \quad f(v)d^3v \tag{8}$$

Then integrating over f ($d^3vf(v) \equiv 1$)

$$F^* = - \frac{4\pi e^4 n^*}{M} \quad d^3v \quad \frac{V - v}{(V-^2v)^3} \quad f(v) \ln\left(\frac{b_x}{b_n}\right) \tag{9}$$

As we will see $\ln (b_x/b_n) \equiv \Lambda$ may vary with v so that it is strictly accurate to write $\Lambda(v)$ under the integral.

3. Let us examine Λ. The maximum impact parameter b_{max} is either the electron beam radius, the b corresponding to transit time through the electron beam ($\sim |V|L/\beta c$ with L = electron beam length), or the Debye length

$$\lambda_D = \sqrt{T_\perp/4\pi e^2 n^*} = 0.051 \sqrt{\beta\gamma T_\perp/J} \tag{10}$$

where J is the beam current density. For cases of practical interest ($J \gtrsim 0.5 A/cm^2$) λ_D < 1mm, which is the least of the above possibilities. Therefore, typically $b_{max} \sim \lambda_D$.

The minimum impact parameter, b_{min}, is determined by the maximum possible e^- - ion momentum transfer, 2m (V-v), kinematically possible. (Note: for much higher e^- densities, e.g., in a solid metal, b_{min} is dominated by other effects):

$$\frac{e^2}{b_{min}} = \frac{m}{2}(V-v)^2 \tag{11}$$

For all practical cooling situations (10) and (11) give $\Lambda \sim$ 10-15.

4. The expression (9) for the total friction drag force is of the same form as that for the electrostatic force of a spatial charge distribution $\rho(\underset{\sim}{r})$ on a test charge e at location R

$$F_{\text{Test Charge}} \propto e \int d^3r \, \frac{\underset{\sim}{R}-\underset{\sim}{r}}{(\underset{\sim}{R}-\underset{\sim}{r})^3} \, \rho(\underset{\sim}{r}) \tag{12}$$

with the net distribution charge Q=1. Ignoring the coulomb log Λ, we see an <u>electrostatic</u> <u>anlogy</u> to the velocity space cooling force. This helps intuitive visualization of the cooling force (e.g. Gauss' Law can be applied). The usual picture[12] (illustrated in Figure 1) is of attraction by a "disc" of electron charge since $T_\parallel << T_\perp$ (as in example 3 above).[4,16] One can show that if, on any planar cut through the origin of the $\rho(r)$ or $f(v)$ distribution, its profile density is monotonically increasing to a maximum at the origin, then $F(V_{ion}) \leq 0$ for all V_{ion}.

5. Two regimes of cooling may be distinguished as,

$$\text{I:} \quad |V| > \sigma_V = \frac{2kT_\perp}{m} \quad \text{(outside)}$$

$$\text{II:} \quad |V| < \sigma_V \quad \text{(inside)}$$

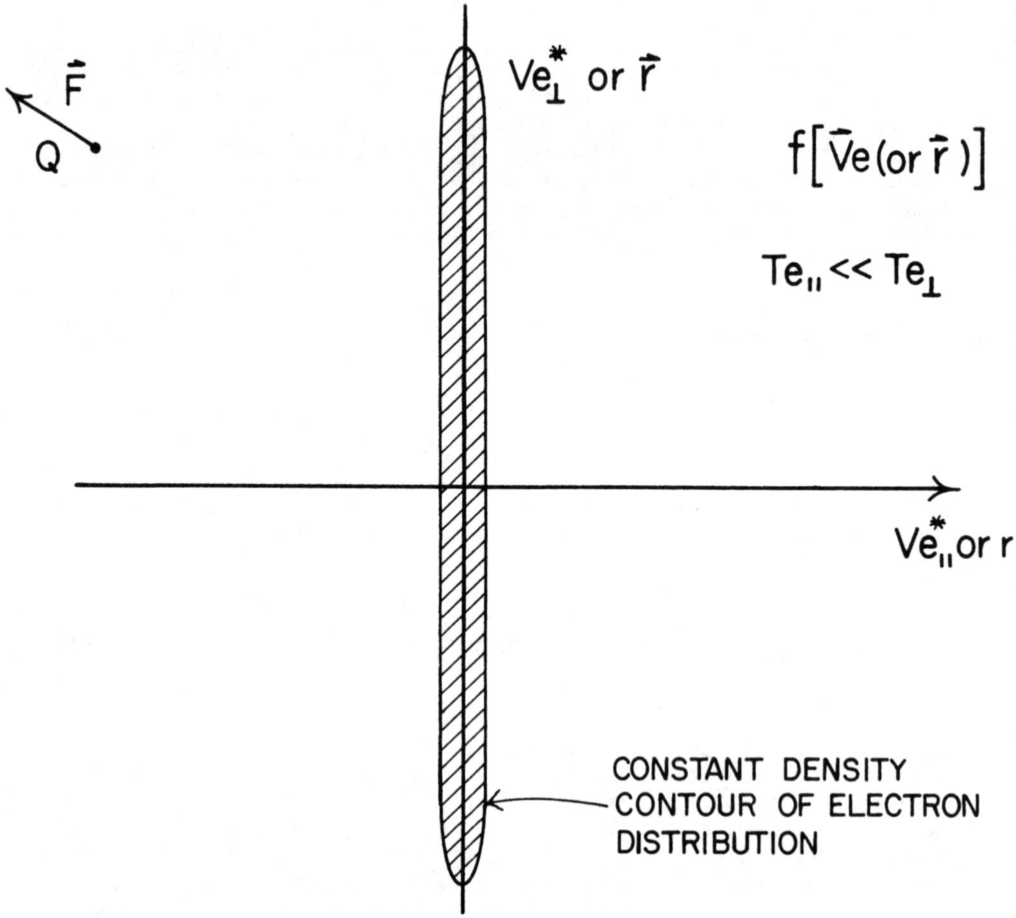

Fig. 1 Force, F, on an ion, Q, from the electron "disc" distribution f.

One refers to the $\bar{p}$'s as being outside/inside the electron distribution. At the transition, $V \sim \sigma_V \sim \sigma_{V\perp}$, the friction force will be maximal (see figure 2)

$$\text{I.}\quad \underset{\sim}{F}^{*}(V) = -K\Lambda_{\perp}\frac{\underset{\sim}{V}}{(V)^3}$$

[f(v) looks like a point charge to to the ion]

(13a)

$$K \equiv 4\pi e^4 n^*/m$$

$$\text{II.}\quad F_{\perp}^{*}(V) = -K\Lambda_{\perp}\frac{V_{\perp}}{(\sigma_{\perp})^3}$$

(13b)

$$F_{\parallel}^{*}(V) = -K\Lambda_{\parallel}\frac{1}{(\sigma_{\perp})^3}$$

[only for $\sigma_{\parallel} < |V_{\parallel}| < \sigma_{\perp}$: then $F_{\parallel}$ is analogous to a test charge attracted to a large $\rho \simeq$ const. disc]

$$F_{\parallel}^{*}(V) \simeq -K\Lambda_{\parallel}\frac{1}{(\sigma_{\perp})^2}\frac{V_{\parallel}}{\sigma}$$

[only for $V_{\parallel} < \sigma_{e\parallel}$ i.e. ion entirely inside the disc]

(13c)

III. PHYSICS OF THE ELECTRON COOLING RATE (example of p^- accumulation)

Further discussion of the net influence of the friction force on the $\bar{p}$ beam requires consideration of a specific scenario for electron cooling use. We first consider direct accumulation of $\bar{p}$'s from a high energy (≳100 GeV) production target in a storage ring incorporating an electron cooler (see figure 3). Once the system parameters (fraction, η, of ring "covered" by electrons; emittances of the $\bar{p}$ beam injected from target; $\bar{p}$ mean momentum (P_0), and $\beta_{x,y}$ functions at the interaction region) are specified we may convert the friction forces (13) into cooling rates, λ. The system performance goal is to accumulate the largest number of $\bar{p}$'s per second (average). Each production target burst fills a very large phase space volume with an essentially uniform density of $\bar{p}$'s

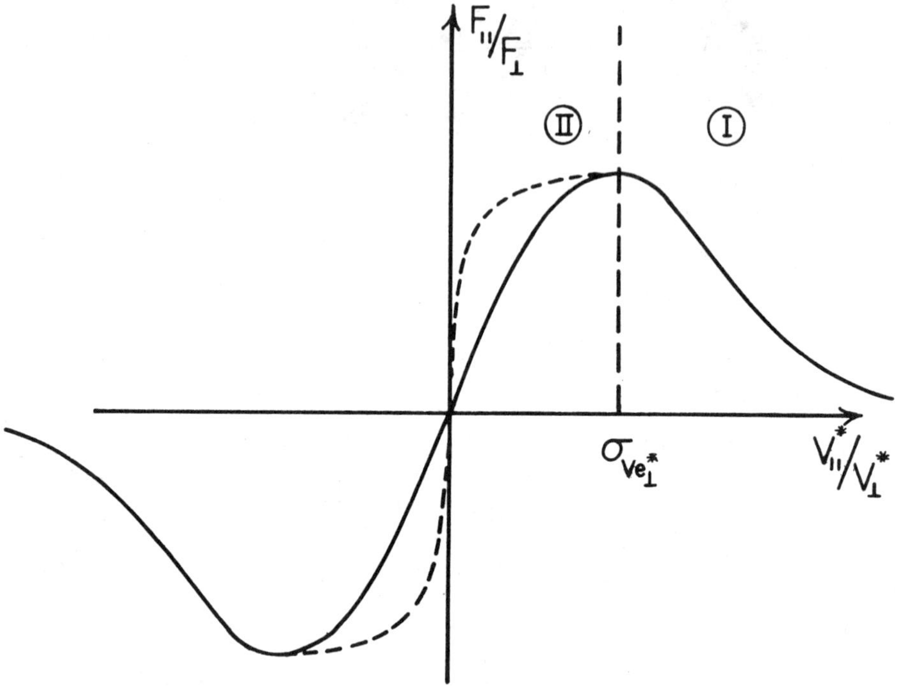

Fig. 2 Friction force on an ion as function of relative ion velocity. Region I: ion outside of electron distribution. Region II: ion "within" electron distribution (.....,$F_{\shortparallel}$, ----$F_{\perp}$).

downstream of the target. The storage ring accepts some bite $\Delta P/P_o$ and $\varepsilon_{\perp o}$ of this phase space.

No further $\bar{p}$'s will fit into it until cooling reduces this volume of $\bar{p}$'s by ~ 1/e. Notice that it is not essential to cool all dimensions - one suffices for this initial step of making room for the next, hot, burst.

In the last section we saw that the maximum friction occurred for matched electron and ion velocities. A typical $T_\perp$(effective) is $\lesssim$ 0.5ev. Then optimum use of the electron beam occurs when

$$T_\perp \sim (T_{\bar{p}\parallel} + T_{\bar{p}\perp}) \frac{m}{M} \tag{14}$$

For example 1, we calculate $T_\perp \sim$ 0.8ev for the Fermilab electron cooling ring. It is therefore well matched to the expected electron beam quality, and the optimum performance is indeed observed.[6] As we saw in example 2, one would like to cool much hotter $\bar{p}$ beams. Equation 13.I shows that the initial cooling friction will be down from the maximum by $(T_\perp/T_{\bar{p}})$ M/m.

2. Since the "hot" $\bar{p}$ beams of interest will violate (14), if we are interested in electron cooling for cyclic $\bar{p}$ accumulation, then the cycle time will be dominated by the initial friction strength given by 13.I. In this case, we can write the beam cooling rate as

$$\lambda_{LAB} = \frac{1}{\gamma} \frac{d|V|}{d\tau} \Big/ |V| = \frac{1}{\gamma M} |F^*(V)| \tag{15}$$

where $|V| \equiv \sqrt{kT_{\bar{p}}/M}$. In terms of LAB quantities (15) is

$$\lambda_{LAB} = - \frac{4\pi e^4}{\gamma m M} \frac{\eta J}{e\beta\gamma c} \left(\frac{M}{kT_{\bar{p}}}\right)^{3/2} \Lambda \tag{16}$$

where η is the storage ring circumference fraction cooled, and J is the electron beam current density. We see that the fundamental energy scaling for electron cooling is $\sim 1/\beta\gamma^2$. However, if we express $T_{\bar{p}}$ in terms of $\varepsilon_\perp \equiv \varepsilon_x + \varepsilon_y$ ($\bar{p}$ beam emittances) and $\Delta p/p_o$

$$T_{\bar{p},} \propto (\beta\gamma)^2 \frac{\varepsilon_\perp}{\pi\beta_\perp} \rightarrow \lambda_{LAB} \sim \frac{1}{\gamma^5\beta^4} \tag{17a}$$

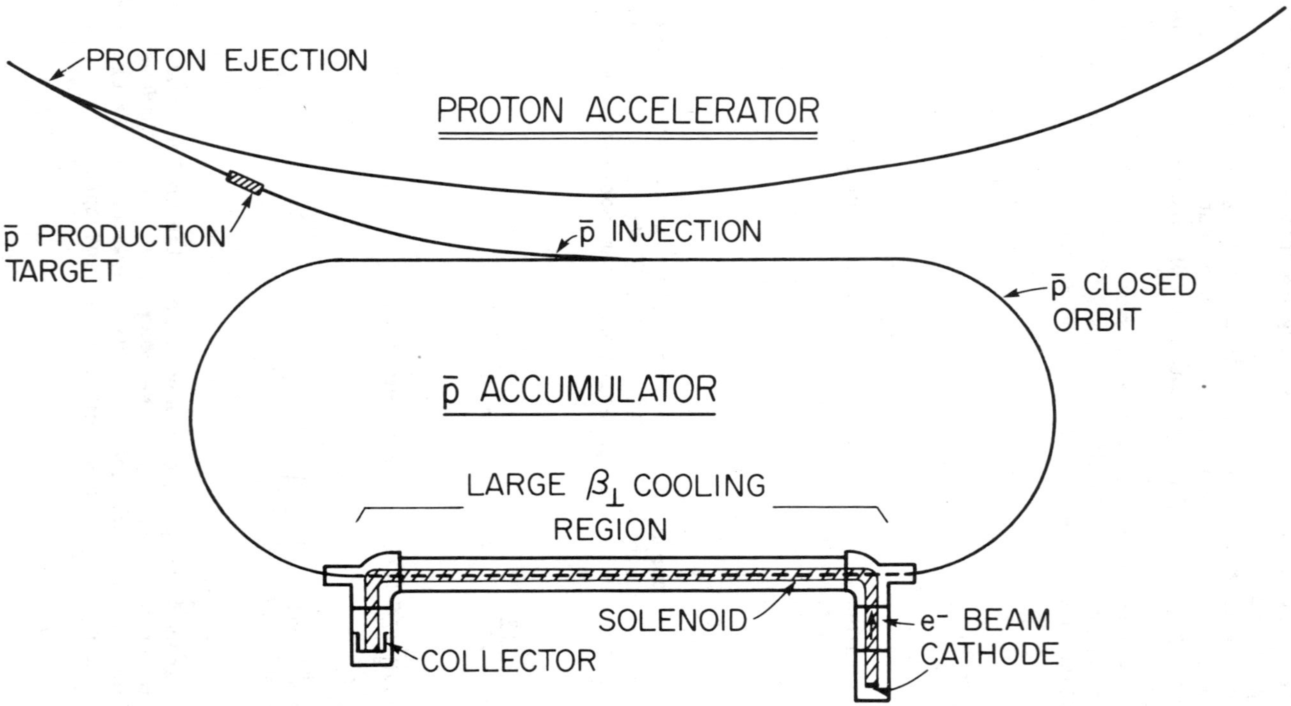

Fig. 3 (a) Schematic electron cooled $\bar{p}$ accumulator. Intense electron beam overlaps with the $\bar{p}$ closed orbit for a significant fraction (η) of the circumference.

$$T_{\bar{p}_\parallel} \propto (\beta\ \Delta p/p)^2 \quad \rightarrow \quad \lambda_{LAB} \sim \frac{1}{\gamma^2\beta^4} \qquad (17b)$$

This case of cooling $\bar{p}$'s far outside the electron distribution is the easiest to fit theoretically (no consideration for the magnetic guide field, distribution shape f(r) or IBS effect enters). Agreement with the data is good (figure 4, also figure 10).

Antiproton production cross sections peak at from 5-8 GeV/c $\bar{p}$ momentum, falling off rapidly below ~4 GeV/c.

$$P_o \sim 5 \text{ GeV/c}$$

A further external constraint is the $\bar{p}$ production cycle time (proton accelerator cycle time). At both CERN and Fermilab this is 2-3 s.[5] Thus

$$1/\lambda_{min} \lesssim 2.5 \text{ s.} \qquad (18)$$

It must be kept in mind that the number N of $\bar{p}$'s into $\varepsilon_\perp$ and $\Delta p/p_0$ per targeted proton scales as

$$N_{\bar{p}} \propto (\beta\gamma)^2\ \varepsilon_\perp\ \Delta p/p_0 \qquad (19)$$

Often the large initial $\Delta p/p_0$ can be reduced by other means (stochastic precooling, pretarget proton bunching/post target $\bar{p}$ debunching), so we assume $T_{\bar{p}\perp}$ dominates. In principal we could always arbitrarily reduce $T_{\bar{p}\perp}$ by increasing the beta function $\beta_\perp$ at the interaction region. Unfortunately there is a practical limit given by $\bar{p}$ beam radius as a function of path length(s) measured from the center of the section

$$a^2(s) = \varepsilon_\perp\ \beta_\perp\ (1+S^2/\beta_\perp^2) \qquad (20)$$

Thus a given electron beam current and current density fix $\beta_\perp$, which also fixes the interaction length to be $\lesssim \beta_\perp$.

To get a feel for these relationships consider the example of figure 5 which shows the result of a full numerical integration of equation (9). Using equations (17) and (19) to scale this to 5 GeV grossly violates (18). Worse, this scaled rate would only represent an ~8πμm x 8πμm x 2.5% $\Delta p/p_0$ beam of $\bar{p}$'s at 5 GeV/c which is a small bite of those available.

Actually it is beyond the state of art to construct an electron beam as in figure 6 but scaled to 5 GeV/c $\bar{p}$'s (electron energy = 3 MeV!). It is doubtful that direct cooling of >5πμm x 5πμm x 2% beams above several GeV/c consistant with (18) is feasible. Several strategies around this are possible:

1. The antiprotons may be decelerated to avoid the large γ at production. In this case, a further scaling relation, assuming adiabatic deceleration must be kept in mind

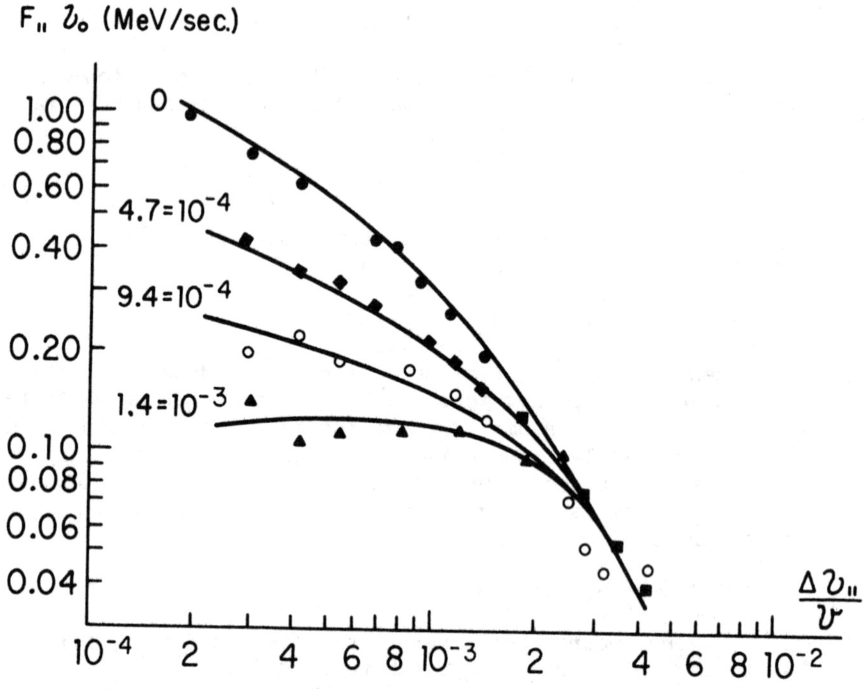

Fig. 4 Novosibirsk data of dependence of longitudinal force on detuning either $\mu_{\parallel}$ or $\mu_{\perp}$ (four different values corresponding to four branches shown).

$$\begin{pmatrix} \varepsilon_\perp \\ \Delta p/p_o \end{pmatrix}_{P_{final}} = \frac{P_o}{P_f} \cdot \begin{pmatrix} \varepsilon_\perp \\ \Delta p/p \end{pmatrix}_{P_o} \qquad (21)$$

This together with (17a), (19) and (20) imply a scaling

$$\frac{N_{\bar{p}}\lambda_{final}}{N_{\bar{p}}\lambda_o} = \frac{\gamma_o}{\gamma_f}\left(\frac{P_o}{P_f}\right)^2 \qquad (22)$$

for η, a, $(\varepsilon_\perp)_o$ and I subject only to the constraint $T_{p\perp} \lesssim T_\perp$, and that $\beta_\perp$ (given by {20}) be the storage ring circumference.

Equation (22) suggests that overall cyclic accumulation rate may be arbitrarily increased by starting with a smaller and smaller $\varepsilon_{\perp o}$ and $\Delta p/p_o$ bite, but recouping this via the nonlinearity of cooling rate with momentum by decelerating to much lower momenta. In practice it is not possible to use (22) to scale so far in momentum:

A. Since $\bar{p}$'s are stored in the electron cooling ring it cannot itself be used for deceleration. A special deceleration ring is required which has a cycling time increasing with $(\varepsilon_\perp)_o$ and p_o/p_f (technical requirements such as RF bunching power grow enormously).

B. The constraint (18) is soon violated. To some extent one can slice up the primary proton cycle into subcycles which can be targeted faster than (18) but this changes the scaling (19) used in (22). As p_o/p_f and the number of such subcycles increases the subcycle time rapidly exceeds the constraint A.

C. Scaling aside, there is the question of absolute rate. We saw in the example (figures 5-6) that an already substantial electron device, working at the validity limit ($T_{p\perp} \simeq T_\perp$) of (22), just gives (18). If P_f is further lowered this current density cannot be maintained for constant $T_\perp$ (see discussion of space charge limitations below).

2. One may push the cooling technology by arriving at cooling times for large $\varepsilon_{\perp o}$ and $\Delta p/p_o$ by brute force increase of η, I and a (implying abnormally large storage ring apertures as well!) some deceleration is necessary so that direct multiplier circuit high voltage supplies may be employed ($\lesssim$1 MV).

For example the Soviet "UNK" project will use electron cooling exclusively

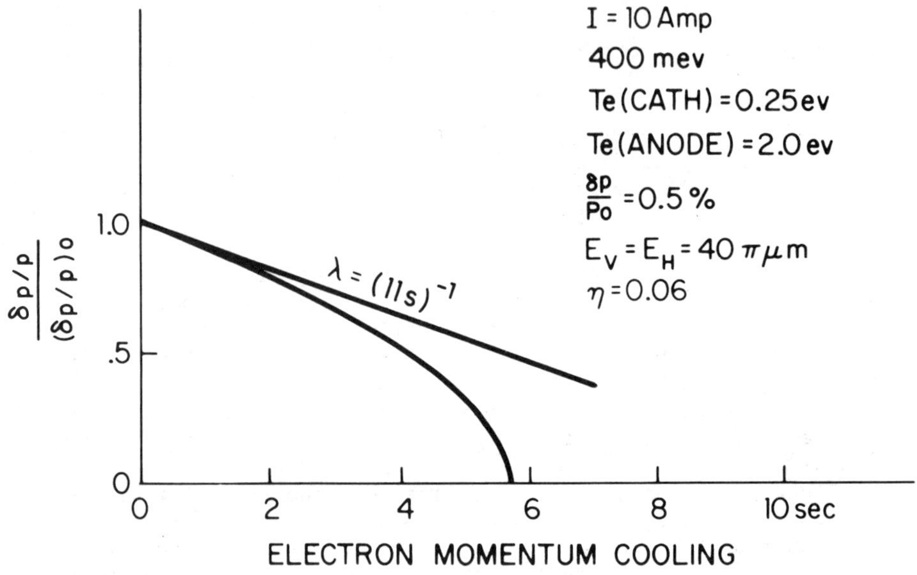

Fig. 5 Full simulation of electron cooling for parameters shown. Note contrast between λ (initial cooling rate) and actual time evolution. Curve is for a <u>single</u> ion with $\delta p/p_o = ½\%$ and $(40\pi)^2$.

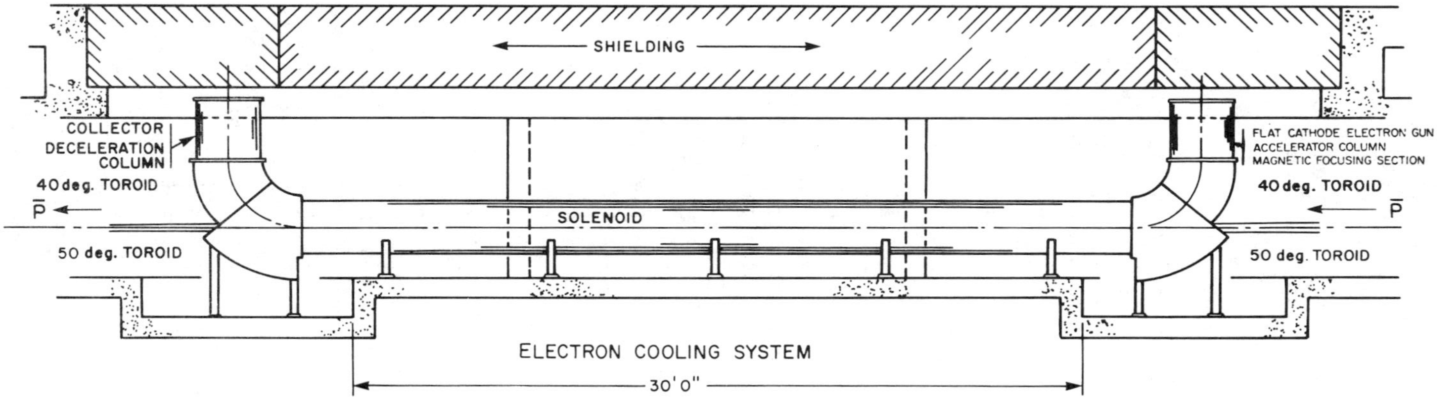

(b) Detail of electron interaction region (corresponding to Fig. 5)

TABLE I

	η	I_e(Amp)	$R_A(\bar{p}/\mathrm{sec})$	a(cm)	Solenoid Power(MW)
FERMILAB	0.05	≤ 10	2.3×10^6	10	0.12
UNK	0.8	~ 265	2.3×10^8	2.5	3.0

(both designs for 400 MeV accumulation; $\beta\gamma = 1.02$)

The use of brute force can be optimized by lattice function ($\beta_x + \beta_y \equiv \beta_\perp$) choice at the cooler location (typically a long straight section). Several interaction sections are necessary (figure 3b represents one). If the electron cooler is made as long as the large $\beta_\perp$ allows ($L \sim \beta_\perp$), then we can write (16) in terms of a,

$$\lambda \propto \frac{1}{\gamma^5 \beta^4} J a^5 \tag{23}$$

Within the limits of lattice design, electron beam design, and the assumption that $T_{\bar{p}\perp} > T_{\bar{p}\parallel} > T_\perp$ (large $I_e \equiv J \pi a^2$ inevitably raises $T_\perp$ and increasing $\beta_\perp$ <u>decreases</u> $T_{\bar{p}\perp}$) λ can be enhanced with an "expanded" electron beam.

The large space charge effects characteristic of the brute force approach are its central design problem. First there is the linear tune shift effect on the circulating $\bar{p}$'s

$$\frac{\Delta\nu}{\nu} = \frac{2JL\,\beta_\perp}{4\pi^2\, 10^7\ \mathrm{Amp}\ (\beta\gamma)^3} \tag{24}$$

where the electron $\bar{p}$ beam size matching relation (20) is assumed. As a rule of thumb one would like $\Delta\nu/\nu \lesssim$ few percent so as to guarantee no problem from (nonlinear) tune spreading. If one is assured of the homogeneity of the electron beam then it's influence is predominately that of a linear lens. In this case the electron beam charge (and current influence) may be considered merely as an additional lattice element and compensated for.

Second, the space charge distorts the electron velocity distribution <u>spatially</u>. The simple theory leading to equation (9) assumes spatial homogeneity. Indeed the confining solenoidal field (figures 3 and 6) is necessary only for space charge confinement. For instance, a beam of current I has a kinetic energy spread (center to edge) of

$$\Delta E(r) = 30\ \mathrm{eV}\ \frac{I\ (\mathrm{amp})}{\beta}\left(\frac{r}{a}\right)^2 \qquad (25)$$

The initial hot $\bar{p}$ beam will be matched in diameter to the electron beam at the interaction straight section. Betatron oscillations cause $\bar{p}$'s to sample the entire e^- beam cross section, therefore seeing the entire spread (25). This effect dominates the effective T

$$T_{\parallel}\ (\text{space charge}) \sim 1/2\ \frac{\Delta E}{E_o}\ \Delta E \qquad (26)$$

= 0.25 eV (I_e = 10 Amp, E_o = 220 kV). Since this is not a spatially homogeneous temperature, it does not imply a large equilibrium $\Delta p/p$. After a betatron cooling time, the beam occupies only a small fraction of (25) so that finally the no space charge limit holds (eq. 9).

The $T_{\perp}$ also gets a (parabolic) spatial dependence across the beam radius due to $B_{\text{axial}} \times E_{\text{space charge}}$ precession of electrons around the beam axis.

These electron temperature effects can be eliminated by allowing the beam to become neutralized by trapping ions formed by beam-gas collisions. Very high current coolers (e.g. for the "UNK" project) require ion neutralization to realize the rates inherent in the basic theory. Such neutralization also cancels the electrostatic lens effect of the electron beam on the circulating $\bar{p}$'s. However, a non-cancellable magnetic action remains.

Calculation of λ requires care under these conditions. The basic formula (9) may be used, however, a integration (average) over betatron phase must be performed since the distribution f(r) is now transverse position (or phase) dependent. The result in figure 5 incorporates such refinements.

A whole host of technical problems arise in the design of such large electron devices, including:

A. H.V. supply. For any (but see below; section VI) electron cooling beam the electrons must be collected by deceleration to recuperate the beam energy (figure 3). Reasonable supplies at ~250 kV can only supply ~10's of milliamps with good regulation (see Section IV). However, some beam is lost to electrodes via incomplete collection (reflection from collector virtual cathode), scraping, or residual gas collisions. Experience indicates that at best $\sim 5 \times 10^{-4}$ will be lost.[8]

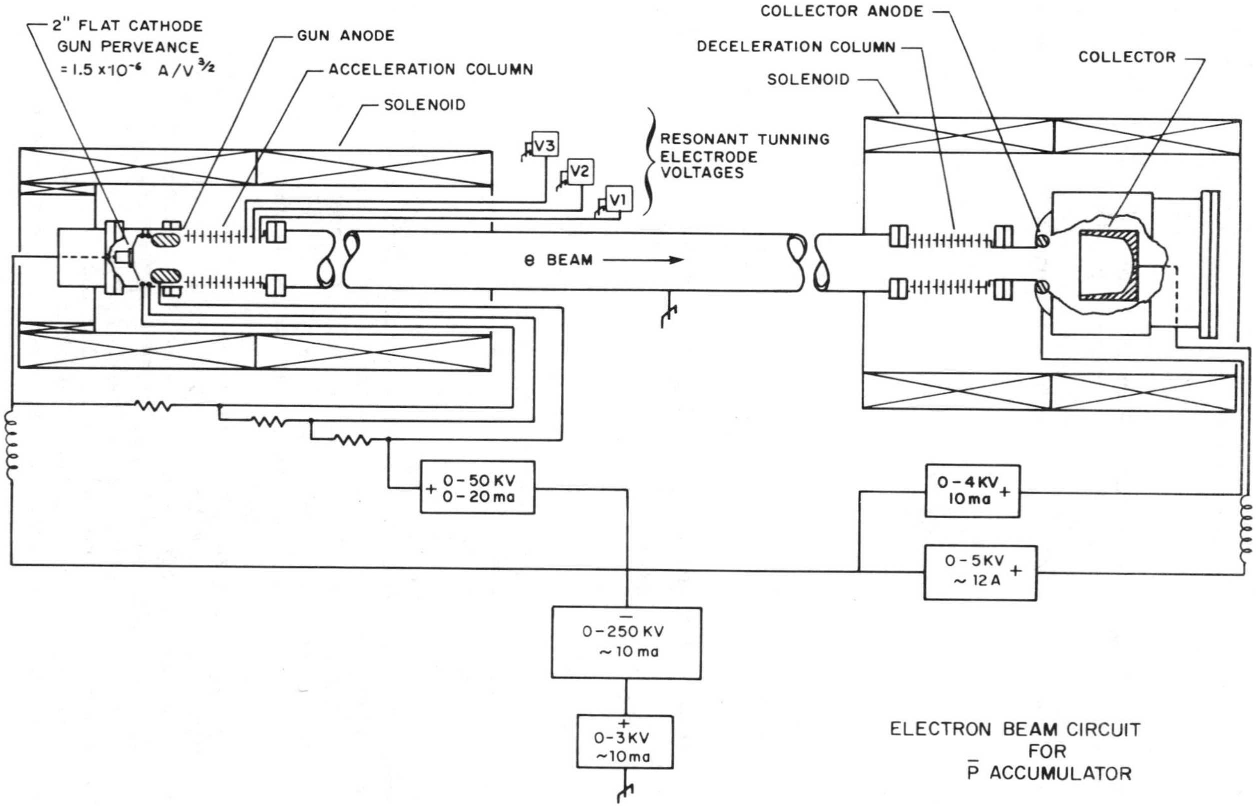

Fig. 6 Schematic of electron beam circuit for design of a cold 10 Amp, 250 KeV electron cooler. Note separately powered accelerating column elements for resonant tuning of T.

B. Vacuum. Unique UHV problems are associated with electron cooler rings all of which are exacerbated by high current and long cooling lengths (large η). Homogeneity conditions for the solenoidal magnetic field (section V) preclude direct pumping access to the interaction regions. The presence of an electron beam inevitably creates a plasma of gas ions, multipactored electrons, etc., which worsens the vacuum roughly in proportion to current. This creates an additional tune shift hazard for the antiprotons as well as a decrease in their lifetime in the ring (dominated by a single Coulomb scatter out of the aperture); given by

$$\tau_{ss} = \frac{4\pi Z^2 r_p^{\ 2}}{\beta^4 \gamma^2 \alpha^2} \qquad (27)$$

where α is the, acceptance angle of the lattice.

C. Gun Perveance.[9] The basic relation for Pierce geometry diode current density is

$$J = \frac{1}{K} V^{3/2} / d^2 \quad (d = \text{diode length}) \qquad (28)$$

where $K = 9\pi \sqrt{m/2e}$. The maximum electric field (E(z)) from such a geometry occurs near the anode $E_{max} = 4/3$ V/d. Reasonable maximum $E_{max} \lesssim 50$ kV/cm and minimum d $\gtrsim$ 2cm gives $J_{max} \lesssim 8\text{A/cm}^2$. Unfortunately good beam quality (low $v_\perp$) is had for small radius to length diode aspect ratios (see Section IV 3). Scaling up all dimensions (constant J and increased cathode diameter) has an absolute perveance ($I/V^{3/2}$) limit. This is plausible since equation (25) tells us that the kinetic energy gradient across the beam grows with I (not J). There is a beam stability threshold for large gradients.

3. A third strategy involves sweeping the electron beam voltage through the $\bar{p}$ energy profile. First a few words about the stacking procedure common to all electron cooling accumulation methods.

It has to be stressed that the accumulation described here is not the close packing of small phase space cells each of which represents the volume filled by each $\bar{p}$ production cycle after it comes into thermal equilibrium with the electron beam. Rather it is an accretion of fresh batches of $\bar{p}$'s into a single fixed, small, phase space volume. It would be technically difficult to move the electron beam spatially (say from "stack" to freshly injected batch), so that momentum stacking is employed. That is, the interaction section is a ~ zero dispersion lattice segment. Beams separated in momentum overlap spatially and with the electron beam. From the electrostatic analogy of the friction force, eq. 7, it is obvious that a fresh batch centered at p_o will be "attracted" to an electron beam centered on a stack at $p_o \pm \delta p$.

It is important to distinguish that the important process for stack accumulation is the <u>accretion</u>. The fact that the fresh batch will be <u>cooled</u> as it is dragged over by ~ δp is not essential. Of course, for complete utilization of the longitudinal phase space of the storage ring $\delta p \sim \Delta p_o$ so that, quantitatively, the time to drag the beam by Δp_o is very nearly equal to the "cooling time" $(1/\lambda)$ of the beam by electrons centered at p_o. In fact, stepping the electron beam to p_o after each injection; cooling, and then dragging back to $p_o \pm \delta p$ (with a larger friction force now: recall the u^{-2} in eq. 8) is faster.

"Sweeping" carries these concepts a step further. Suppose the fresh batch has already been cooled transversely so that $T_{\bar{p}\parallel} >> T_{\bar{p}\perp}$. The dragging or cooling time $1/\lambda$ for assimilating Δp_o will be proportional to $(\Delta p_o)^3$. Clearly the mean electron energy, E, may be swept through Δp_o at a rate given by the maximum friction force (equations 14 and 13 II).

$$c\dot{E}/E = c\beta^2 \dot{P}_o/P_o = \frac{\beta}{M} F_{\parallel \, max} \simeq - \frac{\beta K \Lambda_\parallel}{M} \sigma_\perp^{-3} \tag{29a}$$

which gives a sweep time

$$\tau_{sweep} \equiv \frac{\Delta P_o}{P_o} \frac{P_o}{\dot{P}_o} = - \frac{c\beta M \sigma_\perp^3}{K\Lambda_\parallel} \cdot \frac{\Delta P_o}{P_o} \tag{29b}$$

while (16) gives , for cooling ΔP_o

$$1/\lambda(\Delta P_o) \equiv - \frac{c\beta^2 (v_\parallel)^3}{\gamma K \Lambda_\parallel} \tag{29c}$$

indicating a time reduction of $T_{p\parallel}/T_\perp$ at relativistic energies. If on the other hand, $T_{\bar{p}\perp} > T_\perp$ the reduction will be $T_{\bar{p}\parallel}/T_{\bar{p}\perp}$. Recently, F. Mills has described a scenario in this later regime and has given an elegant application of the electrostatic force analogy to deduce the maximum sweep rate for a given $T_{p\perp}$. For $T_{\bar{p}\perp} > T_\perp$ it is appropriate to approximate the electron velocity distribution as a delta function.[10] Then (9) is equivalent to

$$\dot{u} = \frac{1}{M} \nabla_{\tilde{u}} \frac{K}{U} - \dot{\tilde{v}} = \nabla_{\tilde{u}} \left(\frac{K}{Mu} - \dot{u} \cdot v \right)$$

where $u = V-v$ and $\dot{v}$ is the sweep rate. In configuration space a potential $(K/Mr - \vec{r}\ \vec{E}_0)$ represents the superposition of a point charge K/M and a constant field $-\vec{E}_0$. Gauss' law tells us that infinitely far from the point charge, flux in a tube $r_\perp^2 = 4K/ME_o$ originates from the point. In analogy $T_{p\perp}(\dot{V}) = 1/2\ Mu_\perp^2 = 1/2\ 4K/\dot{v}$ is the maximum transverse temperature beam which will be swept at $\dot{v}$.

In his proposal Mills initially reduces the $\delta P/P_0$ from 5-10% to <1% with an adiabatic debunching technique. The initial ε_x, and ε_y are each reduced ~100 x via transverse stochastic cooling in such a way that several $\bar{p}$ target batches are being successively stochastically and electron cooled simultaneously.

Notice that a lower limit for the sweep time is given by (16) with $T_{\bar{p}} \sim T_\perp$ which has only a γ^{-2} dependence. Also, the reduced ε_x and ε_y of the $\bar{p}$'s at the electron cooling stage allows Small diameter electron beams to "cover" the $\bar{p}$'s even for large $\beta_\perp$ thus allowing relatively large J. These features give sweep times consistent with (18) at P_0 ~8 GeV/c.

The large cooler P_0 is used to advantage here in that the large current densities necessary (~10A/cm^2) present no tune shift problem ($(\beta\gamma)^{-3}$ in (24)). The same momentum scaling holds for the space charge blow up of a beam as a function of z from a focus:[9]

$$\text{electron angle at beam edge} = \sqrt{2\xi \ln r/r_0}$$

$$\text{where } r/r_0 = F^{-1}\left(\sqrt{2\xi}\ z\right)$$

$$\text{with } F(x) = \int_1^x \frac{dy}{\sqrt{\ln y}} \qquad (30)$$

$$\text{and } \xi = \frac{2}{(\gamma\beta)^3}\frac{I}{I_0}$$

Surprisingly, for this design, the blow up is small enough that only a few periodic lenses are required to maintain low $T_\perp$. The following technical problems must be addressed.

A. A ~5 MeV, ~5-10 amp electron beam is a formidable task. Parallel interests exist for developing such beams and supplies for free electron lasers. One is being built. Beam loss is hoped to be $\lesssim 10^{-3}$ which implies a 20-30 kW supply.

B. Present cooling experiments ($\lesssim$100 MeV proton energy) indicate that $T_\perp \lesssim 0.25$ eV can be achieved in the interaction region. At these low energies such temperatures correspond to electron angular deviations of $\lesssim$1 mrad which is an easy precision for the electron guide fields. At ~8 GeV the guide precision must either be ≲0.1 mrad (difficult - through toroids etc.) or the imperfections must be adiabatic on the larmor wavelength (= $2\pi\beta\gamma mc^2/eH$ scale. The latter implies large guide fields and/or large magnet apertures (in ratio to the e^- beam diameter). The small interaction beam diameter ameliorates this situation.

C. As pointed out above (Section 2c) low gun diode voltage allows lowest $T_\perp$. As pictured already for 400 MeV cooling in figure 3, the gun must consist of two sections. First a low voltage, high perveance, diode region which defines the beam current and,second, a uniform gradient acceleration column. It remains to be demonstrated whether such structures can achieve the same low $T_\perp$ obtained from low energy guns. For these small diameter beams the only alternative to higher diode voltage is "convergent diode" optics. By this we mean a large concave cathode with Pierce optics (plus "orthogonal" B field) convergent down to an interaction region value. Both calculation and experience (the Fermilab device has such a gun) show higher $T_\perp$ (compared to asymptotic immersed field cases) for this design.

Finally, it is not necessary to use electron cooling as a cyclic accumulator. Cooling speed at large γ is not a strong point of the method. Low equilibrium $\Delta p/p$ and stack current independence are the true advantages. The ideal arrangement is to initially stochastically stack (a $\bar{p}$ current limited process) but then to periodically (many minutes; hundreds or thousands of accumulation cycles) plunge this partial stack into an electron "freezer".

Scenarios of this kind are similar to that described in 3 above, the difference being that the cooling time is now not directly related to (18). The drawback to this concept is its overall complexity (deceleration is required between prestacking and electron freezer). Since neither fast cooling nor ultimate coldness (see the limitations discussed in Section V) are now required, it is generally possible to find an acceptable trade off between the constraints of damping rate, machine acceptance, and vacuum (equation 27) as functions of ($\beta\gamma$).

In summary we note that e^- cooling for $\bar{p}$ accumulation usually does not require the highest quality electron beam (as measured by $T_\perp$), since the hot $\bar{p}$ batches of interest have $T_{p\perp}$ >> M/m $T_\perp$. The electron gun and transport system and collector designs are compromised between low $T_\perp$ and high current. However, this is not always the case, for instance as in Section 3 where in principal the $\bar{p}$'s could be precooled to the level $T_{p\perp}$ ~ M/m $T_\perp$.

We have not discussed the asymptotic $\bar{p}$ density in any of the above cases. This also is determined by the electron temperature, but in a way which may depend on the "magnetized" cooling which we treat below. A feature of electron cooling is the high phase space density achievable (essentially independent of the initial cooling rate!) and the large damping force it imposes on the cooled stack (independent of number). Thus one expects that the only way to maintain the highest density stacks against IBS, or coherent instabilities is via electron cooling.

IV. FACTORS DETERMINING ELECTRON BEAM TEMPERATURE

1. Equilibrium Distributions

As with any Markovian process there will be diffusion effects. Thermal equilibrium is such an effect. The asymptotic $\bar{p}$ phase space area will be given by diffusional equilibrium between e^- and $\bar{p}$ beams[12]

$$T_{\bar{p}\perp}(\infty) = 1/2\ T_\perp \tag{31a}$$

$$T_{\bar{p}\parallel} = \frac{\pi}{4}\sqrt{T_\perp T_\parallel} \qquad \Lambda_\perp/\Lambda_\parallel \qquad (\ll \Delta T_\perp)$$

This illustrates why low $T_\perp$ is important even when $T_{\bar{p}\perp}$(initial) > $T_\perp$ dominates λ (initial). The flattened electron distribution implies $T_{\bar{p}}\ \infty \ll T_e$ so that extremely small $\Delta p/p\ (\infty)$ are expected with electron cooling. Equation (31) holds only when the ion velocities are well within the electron distribution disc. In this limit the theory is substantially modified by the magnetic guide field (see below)

$$T_{p\perp} = T_{p\parallel} \backsim T_\parallel \tag{31b}$$

Equation (31) indicates one of the experimental methods for <u>determining</u> $T_\perp$. For betatron oscillations a particle's maximum <u>amplitudee</u> $a_{x,z}$ is related to $\Theta_{x,z}$ as $\alpha_{y,z}\ \Theta_{x,z} = \varepsilon_{y,z}/\pi$ and we have

$$T_{p_\perp} = MV^2_{x,z} = Mc^2\ (\beta\gamma)^2\ \Theta^2_{x,z} = M(\beta\gamma)^2\ \alpha^2_{x,z}/\beta^2_{x,z} \tag{32}$$

So a measurement of beam profile rms in x or z (at zero dispersion lattice points) gives T_p.

Another T_p measurement is the production rate of neutral <u>hydrogen atoms'</u> in the interaction region.[11,14] This fascinating phenomena is a dramatic demonstration of the cooling itself but only occurs for <u>single</u> ionized positive ions (i.e. <u>not</u> $\bar{p}$'s, but protons which are universally used to tune up $\bar{p}$ accumulators)

$$\sigma_{capture}(u) \propto \frac{1}{u^2} \tag{33}$$

The neutral beam is separated from the circulating one at the first bend after the cooling region, emerging as a monochromatic proton beam. The exact theory is complex when magnetic guide fields are taken into account, but within the uncertainties observed, hydrogen production is consistent with cooling rates and (32). More directly, the hydrogen spot size can be measured with an external chamber, determining the proton $<\theta^2_{x,z}>$ averaged over the interaction region.

$T_\parallel$ is most easily measured by "Schottky" scan (spectrum analysis of the longitudinal storage ring current).[4] However, all the above temperature measurements only directly measure $T_{p\perp,\parallel}$. Misalignment of e^- beam axis (defined by the solenoidal guide field) or fluctuations of the electron high voltage supply potential can easily introduce "effective" apparent e^- temperatures much higher than the true incoherent one. Let us consider the various practical sources of the effective temperature.

0. Reviewing example 3: we expect ideally $T_\perp \sim 1$ eV and $T_\parallel \sim 10^{-4}$ eV or 10^{-6} eV (31a or 31b), due to the thermal cathode temperature ~0.1 eV. We take these as base values.

1. As we discussed in Section III space charge depresses the kinetic energy of electrons inside the beam. For currents above ~10 amps and at nonrelativistic energies, this can about double the effective temperature. But for cooled ions centered in the electron beam this should not increase the equilibrium $T_{p\parallel}$.

EXERCISE

Show (as shown in figure 7) that if the mean ion closed orbit is coincident with the e^- beam center, and if the interaction region has nonzero dispersion, then for sufficient space charge depression some of the ions may be unstable under the friction force. How can this be remedied? What is the effect of the remedy on T (effective). Note that cooling experiments at both INP, Novosibirsk and at CERN had $\alpha_p \neq 0$.

---

2. The "E x B" drift motion. Quantitatively this contributes to $T_\perp$

$$T_\perp\,(\mathrm{ExB}) = \frac{1}{2} m v_\phi^2 = \frac{1}{2} m \left(\frac{|\vec{E}\ (\mathrm{volt/m})|}{B(\mathrm{gauss})\times 10^{-4}} \right)^2$$

$$= \frac{0.34\, r^2(\mathrm{m})}{\beta^2\, a^4}\ \frac{I^2(\mathrm{amp})}{B^2(\mathrm{gauss})} \qquad (34)$$

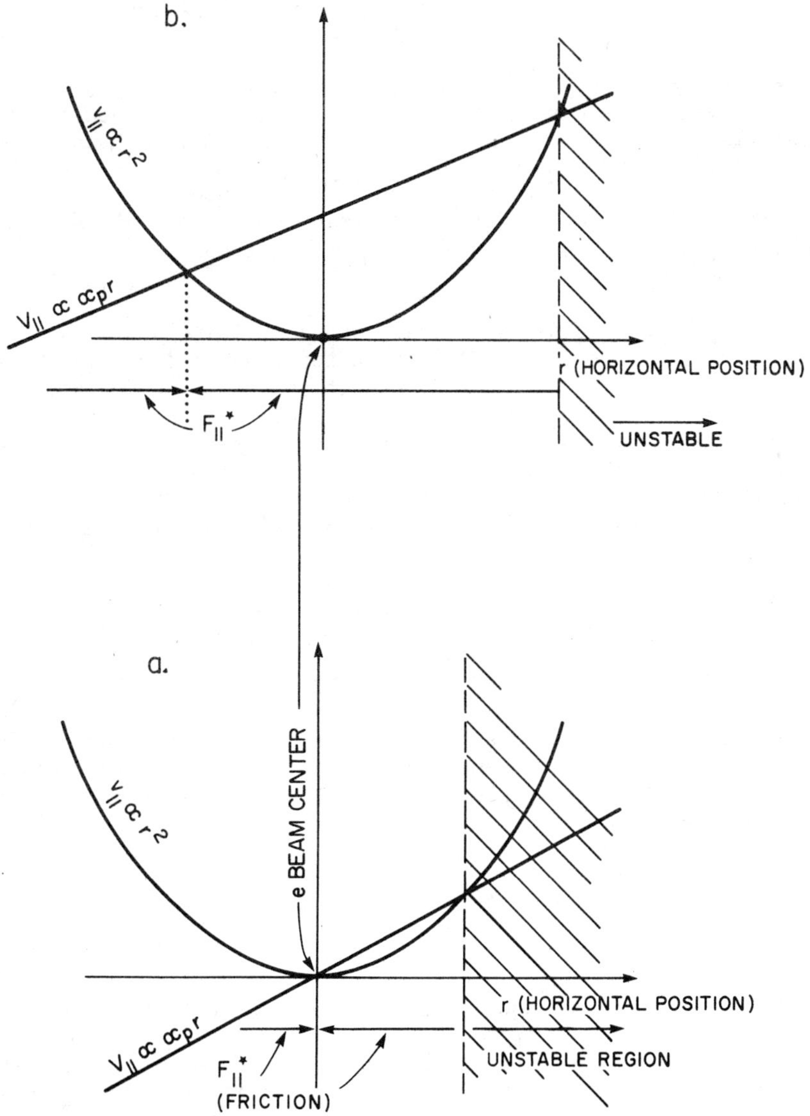

Fig. 7 Relative <u>initial</u> velocities as a function of radial position for $\alpha p \neq 0$ interaction region and for actual T electron progile. (a) Mean velocities matched: large unstable region (b) Mean velocities mismatched: removed unstable region.

= 0.15 eV (at beam edge for figure 4 example)

Brute force confinement eliminates this temperature source!

In many designs (e.g. Section III.3) one would like a physically small e^- gun and accelerating column (ease of H.V. construction which must be inside a solenoid!). Then at the high $\beta_\perp$ interaction region the $\bar{B}$ field tapers off to allow size matching to the $\bar{p}$ beam. This B reduction does not alter $T_\perp$ (EXB) since a^2B is an adiabatic invariant (Bush's theorem).

3. Gun Anode Lens. Near the cathode it is possible to arrange cylindrically symmetric "Pierce" electrodes such that an electron beam of arbitrary I aquires no additional $T_\perp$ above the cathode temperature as it accelerates. The form of the Pierce geometry is of equipotentials cutting the beam. Since we physically cannot have a final V=0 electrode cutting the beam this ideal situation cannot hold everywhere. Some "anode" defining the cathode perveance must exist with an equivalent lens ($\vec{E}_r \neq 0$ fields) hole in it. Such a geometry would induce

$$v_\perp = v_\parallel \, r/3d \tag{35}$$

for electrons at radius r and diode length d. This leads to enormous $T_\perp$ (again $\propto r^2$)!

This situation is saved by magnetic confinement plus "resonant" focusing.[15] In the presence of a solenoidal field B the radial equation of motion for each electron is simple harmonic with a forcing term given by

$$\ddot{r} + \omega^2_{LARMOR}(r - r_o) = e\,E_r(r_o, z(t)) \tag{36}$$

Now $E_r(t)$ = const., such as comes from space charge, just has the effect of shifting the equilibrium r_o (of course it also excites Θ as per eq. 35). But transient $E_r^o(z)$ can start an oscillation (heating), or if the phase is correct could cancel any existing coherent oscillation! Notice that this is true at all r if the $E_r(z)$ perterbations (causing and canceling) are out of phase. For instance an anode of the correct, resonant, length

$$L_{eff} \sim n \, \frac{2\pi \, p_\parallel \, (ANODE)}{Be} = \lambda_{LARMOR}, \qquad n=1,2$$

will impart no coherent $\dot{r}$.

This theory implies that certain "magic" values of B and gun diode voltage ($p_{\parallel}$) will result in low $T_{\perp}$ beams. Figure 8 illustrates this resonance behavior for the CERN cooler. Practically it may be desirable to independently set B and $P_{\parallel}$(ANODE) (and therefore I via Child's law) e.g. to change I without disturbing magnet currents or $p_{\parallel}(\infty)$. For this the anode electrode is not run at ground potential but from a supply referenced from cathode potential (fig. 4). After it a series of resonant tuning electrodes are arranged with spacing $\lesssim \lambda_{LARMOR}/2$ (see figure 4).

4. Imperfections in the solenoidal field homogeneity which are not adiabatic ("bumps") excite similar coherent $v_{\perp}$ excitations. If the bump is cylindrically symmetric with respect to the e^- beam the $T_{\perp}$ induced would be quadratic in r. Typically interfaces between magnetic elements (e.g. gun/toroid; toroid/cooling region) are the culprits. In analogy with the discussion, 3. above, and since the electron cooler (as in fig. 3) magnetic circuit is reflection symmetric about a central plain (or point) we expect resonant behavior similar to (37) but with $P_{\parallel}(\infty)$ instead of $p_{\parallel}$ (anode). This is observed in the great sensitivity of $T_{\perp}$ to $p_{\parallel}(\infty)$ since the dimensions of the magnetic elements are large compared to λ_{LARMOR}.

As a typical example consider the anomaly in axial field in the interaction region caused by the joints between the one meter segments of solenoid winding of the Fermilab cooler (5 segments in all). At radius r the induced $v_{\perp}$ is, for a peak anomaly ΔB of length Δ

$$v_{\perp}/\beta c = \gamma\pi \frac{r\lambda_{LAR}}{\Delta^2} \frac{\Delta B}{B_o} \qquad \Delta >> \lambda$$

$$= \gamma\pi^2 \frac{\Delta r}{\lambda^2} \frac{\Delta B}{B} \qquad \lambda >> \Delta \tag{38}$$

But the field near the center of a loop ($I_{loop} \propto g$) is $\pi I_{loop}/cd$ which gives

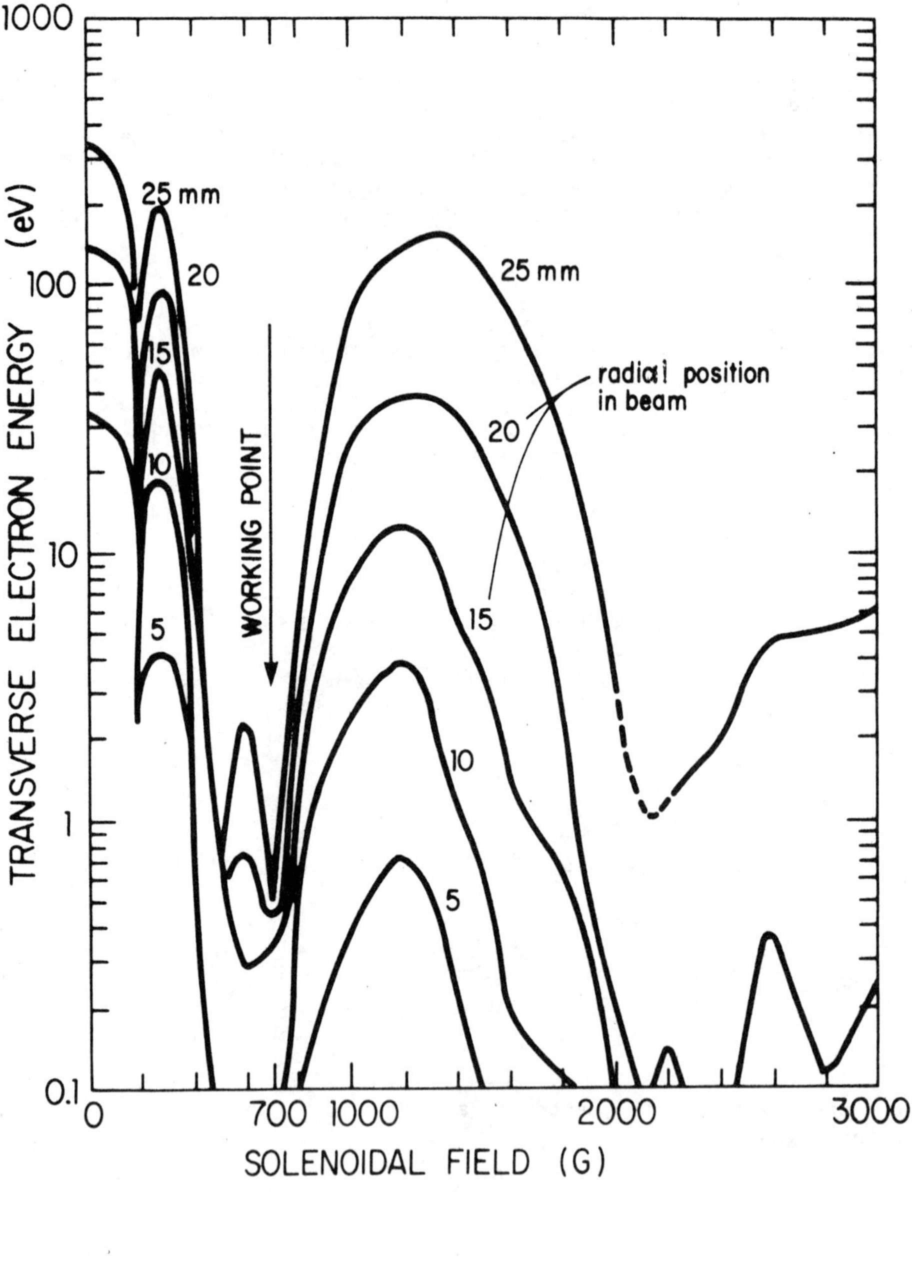

Fig. 8 Curves of T_- as a function of gyro-wavelength (B_o) for various radii ($r_\perp$) within the e^- beam profile.[17]

$$\Delta B/B_o \sim \frac{\pi g}{cd}\ 2\pi/c \sim 1/2\ g/d$$

$$\Rightarrow v_\perp/\beta c = \gamma\ (\pi/2)\ r\ \frac{\lambda g}{\Delta^2 d} \qquad \Delta \ll \Lambda \tag{39}$$

In the Fermilab case $\gamma \leq 1.2$, $r=a \leq 3$ cm, $g \leq 3$ mm, $d \sim 2\Delta = 25$ cm, and $\lambda = 6$ cm ($B_o = 1000$ G), gives $T_\perp < 1$ eV (beam edge).

We have mentioned that beam expansion by adiabatically decreasing B_o, allows lower temperatures in accordance with Bush's theorem. However, viewed as a perturbation this ΔB induces a $T_\perp$ equal to 1/4 of (38). If B_o is, say, halved in the above case the transition region must be $\gtrsim 12\lambda^o_{LAR}$ to have $T_\perp \leq 1$ eV.

5. Misalignment ($\equiv\alpha$) of Average Ion closed orbit direction with the electron beam direction ($\vec{B}_o$ if solenoidal guiding used), will obviously lead to an effective increase in $T_\perp = (mc^2)\ (\beta\gamma)^2\ \alpha^2$. Actually this $T_\perp$ dependence only holds i if $\sigma_\perp \gtrsim \alpha$ (or if $\sigma_\parallel \gtrsim \alpha$ for fully "magnetized" cooling). For the opposite inequalities (i.e. ions outside the friction force maximum) the betatron oscillations are antidamped and grow to a value

$$\langle\theta^2_{ion}\rangle = \alpha^2 \tag{40}$$

So that the equilibrium temperature becomes $T_{\perp ion} = (M/m)T_\perp$(effective). Of course, if $\theta_{ion} \geq \alpha$ initially, then only damping occurs but this same equilibrium holds. For $\beta=1$ a 1 mrad misalignment gives $T_\perp$(effective) $\simeq$ 1ev and $T_{\perp ion} \sim 2000$ev!

EXERCISE

Qualitatively demonstrate this instability by examining the behavior of $\langle F_\perp^*(V)\ V_\perp\rangle$ ($\langle\rangle$ over betatron phase) using the force diagram Figure 2 for $\langle V_\perp\rangle = \alpha \neq 0$.

It is crucial to have some fine tuning of either the closed orbit or the e^- beam direction to line up the effective beam axes. The equilibrium $T_{\perp ion}$ is extremely sensitive to α since equilibrium occurs as the ions cool to the e^- beam center when $\sigma_\perp$ is very small effects 3, 4, 5 vanish). Notice that the misalignment contribution to $T_\perp$(effective) is uniform in r. The necessary fine tuning to $\alpha = 0$ is easily accomplished by incorporating long weak dipole windings along the interaction solenoid length which alter $\vec{B}_o/|B_o|$.

6. High Voltage Supply Regulation and ripple contribute a much higher $T_\parallel$(effective) than the "flattened" cathode temperature

$$T_{\parallel}(\delta V) - \frac{1}{2}\frac{\delta V}{V_0}\delta V \qquad \text{(non-relativisitic)} \qquad (41)$$

where V_0 is the H.V. supply voltage. This result holds for "ripple" on a time scale shorter than $1/\lambda$. Practically, ripple is usually not critical for fast initial cooling. First, because the effect is uniform in r so that it will be very much smaller, due to the suppression factor $\delta V/V_0$, than the large $T_{\perp}(r=a)$ dominating initial cooling. The usual measure of $T_{\parallel,\mathrm{ion}}(\infty)$ is the $\Delta p/p_0$ of the cooled ions. But we see from (31a) that this is $\propto (T_{\parallel}(r=0))^{1/4}$. Table II displays "best" data from Novosibirsk and CERN. The quoted $T_{\perp}$(r=0of Table II,) plus measured H.V. regulation ($\sim 10^{-5}$) would give (via (31a)) $\Delta p/p_0 \sim 4\times10^{-6}$. We discuss this discrepancy further in the next section.

EXERCISE

Calculate: $T_{\parallel}(\delta V)$ relativistically:

Answer: $$T_{\parallel} = \frac{(\delta V)^2}{2m(\beta\gamma)^2\,[1+(\beta\gamma)]^2}$$

In summary, the distinction between the r^2 dependent T(effective) and the uniform (with r) contributions is to be emphasized. The design criteria for a fast initial cooler treats the former while ultimate phase space density is connected with the latter. The $(\beta\gamma)^2$ dependence of $T_{\perp}$ of equations (38) and (39) indicates that for fixed $T_{\perp}$ and fixed coil diameter (d) winding accuracy, segment alignment, flex, etc. all must be held to increased accuracy with momentum. (A way around this was illustrated in Section III.)

TABLE II

	NAPP-M[4] COOLER	ICE[17] DEVICE	FERMILAB[3]
Electron Energy (KeV)	45	25.9	111
Electron Current (A)	0.8	1.3	1-3
Electron Beam Ø(cm)	2	1	5
η/Interaction Length (m)	0.02/1	0.04/3	0.05/5
Solenoidal Field (K Gauss)	1	0.5	$\sim$ 1
"Beam on" Mean Vacuum (T)	5×10^{-10}	2×10^{-9}	1×10^{-9}
HV Supply Regulation	1×10^{-5}	6×10^{-5}	$<3\times10^{-5}$
Betatron e Folding Time (s)	0.19	1.2	15(@ 63 kV)
for Initial FWHM θp =	$\sim 10^{-4}$ (+low $\Delta p/p$)	1.5×10^{-3}	$\sim 1\times10^{-3}$
Momentum e Folding Time (s)	-	0.3-0.4	$\sim$ 2(@ 63 kV)
for Initial FWHM θp =	-	1.5×10^{-3}	1.2×10^{-3}
Equilibrim: $\frac{\Delta p}{p}$(FWHM)/$T_{\parallel}$(effective)	2×10^{-5}/ 10^{-4}ev	4×10^{-5}/-	2.1×10^{-5}/ 3×10^{-4}ev
θp(FWHM)/$T_{\perp}$(effective)	8×10^{-5}/ 0.2ev	2.3×10^{-4}/ 0.2ev	9×10^{-5}/ 0.9ev
Storage Lifetime (s)	3.6×10^{4}	8.4×10^{3}	10^{3}

V. Magnetized Cooling and Maximum Phase Space Density

Applications of electron cooling other than for $\bar{p}$ accumulation often require maximizing the phase space density of an available quantity of ions. Driving high luminosity, high duty cycle studies of resonant nuclear reactions is an example. Production of $\bar{p}p$ (protonium) states by overlaping cold p and $\bar{p}$ beams is another. Also, injecting $\bar{p}$'s into a very low phase space volume Penning trap for g-2 measurement requires "stopping" ultra-cooled stored beams.

If the available quantity is fixed, one must seek low equilibrium temperatures, which the above theory shows to be dominated by the relatively large $T_{\perp}(>>T_{\parallel})$. On the other hand, if cumulative quantities are available (eg. via stacking cycles of production) we have seen that the simple theory allows indefinite particle accretion into the equilibrium Maxwell-Boltzman distribution. This naive picture must be limited by the growing influence of ions on each other as density increases. This can take two basic forms. First, an incoherent IBS which depends on spatial density and relative ion velocity in the same way as the cooling itself. Second, coherent instabilities ("plasma" oscillations of various modes) including the "static" space charge growth. Beyond a certain phase space density reduced $T_{\perp}$ will not in itself permit further cooling. There must be an accompanying increased damping force to offset growing IBS diffusion or incipient instabilities.

Fortunately, there is a correlation between $T_{\perp}(r=0)$ and λ (where now we are "inside" the electron distribution, $V_{\perp} \lesssim \sigma_{\perp}$):

$$\lambda \propto \frac{J}{T_{\perp}(r=0)} \tag{42}$$

At high ion phase space density we no longer achieve equilibria (31). Instead, equilibrium is calculated by setting $-\lambda$ equal to IBS or mode blow up rates calculated from instability theory. The quantitative calculation is difficult (one actually has 3 separate rates for the 3 degrees of freedom connected by non-diagonal diffusion matrices all the terms of which are highly non-linear in $u_{\parallel,\perp}$). Also, the theory of IBS is in an uncertain state even at a qualitative level. No conclusive calculations of $T_{ion}(\infty)$ as a function of N_{ions} has been done. The experimental situation is illustrated in Figure 9.

We have only considered the electron's solenoidal guide field as a container against space charge blow up. Certainly all cooling aimed at maximizing phase space density will be at subrelativistic $\beta\gamma$ and thus require strong solenoidal containment. We now consider the influence of B_z on the cooling process itself.[4,12,13]

The exact electron motion is, of course, a spiral about the solenoidal field lines. $T_{\perp}$ appears as the amplitude of these circles

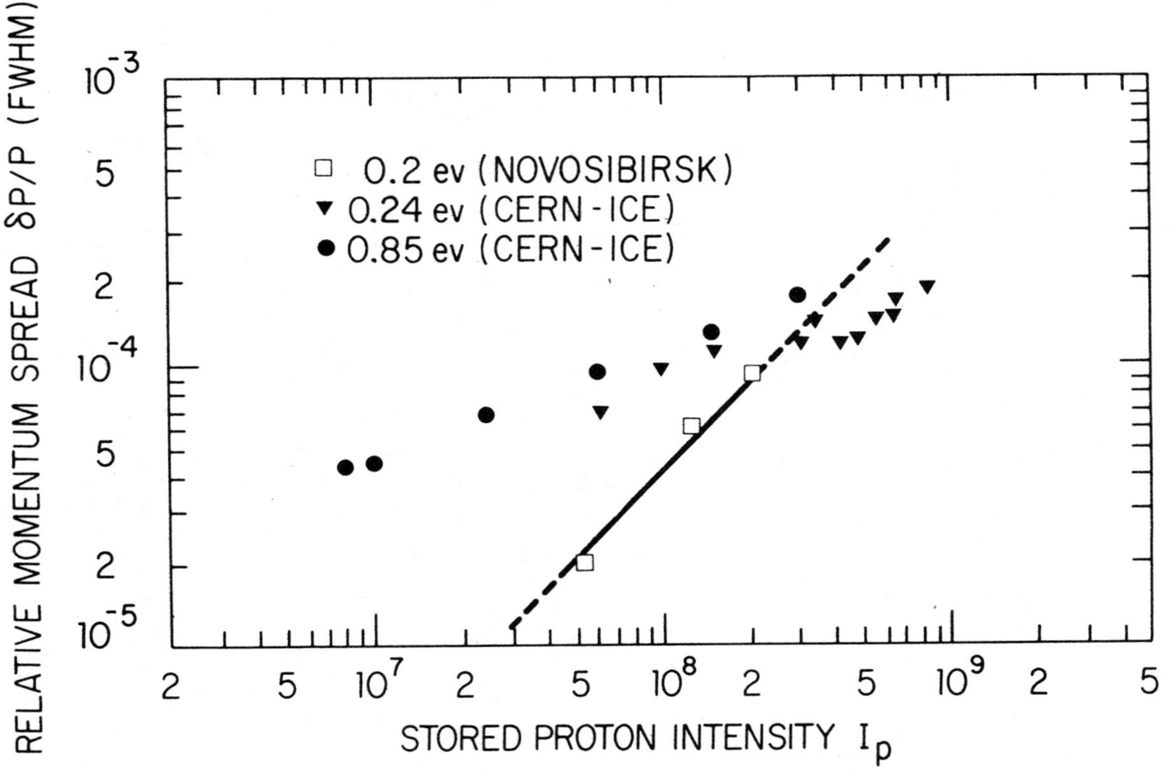

Fig. 9 Longitudinal stored proton equilibrium state as a function of stored current. (See Table II for comparison of other relevant parameters).[4,17]

$$\rho_e = 2.4T_\perp/B = 2\mu m \quad \text{for } T = 0.5 \text{ ev}, \; B = 1\text{K Gauss} \tag{43}$$

We can discuss the physics of $\bar{p} - e^-$ collisions, where the electrons are magnetically constrained by considering three regimes separately. Again a highly flattened disc distribution of electron velocities is assumed.

A. $|V| > \sigma_\perp$. For all impact parameters $b < \rho_e$ we have

$$\tau_{collision} < \frac{b}{|V|} < \frac{\rho_e}{\sigma_\perp} \equiv \tau_{larmor} \tag{44}$$

If we redefine $b_{max} \leq \rho_e$ (which reduces Λ since b_{max} (B = 0) >> ρ_e, typically, see eq. (10)), then cooling is as with no magnetization. Thus, we expect only a slight change in the logarithm for the cooling of initially hot beams.

B. $\sigma_\parallel < |V|\ \sigma_\perp$. As $|V|$ gets smaller, a larger range of impact parameters satisfies $\tau_{collision} < \tau_{larmor}$. Such adiabatic collisions average over several gyroperiods: no transverse momenta can be exchanged between the electron and ion. An ion's energy loss rate is now non-isotropic (not because of the electron temperature anisotropy!) for adiabatic collisions the equivalent of (7) is

$$\dot{E} = \tilde{F}.\tilde{V} = -\frac{2\pi e^4 n^*}{m}\Lambda_{Adiab.}\frac{1}{|V|}\sin^2\theta^* \tag{45}$$

where $\theta^* \equiv V_\perp/|V|$ is the rest frame ion angle with respect to the $\tilde{B}$ direction. Since $v_\perp$ is irrelevant for adiabatic collisions the integral over f(v) (analogous to (9)) becomes a delta function for the very small $v_\parallel$ typical of the flattened electron distribution. The final net drag is then (45). For an isotropic (rest frame) ion distribution the average energy loss rate (integrate (45) over θ^*) can be much larger than for non-adiabadic collisions since $|V|$ in the denominator is now limited by $v_\parallel$ not $\sigma_\perp$.

C. $|V| < \sigma_\parallel$. Adiabatic collisions dominate (Λ_{Ad} comparable to Λ(B=0)). The high transverse electron temperature becomes irrelevant. The transverse damping ($\theta^*=\pi/2$) reaches a maximum (as the ion crosses the disc boundary)

$$\begin{aligned}\lambda_{max} &\simeq \frac{4\pi e^4 n^*}{mM}\Lambda_A(\sigma_\parallel)^{-3} \\ &= \lambda(B{=}0)\Lambda_A/\Lambda(B{=}0)\ \ \sigma_\perp^2/\sigma^2\end{aligned} \tag{46}$$

Notice that Λ_A depends on $\overline{B}$ via (43) and (44).

This predicted fast cooling has been observed in two ways. An already cooled beam (therefore, within regime B-C above) is instantaneously perturbed (e.g. transverse kick) and the redamping rate observed. Second, ion beam lifetime may be observed under conditions (e.g. low ion energy, poor vacuum, low electron beam current) such that the B=0 theory would predict $\lambda <$ the rate of beam diffusional blowup (say, due to Coulomb scatter). In such a situation no persisting beam is expected. The Novosibirsk group has demonstrated dramatic beam lifetime increases (approaching the single scatter limit (27)) at low kinetic energy (1-1.5 MeV) with few milliamp electron beams (space charge limit!).[16] The increased multiple scattering of such a nonrelativistic energy would disallow non-magnetized cooling (see Figure 10).

Since the solenoidal field serves to "freeze" out the electron's transverse degrees of freedom, the ions (low intensity) cool toward an equilibrium $T_{ion} \simeq T_{\parallel}$ (compared to (31a)!). Therefore, a given quantity of ions can attain vastly increased phase space density via the magnetized interaction. As stored intensity increases an actual $T_{ion} > T_{\parallel}$ will be determined by the complicated equilibrium between cooling and ion beam instabilities. However, there will always be a lower magnetized equilibrium value due to the average increased damping (eq. 46).

For example, it is of interest in nuclear physics to upgrade medium energy facilities (cyclotrons) with internal target ion storage rings.[19] At such facilities the magnetized cooling has the potential of providing the highest possible <u>resolution</u> probe for target excitation at energies above those attainable with tandems. In such storage rings ($E_{kinetic} >$ hundred MeV) neither gas scattering nor IBS determines the equilibrium temperatures. Instead, internal target dE/dx (target thickness) heats the beam, while, in a different lattice location, electrons <u>cool</u> the beam. Variation of target thickness t, $\beta_{\perp}$ functions at target or cooler, and cooling strength all control the ion equilibrium temperature through

$$\frac{d}{dt}(T_{ion}) = \frac{4.1\times10^{-9}\ (\text{cgs})\ Z_B^2\ Z_T^2}{\beta^2\ A_B}\ tf\,\frac{\beta_{\perp,T}}{\beta_{\perp}} - \lambda\,(T_{ion} - T_{\perp}) \quad (47)$$

where f is the revolution frequency and λ, the electron (generally magnetized) damping rate depends on $T_{\perp}$ and T_{ion}. A maximum t results from the solution of this where $V_{\perp} \simeq v_{\perp}$. For <u>thin</u> targets, beam equilibrium with the electron temperature ($\rightarrow T_{\parallel}$ for full magnetization!) may be approached. The balance between various time scales involved for the Indiana project is illustrated in Fig. 11.

To not leave the impression that the magnetized limit $T_{ion}(\infty) \simeq T_{\parallel}$ (potentially $\rightarrow$ 2ev for 200 MeV protons - see eq. (41)) can be usefully contemplated, I end by listing some limitations to reaching these extremely low magnetized temperatures:

a. Any solenoid misalignment or imperfection sets an effective $T_{\perp}$ which is not "frozen" out. In practice the solenoid design and

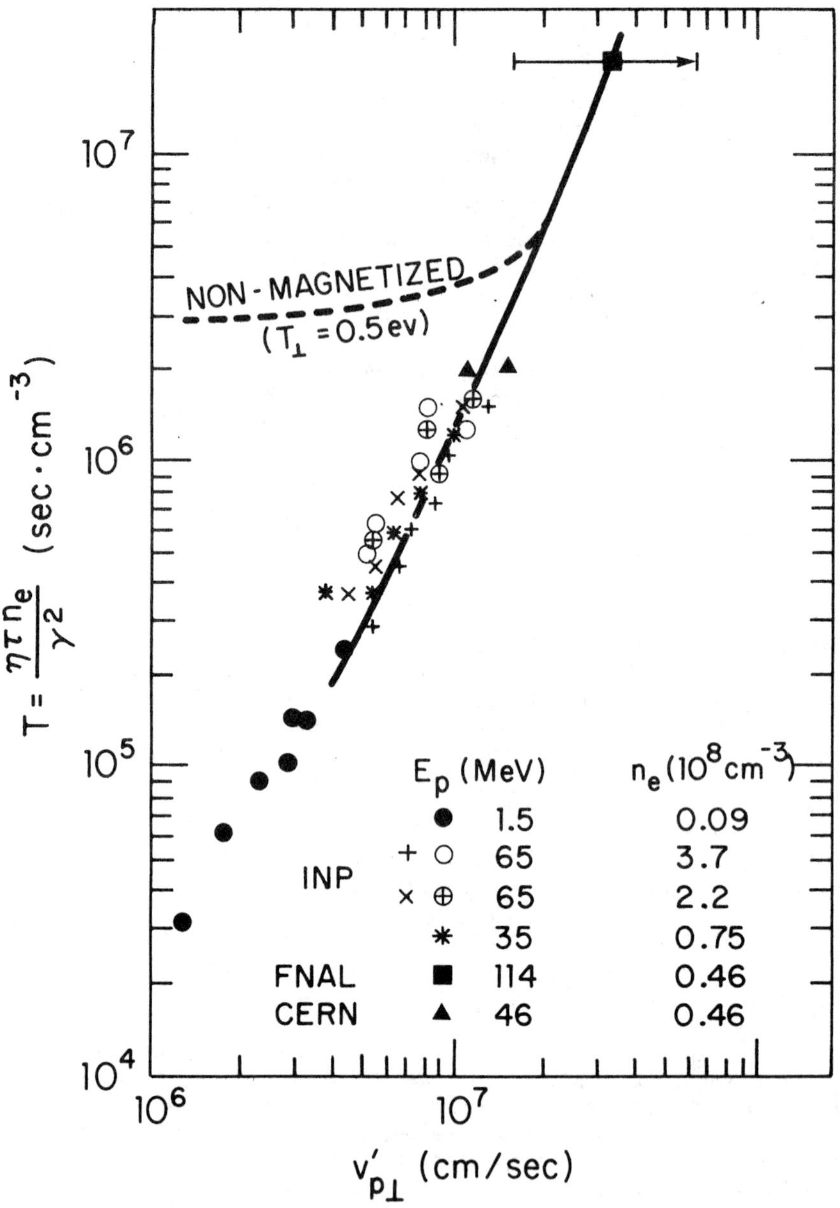

Fig. 10 Comparison of various electron cooling results. FNAL result: non-magnetized. CERN results and INP work at MeV: borderline case (T <0.5 eV!). Note definite magnetization for the 1.5 MeV INP work

solenoid axis - ion closed orbit alignment error probably determines $T_{ion}(\infty)$.

b. Of course, as we discussed above, any large current (ion or electron) effects will increase with phase space density so that the full magnetized limit might only be achieved for extremely low intensity beams. For instance e^--e^- collisions would tend to equilibrate $T_\parallel$ and $T_\perp$.

c. The entire analysis considered here is within the Coulomb log approximation. Before $T_{ion} \simeq T_\parallel$ could be approached higher($\propto \Lambda$ relative to the λ_{max} cooling rate) processes must be considered. be considered.

The most thorough attempt at discussion of these effects appears in the papers of Derbenev and Skrinsky, especially with regard to guide field misalignments (The "monochromatic" instability).[4,12] However as these authors admit, and as the experimental evidence indicates full understanding of electron cooling at low beam temperature and high phase space density is far from complete.

VI. TWO LIMITS: ULTRA-HIGH AND ULTRA-LOW ENERGY COOLING

A final emphasis will now be placed on the storage ring/accelerator system aspect of cooling by considering extreme limits of suggested applications.

Once $\bar{p}$'s are accumulated at low or moderate energies (< 10 GeV) they are actually used in p-$\bar{p}$ colliders (≳300 GeV). The actual beam-beam and beam-gas hard collision loss rates are negligible. However, the luminosity

$$L_u \propto \frac{n_p n_{\bar{p}} f}{a_p^2 + a_{\bar{p}}^2} , \tag{48}$$

where a = rms beam size at collision, will degrade with time as a_p and $a_{\bar{p}}$ grow and eventually as $n_{\bar{p}}$, n_p dwindle. Of course, n_p and a_p can be easily replenished. On the other hand, there is a limit to the allowed strength of the electromagnetic perturbation of protons on antiprotons at the collision point, which may be stated as a tune shift limit

$$\Delta\nu_{max} \propto \frac{n_p \beta_\perp}{\gamma a^2} \quad (\beta_\perp \text{ at collision point}) \tag{49}$$

thus

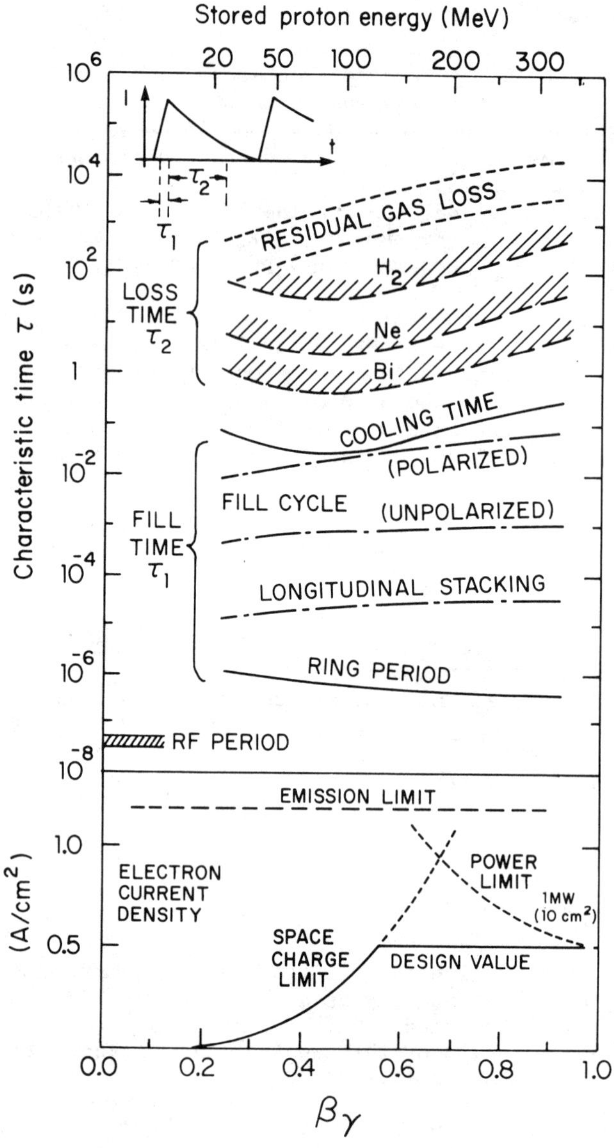

Fig. 11 Cycling time structure of an electron cooled storage ring for internal target nuclear physics. Storage ring is filled in τ, from a cyclotron source. Note how the cooling time becomes dominated by limited electron beam current density due to space charge tune shift.[19]

$$L_{umax} \simeq 5.6\times10^{19} \frac{\gamma\Delta\nu_{max}}{\beta_\perp} n_{\bar{p}} \ \mathrm{cm}^{-2}\ \mathrm{sec}^{-1} \tag{50}$$

Electron cooling in the collider could both increase $\Delta\nu_{max}$ and retard the increase of a with time. The electron energy involved (> 100 MeV) precludes an electrostatic gun-collider design. Instead an electron synchrotron is considered which has a straight section in common with the $\bar{p}$ collider. At first sight (see eq 40) this seems hopeless because of the α^{-2} and η (a practical straight section << circumference of collider) dependence of λ. However, the following points prevail:

1. One is not accumulating $\bar{p}$'s so that the cooling "time" need not be short, ~ 1000 sec is useful.
2. Extremely high electron current densities are possible since $T_\parallel = m\,(\delta p/p_e)^2$ which allows relatively large electron beam momentum spreads (δp).
3. At such large γ space charge blow up (heating) of the electrons is not a problem (see eqn. 30). Thus the cooling straight section may simply be a low $\beta_\perp$ lattice space. The scaling involved is similar to that discussed in Section III.

Since the electrons are not replenished by a gun, synchrotron radiation must be relied on to dissipate the heat transferred from the collider beams. The design of such a device is quite fascinating and largely different from that of the low energy type![20]

An even less explored subject is the production of ultra cold low energy beams ($\leq$ 1 MeV protons ~ 700ev electrons!). Usually such beams are produced by electrostatic ion accelerators. However, some species ($\bar{p}$'s) and higher phase space densities may not be achievable by this classic means. One interesting application is the deceleration of $\bar{p}$'s with the goal of "filling" an ion trap (penning or RFQ type) which has extremely low phase space volume. In this way high precision studies of the $\bar{p}$ are possible.

To do this one starts with a "conventional" electron cooler as a source (eg. the LEAR storage ring ~ 100 MeV $\bar{p}$'s). After cooling asymptotically to $T_{\bar{p}}(\infty)$ one decelerates to the minimum energy possible for the ring (say 5 MeV). The factor 100 MeV/5 MeV = 20 $\equiv \xi$. We may transfer $\bar{p}$'s into a ring of circumference a factor ξ less in some adiabadic manner (e.g. RF bunching). Using equations 20, 21, and 24 we can scale the λ_{max} which an electron cooler in this mini ring can achieve

$$\lambda(\xi)_{max} = \lambda(1)_{max}\, \xi^{-2} \left(\frac{T_\perp(1)}{T_\perp(\xi)}\right)^{3/2} \tag{51}$$

where we have used a non-relativistic limit, and $\beta_\perp(\xi) << \beta_\perp(1)$. That is, the cold $\bar{p}$'s are made to look artificially hot at the cooling region by incorporating a tight focus there. Thus even $T_\perp(\xi) = T_\perp(1)$ will give a further emittence (phase space) reduction upon reaching the asymptotic temperature equilibrium. However, the cooling time is greatly increased, and at the same time gas scattering has increased as β^{-2}. One can expect a net cooling based on the following considerations:

1. Improved vacuum is possible since the e^- beam has very modest I ($\leq$ 10 ma!) and energy.

2. The low cathode emissivity necessary allows lower $T_\perp(\xi)$ ($< T_\perp(1)$).

3. $\beta_\perp(\xi) << \beta_\perp(1)$ implies short cooling regions which, combined with the low electron momentum, allows essentially full magnetization of the electron beam.

4. Long term beam lifetime (single scattering) is not an issue since the $\bar{p}$'s spend only a transient existence in this ring.

References

1. G.I. Budker, in Proc. Intl. Symp. on Electron and Positron Storage Rings, Saclay, 1966, p. II-1-1.

   G.I. Budker et al., in Proc. 4th All-Union Conf. on Charged Particle Accelerators, Vol. II, Moscow, Novk, 1975.

2. F. Krienen, in Proc. 11th Intl. Conf. on High Energy Accelerators, Geneva, 1980, p. 781.

3. R. Forster et al., in Proc. 1981 Particle Accelerator Conf., IEEE Trans. (NS)-28, Part 1, p. 2386 (1981).

4. For an excellent introductory review of cooling methods and for further references, see F.T. Cole and F.E. Mills in Ann. Rev. Nucl. Sci. 31 295 (1981) and for a comprehensive treatment of the electron method see The Physics of Electron Cooling , Ya. Derbenev, A. Skrinski (to be published).

5. Fermilab "TEVI" $\bar{p}$ Source Proposal, Fermilab Report, April 1981.

6. W. Kells et al., in Proc. 1981 Particle Accelerator Conf., IEEE Trans. (NS)-28, Part 1, p. 2583 (1981).

7. "UNK" Antiproton Source Proposal, T. Vsevolozskaja et al., (in Russian-Novosibirsk Preprint 80-182), translated to English in Fermilab $\bar{p}$ Note 150 (1981).

8. L. Oleksiuk, Fermilab $\bar{p}$ Note 124, April 1981 (unpublished).

9. Several good electron optics treatments exist, eg. H. Ivey, Advances in Electron Physics , Vol. XI , p. 158 (1954); Electron Optics and Electron Beam Devices , A. Zhigarev, Mir Publishers, (Moscow, 1975).

10. F. Mills, Fermilab $\bar{p}$ Note 144, September 1981 (unpublished).

11. G.I. Budker et al., Particle Accelerators 7 , 197 (1976).

12. Ya. S. Derbenev, A.N. Skrinsky, Part. Accel. 8, 235 (1978).

13. M. Bell and J. Bell, CERN Preprint TH3017 (1981).

14. M. Bell, J.S. Bell, CERN Report TH-3054 (1981, unpublished).

15. C. Rubbia, CERN Report EP77-2 (1977, unpublished).

16. INP (Novosibirsk) Electron Cooling Group, CERN Report 77-08 (1977).

17. Details of the CERN Ice Ring Electron Cooling experiments appear in M. Be et al., "Electron Cooling in ICE at CERN" to be published. (For a brief summary see: F. Krienen, Proc. 11th Intl. Conf. on High Energy Accelerators, Geneva, 1980, p. 781.

18. A. Piwinski, in Proc. IXth Intl. Conf. on High Energy Accelerators, p. 40 (1974).

19. The IUCF Cooler/Tripler Design Report (1981, unpublished).

20. C. Rubbia, "Relativistic Electron Cooling to Increase the Luminosity of th $\bar{p}p$ Collider", in Proc. Workshop on Cooling of High Energy Beams (D. Cline, Ed.), Madison, Wisconsin (1980).

SUPERCONDUCTING MAGNETS

A. V. Tollestrup
Fermi National Accelerator Laboratory, Batavia, Illinois 60510

TABLE OF CONTENTS

0094-243X/82/870699-106$3.00 Copyright 1982 American Institute of Physics

1. INTRODUCTION

Superconductivity was discovered in 1911 by Omnes, and the BCS theory was developed in 1957. It was 1961 before Kuntzler et al. showed that Nb_3Sn could carry high-current densities at fields up to 10 Tesla, and the theory for cryostatic stability was developed in 1965 by Steckly. Dynamic stability was not understood until 1967 to 1968. One can pose the question: Why 70 years after the effect and 20 years after the theory is superconductivity just now being applied extensively in high-energy physics? The answer is complicated but in essence comes down to a statement that we are still at the edge of the technology. The conductor development has been difficult and slow and has been done by a close collaboration between industry and the magnet builders. In addition, it has been an expensive technology to develop which has undoubtedly hindered its rapid growth. A magnet's design, as we shall see, involves little more than "freshman physics." Nevertheless, production has presented and will continue to present challenges to the technology. Accelerator magnets have very different requirements from the large magnets being developed for magnetohydrodynamics or nuclear energy or that have been successfully used for bubble-chamber applications. The magnets require a very high field, a high accuracy, pulsed fields, small volumes with the concommitant very high current density, and a very extended and diffuse cryogenic system. One can anticipate discovering many systems problems in the next one or two years that still have to be solved.

Nevertheless, there are many projected or partially completed accelerators involving superconducting magnets. In general these magnets are being developed around a conductor that will carry approximately 5,000 amps in a 5 to 6 Tesla field. Such cable in high quality and quantity is now readily available from industry. The following represents a list of such accelerators:

1. Tevatron - Fermilab. Approximately 1-TeV fixed-target machine or 1000 × 1000 GeV $\bar{p}p$ colliding beam.

2. ISABELLE - Brookhaven National Laboratory (BNL). Storage ring of 400 × 400 GeV, pp.

3. UNK-Serpukhov. 3-TeV fixed-target machine.

4. TRISTAN - ep. Phase II proton ring in the range from 300 to 500 TeV.

5. HERA - ep. Phase II proton ring at DESY.

In addition, there are many applications of superconducting magnets in high-energy physics. Some of the more notable are the following:

1. Bubble-chamber magnets at Fermilab, Argonne National Laboratory, and CERN.

2. 8° window frame dipole magnet at BNL.

3. HEUB magnets at Brookhaven National Laboratory.

4. Left bend beamline at Fermilab.

5. Low beta quads at CERN ISR.

6. A superconducting septum magnet at KEK.

7. A solenoid field for a muon channel at KEK.

8. Large superconducting solenoids for detectors such as CELLO.

9. Large analyzing dipoles for high-energy physics.

The following is an exposition of the design principles of such magnets. Emphasis will be placed on accelerator dipoles simply because the author has the most knowledge and experience in this area, and because these magnets have forced the limits of technology and hence furnish a real source of examples to illustrate the problems that have been encountered.

2. MAGNET DESIGN

The design of a superconducting magnet is dominated by the characteristics of the superconductor. There are three important variables: temperature, magnetic field, and current density. For temperatures lower than the critical temperature, there is a region of current density and magnetic field for which a superconductor exhibits zero resistance. A three-dimensional sketch of the surface that separates the region of superconductivity from that of normal resistance is shown in Fig. 1. The problem of the magnet design is to make sure that no point in the magnet coil has combinations of J, B, and T that take it outside the surface shown. The critical temperature, T_c, is completely dependent upon the material that is selected. For NbTi alloy, this is about 10.6° and for Nb_3Sn, a second commercially available alloy, it is about 18° K. The curve in the plane J = 0 tends to be parabolic in shape, and the maximum field attainable is called H_{c2}. The critical field for NbTi alloy at 0° K is about 17.6 Tesla and for Nb_3Sn, is about 35 Tesla.

If one slices the surface shown in Fig. 1 by planes perpendicular to the B-axis, the intercepts tend to be straight lines; however, the projection that we will be most interested in is that obtained by passing planes perpendicular to the temperature axis. These projections are shown for various temperatures in Fig. 2.

The operating temperature is determined by the cooling method; for instance, if the coil is cooled by cool boiling helium, the temperature will be near 4.3° K. If other methods are used, different temperatures may be achieved. It is obvious that lower temperatures will lead to higher values of J and B in an operating magnet. As already mentioned, the maximum critical field is determined by the alloy. The shape of the curves in this projection is complicated and is determined by the metallurgical processes that have been applied to the alloy. Heat treating and cold working, for example, can have an enormous influence on the shape of these curves. The reason that this is important may be seen by considering the magnet load lines that are also plotted on these same coordinates. A load line will be linear if there is no iron that saturates near the field. Different load lines can be drawn for different points in the magnet; for instance, there is a load line for the field on the axis of the magnet, but the one that is important for magnet design is the one

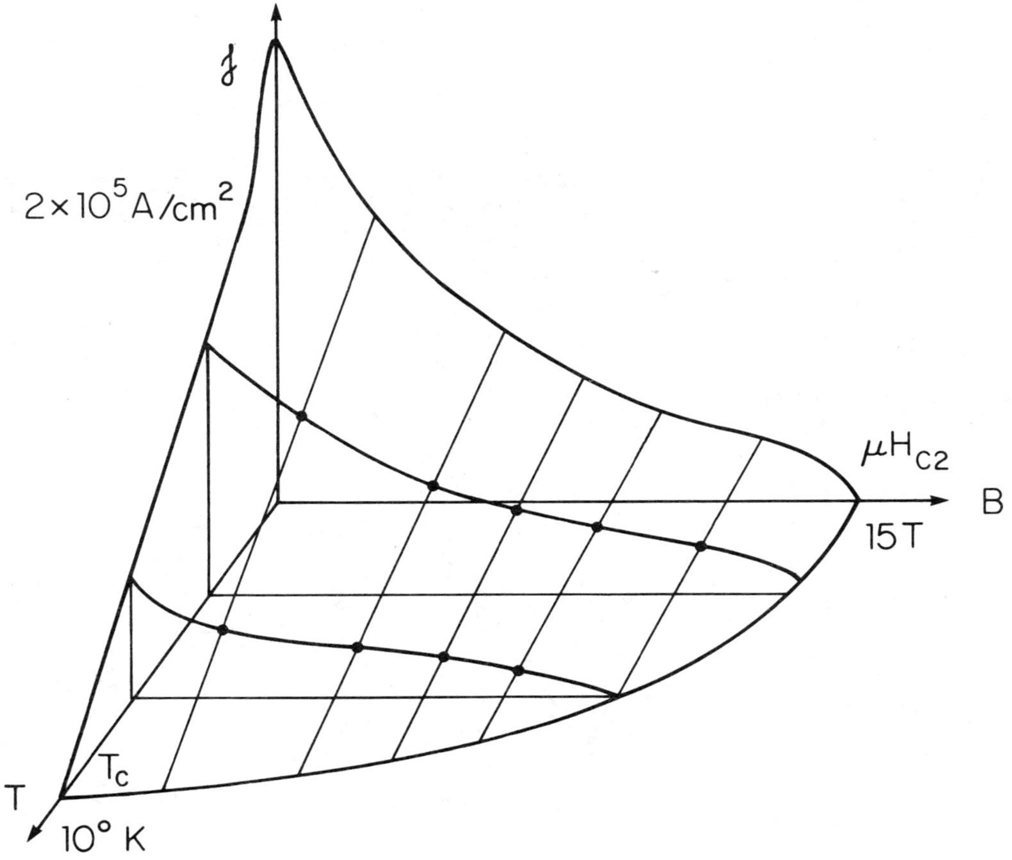

Fig. 1. Typical temperature, magnetic field, current density surface for superconducting wire.

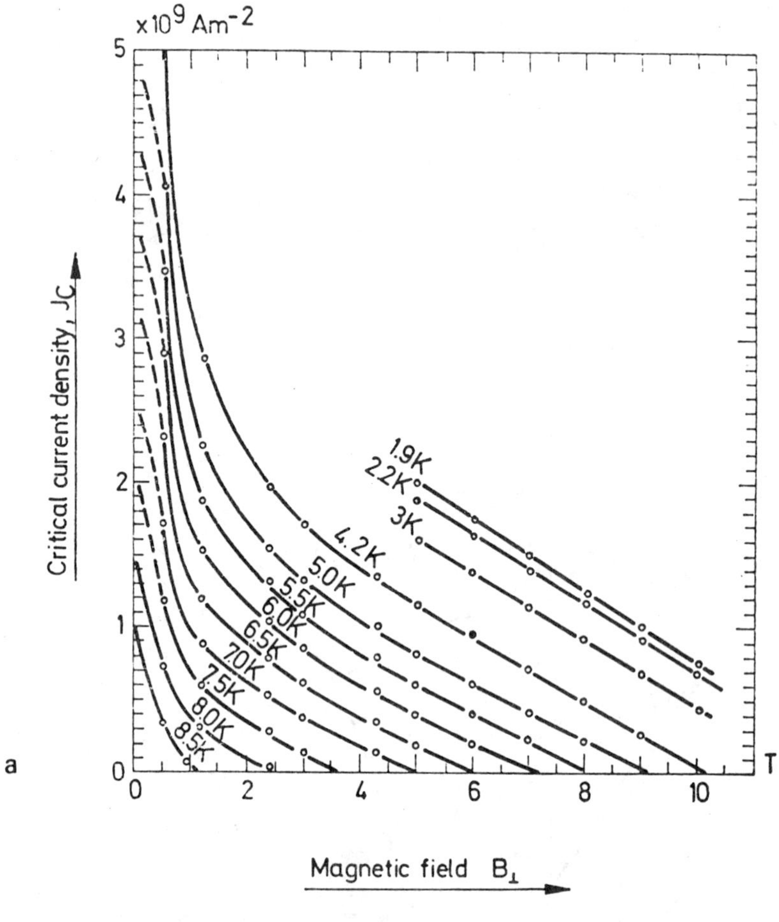

Fig. 2. Temperature dependence of critical current density vs. applied transverse magnetic field for $Nb_{(x)}Ti$ (I. M. I.), taken from Brechna. Load lines are explained in the text.

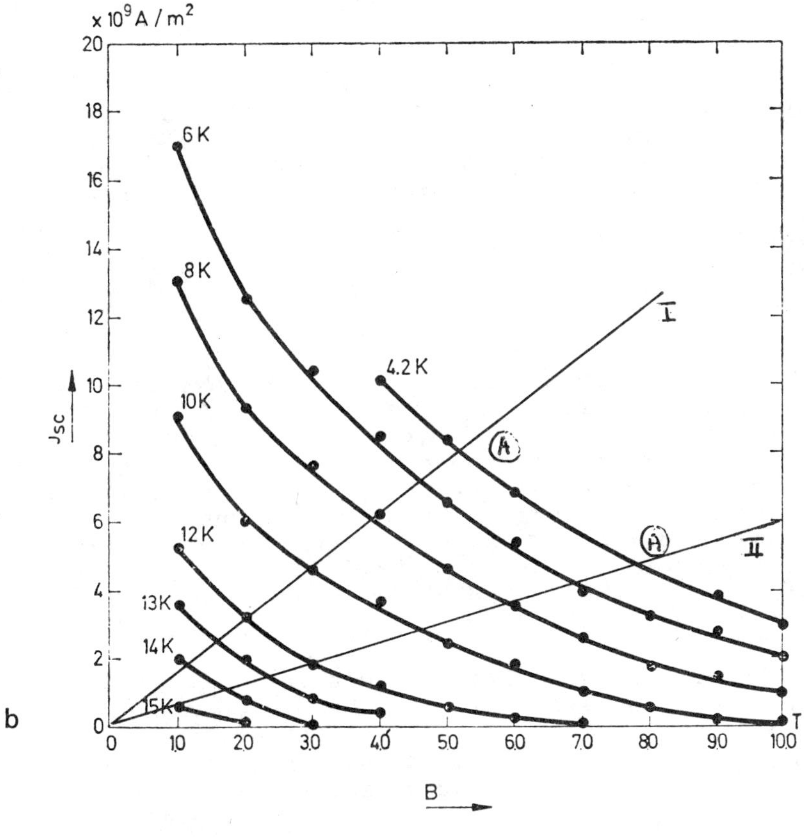

Fig. 2. Temperature dependence of critical current density vs. applied transverse magnetic field for Nb_3Sn (I. G. E.) taken from Brechna. Load lines are explained in text.

determined by the high field point in the magnet winding. The point where this intersects the alloy curve shown as A in Fig. 2 determines the maximum field obtainable in the magnet. A conductor that is optimized for low B and high current may well not be optimized for achieving very high peak fields. (See Load Lines I and II.)

The current densities shown in Fig. 2 are very large, and if these were achievable in practice, we would have much higher magnetic field strengths than are presently available. However, this current density must be reduced in order to achieve space for insulation as well as for additional copper or aluminum that must be used in parallel with the superconductor in order to stabilize it. To see why the copper is necessary, consider the data shown in Fig. 3. The curves show the resistivity of a number of materials as a function of temperature. The three curves at the top of the graph are some common superconductors, and the curve labeled 4. is OFHC copper. At the temperature of liquid helium, there is more than three orders of magnitude difference in the resistivity of these materials. Consider what happens when a magnet quenches, i.e., the superconductor goes normal. If current density in a superconductor is at the order of 10^5 amps per square centimeter and the resistivity is of the order of 10^{-5} watts per cubic centimeter, the conductor would vaporize almost instantaneously.

There are two different classes of magnets. In one, such as are built for bubble chambers, there is enough copper around the superconductor, and it is well enough cooled with helium so that if the current transfers to the copper, the temperaure can well return to a value lower than the critical temperature, and hence, the current will return to the superconductor. Such a magnet is called cryogenically stable. Magnets built for accelerators cannot afford to have this large amount of copper. It reduces the effective current density too much. Hence, if an accelerator magnet quenches, the energy must somehow be either removed or absorbed by the magnet structure, and the power supply turned off until the magnet recovers its low temperature. Thus, the role of the copper is complicated.

To see what effective current densities can be achieved, let us consider the case of the Doubler magnets. The specifications of this magnet are given in Table I. The conductor here is a cable that consists of

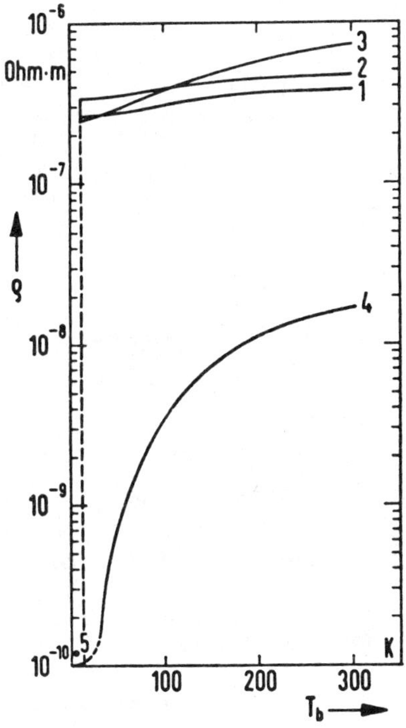

Fig. 3. Resistivity vs. temperature for several materials (from Brechna).

Table I. Dipole Parameters and Specifications

COIL SUPERCONDUCTOR:

Cross Section (In.):

$$\text{Width} - 0.307^{+0.001}_{-0.000}, \quad \text{Thickness} - \begin{cases} 0.055\ ^{+0.000}_{-0.001} & \text{Outer Edg} \\ 0.044\ ^{+0.001}_{-0.000} & \text{Inner Edg} \end{cases}$$

Strand Diameter: 0.0268±0.0003 in.
No. Strands/Cable: 23
Filaments: 8 μm diameter Ni-Ti alloy
2100 filaments/strand
Copper to Superconductor Ratio: 1.8/1 by volume
Strand Twist Pitch: 0.5 in.
Cable Twist Pitch: 2.25 in.
Cable Short-Sample Current (min.): 5350 A @ 50 kG and 4.2 K
Strand Short-Sample Current (min.): 244 A @ 50 kG and 4.2 K
Copper Resistivity Ratio: R(9.5K)/R(273K) = 0.023 ± 0.002
Insulation (IN.): 0.001 thick × 0.375 wide Kapton, spiral wrap, 7/12 lap, plus
0.007 thick × 0.250 wide B-stage/fiberglass, spiral wrap, 1/8 in. gap

RETURN BUS SUPERCONDUCTOR: (Same parameters as above except as follo

Cross Section (In.): 0.286 wide
0.058 outer edge thickness
0.049 inner edge thickness

Insulation: 4 layers - 0.001-in. thick × 0.375-in. wide Kapton, spiral wrap, 7/12 lap plus
0.007-in. thick × 0.250-in. wide fiberglass spiral wrap, butted (no gap) (alternate spirals dry and B-staged)

COILS:

Inner Coil: No. Turns: 2 × 35 Inner Radius: 1.500 in.

Outer Coil: No. Turns: 2 × 21 (including bus 1/2 turn)
Inner Radius: 1.835 in.

Cable Length/Magnet: Inner: 2900 ft
Outer: 1756 ft
Bus: 25 ft

Coil collaring preload requirements 15000 lb/linear inch

YOKE DIMENSIONS (In.):

Inner Radius: 3.765
Width: 15
Height: 10
Maximum Twist: ± 1/32
Sagitta: 0.26

LENGTHS (In.):

Yoke: 235
Inner Coil:
  To Inner Radii: 238.5
  To Outer Radii: 245.115
Outer Coil:
  To Inner Radii: 241.474
  To Outer Radii: 244.635
Cryostat: (to interface): 252

COOLING:

Helium:

1ϕ capacity:	15 ℓ/magnet
2ϕ capacity:	7.1 ℓ/magnet
1ϕ inlet: (1st magnet)	4.5 K
1ϕ outlet: (22nd magnet)	4.6 K
2ϕ inlet: (22nd magnet)	4.47 K
2ϕ outlet: (1st magnet)	4.42 K

Helium flow rate: 20.55 g/s

Nitrogen:

LN capacity:	6.7 ℓ
Max. Temp.: (outlet)	85 K

WEIGHTS:

Collared Coil Assembly	1050 lb
Cryostat	550
Yoke	6800
Total	8400

Quench Current — > 4350 A @ $\geq$ 200 A/s

AC Loss — < 500 J/cycle @ 4000 A and 300 A/

Relative Variation of Integral Field — $<\pm 10^{-3}$ about mean @ 2000 A

Magnetic Vertical Axis — $<1/2\times 10^{-3}$ rad from vertical measured and marked absolute accuracy @ 2000 A

Outside Physical Dimensions

Curvature	± 15 mil from nominal
Flatness and Twist	within 30 mil envelope
Relative Twist	2 mr

Integral Multipole Fields (B_n/B_o at 1 in.) at $\geq$ 2000 A

	Normal	Skew
Quadrupole	$\pm 2.5 \times 10^{-4}$	$\pm 2.5 \times 10^{-4}$
Sextupole	$\pm 6.0 \times 10^{-4}$	$\pm 2 \times 10^{-4}$
Octopole	$\pm 2 \times 10^{-4}$	$\pm 2 \times 10^{-4}$
Decapole	$\pm 2 \times 10^{-4}$	$\pm 2 \times 10^{-4}$

Hipot

Coil, bus, heater to ground — < 5 μA @ 5 kV

Electrical Parameters

(Acceptable tolerance about mean)

R ± 0.3% (dc)
C ± 10%
L ± 2% (at 1 kHz)
Q ± 10% (at 1 kHz)

Vacuum (maximum leak room temp.) 5×10^{-9} atm-cc/s

1000 GeV = 4.420 T for 241 in.
= 27.056 Tm
Our magnets are 6.115 Tm/ka
1000 GeV = 4424 Amps

23 strands. Each strand is 0.0268 in. in diameter. The overall dimension of the cable is shown below. The shape is keystoned so that it will pack densely into the winding.

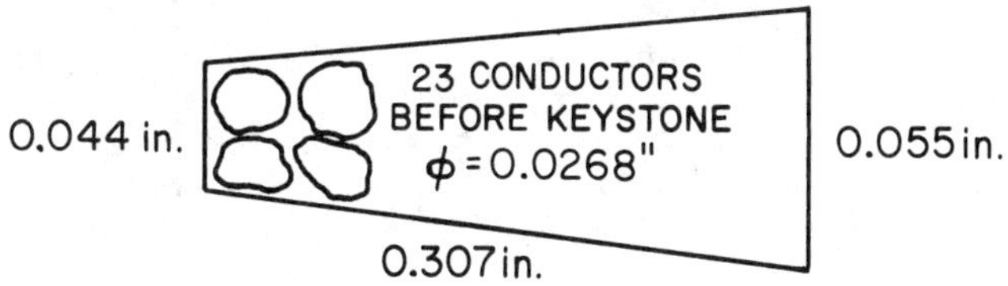

This conductor carries about 4,500 amps at 5 Tesla. It has a copper to superconductor ratio of 1.8. The conductor is insulated with a double overlap wrapping of 1 mil Kapton, and a layer of B stage impregnated glass cloth tape. If one considers the overall packing fraction in a winding, i.e., the ratio of the individual strand area to the area occupied by the coil including insulation, it is 0.71. However, the copper-to-superconductor ratio reduces its effective packing fraction to 0.25. The current density in the inner coil of the magnet is about 3.5 x 10^4 amps per square centimeter averaged over the winding; however, the current in the superconductor is 1.4 x 10^5 amps per centimeter square. The specifications on the superconductor call for a value of 1.8 x 10^5 amps at 5 Tesla.

The ISABELLE cosθ magnet uses braid. It has a dimension of 0.635 in. x 0.024 in. thick. The insulated conductor is 0.640 x 0.27 mils; there are 97 12-mil strands forming the conductor. The superconductor is encased in a copper nickel shell which encloses the pure copper plus the superconductor. The overall ratio of matrix to superconductor is 1.6, and the copper-to-superconductor ratio is 1.3. Table II summarizes the comparison of these two magnets. The current densities at the operating point of both magnets are nearly the same. It is worth noting the enormous reduction of current density that arises from the insulation in stabilizing copper around the superconductor. Since the current density in the

Table II. Comparison of Fermilab and ISABELLE Magnets

	Fermilab Cable	ISA Braid
Dim. (m)	0.307 × (0.044, 0.055)	0.635 × 0.024
Strand	23 in. 0.0268 in. $^{+0.003}_{-0}$	97 0.012 in.
Turn-Turn Insulation	0.010 in.	0.003 in.
Matrix/SC	1.8/1 Cu/Sc	1.6/1 (CuN_L + Cu)/SC
Cu/Sc	1.8/1	1.3/1
Amps	> 4300 4.6° K	4000 A 3.8° K
<u>Packing Fractions</u>		
Strand/Cable	0.854	0.720
Cable/Cable + Insulation	0.832	0.889
Superconductor Cable (No Insul.)	0.305	0.277
<u>Current Density</u>		
Average	35.2 kA/cm^2	35.9 kA/cm^2
Current Density in SC	1.44×10^5 A/cm^2	1.47×10^5 A/cm^2
<u>Field</u>		
Bore	4.3 T	5 T
Peak in Coil	4.7 T	5.3 T

supercondutor in the two magnets is nearly the same and since the BNL magnet is designed to work at lower temperature, it should be more stable than the Fermilab dipole.

We now come to the subject of training. Let's consider the excitation of a newly constructed magnet. The current should increase up the load line until it reaches B_{max}, the point circled A in Fig. 2. When the magnet reaches this point, we should be prepared to reduce the current to zero before the magnet is damaged. However, Nature is not so kind and in general the first excitation of a magnet will not succeed in reaching Point A. The conductor will quench at a considerably lower current. Each subsequent excitation will reach a higher current. A well-behaved magnet may only take one or two quenches to reach this point. A poorly designed magnet may never succeed in reaching Point A. This phenomena is referred to as "training" and is observed to some extent in all superconducting magnets. It is thought to occur from heat generated by the frictional motion of the winding in coming into an equilibrium configuration. We will examine theories of this effect in Section 6.

3. FIELD SHAPE AND FORCES

Dipoles

We will now consider designing a dipole that would be suitable for use in an accelerator. The aperture of the magnet must be big enough to contain the beam, and the field must be uniform to about one part in one thousand across the useful aperture. We will work out the mathematical details of magnets formed by current shells. In this section, we will refer to these results.

There are a number of different designs for dipoles, and they can be classified generally into three types. One is a so-called two-shell design of the Fermilab type. The second is a so-called cosθ magnet being investigated for ISABELLE, and the third is a magnet originally conceived by Gordon Danby at BNL. This geometry is not cylindrical, and the theory developed here does not apply.

For future reference drawings of the Doubler, ISABELLE, and Danby magnets are shown in Figs. 4-8.

The drawing below shows a coordinate system in the general configuration of currents that will be considered here.

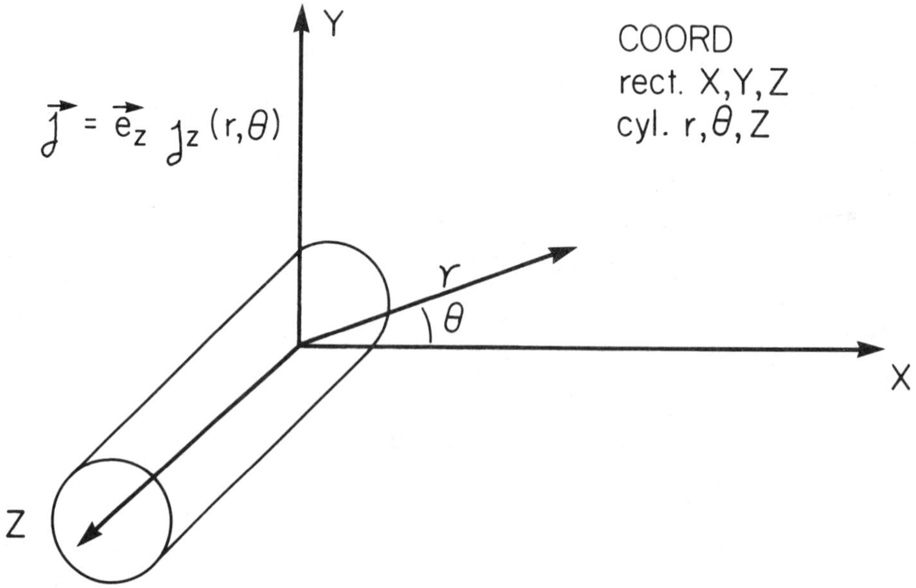

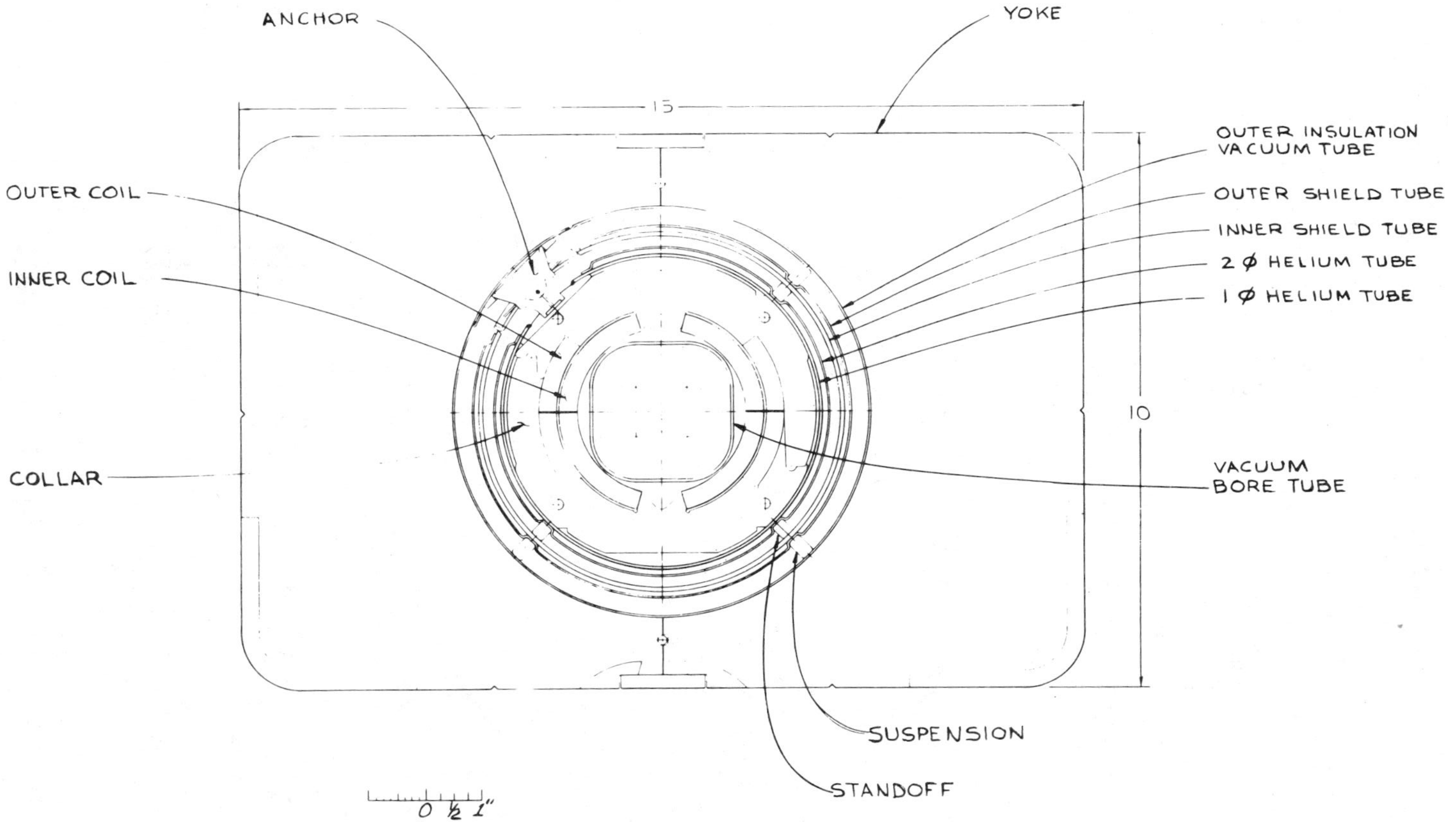

Fig. 4. Cross section of Fermilab dipole magnet (from the Energy Doubler Design Report).

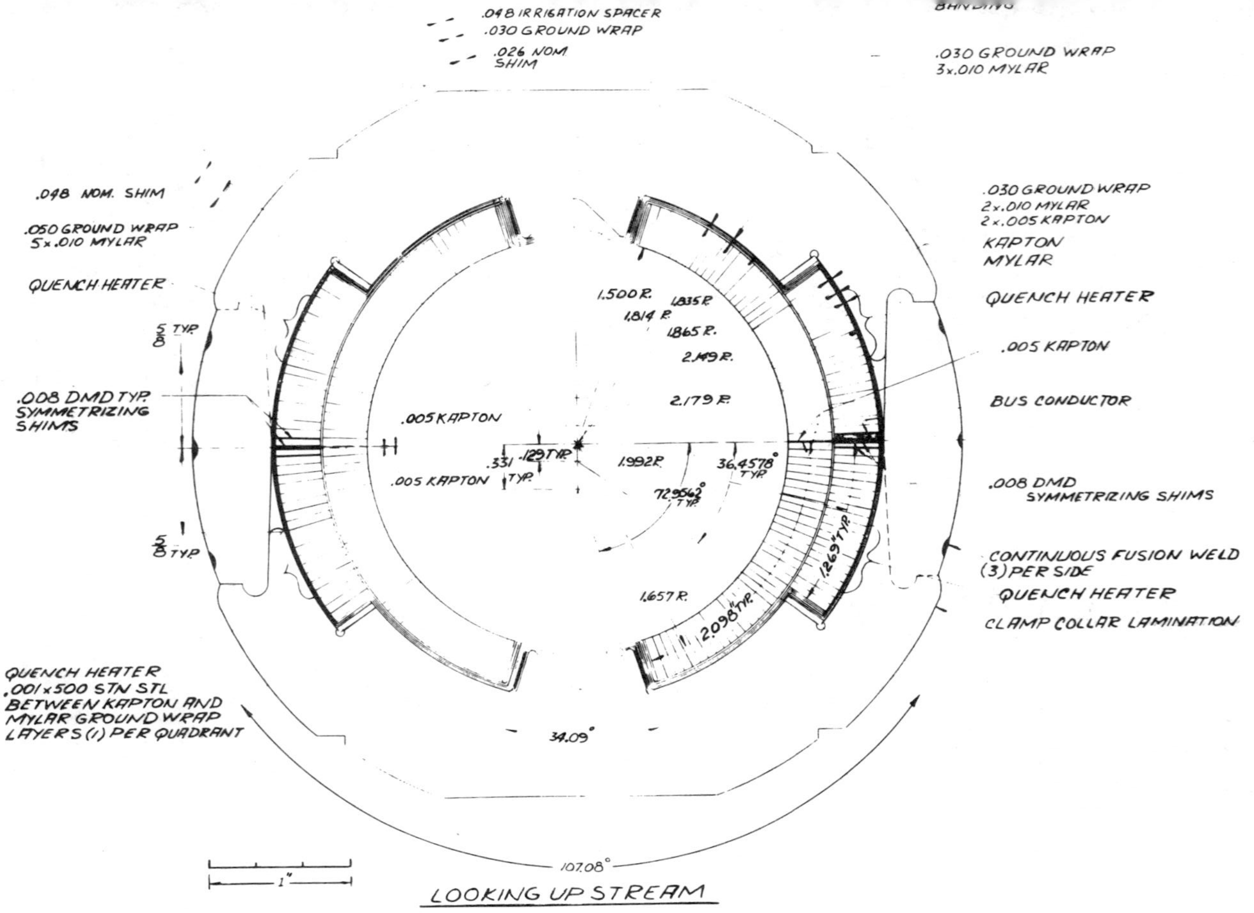

Fig. 5. Cross section of Fermilab dipole collared coil (from the Energy Doubler Design Report).

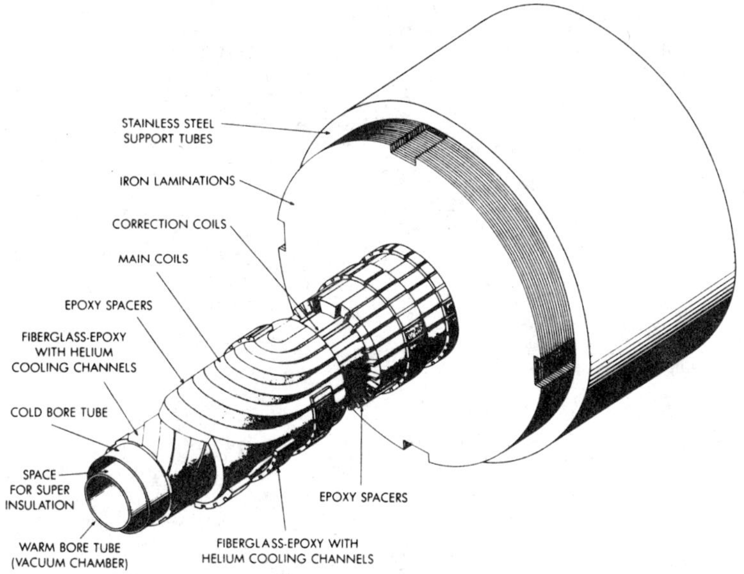

Fig. 6. Isometric view of the ISABELLE dipole magnet, showing the configuration of the coil ends (from the ISA Design Report).

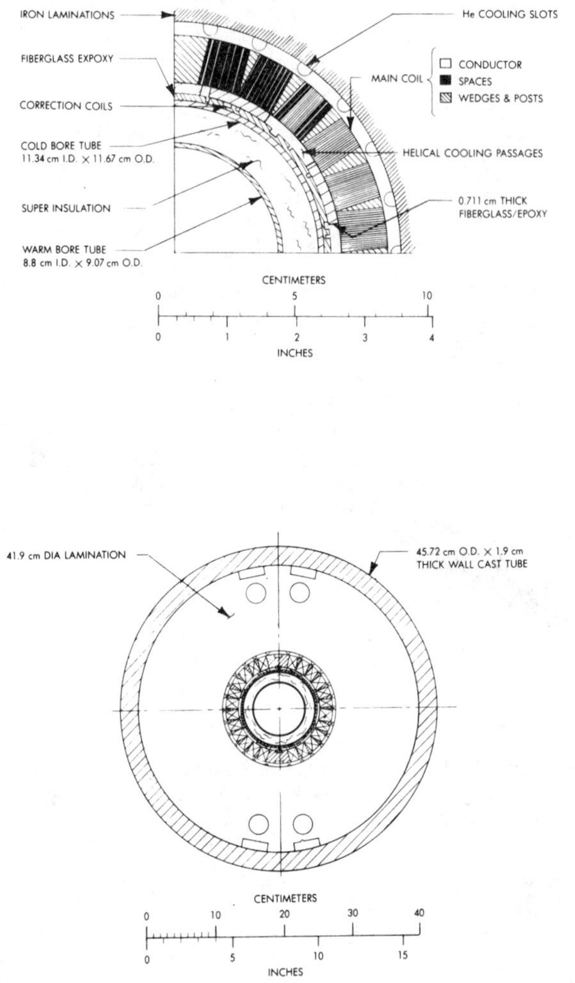

Fig. 7. Schematic of ISABELLE dipole (from the ISA Design Report).

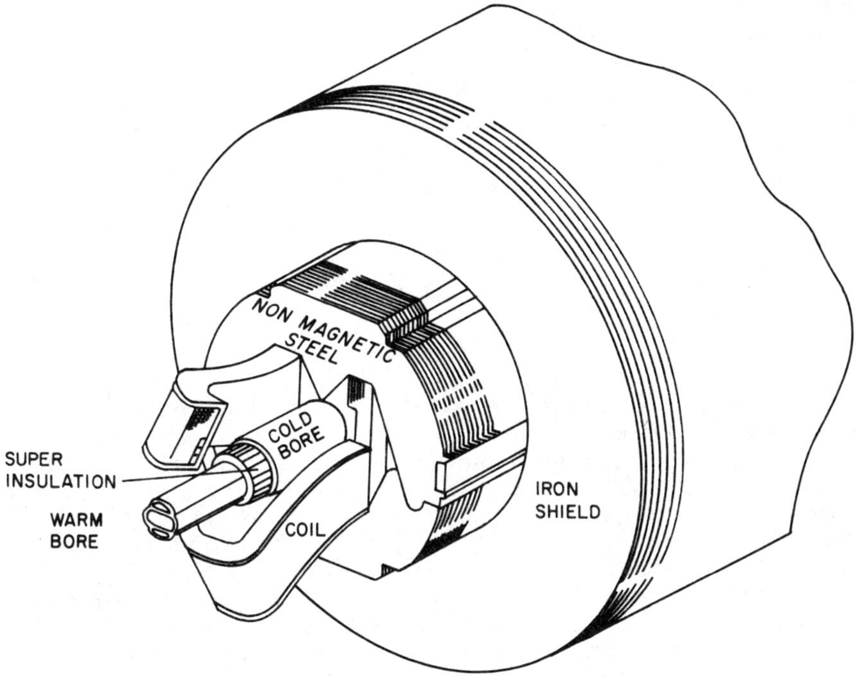

Fig. 8. Danby magnet configuration.

The current distribution flows in the direction of the z axis and is a function of r and θ. This generates an axial magnetic vector potential and leads to a two-dimensional magnetic field through the equations below.

$$\vec{A} = \vec{e}_z \quad \frac{\mu_o}{4\pi}\int_V \frac{J_z\ (r,\theta)\ d\tau}{(r^2 + z^2)^{1/2}}$$

$$\nabla^2 A_z = \mu_o j_z$$

$$\vec{B} = \nabla \times \vec{A} \tag{1}$$

$$B_r = \frac{1}{r}\frac{\partial A_z}{\partial \theta} \quad , \quad B_\theta = - \frac{\partial A_z}{\partial r} .$$

An exact solution is a computer problem. No general rules for achieving the solution of a given field exist. This is because insulation occupies some space. Mechanical support of forces is necessary when the forces are large. The accuracy of placement of the conductor is an important consideration, and the iron return yoke may or may not be saturated. However, the exact computer solution for the cylindrical magnets considered here can be understood from the simpler solution of current shells. If one could achieve a surface current density given by

$$\vec{K} = \vec{e}_z\ K_o \cos\theta\ . \tag{2}$$

This would lead to a vector potential given by

$$A_z = 1/2\ \mu_o K_o\ r \cos\theta = 1/2\ \mu_o K_o\, x \tag{3}$$

which in turn gives a uniform field

$$\vec{B} = \nabla \times \vec{A} = -\ 1/2\ \mu_o K_o\ \vec{e}_y . \tag{4}$$

The ISABELLE magnet has blocks of current with space between them to approximate such a cosθ current distribution as shown in the figure below. The Fermilab magnet has two shells to make this approximation as shown in the left-hand side.

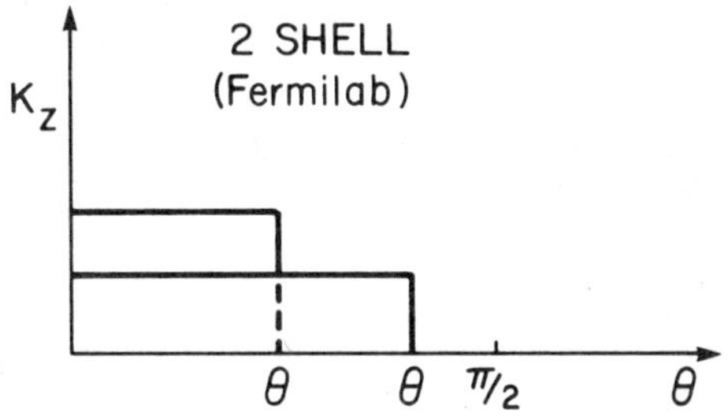

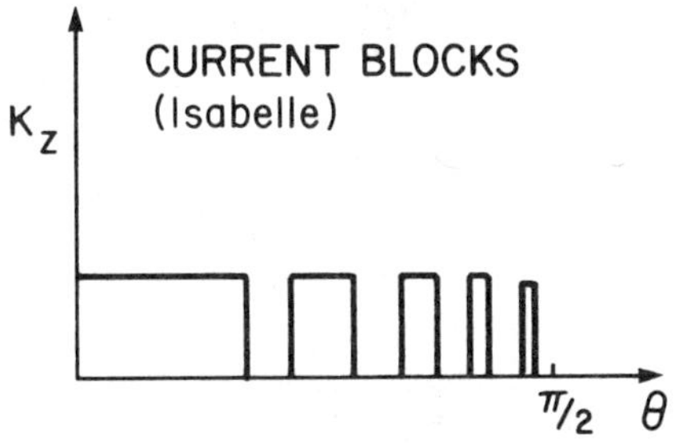

Both magnets use the angles subtended by the coils to control the harmonic content of the field.

More generally, a single current shell at radius a whose current density is given by

$$K_z = K_o \sum_1^\infty (B_n \cos n\theta + A_n \sin n\theta) \qquad (5)$$

will be used to represent one of the windings or a current block in a winding. B_n and A_n are determined by the current distribution. The major terms should be the odd terms in cosθ; all other terms lead to error fields and are unwanted. The vector potential from this current distribution is given by

$$r<a \quad A_Z = 1/2\ \mu_o K_o a \sum_1^\infty \frac{1}{n} (B_n \cos n\theta + A_n \sin n\theta) \left(\frac{r}{a}\right)^n$$

$$r>a \quad A_Z = 1/2\ \mu_o K_o a \sum_1^\infty \frac{1}{n} (B_n \cos n\theta + A_n \sin n\theta) \left(\frac{a}{r}\right)^n$$

from which we find the magnetic field for r less than a.

$$B_\theta = -\ 1/2\ \mu_o K_o \sum_1^\infty (B_n \sin n\theta + A_n \sin n\theta) \left(\frac{r}{a}\right)^{n-1}$$

$$B_r = -\ 1/2\ \mu_o K_o \sum_1^\infty (B_n \sin n\theta - A_n \cos n\theta) \left(\frac{r}{a}\right)^{n-1}$$

Accelerator theory has been developed in terms of an expansion of the field near the x axis. The expansion is given in the following form

$$B_y\ (y=0) = B_o \sum_o^\infty b_n x^n \qquad b_o \equiv 1$$

$$B_x\ (y=0) = B_o \sum_o^\infty a_n x^n \qquad a_o \equiv 0 \tag{8}$$

B_o is the field at the center, x is the distance from the center in the median plane, b_o is identically equal to 1, and a_o is set equal to zero, i.e., the field is aligned with the axis. Note that a_n and b_n have the dimensions $(\text{length})^{-n}$ so the unit of length chosen makes a vast difference. The Saver magnets use 1 inch as a reference which was picked because it is about two-thirds of a, and the harmonics up to 30 pole are all in the range of 0.1 to 20×10^{-4}. If the centimeter unit was chosen, they would be much smaller but would also vary in magnitude by more than four decades. The units used with ISABELLE magnets are generally centimeters.

The a_n and b_n are determined from the A_n and B_n by the following two equations

$$B_o b_n = -\ 1/2\ \mu_o K_o\ B_{n+1}\quad a^{-n}$$

$$B_o a_n = \ 1/2\ \mu_o K_o\ A_{n+1}\quad a^{-n} \tag{9}$$

$$B_o = -\ 1/2\ \mu_o K_o\ B_1 \quad \text{(field on axis)}$$

If the field is to be flat, it is clear that we want $a_n = 0$ and by symmetry about the y axis, all of the odd b^n should also equal 0. For instance, a_1 and b_1 would be skew and normal quadrupole fields, respectively; b_2 represents the sextupole coefficient of the current distribution and can be controlled by the designer. It is clear that the best solution would be to have only b_o; i.e., a pure $\cos\theta$ distribution. This is, however, impractical in practice.

Let us consider now a single shell of arc length $a\theta_1$. We calculate then that

$$b_n = \frac{1}{n+1} a^{-n} \frac{\sin (n+1)\, \theta_1}{\sin \theta_1}, \quad n \text{ even} \qquad (10)$$

If $\theta_1 = 60^\circ$, then we find that $b_2 = 0$ and further

$$B_y\,(y=0) = B_o \left[1 - .04\, x^4 + .012\, x^6 - .0017\, x^8 + .. \right] \qquad (11)$$

Here we have assumed that a is 1.5 in., and the reference radius is 1 inch. There is no way to eliminate any more of the terms in this expression as we have used θ_1 already which is the only variable; however, with two shells we have an additional variable, θ_2. Crudely, we can set the average of θ_1 and θ_2 to 60 degrees to eliminate the sextupole and pick the difference then to eliminate the decapole. A careful calculation shows that if we pick θ_1 to be 78° and θ_2 to be 42°, that then both the sextupole and decapole components will be eliminated, and the first term coming into the expansion will be x^6 for the 14-pole term. [Caution: b_n from different shells, Eq. (10), do not add.] This then illustrates the method that is used to try and control the harmonic content of the field.

As shown in the quadrupole section below, a good approximation to a real magnet can be obtained by using current shells at the average radius of the winding in order to predict the magnetic properties. The angles quoted above become slightly different since the moments of the outer shell are modified by the fact that its radius is slightly larger than that of the inner shell. In summary, one notes that the inner shell has a negative sextupole when its angle exceeds 60° and a positive decapole, while the outer winding has a positive sextupole and a negative decapole. These two terms are balanced against each other to eliminate them. The rest of the harmonics must be accepted, but in general they tend to be small due to the factor of a^{-n} in the expression for b_n, Eq. (9).

Quadrupoles

Quadrupoles can be designed using the same approach. In this case, one wishes to keep the term b_1

and eliminate all the rest. If the symmetry is exact, the next allowed term will be the 12 pole corresponding to the sextupole in the dipole case. This again is eliminated by cutting the shell at an angle equal to 30°. The 20-pole term (corresponding to the decapole in the dipole magnet) is harder to eliminate by balancing against the outside shell. This is because of the a^{-n} dependence in the equations for b_n. The outer radius is sufficiently bigger that it is difficult to get a large enough 20-pole moment to balance the inner winding; hence, the Saver quadrupole resorts to the use of two current blocks on the inner winding to eliminate the 20-pole.

Forces

The forces in the dipoles that we are talking about are large and are scaled by the following equation

$$p_o = B_o^2/2\mu_o \quad , \tag{12}$$

where p_o is the magnetic pressure in a uniform field B. At 5 Tesla, p_o is 10^7 N/m^2 or 1,450 psi. The coil tends to lengthen, radially expand at the equator, and compress azimuthally toward the midplane. These three forces are absorbed differently:

1. Axial stress is generally resisted by tension in the wire itself; for instance, the Saver coils have an axial tension of over 10,000 lbs. and are observed to lengthen by about 0.07 in. at a field of 4 Tesla. In such a long, thin coil, it is difficult to transfer this large force to any structural member other than the wire itself.

2. The radial expansion forces are also large. Roughly they are equal to

$$F \approx 2\ a\ p_o . \tag{13}$$

In the Saver magnet the total force in the x direction is 6,000 lbs. per in. of length and is resisted by stainless-steel collars. In ISA magnets, it is absorbed by the cold iron return yoke. Whatever the mechanism, the support itself must be cold, as the forces are so large that it is not feasible to transfer them to a structural member at room temperature. The heat leak would be enormous due to the large cross-sectional area necessary to keep the stresses at an acceptable level.

3. Finally we come to the azimuthal forces. The calculation for a cosθ distribution gives

$$F_T = \int_0^{\pi/2} F_\theta a d\theta = -2ap_o. \tag{14}$$

This force compresses the winding toward the median plane, and the integrated effect is largest at this point. This force must be resisted by compression in the winding itself and for this reason, to minimize the motion of the turns, a high Young's modulus for the coil package is desirable. We will discuss this in much more detail in Section 4.

Magnetic Field Perturbation

The dipoles for an accelerator should have a field that is uniform to a few parts in 10^4. To achieve this field accuracy the placement of the turns required is of the order of a few thousandths of an inch. The use of superconducting magnets in an accelerator poses a challenge to achieving such accuracy. Magnets using iron to shape the field have achieved a great deal of uniformity through the use of magnet laminations that have been stamped in a precision die; however, superconducting magnets do not have this advantage. Each magnet is slightly different and has its own set of random errors. In addition to these errors there are systematic effects due to the manufacturing process that must be understood. By far the most sensitive measure of these errors is the shape of the magnetic field itself. The field can be measured with great accuracy, i.e., the coefficients a_n and b_n can be determined. One is then faced with the problem of deriving the shape or the change in shape of the magnet coil from the magnetic field measurements. To this end, a simple model is useful.

In order to design a coil accurately, a computer program that integrates over all the current elements is necessary; however, for understanding small displacements of the turn, only a perturbation calculation is needed. A useful approximation of the winding is obtained by replacing each conductor with a line element of current which is located at its center. For instance, the 35 turns of the inner coil of the

Saver dipole can be modeled by 35 line currents located at the conductors' centers. It is then easy to give each of these conductors a small vector displacement and subtract the perturbed field from the original to get the perturbation to the coefficients a_n and b_n for any given distortion of the coil. Such a program is easy to write and has been extensively used in the Saver dipole construction program. Distortions that might be investigated, for instance, are due to the aximuthal compression of the winding from the $\vec{j} \times \vec{B}$ forces, the elliptical distortion of the support or the random displacement of the turns due to insulation thickness errors.

The Effects of Iron and Ends

So far we have neglected end effects. At the ends the current no longer is represented by a two-dimensional distribution, and the field must be integrated in three dimensions. The iron yoke also ends and this generates an additional perturbation to the field. It is easy to saturate the iron near the ends of the yoke. Note that saturation of the iron has two effects. First, it may change the shape of the field with current. The second effect arises because all superconducting machines, for reasons of cooling efficiency, connect the dipoles and quadrupoles in series. If these elements saturate differently, then the tune of the machine will change with excitation even if the field shape remains invariant.

Let us consider the shape of the field at the end of the dipole first. We sketch an end below. The dotted curve indicates where the magnetic field falls to its half value. The integrated length of the magnet will thus be a function of the transverse position at which the particle traverses the magnet. The reason for the change in length with x can be seen in the figure below.

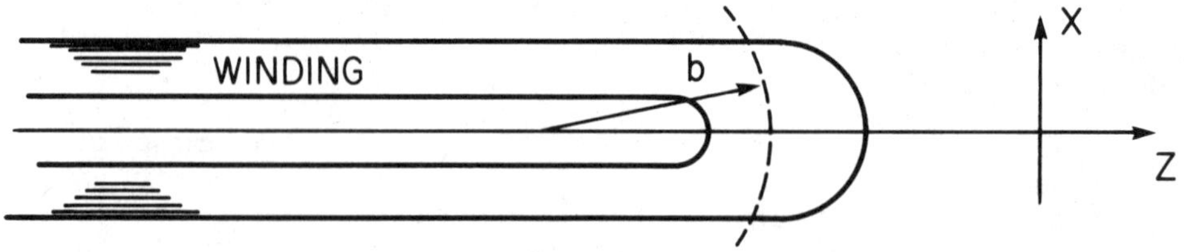

where b is the radius of the dotted curve, and it should be noted that there are two ends. The effective change in length at one end will be given by the equation $\delta z = (\delta x)^2/2b$ Consider a magnet 250 in. long with b = about 4 in. so that $\delta z = x^2/8$. Thus the Bdl as a function of x can be represented as

$$\int B d\ell = B_o L \left[1 - \frac{2}{L} \cdot \frac{1}{8} x^2 \right] = B_o L \, (1 - 10^{-3} \, x^2) \qquad (15)$$

This change in effective length can be compensated by using a positive sextupole within the body of the magnet, i.e., $b_2 = 10 \times 10^{-4}$. This is roughly the case for the Saver magnets. It is very difficult to change the size of this effect in a small diameter magnet. The decapole for the ends is considerably smaller.

Note that we are using the set of coefficients, a_n and b_n, that are averaged over length and, in fact, these are the coefficients generally quoted for a completed magnet.

There is a second effect that arises at the end. For a two-dimensional magnet, the high field point of the winding will generally be located in the turns nearest the poles. Recall that it is necessary to know the value of this field as a function of current in order to pick a safe operating point for the superconductor (Section 2); however, when the whole magnet is considered, the peak field moves to the end. The situation for the Saver magnet is

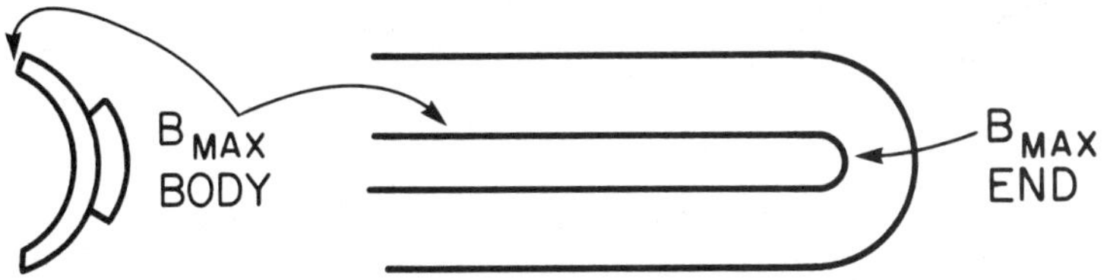

One can adjust the length of the iron yoke so that B_{max} in the body is greater than B_{max} in the end. In order to achieve this result, it may also be necessary to spread out the end turns with spacers. The inner 10 turns of the Saver magnet are spread with such preformed spacers. For magnets with current blocks, the blocks can be spaced or an additional superconductor can be added to shift the critical field point to the body of the magnet.

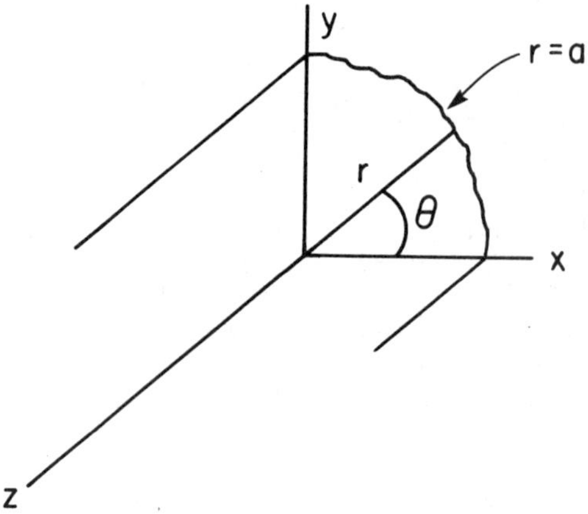

Cyl. Coord. r, θ, z

$$\vec{j} = j_Z \vec{e}_Z \qquad j_Z = \text{function } r, \theta \tag{16}$$

$$\vec{A} = \vec{e}_Z \int \frac{\mu_o}{4\pi} \frac{j_Z d\tau}{(r^2 + z^2)^{\frac{1}{2}}} \tag{17}$$

$$\nabla^2 A_Z = \mu_o j_Z \; ; \; \vec{B} = \nabla \times \vec{A}$$
$$B_r = \frac{1}{r} \frac{\partial A_Z}{\partial \theta} \quad B_\theta = - \frac{\partial A_Z}{\partial r} \tag{18}$$

Consider current on a shell at r = a

$$\vec{K} = \vec{e}_Z K_o \sum_1^\infty (B_n \cos_n \theta - A_n \sin n\theta) \tag{19}$$

For a dipole with $\vec{B} \simeq \vec{e}_y B_o$ the odd B_n will be the major terms.

$$r<a \quad A_Z = \sum_{n=1}^\infty \left[C_n \cos n\theta + D_n \sin n\theta \right] (r/a)^n \tag{20}$$

$$r>a \quad A_Z = \sum_{n=1}^\infty \left[C_n \cos n\theta + D_n \sin n\theta \right] (a/r)^n \tag{21}$$

at r=a Boundary Condition

$$H_{t2} - H_{t1} = K_Z \tag{22}$$

Substituting gives

$$C_n = \frac{\mu_o K_o a}{2n} B_n \tag{23}$$

$$D_n = \frac{\mu_o K_o a}{2n} A_n \tag{24}$$

thus we find for r<a

$$H_\theta = -1/2 \; K_o \sum_{n=1}^{\infty} \left[B_n \cos n\theta + A_n \sin n\theta \right] (r/a)^{n-1} \tag{25}$$

$$H_r = -1/2 \; K_o \sum_{n=1}^{\infty} \left[B_n \sin n\theta - A_n \cos n\theta \right] (r/a)^{n-1} \tag{26}$$

Now it has become common to use the following coefficient to describe the field on the median plane.

$$B_y(y=o) = B_o \sum_{n=0}^{\infty} b_n x^n \qquad b_o \equiv 1 \tag{27}$$

$$B_x(y=o) = B_o \sum_{n=1}^{\infty} a_n x^n \qquad a_o \equiv 0 \tag{28}$$

B_o = Field on axis

a_n , b_n have dimensions L^{-n}

The plane y=o is given by θ=0 so from $H_{\theta 1} H_\rho$ we find:

$$B_y \; (y=o) = - \frac{\mu_o K_o}{2} \sum_{n=1}^{\infty} B_n \; (x/a)^{n-1} \tag{29}$$

$$B_x\ (y=o) = + \frac{\mu_o K_o}{2} \sum_{n=1}^{\infty} A_n\ (x/a)^{n-1} \tag{30}$$

Compare Eq. 29 with Eq. 27 gives

$$B_o b_n = - 1/2\ \mu_o K_o\ B_{n+1}\ a^{-n} \tag{31}$$

$$B_o a_n = 1/2\ \mu_o K_o\ A_{n+1}\ a^{-n} \tag{32}$$

$$B_o = - 1/2\ \mu_o K_o\ B_1 \tag{33}$$

<u>Example 1</u>

Suppose $B_1 = 1$ all other A_n, $B_n \equiv 0$

$$K_Z = K_o \cos\theta$$

The total current down one side is

$$I_T = \int_{-\pi/2}^{+\pi/2} K_o \cos\theta\ \ a d\theta = 2\ K_o a \tag{34}$$

or $$K_o = I_T/2a$$

$$B_o = - 1/2\ \mu_o K_o\ B_1 = \frac{-\mu_o I_T}{4a} \tag{35}$$

A uniform field.

<u>Example 2</u>

Consider a current shell as follows

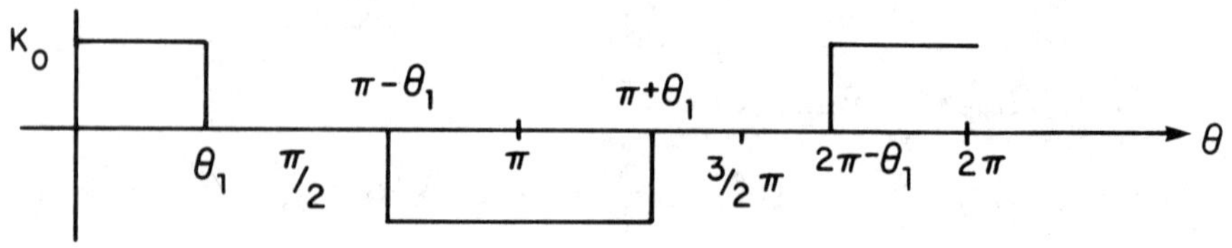

$$K_z(\theta) = K_o \quad \Sigma\, B_n \cos n\theta$$

$$B_n = \frac{4}{n\pi} \quad \sin n\theta_1 \qquad n \text{ odd} \tag{36}$$

so

$$b_n = \frac{-1/2\, \mu_o K_o\, B_{n+1}\, a^{-n}}{-\,1/2\mu_o K_o B_1} \tag{37}$$

$$b_n = \frac{1}{n+1} \quad a^{-n} \quad \frac{\sin\,(n+1)\theta_1}{\sin\,\theta_1} \tag{38}$$

Note $\quad b_2 = 0 \quad \text{For } \theta_1 = 60^\circ$

$b_4 = 0 \quad \text{For } \theta_1 = 36^\circ,\ 72^\circ$ (39)

Next consider the forces on the coil.

$$d\vec{F} = \vec{e}_z K_z \; d\ell \times \vec{B}$$

$$d\vec{F} = d\ell K_z(\theta)\; \vec{e}_z \times \left[\, \vec{e}_r B_r + \vec{e}_\theta B_\theta \,\right]$$

$$= d\ell K_z(\theta)\; \left[\, \vec{e}_\theta B_r - \vec{e}_r\; B_\theta \,\right] \tag{40}$$

$dF_\theta = K_z(\theta)\; B_r d\ell \qquad B_r$ continuous at $r=a$

$d\vec{F}_r$ must be calculated with more care since B_θ is discontinuous.

We calculate F_θ for $\cos\theta$ winding as an example .

If $B_n = 0 \quad n \neq 1 \quad B_1 = 1$ then

$$K = K_o \cos\theta$$

Eq. 28 gives

$$B_r = - 1/2 \ \mu_c K_o \ \sin\theta$$

$$dF_\theta = - 1/2 \ \mu_o K_o^2 \ \sin\theta \ \cos\theta \ a d\theta$$

at median plane the total compressive force is

$$F_T = - 1/4 \ \mu_o K_o^2 \ a \int_0^{\pi/2} \sin 2\theta d\theta$$

But Eq. 32 gives

$$B_o = - 1/2 \ \mu_o K_o \qquad \text{Field on axis}$$

So

$$F_T = - \frac{B_o^2 a}{\mu_o} \int_0^{\pi/2} \sin 2\theta d\theta \tag{41}$$

$$F_T = - \frac{B_o^2 a}{\mu_o}$$

If $P_o = B_o^2/2\mu_o$,

then

$$F_T = - 2a \ P_o \tag{42}$$

$$P_o = \frac{B_o^2}{2\mu_o} = \frac{25}{2 \times 4\pi \times 10^{-7}} = 9.9 \times 10^6 \ \text{N/m}^2$$

P_o	5T	4T
	9.9×10^6	6.4×10^6 N/m^2
	1436	920 psi

In this simple approximation we find

	Fermilab	ISA
B_o	4.3 T	5 T
P_o	1062	1436 psi
a	1.5"	2.5"
F_T	3186	7180 #/linear inch

We now calculate effect of putting a cosθ coil in iron. Consider

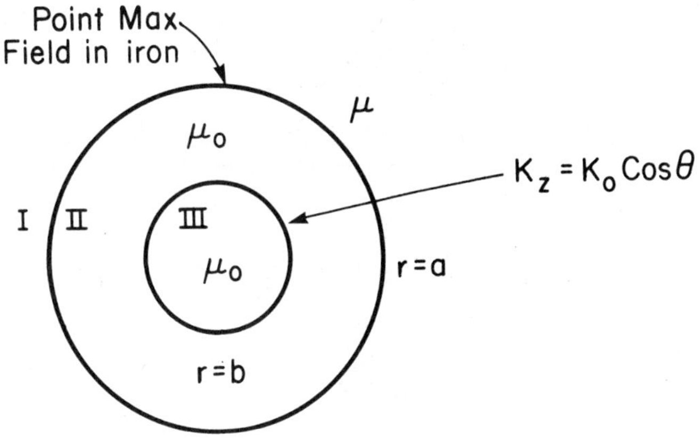

How much does the iron increase the field? There are 3 regions

$$\begin{array}{lll} \text{I} & r>a & A_z = A_1 \left(\frac{a}{r}\right) \cos\theta \\ \text{II} & b<r<a & A_z = A_2 \left(\frac{a}{r}\right) \cos\theta + A_3 \left(\frac{r}{a}\right) \cos\theta \\ \text{III} & r<b & A_z = A_4 \left(\frac{r}{b}\right) \cos\theta \end{array} \qquad (43)$$

The boundary conditions are

$$\text{at } r = a: \quad A_{z1} = A_{z2}$$

$$H_{\theta 1} - H_{\theta 2} = 0$$

$$\text{at } r = b: \quad A_{z2} = A_{z3}$$

$$H_{\theta 2} - H_{\theta 3} = K_z$$

Set B_o is the field in region 3. Then $A_4 = -B_o b$ and we have 3 equations and 3 unknowns.

$$A_1 - A_2 - A_3 = 0$$

$$A_2 \left(\frac{a}{b}\right) + A_3 \left(\frac{b}{a}\right) = - B_o b$$

$$A_1 - \frac{\mu}{\mu_o} A_2 + \frac{\mu}{\mu_o} A_3 = 0$$

Solving gives

$$A_1 = - B_o a \frac{2(b/a)^2 \mu/\mu_o}{D}$$

$$A_2 = - B_o a \frac{b^2/a^2 \ (1 + \mu/\mu_o)}{D} \qquad (44)$$

$$A_3 = - B_o a \frac{b^2/a^2 \ (\mu/\mu_o - 1)}{D}$$

$$D = 1 + (\mu/\mu_o) \left[1 + (b/a)^2 \right] - (b/a)^2$$

as $\mu/\mu_o \to \infty$

$$A_1 \to - B_o a \frac{2 \alpha^2}{1 + \alpha^2}$$

$$A_2 \to - B_o a \frac{\alpha^2}{1 + \alpha^2} \qquad (45)$$

$$A_3 \to - B_o a \frac{\alpha^2}{1 + \alpha^2}$$

$$\alpha = (b/a) \qquad \alpha \leq 1$$

the max field in the iron is at $r = a$, $\theta = \pi/2$

$$B_{max} = B_o \frac{2 \alpha^2}{1 + \alpha^2}$$

If B_o is known, and $B_{max} = B_{SAT}$ then $\alpha = b/a$ determines the size of the hole in the iron by

$$B_o = 1/2\ B_{SAT} + 1/2\ B_{SAT}\ (a/b)$$

the max. increase in the central field from non-saturating iron is $1/2\ B_{SAT} = 1$ T.

If the iron saturates, it can contribute more to the central field, but it makes the field non-linear with current. (i.e. dipoles and quadrupoles do not track when placed in series) and a sextapole moment that is a function of current is induced. Fig. 9 shows this effect in the ISA Dipole.

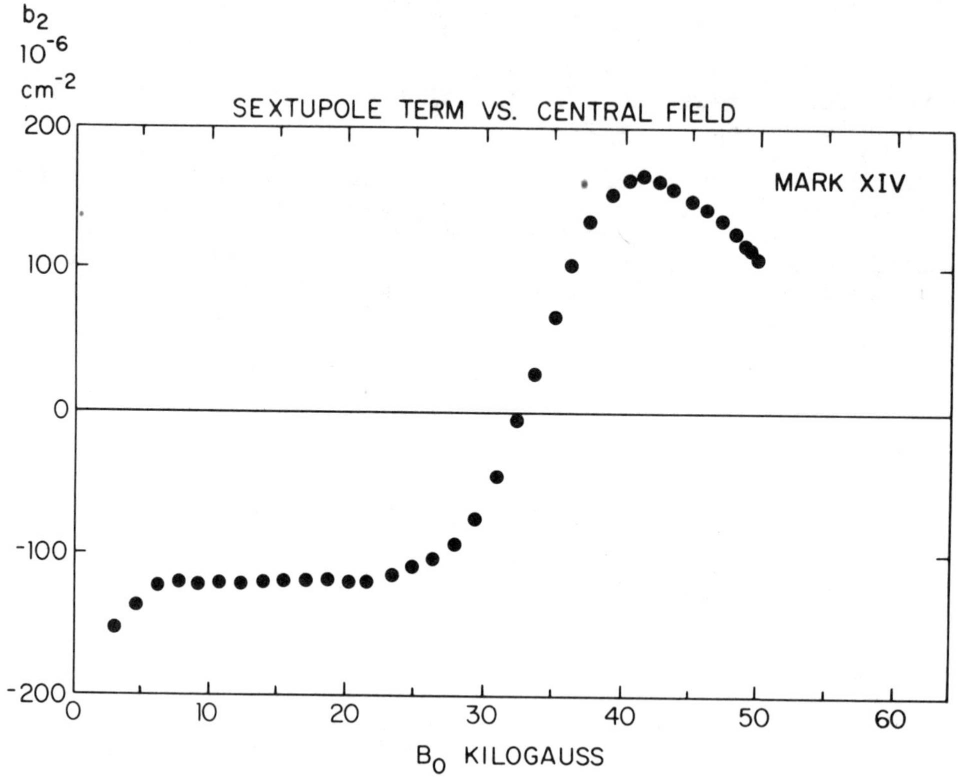

Fig. 9. Sextupole term vs. central field in the ISA dipole.

4. PRELOAD

Azimuthal Coil Deflection

The magnetic forces on individual strands of cable in the magnet winding are rather large. Forces and fields at various points in the winding of the magnet are shown in Fig. 10, based on calculations made by Stan Snowdon. They are shown resolved into radial components and azimuthal components and as a function of wire number. One can see in this figure that the radial forces on the outer coil are rather small. It should be remembered that the radial forces in the inner coil must be transmitted through the outer one to the collars in the central region of the magnet. What we are concerned with here is the azimuthal force. These forces tend to cause a compaction of the coil when the magnet is excited.

Consider for a moment a coil that is restrained in the radial direction but has no azimuthal constraint. Then as the magnet is excited, the topmost conductor on the inner coil experiences a force of almost 120 lbs. per inch of length toward the median plane of the magnet. This force causes a compaction of the coil and hence an azimuthal motion of the conductor. Suppose now that we compress the inner coil so that the mechanical forces on it are greater than the magnetic forces. Under these conditions, there will be no motion of the wire next to the key and by symmetry, there will also be no motion of the midplane. Consequently, the only conductors that can move in an azimuthal direction will be those near 45°. As the coil is excited, pressure on the median plane is increased, and the pressure on the key is decreased. We will now construct a model of the coil that will allow us to get approximately the motions of the wire under these conditions.

We see from Fig. 10 that the azimuthal forces are almost linearly proportional to the conductor number. Furthermore, we have made extensive measurements on the coil package and know its elastic properties. Measurements made by Karl Koepke and John Saarivirta are shown in Fig. 11. Measurements are shown both with the coil at room temperature and cooled to liquid nitrogen temperature. As the coil cools it shrinks, and the Youngs modulus increases. A model of the coil is shown below.

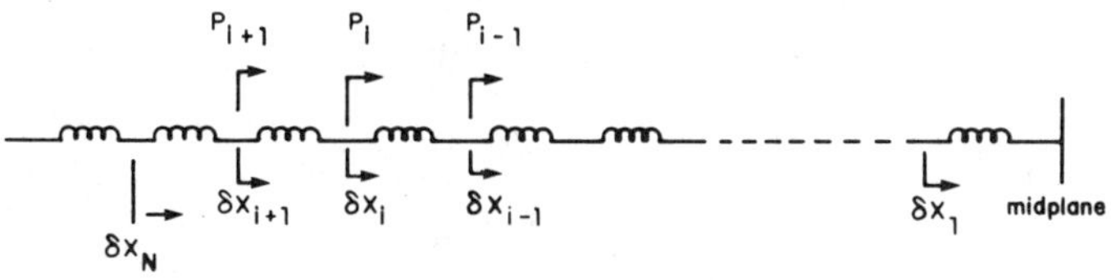

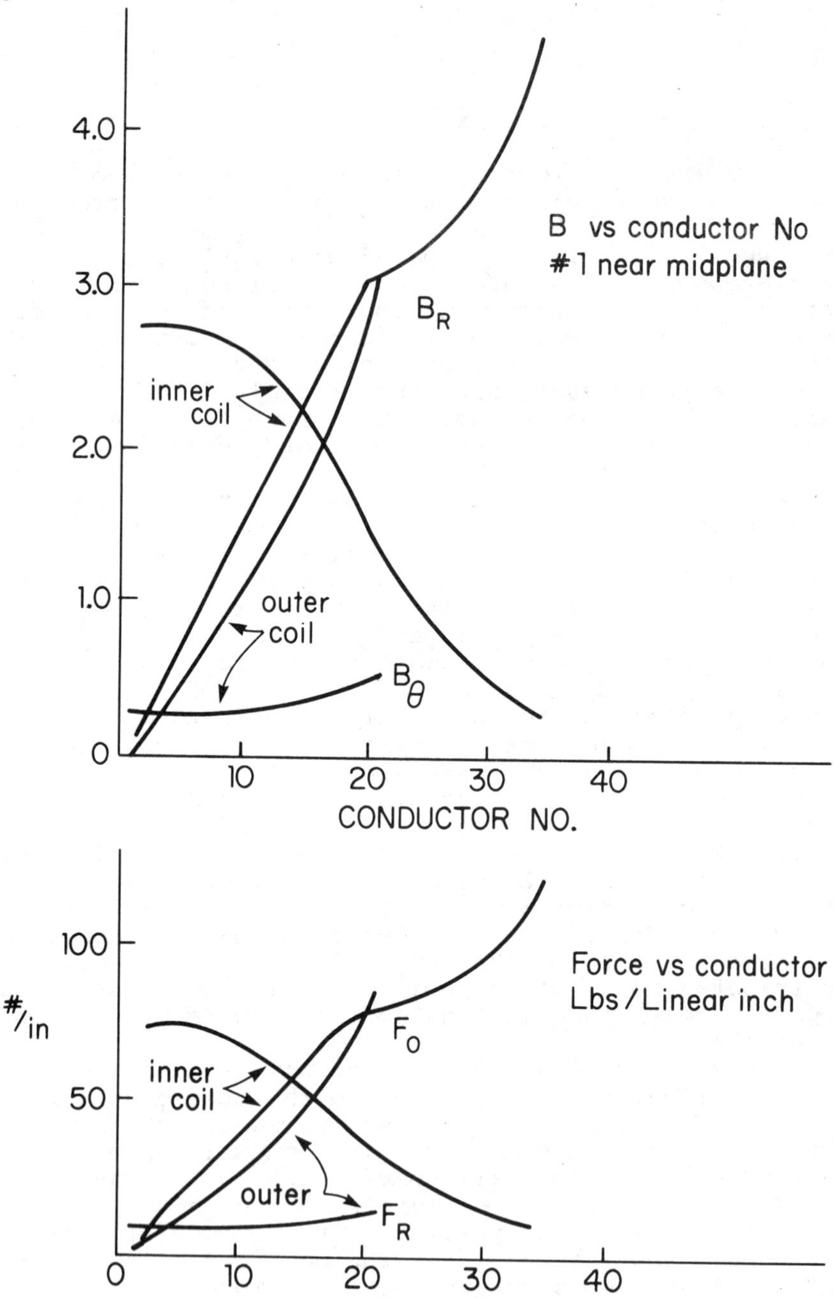

Fig. 10. Magnetic field (tesla) and force vs. conductor number in a Doubler coil.

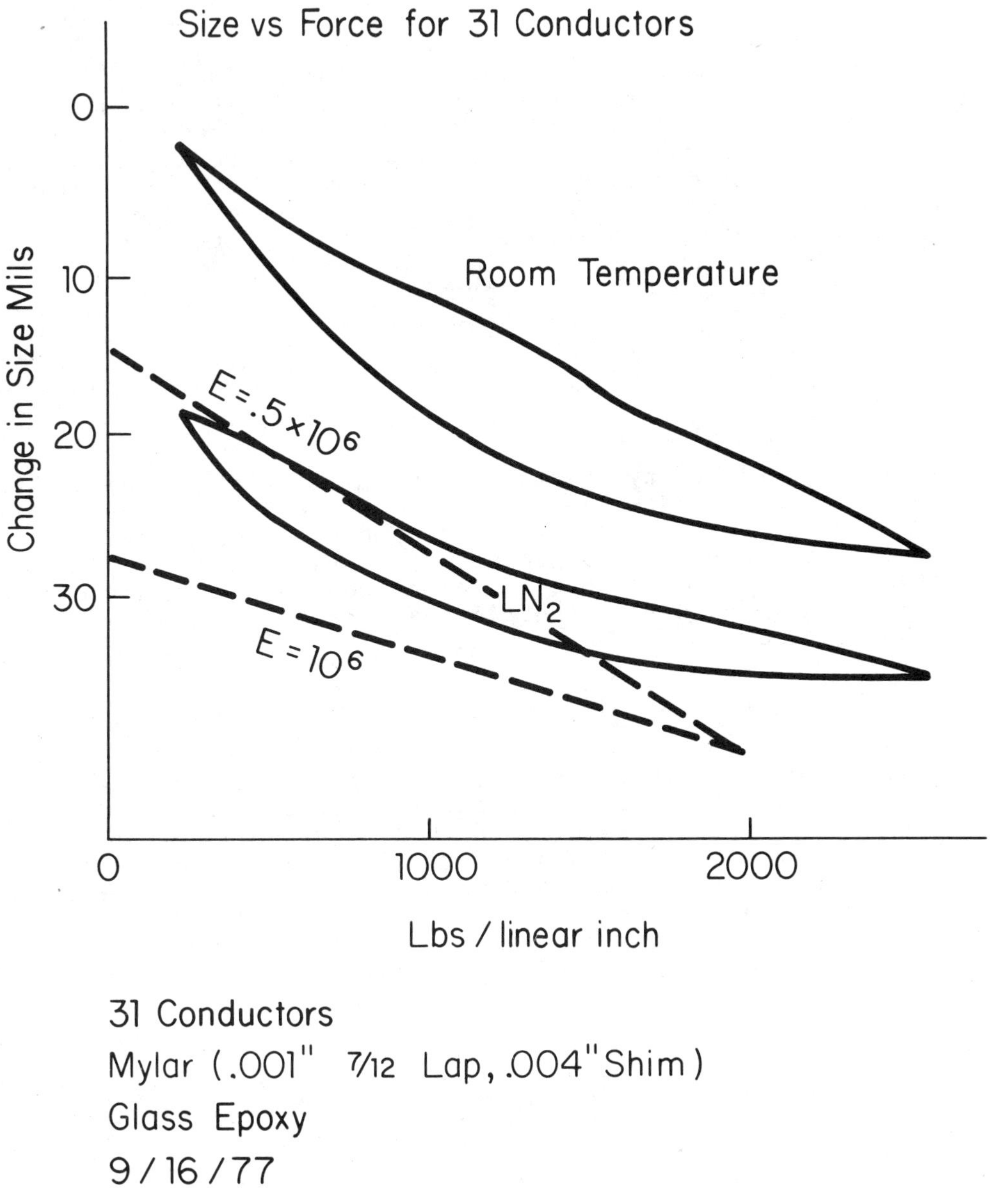

Fig. 11. Size vs. force for a typical Doubler coil.

The coil is a series of springs between conductors with a force applied at each spring junction. The strength of the springs, k, is taken from the measurements in Fig. 10, and the force on the individual conductors is taken from linearizing the forces shown in Fig. 9. The conductor number is designated by i and runs from 1 to N. The boundary conditions on the problem are that we want zero displacement at i = 0 and i = N + 1, the midplane and the key, respectively. Balancing forces, we can write the difference equation:

$$+k(\delta X_{i+1} - \delta X_i) - k(\delta X_i - \delta X_{i-1}) = -P_i = -\alpha i$$
$$\delta X_{i+1} - 2\delta X_i + \delta X_{i-1} = -\frac{\alpha}{k} i \quad (46)$$

The proportionality constant between conductor number and force is designated by α. The solution of this difference equation is a simple cubic. In addition, there are no particular solutions that enable us to fit the boundary conditions. Combining these equations with the boundary condition, we get the solution for the problem

$$\delta X_i = \frac{\alpha}{6k} i \left[(N+1)^2 - i^2 \right] \quad (47)$$

This equation represents the motion of the individual conductors away from the equilibrium position as the magnet is excited. A plot of this equation is shown in Fig. 12 for the inner and outer coils. It is seen that the maximum wire motion occurs about half way through the coil; for the inner coil, it is about 3 mils and for the outer, about 0.6 mil.

The peak deflection occurs at conductor number given by

$$i = \frac{N+1}{\sqrt{3}} \; ; \; \delta X_m = \frac{\alpha}{6k} (N+1)^3 \frac{2}{3\sqrt{3}} = \frac{\alpha (N+1)^3}{9\sqrt{3}k} \quad (48)$$

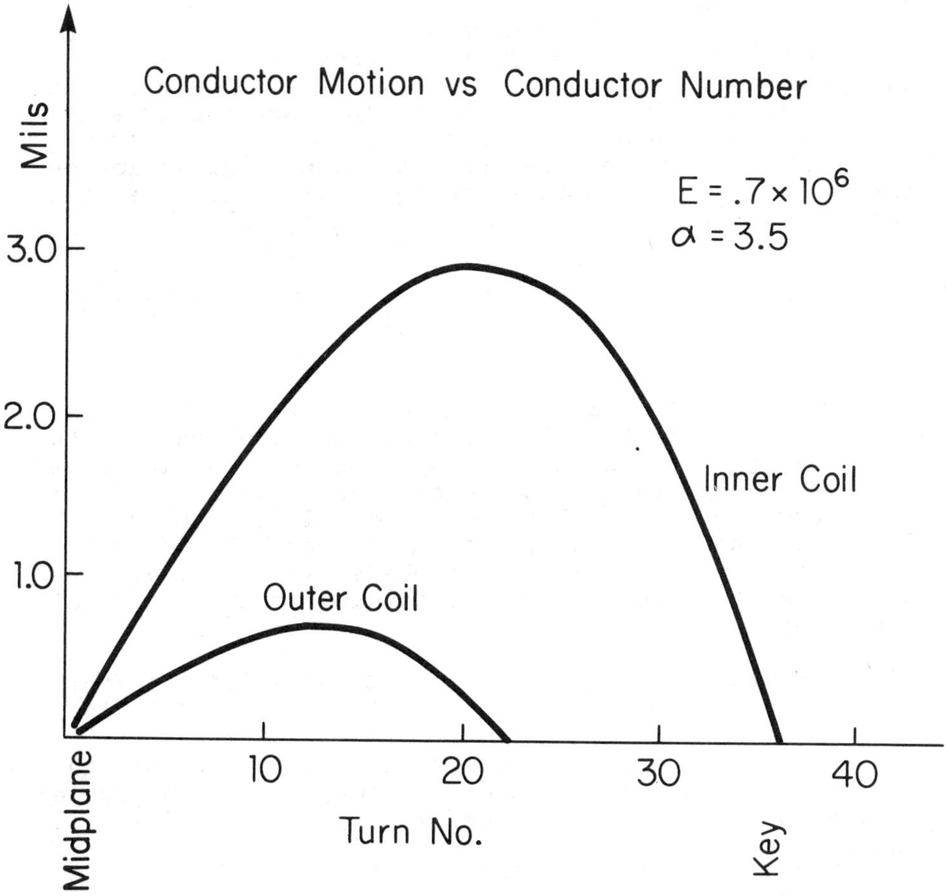

Fig. 12. Conductor motion vs. the conductor number for a typical dipole coil.

For the inner, this gives conductor number 21 away from the median plane and for the outer coil, it gives conductor number 13.

If we insert the constants α and k, we can calculate the maximum deflection, and the forces at the two ends of the coil. In this model, since we have set the deflection at the key equal to zero, the equation will yield a tension. In order to keep the actual physical coil clamped, we must have a precompression on the coil that exceeds this tension. The tension at the key is given by

$$F = k\,\delta X_n = k\,\frac{\alpha}{6k}\,N\left[(N+1)^2 - N^2\right] = \frac{\alpha}{6}\,N(2N+1) \approx \frac{\alpha N^2}{3} \quad (49)$$

The Young's modulus is about 7×10^5 lbs. per square inch (see Fig. 11). Since the wire has a width of 0.315 in. and the average spacing is 0.062 in., we can calculate the spring constant $k \approx 3.5 \times 10^6$. From Fig. 10 we obtain a value of F equal to about 3.5 lbs. per inch per conductor for both the inner and outer coil. Inserting these numbers in Eq. (49), we calculate the necessary preload maximum deflection of the inner and outer coil

Preload = $\frac{\alpha}{6}N\,(2N + 1)$	Inner = 1,450 lb/in.
	Outer = 527 lb/in.

Note that the preload necessary is independent of the spring constant k and is only a function of the force on the conductor due to the magnetic field, as it should be.

The perturbation to the magnetic field caused by the conductor motion that we have calculated is shown in Table III, and the perturbation has been normalized to a peak motion of the wire equal to 10 mils. The displacement, of course, would be proportional to B^2.

Table III

Distortion caused by the elastic compression of the wires in an azimuthal direction due to the magnetic forces.

n	a_n(Inner)	b_n(Inner)	a_n(Outer)	b_n(Outer)
0	0.000	24.166	0.000	6.595
2	0.000	18.040	0.000	8.544
4	0.000	-3.135	0.000	2.622
6	0.000	0.677	0.000	0.094
8	0.000	-0.059	0.000	-0.133
10	0.000	-0.049	0.000	-0.026
12	0.000	0.038	0.000	0.004
14	0.000	-0.017	0.000	0.002
16	0.000	0.005	0.000	0.000
18	0.000	-0.001	0.000	-0.000

Clamping the coil

We come now to to central difficulty of collaring a superconducting coil. We have calculated the forces that are necessary in order to restrict the motion at the end of the coil; however, an unpleasant fact now must be coped with and that is that the coil shrinks more than the collars when it is cooled. The coil, therefore, must be molded considerably bigger than the space in the collars, and the collars must be applied with a correspondingly large pressure. The situation is illustrated in Fig. 13. This measurement is called a split-collar measurement. A 1-in. section was cut out of a finished magnet. Then the collar was carefully measured and then split on one side at the midplane with a saw cut. The force necessary to close this gap back to the original collar size gives the preload. This technique has two advantages:

1. It directly measures the preload in the collared coil. (The mechanical stiffness at the collar is almost negligible, and so the displacement-force curve gives the total elastic constant at the coil package.)

2. It gives a direct measurement of the relative size of the coil to the collar when measurements are made at reduced temperature and hence the preload necessary at room temperature to produce the required preload at low temperature. This is a very difficult measurement to make by any other means.

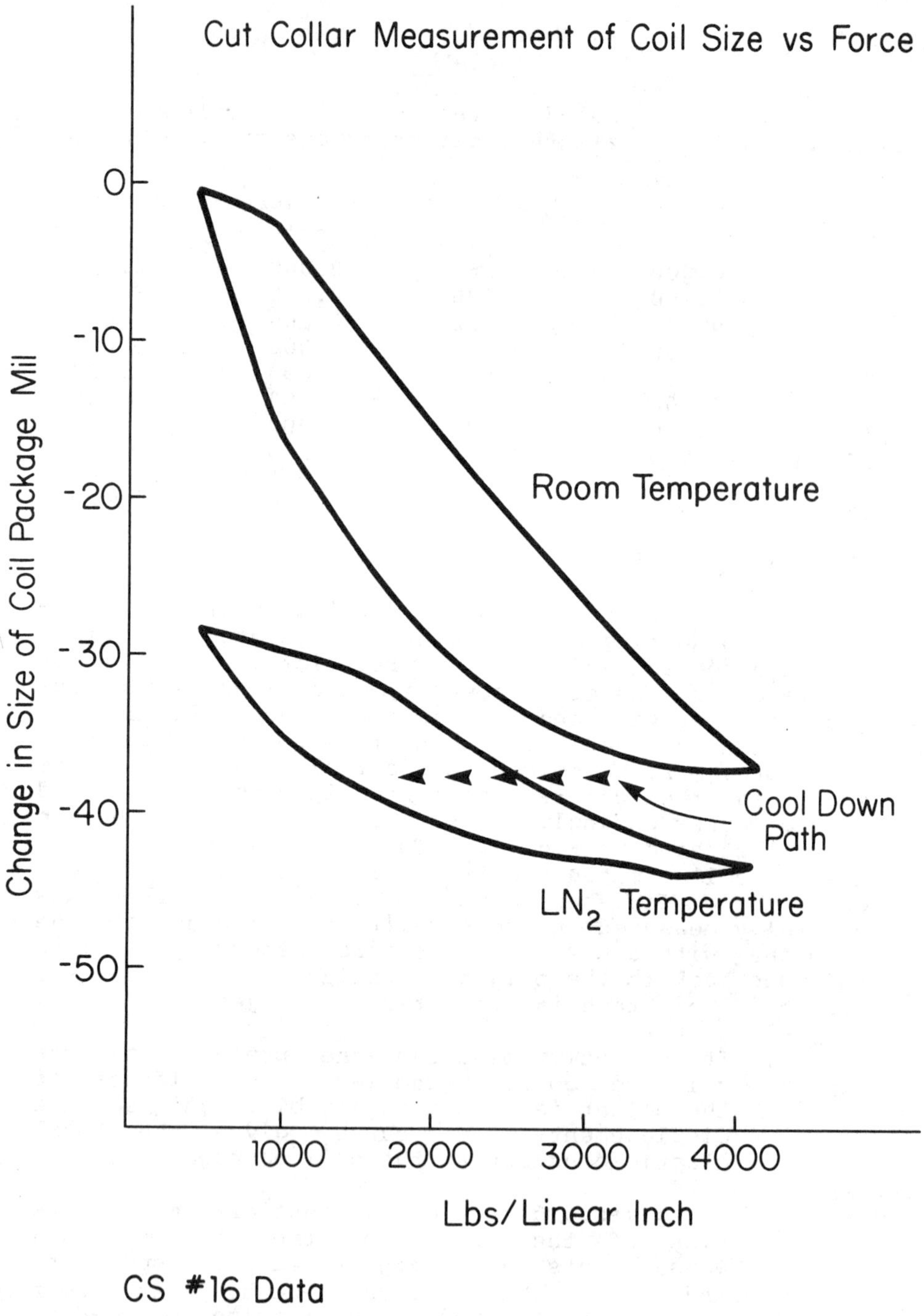

Fig. 13. Cut collar measurement of coil size vs. force.

As a collared coil is cooled, the pressure in it decreases and yet must remain above the minimum force as dictated by the magnetic pressure as calculated above. This forces a design constraint upon the coil matrix. A solution does not have to exist. The necessary room-temperature forces may take the coil out of its elastic range and crush the matrix. The cooldown path is shown as a line of arrows in Fig. 13. The room-temperature force of 4,000 lbs/inch in this case would load to a total force of about 1,400 lbs/inch when cold. This is less than given by Eq. (50) for the total force necessary to clamp the coil. A force at room temperature of 5,000 lbs/inch has been found to clamp the coil.

We now consider how this necessary preload is obtained. It is to be emphasized at this point that the accuracy of the final coil package is determined by the collars. The coil itself is molded oversize and compressed into the collars by means of large presses. We must, however, consider the accuracy of the molding process for the coil and must be sure that even at the minimum fluctuations in its size that the pressures will be large enough so that the coil remains clamped. In the case of the Fermilab coils, each coil is measured by means of a special gauge that applies a high pressure locally to the coil and measures its circumferential size. Ten measurements are averaged for each side of the four coils. The size of the coil is controlled by means of shims at the time the coil is molded. The amount the coils are oversize has been determined so that when the coil is cold and collared, it will still have the necessary preload on it.

A monitor of this preload is furnished by observing the force applied to the coil during the collaring operation. The total force of the press is equal to the sum of the preload necessary on the inner and outer coil plus the safety factor. A typical collaring curve is shown in Fig. 14. The amount that the collars are open is determined by the gaps between scribe marks on some specially selected hardpacks. Before the coil is collared, these hardpacks are assembled, and a fine mark scribed across the interlaced fingers. When these hardpacks are placed around the coil, the gaps between adjacent scribe marks show how far the collars are from being in the closed position. In general it takes about 10,000 lbs per linear inch of coil to close the collars. This is equivalent to 5,000 lbs per linear inch on each side of the magnet and should be compared with about a total of 1,977 lbs. per inch which is the sum of the necessary magnetic preload on the inner and outer coils per inch as given in Eq. (50).

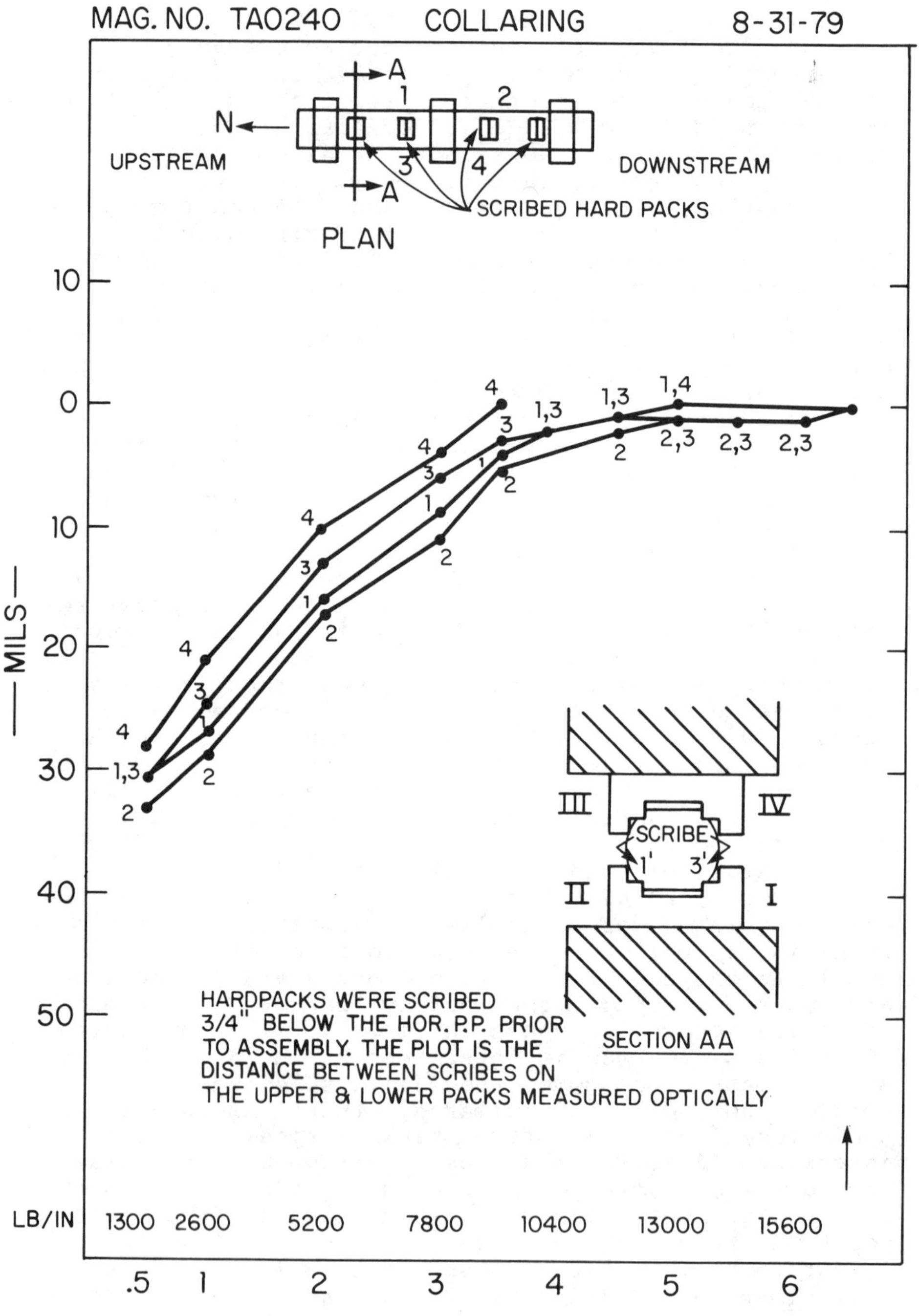

Fig. 14. Typical collar deflections vs. force/in. for Doubler magnet.

Early in the program a great deal of difficulty was experienced in getting adequate preload applied to the coil conductor. In order to study the motion of the wires as the magnet was excited, an instrument called a sissometer was invented which measures the relative deflection between the 32nd conductor on the top coil and the 32nd conductor on the bottom coil. The problem was first observed in the 1-ft magnet series and was not solved until Magnet E22-52. Figs. 15 and 16 show curves from this early series of magnets. The first curve is for Magnet E22-33. The sissometer measurements on this magnet are shown at the bottom of the figure. The vertical axis shows the compaction of the whole coil in mils and hence half this number would correspond to the motion of the conductor away from the key. Two effects are seen in this curve, the first is that there is a large motion in the conductor as the magnet is excited, and second that there is a considerable amount of hysteresis displayed. If the coil is changing its configuration with current, it follows that the multipole moments should also change. The top of the figure shows the change of b_2 and b_4 with field. (The hysteresis displayed in b_2 is partly due to persistent currents.) The upward slope of the curve for b_2 corresponds to the inner coil becoming more compact. This is also verified by the shape of the curve for b_4. As the inner coil becomes more compact, b_4 becomes more negative. The motion of the coil does not correspond to a uniform compaction, however, it is possible to verify the signs for these two effects by looking at Table IV which shows the variation of b_2 and b_4 with key angle.

Table IV

This table applies to the harmonics when the key angles are changed by 10 mils.

n	a_n(Inner)	b_n(Inner)	a_n(Outer)	b_n(Outer)
0	0.000	17.227	0.000	4.577
2	0.000	7.101	0.000	5.845
4	0.000	-4.117	0.000	1.702
6	0.000	1.415	0.000	0.005
8	0.000	-0.012	0.000	-0.105
10	0.000	-0.293	0.000	-0.018
12	0.000	0.046	0.000	0.004
14	0.000	-0.026	0.000	0.002
16	0.000	0.009	0.000	0.000
18	0.000	-0.002	0.000	-0.000

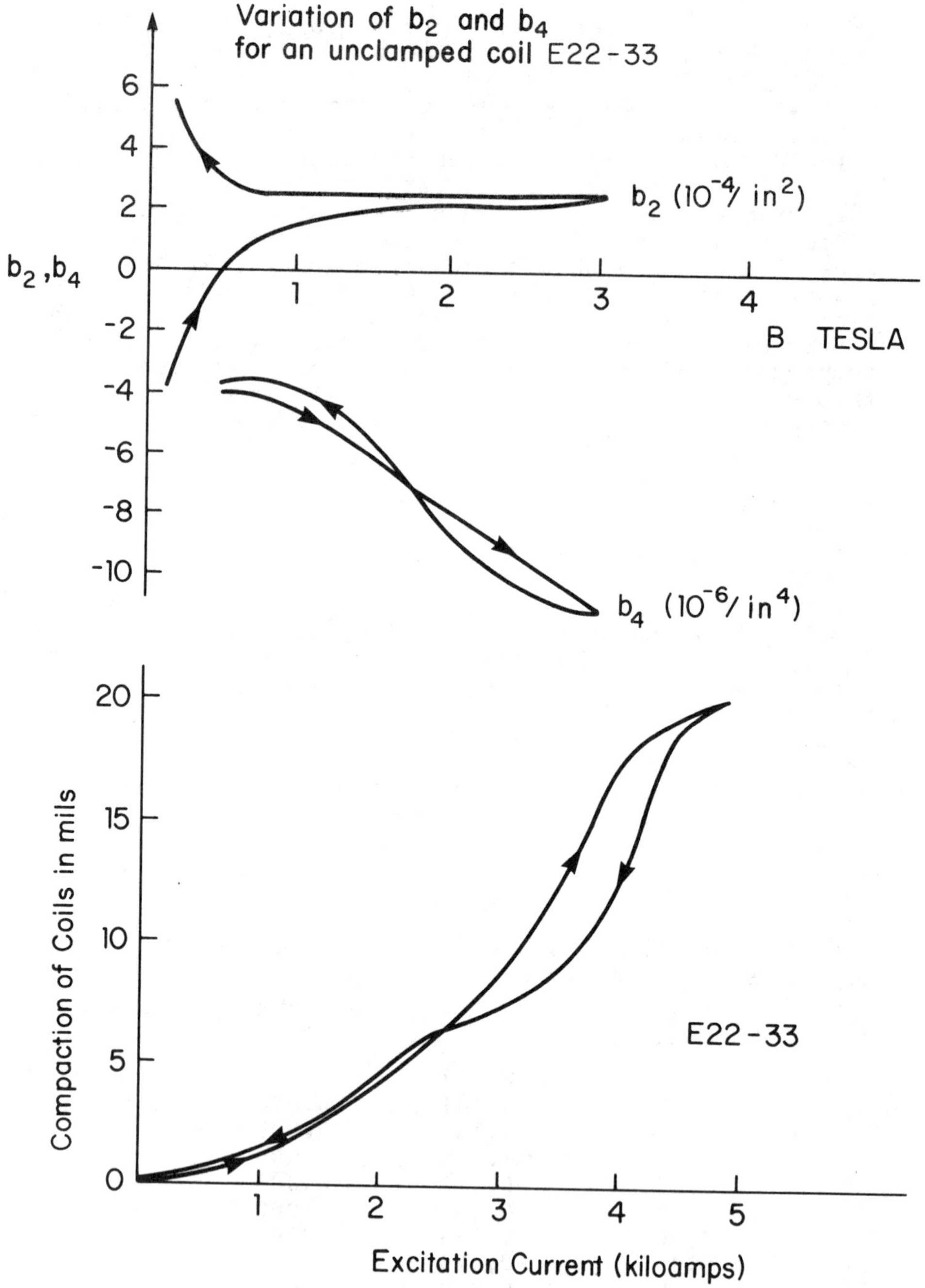

Fig. 15. Change of b2, b4 vs. magnetic field and coil compaction vs. excitation current for an early Doubler test magnet.

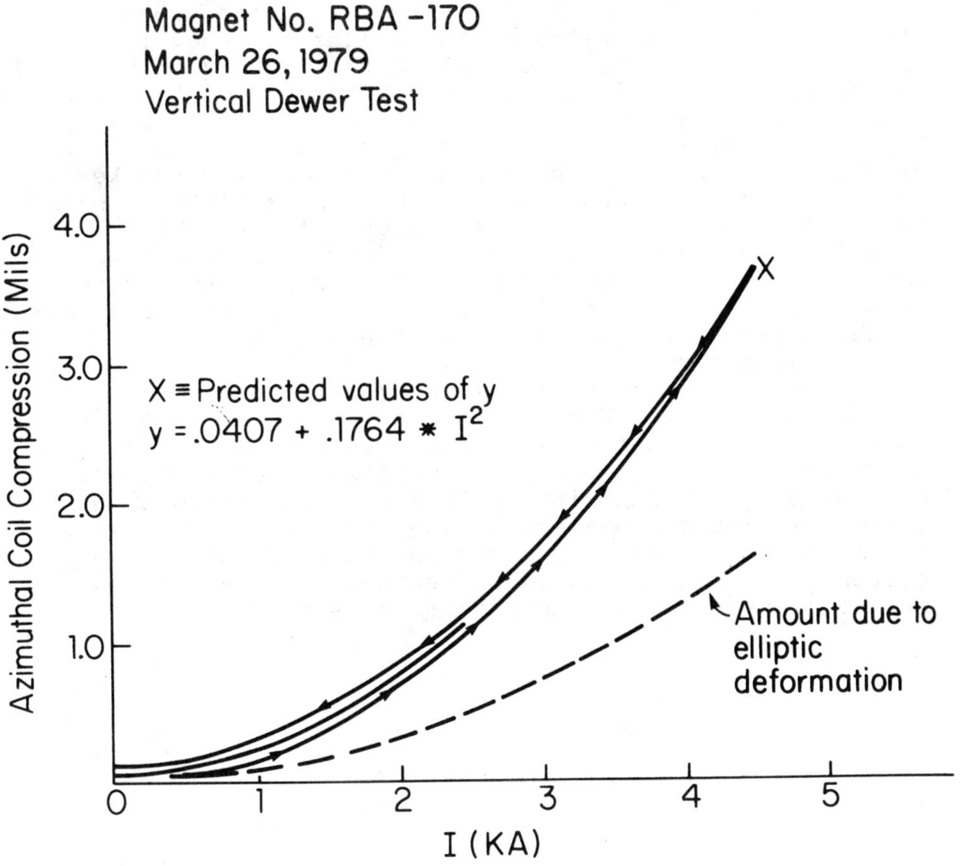

Fig. 16. Coil compression vs. excitation current for an early Doubler test magnet.

Fig. 16 shows the typical sissometer curve for a later magnet. It is noticed that in this case the motion is completely elastic, and no hysteresis is present. The instrument used, the sissometer, responds also to other distortions in the collar as the magnet is excited. The amount of this correction is shown in the dotted curve and should be subtracted from the top curve. The fact that there is any motion at all is because the motion shown is not at the end of the winding but rather three conductors in from the key. In addition to this, the model is a little bit idealized in that in a real magnet there is over 30 mils of compressed Kapton at the key for electrical insulation. Thus the motion that is calculated is not exactly correct but is linearly superimposed upon a small motion at the end of the winding due to the weaker end spring.

We no longer make sissometer measurements on all of the magnets. Fig. 17 shows the histogram of the motion observed in the 22-ft magnets that were tested.

Fig. 18 shows the typical variation of sextupole and decapole moments with field that is observed in our present series of magnets. The vertical scale for these coefficients is the same as Fig. 18 and does not correspond to the standard units that we use in that they are expressed in the centimeter system.

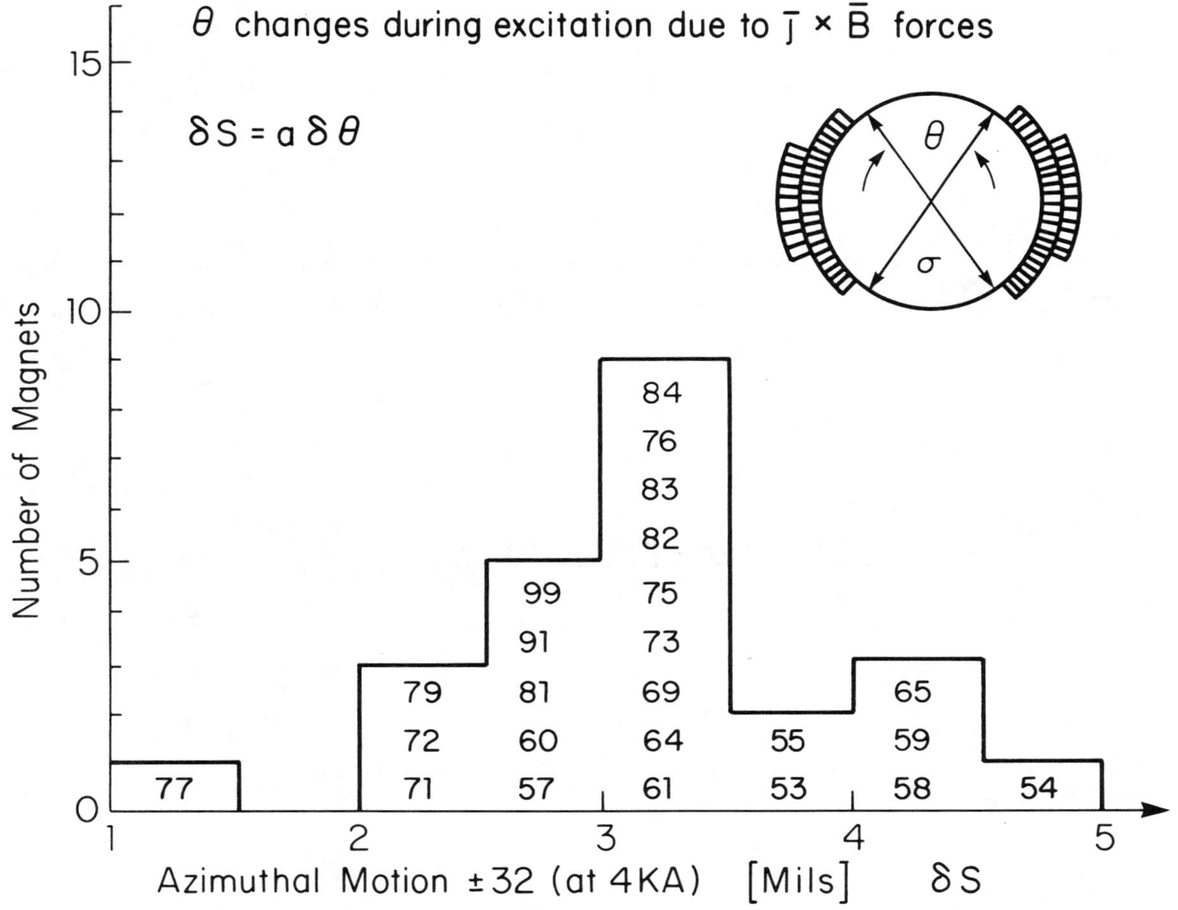

Fig. 17. Histogram of azimuthal motions for a group of typical Doubler magnets.

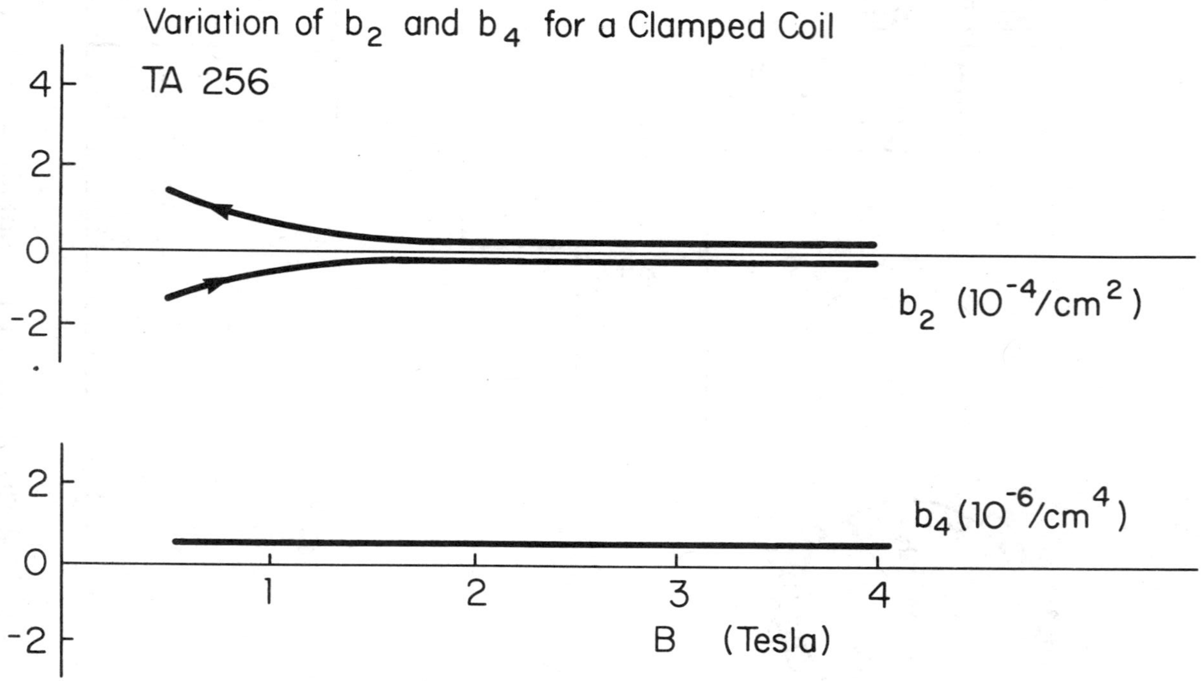

Fig. 18. B2, b4 as a function of magnetic field for recent Doubler magnets.

5. QUENCH WAVES

Formulation of the problem

We consider first a simple description of what happens when a piece of superconducting cable starts to quench. First of all, consider a small piece of the cable length ΔL going normal. Once this happens it be comes resistive. In the Fermilab cable, the resistance is about 1 $\mu\Omega$ per centimeter, and hence, at 5,000 amps through the cable, a power of 25 watts per centimeter length of cable will be generated. The power deposited in a piece ΔL will simply be 25 ΔL. This heat is conducted away by the helium that is in contact with the cable. The conduction of heat by the helium is complicated, and it can take place through forced convection or by means of boiling. There are two regimes for boiling: One is nuclear boiling and the other is film boiling. In general, for a temperature rise of a degree, the heat flux is less than half a watt per square centimeter. In addition, the section L is cooled by conduction to the rest of the cable through its ends. If ΔL is very short, then the power generated becomes small enough so that the cooling exceeds the heat generated, and the region collapses and returns to the superconducting state. If, however, ΔL becomes larger, then the power generated exceeds the cooling of the conductor, and the normal region expands. This expanding front is called a quench wave, and it will spread down the cable with a velocity of the order of 1 to 20 meters per second. It will also spread by conduction through the turn-to-turn insulation into adjacent turns, and thus spread out in an elliptical fashion from the original hot spot. The turn-to-turn velocity is generally much slower than the velocity along the cable. The hottest point in a quenching magnet is in general the point that initially started the quench.

This behavior is governed by the heat equation

$$C(T)\,\frac{\partial T}{\partial t} = \nabla\cdot(k(T)\,\nabla T) + P(\vec{r},T). \qquad (51)$$

Here C(T) is the heat capacity of the material, k(T) is the heat conductivity of the superconducting composite, and P is the power generated per unit volume. If we integrate over the cross section of the conductor, this becomes a one-dimensional heat equation

$$C(T)\,\frac{\partial T}{\partial t} = \frac{\partial}{\partial z}\left(k(T)\,\frac{\partial T}{\partial z}\right) + P(z,T). \qquad (52)$$

The power now becomes $I^2R(T)$, where R is the resistance as a function of temperature of the cable. We will now investigate how these parameters change with temperature.

Figure 19 shows the resistance as a function of temperature of copper and aluminum. Material of different purities follow the same curve at high temperatures. At low temperature, however, impurities introduce scattering centers. The resistance at 4^0 K defines the resistivity ratio of the material. The copper in the Saver cable has a resistivity ratio that is of the order of 100 and with aluminum, one can achieve over 1000. This is complicated by the fact that the resistance is also a function of the magnetic field in which the conductor is immersed. High fields increase the resistivity through the magneto resistance effect. The resistance ratio is also degraded by strain in the material. Since very pure conductors tend to be very soft and have little mechanical strength, the conductor must be supported well if the strain is to be kept to low levels. Finally, we note that the resistance of the superconducting material is so high that it can be effectively neglected for any of our considerations here (see Fig. 3).

The parameter C in the heat equation represents the heat capacity per centimeter of length in the wire. According to the rule of Dulong and Petite, thermal heat capacity should be a constant independent of temperature. However, it is a rapid function of temperature and the superconductor behavior. It is even more complicated by the fact that there is a discontinuity at the transition from the superconducting to nonsuperconducting state. The behavior of some metals and alloys is shown in Fig. 20. In general, at low temperatures the heat capacity can be fit by an equation of the form

$$C(T) = \beta T + \gamma T^3. \qquad (53)$$

The major term in this equation is the term proportional to T^3, and we will use this approximation in studying quench waves.

The other parameter in the heat equation k(T) is the thermal conductivity of the material. Resistivity and thermal conductivity are related and as with the electrical current, we find that the main heat conduction occurs through the copper or aluminum in the matrix. Some curves showing thermal conductivity of various metals are given in Fig. 21. Note that materials like stainless steel are nearly independent of temperature, whereas material like copper can have a very complicated behavior. In the region that we are considering here, we will approximate the thermal conductivity as being proportional to temperature in the low-temperature region.

Static solutions

The small length of cable ΔL that we have been considering as normal can be in equilibrium if its length is exactly right, and this would correspond to a static solution of the heat equation. The length will depend upon the cooling capacity of the helium, the resistance of the wire, and its heat conductivity. It will also depend upon the amount that the temperature must be raised above ambient in order to cause the wire to go normal. For NbTi the critical temperature is of the order of 10^0 K, and the temperature of liquid helium is about 4.3^0 K. However, as has been shown in Fig. 1,

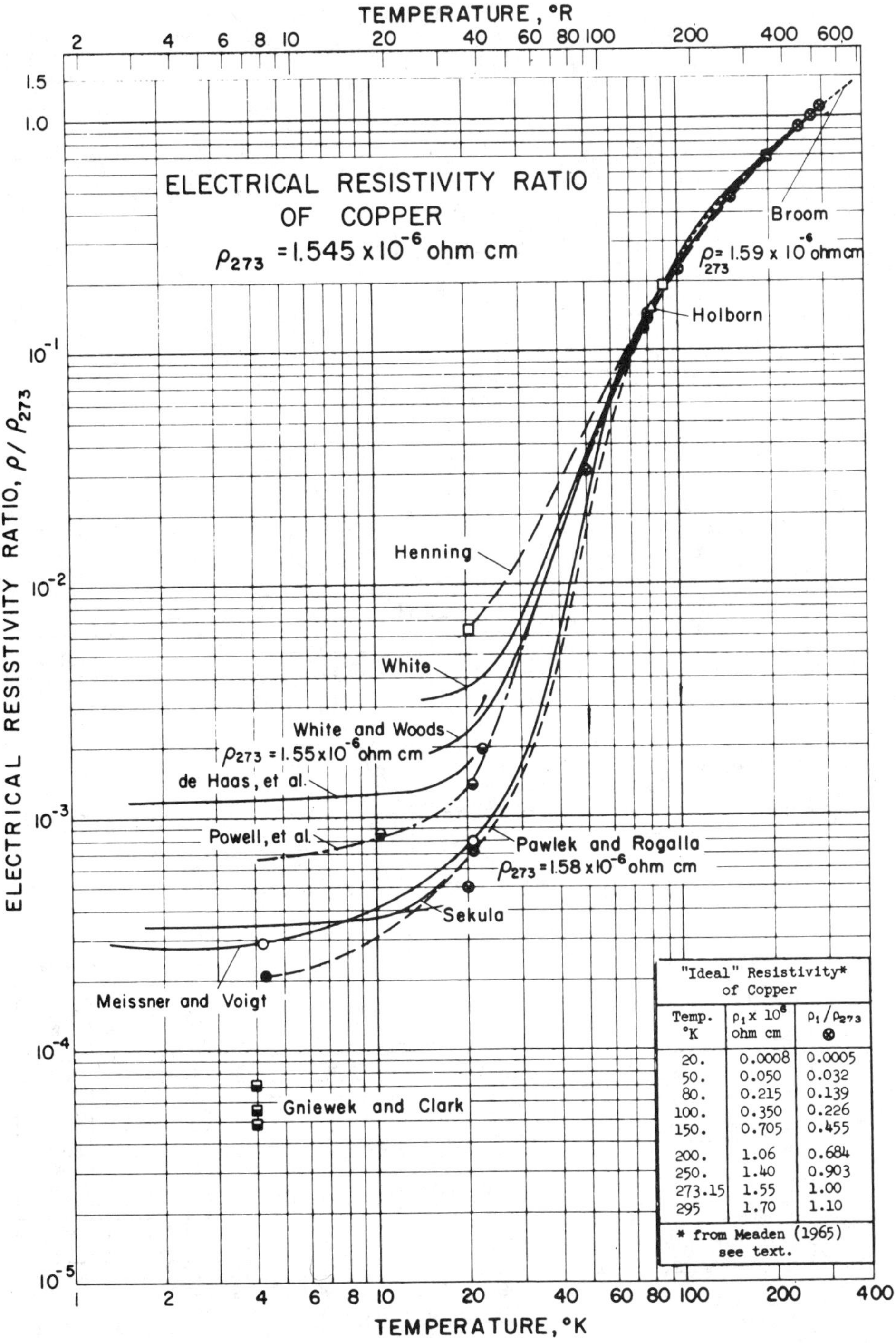

"Ideal" Resistivity* of Copper

Temp. °K	$\rho_i \times 10^6$ ohm cm	ρ_i / ρ_{273} ⊗
20.	0.0008	0.0005
50.	0.050	0.032
80.	0.215	0.139
100.	0.350	0.226
150.	0.705	0.455
200.	1.06	0.684
250.	1.40	0.903
273.15	1.55	1.00
295	1.70	1.10

* from Meaden (1965) see text.

Fig. 19(a). Resistivity as a function of temperature for aluminum [from Cryogenic Data Memornadum - NBS Technical Note 365 (1968)].

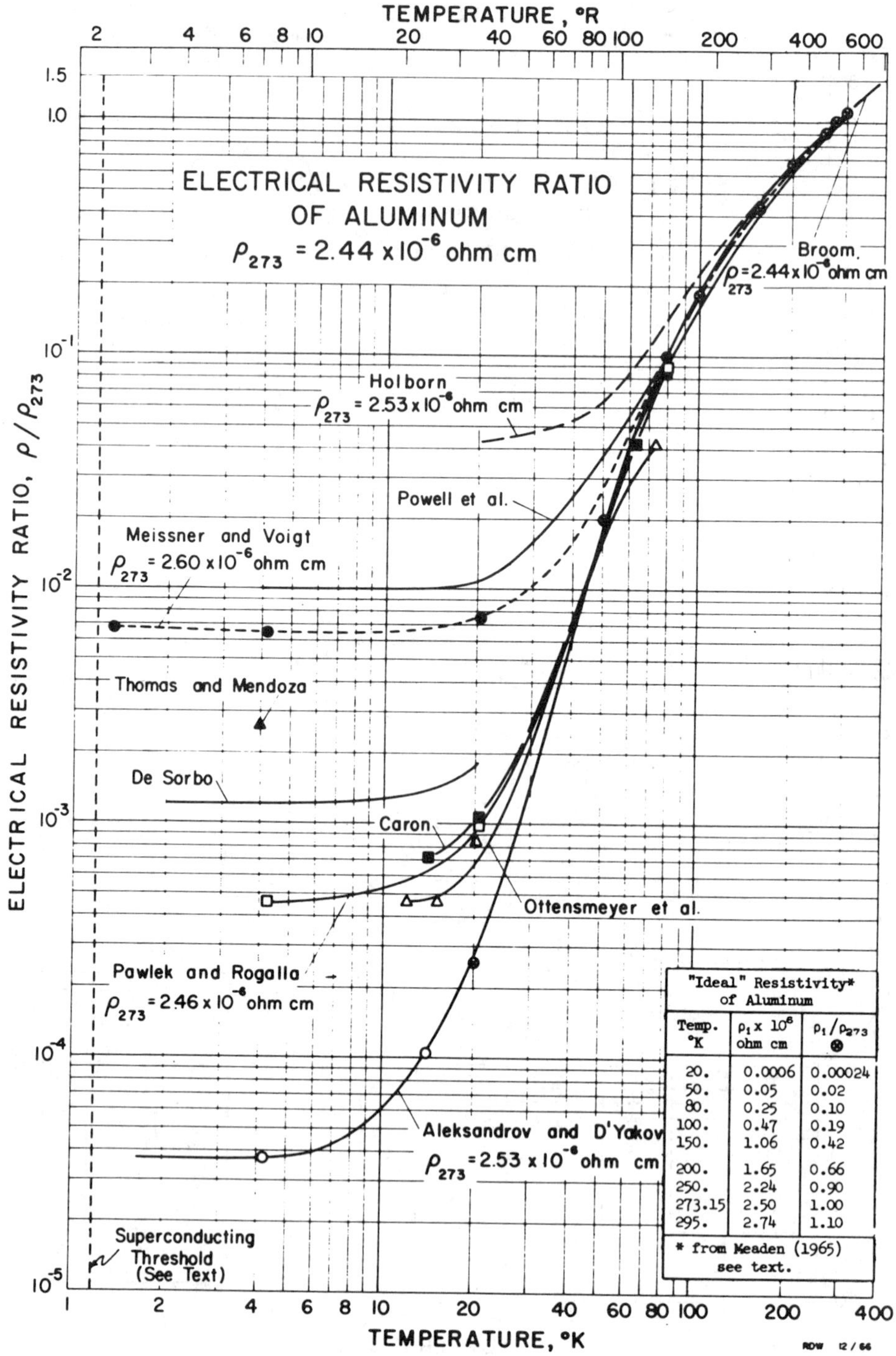

"Ideal" Resistivity* of Aluminum

Temp. °K	ρ_i x 10^6 ohm cm	ρ_i/ρ_{273} ⊗
20.	0.0006	0.00024
50.	0.05	0.02
80.	0.25	0.10
100.	0.47	0.19
150.	1.06	0.42
200.	1.65	0.66
250.	2.24	0.90
273.15	2.50	1.00
295.	2.74	1.10

* from Meaden (1965) see text.

Fig. 19(b). Resistivity as a function of temperature for copper [from Cryogenic Data Memorandum - NBS Technical Note 365 (1968)].

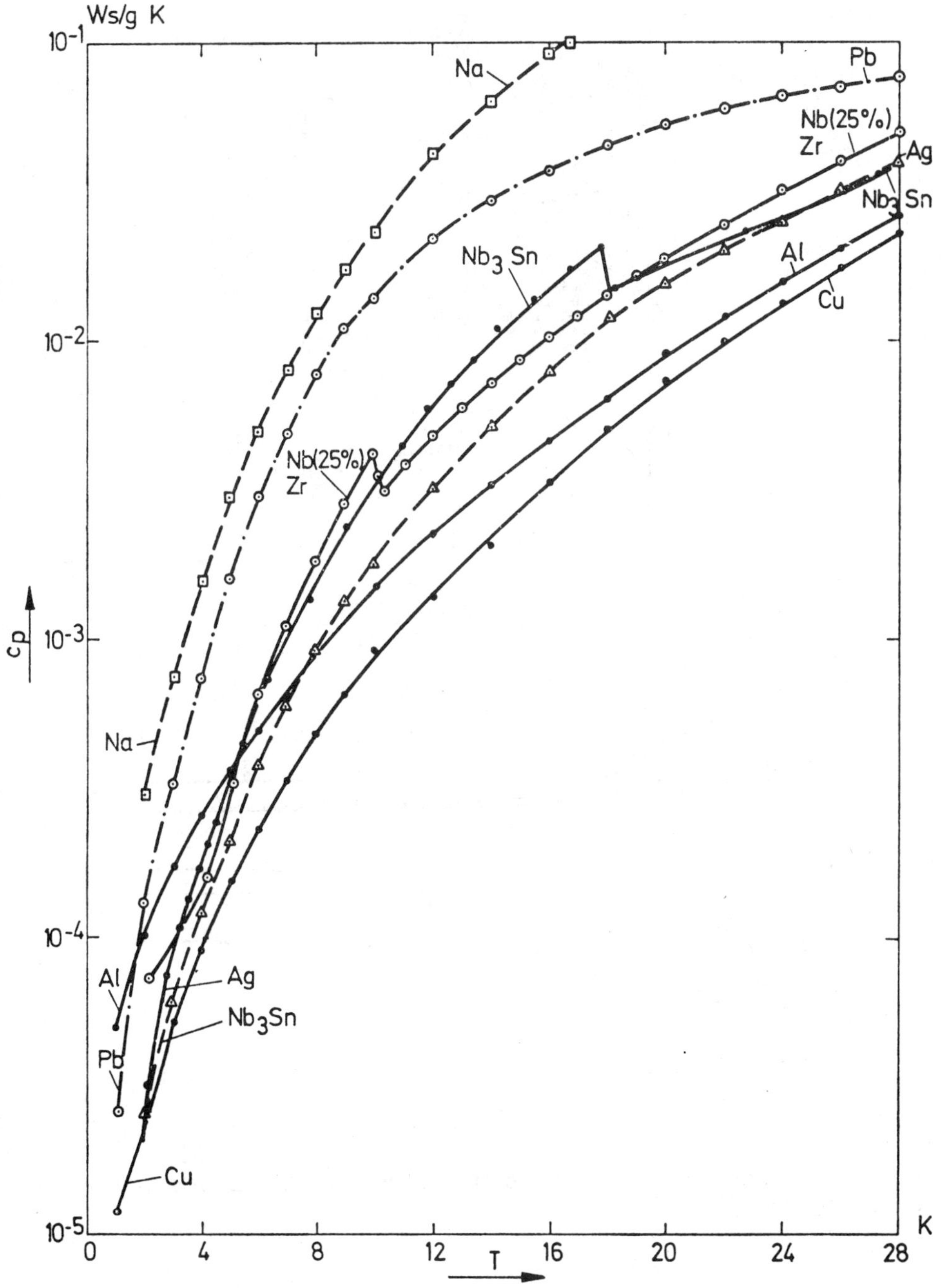

Fig. 20. Specific heat vs. temperature normal metals and Type II superconductors (from Brechna).

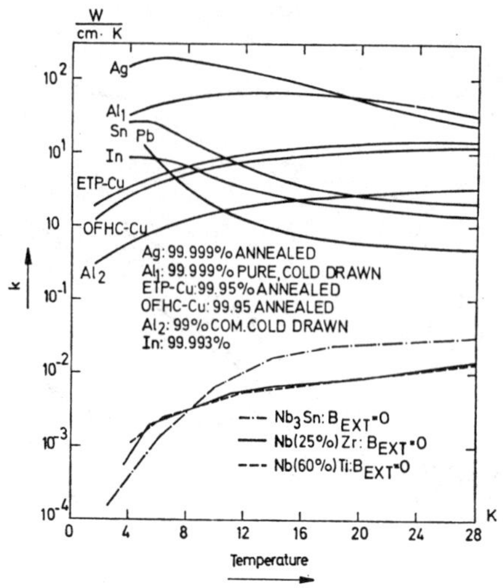

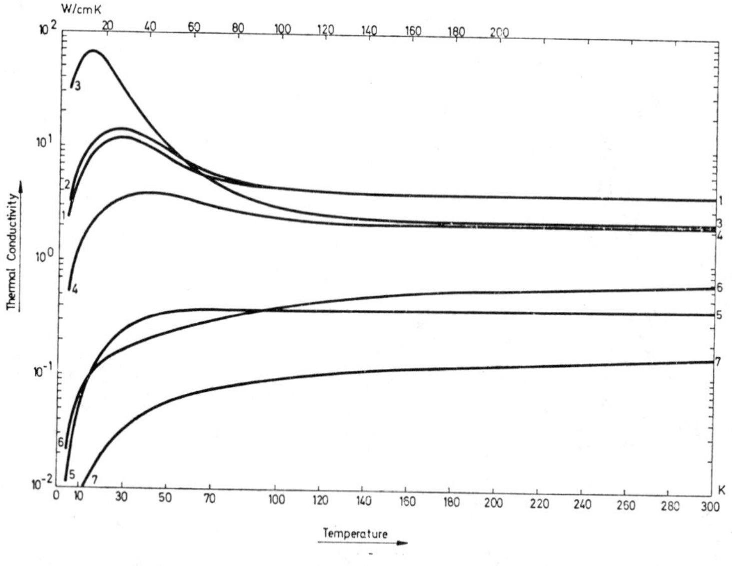

Fig. 21. Thermal conductivity vs. temperature for various materials (from Brechna).

the amount of temperature rise necessary is also dependent upon the magnetic field. The higher the magnetic field, the more sensitive one is to a small temperature rise in the conductor. It is possible to get some idea of the amount of energy necessary to cause a quench by calculating the total energy necessary to establish the state that we have just been discussing. This can be done under several assumptions, the easiest of which is to assume all the parameters in the heat equation are independent of temperature. This is unrealistic, but it leads to simple mathematics. It is also possible to solve the equation when the conductivity is proportional to the temperature and the resistance is independent of temperature. Both of these solutions will be treated later. We will describe only the results here.

Let's consider the case of constant power generation and constant heat conductivity. The cooling is proportional to the temperature difference between the conductor and the bath.

$P = -\alpha(T-T_B)$	$P = I^2R$	$P = \alpha(T-T_B)$	
SC	NORMAL	SC	CABLE

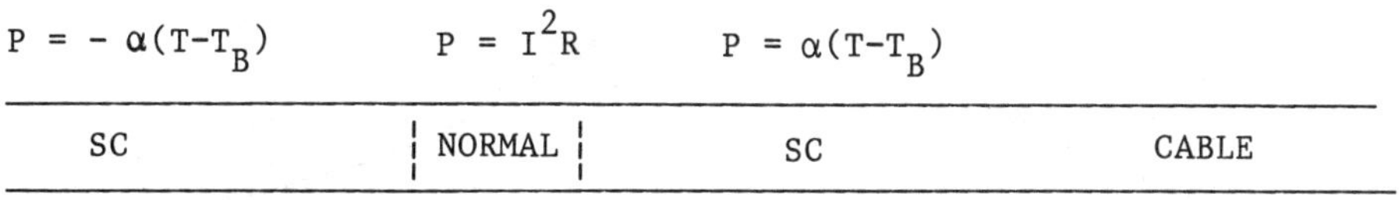

The heat equation for this case becomes

$$k \frac{d^2T}{dz^2} + P(z,T) = 0. \qquad (54)$$

We can now consider two different boundary conditions: The first would be to consider all of the cable on the left-hand side of the origin to be in the normal state, and everything on the right-hand side to be superconducting. On the far left-hand side, the power generation is just in equilibrium with the cooling and on the far right-hand side, the temperature is that of the bath. The solution for this situation is

$T = 2T_C - T_B$
$\lambda = \sqrt{K/\alpha}$
T_C
$T = T_B + (T_C - T_B)e^{-z/\lambda}$
$T = T_B$

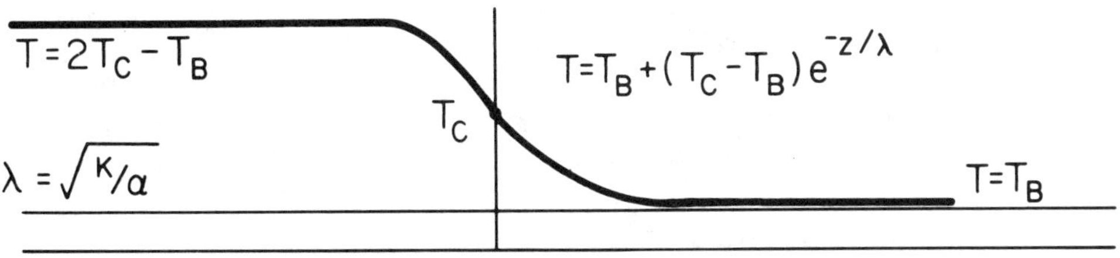

Note that this situation allows us to determine the cooling capacity of the helium for the conductor provided we know the resistance of the conductor at the critical temperature. It is an easy situation to set up in a laboratory. One starts a quench wave and reduces the current to the point where the wave is actually stable. By reducing the current slightly, the wave will collapse; by increasing it, the wave will grow.

A second simple solution is to find a minimum length of cable that can remain normal for a given current. The boundary conditions are shown in the following figure.

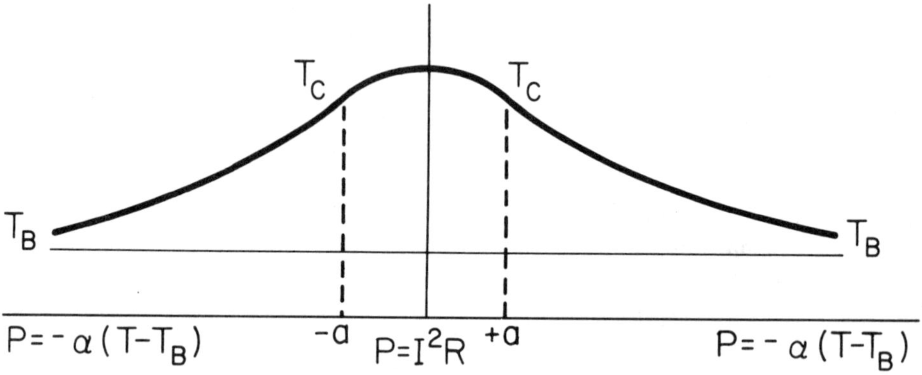

The case we have first considered is the limit of this one, when the length, ΔL, is infinite. As we increase the current above this minimum current, the length ΔL will decrease. The solution to the heat equation in the various regions is given by

$$z<a \quad T = T_C + 1/2\,\frac{P_o}{k}\,(a^2-z^2)$$

$$z<a \quad T = T_B + (T_C-T_B)\,e^{(a-z)/\lambda} \qquad (55)$$

$$\lambda = \sqrt{k/\alpha} \qquad a = \frac{(T_C-T_B)}{P_o}\,\lambda\,.$$

It is now interesting to integrate the total amount of heat contained in the temperature bump that we have just calculated. The thermal relaxation time of the cable is very short. For distances of the order of a centimeter, this time is a few hundred microseconds. Hence, we will make the approximation that if a disturbance occurred and heat was dumped into the cable, the equilibrium shape that we have been calculating is instantaneously established. Thus, by finding how much energy it takes to establish this temperature profile, we will have some measure of the energy necessary to start a quench. Integration of the total heat in the above equations gives us

$$Q = 2 \int_0^{\infty} C(T-T_B)\, dz$$

$$= 2C \left\{ a \left| (T_C-T_B) + 1/3 \frac{P_o a^2}{k} \right| + \lambda\, (T_C-T_B) \right\}$$

T_C = Quench Temperature (56)

T_B = Bath Temperature

$= \sqrt{k/\alpha}$

$$a = \alpha \left| T_C-T_B \right| \cdot \frac{\lambda}{P_o}$$

A numerical example with Doubler cable is instructive. In order to make this calculation, we need a number for α. For a bare surface, this is 0.3 watt per degree K per centimeter square; however, the cable is insulated with a mylar jacket. We will take a = 0.1 watt per centimeter length of cable for this example. The rest of the numbers are taken for the Doubler cable and give the following results:

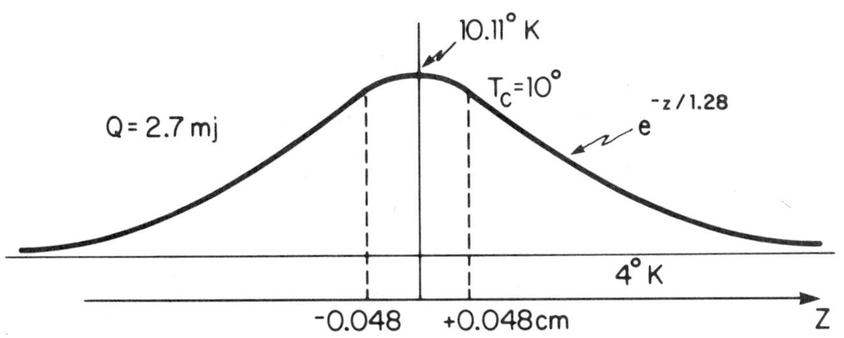

In this calculation, we have assumed that the critical temperature was 6^0 K above the bath temperature which corresponds to NbTi cooled by pool boiling helium. If a magnetic field is present, however, the critical temperature is reduced, and in fact as discussed earlier, this reduction is nearly linear with temperature and goes to zero at about 10.5^0 K. In view of this, we calculate the total heat stored for a difference of $1/2^0$ K between the bath and the critical temperature. This corresponds to operating a magnet within about 10% of its short sample limit. The numbers in this case are

$$a = 0.004 \text{ cm !}$$

$$Q = 0.21 \text{ mj.}$$

The normal region is becoming exceedingly short, and the quantity of energy necessary to cause a quench is going linearly to zero as one approaches the critical temperature. Indeed this quantity of heat is becoming so small that it is probably not the heat capacity of the wire that stabilizes a magnet but rather the heat capacity of the liquid helium that is in immediate contact with the wire. We will come back to these considerations later. At this point it is worth noting that the cable is only compacted to the point where the copper superconducting matrix fills about 90% of the available space. The rest is filled by liquid helium. The heat capacity of this liquid helium is about 5 mJ per centimeter of length of cable, and its latent heat capacity per centimeter length of cable is of the order of 40 mJ. Thus, the solution of the problem that we have given here for the transient state is simple-minded but does serve to focus attention on the extremely small amount of energy necessary to cause a quench in the superconductor magnet.

Quench wave propagation

We now summarize the results of the next section for propagation of a quench wave on a superconducting cable. The physical situation involved here is that the power generation in the normal section of cable is much higher than the cooling in the helium that is available. Hence, heat flows from the hot region into the superconducting region and causes it to go normal, and this transion front propagates down the cable. It is this front that we wish to study. The heat equation for this case can be written

$$C \frac{\partial T}{\partial t} - C\upsilon \frac{\partial T}{\partial \zeta} = \frac{\partial}{\partial \zeta} \left(K \frac{\partial T}{\partial \zeta}\right) + P(\zeta, T) \qquad (57)$$

$$z = \upsilon t + \zeta. \qquad (58)$$

Here the new variable ζ refers to a coordinate system that travels along with the velocity equal to the velocity of the transition. We assume that a solution of the equation exists such that in this system the heat equation loses its time-dependent term and becomes the simple static form that we have been considering previously.

$$\frac{\partial}{\partial \zeta}\left(k \frac{\partial T}{\partial \zeta}\right) + C\upsilon \frac{\partial T}{\partial \zeta} + P(\zeta, T) = 0 \qquad (59)$$

The problem now is to find solutions of this equation subject to the boundary condition that T at $\zeta = 0$ is the critical temperature and that T approaches the bath temperature as $\zeta \to +\infty$. For ζ less than zero, we have power generation of the type that we have been considering previously. Stable solutions of this equation should then lead to the value of the velocity v.

As in the previous case to simplify the mathematics, we consider k, C, and the power generation to be constants as a function of temperature. The solution of the equation with these constants varying with temperature is treated mathematically in the following section. The solutions for this equation to the right of the origin are

$$T = T_B + (T_C - T_B)\, e^{-\zeta/\lambda} \qquad (60)$$

$$\lambda = \frac{k}{C\upsilon}$$

In this equation we have neglected the cooling of the conductor by the helium. The boundary condition has only been that the conductor temperature approach the bath temperature as $\zeta \to \infty$. Thus, the right-hand side of the axis is dominated by the fact that new conductor at bath temperature is continually being fed into the propagating front. The assumption we have made simplifies the mathematics.

To the left of the origin, we have a constant power generation $P_o = I^2 r$, and the equation becomes

$$\frac{\partial^2 T}{\partial \zeta^2} + \frac{C\upsilon}{k}\frac{\partial T}{\partial \zeta} + \frac{P_o}{k} = 0 \qquad (61)$$

This equation can be integrated and has the solutions

$$\zeta < 0 \quad T(\zeta) = A_1\, e^{-\zeta/\lambda} - \frac{P\zeta}{C\upsilon} + A_o. \qquad (62)$$

The right and left side must match in the value of the temperature and its first derivative. Also, the solution that increases exponentially with distance is unphysical, and we set the coefficients of the exponential term equal to zero. Doing this, we obtain the results

$$\zeta < 0 \quad T(\zeta) = \lambda \left[\frac{P_o}{C\upsilon} - \frac{T_C - T_B}{\lambda} \right] e^{-\zeta/\lambda} - \frac{P_o}{C\upsilon} (\zeta - \lambda) + T_o$$

$$0 = \frac{P_o}{C\upsilon} - \frac{T_C - T_B}{\lambda} \qquad (63)$$

$$\upsilon = \sqrt{\frac{P_o k}{C^2 (T_C - T_B)}}$$

Substituting numbers into the equation for the velocity gives values that vary from a fraction of a meter per second to 10's of meters per second. It is obviously exceedinly sensitive to the value of the heat capacity, and the approximation that the parameters of the heat equation are independent of temperature is not very good in this case.

Since the transition temperature is a function of the magnetic field around the wire, the quench velocity changes with field. A set of curves showing the velocity of quench waves in a single strand of the conductor used in the Fermilab cable is shown in Fig. 22. This curve also shows the effect of applying an external magnetic field to the conductor that is carrying current. Figure 22 shows similar measurements, but in zero magnetic field, of quench waves on 23 strand cable of the type used in the Doubler magnets. This curve shows the effect of the velocity going to zero when the cooling by the helium is equal to the heating generated by I^2r.

These data were obtained by putting taps about 1 cm apart on the cable and running the voltage from the taps to a chart recorder as is shown in Fig. 24. A typical output from such a pair of voltage taps is also shown in Fig. 24. The fact that the voltage curve is so flat after the quench front has passed indicates that although the cable is still heating, the resistance is not changing.

Solutions of the heat equation

$$C \frac{\partial T}{\partial t} = \nabla \cdot (\cdot k \nabla T) + P. \qquad (64)$$

C = Heat Capacity

k = Heat Conductivity

p = Power/unit vol (heating or cooling)

Lets now consider the one dimensional case

$$C \frac{\partial T}{\partial t} = \frac{\partial}{\partial z} \left(k \frac{\partial T}{\partial x} \right) + P(x,t) \qquad (65)$$

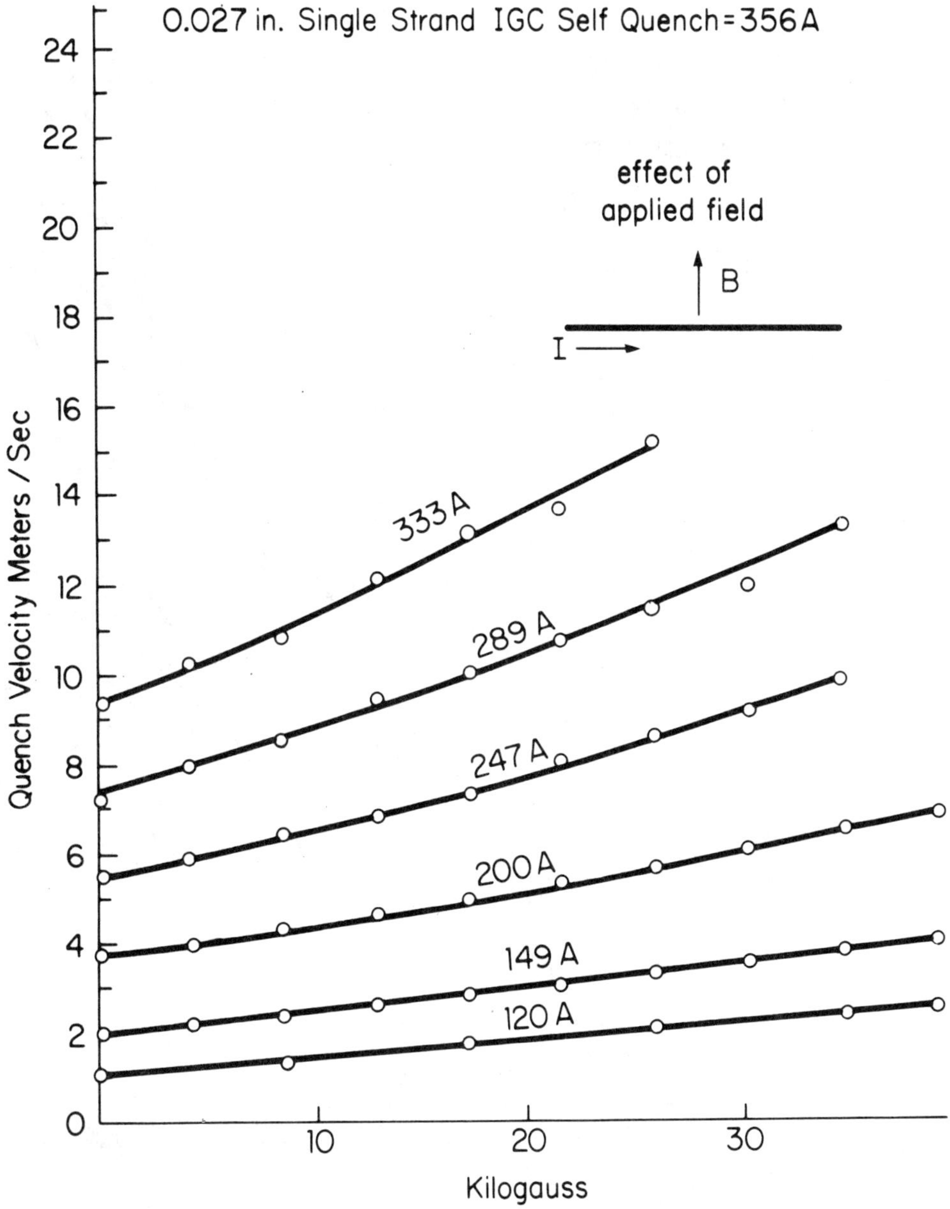

Fig. 22. Quench wave velocity in a single strand of Fermilab cable as a function of magnetic field.

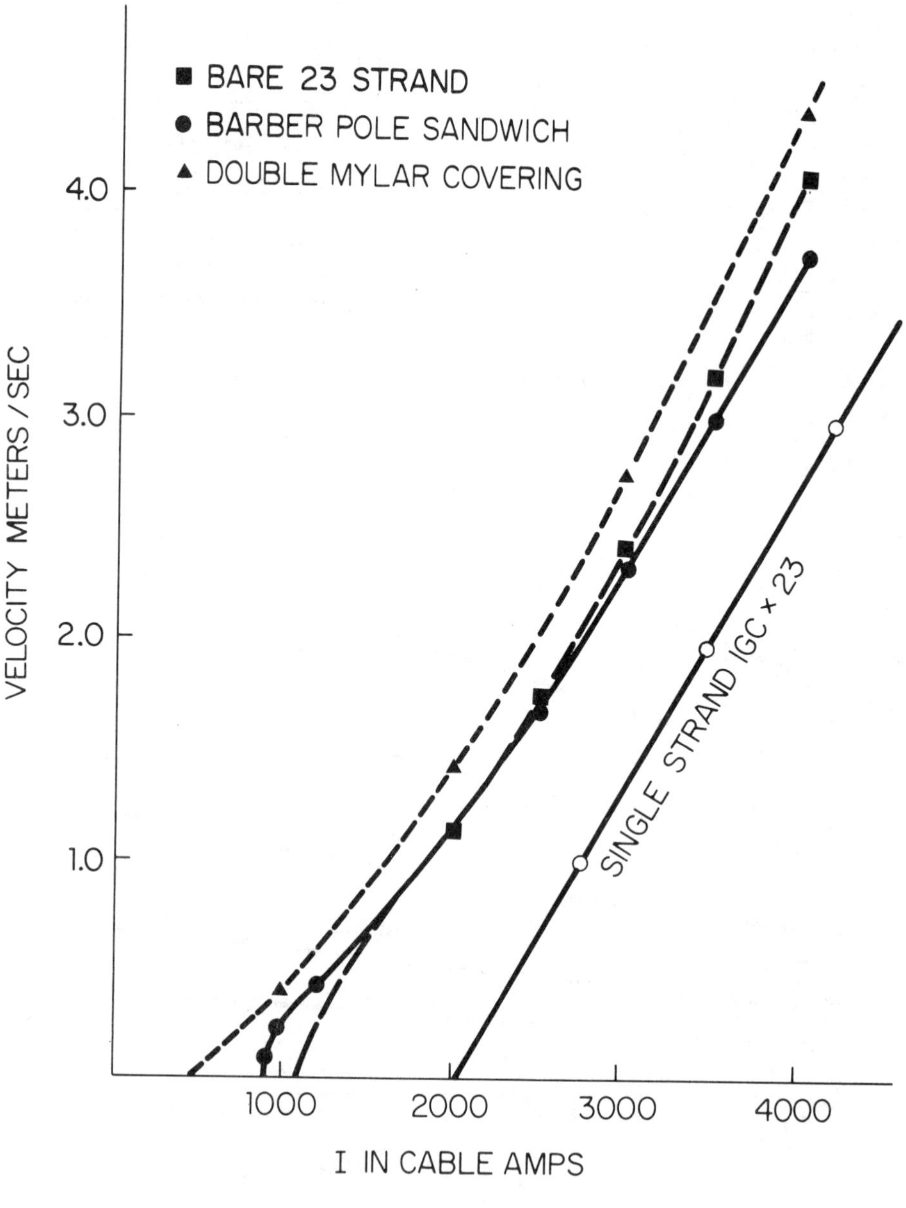

Projection 15

Fig. 23. Quench wave velocity for zero field in 23 strand Doubler cable as a function of current.

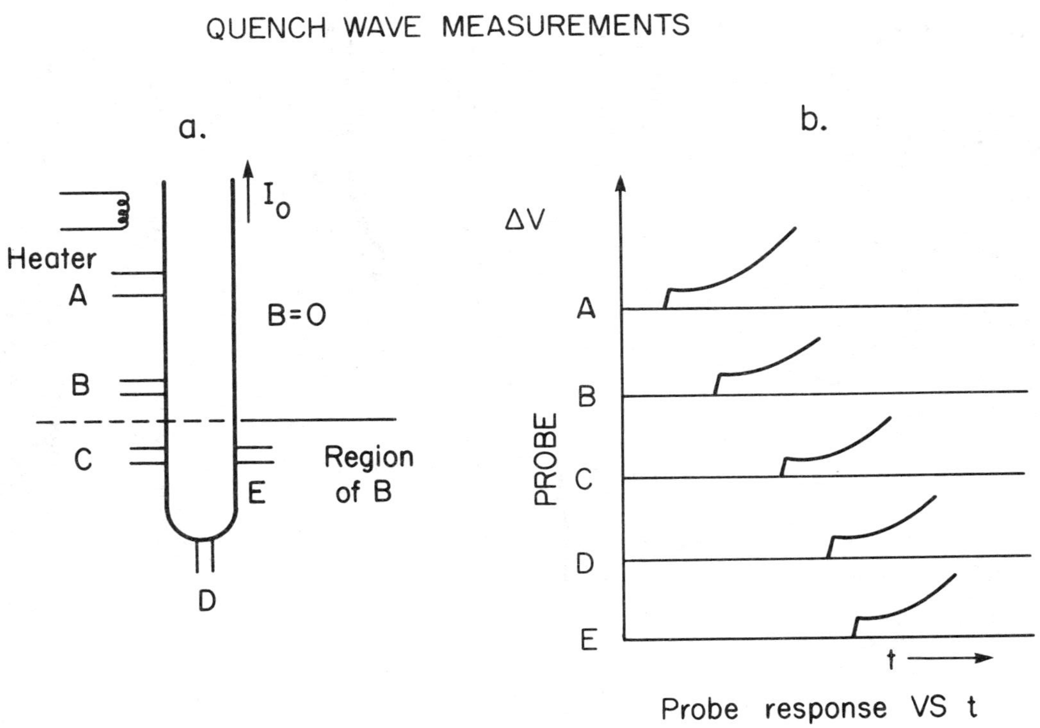

Fig. 24. System for measuring quench waves and a typical voltage output of the probe as a function of time.

At low temperatures, approximately

$$
\begin{aligned}
&C = \alpha T + \beta T^3 \\
&k = \gamma T \\
&P = I^2 R(T) \\
&R(T) \simeq R_o \qquad \text{Heating for } T < T_C \\
&\qquad\qquad\quad R = 0 \quad T < T_C \\
&P = -a(T-T_B) \qquad \text{Bath Cooling}
\end{aligned}
\tag{66}
$$

Consider first the simplest static solution.

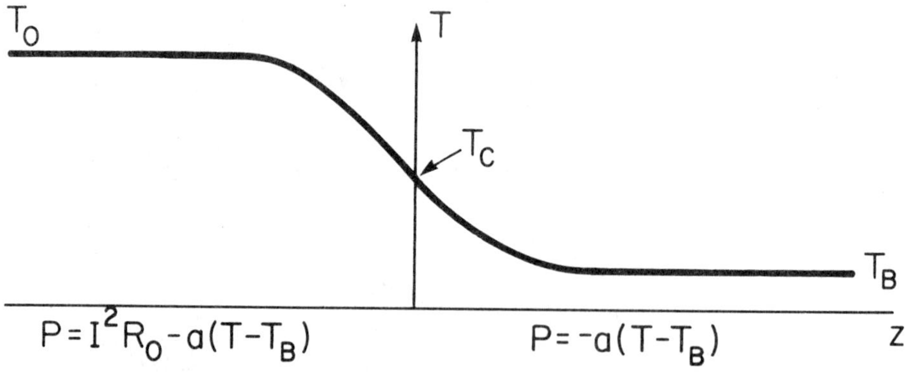

In this case the conductor is normal for $z > 0$ and must be in equilibrium with the cooling. (We assigned $\partial T/\partial t = 0$).

$$
z > 0 \quad \frac{\partial}{\partial z}\left(k \frac{\partial T}{\partial z}\right) - a(T-T_B) = 0 \tag{67}
$$

$$
z < 0 \quad \frac{\partial}{\partial z}\left(k \frac{\partial T}{\partial z}\right) - a(T-T_B) + I_o^2 R_o = 0
$$

The boundary conditions are

$$T \to T_B \quad \text{at} \quad z \to +\infty$$

$$T \to T_B + I^2R_0/a \quad z \to -\infty \qquad (68)$$

$$k+ \frac{\partial T_+}{\partial z} = k_- \frac{\partial T_-}{\partial z} \qquad z = 0$$

$$T_+ = T_- = T_C \qquad z = 0$$

For $z > 0$ $S = t$ $\theta_+ = T - T_B$ (69)

$z > 0$ $\theta_- = T - T_B - I^2R_0/a$

then

$$\frac{\partial}{\partial z}\left(k \frac{\partial \theta_+}{\partial z}\right) - a\theta_\pm = 0 \qquad (70)$$

$$\theta \to 0 \text{ as } z \to \pm\infty$$

The boundary conditions can only be satisfied for one value of I^2R. If I^2R_0 is greater than this value then a wave will propagate to the right and if its less, the cooling will exceed the power generation and the wave will collapse. This is because at = 0 T <u>must equal</u> T_C.

Suppose the solution of Eq. (70) is f( ) then Eq. (69) gives

$$T_+ = f(\ ) + T_B$$

$$T_- = -f(-\ ) + T_B + I^2R_0/a \qquad (71)$$

which satisfy $\left.\frac{\partial T_+}{\partial z}\right|_0 = \left.\frac{\partial T_-}{\partial z}\right|_0$

using $T = T_C$ at $z=0$ gives

$$T_C = f(0) + T_B$$

$$T_C = -f(0) + T_B + I^2R_0/a \qquad (72)$$

then $f(0) = T_C - T_B$

and $I^2R_o = 2a(T_C - T_B)$ (73)

we find $T_0 = 2T_C - T_B$ (74)

If k is a constant then we have

$$k\frac{\partial^2\theta}{\partial z^2} - a\theta = 0$$

$$= A\,e^{\pm z/\lambda} \qquad \lambda = \sqrt{\frac{k}{a}}$$

Cryogenic stability

It is clear from the curves shown in the last section that if the cooling is great enough, the quench wave will not be able to start. A magnet that has been designed so that at its operating current, the cooling is greater than the I^2r loss that would be generated if the cable went normal, is called cryogenically stable. Such magnets have customarily been used for bubble chambers and other large energy storage devices where quenches cannot be tolerated. In order to design a magnet so that it is cryogenically stable, the current density is lower than it would be if the magnet were not stable. This is because of two effects.

The first reason is that one desires a large copper to superconductor ratio so that if the current does transfer to the copper, the I^2r losses are low. The second factor that increases the size of the conductor is the requirement for a large area per unit length of the conductor in contact with helium. Furthermore, there must be adequate paths for the helium to flow in order that it can cool the conductor adequately and return it to the superconducting state.

6. CARE AND TRAINING OF SUPERCONDUCTING MAGNETS

IEEE Trans. MAGN. MAG-17, 863(1981)

Training

Superconducting magnets are notorious for the wide gap between performance that is predicted on paper and that which is achieved for produced magnets. In the early years before multifilament conductors became available, many of the difficulties were associated with flux instabilities. However, the availability of this type of conductor solved this problem, and yet W.P. Smith[1] at RHEL still reported training and degradation in magnets built with such superconductor.

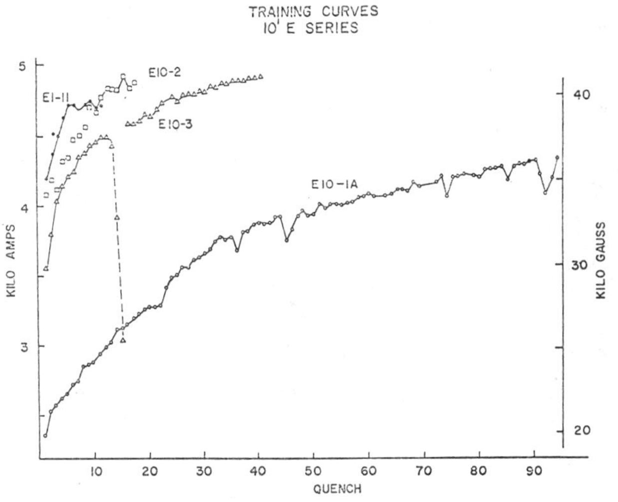

Figure 25

Figure 25 shows some examples of magnets that train. The term training refers to the increase of peak current observed in a magnet when it undergoes a series of tests where the current is ramped up until the magnet quenches. The curves shown in Fig. 25 are for a number of different situations. There is a training curve for a 1-ft long, 3-in. bore magnet from the early Fermilab test program. Also shown is a training curve for a 21-ft long dipole suitable for the Energy Saver.

Training occurs under widely different circumstances. Solenoids in general show a rather small amount of training, whereas, race-track type coils have given a great deal of difficulty in this respect. It should be noted that the training phenomena is most serious in magnet coils where a high current density must be achieved, such as coils for high energy particle accelerators or coils in rotating machinery. In these cases, the conductor cannot be cryogenically stabilized and if a portion of it goes normal, it necessitates shutting off the current through the conductor. While training can be tolerated in some cases as research type magnets, there are other cases where it is completely intolerable. Particle accelerators that are being constructed presently at Fermilab and Brookhaven require over 1,000 superconducting magnets, and each magnet may have of the order of 1 MJ stored in it. It is obviously impossible to construct a machine of magnets that take more than a few quenches to train; or just as importantly, from magnets that do not remember their training after they have been warmed to room temperature. One can easily think of other examples such as coils in rotating machinery or any coil where the energy stored is very large. It is interesting to note that it has been possible to construct field coils for a generator that do not exhibit training. It is also equally true that there are abundant examples where training problems have jeopardized the ultimate success of the program.

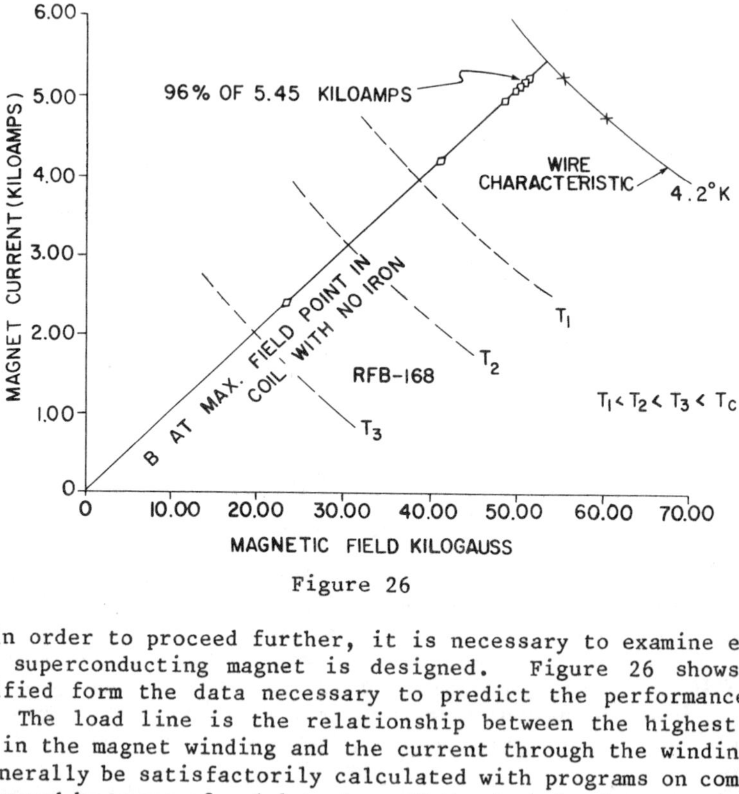

Figure 26

In order to proceed further, it is necessary to examine exactly how a superconducting magnet is designed. Figure 26 shows in a simplified form the data necessary to predict the performance of a coil. The load line is the relationship between the highest field point in the magnet winding and the current through the winding, and can generally be satisfactorily calculated with programs on computers or measured by means of model tests. If the iron in the return yoke is not saturated, this line in general will be linear. The hyperbolic type curves shown are the characteristic curves of the superconductor. Any attempt to operate the conductor above the curve at a fixed temperature will result in the conductor going normal. Hence, Point A shown on the 4° K curve, which is the intersection of the superconductor characteristic and the magnet load line, represents what is called the short sample limit of the magnet. Notice that this prediction requires knowledge of three things:

1. The temperature,
2. The exact "high field point" load line,
3. And the characteristic of the superconductor at that point.

The first two variables can generally be satisfactorily obtained with fair accuracy. However, the characteristic of the superconductor at the high field point in general is not known but must be inferred from measurements done on samples of the conductor taken from the coil of cable from which the magnet is wound. The uniformity of the cable used in the Fermilab magnets is discussed in Ref. 2.

If a magnet trains asymptotically to a value that is less than the predicted short sample limit, it is referred to as degraded performance. Figure 27 shows a histogram of the peak current reached in a series of magnets constructed at Fermilab in terms of the percent of the predicted short sample limit. It is clear that there is a tail on the low current side of the peak. This spread around the peak is consistent with the accuracy of the many measurements that are involved in obtaining the data for a histogram of this type. It is not known what causes the low current tail, but there are many sources that are possible to explain this effect. Reference 3 addresses the question of the performance in terms of the short sample properties in much more detail.

Figure 27

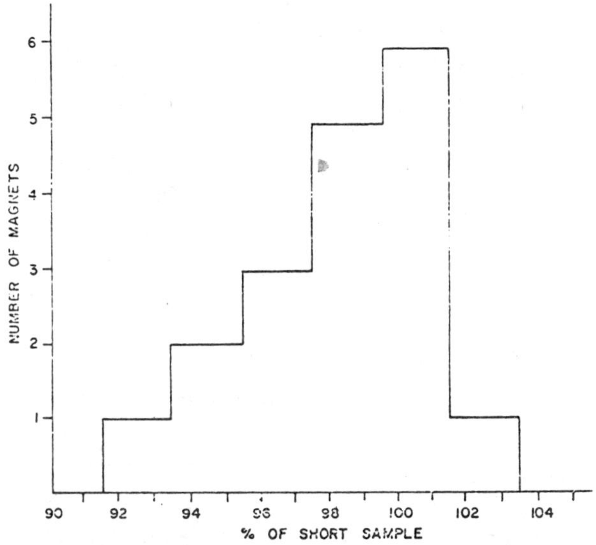

An interesting question to ask is the following: Did the magnet quench start at the predicted high field point? There has been experimental work both at Brookhaven and LBL that narrowed down the quench region to the nearest current block in a race-track winding. In one case at Fermilab, a 1-ft model magnet was instrumented with enough voltage probes so that the quench wave could be observed as it propagated away from the high field point and by means of interpolation it was possible to ascertain the point that the quench started to an accuracy of several millimeters. The majority of the quenches started at the predicted point. However, it was also observed that some of the quenches started elsewhere in the coil.

In the above discussion we have tried to indicate that it is possible to construct a superconducting magnet coil whose performance is governed by the short sample limit. We now address the question of how many quenches does it take to "train a magnet." Solenoids in general show very little or no training. On the other hand, race-track type coils can exhibit a tremendous variability in this parameter. As mentioned above, it has been possible to fabricate field coils in a "race-track" configuration for a generator that shows no training. Some magnets that have been constructed at Fermilab have also exhibited essentially no training. Figure 28 shows a histogram of the number of quenches necessary to train an Energy Doubler magnet to 95 per cent of the short sample limit. It has not been possible to eliminate the training completely, but the number of cycles necessary are small enough so that it can be trained before it is placed in the machine, and indications are that the magnet remembers its training. The above discussion indicates the range of variability observed in the training of superconducting magnets. We now proceed to the theory of this process.

Figure 28

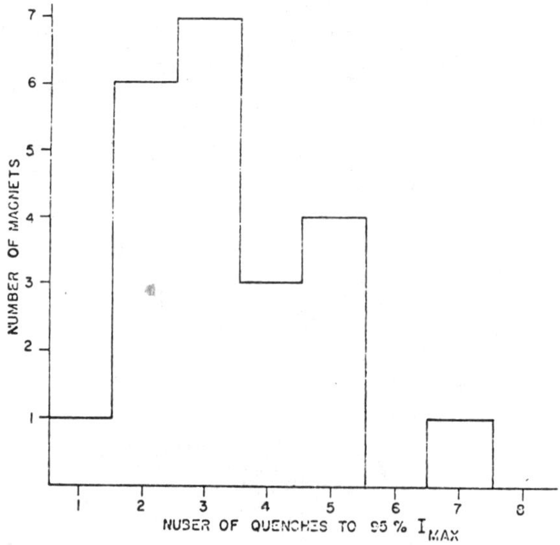

Theories of training

In order to discuss the theory of training, it is necessary to address the questions, "what causes the conductor to become normal before it reaches the short sample limit, and why does the current increase with the quench number"? The theory proposes that the conductor is locally heated by bursts of energy released during excitation of the magnet. There are three sources for this energy. A first source is non-elastic deformation of the conductor material itself. We will come back to this question later. The second is the cracking or failure of the cable support matrix, and the third considers that small frictional motion of the wire due to Lorentz forces can supply the energy. The amount of energy necessary can be estimated as follows.

Figure 29 shows the situation in a piece of conductor when a short section of it has gone normal. In this section, the current is flowing in the copper matrix and generates heat which is removed by

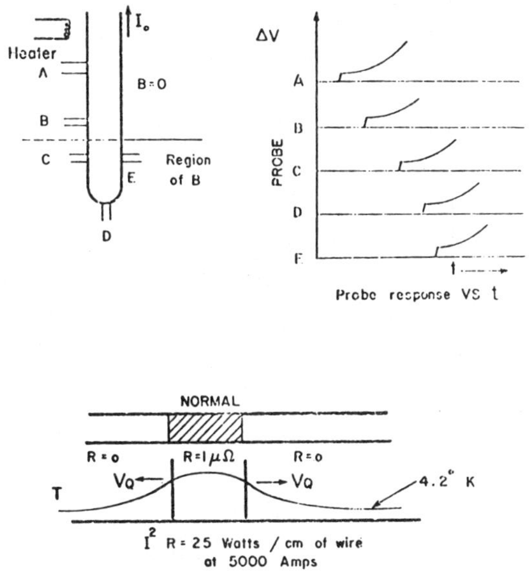

Figure 29

the liquid helium in contact with the cable. If the region that is normal is very short, the cooling can be sufficient to extract the heat produced by the I^2r losses, and the region will collapse and become superconducting again. The numbers shown on the figure are for the Fermilab cable with 5,000 amps through it. If the cable is normal at this current, 25 watts per centimeter of length are generated. This is almost 2 orders of magnitude greater than the ability of the helium to cool the cable. The study of these normal regions has been carried out in great detail. The most complete published work is that of S. L. Wipf at Los Alamos. The zone is governed by the heat equation. At any given current in the magnet, there is a critical length such that if the normal zone is longer than this length, the region will propagate along the cable, and if it is shorter than this length, the cable will recover to the superconducting state. This has been given the name by Wipf of Minimum Propagating Zone (MPZ). In order to obtain an estimate of the amount of energy necessary to start a quench, we calculate the amount of energy that instaneously must be deposited in the cable in order to set up a region equal in length to the minimum propagating zone. Figure 30 shows this energy in a typical conductor.

Suppose now that by some means a delta function of energy in position and time is deposited in the conductor. The thermal relaxation times are exceedingly short, and if the energy is greater than the energy content of the minimum propagating zone, it will trigger a quench. If it is less than this value, the magnet will remain in its superconducting state.

Figure 30, Ref. 5

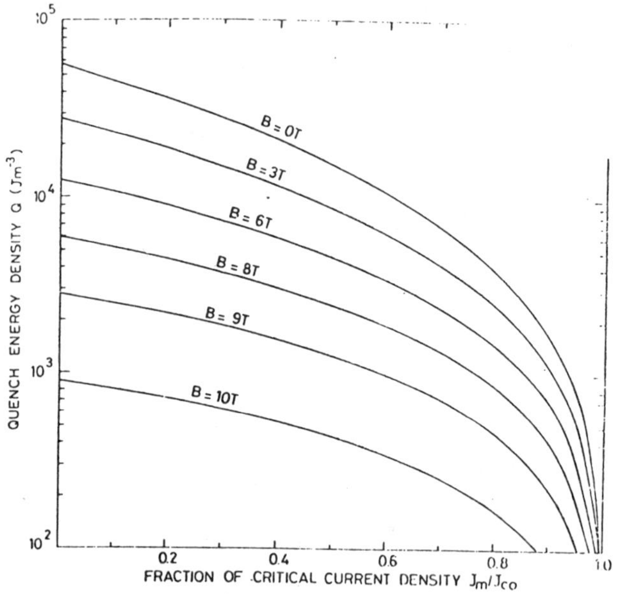

Figure 31 attempts to display the energy balance within a magnet structure. Again for purposes of illustration we have used a Fermilab Energy Saver type dipole. The horizontal scale represents temperature, and the vertical scale represents energy per unit length of the cable. The curve labeled cable enthalpy represents total heat content of the 23 conductor cable as a function of temperature. It is seen at 5° K the energy content of the cable is less than a millijoule per centimeter of length. The slope of this curve at the operating temperature of 4.2° K is about 0.2 mJ per degree K. The critical temperature for NbTi is about 10° K. Hence, it would take about 10 mJ of energy per centimeter of cable in order to raise the temperature to the point where it would no longer be superconducting. However, when the cable is in a magnetic field and carrying a current, a much smaller change in temperature will drive the cable normal. For instance if the cable is at 90 per cent of the short sample limit a change 0.1° K will change the cable from superconducting to normal. In addition to the enthalpy of the cable, there are two other potential heat sinks available. The first is the specific heat of the matrix and any liquid helium contained around the wire. The Fermilab cable has a certain amount of open space available for helium which amounts to about 10 per cent of the cross section of the wire. Hence, an arrow is indicated on the curve showing the heat capacity of the captured liquid helium for 0.1° K change in ambient temperature assuming no boiling. It is equal to 0.65 mJ per centimeter of length of wire. However, the ultimate heat capacity of this captured helium would be represented by the heat of vaporization, and that is shown by an arrow near the top of the graph, and it is 45 mJ per centimeter of length. It is thus clear that the helium represents a major heat sink within the structure of the wire.

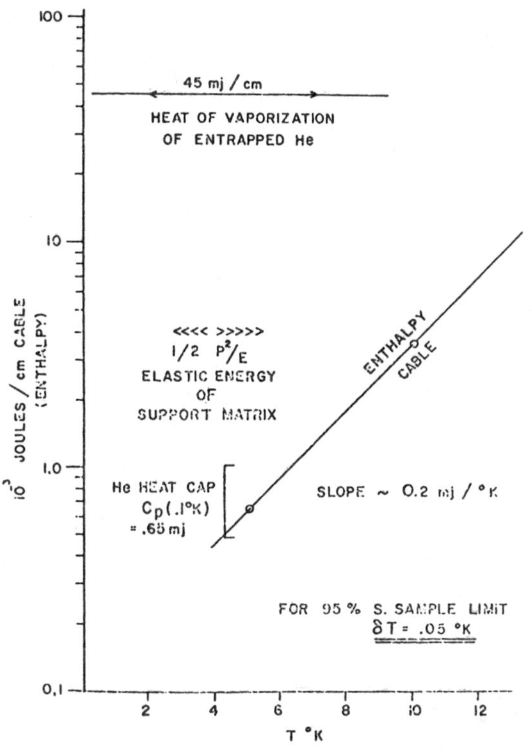

Figure 31

To give some idea of the sources of energy available for initiating a quench, an arrow indicates the elastic energy stored in the matrix. This is calculated from $1/2\ p^2/E$, where p is the pressure within the matrix, and E is the Young's modulus. For the Fermilab magnet, Young's modulus is about 10^6, and p is several thousand pounds per inch. It is clear the energy stored elastically is much larger than the energy needed to drive the wire normal when it is carrying a current close to the short sample limit. However, it is surprising to see that this energy is not enormously large compared to the energy required to initiate a quench, nor is it enormous compared to the heat sink available in the helium.

At this point it should be evident that the situation is extremely complicated. There are two sources of energy available in the magnet to drive the wire normal. The first would involve yielding of the support matrix. The level of energy available for this is represented by $1/2\ p^2/E$. Only a small fraction of this energy would be available if the support yields. On the other hand, it is also possible that frictional forces due to the motion of the superconductor against the supporting structure could also generate heat and drive the conductor normal. A simple calculation would indicate that one of the conductors moving frictionally a distance less than 0.001 cm would generate enough energy to drive the cable normal. It is thus obvious that there are adequate sources of energy to cause a magnet to quench before the short sample limit. In fact it is utterly amazing that magnets are able to operate within 95 per cent of the short sample limit.

The experimental survey: confusion reigns

It is now possible to understand why the experimental situation is so confused. There are a large number of variables available to the magnet builder. He can change the conductor, the support matrix, the shape, the cooling, and the clamping depending on the requirements for the magnet. It would be nice to be able to draw from the literature some general principles that would lead to successful magnet construction. However, this report is little changed from those that have been given over the previous five years. The main challenge of superconducting coil fabrication is to remove the "black magic" and find reproducible principles for coil fabrication. As experience is teaching us, there is still much work to be done. At present each large project invents its own techniques, but these are not readily transferable to other applications. The "freshman physics" is not yet well understood. This will be illustrated in some of the following examples.

Consider now the variables that can be found in various coils reported in the literature.

1. Cooling. The coolant may be subcooled liquid helium, He-II, or subcooled supercritical helium. The coolant can be in contact with the superconductor or isolated from it by the insulation.
2. The support structure. In order to keep the conductors from moving under the Lorentz forces, it is necessary that the mechanical clamping force be greater that the magnetic force when the magnet is cold. This latter condition causes real trouble because of differential shrinking with temperature of the different materials in the magnet.

Three methods are being used at present to clamp the coil:

A. The coil is assembled at room temperature under sufficiently high preload so that the mechanical forces remain greater than the magnetic forces even when the coil is cold. (For externally confined coils, the coil tends to shrink away from the support.) The greater the differential shrinking between the components, the greater the preload necessary at room temperature. In order for this approach to be successful, the magnet constructor must control the elastic properties of the insulation support matrix very carefully. In any given case, a solution does not have to exist. The forces may be so large that they crush the support matrix at room temperature.

B. One can make use of the differential shrinkage in order to clamp the coil. Aluminum in general shrinks more than the coil composite, so if the coil is externally supported and clamped at room temperature, the forces will increase as it is cooled.

C. The last method of clamping the coil involves impregnating it thoroughly with glass-filled epoxy. For this method to be successful, the coil must be vacuum impregnated, and all of the voids must be carefully removed. The glass filling considerably strengthens the epoxy and prevents cracking when thermal and magnetic stresses are applied. A careful stress analysis of the coil must be carried out to ensure that the stress levels do not exceed the strength of the epoxy.

3. Insulation. A number of types of insulation are being used at present. One involves using an open-weave epoxy filled glass tape wrapped around the conductor (BNL). In another (Fermilab) case, the conductor is insulated first by wrapping a mylar film around it and then spiral wrapping the composite with the epoxy filled tape. In still other cases the formvar or some other plastic film type insulation may be applied directly to the superconducting composite. For epoxy impregnated coils, epoxy may also serve the insulation purpose. Insulation has two important effects. First, it can thermally isolate the superconductor from the coolant, but it also is the nearest source of strain energy that the superconducting wire sees, and thus its effect can be very complex.

4. Conductor. The conductor used can be either a simple multifilament single strand composite, or it can have the much more complicated structure of a woven braid or a multistrand cable. In the latter two cases, the strands of the braid of the cable may be additionally fixed with solder filling or left free to move. For braid or cable, one may have an additional source of heat available at a very vulnerable point in the system due to frictional motion of the adjacent strands over each other.

It is now appropriate to examine a number of different experimental programs and see if the results can be understood on the basis of the preceeding considerations.

Example 1. Generator Field Coil, Ref. 6. General Electric has an active program to study superconducting field windings for a generator. During the course of this program, perhaps 10 race-track type windings have been built. These coils are approximately 1-1/2 m long by 1/4 m wide and have a cross section that is of the order of 3 cm × 6 cm. The conductor is approximately 0.05 in. × 0.1 in. monolith, with formvar insulation and a ratio of 1.6 to 1 for copper to NbTi. See Fig. 32. These coils store an energy of the order 1/2 Mj and have a short sample limit of about 7 Tesla. Not only are the

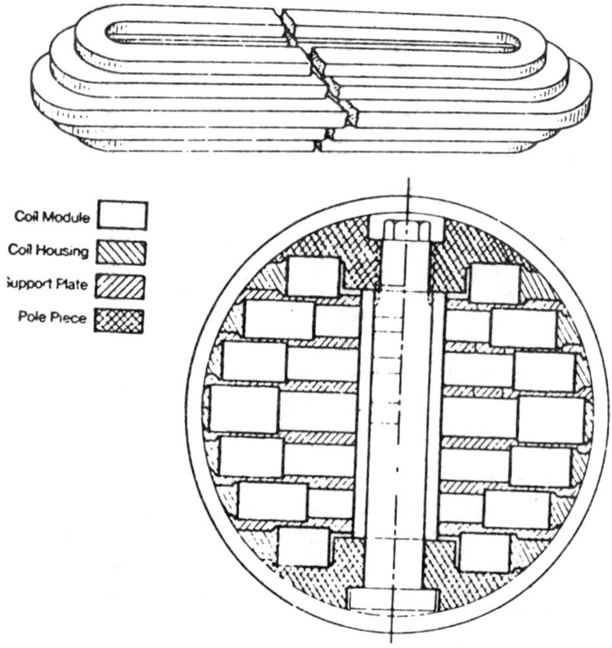

Figure 32, Ref. 6

anticipated forces large, but there is also the possibility of mechanical shock and vibration. Thus, this application represents a real challenge to the support structure. It must absolutely restrain all motion of the conductor. Hence, if one can successfully build this type of winding, one would expect a minimum of training. The results from this program show a series of coils that reach the short sample limit on the first quench. In order to achieve this behavior, the following construction techniques are used:

1. The coils are wound with a great deal of care. The epoxy bonds to the formvar and when the impregnation is done, all voids are carefully filled.
2. Surfaces within the structure that must move tangential to each other are supported by a material with a low shear modulus and a high compressive modulus, thus avoiding any type of stick slip motion.
3. The whole coil assembly has a tapered interface with respect to the torque tube which is shrunk on at room temperature.
4. The moduli of thermal expansion are such that the coil is still preloaded when it is cold. It is interesting to note that only the outer conductors of the winding blocks are in contact with liquid helium. This coil represents an encouraging sign that if one pays enough attention to the support of the individual conductors, the training indeed can be eliminated.

Example 2. Accelerator Type Magnets. BNL, Fermilab, and LBL all have active programs to study saddle type coils for accelerator applications. These coils are several meters long, and 7-1/2 cm to 12 cm i.d. They typically store of the order of 1 MJ and operate at fields of the order of 4 Tesla. This type of magnet must be ramped. The Tevatron goes 4 Tesla in 20 seconds; ISABELLE 5 Tesla in a few minutes. In order to control eddy current losses, it is necessary to use either a Rutherford type cable (Tevatron) or a braid (ISABELLE). The cable has poor structural stability and hence must be embedded in a support matrix. In the past this has been formed by a spiral wrap of B stage glass tape around the conductor. This leaves a porous structure in which helium can circulate and remove the heat generated by ramping. Other solutions to supporting the conductor are being investigated at LBL and Saclay which involve leaving out the B stage glass tape and using plastic film.

Figure 33 shows a cross section of the Fermilab magnet, and Fig. 34 shows the forces on the conductors that must be supported. This particular coil is forced under high pressure into a stainless collar at room temperature. The coefficients of thermal contraction are such that the mechanical forces are greater than the magnetic forces when it is cold. Figure 28 shows a histogram of the number of quenches to short sample limit for these magnets, and Fig. 27 shows the percentage of short sample current achieved. In this magnet, there is no question that the conductors move. The radius changes about 2-1/2 mils during excitation, and there is azimuthal compression of the winding. In fact if the mechanical forces of compression are not greater than the magnetic forces, the last turn next to the key can

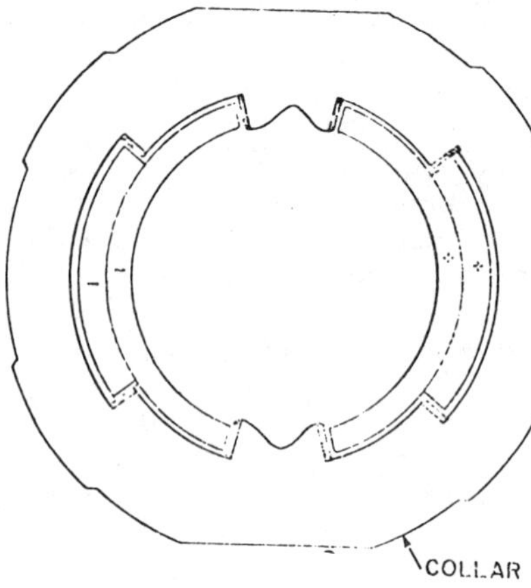

Figure 33

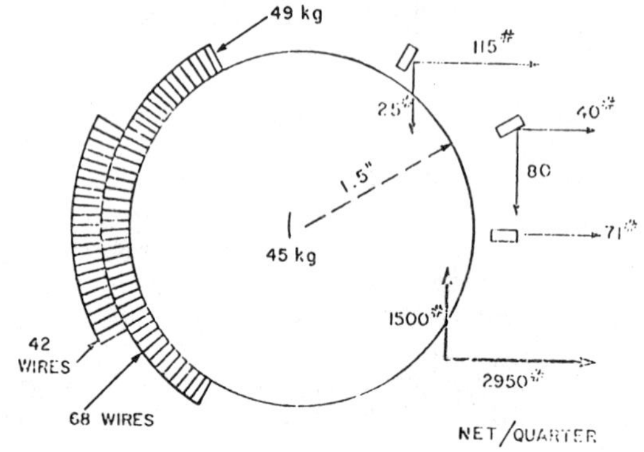

FORCES AND FIELDS IN A
SERIES E MAGNET

$$\text{Axial Force} = \frac{\partial}{\partial z}\left(\tfrac{1}{2} L I^2\right)$$

Figure 34

actually move away from the key during excitation. The change in radius is proportional to I^2 and hence elastic. It shows very little hysteresis. However, the azimuthal motion has given more trouble. Early in the program, magnets were built such that the elastic forces when cold were less than the magnetic forces, and the conductor at the key moved. Figure 35 shows data from 81 magnets whose training took from one quench to over 25. Some magnets in this series had preload small enough so that there was motion of the wire at the collar. Two or three mils of azimuthal motion can be expected on the basis of the elastic properties of the insulation alone. The vertical axis shows the average motion of the cable exhibited by all of those magnets that took a given number of quenches to train. The error bars are just a square root of the number of magnets at each point. It is seen that this motion does not couple into the training until it is large enough so that the conductor is completely unclamped. Why it takes some magnets one quench and others 10 quenches to train when the conductor remains clamped is a mystery.

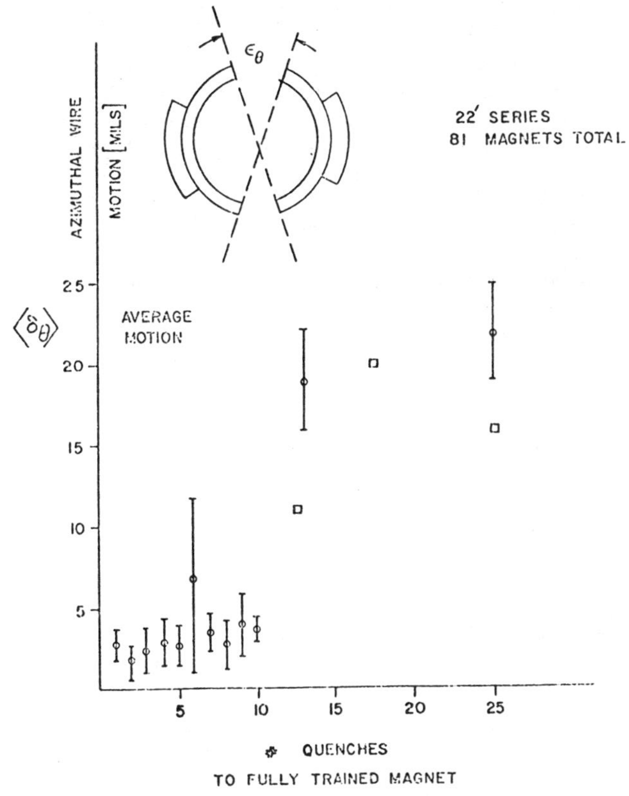

Figure 35

There is another puzzle associated with these magnets. Figure 25 shows the training curve of a 1-ft long model as well as one for a 10-ft long magnet. Both were tested in the vertical dewar and both magnets have the same cross section and were built with substantially the same wire. In this case the high field point is located at the place where the turn next to the key reverses its direction at the end of the coil. The field at this point is about 20 per cent higher than it is on the axis of the magnet. This is true for both the 1-ft magnet and the 10-ft magnet, so if this is the only region of the magnet that needs training, it may not be surprising that the training curves look similar. However, the training curve for a 21-ft magnet in its completed cryostat and yoke looks the same. In this case, the high field point moves away from the very end and is located at any place along the wire next to the key. The field is about 10 per cent higher than it is on the axis. (The length of the iron yoke was carefully adjusted to make this true.) There is now 21 ft of active length of the magnet that can get involved in the training. Early in the development of these magnets, there was a concern that the number of quenches to train a magnet of this type might be proportional to its length. That is clearly not true. A quench apparently relieves strain along the whole length of the magnet.

There is another curiosity in the Fermilab experimental series. Early in the program a great deal of trouble was caused by turn-to-turn shorts. The B stage glass tape could easily be shorted through by a small chip. In an attempt to solve this, several 1-ft model magnets were constructed from cable that had been coated with several mils of epoxy paint. This was over-wrapped with B stage glass tape in the normal manner. Figure 36 shows the training curve of E1-3. Others in this series displayed similar curves, and these were perhaps

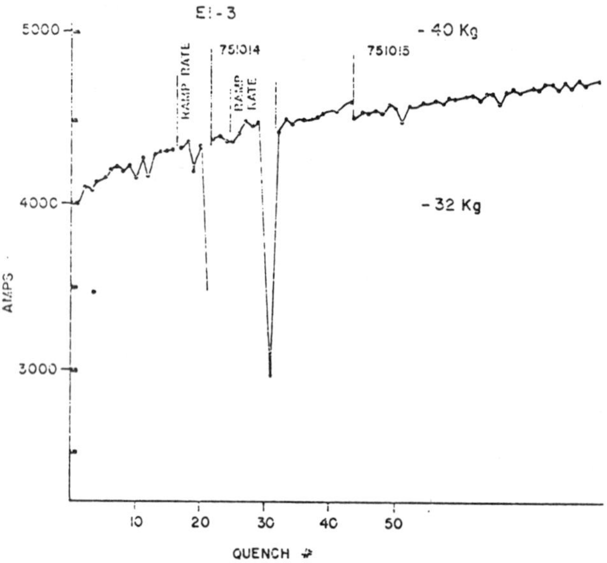

Figure 36

the worst magnets ever made at Fermilab. This led to an early conclusion that due to the thermal stresses that can set up, epoxy was bad if it was in intimate contact with the wire. Consequently, a second solution was adopted which consisted of over-wrapping the Rutherford type cable with 1-mil mylar tape. The B stage glass tape is now wrapped on top of this. Thus, there is no epoxy in contact with the conductor, and there is excellent turn-to-turn insulation. The mylar tape is still porous enough so that helium can enter into the spaces acound the strands. The heat transport through the overall system is clearly decreased. However, it should be noted that there is perhaps another effect that the mylar wrapping can exert; namely, it insulates the strands themselves from heat generated by friction in the coil structure, and hence it may actually have a beneficial effect upon the training properties of the coil.

If a magnet trains badly, it is difficult to isolate and diagnose the cause. At Fermilab, we have used two methods to detect motion of the cable. One is by coupling directly to the winding via a rotary transformer. This method is sensitive to less than 1 mil of motion in the windings but must be corrected for any elastic deformation under the Lorentz forces. We have also verified the conductor motion by observing the behavior of the low order multipole coefficients. In the Fermilab magnets, the key angles effect particularly the quadrupole, sextupole, octapole and decapole moments. It is possible to deduce the motion of the conductors given the variation with current of these lower four moments.

BNL has used a different scheme.[7] Since the critical current is a nearly linear function of temperature, they have plotted the quench current versus temperature for one of their magnets. The quench point is a linear function of temperature until the forces on a current block becomes big enough so that the Lorentz forces exceed the preload. At this point, the quench current does not increase as the temperature is reduced. Frictional motion of the winding is thus indicated. Other techniques involve the use of voltage taps across the coil to find out which parts are going normal and sensitive thermometers embedded in the coil to find out where the quench is starting.

Example 3. Solenoids. Traditionally the easiest coils to build have been solenoids. Symmetry handles the forces well, and many such coils have been built that exhibit little or no training. However, as one goes to higher fields, the forces can become strong enough to cause the layers in the winding to delaminate which then results in training. Reference 6 from the GE field coil program has some examples of solenoids that were built which did not train. A second sample is provided by a solenoid built at the National Magnet Laboratory at MIT, Ref. 8. The structure of this coil has received a comprehensive analysis, both statically and dynamically, and many measurements have been made on the completed coil. Figure 37b shows a typical cross section of the winding; the individual conductors have been modeled as shown in Fig. 37a. The model makes an attempt to include the superconductor, the copper, the formvar insulation, and the epoxy filling. This magnet trained in about five quenches. During training, the coil was well instrumeted with strain gauges, and those

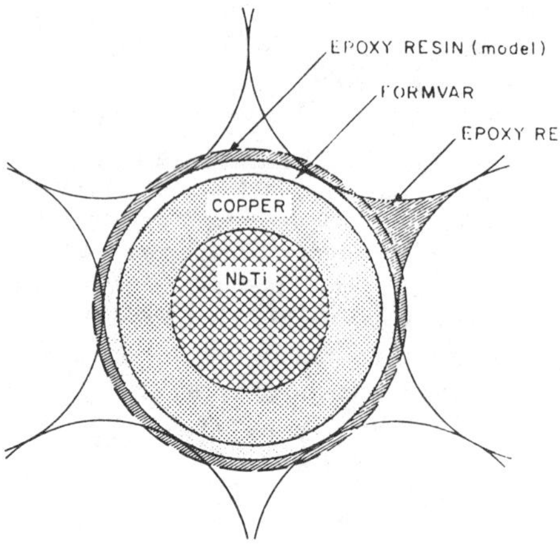

Figure 37a

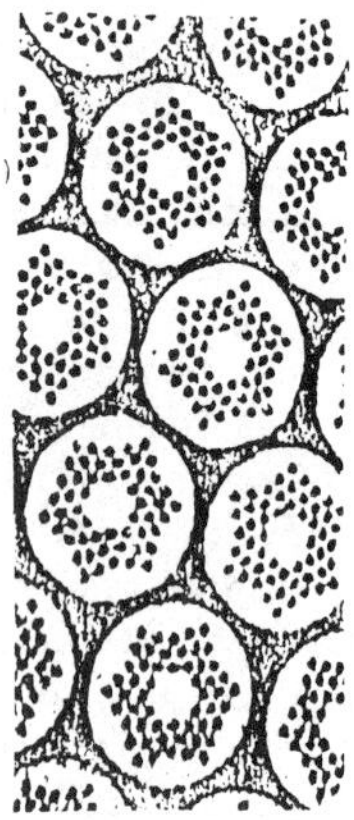

Figure 37b

measurements have confirmed the predictions of the stress analysis. The authors interpret this as strong evidence that the observed training was associated with stress induced cracking before the coil became separated from the coil form. This separation was predicted for the calculations. After it was complete, the coil reached its short sample limit of 95 amps.

In the foregoing an attempt has been made to indicate that superconducting magnets have been built that do not train or at least show a minimum amount of training. It is now useful to look at a series of experiments that were designed to explore why training takes place. These experiments are aimed at understanding the role that the matrix or the conductor may play.

Experiments related to training

A series of experiments at NBS, Ref. 9 is concentrating on effects due to the role the support matrix plays. If cracking of the matrix is responsible for releasing heat, then it should be possible by applying mechanical strain to a coil to cause it to quench at a point below its short sample limit. Figure 38 shows the experimental setup used to apply such a mechanical strain to a small coil that is 18 cm in i.d. Different techniques were used for fabricating these composite rings. The split conical washers can apply a tensile force to the sample which is embedded in an external magnetic field. By controlling the external field and the current through the sample, the fraction of the short sample current carried by the coil can be controlled. Force is applied to the coil until it quenches and the strain at the quench point measured. Figure 38 shows the apparatus used for these experiments. It is seen that a phenomena very much like training is exhibited each time the strain that the coil is able

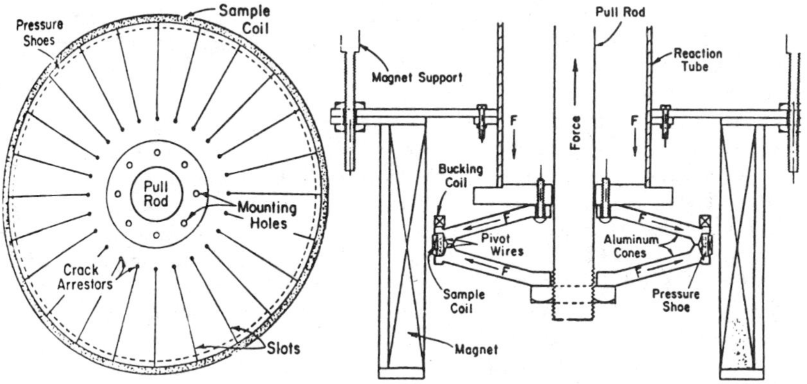

Figure 38

to sustain is larger. Furthermore, as the current carried approaches the critical current, this strain necessary to quench becomes less. Figure 39 shows some of the information that has already been obtained. Two coils, one with the fiberglass and one without fiberglass to strengthen the epoxy are compared. It is found that the

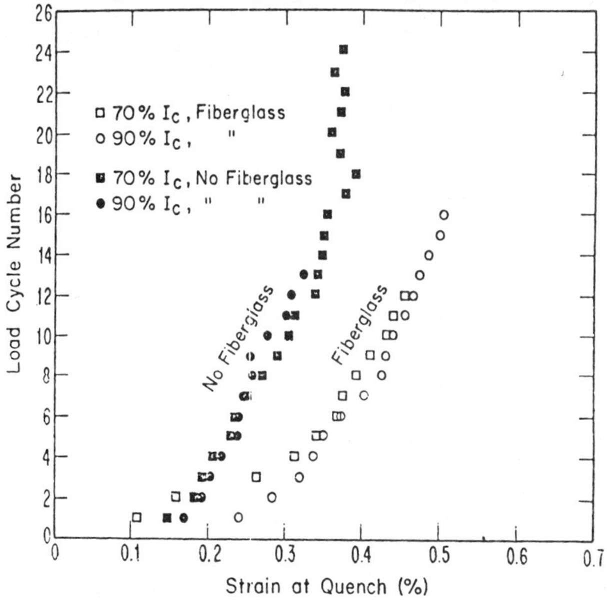

Figure 39

coil with fiberglass is much more resistant to quenching than the one without. A second phenomena similar to real magnets is also observed. If the coil is first strained up to a certain level and then the strain reduced, the first quench is observed to take place at a strain greater than was first applied.

There are two remarks to be made on this experiment. One is that the applied forces are different than in a real coil. In this case the forces are being applied to the matrix, and the matrix is pushing on the wire. A real solenoid has the situation just reversed. A second very important point is that the strains being explored are in the range of a 0.1% to 1% or more. In the previous magnets discussed the strain is much lower than this. For instance, the axial strain on the Fermilab magnets is less than 0.03%. The NMR magnet studied at the National Magnet Lab has a strain of the order of a few tenths of a per cent, and the generator field coil windings have a strain of the order of 0.1%. Thus, actual magnets are operated at strains much below the area that is being explored here. However, it may also be true that there are places in real magnets where the strain is concentrated and hence is much larger than the average values.

A major study by Pasztor and Schmidt, Ref. 10 has indicated that the conductor itself may yield inelastically and that this effect may play some role in the training of magnets. Figure 40 from that paper shows the serrated yielding of a NbTi sample. The theory is that as the sample yields, it heats which softens it and lets it expand locally until it work-hardens. Simple calculations show that the local ΔT could be as high as 60° K due to the very low heat capacity of the material.

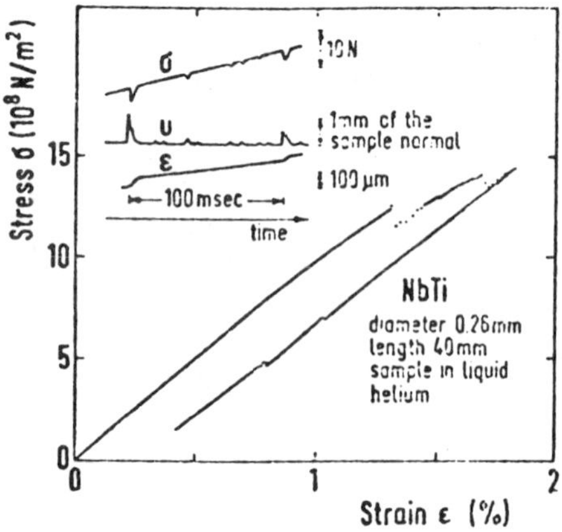

Figure 40, Ref. 10

In their experiment, the conductor was studied under externally applied stress and, as in the preceeding experiment, the wire was immersed in an externally applied field, and the current through it could be varied. Thus, the fraction j/j_c could be varied at will.

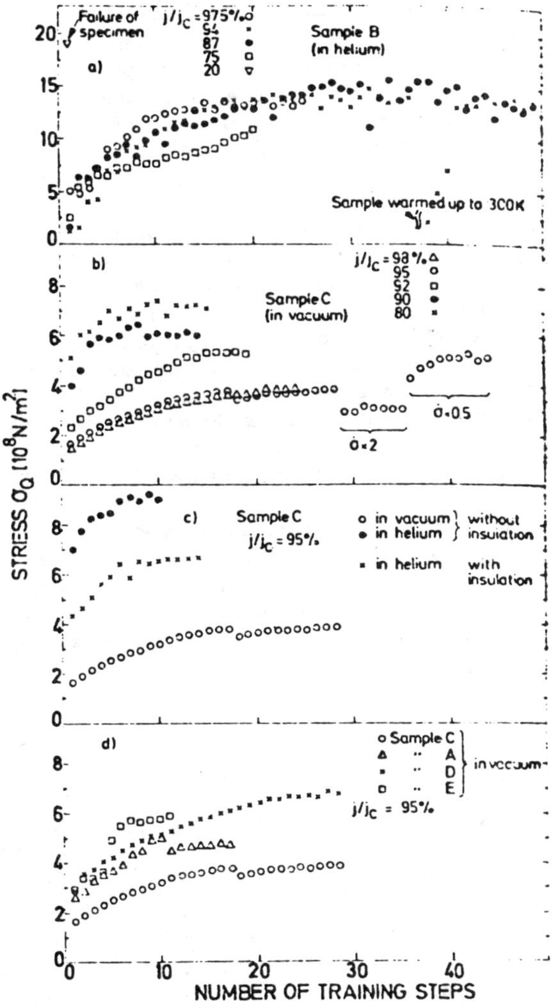

Figure 41, Ref. 10

The sample of wire could also be studied either in vacuum or immersed in liquid helium. Figure 41 shows the results of a number of experiments done during this study. Sample A was a single core conductor 0.4 mm in diameter with a copper to superconducting ratio of 1.3 to 1. Sample B had the copper jacket etched off and hence was a NbTi conductor whose radius was 0.26 mm. Sample C was a multifilament conductor 0.4 mm in diameter with 60 filaments. The copper to superconducting ratio was 1.4 to 1. In the figure, the stress is shown on the vertical axis and as in the case of the previous experiment, we see a phenomena displayed that is very reminiscent of training a magnet. The peak strains here would correspond to almost a per cent. Example C in this figure shows the important role that cooling can play. The top curve was a conductor cooled by liquid helium, and the bottom curve is for the same conductor in vacuum.

The stress strain curve of a sample of conductor was measured, and this curve showed hysteresis. The area contained inside of the loop is a measure of the amount of energy available for directly heating the conductor. Measurement of the actual heating from the inelastic behavior of the wire was made. We cannot go into all of the details of this paper here, but this study does show that applying stress to a conductor can cause it to quench and also training is observed, i.e., greater stress can be applied in subsequent cycles.

The work above stimulated investigation of whether or not winding a coil with prestressed wire would decrease the winding. Reference 11 refers to these experiments. We show here only the result of one such experiment, that of Schmidt and Turck. Figure 42 gives the results of the training of a square shaped solenoid wound with either prestressed or non-prestressed wire. The top two curves refer to an epoxy impregnated magnet, and the lower curve refers to a magnet wound without epoxy. The prestraining resulted in a permanent strain of about 1/4 of a per cent. It is seen that prestraining the wire reduced the training observed in the coil. The wire used in this particular case had a copper to superconductor ratio of 1.9 to 1 and 450 filamets of superconductor. There is not universal agreement that prestraining the wire results in fewer quenches to train. See Ref. 11.

Reference 12 refers to some papers presented at this conference that relate directly to either the source of the energy for causing a quench or the energy required to initiate the quench.

A number of experiments have observed accoustic emission taking place during the training of a magnet. In addition, the previous paper by Pasztor and Schmidt monitored the acoustic emission from the serrated yielding of a single conductor. Figure 43 shows the noise events observed during training of a 1-ft magnet at Fermilab.

Iwasa and Sinclair,[13] have made a very interesting observation. This is illustrated in Fig. 44. Three magnets: a solenoid, a split-pair, and a dipole are plotted. The horizontal scale is I/I_c, and the vertical scale is the acoustic emission rate in arbitrary units. The magnets have been normalized on both the horizontal and vertical axis; however, it is interesting to observe that the shape of these acoustic

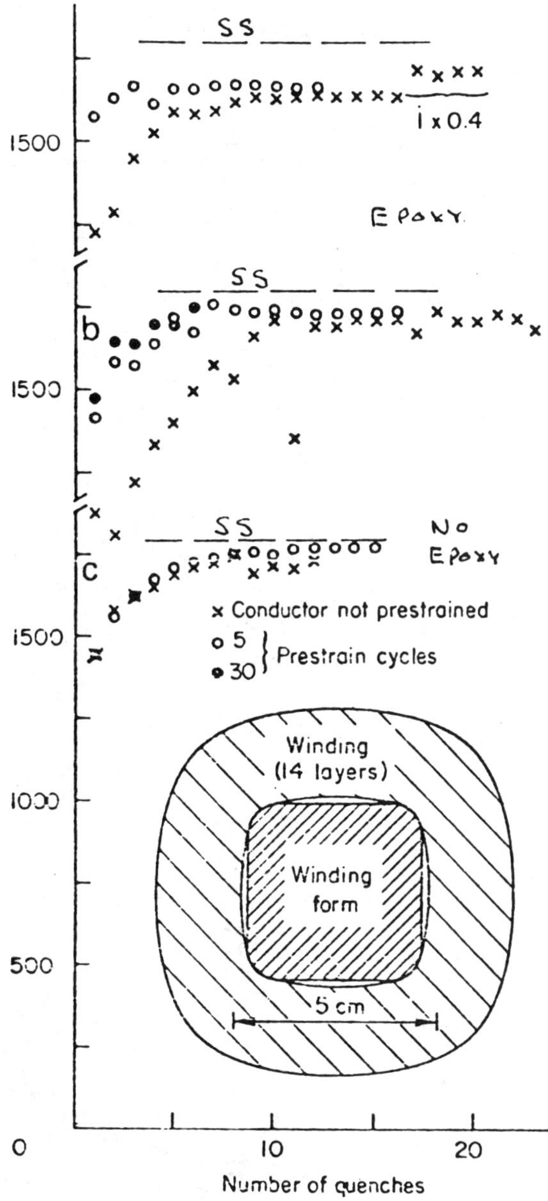
SS
I x 0.4
EPOXY
b
SS
SS
NO EPOXY
c
× Conductor not prestrained
o 5
• 30
Prestrain cycles
Winding (14 layers)
Winding form
5 cm
1500
1500
1500
1000
500
0
10
20
Number of quenches

Fig. 42

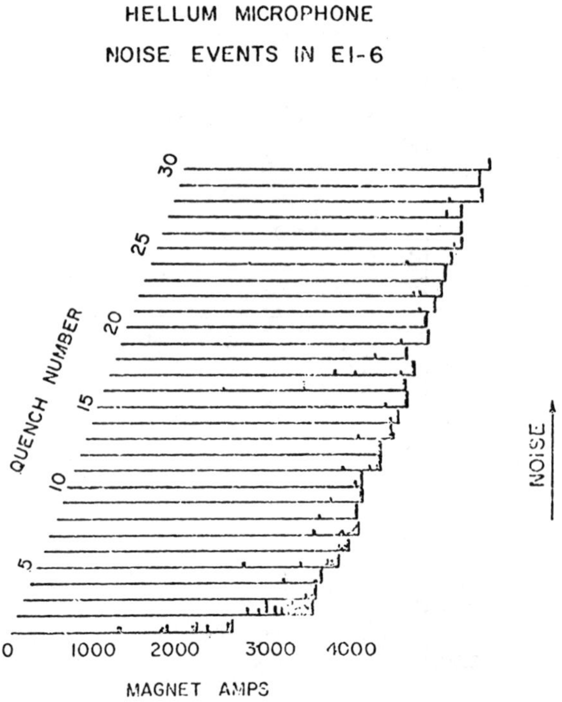
HELLUM MICROPHONE
NOISE EVENTS IN EI-6
QUENCH NUMBER
30
25
20
15
10
5
NOISE
0
1000
2000
3000
4000
MAGNET AMPS

Fig. 43

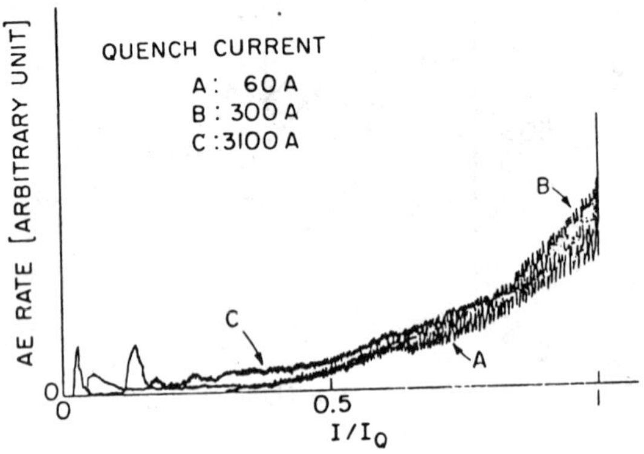
QUENCH CURRENT
A: 60 A
B: 300 A
C: 3100 A
AE RATE [ARBITRARY UNIT]
B
C
A
0
0
0.5
1
I/I0

Fig. 44

emission curves are rather similar. So far, acoustic emission has not been used as a monitor to protect superconducting magnets, and more work in this area may be very interesting, especially if it can be shown that the phenomena has a predictive capability of indicating the quench current of a magnet.

There is another area of study receiving much attention that is related to training. If heat could be removed from the conductor more rapidly, one might expect to affect the training. Superfluid helium is capable of doing this, and the effect on magnets has been studied

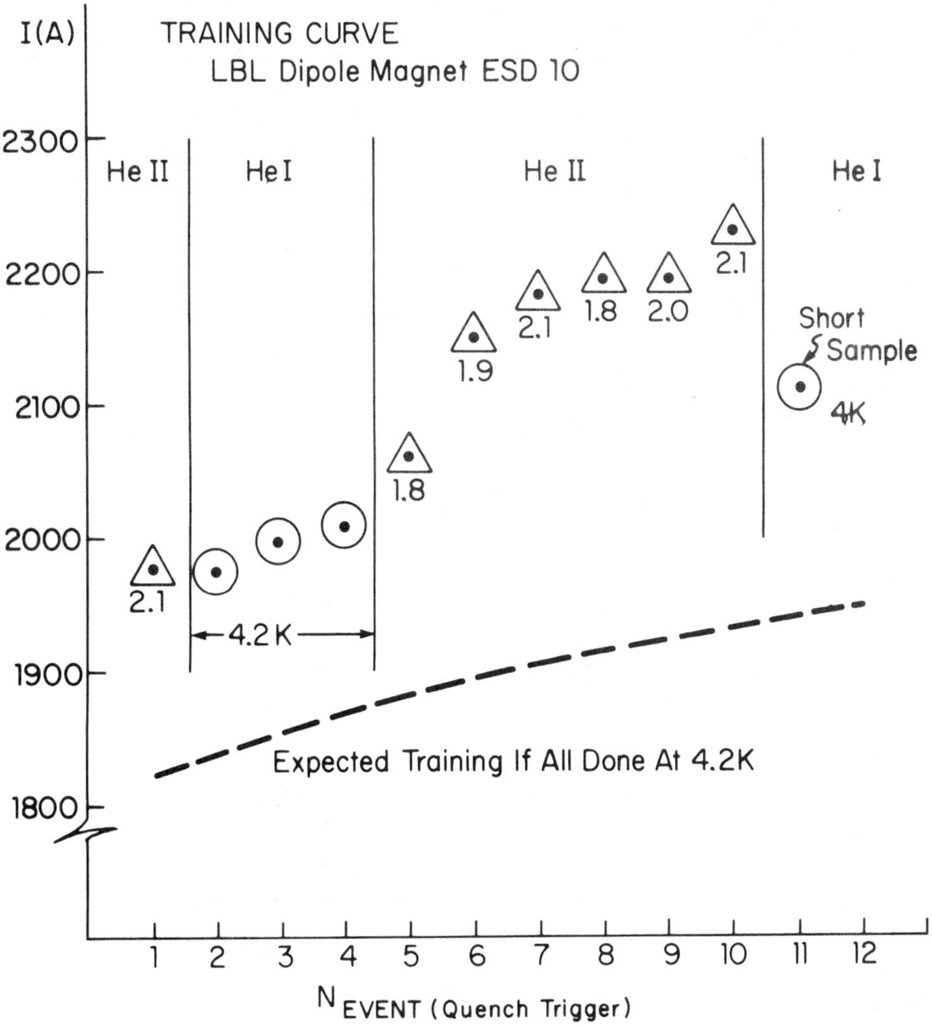

Figure 45

at a number of different laboratories. We pick here an example from an LBL ESCAR dipole. Figure 45 shows these data. The dotted curves along the bottom of the figure show the expected training of this magnet. The points indicate the actual quench level observed as the helium was cycled back and forth between He-I and He-II. The interesting point is made that only a few quenches at low temperature were required in order to train the magnet to its short sample limit at 4.2° K. The expected training if all done at 4.2° K would have been much slower than that observed at the lower temperature.

As a final example of problems caused by training, I would like to refer to the correction coil for the Fermilab Energy Doubler. Originally, it was planned to place dipole trimming coils inside of the focussing quadrupoles. These trimming coils were to operate at only a few hundred amps and are used for removing errors from the closed orbit. Reference 15 is the paper reported at this conference.

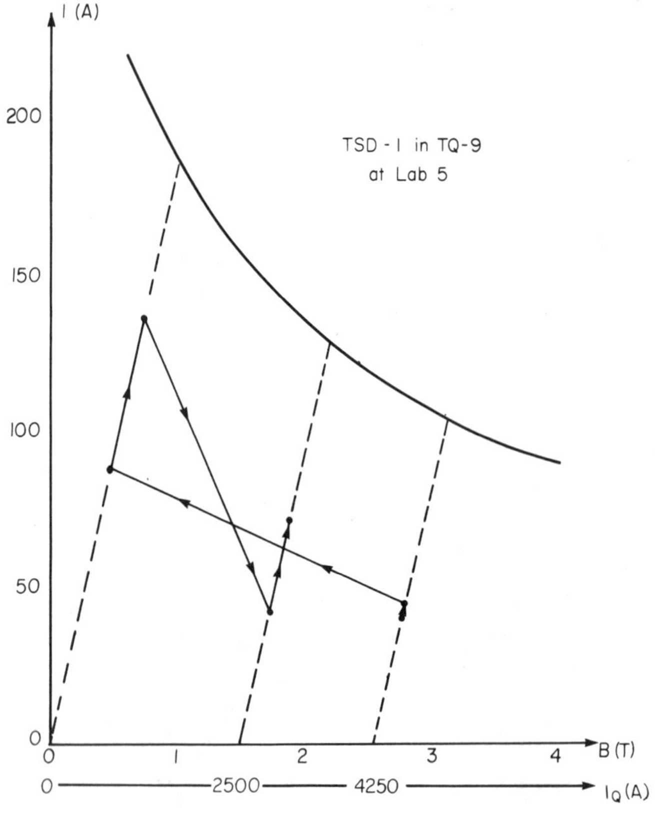

Figure 46, Ref. 15

Figure 46 shows a curve indicating the training history of the dipole TSD-1 in the field of the quadrupole TQ-9. Originally, the trim coil was trained up to the point shown on the right hand load line. The current in the quadrupole was then reduced to zero, and the trim coil trained up the left hand load line. Finally, a field of 1.5 Tesla was then turned on in the quadrupole with the resultant training history shown along the middle load line. Additional experience with this type of phenomena was obtained in a correction coil package containing several different overwound multipoles, such as quadrupole, sextupole, and octapole. Since these coils had to be independently adjusted, it was impossible to predict what the new training history would be for a new field configuration. The coils which were fabricated from a bonded wire were apparently cracking apart under the Lorentz forces. In this particular case, training was completely unacceptable, and the problem was solved simply by going to a larger size wire which would then operate at a smaller fraction of the critical current in a given field.

Summary

In summary, one can say that it is possible to control training. In order to do this, very careful attention must be given to the support of the conductor. However, much additional work in this field would be welcome. Work elucidating the disturbance spectrum of the support matrix and how to control it would be very useful. It is important to understand how to control friction and the heat it generates. Ways to control failure of the matrix at higher force levels become practical. In addition, studies of He-II as a coolant need to be carried out, and finally, the role of acoustic emission relative to the training of magnets needs to be clarified.

I would like to thank my many colleagues that have contributed to the ideas that I have expressed in this talk.

7. ENERGY LOSS IN PULSING MAGNETS

Introduction

When a conventional magnet is pulsed, there are three sources of energy loss: a hysteresis loss in the iron, the eddy current loss in the magnet structure and cable, and finally the I^2r loss in the cable itself. The superconducting magnet has losses that are similar in nature. There is, of course, no I^2r loss except at joints in the cable, either between the magnets or between the coils of an individual magnet. These losses can be kept quite small, for instance, overlapping 3 in. of the Fermilab cable and simply soft soldering the pieces together produces a joint which at helium temperature has a resistance of the order of a few times 10^{-9} ohms. At a current of 4,000 amps this represents a loss of less than 100 mW. This amount of heat is easily removed by the helium if proper precautions are taken.

Eddy currents exist in pulsed superconducting magnets as well as normal magnets. The most serious eddy current losses are within the cable itself since losses at this point will raise the temperature and lower the critical current that the conductor can carry. Making the cable from strands can break up the eddy current loops, and for this reason all magnets that are pulsed must be made of stranded cable. If the pulse rate is very rapid, the strands are insulated from each other, and if they enclose any net flux, the currents in the individual strands will not be equal. In fact, experiments with stranded cables using completely insulated strands have actually shown reverse currents in some of the individual wires. In order to alleviate this effect, an attempt is made to average each of the strands over the field within the coil package as completely as possible. The Rutherford cable as developed at Fermilab does this in the following way. Imagine first of all the 23 strands are wound in a compact spiral around a cylinder with a pitch of about 2 in. This cylinder is then flattened into the cable, and thus each strand averages across the field of a coil every two inches. In the early versions of the Fermilab conductor, the cable was then filled with solder. It was soon discovered that the eddy currents in this structure were excessive. The cable now consists of 23 strands of which half are insulated with a copper oxide covering. This insulation is actually rather poor and allows some current sharing between the strands. On the other hand, its resistance is high enough so that the eddy current loops are effectively broken up.

In a superconducting magnet there are, of course, hysteresis losses in any iron that is located within the field; however, there is another phenomenon that is also referred to as hysteresis. It takes place in the hard Type 2 superconductors. As the current is raised from zero, the field that surrounds the superconducting filaments also increases. At the start the current is all on the surface of the individual filaments within the conductor. As soon as the current density in the filament surface increases to the critical current, however, the field and the current are obliged to penetrate the filament to an increased depth. As the field penetrates the filament, energy is generated which must be removed by the helium coolant. When the cable reaches its critical current, the field has completely penetrated all of the filaments. As the current in the magnet is then reduced to zero, there are persistent currents left within the conductor which generate a persistent magnetic field. For instance, the Fermilab magnets show a residual dipole field of about 10 Gauss due to the persistent currents in the cable. There is also a sextupole moment whose magnitude is such that at 1 in. from the center, it is also of the order of 10 Gauss. The rest of the harmonics generated by the persistent currents are quite small.

This phenomenon of persistent currents leads to a magnetization moment per unit volume of the superconductor that depends upon its history. When the magnet is carried around a complete magnetization cycle, a certain amount of energy is deposited within the conductor. The theory for calculating this energy loss is exactly the same as it is with iron. In the case of iron the persistent currents are atomic currents. It is found empirically that the energy loss per cycle is roughly proportional to the peak field generated by the magnet.

Analysis

In order to calculate the energy loss for a magnet that is being pulsed, we will first make a model for the eddy current losses. In the magnet a large amount of conductor is exposed to changing magnetic field. Inside of the conductor we can visualize little loops in which an induced current flows. Since these loops are resistive (we are not talking about currents in the superconductor), there will be energy loss generated. The circuit diagram for this model is

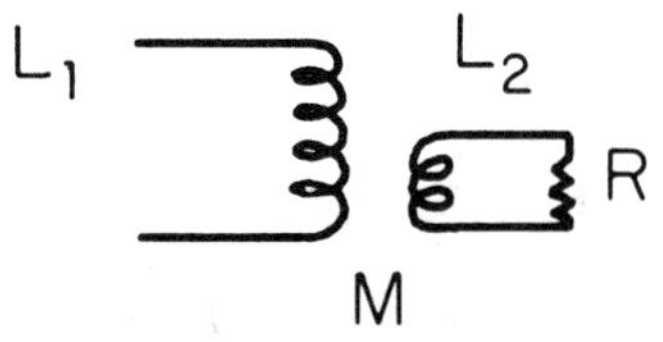

Consider now a magnet that is being pulsed so that the field increases linearly up to a field B_0 in time T_0. The induced current in the loop is illustrated in the figure below.

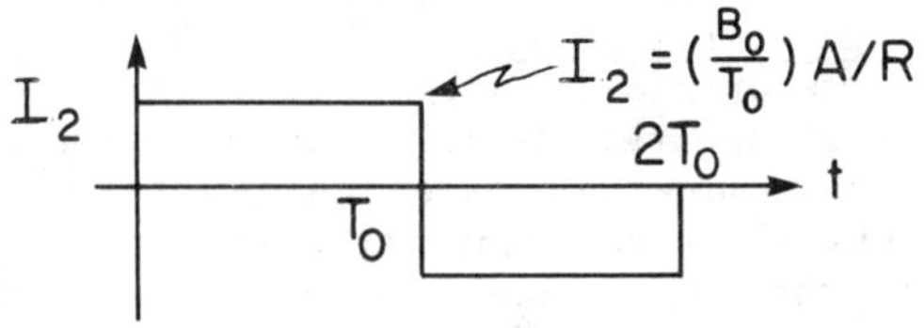

where A is the area of the loop and R is its resistance. The instantaneous power generated in this loop will be I^2R and integrated over the magnet cycle. We find that the energy loss is

$$W = \int_0^{2T_o} I_2^2 \, R dt = 2 B_o \left(\frac{B_o}{T_o}\right)\left(\frac{A^2}{R}\right)$$

We see, therefore, that the energy loss per cycle is proportional to B_o as well as $\dot{B}_o$. The loss can be decreased by increasing the resistance of the loops.

We next consider the hysteresis losses. As mentioned above, these are proportional to the maximum field to which the magnet is pulsed. So we have the total energy loss per cycle as

$$W = \underbrace{\alpha B_o}_{\text{hysteresis}} + \underbrace{\beta B_o \dot{B}_o}_{\text{Eddy Current}}$$

This equation shows that if we hold $\dot{B}$ fixed, the losses are proportional to the maximum field, and that if we hold the maximum field fixed, the losses are proportional to $\dot{B}$. This is exemplified in the two graphs plotted on the following page.

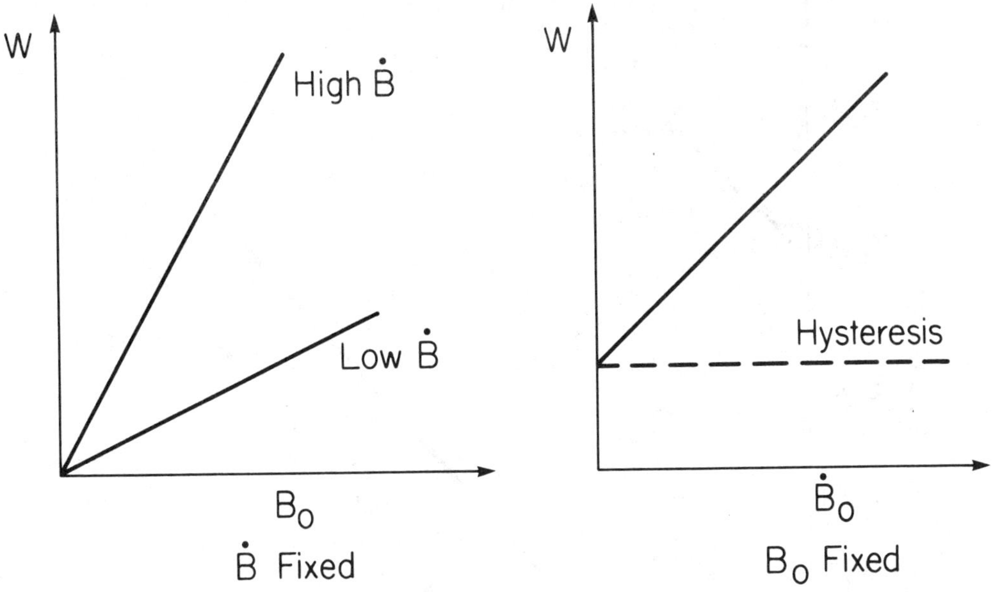

In order to separate these effects, magnet losses can be measured as a function of B_0 and $\dot{B}$, and the two individual losses can then be separately determined. If the eddy current losses are too high, additional resistance must be build into the conductor. If the hysteresis losses are too high, the only recourse one has is to use smaller diameter filament. Here, one is at the mercy of the techniques of conductor fabrication. At present the filaments are of the order of 5 to 10 microns in diameter, and a further reduction in size causes some of them to break.

Examples

Figures 47 and 48 show examples of energy loss in magnets that have been constructed at Fermilab. The one curve shows a magnet that was constructed of conductor which had Sta-Brite cover strands and exhibited very large eddy current losses. (Stay-Brite is a tin silver solder and is a good conductor.) The loss versus B_{max} with fixed $\dot{B}$ is linear. It agrees with the above simple theory. Also, when B_{max} is held fixed and $\dot{B}$ is varied, we see the typical behavior of eddy current losses. The intercept at B = 0 gives the hysteresis losses corresponding to the B_{max}. The second curve shows what happens when the insulation between the strands of the cable is increased somewhat. The cable used here is the so-called "zebra" cable which has every

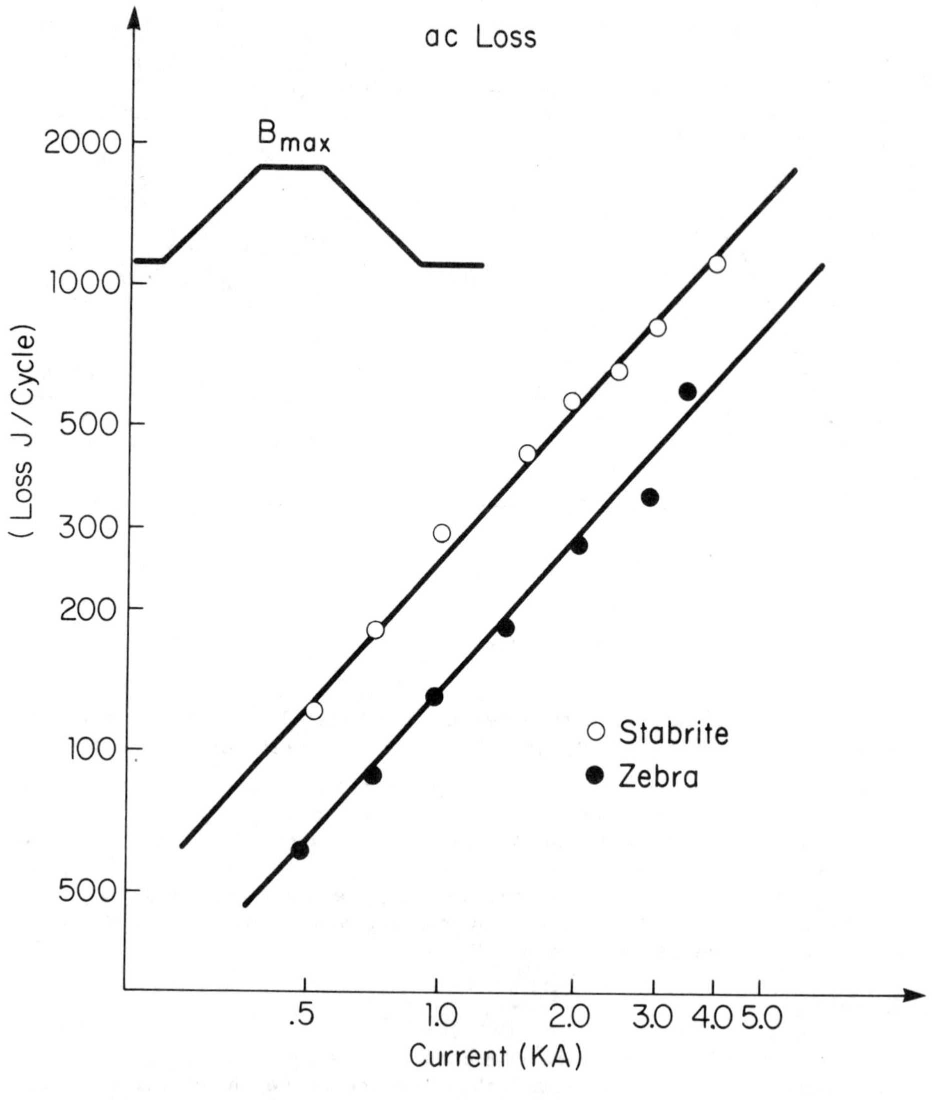

Fig. 47. Energy loss/cycle vs. excitation current for several conductors in magnets from the Doubler program.

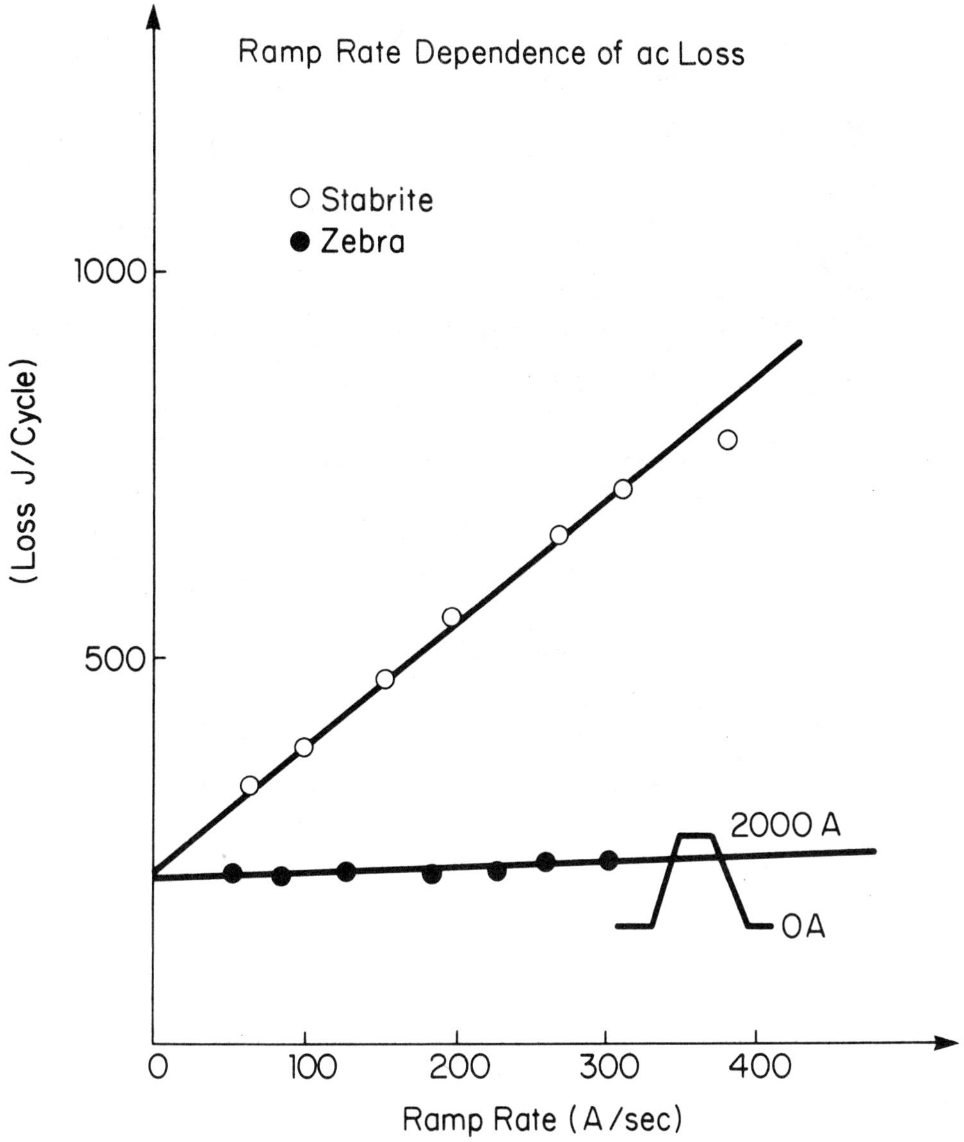

Fig. 48. Energy loss/cycle vs. ramp rate for several conductors in magnets from the Doubler program.

other strand coated with ebonol which is a thin layer of copper oxide. In this case we see that the losses are still proportional to B_{max} for a fixed B, but when we look at the losses as a function of $\dot{B}$, we see that the eddy current losses have essentially disappeared. In this series of experiments, all 23 strands of the cable were insulated. There was severe difficulties in some magnets with current sharing between the strands of the cable. Of the ten magnets built using this cable, some behaved very well, but a few quenched at currents only 75 per cent of the short sample limit. This behavior was attributed to unequal current sharing in the strands. The currents are unequal because although the Rutherford cable is twisted in order to make each strand average over the field, there are obviously small effects causing a net flux linkage between some of the strands. Since the strands are connected together at the ends, currents can flow in these loops. This circulating current will be in the same direction as the transport current in some strands, but opposite in others; hence, some strands will have higher than average current and some strands lower than average. The strands with the high current will quench first and degrade the performance of the magnet. A small resistance between strands aids in current sharing. One is therefore caught in a delicate balance between two extremes. If the insulation is too good, current sharing is poor. If the insulation is too low, the eddy current losses become too high. Similar problems were experienced at Brookhaven with the conductor in the form of a braid. In that case the trouble was not so much with the magnet losses, since the ramp rate of the magnets was very low, but rather with the sextupole field induced by these eddy currents. Attempts were made to solve this problem but were abandoned when the design of the magnet was changed, and a Fermilab-type cable was selected.

REFERENCES

[1] P. F. Smith, M. N. Wilson, and A. H. Spurway, J. Phys. D3, 1561 (1970).

[2] W. B. Fowler, M. Kuchnir, R. H. Flora, and R. H. Remsbottom, IEEE Trans. Magn. Mag-17, 925 (1981).

[3] R. Yamada, M. Wake, W. Fowler, Maximum Quench Current of Energy Doubler Magnets, Fermilab Internal Report TM-969, July 8, 1980.

[4] S.L. Wipf, Stability and Degradation of Superconducting Current-Carrying Devices, Los Alamos Scientific Laboratory Report No. LA 7275.

[5] M. N. Wilson, Degradation and Training: A Review, Rutherford High Energy Laboratory Report No. SMR/41.

[6] S. H. Minnick et al., Design Studies of Superconducting Generators, IEEE Trans. Magn. Mag-15, 703 (1979).
T.E. Laskaris, High-Performance Superconducting Windings for AC Generators, IEEE Trans. Magn. MAG-17, 884 (1981).

[7] W. Sampson, Brookhaven National Laboratory, private communication.

[8] J. E. C. Williams and E. S. Bobrov, The Magnet System of the 500 MHz NMR Spectrometer at the NML, submitted to RSI; See also, E. S. Bobrov and J. E. C. Williams, Direct Optimization of the Winding Process for Superconducting Solenoid Magnets, FA-16. IEEE Trans. Magn. MAG-17, 447 (1981).

[9] J. W. Ekin, R. E. Schram, M. J. Superczynski and D. Taylor, Training of Epoxy-Impregnated Superconductor Windings, IEEE Trans. Magn. MAG-17, 447 (1981).

[10] G. Pasztor and C. Schmidt, J. Appl. Phys. 49, 886 (1978).

[11] C. Schmidt and B. Turck, Cryogenics 17, 695 (1977).
J. W. Ekin, private communication. Apparently the NBS Experiments, Ref. 9, do not show this effect.

[12] D. E. Baynham, V. W. Edwards, and M. N. Wilson, Transient Stability of High Current Density Superconducting Solenoids, IEEE Trans. Magn. MAG-17, 732 (1981).
C. Schmidt, Transient Heat Transfer and Recovery Behavior of Superconductors, IEEE Trans. Magn. MAG-17, 738 (1981).
S. L. Wipf, Some Experiments on Helium Heat Transfer Characteristics Affecting Stability of Superconducting Magnet Operation, IEEE Trans. Magn. MAG-17, 742 (1981).
S. W. Van Sciver, Enthalpy Stability Criterion for Magnets Cooled with Superfluid Helium II, IEEE Trans. Magn. MAG-17, 747 (1981).
T. Okada, S. Nishijima, and T. Horiuchi, Study of Stress/Strain Effects on Superconducting Composite System - A Coil Simulation Experiment on Potting, IEEE Trans. Magn. MAG-17, 1052 (1981).
D. J. Waltman, M. J. Superczynski, and F. E. McDonald, Energy Pulses Required to Quench Potted Superconducting Magnets at Constant Field, IEEE Trans. Magn. MAG-17, 1056 (1981).
K. A. Tepper, Mechanical and Heat Transfer Models for Frictional Heating in Superconducting Coils, IEEE Trans. Magn. MAG-17, 1060 (1981).
R. S. Kensley, H. Maeda, and Y. Iwasa, Frictional Disturbances in Superconducting Magnets, IEEE Trans. Magn. MAG-17, 1068 (1981).

[13] M. W. Sinclair and Y. Iwasa, private communication; see also their

paper Acoustic Emission from Composite Superconductors, in Short Samples and in Magnets, IEEE Trans. Magn. MAG-17, 1064 (1981).

[14] W. Gilbert, private communication.

[15] D. Ciazynski and P. Mantsch, Typical Problems of the Correction Magnets for Fermilab Energy Saver, IEEE Trans. Magn. MAG-17, 165 (1981).

STORAGE RING DESIGN

E. Keil
CERN, Geneva, Switzerland

TABLE OF CONTENTS

0094-243X/82/870806-58$3.00 Copyright 1982 American Institute of Physics

STORAGE RING DESIGN

Part I

The Design of Proton Storage Rings

Eberhard Keil
CERN, Geneva, Switzerland

1. INTRODUCTION

Practical experience with proton storage rings is only available from the Intersecting Storage Rings (ISR) at CERN.[1] In this machine, protons circulate in opposite directions in two independent rings, which intersect in 8 equidistant crossing points. The maximum energy is 31 GeV, and the maximum current is above 50 A in each beam. The ISR were conceived before design principles such as low-β insertions[2] and incorporating of space-charge phenomena were developed. These principles are included in the design of ISABELLE[3] which is currently under construction at the Brookhaven National Laboratory. Design studies for similar machines have been made in several other places.[4,5,6]

In this lecture a design concept is presented for proton storage rings which was developed at the beginning of the LSR study.[6] Most of it was presented[7] at the International Conference on High-Energy Accelerators in 1974. The design and performance of proton-proton colliding-beam storage rings is restricted by two different kinds of limitation: those due to technology and funds, and those due to space-charge phenomena. The design aim is to achieve the best possible luminosity within all these constraints. A procedure is outlined which includes all constraints from the beginning. Sets of machine parameters obtained in this way are given for reference.

The understanding of what follows might be helped by a very qualitative description of the layout of proton storage rings and of the way they are operated.

There are two rings, one for protons circulating clockwise and one for protons circulating counterclockwise. Since the sign of the proton charge is the same for both beams, the protons cannot circulate in one ring. At a few equidistant points around the circumference, the two rings are made to intersect in interaction regions where beam-beam collisions take place. The machine lattices, therefore, consist of arcs and interaction regions. In the arcs the circulating proton beams are bent and focussed in two separated-function FODO lattices which are arranged side-by-side in a narrow tunnel. In the intersection regions, the two rings interchange roles, the outside one becoming the inside one and vice versa. The interaction regions also contain all the special beam transport necessary to match the beam properties between the arcs and the crossing points.

The average circulating proton current in each ring is of the order of several amperes. The beams are not bunched but coasting, i.e. the current line density is constant around the circumference. Since the average current exceeds that from any conceivable injector by a large factor, many injector pulses must be accumulated during a storage ring fill.

Protons obey Liouville's theorem. Hence, the six-dimensional phase-space density cannot be increased during the accumulation, and the phase-space volume occupied by the beam increases at least in the ratio of the currents. For the design procedure, it is assumed that the accumulation is done by the same process as in the ISR, called RF stacking, in which the increase in phase-space volume takes place in synchrotron phase space. This essentially leads to an increase of the momentum spread in proportion to the circulating current.

Liouville's theorem also implies that a beam cannot be injected into a region of phase space already occupied. In practice, beam cannot even be injected into a region of real space already occupied. Therefore, the horizontal aperture must be divided into two parts: a part reserved for injection where the incoming beam can be made to circulate and a stacking region where the coasting beam is gradually built up. The RF stacking moves the injected beam into the stacking region by RF acceleration, thereby liberating the injection orbit for the next injected pulse. Since all this happens at constant magnetic guide field, the energy change during acceleration is of the order of a few percent.

2. INTERACTION REGION DESIGN

2.1. Crossing geometry

Consider two unbunched beams which collide at the center of a straight section of length ℓ with a small crossing angle as shown in Figure 1. To simplify the presentation, it is assumed that the beams have a circular cross section, that the density distribution is Gaussian with standard deviation σ_0, and that the betatron amplitude functions have the same low value β_0 in the horizontal and vertical direction. In contrast to the previous paper,[7] the crossing is now assumed to take place in the horizontal plane.

The betatron amplitude function varies with the distance s from the crossing point like

$$\beta(s) = \beta_0 + s^2/\beta_0 \qquad (1)$$

If the dispersion vanishes in the interaction region then the beam radius varies with s like

$$\sigma(s) = \sigma_0(\beta(s)/\beta_0)^{1/2} = \sigma_0(1 + s^2/\beta_0{}^2)^{1/2} \qquad (2)$$

The crossing geometry shown in Figure 1 can be described in terms of two dimensionless parameters ξ and η defined as follows:[8]

$$\xi = \ell/2\beta_0 \qquad \eta = \alpha\beta_0/\sigma_0 \qquad (3)$$

The parameter $\xi \simeq \sigma(\ell/2)/\sigma_0$ is approximately the ratio between the beam cross sections at the end and the center of the intersec-

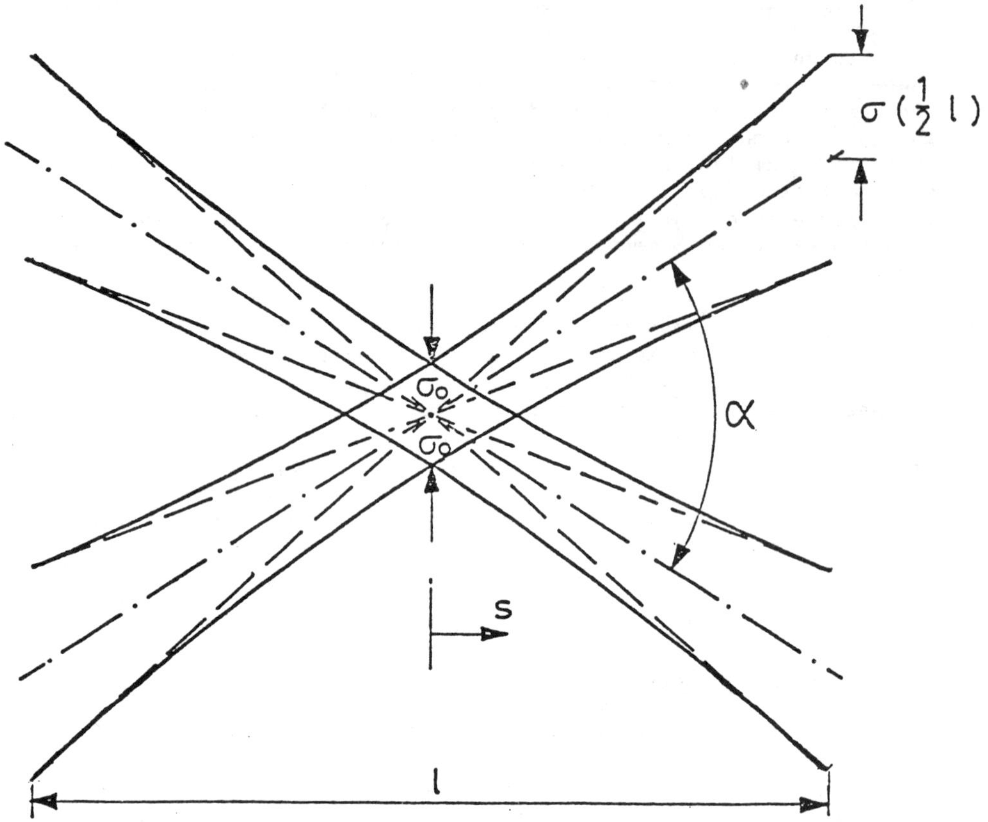

Fig. 1 - Geometry of two coasting beams crossing at an angle α in a low-β insertion of length ℓ.

tion region: ξ<<1 means little variation, i.e. $\sigma(\ell/2) \simeq \sigma_0$ while 1<<ξ means a large variation as typically to be found in a proper low-β insertion with ℓ/β_0>>1. Similarly, the parameter $\eta \simeq \frac{1}{2}\alpha\ell/\sigma$ is approximately the ratio of the beam separation and beam size at s = ℓ/2. When η<<1 the beams collide almost head-on, while with 1<<η the beam separation is large compared to their size at the end of the intersection region.

2.2. Luminosity

The luminosity L obtained in such a crossing point can be calculated from the appropriate overlap integral of the two beam densities. It becomes[9]

$$L = \frac{c\lambda^2\ell}{4\pi\sigma_0^2} \; K\,(\ell/2\,\beta_0,\; \alpha\beta_0/\sigma_0) \tag{4}$$

where the universal function K is defined by

$$K(\xi,\eta) = \frac{1}{\xi}\int_{-\xi}^{+\xi} (1+u^2)^{-1} \; \exp\left(-\frac{1}{4}\,\eta^2u^2/(1+u^2)\right)\, du \tag{5}$$

λ is the (particle) line density and c is the speed of light. For extreme values of the arguments K has asymptotic values:

$$K \to 2 \text{ for } \xi,\ \eta \ll 1 \tag{6}$$

$$K \to 2\pi^{1/2}/(\xi\eta) \text{ for } \xi,\ \eta \gg 1 \tag{7}$$

2.3. Beam-beam tune shift

The beam-beam tune shifts ΔQ_x and ΔQ_y are usually defined for particles with small betatron amplitudes. They can be calculated from the transverse impulses Δp_x and Δp_y which the protons receive when they cross the other beam with lateral displacements x and y, respectively. Define the strength of the kick δ_u where u may either be x or y as follows:

$$\delta_u = \lim_{u\to 0} \frac{\Delta p_u}{p} \tag{8}$$

Here p is the proton mementum. Then the beam-beam tune shift becomes:

$$\Delta Q_u = \frac{\beta_u\delta_u}{4\pi} \tag{9}$$

When this calculation is done for the geometry shown in Fig. 1 the result is[8]

$$\Delta Q_u = \frac{2\lambda r_0 \ell}{Et} I_u \left(\ell/2\beta_0, \alpha\beta_0/\sigma_0\right) \tag{10}$$

Here r_p is the classical proton radius ($r_p = 1.535 \times 10^{-18}$m), $E_t = 4\pi\beta\gamma\sigma_0^2/\beta_0$ is the invariant transverse emittance, and I_u is one of two universal functions

$$I_x(\xi,\eta) = \frac{1}{\xi\eta^2} \int_{-\xi}^{+\xi} (1+u^2) \left[\left(u^{-2} + \frac{\eta^2}{1+u^2}\right) \exp - \frac{1}{2} \frac{\eta^2 u^2}{1+u^2} - u^{-2} \right] du$$

$$I_y(\xi,\eta) = \frac{1}{\xi\eta^2} \int_{-\xi}^{+\xi} (1+u^{-2}) \left[1-\exp\left(- \frac{\eta^2 u^2}{1+u^2}\right) \right] du$$

These functions have asymptotic values for extreme values of their arguments

$$I_x \to 1 \tag{11}$$

$$\text{for } \xi,\eta \ll 1$$

$$I_y \to 1 \tag{12}$$

$$I_y \to \frac{2}{\eta} \left[(\pi/2)^{1/2}/\xi + 1/\eta \right] \qquad \text{for } \xi,\eta \gg 1 \tag{13}$$

When the beams are well separated, $\eta \gg 1$, there is a large cancellation of the forces in the plane of the crossing, the x-plane, since the other beam is on one side of the test particle before the crossing and on the opposite side after the crossing. There is no such cancellation in the y-plane perpendicular to the crossing. Therefore, ΔQ_y is usually more important than ΔQ_x which will not be discussed further.

2.4. Optimum luminosity

It is customary in storage ring design to eliminate one power of λ from (4) by using (10) for ΔQ_y. Doing this, the following expression is obtained:

$$L = \frac{c\gamma\lambda\Delta Q_y}{2r_p\beta_0} \frac{K}{I_y} \left(\ell/2\beta_0, \alpha\beta_0/\sigma_0\right) \tag{14}$$

With the asymptotic expressions (6) or (7), and (12) or (13) this becomes:

$$L \to \frac{c\gamma\lambda\Delta Q_y}{r_p\beta_0} \qquad \text{for } \xi,\eta \ll 1 \tag{15}$$

$$L \rightarrow \frac{c\gamma\lambda\Delta Q_y}{2^{1/2} r_p \beta_o} \frac{1}{1+(\xi/\eta)(2/\pi)^{1/2}} \quad \text{for } \xi,\eta >> 1 \tag{16}$$

These formulae can be interpreted in several ways. The beam-beam tune shift ΔQ_y, can be deduced from it when all the other quantities are measured. This involves no hidden assumptions. If one knows upper limits for the line density λ and for the beam-beam tune shift ΔQ_y one can calculate an upper limit for the luminosity L to be obtained in a future machine. Limits on λ will be discussed in the next section, while the beam-beam effect is the subject of another lecture in this school. This application needs the additional assumption that the beam emittance and/or the crossing angle are adjusted such that (10) is satisfied.

The two asymptotic values for the luminosity, (15) and (16), differ only by a factor $2^{\frac{1}{2}}$ if β_o is the same in the two cases. This result is not surprising. The assumption $\xi,\eta<<1$ implies head-on collisions over a length $\ell \lesssim \beta_o$, while the assumption $\xi,\eta>>1$ implies a variation of the beam size given by (2) and crossing at an angle, such that the luminosity drops by at least a factor of two when going from the crossing point to $s = \beta_o$, i.e. to a similar distance as in the first case.

A superficial look at (16) leads to the conclusion that the luminosity increases beyond bounds when $\beta_o \rightarrow 0$. However, this is not true because the assumptions used in deriving (16) break down below some value of β_o. Instead of growing beyond all limits, the luminosity tends towards an asymptotic value when $\beta_o \rightarrow 0$. It reaches 2/3 of this value when β_o is chosen as follows[9]

$$\beta_o = \left(\frac{\Delta Q_y E_t \ell}{8\pi\lambda\, r_p}\right)^{1/2} \tag{17}$$

with the luminosity

$$L = \frac{4}{3} c\gamma \left(\frac{\pi\gamma^3 \Delta Q_y}{E_t \ell r_p}\right)^{1/2} \tag{18}$$

2.5. Interaction region optics

While the previous paragraph showed that it is not worthwhile to decrease β_o much below (17) it remains to be seen whether this value can actually be reached. Lower limits for β_o are imposed by two optical considerations. The first is related to the strengths of the quadrupoles which focus the beams coming from the crossing point, at a distance $\ell_Q \simeq \ell/2$. Fig. 2 shows several possible arrangements. The same problem arises in the design of e^+e^- storage rings where the geometry is simpler. Therefore, it will be worked out there.

The second limit arises from the fact that quadrupoles focus particles differently depending on their momentum. These chromatic effects are proportional to $\ell_Q/\beta_o \simeq \xi$. Values of $\ell_Q/\beta_o \simeq 50$ have been reached in e^+e^- storage rings which handle beams with a full

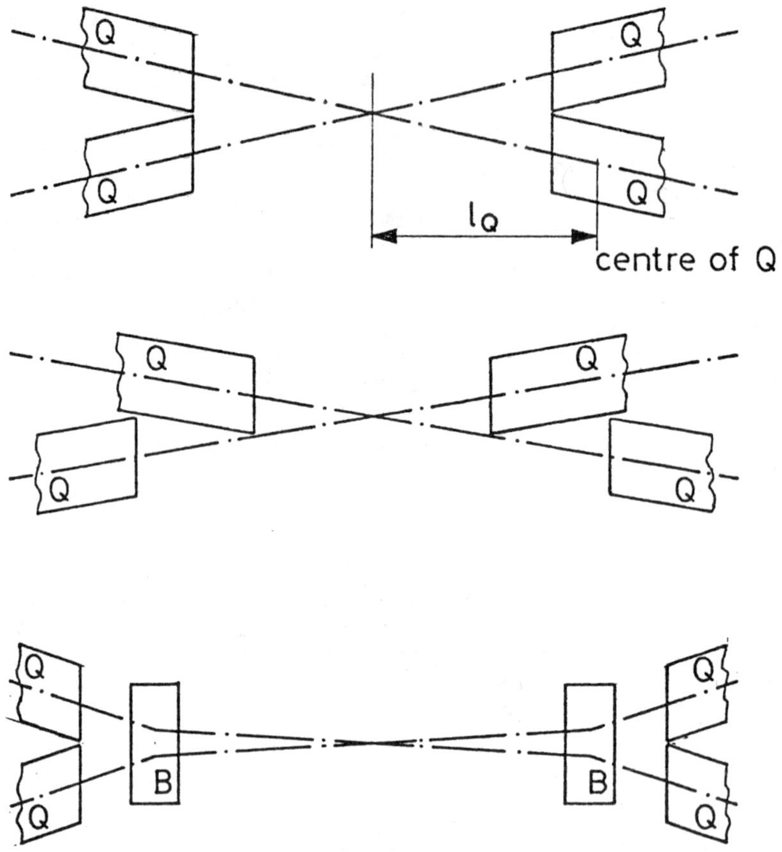

Fig. 2 - Schematic layout of quadrupoles and bending magnets near the crossing points:
(a) Symmetrical layout: crossing angle given by quadrupole diameter and distance
(b) Staggered layout: crossing angle given by quadrupole radius and distance
(c) Layout with separating magnets: no lower limit on crossing angle.

momentum bite of about 1%. A discussion of chromatic effects is beyond the scope of this lecture.

3. SINGLE-BEAM LIMITS

In proton machines with unbunched beams there is no one single-beam phenomenon which is clearly more important than the others. Therefore it is appropriate to include as many of them as one can think of in the design procedure right from the start. In this way, the most important phenomenon will show up automatically.

The discussion given below is not meant to be an up-to-date survey of collective phenomena but rather an application of our present knowledge, suitably simplified, to the design of storage rings. This explains the unconventional way in which some well-known formulae are written. Smooth approximation is used for describing alternating-gradient lattices, i.e. the amplitude function β and the dispersion D are replaced by their average values $\beta \simeq R/Q$ and $D \simeq R/Q^2$. No distinction is made between the radius R of the machine arcs and $C/2\pi$ where C is the machine circumference, i.e. the likely presence of long straight sections is ignored.

3.1. Incoherent tune shift

The incoherent tune shift arises from the space-charge forces of the particles in one beam with each other and with the currents and charges on the vacuum chamber wall. This phenomenon has been found to be quite noticeable in the ISR and had to be compensated.[10] Its most disturbing consequence is a change of the dependence of the betatron tunes Q_x and Q_y on the momentum error which must be kept under control to avoid transverse coherent instabilities as discussed below. To overcome this in the ISR, the tune shift is dynamically compensated during the accumulation of the circulating current. Every few amperes, three operations are performed.

i) The density distribution in momentum is measured by Schottky noise.[11]

ii) The resulting tune shift ΔQ_1 is calculated as a function of $\Delta p/p$.

iii) The poleface windings which allow to modify the radial variation of the magnetic guide field are excited such that the calculated tune shift $\Delta Q_1(\Delta p/p)$ is compensated.

All this is far too complicated for a simple design procedure. Here, only the incoherent tune shift ΔQ_1 for a centered beam in an elliptic vacuum chamber is calculated and used as a yardstick for the difficulties which might arise.

In the highly relativistic limit, $\gamma >> 1$, the direct space-charge effects, due to the repulsion between like charges and the attraction between like currents cancel like $(1-\beta^2) = \gamma^{-2}$, but the effects due to wall currents and charges in the vacuum chamber do not. Neglecting the direct space-charge term, the standard formula[12] may be written as follows:

$$\Delta Q_1 = \frac{N r_p R}{\pi Q \gamma} \left(\frac{\varepsilon_1}{h^2} + \frac{\rho}{R} \frac{\varepsilon_2}{g^2} \right) \tag{19}$$

Here N is the number of protons in the beam, h is the half aperture of the vacuum chamber and g is the half height of the magnet gap. The magnets occupy a fraction ρ/R of the circumference. The image field coefficients ε_1 and ε_2 depend on the shape of the vacuum chamber and magnet pole pieces, respectively. For the parallel-plate geometry of a separated-function lattice one has

$$\varepsilon_1 = \pi^2/48 \qquad \varepsilon_2 = \pi^2/24 \tag{20}$$

Assuming that g and h are equal and calling the half aperture b, (19) can be brought into the form:

$$\Delta Q_1 = \left(\frac{\pi^2}{24} \frac{r_p E_p}{e^2 c^2} \right) \frac{I}{B_M b^2} \frac{R}{Q} \frac{R}{\rho} \left(1 + \frac{2\rho}{R} \right) \tag{21}$$

Here $E_p = m_p c^2$ is the rest energy of the proton, I is the circulating current and B_M is the field strength in the bending magnet, which has been used to eliminate γ and isolated appearances of R. The aim of this manipulation is to write the design equations in a form which links the beam dynamics equations to engineering parameters which by experience only take values within rather narrow limits.

Much of the subsequent discussion will be in terms of round chambers. In this case (21) is pessimistic for a centered beam because the first term in the bracket vanishes. On the other hand, (21) is optimistic because the contributions of variations of the vacuum chamber cross section is neglected. Such variations should be kept small in number and in size to avoid the inductive impedances and cavity resonances associated with them.

There is also a coherent single-beam tune shift[12] which leads to rather similar equations as the incoherent one. Therefore, it is not worked out in this lecture.

3.2. Transverse resistive-wall instability

In the ISR, the transverse resistive-wall instability[13] lead to sudden losses of part or all of the circulating beam up to the time we learned how to control the tune variation with the momentum error (cf. section 3.1) very well and installed feedback systems[14] for the lowest modes. Stabilization by Landau damping is achieved when the following criterion[15] is satisfied

$$|Z_1| \leq \frac{E_p}{e} \frac{\pi Q}{IR} \left| (n-Q)\eta + Q' \right| \frac{\Delta p}{m_p c} \tag{22}$$

Here η is the momentum compaction, $\eta \simeq Q^{-2}$, $Q' = dQ/dp/p$ is the chromaticity and Δp is the momentum spread in the beam (full width at half maximum). The criterion must be satisfied for the total transverse impedance $Z_\perp$ in the machine, for all combinations of I and $\Delta p/m_pc$ which occur during its filling, and for all mode numbers n>Q. In order to avoid a cancellation of the terms between the absolute signs in (22), the chromaticity Q' must be positive. If the storage ring is filled by stacking in momentum space as in the ISR, then I and $\Delta p/m_pc$ grow roughly in proportion and may be interpreted in (22) as the total current and the total momentum spread in the beam.

In the special case of the resistive-wall impedance, the transverse impedance can be written as follows, neglecting the small capacitive impedance:

$$|Z_\perp| = \frac{2}{b^3}\left(\frac{R^3 Z_o}{(n-Q)\sigma}\right)^{\frac{1}{2}} \tag{23}$$

Here Z = 120 πΩ is the impedance of free space and σ is the conductivity of the vacuum chamber material. At low mode numbers n, just above Q, the chromaticity Q' dominates in (22). In order to achieve Landau damping, the tune spread δQ in the beam must satisfy the condition:

$$\delta Q = Q' \frac{\Delta p}{p} \gtrsim \frac{2}{\pi c}\left(\frac{Z_o E_p}{ec}\right)^{\frac{1}{2}} \frac{R}{Q}\left(\frac{R}{\rho B_M}\right)^{3/2} \frac{I}{b^3}\left(\frac{\gamma}{(n-Q)\sigma}\right)^{\frac{1}{2}} \tag{24}$$

In practice, an upper limit for Q is given by the distance between harmful non-linear resonances which the beam must not straddle. Therefore (24) is an equation for the maximum current I_t which is transversely stable.

In an actual machine, there are many more transverse impedances than just the resistivity of the vacuum chamber walls. They all must satisfy the criterion (22). However, the impedance of a few isolated objects can be reduced by careful design and by damping. If a feedback system with bandwidth Δf is used to combat the lowest unstable modes, then Landau damping is still required to damp the modes outside the passband with $(n-Q)f_o > \Delta f$ where f_o is the revolution frequency. Eq. (24) remains valid when the correct n is inserted.

3.3. Longitudinal resistive wall instability

When the momentum spread in a coasting beam is too small, it may bunch itself, i.e. develop a longitudinal instability. The classical example of this is the longitudinal resistive-wall instability.[16] In order to stabilize it by Landau damping, the following criterion must be satisfied:[17,18]

$$\left|\frac{Z}{n}\right| < \left(\frac{E_p}{e}\right)\frac{\eta}{I\gamma}\left(\frac{\Delta p}{m_p c}\right)^2 \tag{25}$$

Here Z is the coupling impedance and n is again the mode number. Δp is the full width of the momentum spread at half height of the distribution function. Eq. (25) must be satisfied for all combinations of I and Δp which occur during the filling process. While stacking in momentum space $I \backsim \Delta p$, and hence the most severe conditions apply to the beginning of the filling.

Let I_p be the current in a single pulse from the injector and let $(\Delta p/m_p c)_i$ be its fully debunched momentum spread. (The phase space volume of the injected beam in the coordinates $(\Delta p/m_p c_1, s)$ is $2\pi R(\Delta p/m_p c)_i$). Then I_p may be used in (25) to find the stable momentum spread, for an injected pulse $(\Delta p/m_p c)_s$. If it is larger than $(\Delta p/m_p c)_i$, the longitudinal phase space volume has to be blown up to meet the stability criterion. It is advantageous to induce this blow-up artificially rather than letting the instability develop, because the instability does not stop right at the threshold but overshoots to a larger momentum spread.[18] If $(\Delta p/m_p c)_i > (\Delta p/m_p c)_s$ there is no instability.

For a centered beam in a circular vacuum chamber of radius b, the coupling impedance for the worst mode n = 1, neglecting the small capacitive contribution, is given by:

$$\left|\frac{Z}{n}\right| = \frac{1}{b}\left(\frac{RZ_o}{\sigma}\right)^{\frac{1}{2}} \tag{26}$$

Combining (25) and (26), and using the current pulse I_p, we find for the stable momentum spread:

$$\left(\frac{\Delta p}{m_p c}\right)_s^2 = \left(\frac{Z_o R}{\sigma}\right)^{\frac{1}{2}} \frac{e}{E_p} \frac{I_p Q^2 \gamma}{b} \tag{27}$$

Next, we calculate the maximum current I_ℓ which can be accumulated from pulses with current I_p each and momentum spread given by (27). In order to do this, we assume that the space available is one vacuum chamber radius b, the other half of the diameter being reserved for injection and betatron oscillations. We convert (27) to relative spread, and use $D = R/Q^2$ for the dispersion to find:

$$I_\ell = \frac{bQ^2 I_p \gamma}{R(\Delta p/m_p c)_s} \tag{28}$$

By combining (27) and (28), and the standard manipulations, this can be brought into the form:

$$I_\ell^2 = \left(\frac{E_p c}{Z_o e}\right)^{\frac{1}{2}} b^3 I_p \left(\frac{Q}{R}\right)^2 \left(\frac{\rho\ \sigma\gamma B_M}{R}\right)^{\frac{1}{2}} \tag{29}$$

As in the transverse case, a real machine has components with coupling impedances other than the resistive-wall impedance, the most conspicuous being the RF system. In the ISR, its shunt impedance is compensated by a feedback system.[19,20] The impedance of all other components must be made small compared to (26) or it must be included in the analysis of longitudinal instabilities.

4. LIMITS DUE TO THE LATTICE OPTICS

From the equations derived so far, in particular (21), (24) and (29), one might conclude that the aperture radius b and the reduced betatron wavelength R/Q can be chosen independently. This is a fallacy as can be seen by looking at a simple FODO lattice.

In a separated function lattice, the bending is done by homogeneous-field dipoles and the focussing by quadrupoles. Assuming that the phase advance in a lattice period of length L_p is 90°, the required focal length of the quadruples is given by

$$f = L_p / 2^{3/2} \tag{30}$$

The focal length which can be achieved in such a quadrupole with gradient dB/dx and length ℓ_Q is in thin-lens approximation with $\ell_Q \ll f$:

$$f = \frac{B\rho}{\ell_Q} \left(\frac{dB}{dx}\right)^{-1} = \frac{B\rho}{\ell_Q} \frac{b}{B_Q} \tag{31}$$

Here $B\rho$ is the magnetic rigidity of the protons, and B_Q is the field at a distance b from the centre of the quadrupole, the "pole tip" field.

The period length L_p and the reduced wavelength R/Q are trivially related

$$L_p = \frac{\pi}{2} \frac{R}{Q} \tag{32}$$

A fraction ρ/R of the length of a cell is occupied by bending magnets. Hence, the maximum length available for a quadrupole in half a cell is $1/2\ L_p(1-\rho/R)$. Introducing a quadrupole filling factor C_Q with $0<C_Q<1$, we find for ℓ_Q

$$\ell_Q = \frac{1}{2} L_p C_Q \ (1-\rho/R) \tag{33}$$

Combining (30) to 33) we obtain the wanted relation between b and R/Q:

$$b = \frac{ec}{E_p} \frac{\pi^2}{2^{\frac{1}{2}} 16} \left(\frac{R}{Q}\right)^2 \frac{B_Q C_Q}{\gamma} \left(1 - \frac{\rho}{R}\right) \tag{34}$$

5. DESIGN PROCEDURE FOR PROTON STORAGE RINGS

The aim of a design procedure is to arrive at a plausible set of parameters such as the aperture b, the tune Q, and lengths of quadrupoles ℓ_Q and lattice cells L_p. This set is known to be free of the beam dynamical problems included in the procedure. It therefore is a good starting point for a detailed engineering study which is likely to lead to some changes in the parameters selected initially.

5.1. Outline of the procedure

The design equations (21), (24), (29) and (34) involve only 15 parameters which can be grouped as follows:

i) Machine energy γ

ii) Injected beam parameters, given by injector: I_p, E_t, $(\Delta p/m_p c)_i$

iii) Storage ring lattice parameters R/ρ, B_M, B_Q, C_Q, b, σ, n-Q

iv) Space-charge parameters δQ_{max}, ΔQ_{1max}

v) Intersection parameters ℓ, ΔQ_{2max}

The most appropriate way of proceeding consists in first choosing these parameters, and then to calculate the maximum currents I_q, I_t and I_ℓ, and the corresponding luminosities. This gives a clear picture of what phenomena are most harmful, and what luminosity is to be expected. Once a likely set of parameters has been found, many more derived quantities can also be calculated.

5.2. Choice of parameters

The coice of the parameters must be guided by extrapolation and scaling from existing proton synchronotrons and storage rings. As examples, machines will be used which were under study at CERN when this procedure was developed. These machines used the CERN SPS as an injector, and the parameters of the injected beam were well known. The choice of the lattice parameters R/ρ, B_M, B_Q, C_Q was guided by a comparison with the large proton synchrotrons at Fermilab and CERN for a conventional copper-steel magnet system. For the superconducting magnet system, parameters in the vicinity of the ISABELLE proposal[3] were used.

The aperture radius b is a variable parameter. The choice of the vacuum chamber material determines σ. It is advantageous to work in the half-integral range of Q just above an integer, with n-Q = 3/4. The total tune spread ΔQ in the beam is approximately given by

$$\Delta Q = \delta Q_{max} + \Delta Q_{1max} + N_x \, \Delta Q_{2max} \tag{35}$$

where N_x is the number of crossing points and ΔQ_2 is the maximum permissible beam-beam tune shift per crossing. The value of ΔQ_{2max} has been the subject of much speculation but choosing $\Delta Q_2 = 0.005$ has become standard practice. Experimental data for the beam-beam limit with bunched beams of protons and anti-protons in the SPS should be coming soon.

The total tune spread ΔQ must fit between harmful nonlinear resonances. For the ISR, two working regions have been used with much success, one with $m + 3/5 < Q < m + 2/3$ of width $\Delta Q = 1/15$ straddling an 8th order resonance at $Q = m + 5/8$, and another one with $m - 0.1 < Q < m$ of width $\Delta Q = 0.1$ with no resonances of order less than 11 where m is an integer. Bearing this in mind, we choose $\delta Q_{max} = \Delta Q_{1max} = 0.02$. The free space around the crossing points was fixed at $\ell = 30$ m.

5.3. Results

Using these parameters, the stored current and the luminosity were calculated. The results are displayed in Figures 3 to 6. The energy γ is used as an abscissa. Each graph contains 4 sets of curves. One has the stored energy W of the circulating beam as a parameter, the other 3 sets the aperture radius b. For smaller values of γ, the longitudinal phase space density limit I_ℓ, becomes independent of b. This means that the momentum spread of the injected beam is large enough for stability and that no aperture dependent blow-up is required. It also means that the circulating current and the luminosity are determined by the density of the injected beam, and that an increase in aperture will not enhance the performance. With the parameters chosen the tightest performance limit is due to the transverse stability, over most of the energy range. For a given energy γ one could change the ratio $\delta Q_{max}/\Delta Q_{1max}$ while keeping their sum fixed to optimise the machine performance at a fixed aperture. For the same luminosity, a superconducting machine needs a smaller aperture.

Table I gives a more detailed comparison between two machines at 400 GeV. Their parameters were adjusted to obtain a luminosity of 10^{33} $cm^{-2}s^{-1}$, and round numbers for the aperture were used, leading to actual values for δQ and ΔQ_1 below the limits imposed. All the derived quantities can be obtained by using formulae in this lecture.

The phase space density D is the number of particles in the beam N divided by the product of all 3 emittances, horizontal, vertical, and longitudinal

$$D = N/(E_x E_y E_\ell) \tag{36}$$

They are defined as follows:

$$\begin{aligned} E_x &= 4\pi\beta\gamma\ \sigma_x{}^2/\beta_x \\ E_y &= 4\pi\beta\gamma\ \sigma_y{}^2/\beta_y \\ E_\ell &= 2\pi R\ \Delta(\beta\gamma) \end{aligned} \tag{37}$$

Because of the way E_ℓ is defined, it has the dimension meters and hence D has dimension m^{-3}. The popular definition of E_ℓ has the dimension of an action (eVs) which is obtained from my definition by multiplication with $m_p c$.

These parameter values were the starting point for more detailed lattice, beam dynamics and engineering studies which were carried out at CERN since 1973.[6] They came to an end in 1976 when we decided to turn our attention to Large Electron-Positron storage rings.

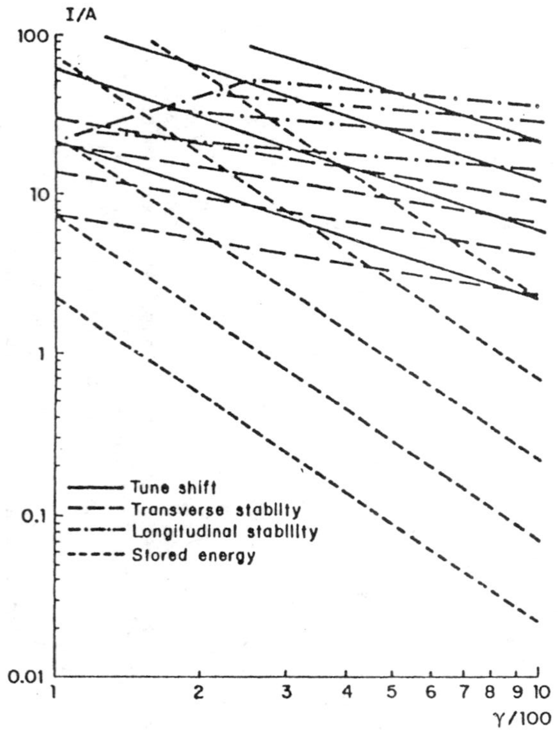

Fig. 3* - Circulating current I vs. energy γ.

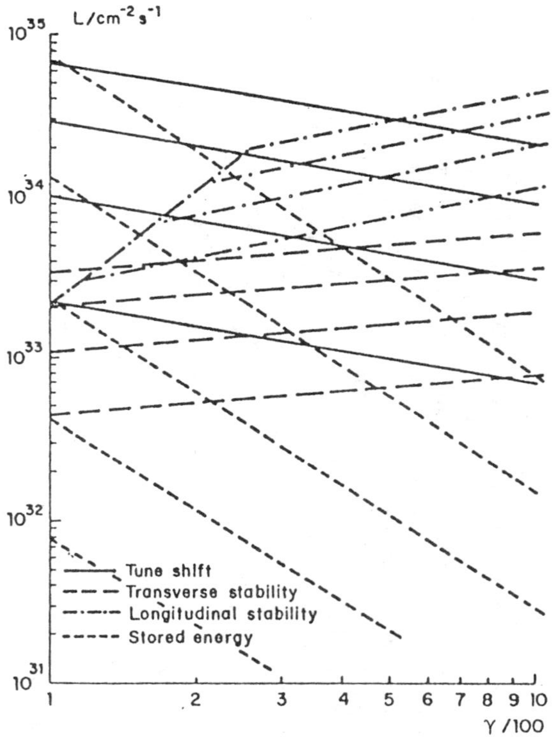

Fig. 4* - Luminosity/crossing L vs. energy γ.

*The parameters used in these figures are: $R/\rho = 1.3$, $B_M = 1.8T$, $B_0 = 0.6T$, $C_0 = 0.5$, $n-Q = 0.75$, $\ell = 30$ m, $\Delta Q_1 = 0.02$, $\Delta Q_2 = 0.005$, $\delta Q = 0.02$, $\Delta(\beta\gamma)_i = 0.013$, $E_t = 29.6\pi \times 10^{-6}$ m, $\sigma = 10^6$ A/Vm, $I_p = 170$ mA.
The curves for tune shift, transverse and longitudinal stability are drawn for an aperture radius b = 20, 30, 40, 50 mm. The curves for the stored energy in the beam are drawn for $W = 1, 10^{\frac{1}{2}}, 10, 10^{3/2}, 100$ MJ.

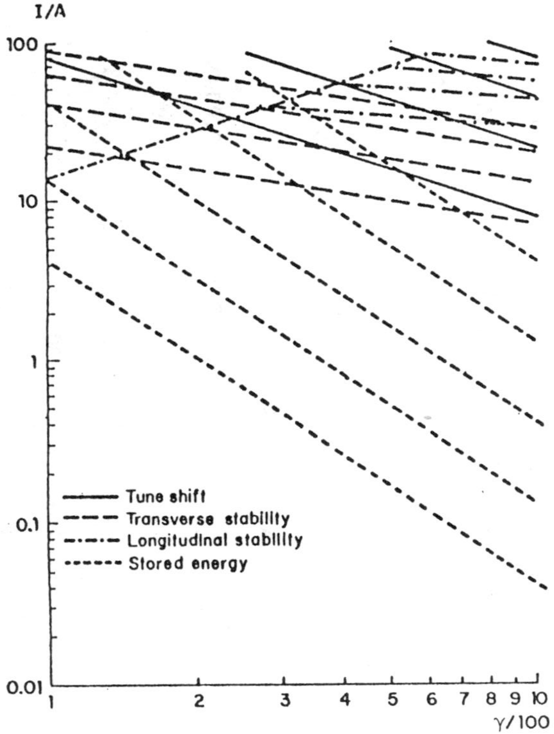

Fig. 5* – Circulating current I vs. energy γ.

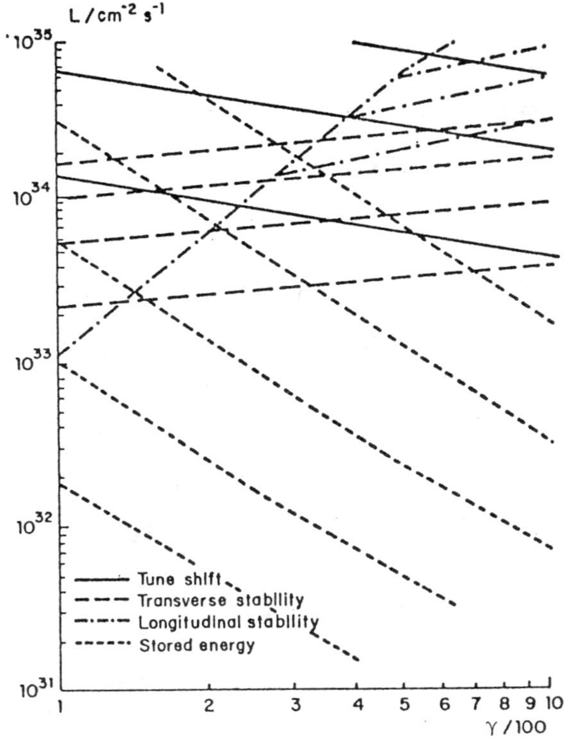

Fig. 6* – Luminosity/crossing L vs. energy γ.

*The parameters used in these figures are: $R/\rho = 1.6$, $B_M = 4T$, $B_Q = 2T$, $C_Q = 0.2$, $n-Q = 0.75$, $\ell = 30$ m, $\Delta Q_1 = 0.02$, $\Delta Q_2 = 0.005$, $\delta Q = 0.02$, $\Delta(\beta\gamma)_i = 0.013$, $E_t = 29.6\pi \times 10^{-6}$ m, $\sigma = 10^6$ A/Vm, $I_p = 70$ mA.
The curves for tune shift, transverse and longitudinal stability are drawn for an aperture radius b = 20, 30, 40, 50 mm. The curves for the stored energy in the beam are drawn for $W = 1, 10^{\frac{1}{2}}, 10, 10^{3/2}, 100$ MJ.

Table 1. Comparison of 400 GeV proton storage ring parameters using conventional and superconducting magnets

		conv.	s.c.
Maximum momentum	p/GeV	400	400
Bending field	B_M/T	1.8	4.0
Quadrupole field	B_Q/T	0.6	2.0
Circumference factor	R/ρ	1.3	1.6
Bending radius	ρ/m	741	333
Average radius	R/m	964	533
Quadrupole factor	C_Q	0.5	0.2
Aperture radius	b/mm	30	20
Wavelength	R/Q/m	36.4	17.9
Wavenumber	Q	26.5	29.8
Phase advance	u	π/2	π/2
Period length	L_p/m	57.2	28.1
Quadrupole length	ℓ_Q/m	3.3	1.1
Current/pulse	I_p/A	0.07	0.07
Density	D/m^{-3}	1.3×10^{20}	1.3×10^{20}
Normalized emittance	E_t/m	30×10^{-6}	30×10^{-6}
Stored current	I/A	5	5
Stored energy	W/MJ	40	22
Tune shift	ΔQ_1	0.015	0.008
Tune spread	δQ	0.007	0.005
Momentum spread	Δp/p	3.6×10^{-3}	4.2×10^{-3}
Free length at crossing	ℓ/m	30	30
Amplitude function	β_o/m	2.0	2.0
Beam-beam tune shift	ΔQ_2	0.005	0.005
Luminosity	L/$cm^{-2}s^{-1}$	10^{33}	10^{33}

REFERENCES

1. K. Johnsen, 9th International Conference on High Energy Accelerators, SLAC, 32 (1974).
2. K. Robinson and G. A. Voss, CEA-TM-149 (1965).
3. 400 x 400 GeV Isabelle Proton-Proton Colliding Beam Facility, BNL 50718 (1978).
4. Proton-Proton Colliding-Beam Storage Rings for the National Accelerator Laboratory, Design Study 1968.
5. A. N. Skrinskij et al., 8th International Conference on High Energy Accelerators, CERN, 72 (1971).
6. D. Blechschmidt et al., Nucl. Instr. Meth. 156, 375 (1978).
7. E. Keil, 9th International Conference on High Energy Accelerators, SLAC, 660 (1974).
8. E. Keil, C. Pellegrini and A. M. Sessler, Nucl. Instr. Meth. 118, 165 (1974).
9. E. Keil, Nucl. Instr. Meth. 113, 333 (1973).
10. P. J. Bryant, 9th International Conference on High Energy Accelerators, SLAC, 80 (1974), and P. J. Bryant et al., CERN-ISR-MA/75-54 (1975).
11. J. Borer et al., IEEE Trans. Nucl. Sci., NS-26, 3405 (1979).
12. L. J. Laslett, BNL 7534, 324 (1963).
13. L. J. Laslett, V. K. Neil, A. M. Sessler, Rev. Sci. Instr. 36, 436 (1965).
14. E. Peschardt, 11th International Conference on High Energy Accelerators, CERN, 506 (1980).
15. W. Schnell and B. Zotter, CERN/ISR-GS-RF/76-26 (1976).
16. V. K. Neil and A. M. Sessler, Rev. Sci. Instr. 36, 429 (1965).
17. V. K. Neil, D. L. Judd and L. J. Laslett, Rev. Sci. Instr. 32, 276, (1961).
18. E. Keil and W. Schnell, CERN/ISR-TH-RF/69-48 (1969).
19. P. Bramham et al., 9th International Conference on High Energy Accelerators, SLAC, 359 (1974).
20. H. Frischholz and W. Schnell, IEEE Trans. Nucl. Sci., NS-24, 1683 (1977).

Part II
The Design of e^+-e^- Storage Rings

1. BASIC DESIGN PRINCIPLES

The design procedure for e^+e^- storage rings presented in this lecture is based on a few basic principles which are enumerated below.

i) We assume that k bunches of electrons and k bunches of positrons circulate in the same magnet lattice in opposite directions. Both the electron and positron bunches are equally spaced around the machine circumference. They collide in 2k intersection (or crossing) points which are all equipped with low-β insertions.

ii) The crossing angle α, i.e. the angle between the electron and positron trajectories, vanishes at the crossing point.

iii) The density distribution of particles inside the electron and positron bunches is Gaussian in all three dimensions which are labelled x for horizontal, y for vertical transverse motion and s along the beam. The rms radii are called σ_x, σ_y and σ_s.

These assumptions are in agreement with those made for a whole series of e^+e^- storage rings built so far and in particular for the three largest storage rings; Table 1 gives a list of these machines and their main characteristics.[1,2,3]

In several machines it had been foreseen that the effective cross section of the colliding beams would be increased by colliding them at an angle α such that $\alpha \gg \sigma_y/\sigma_s$. This would have allowed the colliding current and hence the luminosity to be increased by the ratio of the effective beam areas with and without crossing angle. It turned out in practice that this factor could not be achieved.

Two other machines have used double-ring designs. In the DORIS storage ring[4] the electrons and positrons circulate in two different rings which intersect in two interaction regions. Both beams contain many bunches. In DCI[5] each ring contains one electron and one positron bunch which are brought into collision in two interaction regions such that forces due to the charges and currents of all four colliding bunches cancel to a very large extent. In both machines it was hoped to achieve much higher luminosities than could have been possible with single-ring machines. Unfortunately this hope has not been realized.

2. BEAM-BEAM COLLISIONS

In this section, the basic formulae are presented which describe the effects arising from the collisions of two bunched beams. Experimental observations from existing storage rings are summarized briefly and conclusions for the design parameters of future machines are drawn.

Table I. General parameters of CESR, PEP and PETRA

	CESR	PEP	PETRA	
Maximum energy	8	18	19	GeV
Circumference	768	2200	2304	m
Bending radius	88	165.5	192	m
Number of intersections	2	6	4	
Number of bunches per beam	1	3	2	
Horizontal betatron wave number	9.4	21.3	25.2	
Vertical betatron wave number	9.2	18.2	23.3	
RF frequency	500	353.21	499.67	MHz
Harmonic number	1281	2592	3840	
Length of accelerating structure	8.4	51	90	m
Installed RF power	1.6	6	4.8	MeV
Synchrotron energy loss/turn	4.9	26[1]	58.5	MV
Peak RF voltage/turn	12	50[1]	105	MV

[1]at 15 GeV

2.1. Luminosity

The luminosity L achieved in one crossing point when two beams, each containing a total of N particles in k bunches, collide is given by

$$L = \frac{N^2 f}{4\pi k \sigma_x \sigma_y} \tag{1}$$

Here f is the revolution frequency in the storage ring. The assumption has been made that σ_x and σ_y are constant throughout the crossing region. This requires that $\sigma_s < \beta_y << \beta_x$. When this condition is not satisfied, the variation of the vertical beam size with s due to the variation of β_y has to be taken into account:

$$\sigma_y(s) = \sigma_y(o) \left[\beta_y(s)/\beta_y(o)\right]^{\frac{1}{2}} = \sigma_y(o) \left[1 + \left(s/\beta_y(o)\right)^2\right]^{\frac{1}{2}} \tag{2}$$

Note that in writing down (2) the additional assumption has been made that the beam size is only determined by betatron oscillations. This implies that the dispersion vanishes in the whole crossing region, D = 0. We further assume that it is still possible to neglect the variation of $\sigma_x(s)$ since $\sigma_s << \beta_x$. Under these assumptions the correction factor for the luminosity[6] can be written as follows:

$$F = (a/\pi)^{\frac{1}{2}} \exp(a/2)\, K_0(a/2) \tag{3}$$

where $a = (\beta_y/\sigma_s)^2$.

2.2. Beam-beam tune shift

For particles with betatron amplitudes x and y small compared to the rms beam radii σ_x and σ_y, the linear tune shifts due to the beam-beam interaction in the crossing point can be written as follows, assuming as in (1) that $\sigma_y < \beta_y << \beta_x$:

$$\Delta Q_x = \frac{N r_e \beta_x}{2\pi k \gamma (\sigma_x + \sigma_y) \sigma_x} \tag{4}$$

$$\Delta Q_y = \frac{N r_e \beta_y}{2\pi k \gamma (\sigma_x + \sigma_y) \sigma_y} \tag{5}$$

Here $r_e = 2.817938 \times 10^{-15}$ m is the classical electron radius and γ is the usual relativistic factor.

It is customary to eliminate one power of N from the luminosity equation (1) by using (5). In the limit $\sigma_y << \sigma_x$ one thus obtains

$$L = \frac{N f \gamma \Delta Q_y}{2\, r_e\, \beta_y} = \frac{I\, \gamma\, \Delta Q_y}{2 e r_e \beta_y} \tag{6}$$

This formula can be interpreted in two ways. If it is possible to adjust the beam radii σ_x and σ_y of the two beams in collision such that at a prescribed particle number N (or a prescribed current I), the beam-beam tune shift limit ΔQ_y is just reached then the luminosity will be given by (6). This holds whether or not there is beam blow-up due to the beam-beam effect if values of σ_x and σ_y including the blow-up obtained by simulation are used. This interpretation applies mostly to the design stage of a future machine. The second interpretation of (6) may be used in an operating machine to calculate the beam-beam tune shift from quantities which can all be measured directly. A particularly convenient numerical form of (6) is

$$\Delta Q_y = 4.6 \, \frac{\beta_y(\mathrm{m}) \; L \; (10^{30}\mathrm{cm}^{-2}\mathrm{s}^{-1})}{I \; (\mathrm{mA}) \; E \; (\mathrm{GeV})} \qquad (7)$$

The effect of a variation of the vertical beam size σ_y (s) as given in (2) can again be taken into account by applying a correction factor[7] to (5):

$$F_{BB} = \left(\frac{a}{2\pi}\right)^{\frac{1}{2}} \; \exp(a) \; \left[K_1(a) + K_0(a)\right] \qquad (8)$$

The small-amplitude tune shifts ΔQ_x and ΔQ_y are the maximum values reached in the beam-beam collisions. The distribution functions of ΔQ_x and ΔQ_y as shown in Figs. 1 and 2 are obtained by a Monte-Carlo calculation.

2.3. Experimental observations in existing machines

It is appropriate to review the experimental observations in existing machines in order to see how their design procedures have to be modified before they are applied to yet another generation of e^+e^- storage rings.

The luminosities achieved in the latest generation of e^+e^- storage rings, CESR, PEP and PETRA, fall short of the design values. Part, but not all of this, may be attributed to the beam-beam limit being lower than expected. The discussion is restricted to these three machines because several reviews of the behavious in the older machines are already available.[8,9,10]

Table II shows a comparison between the most important machine parameters in the design proposal (D) and actually measured (M). In order to obtain comparable figures it was necessary to change the energy occasionally, and to scale the other parameters appropriately. The table describes the machine performance at the end of January 1981. Since that time, mini-β insertions have been installed in PETRA,[11] and the luminosity in PEP has been doubled.[12]

The following observations can be made from a close inspection of Table II.

i) The horizontal emittance E_x is always smaller than that used in the design, the factor varies between 6 and about 1.

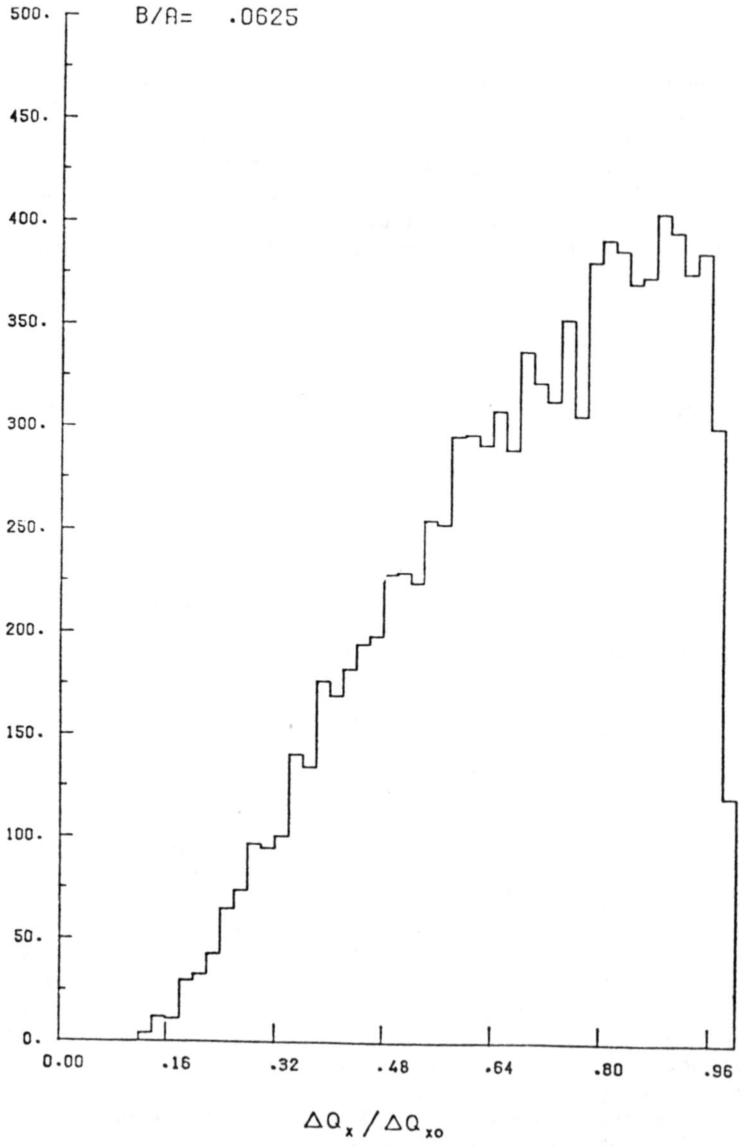

Fig. 1 - Distribution function of the horizontal beam-beam tune shift $\Delta Q_x/\Delta Q_{xo}$ for an axis ratio $\sigma_x/\sigma_y = 16$; ΔQ_{xo} is the horizontal beam-beam tune shift at vanishing betatron amplitudes.

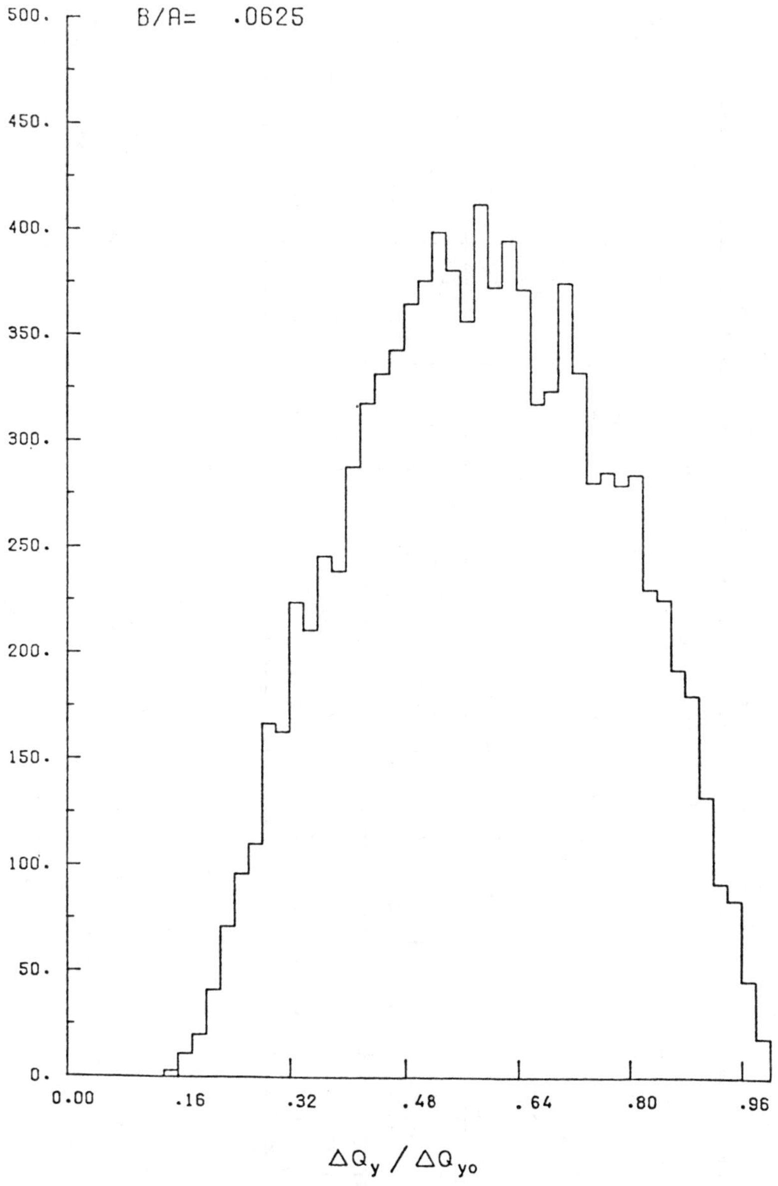

Fig. 2 - Distribution function of the vertical beam-beam tune shift $\Delta Q_y/\Delta Q_{yo}$ for an axis ratio σ_x/σ_y = 16; ΔQ_{yo} is the vertical beam-beam tune shift at vanishing betatron amplitudes.

Table II. Summary of CESR, PEP and PETRA performances

D(M) indicates the design (measured) value.

Machine		GeV	E_x μm	β_y m	k	I/k mA	L 10^{30} $cm^{-2}s^{-1}$	σ_x mm	σ_y mm	ΔQ_x	ΔQ_y
CESR	D	5.0	1.8	0.10	1	84	59	1.40	0.09	0.061	0.06
	M	5.3	0.28	0.13	1	6.6	1.5	0.86	0.032	0.019	0.025
PEP	D	14.5	0.12	0.11	3	11	60	0.67	0.02	0.06	0.06
	M	14.5	0.11	0.23	3	5	3.8	0.7	0.06	0.03	0.018
PETRA	D	14	0.21	0.15	4	17	86	0.79	0.039	0.06	0.06
	M	14	0.10	0.21	2	3.3	3.3	0.57	0.027	0.024	0.034

ii) The vertical amplitude function β_y is always larger than in the designs, the increase varies between 30 and 110%.

iii) CESR and PEP are operated with the designed number of bunches, PETRA is not. A few attempts at operating PETRA with k = 4 did not achieve a higher luminosity than operation with k = 2. Since there are only four equidistant experiments in PETRA the incentive to operate it with k = 4 is not very high.

iv) The beam-beam tune shifts are smaller than the design value $\Delta Q = 0.06$ by about a factor of two.

v) The circulating current per bunch is always smaller than in the design, but the ratio is about $E_x \Delta Q$ as one would expect.

vi) The luminosity is always much smaller than in the design because of all the above reasons combined. The ratio is about $\Delta Q^2 k E_x / \beta_y$ as expected.

vii) The horizontal beam size σ_x behaves as expected.

viii) The vertical beam size σ_y varies significantly; it is smaller than the design figure in CESR which is fortunate for its performance; it is as expected from the design in PETRA; and it is larger than the design figure in PEP.

The only conclusions that emerge clearly from this discussion are that the beam-beam tune shift limit should be lowered to $\Delta Q = 0.03$, and that ratios of $\ell_x/\beta_y > 50$ should be avoided where ℓ_x is half the free space around the crossing points. Reducing ℓ_x in a mini-β insertion is a better way of achieving the latter than increasing β_y.

The data on E_x and σ_y, taken together with the experimental observation that attempts at increasing them led to intolerable reductions in lifetime, suggest that a larger ratio between radial and vertical apertures and beam radii might improve the performance. On the whole, a better understanding of the beam dynamics in the existing machines would help in the design of future machines.

2.4. Computer simulation

Taking an experimental figure for the beam-beam limit from CESR, PEP or PETRA and applying it to a new machine ten times their size needs justification, in particular since this process has failed in the extrapolation from ADONE and SPEAR to the above three machines. In the absence of analytical scaling laws, computer simulation is the only choice. A program may safely be used to predict the performance of future machines if it agrees with the observations in at least one existing machine.

Computer simulation has been attempted for a long time, but only recently in quantitative agreement with experimental data.[13,14,15] The reasons for this success are more powerful computers and the smaller number of turns in the larger storage rings which must be included in a simulation to cover a few damping times. In order to obtain quantitative agreement with experimental data many more phenomena must be included than just the transverse kicks due to the beam-beam forces, linear transformations through the machine arcs, radiation damping and quantum excitation. This includes the synchrotron motion and its coupling mechanisms to the transverse motion, e.g. dispersion in RF cavities and the betatron phase differences at the crossing points due to the longitudinal position, errors in the machine parameters due to closed-orbit distortions, e.g. phase differences along the arcs, differences in amplitude functions and dispersion, and small beam offsets at the crossing points.

Another recent innovation is the simulation of the collision between two "strong" beams, which includes the action of the first beam on the second as well as that of the second beam on the first, in contrast to earlier simulations which treated the case of a "weak" beam colliding with an unperturbed "strong" beam. The results of such a simulation of two strong beams for LEP[16] are shown in Fig. 3.

3. SIZE OF e^+e^- STORAGE RINGS

The aim of this section is finding a method to determine the size of an e^+e^- storage ring. As can be seen from eq. (7) the machine size is irrelevant for the expected luminosity provided all the parameters on the right-hand side are independent of the machine radius.

Clearly, the cost of the civil engineering for the machine tunnel and also of the machine components inside the tunnel increase with the machine size. This is a good reason for keeping the size small. On the other hand, the energy loss due to synchrotron oscillations which must be made up for by an RF system is inversely proportional to the machine radius, as will be shown in the next section. Hence the size and cost of the RF system will be inversely proportional to the machine radius, and the optimum machine size will be the result of a balance between these two effects.

3.1. Synchrotron radiation and its effects on beam dynamics

The synchrotron radiation emitted by electrons and positrons when they are transversely accelerated in a magnetic field has a very profound effect on the beam dynamics in an e^+e^- storage ring.

To keep the formulae simple, an isomagnetic machine is assumed with a constant bending radius ρ in all the dipoles. It is also assumed that the reduced betatron wavelength can be approximated by R/Q and the dispersion by R/Q^2. Here R is the average radius of the machine arcs, and Q is the contribution of the arcs to the tune.

The energy loss per turn U_s is given by

$$U_s = \frac{4\pi r_e E_e}{3} \frac{\beta^3 \gamma^4}{\rho} \tag{9}$$

Here E_e is the rest energy of the electron. The strong energy dependence of U_s has the consequence that the horizontal and vertical betatron oscillations and the synchrotron oscillations are all damped when the machine lattice is chosen correctly. The damping rate for the uth mode is

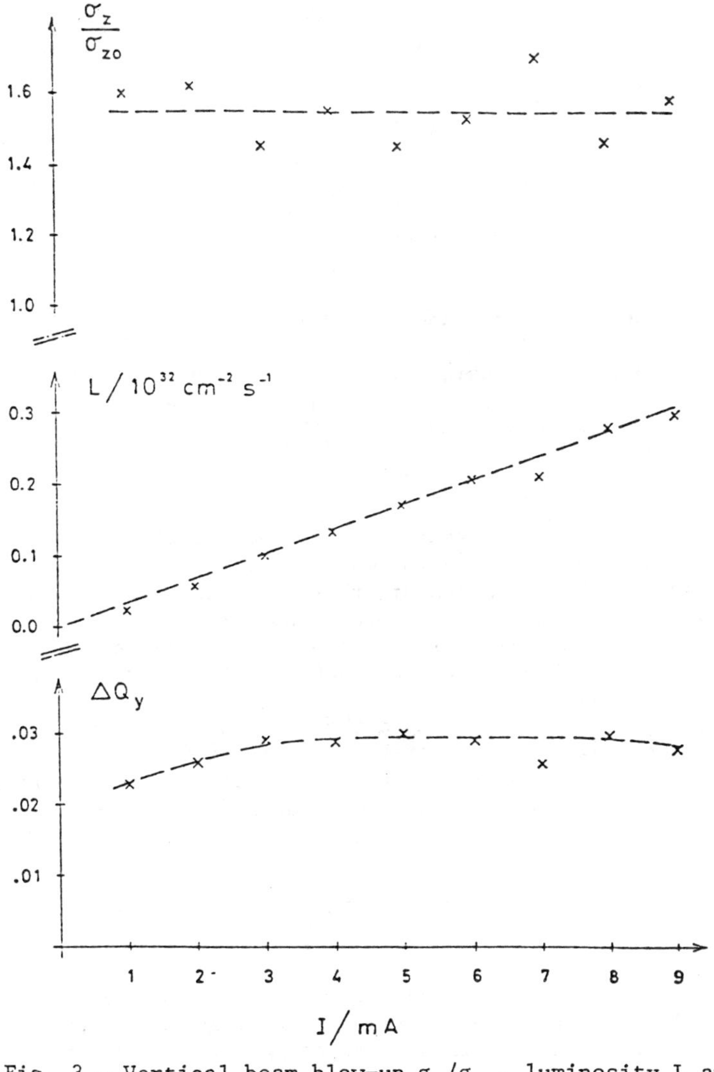

Fig. 3 - Vertical beam blow-up σ_z/σ_{zo}, luminosity L and vertical beam-beam tune shift ΔQ_y vs. circulating current I. The parameters are for LEP at 51.5 GeV and four bunches in each beam. The unperturbed beam-beam tune shift ΔQ_y = 0.06 and the unperturbed axis ratio σ_{xo}/σ_{yo} = 16 are held constant by varying the horizontal rms beam radius σ_{xo} in proportion to I. Typical errors are also included.

$$\tau_u^{-1} = \frac{1}{2}\frac{U_s f J_u}{E} \tag{10}$$

Here J_u is the damping partition number for the uth mode. The sum of the three J_u's is fixed for any machine:

$$J_x + J_y + J_s = 4 \tag{11}$$

If the machine has a median plane, the following stronger relations hold:

$$J_x + J_s = 3 \quad \text{and} \quad J_y = 1 \tag{12}$$

For an orbit in the machine on which all the bending happens in homogeneous-field dipoles and all the focusing in separated quadrupoles, the partition numbers are:

$$J_x \simeq 1 \qquad J_y = 1 \qquad J_s = 3 - J_x \tag{13}$$

These stringent conditions are only satisfied for the equilibrium orbit in an ideal separated-function machine without errors. Even in such a machine the off-momentum orbits are bent and focused in the quadrupoles, and hence only the weaker conservation law (12) applies. In a real separated-function machine with closed-orbit distortions, no orbit is in a plane and hence only the weakest conservation law (11) applies.

All modes of oscillation are damped when all three J_u's are positive. From the above it follows that this condition is satisfied for only a small momentum bite, the damping aperture, determined by the rate of change of J_x with the momentum error $\Delta p/p$.

$$dJ_x/dp/p = -\frac{4\ell_B}{\ell_Q}\,\frac{2 + \frac{1}{2}\sin^2\frac{\mu}{2}}{\sin^2\frac{\mu}{2}} \tag{14}$$

This formula applies to a machine consisting of simple FODO cells with two quadrupoles of length ℓ_Q and two bending magnets of length ℓ_B, and a phase advance μ in a period, i.e. the effect of the insertions is ignored.

In a real machine damping of all three modes of oscillation is only assured if the closed-orbit distortions remain below some small value.

The sudden energy changes of the electrons and positrons due to the emission of synchrotron radiation quanta cause a diffusion-like growth of the energy spread which together with the damping discussed above leads to a relative equilibrium energy spread σ_e:

$$\sigma_e^2 = C_q\gamma^2/(\rho J_s) \tag{15}$$

The constant $C_q = 55\hbar/(32\sqrt{3}mc) = 3.84 \times 10^{-13}$ m is equal to the Compton wavelength of the electron, apart from the numerical factor,

C_q. When the synchrotron radiation photons are emitted at a position where the dispersion does not vanish they cause sudden changes of the equilibrium orbit while the actual particle trajectory is continuous. This causes a diffusion-like growth of the betatron oscillation amplitude which is balanced by the radiation damping. The resulting equilibrium horizontal beam emittance E_x is

$$E_x = \sigma_x^2/\beta_x = \frac{C_q \gamma^2}{J_x \rho} \frac{R}{Q^3} \tag{16}$$

Since J_s appears in (15) and J_x appears in (16), the variation of J_x with $\Delta p/p$ can be used to adjust σ_e and E_x. This has been done[17] in PETRA to increase the bunch length at injection to overcome a transverse single-beam instability, and to increase the emittance of the colliding beams to gain luminosity. The required change in equilibrium momentum is obtained by a small change in the RF frequency.

The synchrotron radiation power for both beams P, assuming equal currents, is obtained by multiplying (9) with twice the circulating current.

$$P = 2IU_s = \frac{8\pi}{3} \frac{E_e r_e}{e} \gamma^4 I \rho^{-1} \tag{17}$$

Using (17) to eliminate I from (6) yields a relation between luminosity L and $P\rho$.

$$L = \frac{3}{16\pi} \frac{\Delta Q \rho P}{r_e^2 E_e \gamma^3 \beta_y} \tag{18}$$

In order to keep the luminosity constant, it is necessary to scale the product $P\rho$ like γ^3, if the other parameters are left unchanged. For the design of a storage ring, separate scaling laws for P and ρ would be more useful.

The critical energy of the synchrotron radiation photons E_c, defined such that half the radiated energy is in photons with energies smaller than E_c, is given by

$$E_c = \frac{3}{2} \hbar c \gamma^3 \rho^{-1} \tag{19}$$

The critical energy is an important design parameter of e^+e^- storage rings since MW's of synchrotron radiation power are generated. Because of the risk of radiation damage it must be absorbed in materials with high radiation resistance such as pure metals, and not in more easily damaged components such as plastic water hoses and cable insulation. In machines such as CESR, PEP and PETRA the aluminum vacuum chamber is therefore surrounded by a lead shield. Even more lead thickness is required for LEP since the critical energy is higher and since the photon absorption cross-section has a minimum around 0.5 MeV where practically only the Compton effect contributes to it.

3.2. Power dissipation in RF cavities

Up to now, the most economic way of making up the synchrotron radiation losses has been RF systems consisting of copper or aluminum cavities fed by RF power sources in CW operation. A superconducting RF cavity has been tested in the Cornell synchrotron,[18] and tests are foreseen in several storage rings. The use of pulsed RF power sources[19] has been studied. The LEP design foresees a combination of accelerating and storage cavities, both made of copper, in order to reduce the RF losses by almost a factor of two.[20] For the discussion below, room-temperature RF cavities and CW power sources are assumed.

The peak voltage V supplied by the RF system must be larger than the losses U_s due to synchrotron radiation in order to have RF buckets whose energy width is much larger than the energy spread in the beam. Hence, the probability is low that an electron jumps out of the bucket and gets out of synchronism with the RF accelerating field when it emits a very high energy synchrotron radiation photon. This results in a quantum lifetime of many hours. Introducing the stable phase angle ϕ_s, counted from the last zero crossing of the RF voltage before the bunch passage we have

$$V = U_s/\sin\phi_s \tag{20}$$

The RF power P_d necessary to build up this voltage is given by:

$$P_d = V^2/(ZL_c) \tag{21}$$

Here, Z is the shunt impedance per unit length, a parameter determined by the shape and material of the cavities, and L_c is the total active length. Note that the definition of Z in (21) differs from the convention of electrical engineering by a factor of two. Combining (9) and (21), the following scaling law is obtained:

$$P_d L_c \rho^2 \simeq E^8 \tag{22}$$

It is logical to vary P_d and L_c in proportion because this just means that in scaling from one machine to another one keeps the dissipated RF power per unit length constant. Similarly, varying L_c and ρ in proportion implies "scaling machines" in which the fraction of circumference occupied by the RF system remains the same. In order to be compatible with (22) these two scaling laws imply:

$$P_d \simeq L_c \simeq \rho \simeq E^2 \tag{23}$$

In this discussion, the synchrotron radiation power P has been neglected. This is a good approximation for machines much larger than PEP or PETRA.

3.3. Determination of the storage ring size by cost optimization

After the preparatory sections on synchrotron radiation and RF power, the problem of determining the storage ring size can be tackled by cost optimization as proposed by B. Richter.[21] The basic assumption is that the total construction cost and the capitalized operating cost can be grouped under a small group of headings and expressed in the form of unit prices multiplied by appropriate factors such that the unit prices have little variation over the range of parameters considered. Quantitatively, this is expressed as follows:

$$\mathrm{Cost} = 2\pi k_1\rho + k_2(P + P_d) + k_3L_c + k_4(P + P_d)/\varepsilon + k_5(P + P_d) + k_6 \qquad (24)$$

The coefficient k_1 describes the unit cost of the machine arcs and includes all components in the arcs such as the tunnel, the magnet and vacuum systems, beam observation etc. The coefficient k_2 describes the unit cost of the RF power installation, i.e. the power supplies, RF generators (klystrons), waveguides etc., and the buildings to house them. The coefficient k_3 describes the RF cavity system, including the cavities proper, their control and vacuum systems and the piece of tunnel housing them. The coefficient k_4 describes the capitalized power from the mains. It includes an efficiency $\varepsilon < 1$ between mains and RF power and an assumption about the total operating time. Similarly, the coefficient k_5 describes the capitalized klystron replacement. In order to obtain it one needs an estimate of the klystron lifetime. The last coefficient k_6 describes that part of the storage ring which is not a strong function of the four variables ρ, L_c, P_b and P_d, e.g. the experimental areas. The four variables are limited by two constraints (17) and (18) and by (20) and (21). By substituting them into (24) an equation in the two unknowns L_c and ρ is obtained. The cost optimum is found by imposing the condition that the two partial derivatives with respect to L_c and ρ vanish.

The results for L_c and ρ are quite lengthy but easy to derive and hence not given here. It can be shown in general that L_c is proportional to V at all energies. This means that the RF system is optimized at a fixed voltage gradient and hence at a fixed power dissipation per unit length. In the limit $P \ll P_d$ the scaling law (23) is found.

For numerical work, the unit prices k_i must be known. Table III shows two columns, one derived from PEP costs by B. Richter,[21] and a second one which was popular during the early stages of the LEP study.[22]

Table III. Unit prices for cost optimization

Item		$	SF
Arcs	k_1	12.8 M	36.2 M km^{-1}
RF power installation	k_2	0.58 M	2.3 M MW^{-1}
RF cavities	k_3	81.00 M	126.0 M km^{-1}
RF power	k_4	30	55 MWh^{-1}
Klystron replacement	k_5	-	53 MWh^{-1}

Parameters for storage rings with design energies between 70 and 100 GeV are shown in Fig. 4. Assuming that the average RF power is just the sum $(P + P_d)$ is an oversimplification, good enough for a cost optimization, because it neglects transient beam loading and higher-mode losses. These phenomena will be included in the RF system design later on.

4. DESIGN OF e^+e^- STORAGE RING LATTICES

The starting points for the design of a storage ring lattice are the operating energy E, the bending radius ρ and the active cavity length L_c obtained above by cost optimization. In this process, a few parameters related to the design performance, such as the luminosity L, the number of bunches k, the circulating current I, the beam-beam tune shift ΔQ and the vertical amplitude function β_y at the crossing points, have already been chosen. They are related by eq. (4) to (6).

Finally, a lattice layout must be adopted. The one assumed here is schematically shown in Fig. 5 for half the space between two crossing points. Starting at a crossing point, there is one half of a low-β insertion, consisting of a few quadrupoles and straight sections. This is followed by an RF section containing mainly quadrupoles and RF cavities. Next comes a dispersion suppressor consisting of quadrupoles, dipoles and strategically placed straight sections. The rest of the lattice is occupied by closely-packed regular lattice cells, containing sextupoles, quadrupoles, dipoles and very short straight sections.

The amplitude functions β_x and β_y are assumed to be periodic but not necessarily the same in the RF section as in the regular cells. They are matched between these sections by the dispersion suppressor. The low-β insertion matches them from the values in the RF section to the small values at the crossing points. The dispersion D_x vanishes in the low-β insertion and the RF section; it is matched to its regular lattice value by the dispersion suppressor as implied by its name. There are several reasons for this choice:

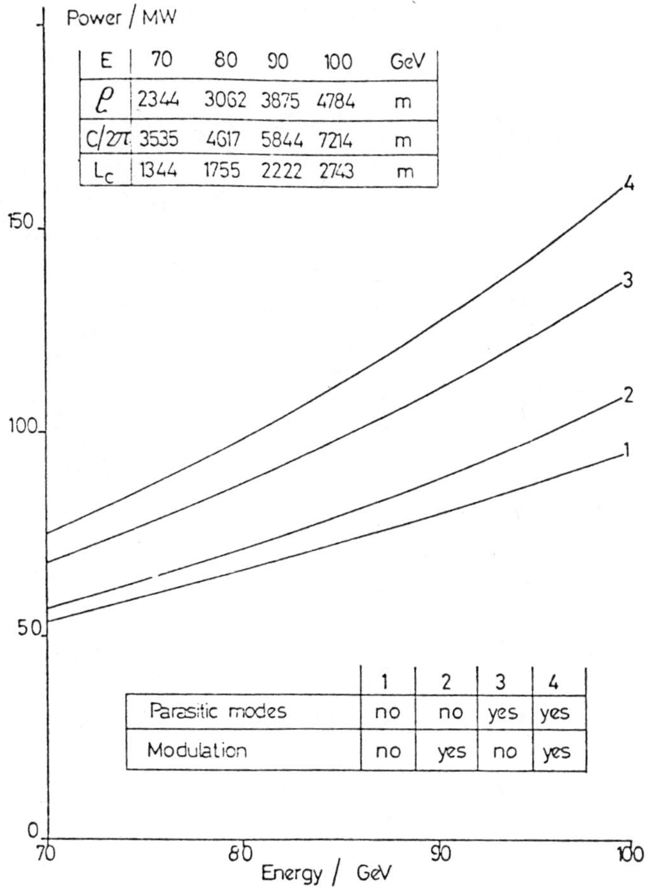

Fig. 4 - RF generator power vs. energy for machines whose size scales like E^2, with and without excitation of parasitic modes, and with and without transient beam loading taken into account. The parameters of the machines are k = 4, β_y = 0.1 m, β_x = 1.6 m, ΔQ= 0.03, L = 2.5 x 10^{31} $cm^{-2}s^{-1}$.

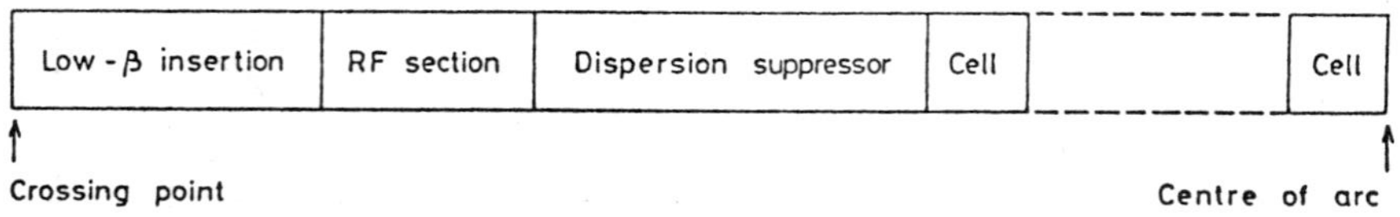

Fig. 5 - Schematic layout of half an octant of machine lattice, starting at a crossing point.

i) The energy spread does not influence the beam size at the crossing points.
ii) Experiments in VEP-2M[23] and computer simulations[14] show that the beam-beam limit ΔQ has a maximum at $D_x = 0$ and decreases when $D_x \neq 0$.
iii) One driving mechanism for synchro-betatron resonances,[24] dispersion in the RF cavities, is absent.

4.1. Choice of the tune

The choice of the tune is straightforward when the storage ring is to be operated at a fixed energy and a fixed current. In this case the tune can be used to adjust the beam size such that the beam-beam limit ΔQ is just reached at the operating energy and the operating current. Too large a beam size would mean that the available current is not used to achieve the maximum luminosity. When the beam size is too small, the beam-beam limit is reached at a lower current than the design current. The desired horizontal beam emittance E_x is already implied in (4) which for $\sigma_x >> \sigma_y$ can be written as follows:

$$E_x = \frac{\sigma_x^{\,2}}{\beta_x} = \frac{Nr_e}{2\pi k\gamma\Delta Q} = \frac{r_e^2\beta_y\ L}{\pi f\gamma^2 k(\Delta Q)^2} \qquad (25)$$

The horizontal emittance E_x obtained in a lattice due to quantum excitation was given in (16). In order to have equal emittances in (16) and (25) the contribution Q of the regular lattice to the tune must satisfy the relation:

$$Q^3 = \frac{2\pi C_q}{r_e}\ \frac{k\gamma^3\Delta Q}{J_x\ N}\ \frac{R}{\rho} = \frac{C_q c}{2r_e^{\,2}}\ \frac{k\gamma^4\Delta Q^2}{J_x\rho\ \beta_y L} \qquad (26)$$

In this simplified presentation the contribution of the dispersion suppressors to the quantum excitation is neglected (the low-β insertion and the RF section do not contribute to the quantum excitation since $D_x = 0$). It follows from (26) that Q is a function of γ unless N varies like γ^3 or, equivalently, L varies like γ^4, when all the other parameters are fixed. This is called their natural variation.

This steep luminosity variation is too restrictive on the energy range in which the storage ring operates at good luminosity. In order to expand the energy range other methods of beam-size adjustment must be introduced.

The most obvious solution would be operating the storage ring at different tunes, depending on the energy. The tune cannot be a continuous variable because of non-linear resonances. Still, a given lattice can be matched in principle to a large number of tunes. A table containing 33 lattices was published[25] before PETRA came into operation. However, it turned out in practice that setting up PETRA at one tune and bringing it to an acceptable performance level is so lengthy and laborious that it was not repeated several times. Hence, in the design of a future machine only very few different tunes should be assumed. In this manner, the performance is optimized at the energies corresponding to those tunes. Other methods of beam-size adjustment must be used for energies falling between those values. They will be discussed later.

4.2. Lattice period design

The schematic layout of a lattice period is shown in Fig. 6. A period contains two quadrupoles QF and QD, focusing and defocusing in the radial plane, two sextupoles SF and SD next to the quadrupoles for chromaticity correction, dipoles B and a few straight sections. The betatron wavelength λ is approximately known from the tune Q just calculated:

$$\lambda = \frac{R}{\rho} \frac{2\pi\rho}{Q} \tag{27}$$

Here $R/\rho > 1$ is the ratio between period length and the length of the bending magnets in a period. The next parameters to be fixed are the phase advance μ and the period length L_p which related by:

$$L_p = \frac{\lambda\mu}{2\pi} = \frac{R}{\rho} \frac{\mu\rho}{Q} \tag{28}$$

Because of synchrotron radiation as much space as possible should be occupied by dipoles, i.e. R/ρ should be close to unity. At constant aperture the quadrupole length is almost independent of μ because λ is held constant. Hence the best R/ρ is obtained by making μ and L_p high. The theory of AG focusing shows that $0 < \mu < \pi$.

Phase advances in the neighborhood of $\pi/4$ have been adopted for SPEAR, CESR, PEP and PETRA. It is intended to operate LEP with phase advances close to either $\pi/3$ or $\pi/2$ which is more in keeping with the choices for the large proton synchrotrons at CERN and FNAL. All the operating machines have phase advances which are not simple fractions of 2π. It has been shown[26] that some of the undesirable side effects of the chromaticity correction by sextupoles, such as the excitation of systematic third and fourth order resonances, can be avoided by grouping SF or SD sextupoles in different cells into pairs with an odd number of half betatron wavelengths between them and exciting them together. This obviously implies phase advances per cell which are simple rational fractions of 2π.

At this point all parameters needed for the calculation of the lattice parameters such as the amplitude functions β and the dispersion D are available. In thin lens approximation this can be done using the formulae given by Collins.[27] It is hardly more difficult with lenses of finite length.

The half apertures A_x and A_y in the lattice cell are determined by the beam sizes σ_x and σ_y, multiplied by aperture allowance factors F_x and F_y. Sometimes closed orbit allowances C_x and C_y are added:

$$A_x = \sigma_x F_x + C_x \qquad A_y = \sigma_y F_y + C_y \tag{29}$$

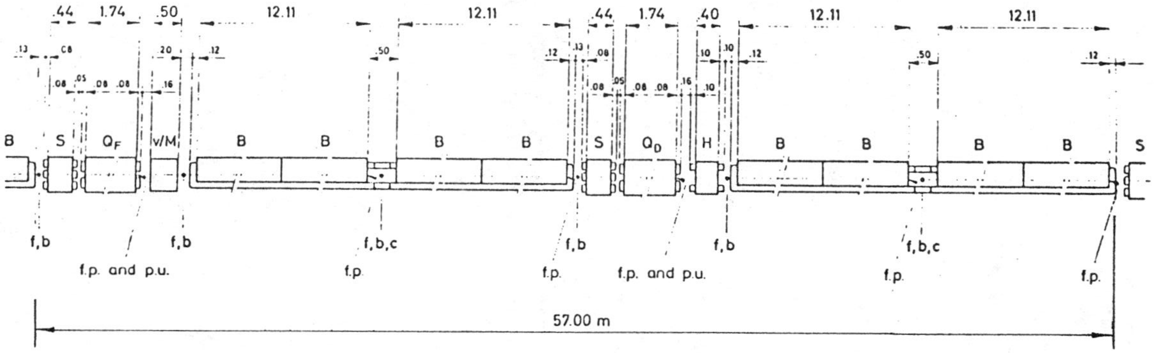

B	- bending magnet	f	- flange
Q_F, Q_D	- quadrupoles	b	- bellows
S	- sextupole	c	- connections
H	- horizontal - field dipole	f.p.	- fixed point
M	- multipole	p.u.	- pick - up
		v	- valve

Fig. 6 - Schematic layout of a lattice cell in LEP-70.

The horizontal beam size σ_x follows from the beam emittance E_x (25) the momentum spread σ_e (15) and the cell parameters β_x and D_x taken at an F quadrupole where they are largest:

$$\sigma_x = \left(E_x\beta_x + (\sigma_e D_x)^2\right)^{\frac{1}{2}} \tag{30}$$

The vertical beam size σ_y vanishes in a perfect machines with a median plane, and without emittance exchange due to coupling between horizontal and vertical betatron motion and without beam-beam collisions. In order to be safe when coupling occurs accidentally, the recent machines have been designed for a vertical emittance half as large as the horizontal one, leading to

$$\sigma_y = \left(\frac{1}{2} E_x\beta_y\right)^{\frac{1}{2}} \tag{31}$$

Here, β_y is taken at a D quadrupole where it is largest.

The aperture allowancee factors F_x and F_y are typically around 10. This may seem generous remembering that factors between 6 and 7 are adequate to ensure a quantum lifetime of the order of a day. On the other hand, a reduction of the beam-beam limit ΔQ was observed[28] when the vertical aperture was artificially reduced by scrapers. Computer simulations have shown[29] that, close to the beam-beam limit, the vertical distribution function changes and develops tails which are much higher than those expected for a Gaussian with the same standard deviation. This invalidates the calculation of the quantum lifetime based on a Gaussian distribution and is another reason to make F_y not too small. The closed orbit allowances are at most about 10 to 20 mm.

The vacuum chamber cross-section is kept constant through a lattice period. This allows the chamber to be fabricated by extrusion out of aluminum. It also avoids collective phenomena due to electromagnetic fields arising when the bunches pass variations of the vacuum chamber cross-section. Therefore the chamber half radii are given by A_x and A_y calculated above.

The lattice cell design is completed by determining the lengths of the quadrupoles. Their pole tips must fit around the vacuum chamber; their focal length is known from L_p and μ. Their length ℓ_Q is determined by the "poletip field" B_Q and the energy E as discussed in the p-p storage ring lecture. For the lattice quadrupoles which must be mass-produced cheaply, an economic optimum for B_Q is about 1 T. Up to this field, quadrupoles can be built with straight poles and simple coils without too much saturation where the poles join the yoke. This argument will be developed further while discussing the insertion design. Similar arguments apply to the choice of the sextupole length.

Adding up all the lengths of the magnetic elements in a lattice cell, and the lengths of all the straight sections, we obtain the final values of L_p and hence of R/ρ.

4.3. Dispersion suppressors

Dispersion suppressors are inserted between the regular lattice of the arcs and the RF section in order to match the dispersion from its value in the arc to zero. They all work by launching a forced oscillation of the dispersion from its normal value to zero, by adjusting the dispersion in the dispersion suppressor to about half the lattice value. This can be achieved by reducing the bending at constant focusing, by increasing the focusing at constant bending, or by hybrid schemes. In the first case, the dispersion suppressor consists of two lattice cells with standard focusing, but with the bending angles changed to ϕ_1 and ϕ_2, as shown in Fig. 7, which are only functions of the phase advance μ and given by

$$\phi_1 = \phi\left(1 - \frac{1}{4\sin^2\frac{\mu}{2}}\right) \qquad \phi_2 = \frac{\phi}{4\sin^2\frac{\mu}{2}} \tag{32}$$

Here ϕ is the bending angle in a normal cell. It can be seen that $\phi_1 + \phi_2 = \phi$, i.e. half the normal value.

The angles ϕ_1 and ϕ_2 take particularly simple values
when $\mu = \pi/2$ with $\phi_1 = \phi_2 = \phi/2$
and when $\mu = \pi/3$ with $\phi_1 = 0$, $\phi_2 = \phi$.

The behavior of the dispersion is also shown in Fig. 7. In normalized betatron phase space where betatron oscillations are represented by circles. Although this scheme is conceptually very simple it has two disadvantages. It only works for one phase advance, because the machine geometry depends on ϕ_1 and ϕ_2. The quadrupole spacing and excitation in the lattice and the RF sections are the same.

Changing the focusing does not have these drawbacks. In order to halve the dispersion, the product of period length L_p and bending angle ϕ must be halved. In LEP, this is achieved by reducing the number of bending magnets in a cell from 12 to 8. When the focusing is also changed, the β-functions must be matched between their values in the arc and in the RF section. The matching of dispersion and β-functions is achieved by adjusting the gradients of six quadrupoles in the dispersion suppressor, using general-purpose matching programs such as AGS,[30] MAGIC,[31] SYNCH[32] or TRANSPORT.[33] The matching is straightforward in principle, but finding a smooth transition of the β-functions without too pronounced peaks and valleys may require some work.

4.4. RF section design

In the early e^+e^- storage rings the RF system consisted of a few cavities which could be housed in straight sections left between magnetic elements. This was done in SPEAR, CESR and PEP. In all these cases no particular beam-optical requirements arise. In PETRA the RF system is housed in special lattice sections which include quadrupoles between strings of RF cavities. This is even more true in LEP. In either machine, some thought must be given to the way

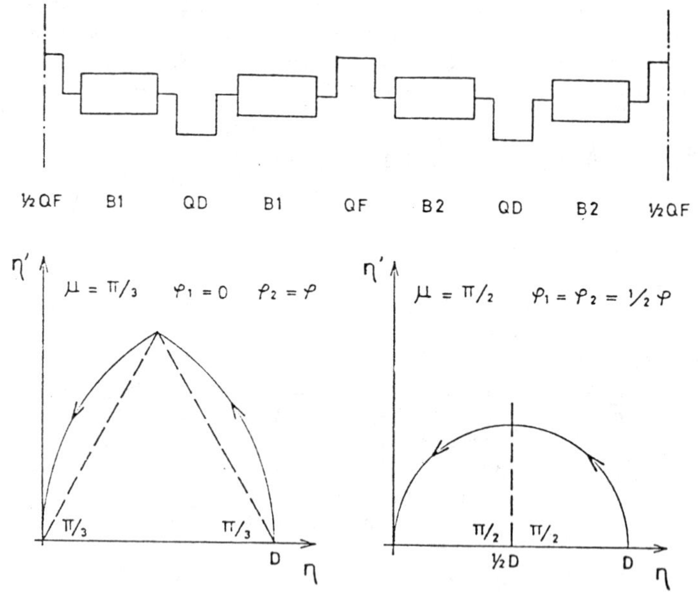

Fig. 7 - Schematic layout of a dispersion suppressor with nonstandard bending magnets B1 and B2 and deflection angles ϕ_1 and ϕ_2. Transformations of the dispersive trajectory through the dispersion suppressor in normalized phase space for phase advances $\mu = \pi/3$ and $\pi/2$.

the lattice is arranged. It is attractive to adapt the period length to the length of a number of cavities which should be considered as an entity, e.g. because they are fed from the same power source. In LEP, groups of eight cavities which are fed from one klystron fit between two quadrupoles.

The minimum aperture of the vacuum chamber is obtained when the betatron amplitude functions are matched to the periodic solution of the quadrupole lattice. This has the advantage of minimizing the aperture of the quadrupoles and hence their length and cost, and also the diameter of the beam hole in the RF cavities. In this way their shunt impedance is maximized and the RF power needed for a given accelerating voltage is minimized. On the other hand, the wake fields due to the excitation of the RF cavities by the bunches are also largest; this enhances the higher-mode losses and collective phenomena due to the wake fields. An optimization of the RF cavities taking all these phenomena into account has not even been attempted.

4.5. Low-β insertion design

The final insertion element between the RF section and the crossing point is the low-β insertion proper. It only has to satisfy four constraints, that the amplitude functions β_x and β_y at the crossing point be small and that α_x and α_y vanish. They can in principle be satisfied with four variable parameters, e.g. the strengths of four quadrupoles. They are usually arranged as one quadrupole doublet which does most of the focusing and two quadrupoles for fine tuning.

In an e^+e^- storage ring the vertical emittance $E_y = \sigma_y^2/\beta_y$ is usually much smaller than the horizontal emittance $E_x = \sigma_x^2/\beta_x$. It follows from (4) and (5) that the beam-beam tune shifts ΔQ_x and ΔQ_y are about equal when β_x and β_y have the same ratio as E_x and E_y, hence $\beta_y \ll \beta_x$. Because of this β_y rises more steeply with the distance s from the crossing point as shown in (2). It is therefore natural that the first quadrupole from the crossing point is vertically focusing, and the second one vertically defocusing. The opposite holds in the horizontal plane.

The luminosity equation (6) shows that β_y should be small in order to have a good luminosity. How small β_y can be made depends on several things:

i) It was already mentioned in Section 2 that one does not gain as much as one might hope from reducing β_y when it becomes comparable to the bunch length σ_z.

ii) Beam-beam simulations show that there is an even more rapid drop in luminosity in the region where $\beta_y \simeq \sigma_y$, because of the rapid change of the betatron phase at the crossing points $d\phi \simeq ds/\beta$ associated with the synchrotron oscillations.

iii) The contribution to the vertical chromaticity $\xi_y = dQ_y/dp/p$ due to the quadrupole doublet is approximately $\Delta\xi \simeq \ell_x^2/(4\pi\beta_y f)$ which becomes with (33) $\Delta\xi_y \simeq \ell_x/(2\pi\beta_y)$. The chromaticity $\Delta\xi$ can be used as a quantitative measure of the difficulties associated with the geometrical and chromatic aberrations of the storage ring lattice, to use terms of geometrical optics. A treatment of the chromaticity is beyond the scope of this lecture; useful information can be found in references 34 and 35. Since the chromaticity ξ_y depends on the ratio ℓ_x/β_y it limits β_y only when ℓ_x is also limited. This is the reason why mini-ℓ_x insertions have recently been installed in PETRA, PEP and CESR.

iv) The final limit on the smallness of β_y arises from the simple laws of geometrical optics.[36]

In thin-lens approximation the focal length d of the first quadrupole must fall into the range

$$\frac{1}{2}\,\ell_x \le d \le \ell_x \tag{33}$$

Here ℓ_x is the distance between the centre of the first quadrupole and the crossing point. The left inequality means that a vertical ray originating at the crossing point will be point-focused a distance ℓ_x behind the quadrupole (if the second quadrupole is not in the way). The right inequality means that all vertical rays originating at the crossing point are made parallel behind the first quadrupole, i.e. there is not enough focusing because the second quadrupole is vertically defocusing. In the following, d shall be approximated by $\ell_x/2$. It also follows from geometrical optics that the distance between the two quadrupole centres must be smaller than ℓ_x. Otherwise the vertical ray will cross the axis between the two quadrupole centres and never be focused by the doublet.

The approximate relation between the focal length d and the quadrupole parameters length ℓ_Q and gradient dB/dx and the rigidity $B\rho$ of the beam is (valid for $d \gg \ell_Q$):

$$d = \frac{B\rho}{\ell_Q}\left(\frac{dB}{dx}\right)^{-1} = \frac{B\rho}{\ell_Q}\,\frac{A_x}{B_Q} \tag{34}$$

Since the field in a quadrupole is linear, its gradient may be expressed as the ratio of the "poletip field" B_Q and its aperture radius A_x. It is appropriate to allow as much aperture in the low-β quadrupoles as in the normal lattice cells, and hence to use (2) for the variation of β_x with s, (25) for the emittance, (29) for the aperture calculation, and (30) for the beam size calculation, with $D_x = 0$. Thus the aperture A_x becomes for $\ell_x \gg \beta_x$:

$$A_x = F_x\ell_x\left(\frac{Nr_e}{2\pi k\gamma\Delta Q\beta_x}\right)^{\frac{1}{2}} + C_x = \frac{F_x r_e \ell_x}{\gamma\Delta Q}\left(\frac{L\,\beta_y}{fk\beta_x}\right)^{\frac{1}{2}} + C_x \tag{35}$$

An analogous calculation for A_y with (31) instead of (30) yields, for $\ell_x \gg \beta_y$:

$$A_y = F_x\ell_x\left(\frac{Nr_e}{4\pi k\gamma\Delta Q\beta_y}\right)^{\frac{1}{2}} + C_y = \frac{F_y r_e \ell_x}{\gamma\Delta Q}\left(\frac{L}{2fk}\right)^{\frac{1}{2}} + C_y \tag{36}$$

It may be seen that A_x and A_y differ mainly by β_x appearing in A_x and $2\beta_y$ in A_y, neglecting the coefficients C_x, C_y, F_x and F_y. One can also show that the essential parts of A_x and A_y are equal when the emittances have the same ratio as the β's at the crossing point. In other words, a beam with optimum emittance ratio has a circular cross-section at the first quadrupole. For the following, (36) is used since it yields the larger aperture.

By eliminating A_y and d from (33), (34) and (36) an equation for ℓ_Q is obtained, assuming for simplicity $C_y = 0$:

$$\ell_Q = \frac{eZ_oF_y}{4\pi B_Q}\left(\frac{N\gamma}{\pi k r_e \Delta Q \beta_y}\right)^{\frac{1}{2}} = \frac{eZ_oF_y}{4\pi B_Q \Delta Q}\left(\frac{2L}{fk}\right)^{\frac{1}{2}} \tag{37}$$

Here the impedance of free space $Z_o = 120\pi\Omega$ has been introduced. The desired relation for β_y is obtained by expressing the ratio ℓ_x/β_y by the contribution of the doublet to the chromaticity $\Delta\xi_y$

$$\ell_x/\beta_y = 2\pi\ \Delta\xi_y \tag{38}$$

and by introducing another law of geometrical optics

$$\ell_x = G_Q\ell_Q \tag{39}$$

where the coefficient G_Q is typically 2. This permits ℓ_Q and ℓ_x to be eliminated completely and the following lower limits for β_y to be obtained:

$$\beta_y = \frac{eZ_o}{8\pi^2}\ \frac{G_QF_y}{\Delta\xi_y B_Q}\left(\frac{N\gamma}{\pi k r_e \Delta Q\beta_y}\right)^{\frac{1}{2}} = \frac{eZ_o}{8\pi^2}\ \frac{G_QF_y}{B_Q\Delta Q\Delta\xi_y}\left(\frac{2L}{fk}\right)^{\frac{1}{2}} \tag{40}$$

These limits are satisfied by the existing mini-ℓ_x insertions. It can be seen from the equation containing L and f which is proportional to E^{-2} from (23) that, with the other factors remaining constant, β_y is proportional to the energy. It can also be seen that β_y is proportional to B^{-1}. This is one of the reasons why superconducting quadrupoles are seriously considered for the mini-ℓ_x insertions of LEP.

The exact calculation of the quadrupole gradients is again the realm of the general-purpose matching programs mentioned earlier, once an arrangement of quadrupoles and straight sections following the above principles has been found.

Two other phenomena entering into the design of the interacting regions are related to synchrotron radiation. They may become important for e^+e^- storage rings beyond LEP energies. An electron which is accelerated transversely by a bending field B emits synchrotron radiation corresponding to an energy loss per unit length given by

$$\frac{dU}{ds} = \frac{2}{3}\ \frac{r_e e^2 c^2}{E_e}\ \gamma^2 B^2 \tag{41}$$

Here E_e is the electron rest energy. This also holds for electrons which pass a quadrupole off-axis. In a low-β insertion, this happens in two places, in the field of the opposite beam[37] and in the quadrupoles next to the crossing points. Since the bending field in a quadrupole is proportional to the distance from the axis, these phenomena depend on the square of the particle amplitude. They are calculated here for one standard deviation.

The first phenomenon was dubbed "beamstrahlung."[38] The bending field is typically 1 T, much higher than in the dipole magnets, but the interaction length is comparable to the bunch length σ_z. Hence the critical photon energy is high, as can be seen from (19), but the number of photons emitted by one electron in a collision is low. Therefore the dominant effect of beamstrahlung is a contribution to the energy spread of the beam.

The contribution of a length ds to the quantum excitation is proportional to $\kappa^3 ds$ where κ is the local curvature. If the condition is imposed that the contribution of the beam-beam collisions with curvature ρ_b^{-1} to the energy spread in the beam be not larger than that of the machine lattice the following condition must be satisfied:

$$\frac{n_x \sigma_z}{\rho_b^3} \lesssim \frac{2\pi}{\rho^2} \tag{42}$$

Here n_x is the number of crossings and σ_z the bunch length. The curvature ρ_b^{-1} due to the field of the other beam is approximately given by[38]

$$\frac{1}{\rho_b} = \frac{4\pi \Delta Q \sigma_y}{\beta_y \sigma_z} \tag{43}$$

Eliminating ΔQ by using (5) and introducing the luminosity L from (1) yields the condition:

$$\left(\frac{L}{kf}\right) \frac{n_x}{\gamma^2}^{2/3} \left(\frac{\rho}{\sigma_z}\right)^{4/3} \leq \left(\frac{2}{\pi}\right)^{1/2} \frac{1}{r_e^2} \left(\frac{\sigma_x}{16\pi\sigma_y}\right) \tag{44}$$

Here the right-hand side depends only on the axis ratio σ_x/σ_y which should be relatively independent of energy. Remembering from the scaling laws that $\rho \backsim \gamma^2$ and $f \backsim \gamma^{-2}$ shows that $L \backsim E^{-8/3}$. In a machine such as LEP with k = 4, f = 11.25 kHz, n_x = 8, E = 85 GeV, ρ = 3104 m, σ_z = 12 mm and σ_x/σ_y = 16, the luminosity limit is L = $2 \cdot 10^{32}$ cm^{-2}s^{-1}, i.e. more than an order of magnitude above the design luminosity.

The quadrupole radiation as it shall be called here is mainly generated in the insertion quadrupoles. Its characteristics differ from that of the beamstrahlung. The bending radius and hence the critical photon energy are more comparable to those in the dipoles. In addition the length of the quadrupole is much higher than the bunch length and hence the number of photons emitted by one particle when passing the quadrupole is higher. Therefore the dominant effect of quadrupole radiation is its contribution to the energy loss, while its contribution to the energy spread is negligible.

By arguing about the quadrupole aperture and strength in the same way as above one can obtain an expression for the ratio of the synchrotron radiation losses in an interaction region quadrupole U_Q to those in the bending arcs U_s:

$$U_Q/U_s = \frac{r_e ce}{\pi E_e} \frac{B_Q \rho}{F_x \gamma^2 \Delta Q} \left(\frac{L}{2\pi fk}\right)^{1/2} \tag{45}$$

If the betatron amplitudes of the electrons were constant then quadrupole radiation would cause a variation of the stable RF phase angle with the betatron amplitude.[39] Since the betatron amplitudes vary continuously due to quantum excitation and radiation damping, quadrupole radiation is a new coupling mechanism between betatron and synchrotron oscillations. Its effect on both distribution functions is being investigated.[40] With all parameters except f fixed and with $f \sim E^{-2}$, $U_Q/U_s \sim E$.

Both beamstrahlung and quadrupole radiation are potential sources of background for experiments installed in the interaction regions, and as such deserve appropriate studies in collaboration with the detector groups.

4.6. Beam size variation

In a fixed lattice with constant J_x the natural luminosity variation is given by (26): the luminosity is proportional to the fourth power of the energy. If one were to operate a machine over an energy range from 20 to 120 GeV this would imply a luminosity variation by a factor $6^4 = 1296$. The reason for this variation is the fact that the emittance of the beam varies like the square of the energy as shown in (16). Clearly, the machine aperture must be chosen such that it is large enough for the highest energy. Therefore it would permit a constant emittance over the whole energy range. It follows from (1) and (5) that this leads to a luminosity variation like the square of the energy. In the above example this corresponds to a luminosity variation by a factor $6^2 = 36$, i.e. to a luminosity at 20 GeV which is a factor of 36 higher than with the natural variation.

One way of achieving a constant emittance over a range of energies is suggested by (16): varying the damping partition number J_x as a function of the energy. In the example, the energy range from 53.7 to 120 GeV can be covered when J_x is varied from 0.5 at 53.7 GeV to 2.5 at 120 GeV. According to (12) this implies varying J_x from 2.5 to 0.5. The beam emittance is enlarged from 53.7 to 75.9 GeV, has its natural value and $J_x = 1$ at that energy, and is reduced from 75.9 GeV to 120 GeV. In order to apply this principle, the lattice tune Q must be determined from (26) by using the appropriate combination of J_x and γ.

For energies below 53.7 GeV the same principle may be repeated at a lower tune Q, thus covering another factor $\sqrt{5}$ in energy, i.e. down to 24 GeV. According to (26) this requires changing J_x from 0.5 to 2.5 at 53.7 GeV and changing the tune Q by a factor $0.2^{1/3} = 0.585$. It was mentioned in section 4.2 that the phase advance per cell can only take values which are simple fractions of 2π, e.g. $\pi/2$ or $\pi/3$. This means that the tune changes in the ratio 2/3 which differs from the above "ideal" ratio, and reduces the energies which can be covered in this manner, to a range from 29.2 to 53.7 GeV. Still, the luminosity at 20 GeV is only about a factor of two smaller than in a perfect constant-emittance scheme.

A method for beam enlargement is exciting wiggler magnets.[41] They are strings of dipole magnets with fields of polarity +--+. They cause no net deflection of the beam, and can be used at all field levels between zero and maximum without distorting the closed orbit. Their field is higher than in the lattice dipoles. Hence they increase the synchrotron radiation losses and the energy spread in the beam, and reduce the damping times. They increase the beam emittance when they are installed where the dispersion does not vanish. By varying the wiggler field the emittance can be varied continuously between the natural and a larger value. Wigglers are not an economic method of reducing the emittance at high energy since the reduction factor is at best equal to the ratio of the synchrotron radiation losses with and without wigglers. Wigglers have been operated successfully in ADONE[42] and in PEP.[43]

A variant of wigglers is "kinks"[44] which do have a net deflection. A way of installing them in a lattice with several dipoles in a row is to excite only part of the dipoles. This decreases the bending radius ρ and increases the emittance according to (16). Such a scheme changes the lattice geometry and has not been realized in practice.

In principle, the beam size at the crossing points can also be increased by making the dispersion there different from zero. Because of the reasons enumerated at the beginning of Chapter 4 this possibility is not incorporated in the design procedure.

All the previous methods involve a change of the horizontal emittance E_x from its natural value while keeping the emittance ratio E_y/E_x constant and equal to β_y/β_x. An alternative is reducing the emittance ratio E_y/E_x and thus reducing the beam cross-section $\sigma_x\sigma_y$, while letting E_x follow its natural variation. It is used in CESR, but not in PEP and PETRA (cf. Table II). Adopting this seemingly simple method in the design of even bigger machines is dangerous for two reasons. Computer simulations have shown that emittance ratios E_y/E_x much smaller than β_y/β_x cannot be reached in a machine with realistic magnet alignment errors and closed orbit correction methods.[45] Other computer simulations show that a higher luminosity is obtained when E_x is varied in proportion to the current rather than fixed, the vertical emittance E_y adjusting itself by the beam-beam effect.[16]

4.7. Conclusions for lattice design

The techniques described above may be used to arrive at an approximate e^+e^- storage ring lattice with a good performance over a wide energy range. The formulae used (or more accurate ones not given here) have been incorporated in the DESIGN code which computes surprisingly accurate parameter lists. The aperture determines the

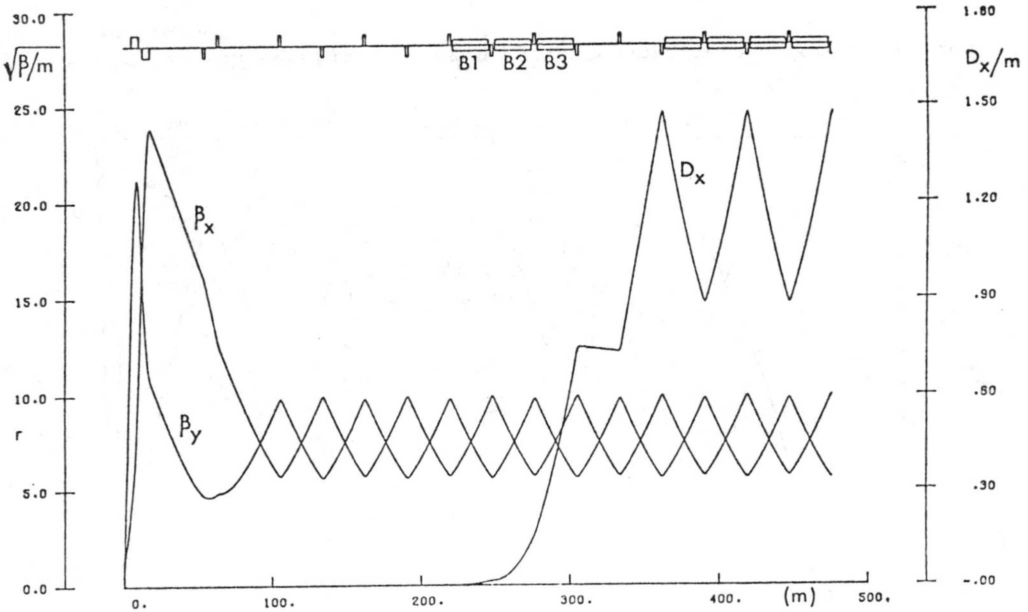

Fig. 8 - LEP lattice version 1 as reported in LEP-70, the dispersion suppressor is achieved by missing dipoles.

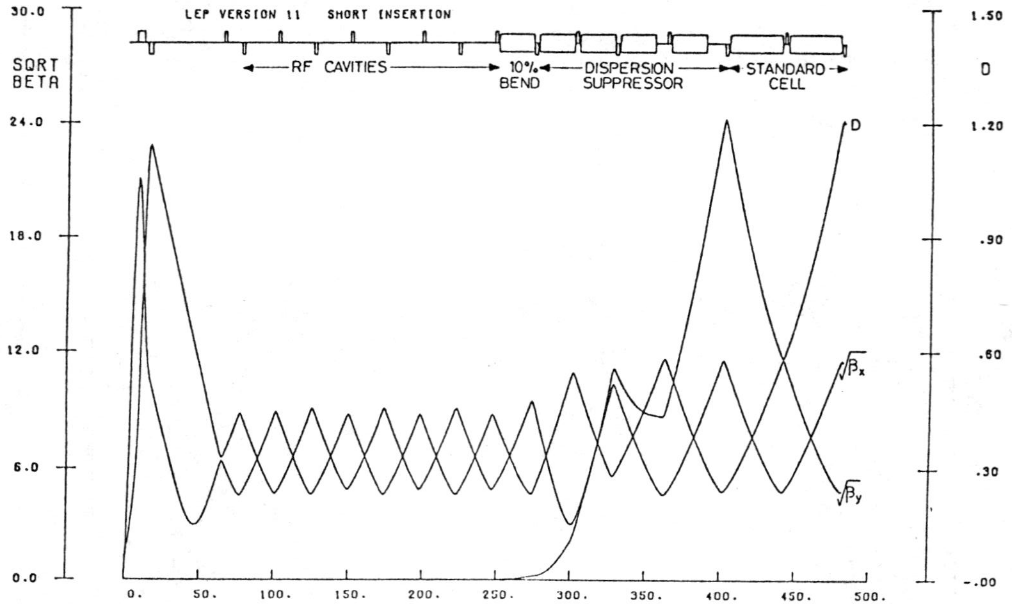

Fig. 9 - LEP lattice version 2 using focusing change method for dispersion suppressor.

performance in the lower part of the energy range while the focusing determines the performance in the upper part. The beam size must be adjusted artificially by varying the damping partition numbers, by wiggler magnets and/or by changing the tune in a few large steps in order to approach the quadratic variation of the luminosity with energy resulting from the aperture of the machine.

Turning an approximate lattice into a final one involves considerations which are beyond the scope of this lecture. The total cost of a machine is only an indication for its final size. Considerations concerning the construction site and, in the case of circular injectors, simple circumference ratios also play an important role. The only collective phenomenon included so far is the beam-beam effect. Phenomena limiting the current in each beam will be discussed briefly in Chapter 6. Incorporating them into the design involves modifications of the procedure described.

Examples of final lattices are shown in Figs. 8 and 9. In the LEP-70 lattice[46] the dispersion suppressor design is based on leaving out bending magnets. This leads to the regular behavior of the β-functions shown in Fig. 8. The LEP Version 11 lattice[47] is shown in Fig. 9. It has the other dispersion suppressor design, and different periodic β-functions in the RF section and the lattice.

5. RF SYSTEM DESIGN

In the discussion of RF systems in Section 3.2 the RF power absorbed by the beams was neglected; however, it was included in the overall power considerations in Section 3.3. It was shown in Fig. 4 that this is an oversimplification.

5.1. Energy conservation

The power absorbed by the beam is only one aspect of beam loading in large machines. More important is the energy which has to be transferred from the electromagnetic field in the RF cavities to a bunch of electrons or positrons in a single traversal. The stored energy W_s is given by:

$$W_s = \frac{Q\,P_d}{2\pi f_{RF}} = \frac{Q}{2\pi f_{RF}} \cdot \frac{V^2}{ZL_c} \tag{46}$$

where Q is the quality factor of the cavity, L_c the total cavity length, P_d the power dissipated in the cavity walls and Z the shunt impedance per metre. The energy W_e to be delivered to each bunch at each passage through the cavities is

$$W_e = \frac{2P_b}{k\,f_{rev}} \tag{47}$$

where P_b is the power delivered to one beam, and f_{rev} is the revolution frequency. Introducing the harmonic number $h = f_{RF}/f_{rev}$ yields for the ratio W_e/W_s:

$$W_e/W_s = \frac{P_b}{P_d} \cdot \frac{4\pi h}{kQ} \tag{48}$$

Energy conservation implies $W_e/W_s \leq 1$, and (48) shows that this condition becomes more difficult to satisfy with increasing h, i.e. increasing machine size.

A tighter criterion is obtained by the following argument, based on the superposition of cavity fields which are represented in Fig. 10 by complex rotating vectors. The voltages V^+ and V^- represent the cavity field immediately before and after the bunch passage, and the voltage V_{bo} the voltage excited by the passage of a bunch in the fundamental mode of a cavity which is not driven by an RF generator. ϕ^+ is the phase just before the bunch passage, counted from the peak of the RF wave. Using the law of cosines on the triangle in Fig. 10 and energy conservation which implies that the energy W_e transferred to the bunch must be equal to the difference in the energy stored in the cavity field before and after the bunch passage, yields the relation:

$$W_e = \frac{Q}{2\pi f_{RF} Z L_c} \left[(2V^+ V_{bo} \cos\phi^+ - V_{bo}^2) - (B-1)V_{bo}^2 \right] \tag{49}$$

The first term on the right-hand side refers to the driven fundamental mode of the RF cavity. The second term refers to the higher modes in the cavity which are only excited by the bunch passage, and are assumed to decay between bunch passages. They are globally described by the beam loading enhancement factor B, the ratio between the energy radiated into all modes to that radiated into the fundamental mode of a cavity when it is not excited by a transmitter. B is only a function of the bunch and cavity geometry. V_{bo} is proportional to the bunch charge q, and W_e reaches a maximum at $V_{bo} = (V^+/B)\cos\phi^+$ where the ratio W_e/W_s becomes

$$W_e/W_s = \frac{\cos^2\phi^+}{B} \tag{50}$$

For two beams circulating in opposite directions the following criterion applies:

$$W_e/W_s = \frac{2\cos^2\phi^+}{B+1} \tag{51}$$

In its derivation use has been made of the fact that for higher modes the bunches of the two beams do not pass the cavities at intervals which are multiples of their periods. Therefore the random phase approximation may be used.

Equations (49) and (50) only indicate whether or not a set of RF parameters violates energy conservation. They do not say how to choose a parameter set which works.

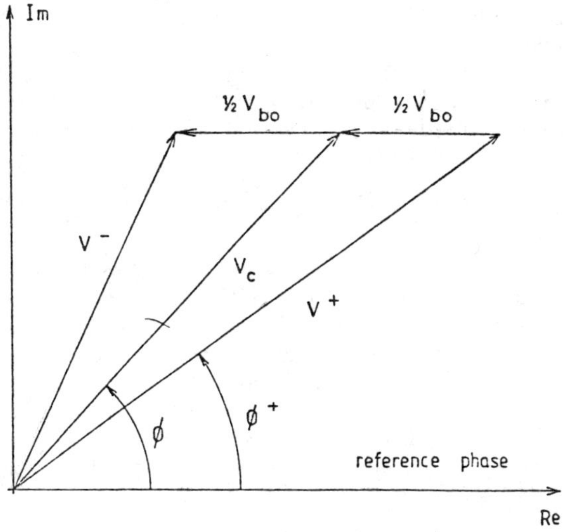

Fig. 10 - Phasor diagram for beam loading analysis.

5.2. Transient beam loading

The transient beam loading due to a few intense bunches traversing RF cavities periodically was analyzed by P.B. Wilson,[48] mainly with the help of complex vector diagrams. Only a brief summary of this work is given here.

The beam-induced voltage V_{bo} will usually not decay completely between two successive bunch passages. In this case the contributions from all previous passages will build up to a voltage V_b which can again be obtained by vector addition and is given by

$$V_b = \frac{1}{2} V_{bo} \frac{\left[1 - \exp(-2\tau)\right] + i \exp(-\tau)\sin\delta}{1 - 2\exp(-\tau)\cos\delta + \exp(-2\tau)} \tag{52}$$

Here $\tau = T_b/T_f$ is the ratio between bunch spacing $T_b = h/(kf_{RF})$ and filling time of the cavity $T_f = Q/(\pi f_{RF})$, and $\delta = 2\pi T_b(f_o - f_{RF})$ is the phase difference between successive passages due to the small difference between the frequency of the transmitter f_{RF} and the resonant frequency f_o of the lowest mode.

The next thing to be considered is the coupling between the RF cavity and the transmitter. It is described by a coupling parameter β. The result of all these considerations is an expression for the generator power P_g:

$$\begin{aligned} P_g = \left(\frac{1+\beta}{\cos\psi}\right)^2 \frac{1}{4\beta Z L_c} \Big\{ &\left(V_c\cos\phi + IZL_c\tau\, F_1(\beta,\psi)\right)^2 \\ &+ \left(V_c\sin\phi + IZL_c\tau\, F_2(\beta,\psi)\right)^2 \Big\} \end{aligned} \tag{53}$$

Here $\psi = \tan^{-1}(\delta/\tau)$ is the tuning angle which together with β can be chosen freely to minimize P_g. V_c is the peak cavity voltage and ϕ is the stable phase angle counted from the peak of the RF voltage. The energy supplied to the beam in one turn is $V_c\cos\phi$. Both V_c and ϕ are chosen such that an adequate bucket height and therefore quantum lifetime is obtained. I is the circulating current, summed for the two beams if necessary; $\tau_o = T_b/T_{fo}$ where $T_{fo} = Q_o/(\pi f_{RF})$ is the unloaded filling time of the cavity when it is not connected to the generator and has a quality factor Q_o. The functions F_1 and F_2 are defined as follows:

$$\begin{aligned} F_1 &= \left[1 - \exp(-2\tau)\right]/D \\ F_2 &= \exp(-\tau)\sin(\tau\tan\psi)/D \\ D &= 1 - 2\exp(-\tau)\cos(\tau\tan\psi) + \exp(2\tau) \\ \tau &= (1+\beta)\tau_o \end{aligned} \tag{54}$$

The formalism described includes energy conservation and transient beam loading. It therefore automatically satisfies the criterion (50).

5.3. Calculation of the loss parameters

In the above analysis, quantities like V_{bo} or U_{hm} are needed. The voltage V_b is excited in the fundamental mode by the passage of a bunch through an RF cavity which is not driven by a generator. It was mentioned that higher modes of the cavity are also excited by the passage of a bunch. The average energy loss of the particles in a bunch can be described by the voltage U_{hm}. There are two notations, using impedances Z_{hm}, or k-factors k_{hm}:

$$U_{hm} = IZ_{hm} = q\, k_{hm} \tag{55}$$

Here I is the total circulating current (in all bunches) while q is the charge per bunch. Since the two expressions must be the same, Z_{hm} and k_{pm} are related

$$k_{pm} = Z_{pm}\, k\, f_{rev} \tag{56}$$

where k is the number of bunches and f_{rev} is the revolution frequency.

Similarly, the excitation of the fundamental mode is given by

$$V_{bo} = 2IZ_{fm} = 2q\, k_{fm} \tag{57}$$

The appearance of a factor of two in (57) is called the "fundamental theorem of beam loading" by Wilson. It expresses the fact that the excitation of a mode in a cavity is exactly twice the average energy loss it causes for the particles in a bunch.

The loss factors k_{fm} and k_{pm} are calculated by computer programs which solve Maxwell's equations. Programs like KN7C[49] and SUPERFISH[50] find the resonant frequencies f_i and the factors $(R/Q)_i$ associated with them for the i-th mode. The loss factor k_i can then be calculated from

$$k_i = \frac{\pi f_i}{2}\left(\frac{R}{Q}\right)_i \tag{58}$$

The program BCI[51] operates in the time domain and calculates directly the total loss parameter for bunches of rms length σ_s.

6. CONCLUSIONS

The size of an e^+e^- storage ring scales roughly like the square of its design energy. This follows from a cost optimization. The design performance, the circulating current, and the beam size are related by the maximum tolerable beam-beam tune shift. The same beam size must be arrived at by the balance between synchrotron radiation, quantum excitation and damping. This requirement fixes the lattice parameters to a very large extent. If the machine is to be operated over a large energy range the natural beam size variation must be modified. Several possibilities for this are discussed.

The design of the RF system is substantially modified by the transient beam loading due to the passage of widely-spaced intense bunches through the RF cavities. In these considerations the energy losses due to the excitation of higher modes of electromagnetic cavity fields must also be included.

It must not be forgotten that limits on the circulating current arise from collective phenomena other than the beam-beam effect. The most important ones are longitudinal and transverse coherent instabilities of bunched beams[52] which are mostly driven by electromagnetic accelerating and deflecting fields excited by the passage of the bunches through the RF cavities. These instabilities have been observed in the operating machines, but have not been a drastic performance limitation. When the design energy increases, these instabilities become more serious. Their scaling laws with energy and other machine parameters have been discussed elsewhere.[53]

The rapid increase in both the cost and the severity of collective instabilities leads to an upper limit of the energy up to which the design concept presented here can be applied.

REFERENCES

1. G. A. Voss et al., IEEE Trans. Nucl. Sci. NS-26, 2970 (1979).
2. M. Billing et al., Proc. 11th Internat. Conf. on High Energy Accelerators, Geneva 1980, 26 (1980).
3. PEP Conceptual Design Report, SLAC-189 (1976).
4. The DORIS Storage Ring Group, DESY 79-08 (1979).
5. P. Marin, Proc. 9th Internat. Conf. on High Energy Accelerators, Stanford 1974, 49 (1974).
6. G. E. Fischer, SPEAR-154 (1972).
7. SPEAR Storage Ring Group, IEEE Trans. Nucl. Sci. NS-20, 838 (1973).
8. S. Tazzari, IEEE Trans. Nucl. Sci. NS-28, 2420 (1981).
9. P. Marin, Proc. 11th Internat. Conf. on High Energy Accelerators, Geneva 1980, 742 (1980).
10. F. Amman, IEEE Trans. Nucl. Sci. NS-20, 858 (1973).
11. J. Rossbach, IEEE Trans. Nucl. Sci. NS-28, 2025 (1981).
12. A. W. Chao and H. Wiedemann, private communication (1981).
13. S. Peggs and R. Talman, Proc. 11th Internat. Conf. on High Energy Accelerators, Geneva 1980, 754 (1980).
14. S. Myers, IEEE Trans. Nucl. Sci. NS-28, 2503 (1981).
15. A. Piwinski, IEEE Trans. Nucl. Sci. NS-28, 2440 (1981).
16. S. Myers, LEP Note 310 (1981).
17. D. Degèle, 11th Internat. Conf. on High Energy Accelerators, Geneva 1980, 16 (1980).
18. J. Kirchgessner et al., IEEE Trans. Nucl. Sci. NS-22, 1141 (1975).
19. P. B. Wilson, IEEE Trans. Nucl. Sci. NS-26, 3255 (1979).
20. W. Schnell, CERN/ISR-LTD/76-8 (1976).
21. B. Richter, Nucl. Instr. Methods 136, 47 (1976).
22. J. R. Bennett et al., CERN 77-14 (1977).
23. I. B. Vasserman et al., All-Union Conf. Charged Particle Accelerators, Dubna 1978 [In English CERN Trans. Int. 78-01 (1978)].
24. A. Piwinski and A. Wrulich, DESY 76/07 (1976).
25. S. Kheifets and E. Messerschmid, DESY/PET-77-18 (1977).
26. D. A. Edwards and L. C. Teng, in CERN/ISR-AS/74-64 (1974).
27. T. Collins, this school.
28. H. Wiedemann, Proc. 11th Internat. Conf. on High Energy Accelerators, Geneva 1980, 744 (1980).
29. S. Myers, LEP Note 327 (1981).
30. E. Keil, Y. Marti, B. W. Montague and A. Sudboe, CERN 75-13 (1975).
31. A. S. King, M. J. Lee, W. W. Lee, SLAC-183 (1975).
32. A. A. Garren and J. W. Eusebio, UCID-10153 (1965).
33. K. L. Brown, D. C. Carey, Ch. Iselin and F. Rothacker, CERN 80-04 (1980).
34. M. H. R. Donald, P. L. Morton and H. Wiedemann, IEEE Trans. Nucl. Sci. NS-24, 1200 (1979).
35. P. Faugeras, A. Faugier, J. Gareyte, A. Hilaire, IEEE Trans. Nucl. Sci. NS-26, 3577 (1979).

36. E. D. Courant and E. Keil, Proc. Workshop on Possibilities and Limitations of Accelerators and Detectors, Fermilab October 1978, 135 (1979).
37. A. Hofmann and E. Keil, LEP-70/86 (1978).
38. J. E. Augustin et al., Proc. Workshop on Possibilities and Limitations of Accelerators and Detectors, Fermilab October 1978, 87 (1979).
39. A. W. Chao and A. A. Garren, PEP Technical Memo 141 (1978).
40. J. M. Jowett, private communication.
41. J. M. Paterson, J. R. Rees and H. Wiedemann, SLAC Report SPEAR-186, PEP-125 (1975).
42. Frascati Laboratory Internal Memorandum RM-13 (1979).
43. J. M. Paterson, Proc. 11th Internat. Conf. on High Energy Accelerators, Geneva 1980, 7 (1980).
44. A. Hutton, CERN/ISR-TH/77-48 (1977).
45. The LEP Study Group, CERN Report ISR-LEP/79-33 (1979).
46. The LEP Study Group, CERN Report ISR-LEP/78-17 (1978).
47. A. Hutton, LEP Note 289 (1981).
48. P. B. Wilson, 9th Internat. Conf. on High Energy Accelerators, Stanford 1974, 57 (1974).
49. E. Keil, Nucl. Instr. Methods 100, 419 (1972).
50. K. Halbach, Particle Accelerators 7, 213 (1976).
51. T. Weiland, 11th Internat. Conf. on High Energy Accelerators, Geneva 1980, 570 (1980).
T. Weiland, CERN/ISR-TH/80-07, 45 and 46 (1980).
52. C. Pellegrini, this school.
53. E. Keil, Proc. 1979 Internat. Symposium on Lepton and Photon Interactions at High Energies, Fermilab, 305 (1979).

INTERACTIONS BETWEEN ELECTROMAGNETIC FIELDS AND ELECTRONS

R. H. Pantell
Stanford University, Stanford, California 94305

TABLE OF CONTENTS

0094-243X/82/870863-56$3.00 Copyright 1982 American Institute of Physics

INTERACTIONS BETWEEN ELECTROMAGNETIC FIELDS AND ELECTRONS

R. H. Pantell
Stanford University, Stanford, California 94305

1. INTRODUCTION

Various types of interactions between electromagnetic waves and electrons will be discussed, with emphasis on radiation that is in the optical and near-optical portions of the spectrum. There is increasing interest in this subject because of the recent development of the free-electron laser[1] (FEL). Related areas of application include optical klystrons, laser-driven electron accelerators, and harmonic generators.

Energy and momentum conservation requirements will be considered first, since these conditions determine what is and what isn't possible regarding energy exchange between waves and electrons. Next, several examples will be considered in some detail including the FEL, energy transfer by a limited interaction length, and the stimulated synchrotron effect. Laser-driven particle accelerators will then be described and estimates will be made for possible acceleration gradients and total energy transfer. Finally, experimental results will be presented and compared to predictions obtained from theoretical models.

2. ENERGY AND MOMENTUM CONSERVATION

If an electron is to emit or absorb electromagnetic radiation then certain conditions must be satisfied: (a) Energy must be conserved from the initial to final state, (b) Momentum must be conserved from the initial to final state, and (c) The dispersion relationships for the interacting particles must be satisfied. These requirements determine what is or isn't possible regarding schemes for amplification of waves or acceleration of electrons.

Since the conservation conditions involve energy and momentum it is helpful to study the interactions in energy-momentum space.[2] As an example, Fig. 1 illustrates photon emission by a free electron. The upper surface represents the electron dispersion relation

$$E = [m^2c^4 + c^2p^2]^{1/2}, \tag{1}$$

where E is the electron energy, p is the magnitude of momentum, m is rest mass, and c is the velocity of light. The lower surface, a cone, is the dispersion relation for photons in vacuum

$$E_\gamma = cp_\gamma, \tag{2}$$

where the subscript γ has been added to designate photon energy and momentum.

The final-state energy and momentum lie on the surface of the cone drawn with dashed lines in Fig. 1. This cone is the photon dispersion surface subtracted from the energy and momentum of the incident electron. Since this energy and momentum must be carried off by the final-state electron, the dashed-line cone must intersect the electron dispersion surface if this process is to occur. From Fig. 1 it is seen that photon emission by a free electron is not possible.

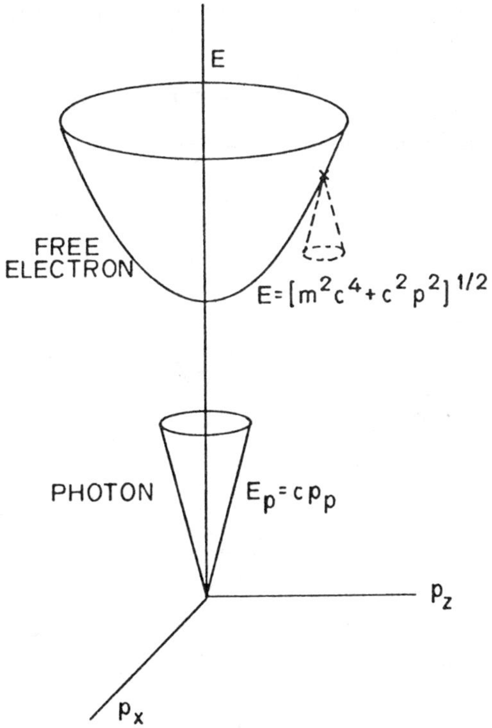

Fig. 1. Photon emission by a free electron, as pictured in energy-momentum space.

Similarly, photon absorption cannot occur. This process inverts the dashed cone, which also does not intersect the electron dispersion surface. Therefore, neither emission nor absorption can occur from a free electron.

A. The Cherenkov effect[3]

Interaction in a medium, where the effect of the medium is to introduce an index of refraction n, causes a widening of the photon cone as shown in Fig. 2. For the photon the energy-momentum relationship is now

$$E_\gamma = (\frac{c}{n})\ p_\gamma. \qquad (3)$$

It is possible for the final-state energy surface, represented by the inverted cone, to intersect the electron dispersion surface. The line of intersection is shown in Fig. 2(a) connecting points (E_1, p_1) and (E_2, p_2), where (E_i, p_i) are electron energy and momentum.

Figure 2(b) is the projection of Fig. 2(a) on the momentum plane. The angle of photon emission θ relative to the direction of incident electron motion can be calculated from the conservation conditions. From energy conservation

$$E_2 + \hbar\omega = E_1 , \tag{4}$$

where $\hbar\omega$ is the photon energy, and from momentum conservation

$$\underline{p}_2 + \hbar\underline{k} = \underline{p}_1 , \tag{5}$$

where $\underline{k}$ is the light wave vector and

$$|k| = \frac{\omega n}{c} . \tag{6}$$

If the photon energy is much less than the particle energy then initial and final electron speeds are approximately equal: $\beta_1 c \cong \beta_2 c$. The dispersion relation (1) can be used to obtain

$$E_1^2 - E_2^2 = c^2\left(p_1^2 - p_2^2\right), \tag{7}$$

implying

$$E_1 - E_2 \simeq \beta c \, (p_1 - p_2) \tag{8}$$

Equations (4) and (8) can be combined

$$E_1 - E_2 = \hbar\omega \simeq \beta c \, (p_1 - p_2). \tag{9}$$

Momentum conservation in the direction of electron motion requires

$$\left(\frac{\hbar\omega n}{c}\right) \cos\theta \simeq (p_1 - p_2), \tag{10}$$

where $\beta = v/c$ is the ratio of electron velocity to the velocity of light. The ratio of Eq. (10) to Eq. (9) gives the Cherenkov condition

$$\cos\theta = (n\beta)^{-1}. \tag{11}$$

Energy and momentum conservation also require that the phase velocity of the electromagnetic wave along the particle trajectory equals the particle velocity. This means that the particle remains in a constant phase of the field. From Fig. 2(b), the phase velocity v_p along the particle direction is

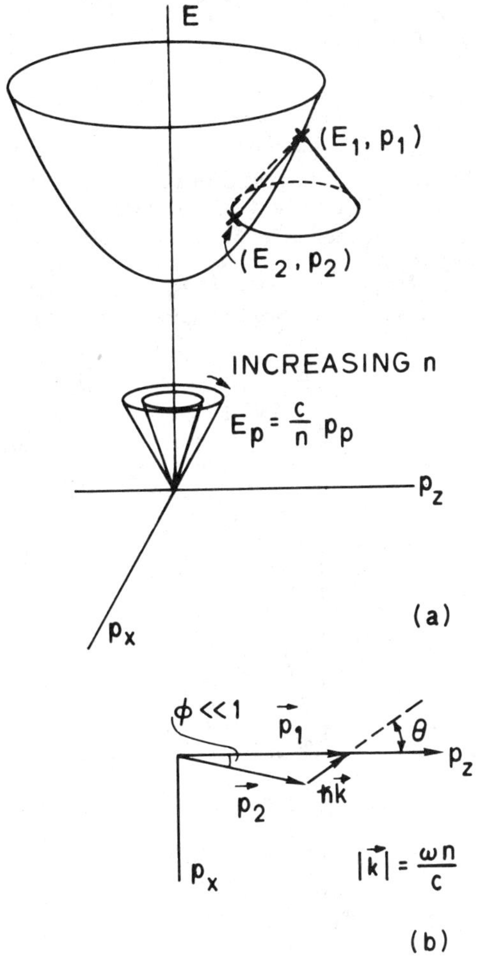

Fig. 2. The Cherenkov effect.

$$v_p = \frac{c}{n} (\cos\theta)^{-1} , \qquad (12)$$

and from Eq. (11) we find that $v_p = \beta c = v$.

B. Compton scattering[4]

In vacuum a free electron can absorb one photon and emit another. This is Compton scattering, which is illustrated in Fig. 3 for the collinear case. Starting with an electron characterized by E_1 and p_1 a backward-directed photon of energy $\hbar\omega_1$ is absorbed, a forward-directed photon of energy $\hbar\omega_2$ is emitted, and the final-state electron is specified by the point (E_2, p_2). Both dashed lines in Fig. 3 have slopes with magnitudes equal to c.

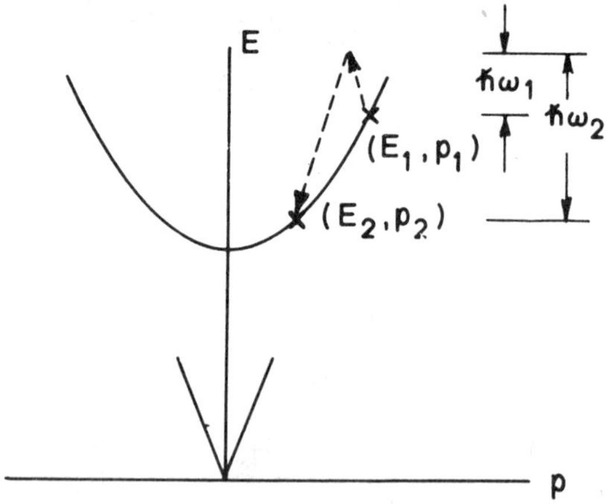

Fig. 3. Collinear Compton scattering.

Energy conservation is

$$E_1 + \hbar\omega_1 = E_2 + \hbar\omega_2 . \qquad (13)$$

Momentum conservation is

$$p_1 = p_2 + \left(\frac{\hbar\omega_2}{c}\right) + \left(\frac{\hbar\omega_1}{c}\right), \qquad (14)$$

and with the dispersion relationship

$$E_1 - E_2 \simeq \beta c \, (p_1 - p_2) , \qquad (15)$$

the frequency ratio is found to be

$$\left(\frac{\omega_2}{\omega_1}\right) \simeq 4\gamma^2 \qquad (16)$$

for $\gamma \gg 1$, where $\gamma = (1 - \beta^2)^{-1/2}$. For example, if the absorbed radiation wavelength is 1.0 cm, then for 4.5 MeV electrons the emitted radiation wavelength is 25 μm.

C. Bremsstrahlung[5]

In bremsstrahlung, recoil momentum of an atom is used to bridge the momentum gap, as illustrated in Fig. 4. Relatively little energy is involved in the scattering of an atom, and so the recoil is shown as a horizontal vector in the E-p plane. In Fig. 4 both the recoil momentum and emitted photon are in the same direction as the incident electron.

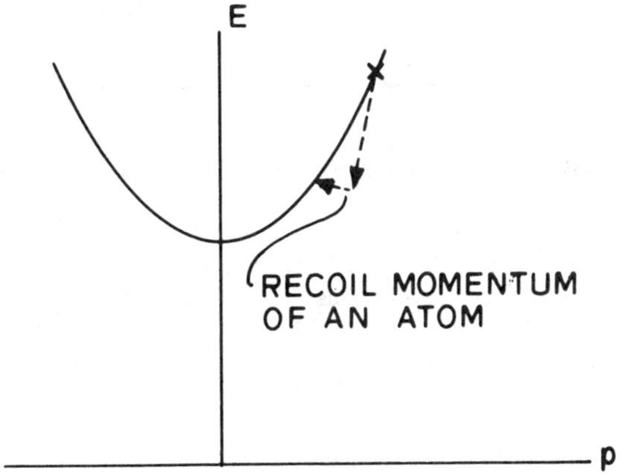

Fig. 4. Bremsstrahlung.

D. Static fields

Static electric and/or magnetic fields can permit photon absorption or emission by providing a multiplicity of electron E-p dispersion curves, corresponding to the energy eigenstates of the electron-field interaction. Figure 5 illustrates the case of a constant magnetic field applied along the z direction, with the electrons following helical trajectories centered about the z axis. The electron energy is given by[6]

$$E = (m^2c^4 + c^2p^2 - 2\ell\ c^2|e\beta|)^{1/2}\ , \tag{17}$$

where $\ell = 0, 1, 2 \ldots$

As shown in Fig. 5 it is possible for a photon to be emitted, with the vector representing the photon in the E-p plane terminating on an electron dispersion curve corresponding to a different ℓ-value than the ℓ-value for the initial electron state.

In the laboratory frame the change in electron energy for single-photon emission is

$$\Delta E = \hbar\omega \simeq \left(\frac{\delta E}{\delta p}\right) \Delta p + \left(\frac{\delta E}{\delta \ell}\right) \Delta\ell, \tag{18}$$

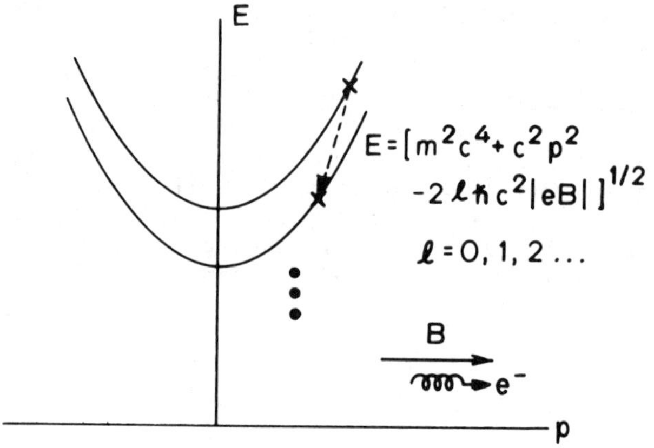

Fig. 5. Photon emission from an electron in a static, uniform magnetic field.

where $\delta E/\delta p = c^2p/E$, $\delta E/\delta \ell = |e\beta|\ (c^2/E)$, $\Delta p = \ \omega/c$, and $\Delta \ell = 1$. Substituting the above expressions into Eq. (18) and using $E = \gamma mc^2$ and $p = mc[\gamma^2 - 1]^{1/2}$, then for $\gamma >> 1$ it is found that

$$\omega = 2\gamma^2\omega_c, \tag{19}$$

where ω_c, the relativistic cyclotron frequency, is defined as $\omega_c = |eB|\ (\gamma m)^{-1}$. Equation (19) applies to the collinear case, and if there is an angle θ between the direction of motion and the direction of emission Eq. (19) becomes

$$\omega = \frac{\omega_c}{1 - \beta\cos\theta}\ . \tag{20}$$

The electron remains in a constant phase of the electromagnetic field, as is now shown. For a circularly polarized wave, the phase ψ is

$$\psi = \omega t - \left(\frac{\omega}{c}\right)z - \phi\ , \tag{21}$$

where, for the electron

$$z = vt \qquad \phi = \omega_c t, \tag{22}$$

assuming that the transverse component of velocity is small. Therefore, the phase seen by the particle is

$$\psi = \omega t(1 - \beta - \omega_c/\omega) \tag{23}$$

and substituting $1 - \beta \cong (2\gamma^2)^{-1}$ and $\omega_c/\omega \cong (2\gamma^2)^{-1}$ into Eq. (23) gives $\psi = 0$.

Photon emission from an electron can also be obtained by replacing the static magnetic field with other electrostatic field arrangements. For example, crossed electric and magnetic fields can be used as in a magnetron to give the electron cycloidal motion, or a periodic magnetic field (a "wiggler") can be used to provide undulating motion.

E. Limited interaction length[7]

As shown in Fig. 6, if a free electron emits a photon in the forward direction, energy conservation requires a momentum gap δp equal to

$$\delta p \simeq \hbar\omega \left[\frac{1}{v} - \frac{1}{c}\right] \simeq \frac{\hbar\omega}{2c\gamma^2}, \tag{24}$$

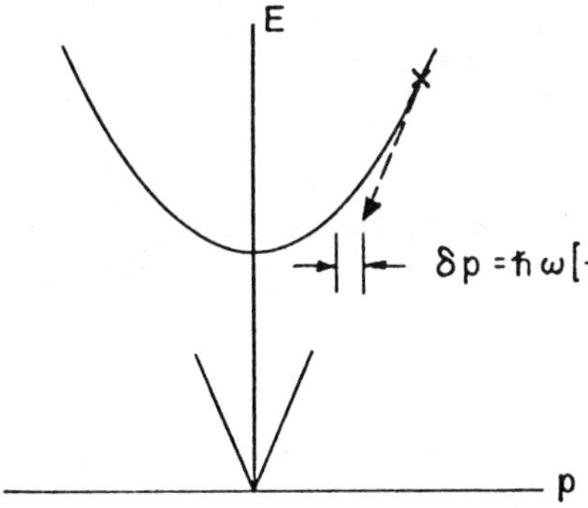

Fig. 6. Photon emission over a limited interaction length.

where it has been assumed that the photon energy is much less than the electron energy. If the interaction length δz is limited so that

$$(\delta z)\,(\delta p) \leqslant \hbar/2 \tag{25}$$

then an uncertainty exists in the momentum that could allow single photon emission from an electron. From Eqs. (24) and (25) it is found that the required condition is

$$\frac{\delta z}{\lambda} \leqslant \frac{\gamma^2}{2\pi}. \tag{26}$$

If γ is sufficiently large then the interaction length can be many wavelengths long.

From a classical viewpoint, the interaction region is limited so that the phase slippage of the particle in the field is small compared to 2π. For high γ, v_p and v are almost equal so that δz can be thousands of wavelengths long.

3. SOME EXAMPLES

A. The free-electron laser

The Free-Electron Laser (FEL) is the adaptation of the microwave frequency "traveling wave tube" principle, to the optical and infrared parts of the spectrum.[8] An FEL consists of three components: an electron beam, an electromagnetic field, and a magnetostatic wiggler field. This device can be tunable from the ultra-violet through the infrared; it has the potential for high power, and someday it may be operated at respectable efficiencies.

The magnetic flux density of the wiggler field, B, is oriented transverse to the z-direction, the direction of relativistic motion of the electron beam, and the orientation of B rotates as one follows the z-direction. Flux density B can be written as

$$B = \hat{u}_x B_0 \cos\kappa z + \hat{u}_y B_0 \sin\kappa z \;, \tag{27}$$

where $\kappa = 2\pi/s$ and s is the periodicity of the magnet. The electron follows a helical trajectory given by

$$\begin{aligned} r &= \text{radius of the helix} = \frac{|e|B_0}{\gamma m\kappa^2\beta_z c} \\ \phi &= \left(\frac{2\pi}{s}\right)z + \phi(0) \\ z &= \beta_z ct, \end{aligned} \tag{28}$$

where β_z is the ratio of the z-component of velocity to the velocity of light. The electron velocity in the x-y plane, v_t, is a constant given by

$$v_t = \frac{|e|B_0}{\gamma m\kappa} \;. \tag{29}$$

Figure 7 shows the FEL interaction in energy-momentum space. The magnetic wiggler field provides a momentum p_m given by

$$p_m = \frac{2\pi\hbar}{s}. \tag{30}$$

A synchronism condition can be derived by using energy conservation and the dispersion relationship given by Eq. (9), and from momentum conservation

$$p_1 - p_2 = \frac{\hbar\omega}{c} + \frac{2\pi\hbar}{s} \;. \tag{31}$$

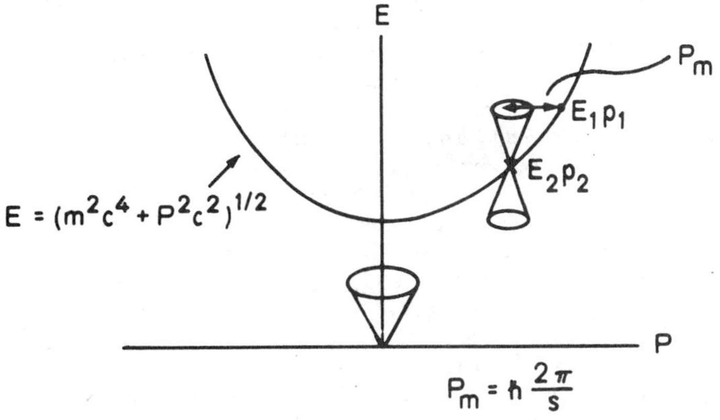

Fig. 7. Free-electron laser interaction.

The simultaneous solution of Eqs. (9) and (31) gives

$$\frac{\lambda}{s} = \frac{1 - \beta}{\beta} \tag{32}$$

For the non-collinear case, $\beta \to \beta_z$ and Eq. (32) becomes

$$\frac{\lambda}{s} = \frac{1 - \beta_z}{\beta_z} \tag{33}$$

which is the synchronism condition.

Equation (33) may be rewritten in terms of the parameters of the wiggler field. Since

$$\gamma^2 = \frac{1}{1 - \beta^2} = \frac{1}{1 - \beta_z^2 - v_t^2/c^2}\ , \tag{34}$$

from the expression for v_t given by Eq. (29) we have

$$\beta_z \simeq 1 - \frac{1}{2\gamma^2}\left[1 + \left[\frac{eB_0}{mc\kappa}\right]^2\right] \tag{35}$$

for β_z close to 1. The combination of Eqs. (33) and (35) then gives

$$\frac{\lambda}{s} = \frac{1}{2\gamma^2}\{1 + [0.093\ sB_0]^2\}\ , \tag{36}$$

where s is in cm and B_0 is in kG. For example, with s = 5 cm, B_0 = 1 kG, and $\gamma = 98.8$ (E = 50 MeV), the wavelength is $\lambda = 3.1$ μm.

Equation (33) implies that the electron remains in a constant phase of a circularly polarized electromagnetic wave propagating in the z-direction. This can be shown by substituting the particle coordinates, Eq. (28) into the expression for the phase of an electromagnetic wave, and requiring synchronism.

If there was no bunching of the electron beam then there would be no net energy transfer from the electrons to the light wave, since there would be an equal amount of acceleration and deceleration. However, from Eq. (35) it is seen that as γ decreases β_z decreases, and as γ increases β_z increases. This means that the particles bunch along the z-axis around particles that are in phase with zero amplitude field. By having the phase velocity of the wave slightly below the value determined from the synchronism condition the bunch of electrons will be in a retarding field and therefore be delivering energy to the light wave. If a laser-driven electron accelerator is desired, then the phase velocity should be slightly greater than the synchronous velocity to place the particles in an accelerating field.

The bandwidth of an FEL is primarily determined by the number of periods in the wiggler magnet. If N_p is the number of periods, then

$$\frac{\Delta\lambda}{\lambda} \simeq \frac{1}{N_p} , \tag{37}$$

where $\Delta\lambda$ is the full-width, half-maximum (FWHM) linewidth.

The efficiency of the FEL is largely a function of the linewidth. As energy is removed from the electrons, γ decreases causing the wavelength, as given by Eq. (36), to increase. When the wavelength changes sufficiently so as to fall outside the linewidth expressed by Eq. (37) then the gain of the amplifier decreases. This places an upper limit on the amount of energy that can be extracted from the electrons. From Eq. (36) we have

$$\frac{\Delta\lambda}{\lambda} = -2\,\frac{\Delta\gamma}{\gamma} , \tag{38}$$

so that for $N_p = 100$, $\Delta\gamma/\gamma$ is limited to 0.5%.

A number of modifications can be made to improve the efficiency. From Eq. (36) it can be seen that s and/or B_0 can be decreased to compensate for decreasing γ. An FEL with decreasing s is described as a "tapered-wiggler" magnet. Alternatively, a storage ring can be employed so that the electrons have their energy losses restored and are reused. A technique used for microwave tubes and that can be applied to the FEL with a single-pass electron beam is to accelerate electrons to a high voltage at the amplifier section relative to the electron gun, but to depress the collector potential so that it is not much greater than that of the emitter, so that most of the electron energy is recovered by the collector.

Finally, it has also been proposed that the electrons passing through the FEL can be used as in a klystron to generate microwaves that are then fed back into the linear accelerator. Free-electron lasers that have operated thus far have had efficiencies below 1%, but it is anticipated that future FEL's will operate at several per cent efficiency.[9]

B. Limited interaction length

Another example of energy exchange between electrons and radiation is the limited interaction approach illustrated in Fig. 6. As shown in Fig. 8, it is assumed that a focused Gaussian laser beam is propagating in the z direction, polarized in the x direction at z = 0, and uniform in the y direction. An electron moving in the x-z plane intersects the z axis at an angle ψ.

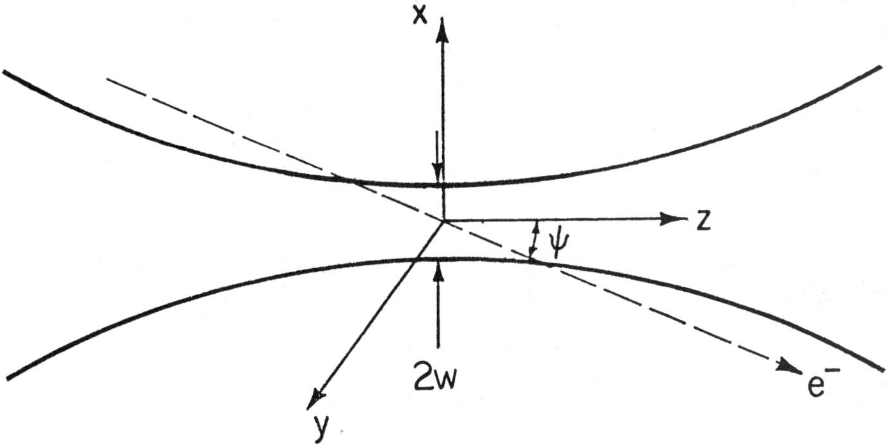

Fig. 8. Electrons move at an angle c relative to the z axis through a focused light beam that has a Gaussian variation in the x direction and is uniform in the y direction. The field is polarized in the x direction at z = 0.

The electromagnetic field components may be represented as the real part of a superposition of plane waves traveling in different directions,

$$\begin{aligned} E_x(x,z,t) &= \exp(-i\omega t) \int_{-\pi/2}^{\pi/2} d\theta\, a(\theta) \cos\theta \\ &\quad \times \exp(ikx \sin\theta + ikz \cos\theta), \\ E_z(x,z,t) &= -\exp(-i\omega t) \int_{-\pi/2}^{\pi/2} d\theta\, a(\theta) \sin\theta \\ &\quad \times \exp(ikx \sin\theta + ikz \cos\theta)\,, \end{aligned} \tag{39}$$

where ω is the frequency, $k = 2\pi/\lambda$, and λ is the wavelength. This angular spectrum is the Fourier transform of the transverse spatial distribution of the laser light.

Assuming a Gaussian-shaped line focus located at $z = 0$, then

$$E_x(x,0,t) = E_0 \exp(-i\omega t)\, \exp(-x^2/w^2) \, , \tag{40}$$

where w is the beam spot size and $E_0 = |E_0| \exp(i\phi)$ is the complex field amplitude, the expression for $a(\theta)$ becomes

$$a(\theta) = \left(\frac{kw}{2\pi^{1/2}}\right) E_0 \, \exp - \left[\left(\frac{wk \sin\theta}{2}\right)^2\right]. \tag{41}$$

The expansion, as given by Eqs. (39) and (41), closely represents the field if $2\pi w/\lambda >> 1$, which is satisfied for the considered range of parameter values.

The change in particle momentum in the direction parallel to its initial motion is given by the force equation

$$\frac{dp_{\parallel}}{dt} = -qE_x \sin\psi + qE_z \cos\psi, \tag{42}$$

where ψ is the angle in the x-z plane at which the electron is traveling, as shown in Fig. 8, where q is the charge of the particle and $p_{\parallel}$ is the momentum change produced by the field along the direction of initial motion. For momentum changes small compared to the initial momentum, the energy change ΔW is given, from Eq. (8), by

$$\Delta W \simeq \beta c \Delta p_{\parallel}, \tag{43}$$

where c is the velocity of light and β is the ratio of particle velocity to the velocity of light.

To determine ΔW the Born approximation is used, i.e., the particle trajectory in the interaction region is assumed to be unaltered by the presence of the field. Taking $z = 0$ at $t = 0$, we have

$$z = \beta ct \cos\psi \, , \tag{44}$$

$$x = x_0 - \beta ct \sin\psi \, , \tag{45}$$

where x_0 is the x position of the particle at $z = t = 0$. The simultaneous solution of Eqs. (39) - (45) gives the expression for ΔW

$$\Delta W = \frac{-qw|E_0|}{2\pi^{1/2}} \int_{-1}^{1} du \exp(-a^2u^2) \frac{d[\ln[f(u)]\}}{du}$$
$$\times \sin\left[\frac{2\pi s}{\beta\lambda} f(u) - \frac{2\pi x_0}{\lambda} u - \phi\right], \tag{46}$$

where $a = \pi w/\lambda$, $s = z/\cos\psi$ is the distance measured along the direction of particle motion, and

$$f(u) = 1 + \beta u \sin\psi - \beta(1 - u^2)^{1/2}\cos\psi . \tag{47}$$

Equation (46) has been integrated numerically to determine ΔW for various sets of parameters. Figure 9 shows the ratio $\Delta W/P^{1/2}$, where P is the laser power per unit length in the y direction, as a function of the distance s along the direction of electron travel. Curve (a) is $\Delta W/P^{1/2}$ for phase $\phi = 0$, and curve (b) is for $\phi = \pi$. Curve (a) is an odd function of s and curve (b) is an even function of s. The dashed line is the envelope of these curves and therefore shows the maximum $\Delta W/P^{1/2}$ for any given value of s. Input parameters for Fig. 9 are the laser wavelength $\lambda = 1.06$ μm and the particle energy $\gamma = (1 - \beta^2)^{-1/2} = 200$ (1012 MeV). The angle of intersection $\psi = 8.5$ mrad, the laser spot size w = 55 μm, and $x_0 = 0$ were chosen to maximize $\Delta W/P^{1/2}$. It is seen from Fig. 9 that, if the interaction length is unlimited, $\Delta W = 0$.

To calibrate the ordinate in Fig. 9, consider a laser providing 10^{12} W/cm corresponding to a field strength, for w = 55 μm, of 3.7 $\times 10^8$ V/cm. This field strength is easily within the capability of lasers. Then, at an ordinate value of unity in Fig. 9 the energy transfer ΔW is 1.0 MeV.

Figure 10 illustrates the maximum energy exchange (i.e., the value of the dashed curve in Fig. 9 as a function of the angle ψ between the electron velocity and the light wave vector, and Fig. 11 illustrates the maximum energy exchange as a function of the laser beam spot size w.

The effect of displacement of the electron along the x direction, so that it misses the center of the light beam focus, is illustrated in Fig. 12. Numbers written adjacent to points on the electron trajectories give the maximum values for $\Delta W/P^{1/2}$ at the corresponding positions. It is seen that ΔW falls off fairly rapidly at z = 0 as the electron is displaced in x, but the falloff is much slower along a path near the z axis. The values of $\Delta W/P^{1/2}$ shown near the z axis are at the position of maximum $\Delta W/P^{1/2}$.

A calculation has been performed without the Born approximation, allowing the particle trajectory to be altered by the electromagnetic field. The Born approximation becomes invalid when the laser power is sufficiently high so that an electron receiving

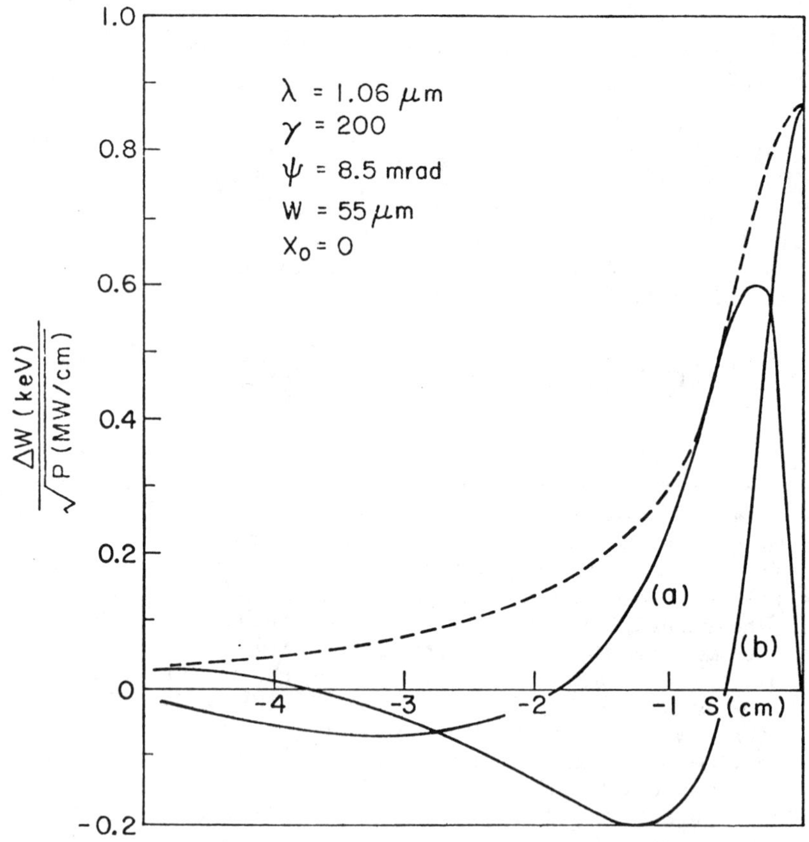

Fig. 9. The ratio $\Delta W/P^{1/2}$ as a function of distance along the direction of electron motion, where ΔW is the energy exchange and P is the power per unit length in the y direction in the laser beam. For curve (a) $\phi = 0$ and for curve (b) $\phi = (1/2)\pi$, where ϕ is the phase constant that appears in Eq. (46). The dashed line is the envelope of the curves for different phase constants and therefore is the maximum value of $\Delta W/P^{1/2}$ at any given position.

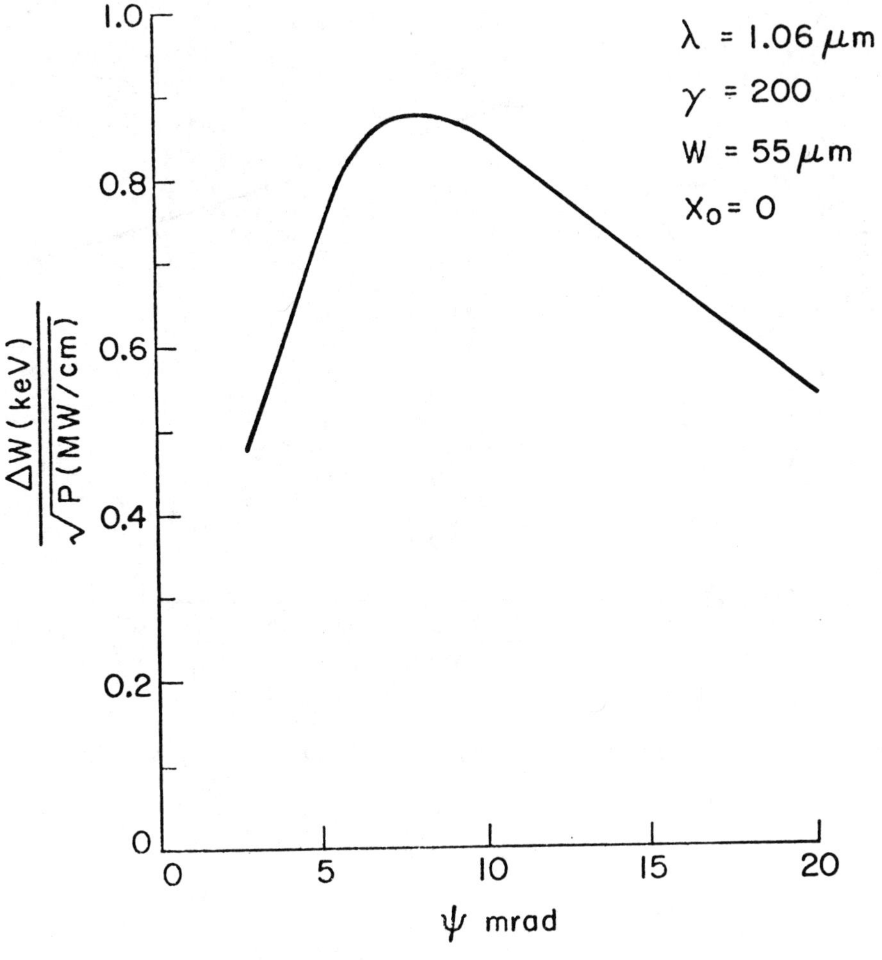

Fig. 10. The maximum value of $\Delta W/P^{1/2}$ as a function of the angle ψ between the electron velocity and the light wave vector.

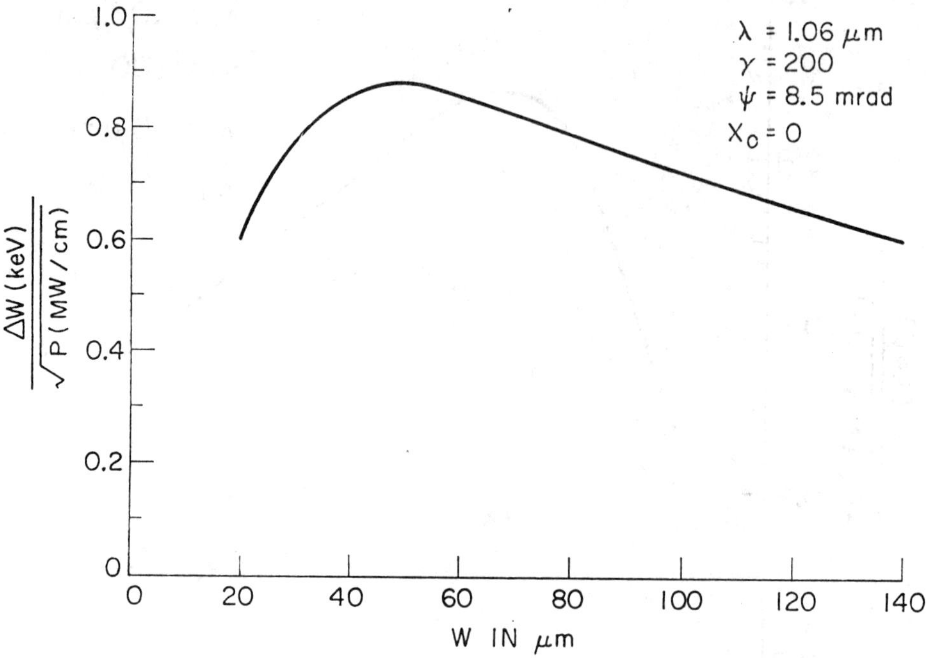

Fig. 11. The maximum value of $\Delta W/P^{1/2}$ as a function of the laser beam spot size.

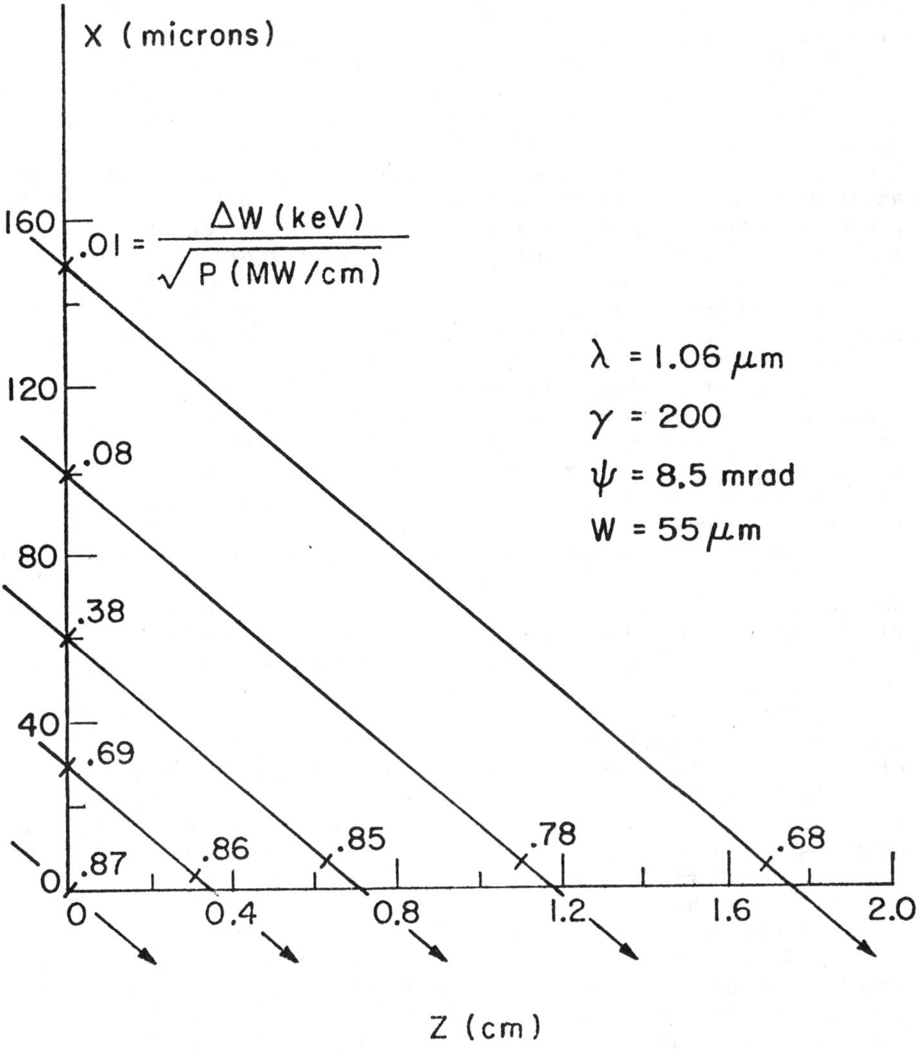

Fig. 12. The effect of displacement of the electron along the x direction. Numbers written adjacent to points on the electron trajectories give the maximum values for $\Delta W/P^{1/2}$ at the corresponding positions.

maximum ΔW gains in position about one-quarter of a wavelength within the interaction region on an electron receiving zero ΔW. For the parameter values given in Fig. 9, this requires an energy change of 10 MeV or greater. The energy exchange of interest is on the order of 20 keV, in which case the exact solution for ΔW differs from the Born approximation solution by one part in 10^4.

A light beam focused in two dimensions has also been considered, in which the field has a Gaussian variation in the x and y directions. (Refer to Fig. 8.) The energy exchange ΔW for a particle passing through the focus is approximately the same for both two-dimensional and one-dimensional focused light beams if the axial electric field is the same in both cases. This means that the power required to obtain a specified ΔW for a one-dimensional beam must be $\simeq w_y/w_x$ times the power in a two-dimensional beam, where w_x is the spot size along the x direction and w_y is the spot size along the y direction. The advantage in using a one-dimensional light beam is that the y dimension of the particle beam can be increased, thereby allowing a greater electron current.

One method for obtaining energy exchange between focused laser light and free electrons over a limited interaction length is to terminate the electromagnetic field in the region of intense field strength by a conducting barrier. Since the field strengths of interest are on the order of 10^6 V/cm or larger, most materials will be damaged or vaporized.

A possible technique for terminating the high fields of a pulsed laser is to use a moving thin foil which presents a new surface for each pulse. For a foil that is 10 μm thick and with a field strength of $\simeq 10^6$ V/cm, the burn-through time for light is in excess of 8 nsec[10] so that this does provide a means for stopping the radiation.

If the purpose of the energy exchange is to produce bunching of the particles so that the electrons can reradiate in the manner of a klystron, then the energy ΔW must exceed the random energy of the particle beam.[11] There will be two sources of random energy: the intrinsic energy spread of the incident particle beam, and that which is introduced by the foil used to terminate the light. For the superconducting accelerator at Stanford the intrinsic energy spread at $\gamma = 200$ is $\simeq$ is 20 keV, and a 20-μm Al foil introduces $\simeq$ 10-keV random energy. This means that the total rms energy spread is $\simeq$ 22 keV. From Fig. 9 it is seen that the laser power per unit length must be > 640 MW/cm for significant bunching to occur, and this power requirement is well within the capability of pulsed Nd lasers. The corresponding field strength at the focus is 6.0×10^6 V/cm.

The alignment of the foil is important for this determines the amount of energy exchange for particles that do not pass through the center of the light-beam focus (refer to Fig. 12) and also affects the thickness of the foil presented to the electrons. To minimize the random energy on the particle beam the thickness should be as small as possible, which suggests that the foil should be normal to the electron velocity. On the other hand, from Fig. 12 it is seen that the foil should be nearly parallel to the particle direction to maximize ΔW for particles that do not pass through the focus. There will be an optimum foil position, between the normal and parallel positions, which will depend upon the x dimension of the electron beam.

The foil can be eliminated by altering the electron's trajectory as it passes through the interaction region. From Figs. 9 and 10 we see that a change in direction of the particle by $\simeq 8$ mrad in a distance less than the distance over which most of the energy exchange occurs (i.e., less than 1 cm) is equivalent to an abrupt termination of the field. For example, bending the electron orbit with a transverse magnetic field of 10 kG over a distance of 0.5 cm at the parameter values given in Figs. 9 and 10, obtains 30% of the peak energy exchange without the use of a foil.

C. Stimulated synchrotron effect

In this section an analysis is presented of energy exchange by the stimulated synchrotron effect,[12] in which an electron follows a circular orbit in a constant magnetic field, and passes through a focused laser light beam. There are several advantages to this approach over alternative approaches. Unlike the free-electron laser, it is not necessary to operate at a specified light frequency and a periodic magnetic circuit is not required. Stimulated synchrotron interaction is not as strong as the stimulated Cherenkov effect, but no medium is required for wave-vector matching, eliminating problems associated with electron scattering in the medium. Another approach, the Smith-Purcell effect,[13] involves propagating an evanescent wave along a boundary, and the laser light field strength is limited to values that are orders of magnitude below the field strengths attainable in vacuum. Limited interaction length in vacuum, discussed above, requires the presence of a light reflector or scatterer, and does not provide as much energy exchange as the stimulated synchrotron effect.

Figure 13 illustrates the interaction. The light beam is propagating in the z direction, is uniform in the y direction, and is brought to a focus at $z = 0$ with the electric field polarized in the x direction. Electrons move in a circular orbit with radius R and velocity v in the x-z plane, and the magnetic flux vector B is in the y direction. The electromagnetic field components can be represented as a superposition of plane waves as given by Eqs. (39)-(41).

It is assumed that the particle trajectory is unaltered by the presence of the electromagnetic wave. This condition may be violated for sufficiently high field strengths, as discussed later in this paper, but it is a valid assumption for the range of parameter values considered in the present case. Taking $z = z_0$ at $t = 0$, the electron position is given by

$$z = z_0 + R \sin\left(\frac{vt}{R}\right) \tag{48}$$

$$x = x_0 - R \cos\left(\frac{vt}{R}\right), \tag{49}$$

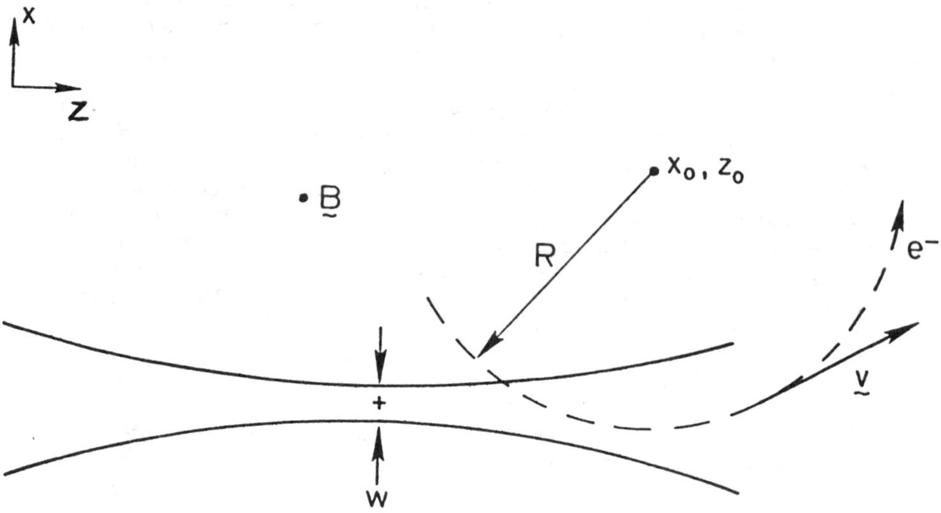

Fig. 13. Electrons traverse a circular orbit in a magnetic field and pass through a focused light beam that has a Gaussian variation in the x direction and is uniform in the y direction.

where (x_0, z_0) are the coordinates of the center of the electron orbit.

We wish to determine the change in particle energy ΔW that results from the field. Assuming that this change is small compared to the initial particle energy, we have that

$$\Delta W \simeq v p_{\parallel}, \qquad (50)$$

where $p_{\parallel}$ is the momentum change along the direction of the circular orbit. From the force equation,

$$\frac{dp_{\parallel}}{dt} = -|e|E_x \sin\left(\frac{vt}{R}\right) - |e|E_z \cos\left(\frac{vt}{R}\right), \qquad (51)$$

where $|e|$ is the magnitude of the electron charge.

The simultaneous solution of Eqs. (39)-(41) and (48)-(51) gives

$$\Delta W(\mathrm{eV}) = -2\pi^{1/2}\alpha R\, |E_0|\, \frac{dJ_n(\beta N)}{d(\beta N)}\, I, \qquad (52)$$

where

$$I = \int_{-1}^{1} du \frac{\exp(-\alpha^2 u^2)}{(1 - u^2)^{1/2}}$$
$$\times \cos[kx_0 u + kz_0 (1 - u^2)^{1/2} - N \sin^{-1} u + \phi], \tag{53}$$

$\alpha = w\pi/\lambda$, J_N is the Bessel function of integral order N, $N = 2\pi R/\beta\lambda$ (taken to be an integer), $\beta = v/c$, and ϕ is a phase factor that can have any value between zero and 2π. For the condition $N/2\gamma^2 >> 1$, where $\gamma = (1 - \beta^2)^{-1/2}$, and with $\beta \simeq 1$, Eq. (52) becomes

$$\Delta W(eV) = - \frac{2^{5/3}\pi^{1/2}\alpha|E_0|RI}{N^{2/3}} A_i'\left(\frac{N^{2/3}}{2^{2/3}\gamma^2}\right), \tag{54}$$

where A_i' is the derivative of the Airy function. Equation (54) is a good approximation in the range of parameter values used to obtain the curves shown in Figs. 14-19.

Equation (54) has been evaluated for various sets of parameters. For the solid curves in Fig. 14, the ratio $\Delta W/P^{1/2}$, where P is the laser power per unit length in the y direction, is plotted as a function of the orbit radius R for different values of the spot size. The center of the orbit is chosen to be at $x_0 = R/\beta$, $z_0 = 0$, which causes the particle to pass slightly above the center of the light beam focus. Maximum energy exchange occurs for $\phi = 0$. The wavelength is 1.06 μm, corresponding to the Nd laser, and for Fig. 14(a) $\gamma = 100$, and for Fig. 14(b) $\gamma = 200$. It is seen from Fig. 14 that the effect of increasing γ is to increase ΔW, and also to increase the values of R and w at which the maximum exchange occurs. For the same γ, the energy exchange by the stimulated synchrotron effect is about three times as much as by the limited interaction length.

If the energy exchange is to produce bunching of a particle beam, as in a klystron, then ΔW from Fig. 14 must equal or exceed the energy spread of the beam. At $\gamma = 200$, for example, the energy spread for the Stanford Superconducting Accelerator beam is $\simeq$ 20 keV. With an orbit radius R = 100 cm and spot size w = 30 μm, the laser power required for bunching is $\geqslant$ 59 MW/cm, corresponding to a field strength at the focus $\geqslant 2.7\times10^6$ V/cm. This power requirement is readily met by pulsed Nd lasers.

Also in Fig. 14, curves of maximum energy exchange as a function of orbit radius, allowing x_0 and w to vary, are shown as dashed lines. The maximum ΔW occurs when x_0 is greater than R, which means that the electron receiving the greatest energy from the wave passes above the focal point. This effect is shown in Fig. 15.

Figure 15 illustrates the variation of the ratio $\Delta W/P^{1/2}$ with changes in x_0-R. This dependence is important because it determines the allowed x dimension of the electron beam as it passes through

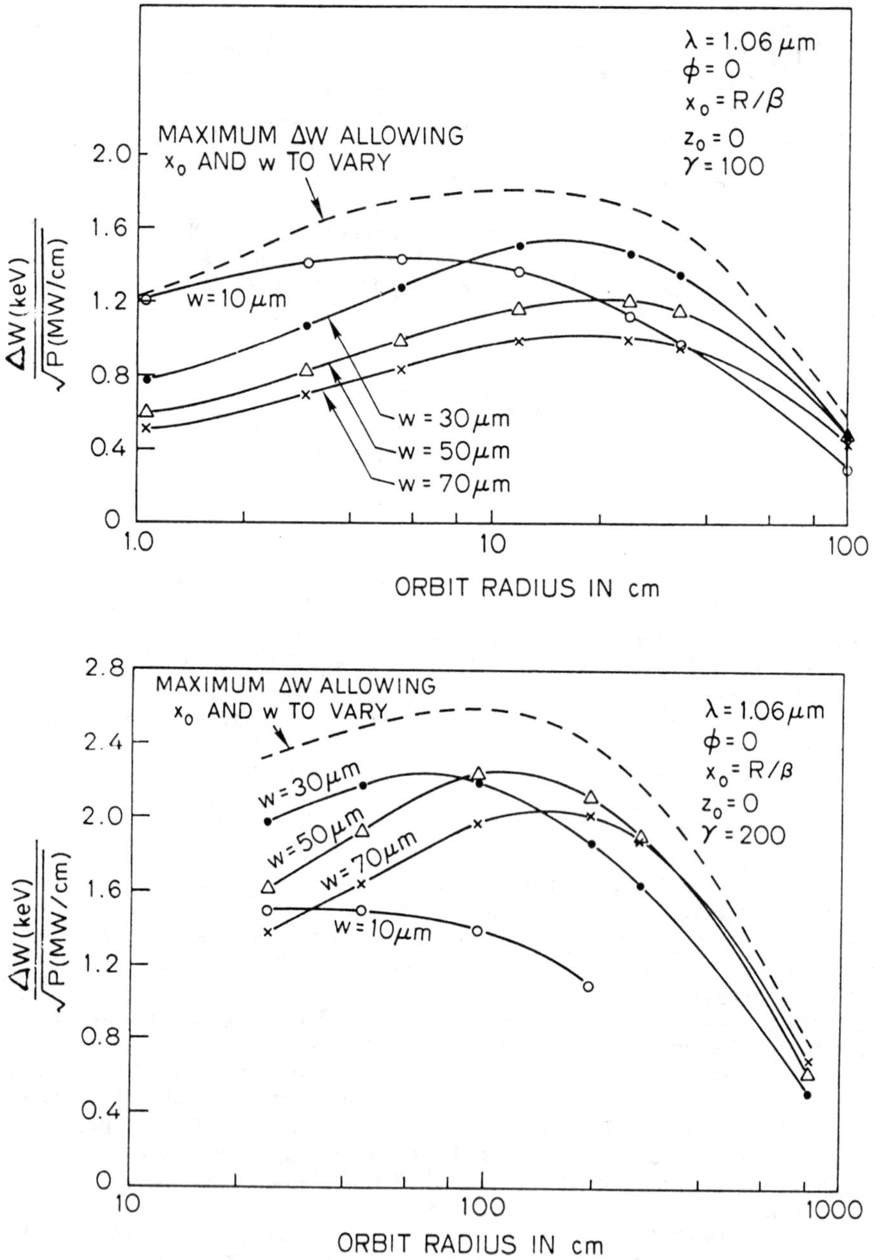

Fig. 14. Curves showing the dependence of the energy exchange on the radius of the electron orbit. For the solid curves the light beam spot size and the center of the orbit are fixed. The dashed curves are obtained by allowing these parameters to vary and maximizing the energy exchange. (a) is for γ = 100 and (b) is for γ = 200.

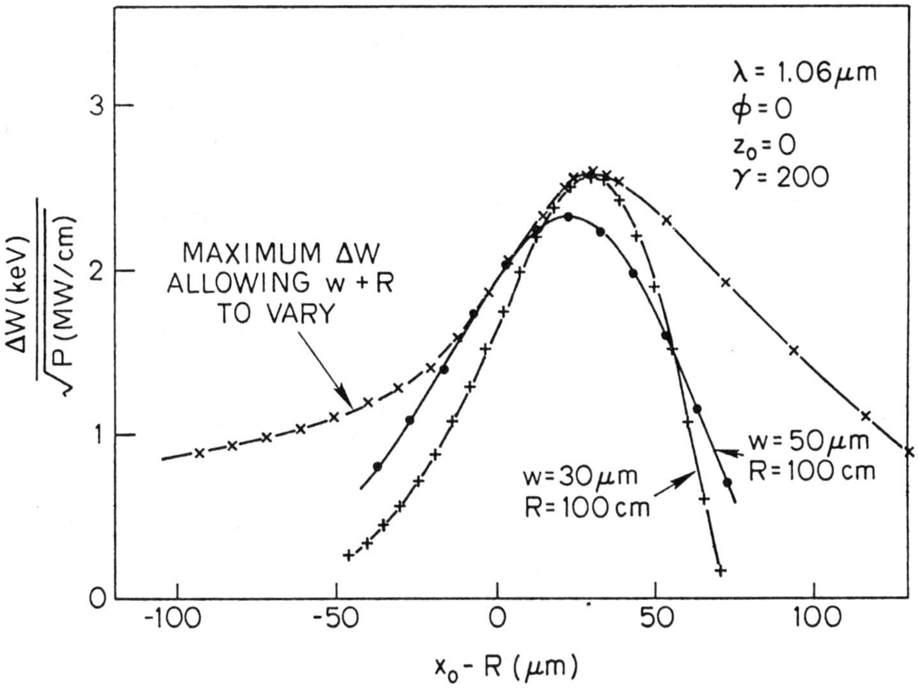

Fig. 15. Dependence of the energy exchange on the x coordinate of the center of the electron orbit. The solid curves are for fixed value of spot size and orbit radius, and the dashed curve is the maximum energy exchange when these parameters are allowed to vary.

the laser light. For example, for the parameter values used for the solid curve in Fig. 15, if ΔW is not to vary more than ± 5% for different electrons, then the x dimension must not exceed 35 μm. The dashed line gives the maximum ΔW with w and R allowed to vary. The curves in Fig. 15 show that the energy exchange is greatest at values of x_0 greater than R.

Figure 16 illustrates the variation of energy exchange with the spot size of the light beam with λ, ϕ, and z_0 held constant, and x_0 and R varied to optimize ΔW. It may be desirable to operate at a spot size larger than the optimum w, to allow a larger-diameter electron beam. For example, at $\gamma = 200$ a spot size equal to 100 μm gives only 65% of the maximum possible energy exchange, but with ΔW limited to ± 5%, the electron beam can be approximately three times the size of the beam when ΔW is optimized.

Figure 17 shows the dependence of $\Delta W/P^{1/2}$ on z_0. For this case the phase is allowed to vary as $\phi = -kz_0$, because this is the phase that gives the maximum value for ΔW. Since the z dimension for which the spot size of the light beam is within a factor of 2 of the spot size at the focus (i.e., the Rayleigh range) is several centimeters long, one would expect the dependence on z_0 to be much more

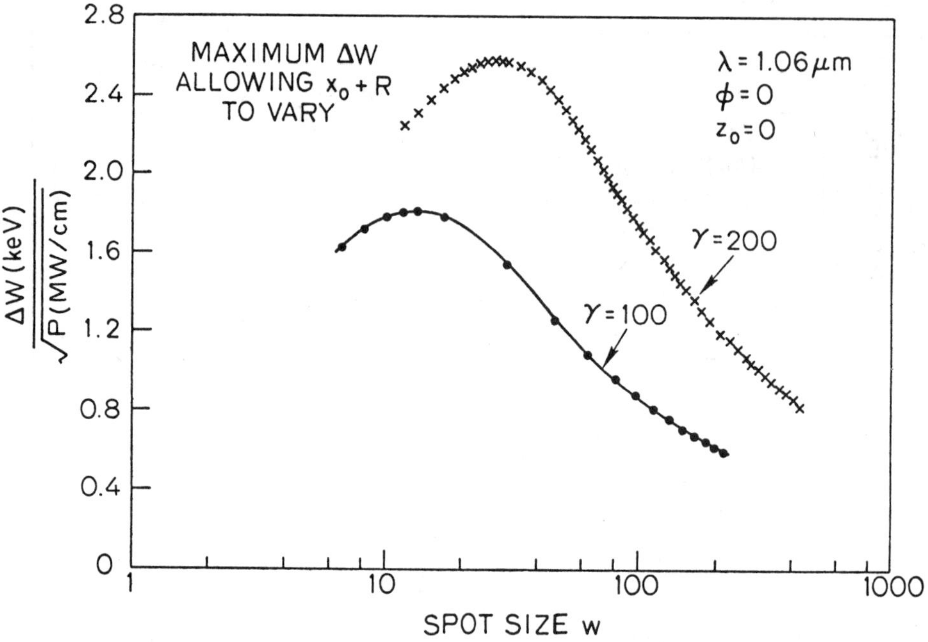

Fig. 16. Energy exchange as a function of the spot size of the light beam.

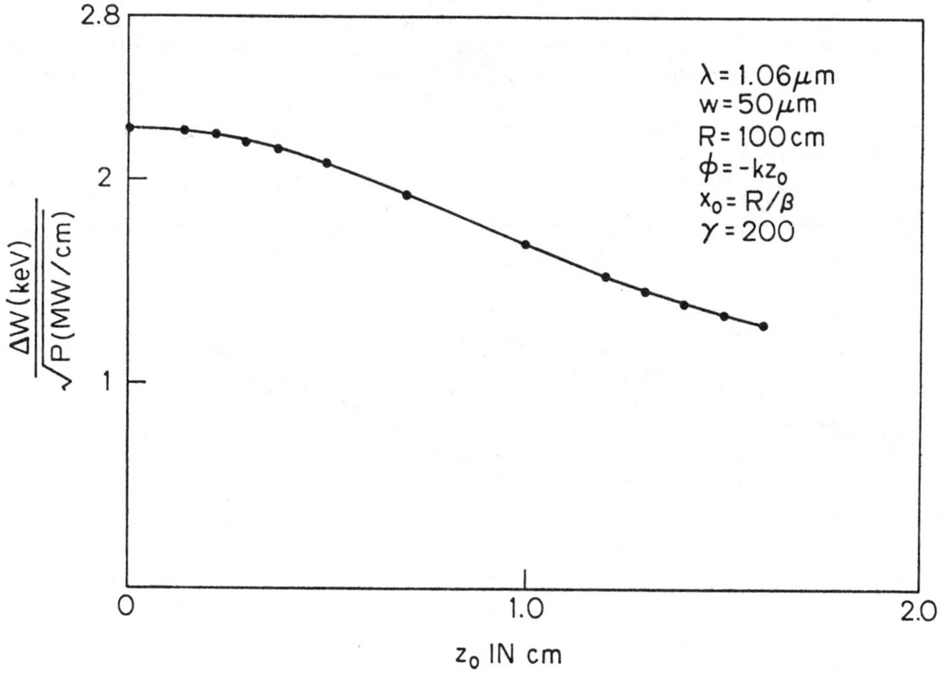

Fig. 17. Energy exchange as a function of the z coordinate of the center of the electron orbit.

gradual than the dependence upon x_0. Figure 17 confirms this, since the decrease of ΔW with z occurs over a distance on the order of 1 cm.

Figure 18 gives the dependence of the energy exchange on the particle energy parameter γ. For this curve λ, ϕ, and z_0 are held constant and x_0, w, and R are allowed to vary to optimize ΔW. It is seen that the energy exchange is very small for $\gamma < 20$ and increases rapidly between $\gamma = 20$ and 50. As γ continues to increase, the energy exchange also increases, with approximately a logarithmic dependence.

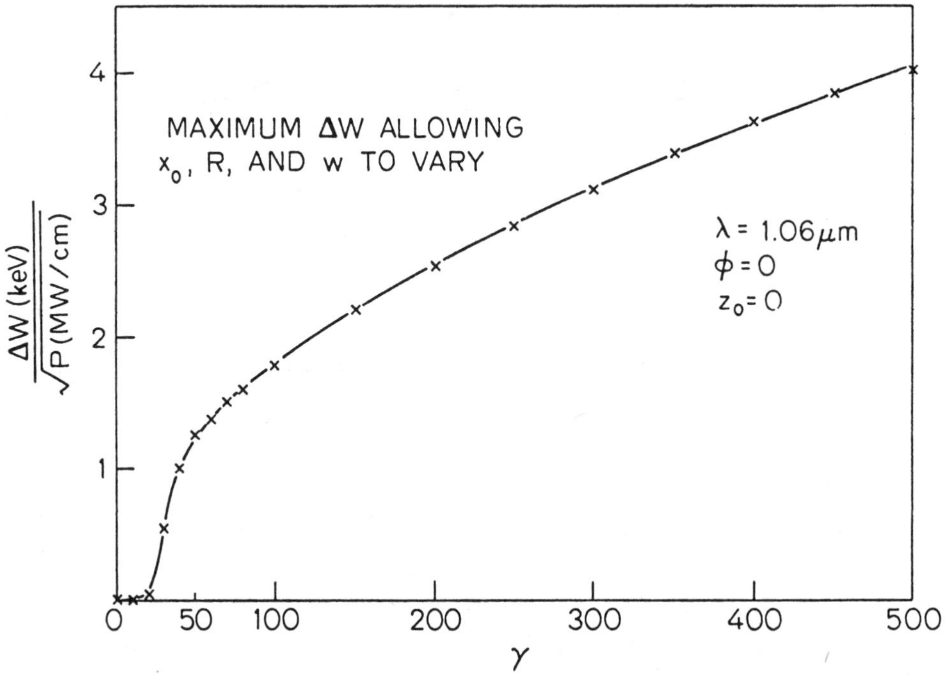

Fig. 18. Energy exchange as a function of the electron energy parameter γ.

The relative contributions to ΔW from E_x and E_z can be seen in Fig. 19 at the top of the next page. The two curves give the projection of the fields on the direction of electron motion (for $\phi = 0$) as a function of s, the distance measured along the particle trajectory ($s = 0$ is taken to occur at $z = z_0 = 0$). Energy exchange is proportional to the integral of these curves, so that the primary source of energy exchange results from E_x. This is a consequence of the fact that the particle sees an E_x field of constant sign over a rather long region about the center of the orbit.

Some important considerations are the magnetic field requirement, the synchrotron radiation energy loss, and the degree of

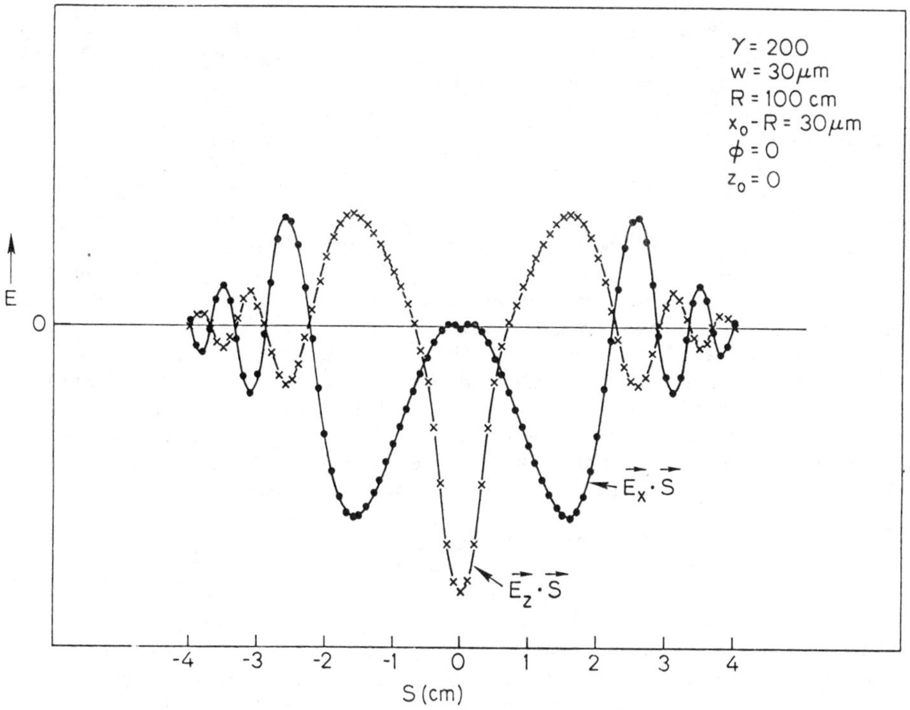

Fig. 19. Electric field (in arbitrary units) projected along the particle trajectory as a function of distance along the trajectory.

validity of the assumption that the particle trajectory is unaltered by the field within the interaction region.

At a given particle energy γ and orbit radius R in centimeters, the required magnetic flux density B in kilogauss is

$$B \simeq 1.72\ \gamma/R. \tag{55}$$

From Fig. 14(b) we find that at $\gamma = 200$ the optimum R is approximately 100 cm. Substituting these numbers in Eq. (53) gives B = 3.44 kG, which is readily attainable.

For one orbit of the electron, the synchrotron radiation loss is

$$\Delta W\ (\text{keV}) \simeq 6.0\times10^{-10}\gamma^4 R^{-1}, \tag{56}$$

where R is measured in centimeters, using $\gamma = 200$ and R = 100 cm, $\Delta W = 9.6\times10^{-3}$ keV. Since this number is about three orders of magnitude less than the energy exchange of interest, this loss would not be significant.

If the electric field is sufficiently large, then the assumption that the particle trajectory for a single transit is unaltered by the wave would not be valid. An upper limit on the field occurs when the particle moves within the strong interaction region approximately one-fourth of an optical wavelength relative to its position in the absence of the field. For the parameter values $\gamma = 1.06$ μm, R = 100 cm, w = 50 μm, and $\gamma = 200$, this limit is in excess of 10^{10} V/cm, which is three to four orders of magnitude larger than the field required to produce bunching.

The analysis presented above is for a single transit of the electrons through the light beam focus. If the time of flight of a particle in its orbit is adjusted so that it experiences the same phase on successive transits, then the energy exchange can be increased in proportion to the number of transits. Alternatively, if the bunching distance (i.e., the distance for which the electron beam current at the laser frequency is maximum) is arranged to equal the orbital distance, then the particles can deliver energy to the electromagnetic field in the manner of a klystron.

4. LASER-DRIVEN ACCELERATORS

There have been a number of proposals to use lasers for particle acceleration[14–17] because of the high field strengths available from laser beams. Existing machines have maximum acceleration gradients of 10^7–10^8 eV/m, whereas calculated gradients using lasers are in the range from 10^9 to over 10^{10} eV/m. This would mean that for a given output energy the acceleration length could be reduced by about two orders of magnitude.

Several types of laser devices have been suggested: those based upon direct acceleration by the electric field in the light wave, those utilizing the ponderomotive force, and a third type in which the laser introduces a disturbance in a medium which then exerts a force on the particle. In the first case the energy exchange is linearly proportional to the electric field and it is necessary that the particle remain in a constant phase of the wave over the interaction region, whereas in the second case phase synchronization is not critical. However, calculations performed by computer simulations of electron trajectories through the focus of intense laser beams[18] indicate that the ponderomotive acceleration is small and results in highly divergent electron paths. For example, with a field strength of 1.6×10^{11} V/cm at the laser focus and 2.5 MeV incident electrons, the maximum energy gain is only 1.8 MeV with a 30° angular spread of the exiting particles.

The third case is exemplified by the plasma-wave accelerator proposed by Tajima and Dawson.[19] An intense laser beam excites space charge waves in a plasma, which result in fields along the direction of propagation with a phase velocity less than the velocity of light. These fields can be used to accelerate electrons, and for the chosen parameter values (a plasma with 10^{18} electrons/cm^3. and a $\lambda = 1$ μm wavelength laser focused to a power density of 10^{18} W/cm^2) an electron energy of 10^9 eV can be produced in a length of 1 cm. However, at the assumed laser-light field intensities in the plasma non-linearities and instabilities (e. g., Raman and Brillouin scattering, breakdown, parametric oscillations) would occur which would deplete the energy in the laser beam and alter the expected field configurations.

In view of these considerations, the discussion will concentrate on the first category of laser accelerators in which energy transfer is accomplished by the force exerted by the electric field on the particle. This interaction does not occur in vacuum since energy and momentum cannot be conserved. Therefore, the accelerator must include a medium, a nearby surface, a static field, or an additional electromagnetic wave as in the Compton effect.

Several characteristics are shared by all accelerators of this type regardless of the particular environment that is provided. The electric field along the direction of acceleration is small compared to the total field unless the transverse dimensions of the light wave are on the order of the light wavelength. This conclusion follows directly from the condition

$$\nabla \cdot \underline{E} = 0, \tag{57}$$

where $\underline{E}$ is the electric field. This means that there is a tradeoff between the size of the electron beam that can be accommodated and the fraction of the total field contributing to acceleration. If the transverse dimensions are limited to the size of the wavelength then the CO_2 laser at $\lambda = 10.6$ μm would be preferable to a shorter wavelength Nd or dye laser.

Secondly, the duty cycle will be low. Peak laser power should be on the order of 10^{12} W for high acceleration gradients and large energy transfer. If the average power for a CO_2 laser is 100 kW, the duty cycle would be $\simeq 10^{-7}$. Beam emittance and energy spread would be appreciably greater for the laser accelerator than for its microwave counterpart, because there will be significant variations in field strength across the area of the light beam, and there will also be strong field components in the direction normal to the direction of primary acceleration. Large emittance and energy spread make phase matching more difficult.

Periodic refocusing of the light, by lenses or mirrors, will be necessary. For example, a Gaussian-focused light beam has a Rayleigh range, R, in which the radius doubles from its minimum value given by

$$R = \frac{\pi w^2}{\lambda}, \tag{58}$$

where w is the laser beam spot size at the focus. For $\lambda \simeq 1$ μm and $w \simeq 1$ mm, $R \simeq 3$ m. This means every few meters it will be necessary to insert a focusing element to increase the power density for high acceleration gradient.

The peak and average current available from a laser accelerator can be estimated from power considerations. For an acceleration gradient of 10^9 eV/m, the power absorbed per electron is 4.8×10^{-2} watts. If the accelerator operates at 1% efficiency and the peak laser power is 10^{12} watts, then $\simeq 2\times10^{11}$ electrons can be accelerated during the laser pulse. If the light pulse is 10 psec in duration, then the average current within that time period is 3.2×10^3 A. With a duty cycle of 10^{-7}, the time-averaged current is 320 μA. Currents derived on the basis of power estimates give no consideration to possible beam instabilities, wake fields excited by the particles, and space-charge forces, all of which could limit the current to lower values.

In the following sections there are discussions of the advantages, disadvantages, and acceleration gradients available from various laser acceleration configurations.

A. The Smith-Purcell accelerator

There have been several proposals[15,20] to use the Smith-Purcell effect for laser electron acceleration, in which a grating excites an electromagnetic wave from an incident wave that propagates with a phase velocity less than the velocity of light, so that the electron can remain in a field of constant phase. The primary advantage of this type of accelerator is that a significant fraction of the total field can be in the direction of acceleration. As pointed out previously, this means, however, that the transverse size available for the particle beam is limited to dimensions comparable to the wavelength. Another difficulty is that, because of the high local fields existing at the surface of a grating, the mean accelerating field may not be too large without causing breakdown and destruction of the grating.

In the first attempt to use a grating for acceleration,[15] the grating was destroyed by a millisecond-duration laser field which was on the order of 10^4 V/cm. A recent proposal by Palmer[20] suggests using 30 psec pulses, which allows higher laser fields, because a buildup time is required for breakdown. There have not been measurements to determine breakdown field strengths at grating surfaces for 30 picosecond-duration pulses at λ = 10.6 μm, but estimates can be made. Saxman has performed measurements on the damage threshold of gratings using a CO_2 laser,[21] and found that gratings could withstand a few 1.0 nsec duration pulses up to intensities of 10^{10} W/cm^2, corresponding to fields of 2.7×10^6 V/cm. Avalanche breakdown theory indicates that the threshold field should vary as $\tau_p^{-1/4}$, where τ_p is the pulse duration. Using this dependence, the breakdown field for a 30 psec pulse is 6.5×10^6 V/cm. This means that the acceleration gradient would be limited to $\simeq 0.65$ GeV/m. However, breakdown field is a function of the shape of the grooves in the grating, and it would be necessary to make an experimental determination of the breakdown field for a specific grating design.

Palmer points out that much higher gradients could be achieved by allowing destruction of the grating. The proposed grating is 1 Km long with a 10 μm line spacing. It would be difficult and costly to constantly adjust and replace such a grating. Allowing for grating destruction, an acceleration gradient of 20 GeV/m is calculated for a CO_2 laser with a peak power of 2.5×10^{15} watts occurring in 30 psec pulses. A laser of this magnitude has never been constructed and would probably have a very low duty cycle because of system misalignment introduced by thermal and shock effects.

The transverse size of the electron beam is limited. In the Palmer design the electric field falls to $\simeq e^{-1}$ of its surface value at $\lambda/2$, which is 5 μm for a CO_2 laser. The other transverse dimension is calculated to be 25 μm. If an energy spread on the order of e is acceptable, then the electron beam dimensions are $\simeq 5 \times 25$ μm over a 1 Km distance. It would be difficult to obtain significant current with these size constraints.

McIver and Lubin[18] showed that a plane-wave approximation for the field is not appropriate in the focal region for a spot size on the order of the wavelength. With the parameters of the system they analyzed, the energy gain was overestimated by an order of magnitude in the plane-wave approximation. It would be necessary to perform a "McIver and Lubin" type of analysis for the grating accelerator to obtain a good estimate of the acceleration gradient.

Finally, the laser fires in 30 psec bursts in which time the electron travels approximately one centimeter. Therefore, over a 1 Km length it would be necessary to have $\simeq 10^5$ bursts, with phase synchronism maintained for all the bursts. This would be difficult to accomplish.

B. Free-electron laser acceleration

Certain disadvantages of the Smith-Purcell accelerator can be eliminated by alternative acceleration schemes, such as using the principle of the free-electron laser.[17,22,23] In particular, the transverse dimensions can be large compared to the optical wavelength, a single traveling-wave pulse can be used to accelerate the electrons, and no medium is present so that higher field strengths can be obtained. On the other hand, the field component accelerating the electrons is small compared to the total field.

In the free-electron laser a periodic magnetic field forces helical or undulating motion on the particle. Along the particle trajectory a plane electromagnetic wave has a phase velocity less than the velocity of light, so that an electron can remain in a field of constant phase. The acceleration gradient, averaged over twice the Rayleigh range of a Gaussian-focused laser beam, is

$$\frac{dW}{dz} = 13.6\,\frac{P^{1/2}}{w}\,\sin\theta\,\sin\psi \quad \text{eV/m}, \tag{59}$$

where P is the laser power in watts, w is the minimum spot size of the laser beam in meters, ψ is the phase of the electromagnetic wave for a given electron, and θ is the pitch angle of the helical trajectory. If s is the periodic spacing of the magnets,

$$\sin\theta \simeq \left(\frac{2\lambda}{s}\right)^{1/2}. \tag{60}$$

With a 10^{12} watt laser focused to a 1 mm spot, $\lambda = 10$ µm and $s = 5$ cm, the gradient for $\sin\psi = 1$ is 0.27 GeV/m. Twice the Rayleigh range is 63 cm, so that the light would have to be refocused in about that distance. The required magnetic field B is given by

$$B = 0.153 \frac{\lambda^{1/2}}{s^{3/2}} \gamma \text{ kilogauss,} \tag{61}$$

where distances are measured in meters. As the electron energy increases with distance the magnetic field must also increase. At $\gamma = 10^3$, for example, and for the dimensions given previously, B = 43 kilogauss. For GeV particle energies, large magnetic fields are required, necessitating supercooled or pulsed magnets.

It is possible to increase the Rayleigh range and thereby reduce the number of focusing elements by altering λ or w, but this reduces the acceleration gradient. The product of the electron energy increase within twice the Rayleigh range, ΔW, and the acceleration gradient in $(\text{eV})^2/\text{m}$ is

$$\Delta W \frac{dW}{dz} = 2.4 \times 10^3 \frac{P(\text{watts})}{s(\text{meters})} \sin^2\psi. \tag{62}$$

To increase this product the laser power must be increased or the periodicity must be decreased. The latter, however, raises B and enhances the synchrotron radiation loss.

C. Stimulated Cherenkov accelerator[24]

A laser beam can transfer energy to electrons if a medium is provided to retard the phase velocity of the wave below the electron velocity. Then, with an intersection angle θ_c between the directions of wave and particle propagation, where

$$\theta_c = \cos^{-1}\frac{1}{n\beta} \tag{63}$$

the electron remains in a field of constant phase. Equation (63) is the Cherenkov condition, θ is the Cherenkov angle, and n is the index of refraction of the medium. [Refer to Section 2A for the derivation of Eq. (63).] The advantages of this interaction are that the transverse dimensions of the electron beam can be much larger than the light wavelength and no magnetic field is required. On the other hand, the presence of a medium means that voltage breakdown can occur, and that electrons experience multiple scattering. With multiple scattering there is some energy loss due to the scattering, and the particles change direction so that the Cherenkov condition is no longer satisfied. In addition, the component of electric field along the direction of acceleration is small compared to the total electric field.

The acceleration gradient, averaged over a single transit of a particle through the beam, is

$$\frac{dW}{dz} = 38.8 \frac{P^{1/2}}{\sqrt{n}\, w} \sin\theta_c \sin\psi, \tag{64}$$

where ψ is the phase of the field and the total energy change for $\sin\psi = 1$ is

$$\Delta W = 38.8 \frac{P^{1/2}}{\sqrt{n}}, \tag{65}$$

where W is electron energy in eV, the laser power P is in watts, and the minimum spot size w of the laser beam is in meters. Equation (64) is very similar to Eq. (59) for the free electron laser, and so the acceleration gradients are comparable.

If hydrogen at 1 atm is used to provide the index of refraction, then for large γ

$$\theta_c \simeq 1.7\times10^{-2} \text{rad}. \tag{66}$$

From Eq. (64) the acceleration gradient for a 10^{12} watt laser focused to a 1 mm spot (taking $\sin\psi = 1$) is 0.66 GeV/m and the energy increment for a single pass is 38.8 MeV. The interaction region is approximately 10^{-5} radiation lengths, which means that at $\gamma = 10^4$ the angular divergence due to scattering is $\simeq$ 10 μrad and the energy loss is $\simeq$ 50 keV, which is much less than the energy gain from the laser.

Angular divergence from scattering is important when it becomes comparable to the divergence of the laser beam, given by $\lambda/\pi w$. Since scattering angle varies as the square-root of the number of crossings of the laser and electron beams, $\simeq 10^5$ crossings could occur at $\gamma = 10^4$ without appreciably reducing the inverse Cherenkov interaction. (This is for λ = 10 μm and w = 1 mm.) As γ increases, electron beam divergence from multiple scattering decreases.

It should be noted that the expressions for acceleration gradient and energy exchange for both the free electron laser and the inverse Cherenkov case apply when there is circular focusing of the laser light. If cylindrical focusing is used, then Eqs. (59), (64), and (65) must be multiplied by $1.12\,(w/d)^{1/2}$, where w is the spot size in the plane formed by the momentum vectors of the photon and electron, and d is the spot size in the direction normal to this plane.

An advantage of the inverse Cherenkov interaction is that for high γ the Cherenkov angle is independent of the particle energy. This means that the gas pressure remains constant as the electron is accelerated. In contrast, the magnet wiggler field used for the free electron laser must increase as γ increases.

A limitation on acceleration gradient may result from breakdown in the gas. Picosecond pulses of Nd laser light in nitrogen and oxygen gases at 1 atm have a breakdown field of $\simeq 3\times10^8$ V/cm.[25] Electric field E_0 is related to power by the expression (mks units)

$$E_0 = \frac{21.9\ P^{1/2}}{\sqrt{\pi}\ w} . \tag{67}$$

For $w = 1$ mm and $E_0 = 3\times10^8$ V/cm, $P = 1.9\times10^{12}$ watts. From Eqs. (64) and (65), the acceleration gradient is $\simeq 0.9$ GeV/m and the energy increase per pass is $\simeq 54$ MeV. It is seen, therefore, that reasonably high gradients can be obtained without breakdown, assuming that the breakdown field in hydrogen is the same as for nitrogen and oxygen.

Table I lists important criteria and summarizes the performance properties of the approaches considered in the previous section. The estimates of problem-free, moderate-problem, and high-difficulty features of the different approaches are indicated.

Table I Performance criteria for laser accelerators

√ means no difficulty
0 means some difficulty
x means great difficulty

	Type of Accelerator		
	Smith-Purcell	FEL	Inverse Cherenkov
Free from breakdown	0	√	0
Large accelerating field component	√	x	x
Large transverse area	x	√	√
Ease of maintaining phase synchronism	x	√	0
Simplicity of design	x	x	√
Straight line trajectory (no need for B-fields)	√	x	√
Collision-free motion	0	√	x
Requirement for light refocusing	x	√	0

It is possible that different accelerator types could be used for different ranges of particle energy. For example, the FEL may be appropriate for low γ where large magnetic fields are not required, and the inverse Cherenkov accelerator could be used for high γ where scattering effects in the index-matching gas are less important. Also, different types of accelerators could be combined A gas could be placed above a grating or in an FEL to modify the performance characteristics.

There are other categories of laser accelerators which have not been considered and which might be useful in certain energy ranges. An inverse synchrotron accelerator does not involve a medium or a grating, but does require a magnetic field. The limited interaction accelerator does not have a medium, grating, or a magnetic field, but does need to have the field terminated periodically.

5. EXPERIMENTAL RESULTS ON THE STIMULATED CHERENKOV INTERACTION

Three experiments have been performed involving the stimulated Cherenkov interaction at optical wavelengths:[26] a measurement of the change in the momentum spectrum of an electron beam passing through a laser field; the dependence of this change on the refractive index of the medium in which the interaction occurs; and the observation of second harmonic radiation from the resulting bunched electron beam. In each case the results predicted from analysis and Monte Carlo simulations are in good agreement with the data. The interaction takes place in a gas which provides a sufficiently high index of refraction to bring the phase velocity of the wave below the particle velocity.

Figure 20 shows the essential features of the stimulated Cherenkov interaction. The electron velocity is $v_0 = \beta c$ and the light wave phase velocity is c/n. When the angle of intersection of the particle and light beams equals the Cherenkov angle θ_c, then the electron remains in a constant phase of the field.

Equation (65) for the amount of energy transfer for the stimulated Cherenkov effect applies to an idealized situation in which every electron intersects the laser beam at the Cherenkov angle. Practical considerations which modify this ideal situation are the divergences of the electron and light beams, collisions of the particles with the gas molecules of the index-matching medium, transverse dimensions of the two beams, and the random energy distribution of the incident electrons. A Monte Carlo computer simulation has been developed to include all these effects, using the Rutherford scattering formula to account for collisions in the gas. The computer results will be presented and compared with the experimental data.

A. Momentum modulation of the electron beam

1. Experimental arrangement

Figure 21 illustrates the experimental arrangement for observing changes in the energy spectrum of an electron beam passing

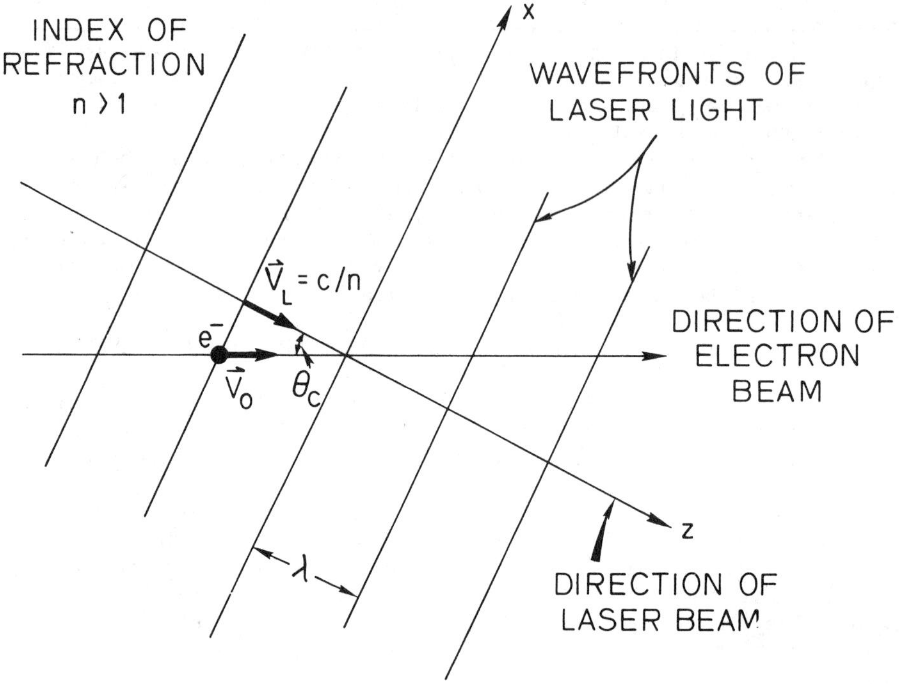

Fig. 20. The stimulated Cherenkov interaction.

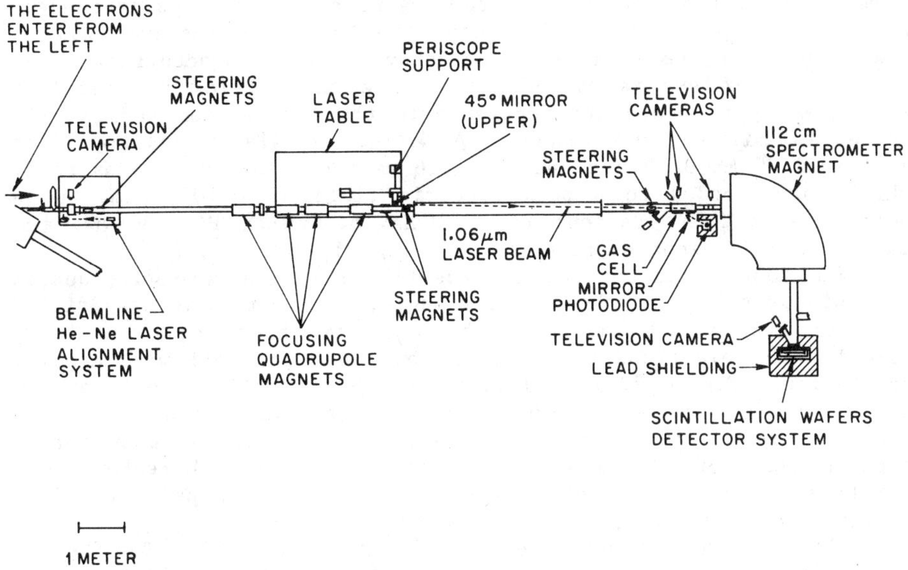

Fig. 21. Experimental Setup. Simplified schematic depicting the top view of the experimental system. The electrons enter the beamline from the left, interact with the laser in the gas cell on the right, and are detected at the end of the system. The entire experiment is remotely controlled.

through an electromagnetic wave. The electrons, generated by Stanford's Superconducting Accelerator (SCA), entered the evacuated beamline pipe from the left. Steering magnets and focusing quadrupole magnets positioned and focused the beam into the gas cell on the right where the stimulated Cherenkov interaction occurred.

The laser system was situated on a table below the beamline with the light guided and focused into the interaction region by a series of mirrors and lenses. A Nd:YAG oscillator and amplifier delivered 30 MW of 8 nsec pulsed 1.06 μm radiation to the gas cell, which corresponded to a peak electric field of $\simeq 10^6$ V/cm, and a peak acceleration gradient of $\simeq$ 2 MeV/m. With P = 30 MW, the maximum energy exchange from Eq. (65) was 213 keV.

The laser beam entered the gas-filled region through a quartz disk window tilted at the Brewster angle. A 1.6 mm hole was drilled in this window and covered with 0.025 mm thick beryllium foil, allowing the electrons to enter and exit the gas cell with minimal scattering. Figure 22 is a diagram of the gas cell. The laser and the electron beams were focused to $\simeq$ 1.2 mm diameter and $\simeq$ 1.6 mm diameter, respectively, at the center of the cell. With an intersection angle of 18 $\pm$ 1 mrad, corresponding to the Cherenkov angle for this system, the electron and laser beams overlapped for $\simeq$ 7 cm or approximately 10^5 optical wavelengths.

The experimental parameters are given in Table II. The SCA was chosen because it provided a multi-MeV electron beam of high current and low energy spread. A Nd:YAG laser was used as the light source so as to obtain near-visible radiation at high field strengths and high repetition rate. Hydrogen gas was used as the Cherenkov medium because of its low atomic number, which provides the least scattering for a given index of refraction.

Electron energy spectra were measured with a 90° spectrometer magnet, using a position detector at the exit focus. The detector consisted of 18 scintillation wafers, each 0.25 mm thick, with an energy resolution of 10.5 keV per wafer.

Figure 23 illustrates the timing arrangement for the laser pulses, the electron beam pulses, and the detector sampling gate. Laser light is emitted on every other electron beam spill, and the change in the electron energy spectrum due to the laser is measured by subtracting spectra from alternate spills. As indicated in Fig. 23, only about one-ninth of the electrons sampled by the detector are modulated by the laser light.

Figure 24 shows sketches of the unmodulated, modulated, and difference spectra. The incident (unmodulated) energy distribution [Fig. 24(a)] is approximately Gaussian. With the laser present [Fig. 24(b)] the spectrum is broadened and reduced in amplitude at the center. Since the particles are uniformly distributed in the phase of the field an equal number gain and lose energy. Subtracting the unmodulated from the modulated spectrum generates the difference spectrum of Fig. 24(c).

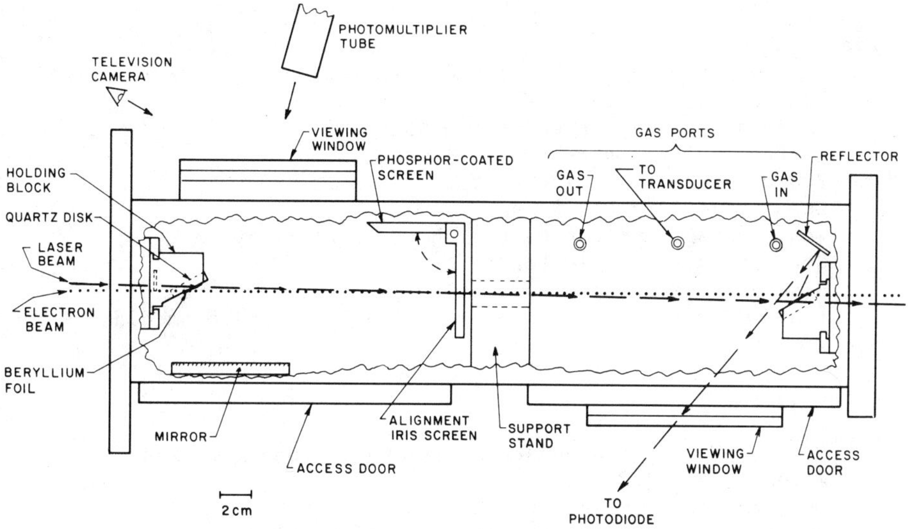

Fig. 22. Diagram of the gas cell in which the stimulated Cherenkov interaction occurs.

Table II. Experimental parameters

Electron beam

Source: Stanford's Superconducting Linear Accelerator (SCA)
Beam energy: 101.8 MeV
Intrinsic energy spread: ≃ ± 15 keV
Width of electron bunches: 9 psec
Bunch separation: 798 psec
Beam spill length: 3 msec
Beam average current (during beam spill): 60 μA
Repetition rate: 10 Hz
Beam divergence in interaction region: ≃ 1.0 mrad
Focused spot size at interaction region: ≃ 1.6 mm diam

Laser beam

Laser type: Nd:YAG unstable resonator configuration
Wavelength: 1.064 μm
Q-switched pulse length: 8 nsec
Power delivered to interaction region: 30 MW
Beam divergence (half-angle): 0.56 mrad
Focused spot size at interaction region: 1.2 mm diam
Linewidth: 0.4 cm^{-1}
Multimode operation

Interaction Region (Gas Cell)

Phase-matching mediums: Hydrogen gas (99.999% pure)
Methane gas (99.97% pure)
Interaction angle: 18 ± 1 mrad
Temperature: 19.5°C
Length of gas cell: 41 cm
Length of electron/laser beam overlap: ≃ 7 cm

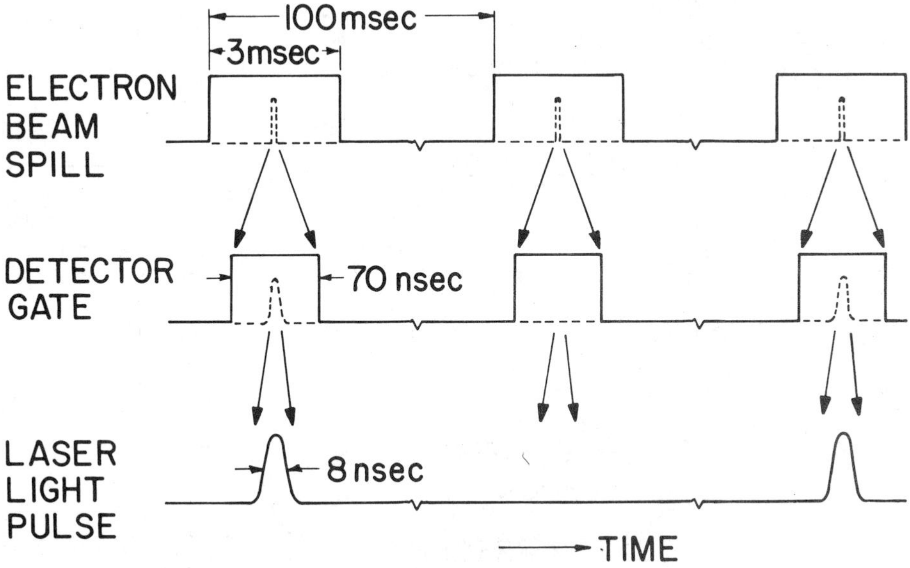

Fig. 23. Electron and Laser Timing. Temporal relationship between the electron beam pulses, detector sampling gate, and laser pulses. Not shown in the diagram is the train of 9 psec long electron bunches, separated by 798 psec, contained within the 3 msec long beam spill. Thus the 8 nsec laser pulse intersected ≃ 10 bunches; however, the detector gate sampled ≃ 90 bunches which reduced the signal-to-noise ratio.

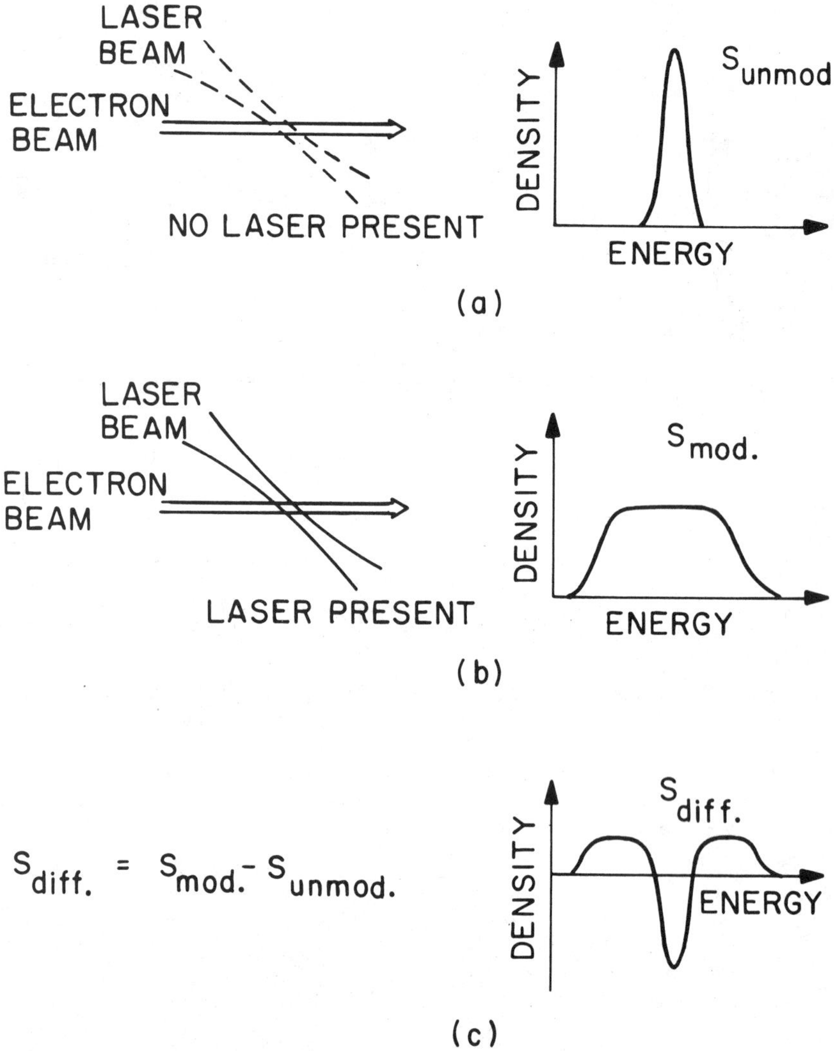

Fig. 24. Electron Energy Spectrums. Schematic of electron energy spectrums. (a) With no laser interaction the unmodulated electron energy spectrum is approximately Gaussian; (b) With laser interaction the modulated electron energy spectrum becomes flat and broad; (c) Subtracting the spectrum in (a) from (b) produces a different spectrum.

2. Experimental observation of energy modulation

A measured difference spectrum, generated by subtracting an unmodulated spectrum from a modulated one, is shown in Fig. 25. The pressure was set to 1.28 atm in hydrogen, which was the pressure that maximized the difference spectrum. Each data point was obtained from 2500 subtractions, and the error bars correspond to one standard deviation. Fluctuation in the electron beam current on a pulse-to-pulse basis was the primary noise source. The solid curve in Fig. 25 represents the computer simulation, where the peak is scaled to fit the data. The fact that the data points show a somewhat narrower energy spread than the simulation suggests that the intrinsic energy spread of the SCA beam is closer to 20 keV (FWHM) rather than the 30 keV assumed in the simulation.

To verify that the energy exchange resulted from the stimulated Cherenkov effect, the dependence of the interaction on the index of refraction of the phase-matching medium was measured. The Cherenkov angle varies approximately as the square root of the gas pressure, so that for a fixed angle of intersection between the electron and laser beams there is a single pressure for which the Cherenkov condition is satisfied. Since, in this experiment the laser beam had a divergence of 0.54 mrad and the electron beam had a divergence of $\simeq$ 1.0 mrad, there is a range of pressures over which portions of the electron and laser beams intersected at the Cherenkov angle.

The experimental results are depicted in Fig. 26 which shows the peak of the difference spectrum, obtained from an average of the two peak channels in the spectrum, plotted as a function of gas pressure. Each data point represents between 500 and 1500 samples. The maximum dip in the pressure curve occurred at 1.28 ± 0.04 atm, corresponding to a Cherenkov angle of 17.4 ± 0.3 mrad. This value agreed within experimental uncertainty with the measured intersection angle of 18 ± 1 mrad, which is shown as the pressure range ΔP in Fig. 26. The solid curve is the pressure dependence computed from the Monte Carlo simulation for the parameters given in Table II. The narrower width of the simulation curve is probably a consequence of assuming a smaller electron beam divergence than the actual value.

Similar results were obtained using methane gas as the phase-matching medium rather than hydrogen. The data are shown in Fig. 27, with energy change plotted as a function of pressure. Energy exchange is less in CH_4 than in H_2 since the larger molecular weight of the former molecule produces more scattering. Scattering is detrimental primarily because it changes the direction of the electron so that the Cherenkov condition is not satisfied and phase slippage occurs. The peak dip in the methane pressure curve was at 0.45 ± 0.03 atm, corresponding to a Cherenkov angle of 17.7 ± 0.6 mrad. Again, this result agreed, within experimental error, with the measured intersection angle of 18 ± 1 mrad.

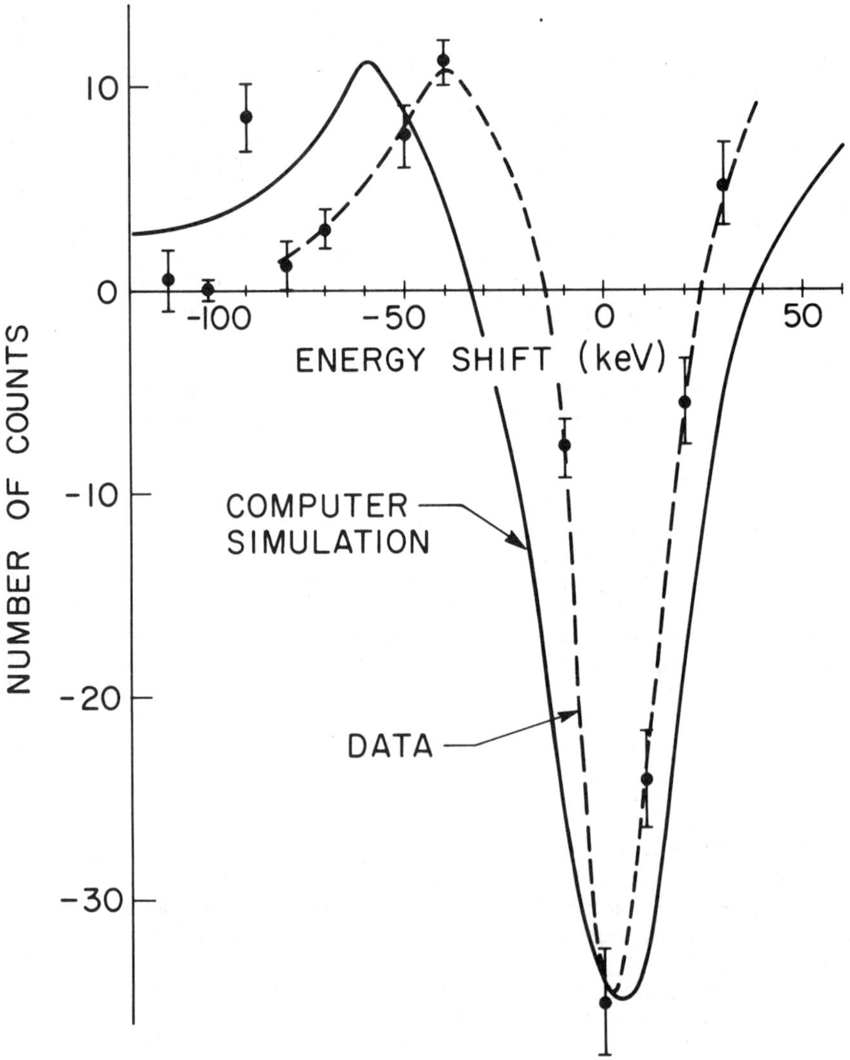

Fig. 25. Difference Electron Energy Spectrum. The dashed curve is the data for hydrogen under the conditions given in Table I. The hydrogen pressure was 1.28 atm and corresponded to the Cherenkov angle equaling the electron-laser intersection angle. The solid curve represents the computer simulation for the same experimental conditions; the peak was scaled to fit the data.

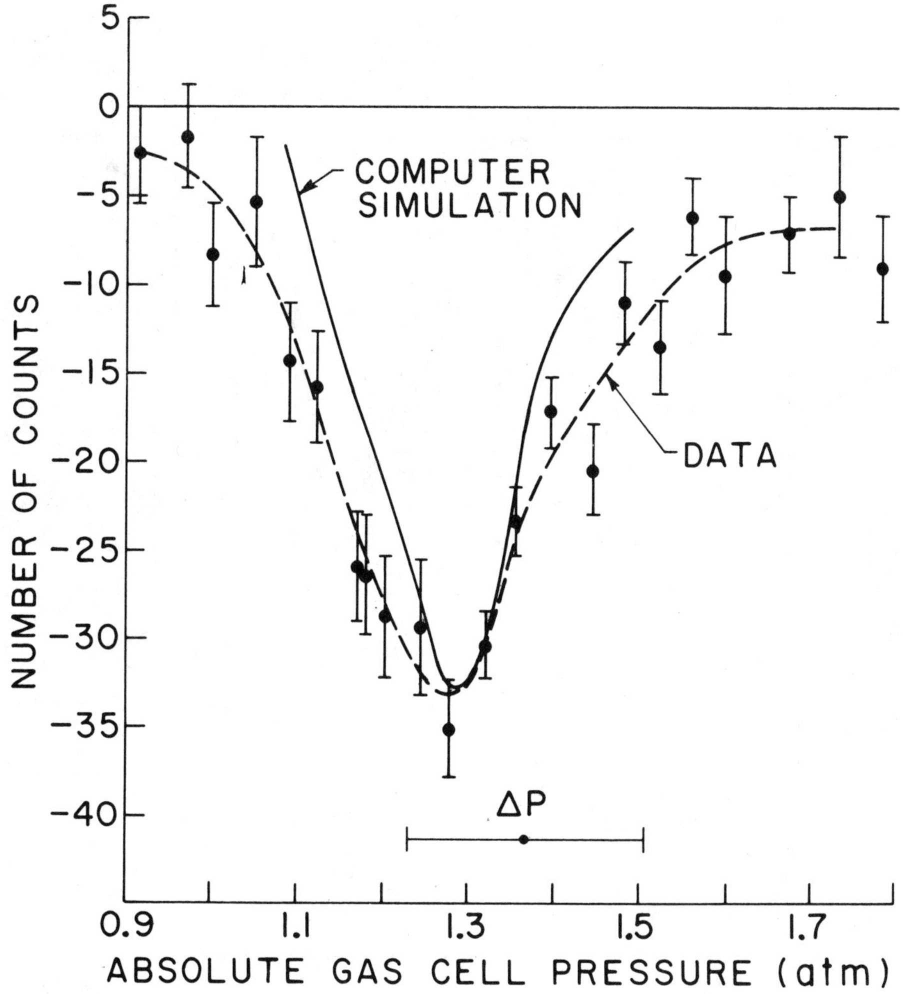

Fig. 26. Pressure Dependence using Hydrogen. Plotted is the average of the peak two channels of the difference spectra at each pressure. The pressure range, ΔP, shown on the bottom of the graph corresponds to an 18 ± 1 mrad intersection angle as determined by the Cherenkov condition. The solid curve is the computer simulation of the dependence; the peak was scaled to fit the data. The electron beam energy was 101.8 MeV.

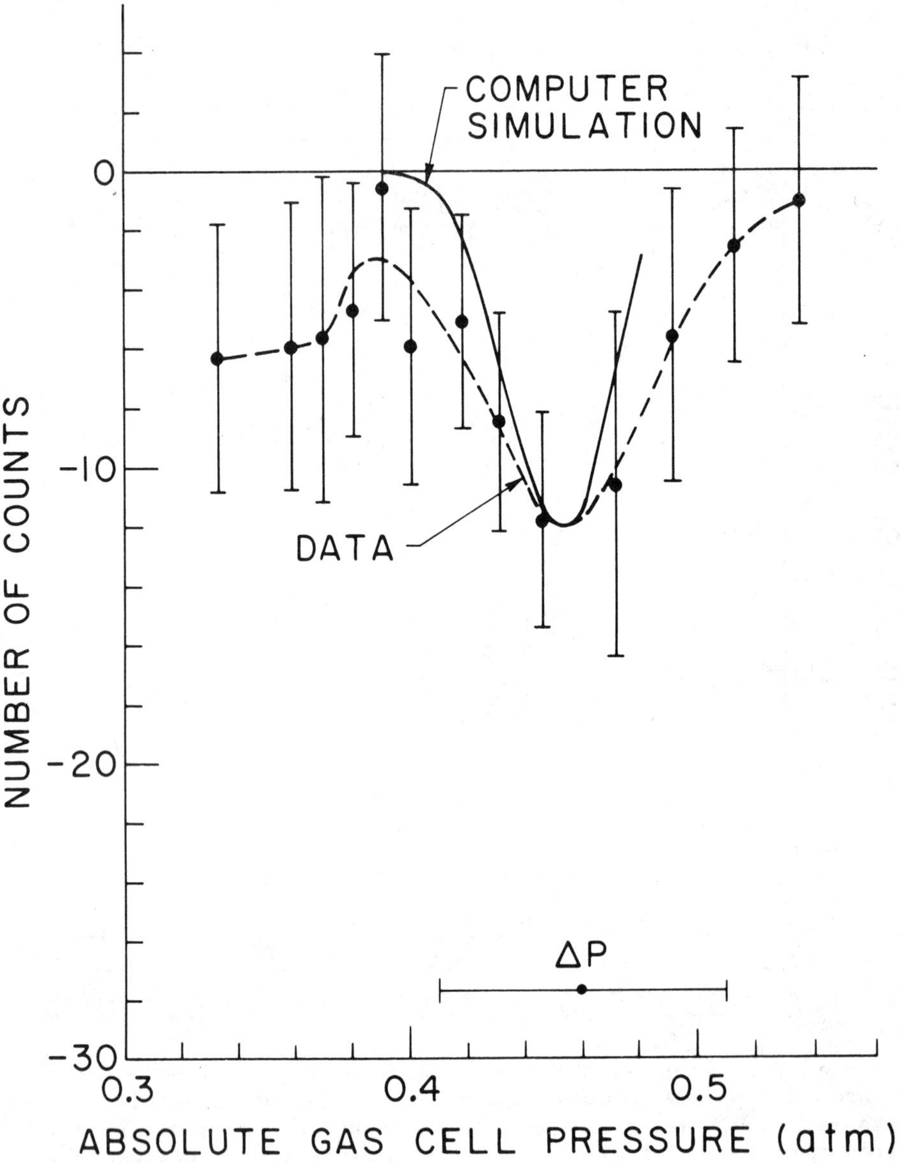

Fig. 27. Pressure Dependence using Methane. This was plotted in the same manner as Fig. 26. The pressure range, ΔP, indicated on the bottom of the graph, corresponds to an 18 ± 1 mrad intersection angle as determined by the Cherenkov condition. The computer simulation of the dependence is given by the solid curve; the peak was scaled to fit the data. The electron beam energy was 101.8 MeV.

B. Harmonic generation

1. Analysis

Electrons subjected to the energy modulation described in the previous section will, if allowed to drift, bunch in the manner of a klystron. With the arrangement shown in Fig. 8, the z-component of the current density in the electron beam I_z is

$$I_z = -|e|v_0\cos\psi\ \delta\{x - x_0 - v_x(t - t_1)\} \times \delta\{z - v_z(t - t_1)\}$$
$$\times\ \delta\{y - y_0 - v_y(t - t_1)\}\ , \tag{68}$$

where δ is the Dirac delta function; $|e|$ is the magnitude of electron charge; v_0 is the dc particle velocity; ψ is the angle between v_0 and the z axis; v_i is the particle velocity in the i-direction; and $x = x_0$, $y = y_0$, and $z = 0$ at $t = t_1$.

For a light beam with a Gaussian intensity variation, the velocity modulation of the electrons is found from the force equations to be

$$\Delta v_{\parallel}/c \simeq \Delta\exp\left\{-\left(\frac{\psi - \theta_c}{\theta_L}\right)^2\right\}\cos\left\{\omega t_1 - \frac{\omega n x_0}{c}\sin(\psi - \theta_c)\right\}$$
$$\Delta v_{\perp}/\Delta v_{\parallel} \simeq -\gamma^2\tan\theta_c, \tag{69}$$

where $\Delta v_{\parallel}$ is the velocity modulation due to the laser field in the direction of initial motion; $\Delta v_{\perp}$ is the modulation normal to the direction of initial motion; θ_L is the divergence angle of the laser light; and

$$\Delta = \frac{7.59\times10^{-5}}{\gamma^3}\ \frac{P^{1/2}}{n^{1/2}}\cos\theta_c\ , \tag{70}$$

where P is the laser power in watts.

Expressing the current as a Fourier series

$$I_z = \sum_m a_m e^{-imt} \tag{71}$$

the substitution of Eq. (69) into Eq. (68) gives for the Fourier coefficients

$$a_m \simeq I_0(-i)^m J_m(A) \exp\left\{-\frac{(x - z\tan\theta_c)^2 + y^2}{a^2}\right. \tag{72}$$
$$\left. \times \exp\frac{im\omega n}{c}\left[z\cos(\psi - \theta_c) + x\sin(\psi - \theta_c)\right]\right\},$$

where "a" is the electron beam spot size for a Gaussian distribution of current, I_0 is the dc current density, J_m is the Bessel function of first kind and order m, and

$$A = 0.477\times10^{-3}\,\frac{mz}{\lambda}\,\frac{P^{1/2}}{n^{1/2}}\,\frac{1 + \gamma^2 \tan^2\theta_c}{\gamma^3}\exp\left\{-\frac{(\psi - \theta_c)^2}{\theta_L{}^2}\right\}. \tag{73}$$

The Cherenkov radiation from a bunched electron beam can be calculated.[4] System parameters for the experimental observation of second harmonic generation, i.e., m = 2, are listed in Table III, and for these parameters the second harmonic power is predicted to be ≃ 8 μW. Allowing perfect collimation of the electron beam, i.e., $\psi = \theta_c$ for all particles, and keeping the other parameters fixed, the calculated power is ≃ 2W. The reason for the large reduction in radiation with an electron beam divergence of 1.6 mrad is that particles moving in different directions have random z-component velocities which tend to eliminate the bunching.

2. Experimental arrangement

The purpose of the experiment was to observe emission at the second harmonic (λ = 0.532 μm) of the laser frequency from the bunched electron beam. Second harmonic power directly from the laser was too low to be observed and did not interfere with the measurements.

Figure 28 shows the arrangement used to extract the second harmonic. The electron and laser beams intersected at the Cherenkov angle at the center of the gas cell, and 65 cm downstream and still within the cell a narrow (1.6 mm) mirror was positioned to reflect the Cherenkov light from the electron beam at a distance of 19 to 28 cm from the center of the cell. Optimum bunching for the second harmonic occurs ≃ 23 cm from the intersection point. At λ = 0.532 μm the index of refraction of hydrogen gas is nearly identical to its value at 1.06 μm, so that the second harmonic is emitted parallel to the incident laser beam. Extracted light was then sent through filters to remove the 1.06 μm background radiation, and a 10 Å bandwidth filter centered at 0.532 μm was placed in front of the photomultiplier to eliminate most of the broadband spectrum of the spontaneous Cherenkov emission.

As in the momentum modulation experiments described in the previous section, the laser was on for alternate pulses of the

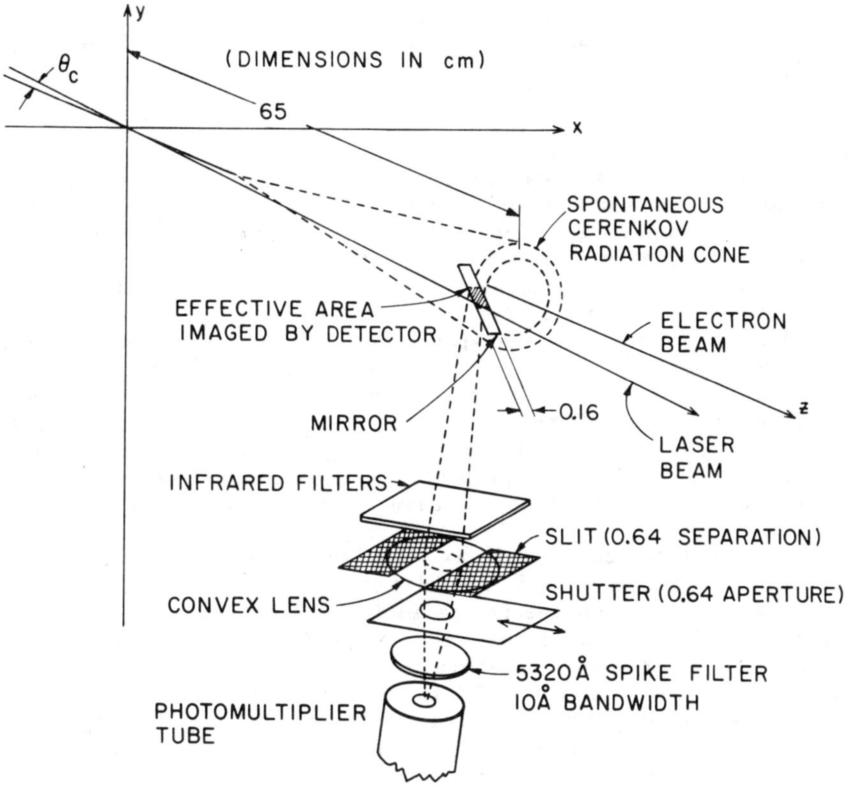

Fig. 28. Experimental arrangement for measuring the coherent Cherenkov radiation.

electron beam. However, a different gun was used so that all electrons were subjected to the laser pulse. (In the momentum modulation experiments only about one detected electron in nine passed through the laser light.) Measurements were made of the difference in the second harmonic radiation with the laser on and with the laser off.

3. Results

Figure 29 is a presentation of the experimental results, with the number of counts from the detector plotted as a function of hydrogen gas pressure. Count number is proportional to the optical radiation intensity, with each count corresponding to $\simeq 0.17\ \mu W$. Each data point is an average over 1000 samples, and the error bars represent one standard deviation.

The upper curve is the spontaneous Cherenkov radiation measured with the laser off. The falloff in emission at both the high and low ends of the pressure scale is due to clipping of the light by apertures in front of the photomultiplier as a result of having the emission angle change with pressure. Spontaneous Cherenkov emission intensity is linearly proportional to electron current, and fluctuation in the electron beam current is the primary noise source.

Data points on the lower portion of Fig. 29 represent the emission intensity at the second harmonic with the laser on after subtracting the spontaneous radiation and other background noise signals. The solid curve is the predicted radiation from the bunched beam for the parameters in Table III. Comparing the data with the theoretical curve it is seen that the radiation falls to zero below the optimum pressure; that the measured position of the peak as a function of pressure is near the predicted value; that the emission decreases above the optimum pressure; that the intensity of radiation is close to the calculated value; and that the measured pressure width is in reasonable agreement with the theory. The gas cell was not designed to withstand pressure greater than 2 atm, thus limiting the maximum pressure that could be applied.

Each data point for the coherent emission is from 1000 samples, with electron beam current fluctuation the major noise source. The second harmonic of the bunched beam is proportional to the square of the current.

The direction of the enhanced emission resulting from bunching does not change as pressure is varied, but rather remains parallel to the direction of the incident laser light. This means that the variation in intensity associated with direction in spontaneous light does not apply to the bunching radiation.

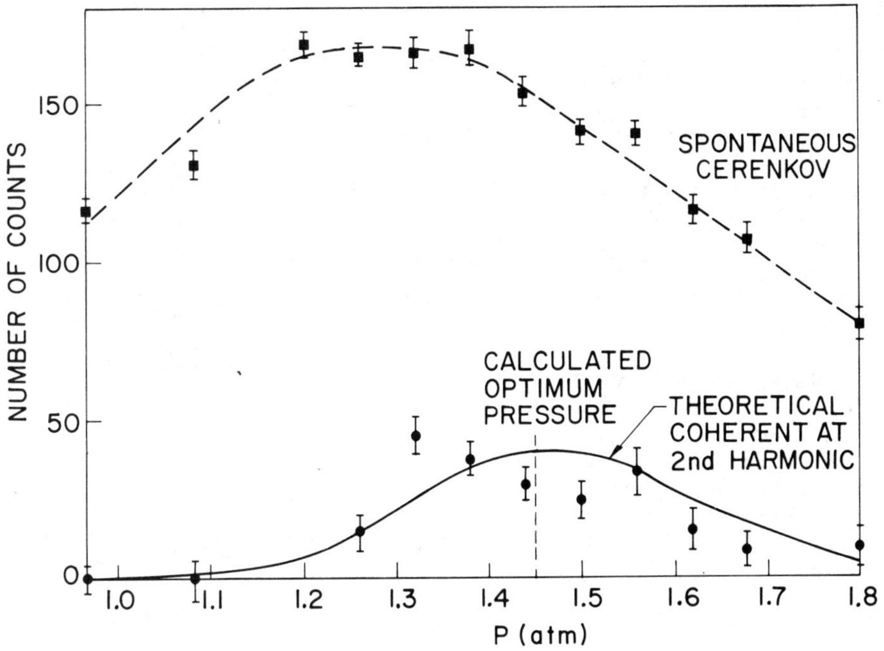

Fig. 29. Measurement of the second harmonic generation radiated from an electron beam bunched by passing it through the light from a Nd:YAG laser.

Table III System parameters for the coherent Cherenkov radiation experiment

Electron Beam
- Source: Stanford's Superconducting Linear Accelerator (SCA)
- Beam energy: 55.9 MeV
- Intrinsic energy spread: ≃ ± 8 keV
- Width of electron bunches: 5 psec
- Bunch separation: 85 nsec
- Beam spill length: 1 msec
- Peak beam current: ≃ 1 A
- Repetition rate: 10 Hz
- Beam divergence in interaction region: ≃ 1.6 mrad
- Focused spot size at interaction region: 1 mm diam

Laser beam
- Same as given in Table II

Interaction region (gas cell)
- Phase-matching medium: hydrogen gas (99.999% pure)
- Interaction angle: 17 ± 1 mrad
- Temperature: 18.7°C
- Length of electron/laser beam overlap: ≃ 7 cm

C. Discussion

It has been noted that the radiation is greatly reduced when there is divergence of the electron beam. A magnetic field can be introduced which restores the bunching and therefore the harmonic current, by making the path length in the field a function of the angle of incidence of the particle. Figure 30 illustrates the arrangement, with the electrons entering the magnetic field at point A at an angle ψ to the laser light propagating in the z-direction, and exiting at point C also at an angle ψ to the z-axis. Flux density integrated over the path length is zero. The Fourier coefficient a'_m with the magnetic field present is

$$a'_m = a_m \frac{J_m(A+\xi)}{J_m(A)} \exp\left\{-i\frac{m\omega n}{c}x'\sin(\psi-\theta_c) + i\rho\right\}, \qquad (74)$$

where a_m is given by Eq. (72),

$$\begin{aligned} \xi &= 2\pi m\Delta(\sec\theta_c)\frac{2\ell}{\lambda}\left[1+\frac{\gamma^2}{3}\frac{\ell^2}{R^2}\right]\times\exp\{-(\theta-\theta_c)^2/\theta_L{}^2\} \\ x' &= 2\ell\left[\frac{\ell}{2R}-\tan\theta_c\right]. \end{aligned} \qquad (75)$$

R is the radius of curvature of the electron in the field, ℓ is the magnet width, ρ is a constant phase factor, and point C is taken as the origin of the coordinate system. If, for example, the radiation from the beam near point C is to be maximized, then the magnetic field parameters ℓ and R should be chosen to maximize J_m and to have $x' = 0$. Typical values for our experiment are $\ell = 10$ cm and B = 3.0 kG.

A klystron oscillator could be constructed by adding two mirrors to Fig. 30, one to the left of point A, one to the right of point C, and with the mirror planes normal to the z-axis. Velocity modulation of the electron beam is at point A and emission is at point C. The efficiency of such a klystron, with reasonable parameter values, would be on the order of one per cent. An advantage of a stimulated Cherenkov oscillator is its simplicity, and a disadvantage is that the interaction occurs in a gas with the possibility of breakdown at high field strengths. For 10-picoseconds duration electron pulses, typical of a linear accelerator, the breakdown field[25] would be about 10^8 V/cm. This corresponds to a power density in the medium greater than 10^{13} W/cm^2, which is sufficiently large for most applications.

This work was supported by the Office of Naval Research under Contract No. N00014-78-C-0403, by the National Science Foundation under Grant No. NSF ECS-7901743, and by the U.S. Department of Energy under Contract No. DE-AT03-76ER71042.

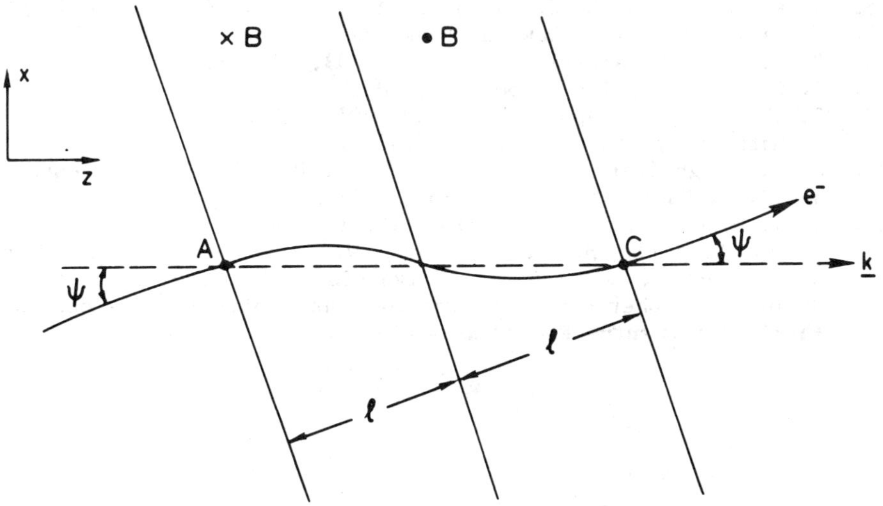

Fig. 30. An arrangement, using a magnetic field, to increase the radiation from a bunched electron beam with angular divergence.

REFERENCES

1. L. R. Elias et al., Phys. Rev. Lett. **36**, 717 (1976).
2. A. Gover and A. Yariv, Appl. Phys. **16**, 121 (1978).
3. J. V. Jelley, **Cherenkov Radiation and Its Applications**, (Pergamon, London, 1958).
4. J. D. Jackson, **Classical Electrodynamics** (John Wiley and Sons, New York, 1962).
5. Ibid, p. 505.
6. J. J. Sakurai, **Advanced Quantum Mechanics** (Addison-Wesley Publishing Company, London, 1973).
7. J. A. Edighoffer and R. H. Pantell, J. Appl. Phys. **50**, 6120 (1979).
8. R. M. Phillips, I.R.E. Trans. E.D. **7**, 231 (1960).
9. C. A. Brau, Laser Focus **17**, 48 (1981).
10. M. A. Gusinow, J. P. Anthes, M. K. Matzen, and D. Woodall, Appl. Phys. Lett. **33**, 800 (1978).
11. C. K. Chen, J. C. Sheppard, M. A. Piestrup, and R. H. Pantell, J. Appl. Phys. **49**, 41 (1978).
12. R. H. Pantell and J. A. Edighoffer, J. Appl. Phys. **51**, 1905 (1981).
13. S. J. Smith and E. M. Purcell, Phys. Rev. **92**, 1069 (1953).
14. K. Shimoda, Appl. Opt. **1**, 33 (1962).
15. Y. Takeda and I. Matsui, Nucl. Instrum. Meth. **62**, 306 (1968).
16. J. D. Lawson, Report #RL-75-043 (Rutherford Labs, Chilton, U.K., March 10, 1975).
17. W. B. Colson and S. K. Ride, Appl. Phys. **20**, 61 (1979).
18. J. K. McIver and M. J. Lubin, J. Appl. Phys. **45**, 1682 (1974).
19. T. Tajima and J. M. Dawson, Phys. Rev. Lett. **43**, 267 (1979).
20. R. B. Palmer, Particle Accelerators **11**, 81 (1980).
21. A. C. Saxman, private communication.
22. R. B. Palmer, J. Appl. Phys. **43**, 3014 (1972).
23. H. Motz, Cont. Phys. **20**, 547 (1979).
24. J. A. Edighoffer, W. D. Kimura, R. H. Pantell, M. A. Piestrup, and D. Y. Wang, Phys. Rev. **23A**, 1848 (1981).
25. R. J. Dewhurst, J. Phys. **11D**, L191 (1978).
26. R. H. Pantell, M. A. Piestrup, W. D. Kimura, D. Y. Wang, A. M. Fauchet, and J. A. Edighoffer, "Measurements of the Stimulated Cherenkov Interaction at Optical Wavelengths," **Physics of Quantum Electronics** (1981).

COLLECTIVE FIELD ACCELERATORS

Andrew M. Sessler
Lawrence Berkeley National Laboratory, University of California,
Berkeley, California 94720

TABLE OF CONTENTS

0094-243X/82/870919-42$3.00 Copyright 1982 American Institute of Physics

COLLECTIVE FIELD ACCELERATORS

Andrew M. Sessler
Lawrence Berkeley National Laboratory,
University of California, Berkeley, California 94720

1. INTRODUCTION

For fifteen years collective-field accelerators have been actively pursued, and suggestions for the use of collective fields can be traced to Alfven and Wernholm,[1] Harvie,[2] and Randorf[3] in the early 1950's. The field was given a considerable boost by the work of Veksler,[4] Budker[5] and Fainberg[6] which was initiated in the mid-1950's and then by the actual observations of collective acceleration by Graybill and Uglum[7] and by Plyutto[8] in the 1960's. By 1970 there was activity in many laboratories, in many countries of the world, and this work continued--with ups and downs--throughout the decade of the 1970's. Yet, in 1981 there is still no practical collective accelerator, i.e., no device which is routinely employed for its accelerated particles.

Why? Clearly, and we shall see this in detail, the collective acceleration of ions is a difficult thing to achieve in a controlled, rapidly repeatable, reliable, and inexpensive manner. Yet, the promise is there. If we look to the future, and we shall in Section 6, we see there are a number of approaches which would appear to be capable of producing a practical accelerator.

But, the effort of past years has been valuable not only in that it makes possible the work which, hopefully, will produce practical collective accelerators in the future, but because the pursuit of collective accelerators has been a driving force in both our understanding of beam behavior and in our development of the technology with which we control and manipulate particle beams.

The subject of collective ion acceleration has been reviewed this very year (Refs. 9 and 10) and has been treated in a compilation of conference papers[11] and in a very good textbook.[12] The field does not need another review article, as the above-mentioned two review articles are quite excellent for learning the current status of work, nor does the field need a discussion of the basic physics, which is quite excellently discussed in the textbook by Olson and Schumacher.[12] In fact, what the field needs is more work with associated technical publications and not more summaries of other people's work.

Nevertheless, we shall in this article present still another overview of collective acceleration. We will, however, lean heavily on the existence of the two recent reviews and the text so that the present treatment will not be at all comprehensive. We hope that the reader will be motivated to read further and, if so, we will feel that our efforts have been rewarded. In particular, we will, where possible, use the notation of the text so that the transition to Olson and Schumacher will be relatively easy.

A. Motivation

In non-collective accelerators, the charge and current of the accelerated particles is small, or at best a restriction on the performance of the device. Thus, to fair approximation,

$$\nabla \cdot \mathbf{E} = 0,$$

$$\text{and} \quad \nabla \times \mathbf{B} - \frac{1}{c}\frac{\partial \mathbf{E}}{\partial t} = 0. \tag{1.1}$$

In collective devices, on the other hand,

$$\nabla \cdot \mathbf{E} = 4\pi\rho,$$

$$\text{and} \quad \nabla \times \mathbf{B} - \frac{1}{c}\frac{\partial \mathbf{E}}{\partial t} = \frac{4\pi}{c}\mathbf{J} \tag{1.2}$$

which opens up a world of possible configurations. In fact, as we shall see, there are a great many configurations which have been proposed for collective accelerators and one of the problems is to categorize the various approaches and to limit activity to those few which appear easiest to achieve or most advantageous if realized.

In conventional accelerators the E and B fields are produced by external conductors and hence are limited by the properties of those materials. Thus the performance of these accelerators (accelerating gradients and/or bending radius; i.e., size) is limited. Collective accelerators, on the other hand, can have much higher fields than conventional accelerators [because of Eq. (1.2)] and hence give the promise of being compact and, possibly, cheap.

Of course, one must make the ρ or J which is employed to create the E and B fields of collective accelerators. As we shall see, the production of ρ or J requires a significant device and hence--as presently envisioned--collective devices are not as attractive as one would at first think.

Nevertheless (ignoring collective instabilities) one cannot help but feel that the removal of the constraint of Eq. (1.1) should allow the design and development of very attractive devices. This has been, and remains, the fascination of collective-field accelerators.

B. Physical principles

Collective accelerators either generate the collective E and/or B field from an approximately stationary source of charge (such as in the electron ring accelerator or in the collective focusing accelerator) or from a streaming source (such as in the ionization front accelerator or in the autoresonant accelerator). In either case the source of charge is an intense relativistic electron beam (IREB). Such beams can be generated in a variety of ways and parameters vary considerably. A "typical" machine has a pulsed diode with a field emission cathode and a foil anode so that the beam has:

τ = pulse length: 10 nsec - 100 nsec
I_e = beam current: 10 kA - 100 kA
W_e = beam energy: 1 MeV - 10 MeV.

Such beams have electron densities, n_b, in the range of 10^{11} - 10^{13} cm^{-3} and radii, r_b, of 1-10 cm. Hence a typical electric field

$$E_r = 2\pi n_b e r_b \tag{1.3}$$

$$\text{or } E_r\left(\frac{MV}{meter}\right) = \frac{6.0\ I_e(kA)}{\beta r_b(cm)}\ , \tag{1.4}$$

where β is the electron speed in units of the velocity of light. For I_e = 10 kA, β = 1, r_b = 1 cm we have E_r = 60 MV/m. This might be compared to SLAC which produces 30-GeV electrons in 3 km; i.e., E_z = 10 MV/m or to ion linacs (such as LAMPF) which typically have $E_z \simeq$ 1.MV/m.

For electron streams there are two methods which appear likely to give collective acceleration. The first involves localized space charge. If for some reason a potential well is created and moved, there can be acceleration produced by the moving potential well providing the motion is at first slow and then faster and faster. We shall see that space charge wells appear to occur naturally and that their motion can be controlled so that this may be a practical way to make a collective accelerator. Section 2 is devoted to the physics of electron beams, the control of beam front velocities (the potential wells which do the acceleration are at the beam front) and the experimental situation in regard to localized space charge accelerators.

A second way in which electron streams may lead to collective acceleration is by space charge waves; that is, if a wave is created on an electron stream then the associated fields can be used to accelerate ions. Of course, the wave speed must be appropriate (i.e., varying in time or space) but there are many modes of oscillation of an electron beam and some of these modes, with external parameter variation (such as the applied longitudinal magnetic field), can be made to vary in speed in the appropriate manner. The third section, Section 3, is devoted to this subject.

Acceleration by the use of charge clusters (as contrasted with electron streams) has been limited to electron rings. In this concept, due to Veksler,[13] the electrons are made to go in a small circle of a few cm radius, and are relativistic so that the electrostatic repulsion of electrons is almost balanced by the magnetic attraction of moving electrons. Motion of the ring as a whole then provides ion acceleration. In order to obtain good ion acceleration the electron ring must have a high density [$\sim 10^{12} cm^{-3}$, see Eq. (3)] while, of course, there are instabilities which limit the electron density. Section 4 is devoted to electron ring configurations, instabilities, and the experimental situation of electron ring accelerators.

Another use of charge clusters is in the collective focusing accelerator. In this device, a large electron gas is built up, by inductive injection, and then employed to provide focusing for ions.[14] The ions are accelerated by conventional means (a betatron) while the electrons are not accelerated because they are magnetically trapped in mirror regions. Section 5 is devoted to this topic.

Finally, there are other physical effects which have been considered for acceleration, but which have not yet been employed in an experimental device. These include impact acceleration, inverse Cherenkov effect and the two stream effect. We will not go into these in any detail.

C. An historical survey

Although the idea of collective acceleration had been around for some years, it was the reports from the Soviets at the first international accelerator conference in 1956 that stimulated interest in the subject.[4-6]

Veksler pointed out that by "impact acceleration" one could reach very high energies. In fact, Veksler believed this would be the only way to reach very high energies (and he may have been correct!). Consider a bunch of N_1 electrons of mass m, moving relativistically (γ>>1) which collide with a stationary light bunch of N_2 ions having mass M (see Fig. 1). We must have

$$N_1 m >> N_2 M\gamma; \tag{1.5}$$

i.e., the stationary bunch must be much less massive than the electron bunch. In a head-on collision each ion will receive the energy

$$W_i \simeq 2\gamma^2 Mc^2 , \tag{1.6}$$

which can be very large indeed.

Now, no one knew then (or now even!) what to do with Veksler's observation. Namely, no one knew how to keep bunches of particles together (also one needs very large currents, as is shown by Olson,[12] but his remarks set lots of physicists thinking.

At the same conference, Budker first discussed "Budker beams" (collective focusing), and Fainberg introduced the idea of wave acceleration. This work precipitated theoretical work and thought, but no experimental programs were initiated.

Then, in 1967, Veksler's group (at Dubna) reported on their work on electron ring accelerators (which had been in progress for quite some years)[13] and this report stimulated experimental programs at Berkeley, Garching, Karlsruhe, and Maryland. Considerable work on this approach has been done and in 1978 it was reported by the Dubna group[11] that they had accelerated nitrogen and heavier ions, at 2-4 MeV/amu-meter, to a few MeV. Other groups were started during the 1970's and successful acceleration of ions was also achieved by the Garching group,[15] as early as 1974. However, the low gradients achieved and the complexity of electron ring accelerators was

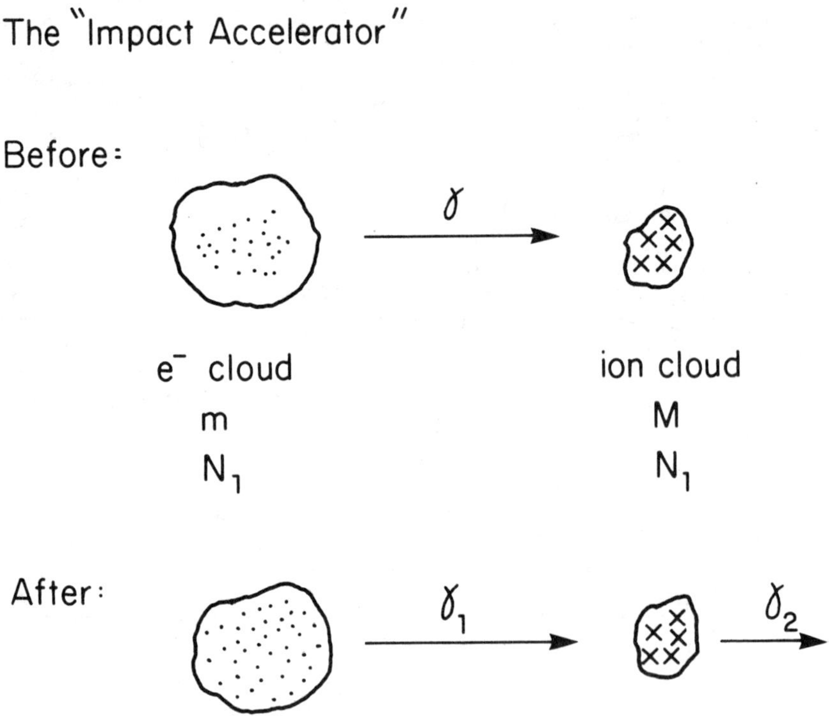

Fig. 1. The "impact accelerator."

Momentum Conservation:

$$N_1 m\gamma\beta = N_1 m\gamma_1\beta_1 + N_2 M\beta_2\gamma_2$$

$$(\beta \simeq \beta_1 = 1)$$

Energy Conservation:

$$N_2 M + N_1 m\gamma = N_1 m\gamma_1 + N_2 M\gamma_2$$

discouraging so that at present all activity, except that at Dubna, has been terminated.

In 1970, Graybill and Uglum[7] reported observations of collective acceleration of protons to 5 MeV and nitrogen to 20 MeV in an intense electron beam (W_e = 1.6 MeV, I_e = 30 kA). Although Plyutto[8] had been observing collective acceleration for a decade, the reported observation of "naturally occurring" acceleration to considerable energies by Graybill and Uglum stimulated much activity.

Work on electron stream acceleration has been done at many laboratories--Physics International, Air Force Weapons Laboratory, Sandia, Irvine, Cornell, Boeing, and Maryland (and no doubt at other locations also)--and resulted in the routine production of very energetic ions. For example (see Refs. 9 and 10), acceleration of protons to 60 MeV has been achieved and, at Maryland, pulses of 10^{12} Xenon ions have been accelerated to an energy of 900 MeV. In this work, the accelerating gradient is about 100 MV/m but only extends over about 10 cm.

In 1973, it was observed by Luce that a vacuum diode with a plastic anode would routinely produce large numbers of accelerated protons.[16] The use of "Luce diodes" was quickly taken up by the groups mentioned above.

It became clear that one must understand the mechanism of "natural acceleration" and then somehow control and (possibly) stage the acceleration mechanism. Efforts in this direction were made by many workers and in Section 2 we shall discuss the theoretical and experimental work on controlled acceleration. One of the most interesting ideas, based upon the developing understanding of "natural acceleration," was the proposal made by Olson in 1974[17] to control the beam front by means of properly-timed laser pulses. This will be discussed in Section 2, as well as other proposals to control the beam front such as Rostoker's proposal[18] to employ a properly-pulsed wall plasma.

Although Fainberg's group had been investigating wave acceleration for many years, it was after the proposal of Drummond and Sloan[19] in 1973 that considerable interest developed in wave accelerators. A large program was initiated at Austin to investigate this concept (the Autoresonant Accelerator) and they have now (1981) demonstrated that they can enhance the wave in which they are interested (the lower-hybrid Doppler-shifted cyclotron mode) while not enhancing other modes.

Other groups have also started to study wave acceleration. In 1976, Sprangle et al.[20] proposed that a slow space charge wave could have its velocity controlled and hence, also, be employed for ion acceleration. Such an activity (the Converging Guide Accelerator) is now underway at the Naval Research Laboratory. Research is also being performed on a wave accelerator, employing a slow-wave structure, at Cornell.

In Section 3 we will discuss the various approaches to wave acceleration. Recently (1979) Friedman has suggested a collective accelerator which is sophisticated in concept and has many controllable features.[21] In this approach (the Collective Particle Accelerator), a hollow electron beam is chopped into a sequence of short cylinders and then sent into a guide field which consists of short solenoids. The electron segments throb radially and, hence, produce an axial electric field whose phase velocity can be varied by changing the magnetic field strength or electron beam current. Experimental work has proceeded to the point of generating electron segments and propagating them some meters.

There have also been, through the years, some other approaches to collective field acceleration and many other researchers which we have not had room to mention. We have attempted in this brief survey to show only the main trends of development. The field has attracted considerable attention and we believe that some of its promise should soon be realized.

D. Related work

There are some closely related devices which although not collective accelerators involve similar theoretical understanding of particle behavior or similar experimental techniques. Of course, most conventional particle accelerators and controlled fusion devices fall into this category, but will not be mentioned further other than to note that many of the experimental and theoretical techniques employed for collective accelerators are borrowed from these two fields.

Special mention needs to be made, however, of the electron beam ion source,[22] the Gabor lens,[23] and the autoaccelerator.[24]

The electron beam ion source (EBIS) employs a dc electron beam, of energy $W_e \simeq 10$ keV and current $I_e \simeq 1$ A, to confine ions radially by its electrostatic field. The electron beam is made to propagate in a uniform longitudinal magnetic field and there are drift tubes about the e^- beam that contain the ions longitudinally by an externally applied potential. The vacuum must be good, so that the beam becomes only neutralized by the injected (desired) ions. The continuous electron bombardment causes the ions to become highly ionized. The potential well is then changed so as to spill the ions out and the process repeated. Performance indicates that the ionization process is very efficient, in fact more so than one would expect from an equilibrium unneutralized electron beam. It is thus believed that the electron beam "collapses," i.e., is in an equilibrium corresponding to almost complete neutralization with radius $r_b \simeq 0.01$ cm. Much work is being done on the EBIS (because of its use as an ion source for nuclear physics) and the mechanisms of collapse and ionization can be expected to be better understood in the future. It is interesting to note that the device is already working better than expected.

A "Gabor lens" is a lens in which the focusing of ions is primarily supplied by a spacially distributed electron cloud. The focusing can be very strong (much greater than with external fields) and the lens quite linear. The electrons, in one version, are held radially by a longitudinal magnetic (solenoid) field and held longitudinally by an electrostatic potential. Such a device has, in fact, been operated.[25]

The autoaccelerator is a device which exploits the existence of a mature technology for producing intense beams of electrons of a few MeV in order to accelerate a less intense beam to considerably more energy (perhaps $\simeq$ 300 MeV). The concept, which is being pursued at the Naval Research Laboratory,[24] was originally developed by Soviet workers.[26] It is well-described in the review paper by Keefe[9] and will only be outlined here. The essential idea is to send an intense beam through a linear array of cavities with filling time such that a small number of electrons near the end of the current pulse are subjected to an accelerating field of considerable magnitude. (The majority of electrons lose energy in this process so that the cavities are a "transformer"; i.e., a means of taking a bit of energy from lots of electrons and giving a few electrons a large energy.) The device not only uses existing IREB's, but it avoids the problems associated with insulation and switches (as does any collective device). To date, electrons have been "autoaccelerated" from 0.3 MeV to 3.0 MeV.

2. LOCALIZED SPACE CHARGE

In this section we shall develop the ideas necessary to understand and, hence, construct a controlled space-charge-well accelerator. It is possible that the "naturally-occurring acceleration" of particles (now seen in more than 30 different experiments) has some other explanation, but that is unlikely (except for the Cornell work which seems to imply an ion-electron wave mechanism) and we restrict ourselves, here, to the space-charge-well explanation. Other theories have been suggested to explain the observations and the interested reader is referred to the textbook by Olson[12] where an exhaustive treatment of a dozen or so different approaches is given.

A. Electron beam physics

Electron beam physics is at the crux of all collective acceleration methods and whole books have been devoted to the subject,[27] as well as hundreds of original papers and review papers. Here, we only touch on the essentials for our purpose.

Consider an intense relativistic electron beam (IREB) which is made up of electrons of mass, m, moving with speed, v_e, so $\beta_e = v_e/c$ and $\gamma_e^{-2} = 1-\beta_e^2$, and of uniform intensity so that the density is n_b (in the laboratory frame) out to a radius r_b and zero outside that radius. At the beam edge the electric field due to the electrons is radial and

$$E_r(r_b) = \frac{2I_e}{\beta_e c r_b} , \tag{2.1}$$

where I_e is the beam current. $E_r(r)$ varies linearly with r inside the beam and drops off inversely with r outside the beam. The magnetic field varies the same way with r as the electric field and, at the edge,

$$B_\theta(r_b) = \frac{2I_e}{cr_b} \,. \tag{2.2}$$

In practical units (and taking β_e=1),

$$E_r(r_b) = 6.0 \, \frac{I_e(A)}{r_b(cm)} \left(\frac{kV}{m}\right) , \tag{2.3}$$

which is just Eq. (1.4).

For an electrically neutral beam a particle at the beam edge has a cyclotron radius, r_{ce}, given by

$$r_{ce} = \frac{\beta_e \gamma_e mc^2}{e B_\theta(r_b)} \tag{2.4}$$

and when $r_{ce} > r_b/2$ the beam will not propagate. Setting $r_{ce} = r_b/2$ and solving for the beam current at which propagation stops, we find

$$I_A = \frac{\beta_e \gamma_e mc^3}{e} , \tag{2.5}$$

where this critical current is known as the Alfven, or Alfven-Lawson, current. In practical units (and, again, setting β_e=1)

$$I_A = 17 \gamma_e \text{ (kA)}. \tag{2.6}$$

However, beams with currents larger than the Alfven limit can propagate or be propagated. This can occur if the beam is profiled in shape (hollowed-out), or has a return current (usually of low-energy electrons) so that the net current in less than I_A, or rotates so that the centrifugal force balances the magnetic force, or is electrically unneutral so that the electric force is also important, or is propagated in an axial field, B_z. Generally, in collective acceleration schemes, one has $I_e < I_A$. The Alfven limit is important in the beam-generation process in a diode and in naturally occurring acceleration.

The space-charge limiting current, on the other hand, is crucially important to collective acceleration schemes. Consider a beam which is launched into a grounded tube of radius R. Because the beam, which is taken to be fractionally neutralized to degree f, has an associated electric field there is a potential difference between the tube and the beam center of

$$\Delta\phi = \left(\frac{I_e}{\beta_e c}\right) (1-f) \left[1+2 \ln \frac{R}{r_b}\right] . \tag{2.7}$$

The electrons have a kinetic energy of $(\gamma_e-1)\,mc^2$ and if we equate this to the potential difference of Eq. (2.7) and then solve for the current we obtain

$$I_1 = \beta_e\,(\gamma_e-1)\left(\frac{mc^3}{e}\right)[1+2\ln R/r_b]^{-1}\,(1-f)^{-1}. \qquad (2.8)$$

This space charge limiting current, defined for $f = 0$, was first derived by Olson and Poukey.[12] For $I_e > I_1$ a beam will not propagate, but the beam can "get around" this limit by (for example) becoming neutralized. We shall see that this is precisely the mechanism behind space charge acceleration. Note that, for $f = 0$, $I_1 < I_A$.

The literature abounds with other space charge limiting currents, but for our purposes I_1 is sufficient. Of course, one could do a better job by including self-consistency, etc., but that is not necessary to understand the physics of IREB. The reader need only remember that we can have charge neutralization (the factor, f, above) and/or current neutralization. In short, we can introduce two phenomenological constants, $f = f_e$ and f_m, one on the electric force and one on the magnetic force. Finally, the reader may know about the Budker parameter ν_e which is defined by

$$\nu_e = \gamma_e\,\frac{I_e}{I_A}, \qquad (2.9)$$

so that (ν_e/γ_e)--a figure of difficulty for propagating IREB--is just I_e/I_A.

B. Naturally occurring acceleration

This acceleration of ions has been observed in two rather different situations. The first is when an IREB is injected into a gas (see Fig. 2) and the second is within the diode of an IREB.

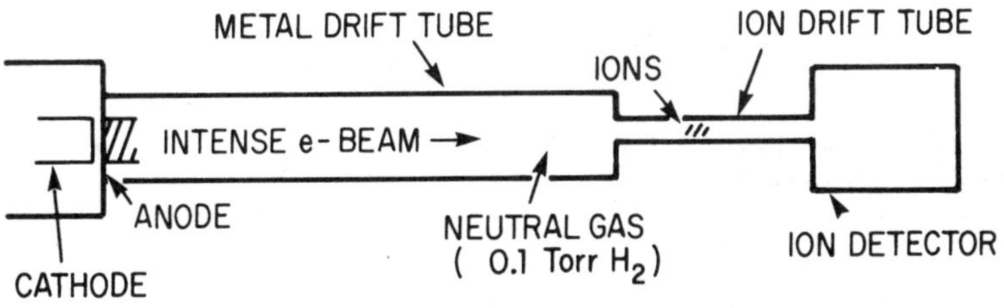

Fig. 2. Basic apparatus for production of IREB and observation of "natural" collectively accelerated ions produced from a neutral gas.

The first situation was that of Graybill and Uglum[7] in their pioneering observation of collective acceleration in 1968. Since then there has been a great deal of work on the subject and people have varied the gas (its pressure and type), the IREB (its energy, current, radius, on-time, and rise-time), the tube (its radius and length), and the diode (its cathode-anode spacing and cathode radius). Suffice it to say, that quite a lot is now known about this process.

The second situation was first studied by Plyutto[8] in 1960 who examined plasma-filled diodes. Much later, Luce[16] worked with a vacuum diode in which the anode was made of plastic (a source of ions) and had a hole (through which the IREB did not quite fit). These "Luce diodes" have subsequently been employed by many workers and proven very effective in accelerating ions.

The experimental situation is, of course, complicated and the degree of ion acceleration does depend upon the parameters listed above. Roughly, one observes perhaps 10^{13} protons, accelerated to 10 MeV, in a pulse length of 10 nsec. The variation in these parameters is large, but the above are typical parameters for an IREB of about 50 kA and a pulse length of 50 nsec. Figure 2 outlines a particular "Luce diode" experiment.

C. Space charge wells

Let us focus upon an IREB injected into a gas contained in a tube. For $I_e > I_1$ the beam stops near the anode forming a deep potential well of the order of 2 or 3 times the beam kinetic energy. The beam then becomes neutralized, by beam-induced ionization processes in the gas, and hence is able to move on. The process of "moving on" is the acceleration process, although the existence of the initial potential well explains many of the observations of naturally occurring acceleration.

Why does the beam stop initially and why is the potential well two or three times as deep as the beam kinetic energy? That the beam should stop can be seen from a one-dimensional model. Consider a beam of electrons which penetrate a conducting plane located at x = 0. Poisson's equation for the potential ϕ is

$$\frac{d^2\phi(x)}{dx^2} = 4\pi e\, n(x) \ , \qquad (2.10)$$

where n(x) is the electron density and the electron charge is (-e). The equation of motion is

$$\frac{d}{dt}\left(m\, \gamma_e(t)\frac{dx}{dt}\right) = e\,\frac{d\phi}{dx} \ , \qquad (2.11)$$

with boundary conditions that at t = 0 the particle is at x = 0 and has speed v_e. The Poisson equation has the boundary conditions that $\phi(x=0)=0$ and $d\phi/dx(x=\infty)=0$.

Solution of this relativistic Child's law problem was first given by Jory and Trivelpiece[28] who found a potential variation out to a distance x_m (and not beyond that point) corresponding to electrons being just reflected at x_m, with x_m given by

$$x_m = \frac{c}{2\omega_b \beta_e^{1/2} \gamma_e} \int_1^{\gamma_e} \frac{dy}{(y^2-1)^{1/4}} , \qquad (2.12)$$

where $\omega_b^2 = 4\pi n e^2/m\gamma_e$ is the beam plasma frequency. For $\gamma_e \gg 1$, $x_m \simeq (c/\omega_b)(2\gamma_e^{-1/2})$.

This static solution is not realized in a time-dependent situation and Poukey and Rostoker[29] used a numerical simulation technique to obtain a better idea of what really occurs in the one-dimensional case. They found that the potential will be deeper than the beam kinetic energy $(\gamma_e-1)\, mc^2$ by a factor between 2 and 3, essentially because in a time-dependent problem electrons keep moving into the region $x > 0$ even when the potential is large enough to stop them at some distance and, hence, they contribute to the potential. This is not a static solution and Poukey and Rostoker study the resulting oscillations in the potential. At a time, t, given by

$$t = 4 \frac{(\gamma_e-1)^{1/2}}{\beta_e\gamma_e} \left(\frac{1}{\omega_b}\right)$$

there is a pile up of density close to x_m, where the oscillation in potential is only about 10%. At the same time, electrons which were injected early are at $x = 4x_m$ where the well depth is about two times the beam kinetic energy. Thus, most of the beam is stopped at a distance x_m [given by Eq.(2.12)], but some of it, near the head of the pulse, moves on in an associated deep potential well. For $\gamma_e \gg 1$ they find, analytically, that the potential well is two times the beam kinetic energy.

Olson[12] has extended the work of Poukey and Rostoker to include finite (rather than zero) current rise-time and also to include two dimensional effects. He concludes that a deep well (2-3 times the beam kinetic energy) forms from the anode out to a distance of (say) twice the pipe radius; i.e., that, as in the one dimensional case, there is beam stopping, but that two dimensional effects (especially on well-shape and location) are important.

D. Well motion

The potential well will move when the beam becomes charge neutralized. Many processes contribute to beam neutralization; namely electron impact, ion impact, and charge-exchange scattering. These have been considered by a number of workers and are summarized by Olson.[12] Olson has emphasized the importance of ion impact and gives a "handy formula" which is

$$\tau \simeq \frac{1.0}{p(\mathrm{Torr})}\mathrm{nsec}, \qquad (2.13)$$

for the charge neutralization time, τ.

At high pressures there is "run-away" of the beam front; i.e., neutralization occurs very quickly and other phenomena limit the beam propagation rate. One of these phenomena is simply power balance between the incoming beam and the power required to establish electric and magnetic fields and to accelerate ions.

At very low pressures the ions created in the deep well of a stopped beam will, according to Olson, determine the beam front velocity.

At intermediate pressures the front velocity, $\beta_f c$, is determined by the ionization time, τ, so that

$$\beta_f \simeq \left(\frac{1}{15}\right) R(\mathrm{cm})\ p\ (\mathrm{Torr}). \qquad (2.14)$$

Thus the beam front has a velocity which at low pressures is the ion velocity, at intermediate pressures is given by Eq. (2.14), and at high pressures is limited (say) by power balance. For many experiments one is in the intermediate regime, $p \simeq 1/10$ Torr, and $R \simeq 10$ cm so the beam front velocity $\beta_f c \simeq (1/15)c$; i.e., is very much less than the electron velocity. This steady state does not lead to further ion acceleration, but the transition may lead to trapped ions having a speed of $\beta_f c$. This is not a large energy and, hence, control of the beam front velocity is the key element in all of the proposed space charge well accelerators.

E. Control of beam front velocity

There are three methods which have been proposed and which are being pursued experimentally for beam front velocity control. These are the use of (1) slow wave structures (Maryland), (2) pulsed wall plasma (Irvine), and (3) laser initiation of ionization (Sandia).

In a way these approaches are not very profound theoretically; put another way, these approaches should work. They do have different experimental requirements and their experimental realization is non-trivial in most cases.

Taking these approaches in order, we start with the research at Maryland on slow wave structures. Without such structures, and employing Luce diodes, they obtained 8-10 MeV protons; with the structures they have produced a high-energy component with 16±1 MeV.[10] They plan to continue work on this approach. Rostoker has proposed (private communication) the use of ions injected from a pulsed wall plasma as a method for time-controlling the beam front. He hopes in this way to routinely produce 10-MeV protons and then extend this approach to produce 100-MeV protons.[9]

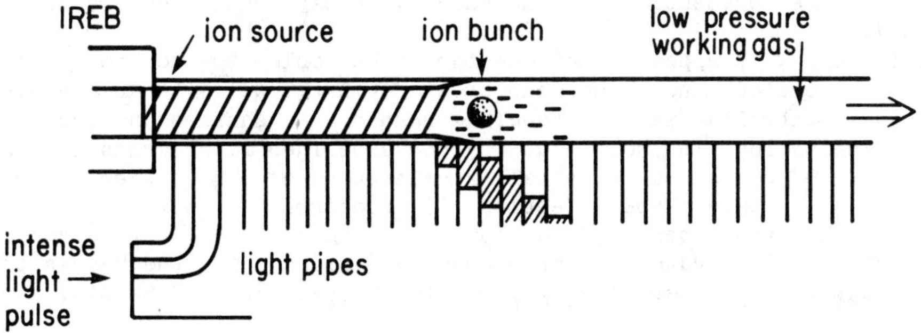

Fig. 3. Ionization front accelerator (IFA).

Without question, most effort has been put into the approach which uses laser initiation of ionization or the ionization front accelerator (IFA) (see Figure 3). This approach, initiated by Olson,[12] employs light pipes of various lengths to control beam front velocity. Olson has been able, in this manner, to control beam front speeds, but has not yet definitively accelerated ions. These data suggest that an accelerating gradient of 50 MV/m has been achieved.

Finally, it should be mentioned that proposals have been made to stage the naturally occurring acceleration process. This is being pursued by Adamski.[11] Also, the Maryland people[10] have been exploring laser-produced plasmas and localized gas clouds as sources of ions for collective acceleration. In this last approach they have obtained the present record in energy, namely Xenon accelerated to 900 MeV.

In short, it appears that one should be able to control beam front velocities and hence employ space-charge wells to accelerate ions in a controlled and reproducible manner. We shall soon see if this expectation is born out by the experimental programs now in progress. Of course, a practical accelerator of the space-charge type or the wave type (Section 3) requires an inexpensive and reliable IREB generator that can be pulsed many times per second. The rapid pulse rate is not presently available but the requisite technology is being developed vigorously at Livermore and Sandia.

3. WAVE ACCELERATOR

Intense relativistic electron beams provide a plasma (non-neutral) which can support waves and if these waves are appropriate they can be employed to accelerate particles. Furthermore, if the waves are unstable they will grow, at the expense of the electrons kinetic energy (which is large and relatively easy to obtain), and thus develop an accelerating field for ions. In fact, this is the basis of wave accelerators.

In order to be effective for acceleration the wave needs to have a controllable and variable phase velocity. The wave must be coherent over scales which greatly exceed the wavelength and wave period. Furthermore the wave that does the acceleration must grow while not affecting other waves, i.e., mode coupling, spectral broadening, and wave-particle effects (on all but the desired wave) must be small. All of this must be true not only while the wave grows, but also while the desired wave has its phase velocity increased and significant energy removed from it by ions. Truly, this is one of the most demanding applications of IREBs. It is, perhaps, remarkable that it appears possible to do all of this! (We have to say "appears" since acceleration has not yet been achieved, but all the requirements for acceleration have already been achieved.)[30]

A. Waves on beams

In most of the wave accelerators, and, in particular, in the autoresonant accelerator (ARA) of Drummond and Sloan,[19] the IREB is quite intense. (In recent work in Austin, the beam was $I_e = 20$ kA and $W_e = 2.25$ MeV.) For this reason, and also because it is essential for the desired mode for the ARA, a longitudinal magnetic field is put on the beam (in the Austin case, this field was 3.5 kG.).

The electron beam may be stably propagated in an evacuated conducting pipe if the following conditions for an equilibrium are met

$$\omega_p^2 < \frac{\gamma_e \Omega c}{r_b} , \tag{3.1}$$

$$2\omega_p^2 < \gamma_e^2 \Omega^2 , \tag{3.2}$$

$$\omega_p^2 r_b^2 < 4c^2 , \tag{3.3}$$

where r_b is the beam radius, ω_p is the plasma frequency

$$\omega_p^2 = \frac{4\pi n e^2}{\gamma_e m} , \tag{3.4}$$

where n is the laboratory beam density, and Ω is the cyclotron frequency,

$$\Omega = \frac{eB}{\gamma_e mc} , \tag{3.6}$$

where B is the applied field.

The condition of Eq. (3.1) is simply that the self electric field is smaller than the confining magnetic field. The condition of Eq. (3.2) is the relativistic form of the condition for (Brillouin) equilibrium. The condition of Eq. (3.3) is simply Eq. (2.6) put into different notation.

Stable propagation requires, according to Davidson and Krall,[31] that

$$\omega_p^2 < \frac{\gamma_e \Omega c}{r_b} , \tag{3.7}$$

which is simply the condition that the self electric field is less than the magnetic field as in Eq. (3.1). Thus, for a beam below the Alfven current, simply making the applied field large enough is sufficient for stable propagation. A model of the beam can be made in which the system is longitudinally homogeneous and axial symmetric. In this case, modes of the system can be found,[19] for the electric field components vary as

$$E_r,\ E_\theta \sim J_1(\alpha r)e^{i(kz-\omega t)}\ ,$$
$$E_z \sim J_0(\alpha r)e^{i(kz-\omega t)}\ , \qquad (3.8)$$

where J_0 and J_1 are Bessel functions and k, ω, α are good mode numbers. One finds the dispersion relation by using Maxwell's equations and the electron dynamical equations. One obtains for an unlimited beam in the radial direction

$$(\omega-kv_e)^2\ [\omega^2-\omega_p{}^2-(\alpha^2+k^2)c^2]^2\ [(\omega-kv_e)^2-\omega_p{}^2/\gamma_e{}^2]$$
$$-\Omega^2\ [\omega^2-(\alpha^2+k^2)c^2]\ \Big[(\omega-kv_e)^2\ [\omega^2-(\alpha^2+k^2)c^2]$$
$$-\ \frac{\omega_p{}^2}{\gamma_e{}^2}\ (\omega^2-k^2v_e{}^2)\Big] = 0 \qquad (3.9)$$

Sloan and Drummond have studied this dispersion relation and find that there is only one mode with a phase velocity which is very small and this is the Doppler-shifted lower cyclotron mode for which approximately

$$\omega = k\ v_e - \Omega. \qquad (3.10)$$

The phase velocity is

$$v_{ph} = \left(\frac{\omega}{\omega+\Omega}\right) v_e. \qquad (3.11)$$

Now by varying B as a function of distance down the accelerator one can control v_{ph}. One simply wants B to drop off with distance, and since electrons follow field lines one must also flare the beam pipe so as to have it large enough to contain the IREB. Now it is interesting to note that in all presently contemplated single wave accelerators one has a fixed frequency, ω, and hence in order to make the phase velocity increase (towards c) one must make the wave number, k, decrease; i.e., make the wavelength of the disturbance increase. This means that the accelerating field decreases as the ions are accelerated to higher and higher energies. Of course the initial field may be large enough that this is no problem, or the wave may be growing with distance so as to compensate the variation of k, or one may design frequency jumps (as in LAMPF) to avoid this problem.

There are other modes in Eq. (3.9) than the Doppler shifted cyclotron mode--seven more, to be precise. Sprangle et al.[20] have focused attention on the space charge mode. It is, like the Doppler shifted cyclotron mode, also a negative energy wave. The space-charge wave has phase velocity given by

$$v_{ph} = \left(\frac{\omega}{\omega+F\omega_p}\right) v_e \ , \qquad (3.12)$$

where F is a form factor which depends upon the ratio of beam radius to pipe radius. (For an unlimited plasma $F \equiv 1$.) By making the walls come closer to the beam, the converging guide accelerator, one can make v_{ph} increase, but the requirements on the guide are rather severe (more later) and it seems unlikely that v_{ph} can be made very small so that such an accelerator will probably need an ion injector.

Perhaps the obvious should be noted, namely that a wave accelerator works by producing a spacial variation of charge and hence an accelerating field. [In the ARA the desired mode is an axisymmetric (m=0) moving "string of sausages", i.e., a modulation of the beam envelope.] Thus wave accelerators develop--at best (if the wave length ≃ beam radius)--accelerating fields limited by Eq. (1.4).

B. The autoresonant accelerator

As we have already stated, this accelerator [see Fig. 4(a)], proposed by Drummond and Sloan,[19] works by growing a Doppler shifted, axisymmetric, cyclotron wave and then varying this wave's phase velocity so as to accelerate ions. Because the wave train is long the ARA can be a quasi-continuous accelerator and the duty factor should be limited by the length of IREB pulses (which so far have been very limited in practice).

The ARA people have had an experimental program since the early 1970's. They have done extensive studies on the condition for propagation of an IREB and Miller et al.[32] have shown that the conditions of Eqs. (3.1-3.4) are experimentally verified.

Most importantly, they have been able to show[30] the growth of the desired mode to large-amplitude, and with low, and variable, phase velocity. Thus they have taken an IREB (W_e = 2.25 MeV, I_e = 20 kA, τ = 80 nsec, ($\Delta W/W$) = 5%) and sent it through a helical structure which is driven by an external oscillator tuned to 239 MHz. The external magnetic field is 3.5 kG which implies that the desired mode has a wavelength of 13 cm. They find that the beam develops just this wave (λ=12.5±1 cm) over the voltage flattop. The antenna which drives the desired mode is put 1 meter from the diode and measurements are made 1 meter from the antenna. The wave develops an accelerating field of 10 MV/m and is very pure (i.e., it is more than 90% symmetric and quite sinusoidal in appearance).

The ARA group has varied the wave phase velocity by varying the guide magnetic field and found that it is in accord with theory [Eq. (3.11)]. They have reported velocities from 0.06 c to 0.20 c which is adequate for first ion acceleration experiments.

Thus everything seems to be in order for actual ion acceleration, and that should be achieved very soon.

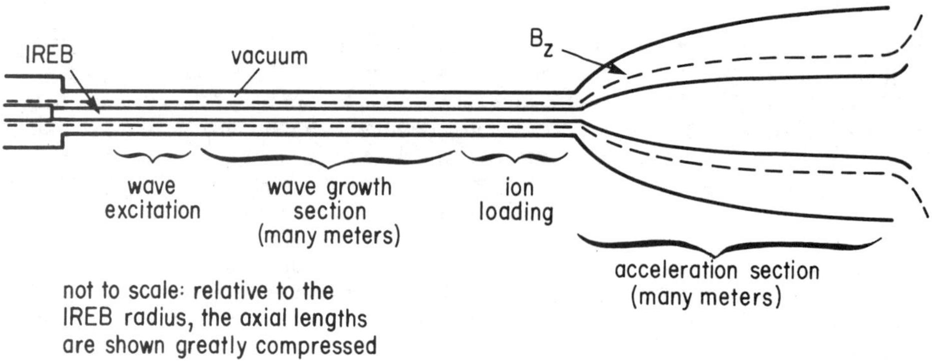

Fig. 4(a). Autoresonant accelerator (ARA).

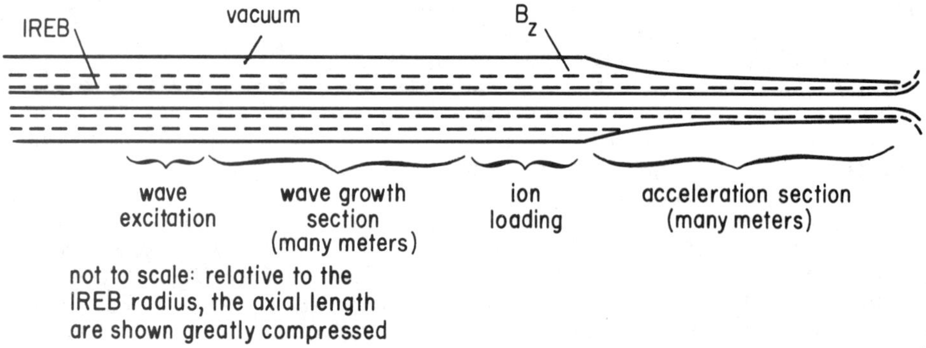

Fig. 4(b). Converging guide accelerator (CGA).

C. The converging guide accelerator

The Converging Guide Accelerator (CGA) [see Fig. 4(B)] employs, as we have said, a space-charge wave which depends logarithmically on the guide pipe radius and, hence, its phase velocity can be controlled. Introduced by Sprangle et al.,[20] experiments have been done by Nation et al.[33]

First it was noted by Briggs and by Godfrey (see Ref. 12), that the phase velocity can only be made small if the beam current is large. In fact, the electron current must be made to approach the space charge limiting current I_1 [Eq. (2.8)] and IREB propagation is, at best, intermittent and unreliable. Thus, Nation et al. have suggested the use of an ion injector so that $v_{ph} \gtrsim 0.2$ c.

Second, the dependence upon the wall spacing is very weak so that going from $\beta_{ph} = 0.2$ to $\beta_{ph} = 0.40$ (W_i = 19 MeV to W_i = 85 MeV) the wall must vary from 2.97 cm to 2.73 cm (i.e., 45 microns/cm!) and up at E_i = 200 MeV one needs a wall radius that varies by just 13 microns/cm. A more rapid variation will lose the ions if the accelerating field isn't large enough.

Nevertheless, Nation et al. have employed a 250 keV beam in an iris-loaded structure to grow the desired mode at 1.1 GHz. The growth was rapid and they achieved an accelerating field E_z = 6 MV/m. Operation with v_{ph} below 0.2 c was found to be marginal, but above this phase velocity the operation was stable and the Cornell group now plans to inject protons of energy ∽20 MeV and thus accelerate them in the CGA.

D. Ion-electron waves

The Cornell group has studied the acceleration of protons in evacuated drift tubes.[34] They employ a Luce diode with an IREB having W_e = 500-700 keV, and $I_e \leq 65$ kA, and a pulse duration τ = 100 nsec. The acceleration of protons is <u>not</u> near the pulse head (as in space-charge well models) but is throughout the pulse. They conclude that the acceleration is by a wave mechanism and, probably, the wave is an electron-ion two-stream mode. It is not clear how to employ this mode to build a scalable accelerator, but the observations motivate presentation of some formalism.

Davidson[1] has given a linear fluid analysis for a beam of small radius, r_b, compared to a guiding pipe radius, R. The two-stream instability is largest for

$$\frac{\mathrm{Re}(\omega)}{k} = \frac{1}{2}\left(\frac{Zm}{2M}\right)^{1/3} \gamma_e v_e, \tag{3.13}$$

where we have assumed the ion velocity is negligible compared to the electron velocity. At this frequency, the growth rate is

$$(\mathrm{Im}\omega) = \frac{\sqrt{3}}{2}\left(\frac{Zm}{2M}\right)^{1/3} \gamma_e^{1/2} \omega_p , \tag{3.14}$$

provided the electron current is greater than a critical current, I_{crit}, where

$$I_{crit} = \frac{I_A\ (v_e/c)^2\gamma^2}{2\ \ln\left(\frac{R}{r_b}\right)\left[1 + \frac{ZM}{m}^{1/3}\ \gamma_e\right]^3}, \qquad (3.15)$$

and the Alfven current, I_A, is given by Eq. (2.6).

Comparison between these formulas and the experimental observations is given by the Cornell group who find reasonable agreement and conclude that the observations are probably explained by a two-stream mechanism.

Before leaving this section, we note that the Collective Particle Accelerator of Friedman,[21] discussed in Section 1, is also a non-linear wave accelerator. Since his accelerator has many waves (the rings in his concept must be described by many Fourier components), it need not be constrained to a decreasing accelerating field with increasing phase velocity, and may function well even relativistically.

4. ELECTRON RING ACCELERATOR

An electron ring accelerator (ERA) works by making a compact cluster of electrons, imbedding ions in the cluster, and then accelerating the cluster (see Fig. 5). There have been a number of different ways proposed for creating electron rings: the Maryland scheme in which a hollow electron beam is fired into a cusp field, the static-field compressor scheme of Christofilos,[35] Laslett and Sessler,[36] and the dynamic-field compressor. The last approach has been employed by the groups at Dubna, Garching, and Berkeley. It is the only approach described here, but interested readers can consult the text of Olson and Schumacher.[12] This text gives an excellent treatment of the whole subject of ERAs. There has also been research by the group at Bari and Lecce[37] on the transverse ERA. This will not be discussed in the present review.

A. Compression single-particle dynamics

In order to obtain a large field gradient, one must make rings of small dimensions and large current. The maximum field of a ring, E_{max}, is given (roughly) by Eq. (1.4). Expressing the field in terms of the number of electrons in a ring, N_e, we have

$$E_{max}\ (\mathrm{MV/m}) = \frac{(4.58\times10^{-12})\ N_e}{R(\mathrm{cm})\ a\ (\mathrm{cm})}, \qquad (4.1)$$

where R and a are the ring major and minor radii. Thus for 10^{13} particles, R = 4 cm, and a = 0.3 cm, we have E_{max} = 38.MV/m.

Fig. 5. The electron ring accelerator.

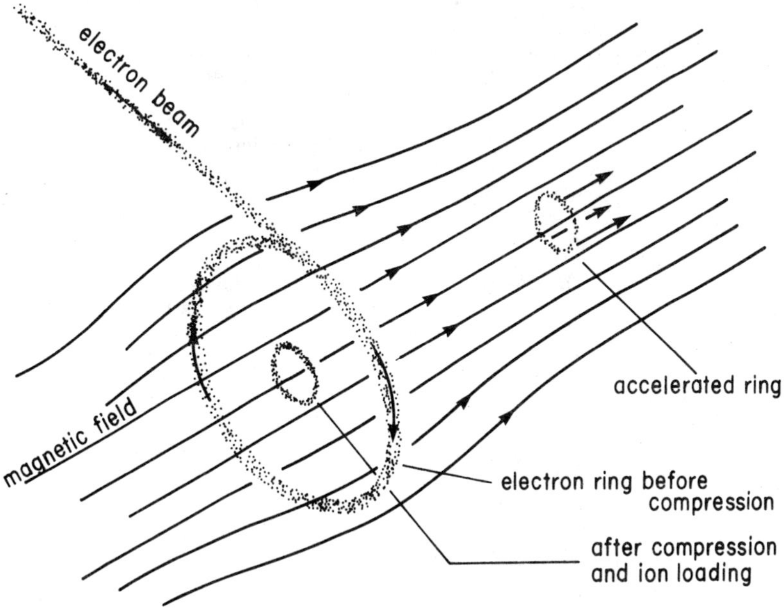

(a). Schematic of an ERA using ring compression.

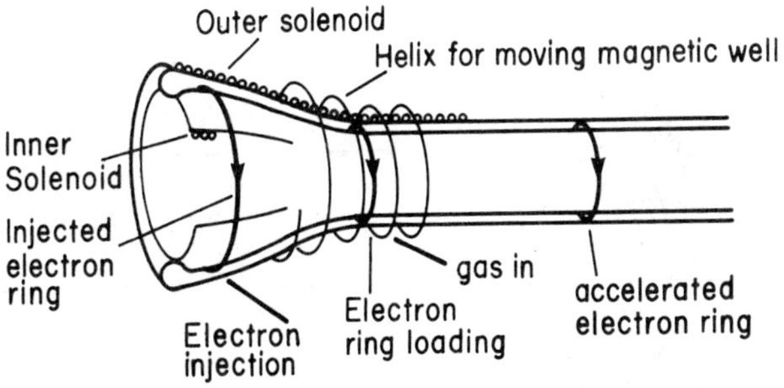

(b). The ERA static-field compressor.

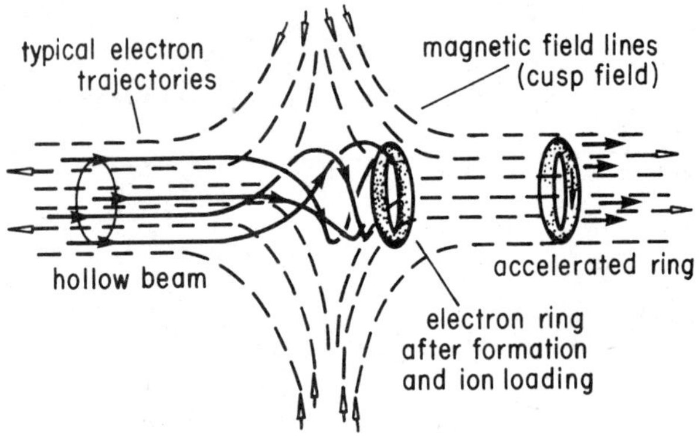

(c). The Maryland Cusp-field scheme.

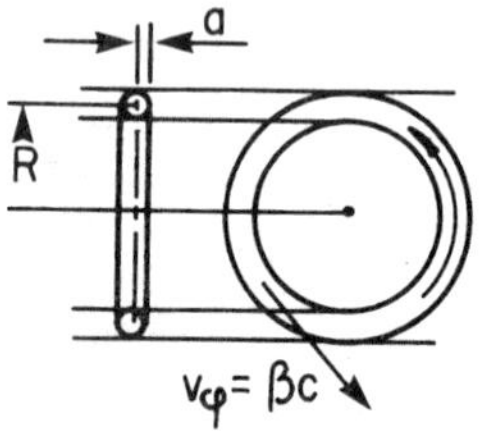

(d). Electron ring geometry.

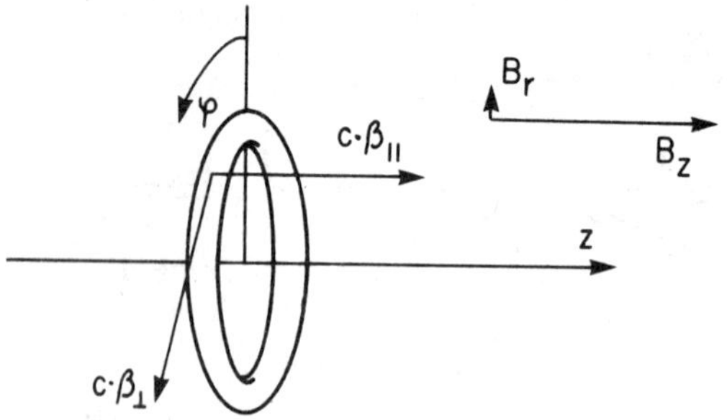

(e). Electron ring with velocity and magnetic field components.

Production of such a ring can be achieved (in fact, has been achieved by the Dubna group) by a dynamic compressor in which the electrons are injected at a large radius R_1 and then compressed to a small radius R_2. At the same time transverse dimensions are also reduced so that the maximum electric field is greatly increased by the compression process.

To a good approximation the magnetic field of a compressor can be considered as spacially uniform, but varying (slowly) in time. In an axially symmetric field the azimuthal component of the canonical momentum is conserved

$$Rp_T - eRA_\theta = \text{const}, \tag{4.2}$$

where p_T is the electron transverse momentum and the vector potential is A_θ. For a uniform field of magnitude B,

$$eRA_\theta = \frac{1}{2}e\, BR^2. \tag{4.3}$$

Since always

$$eBR = p_T \tag{4.4}$$

BR^2 is a constant during compression. Thus we have (injection is labeled by 1, the compressed ring by 2)

$$R_2 = R_1\left(\frac{B_1}{B_2}\right)^{1/2}, \tag{4.5}$$

$$p_{T2} = p_{T1}\left(\frac{B_2}{B_1}\right)^{1/2}. \tag{4.6}$$

During compression the ring current will, of course, increase by the (R_1/R_2) ratio. For the Berkeley compressor the initial current was about 1/2 kA, while the compressed current was about 2 kA.

The minor ring dimensions are determined by conservation of the adiabatic invariant

$$\frac{p_T}{R}\nu a^2 = \text{const}, \tag{4.7}$$

where ν is the betatron tune. Thus,

$$a_2 = a_1\left(\frac{B_1\nu_1}{B_2\nu_2}\right)^{1/2}. \tag{4.8}$$

The initial tune, ν_1, is close to that for an external field gradient of 0.5, i.e., $\nu_1{}^2 \simeq (1/2)^{1/2}$ or $\nu_1{}^2 \simeq (1-1/2)^{1/2} = (1/2)^{1/2}$. The final tune is determined by ion-loading, images, and curvature in a manner which we must discuss. One can see that the field is not really uniform in space, and a proper calculation of compression has, of course, not been performed. The above is a good approximation for compression dynamics.

A compressor, then, has injection at a large radius in a field with gradient close to 0.5, but compresses to a rather uniform field. Thus the electrons are taken through a great many resonances for, roughly,

$$\nu_r^{\,2} = 1-n, \qquad (4.9)$$
$$\nu_z^{\,2} = n$$

and resonances occur, for example, at

$$2\nu_r - 2\nu_z = 0, \; n = 0.5$$
$$\nu_r + 2\nu_z = 2, \; n = 0.36$$
$$2\nu_z = 1, \; n = 0.25$$
$$\nu_r - 2\nu_z = 0, \; n = 0.2$$
$$\nu_r = 0, \; n = 0.0. \qquad (4.10)$$

Much consideration has been given to resonance crossing and the effect of crossing on the electron ring. We shall not go into this subject here (the reader may wish to study Ref. 12 and the works cited therein), but resonance crossing has required the construction of very good fields (so as to reduce the driving terms) and/or rapid motion through the resonances.

We turn, now, to the subject of electron tune. Clearly, for an electron ring accelerator to work there must be focusing of the electrons while the electron ring is accelerated. Such focusing may be supplied by ions and/or by images. Thus, when the external field has a gradient in the range of 0.05 to 0.95 then Eq. (4.9) is adequate, but outside of that range one must take into account the above-mentioned effects. Laslett and Ivanov have done just that[38] and obtain

$$\nu_r^{\,2} = (1-n) + \mu\left\{\frac{2R^2}{\sigma_r(\sigma_r+\sigma_z)}\left(f - \frac{1}{\gamma^2}\right) + (1-f)\,\frac{P}{2}\right.$$
$$-4\left[\frac{(1-f)\varepsilon_{1,E}}{(S_E-1)^2} - \frac{\beta^2\varepsilon_{1,M}}{(S_M-1)^2}\right]$$
$$\left.-n\left[(1-f/2)\,P + \frac{(1-f)}{(S_E-1)} - \frac{\beta^2}{(S_M-1)}\right]\right\} \qquad (4.11)$$

and

$$\nu_Z^2 = n+\mu \left\{ \frac{2R^2}{\sigma_Z(\sigma_r+\sigma_Z)} \left(f - \frac{1}{\gamma^2}\right) - (1-f)P/2 \right.$$

$$+4 \left[\frac{(1-f)\varepsilon_{1,E}}{(S_E-1)^2} - \frac{\beta^2\varepsilon_{1,M}}{(S_M-1)^2} \right]$$

$$\left. +n \left[(1-f)\ P/2 + \frac{(1-f)}{(S_E-1)} - \frac{\beta^2}{(S_M-1)} \right] \right\}, \qquad (4.12)$$

where σ_r and σ_z are the beam minor radii (rms), f is the fraction of neutralization,

$$\mu = \frac{N_e\, e^2}{mc^2 2\pi R\gamma}, \qquad (4.13)$$

$$P = 2\ \ln\left(\frac{8\sqrt{2}\ R}{\sigma_r+\sigma_z}\right), \qquad (4.14)$$

RS_E = radius of tube giving electric images,

RS_M = radius of tube giving magnetic images,

ε_{1E} = electrostatic image coefficient,

ε_{1M} = magnetostatic image coefficient.

One can readily see that ions ($f \neq 0$) give focusing, and one can see that if image currents can be reduced ($\varepsilon_{1M} \simeq 0$) then axial image focusing can be obtained. (The radial focusing when n = 0 is, of course, strong.) Image cylinders that do this are routinely used ("squirrel cages") and thus provide external electrostatic image focusing to the ring. To a good approximation (ignoring magnetic images and curvature effects) one has at the end of compression (n = 0)

$$\nu_r \simeq 1,$$

$$\nu_z \simeq \mu \left[\frac{2R^2}{\sigma_r(\sigma_r+\sigma_z)} \ (f - \frac{1}{\gamma^2}) + \frac{4\varepsilon_{1,E}}{(S_E-1)^2} \right]. \qquad (4.15)$$

Thus axially focusing is provided by ions and electrostatic images. These values of ν can then be employed, with Eq. (4.8), to determine the compressed ring minor dimensions.

B. Collective efforts

Because in an ERA one wants to obtain a large ring electric field it is not surprising that collective effects are the phenomena which limit ERA performance. There are three phenomena which we know to be important: the negative mass effect, the resistive wall effect, and ion-electron transverse instabilities. We only are considering instabilities here, and not collective equilibrium effects which have already been considered in the last section.

The negative mass instability[39] gives a limit on the number of ring particles which can be written in the form

$$N_e \leq \left(\frac{1}{\nu_r^{\,2}} - \frac{1}{\gamma^2}\right) \frac{\gamma R Z_o m c^2}{2\beta^3 e^2 \left|\frac{Z_{\tilde{m}}}{\tilde{m}}\right|} \left(\frac{\Delta E}{E}\right)^2, \qquad (4.16)$$

where β and γ are electron relativistic factors, Z_o is the impedance of free space (377 ohms), ΔE is the ring energy spread (FWHM) and $Z_{\tilde{m}}$ is the impedance which the ring "sees" at frequency $\tilde{m}f_o$ where f_o is the electron cyclotron frequency.

In fact, the negative mass instability limit is one of the most serious limits for an ERA and much effort has been put, both theoretically and experimentally, into reducing the coupling impedance $Z_{\tilde{m}}$. Approximately (ignoring the radiative contribution) we have

$$\frac{Z_{\tilde{m}}}{\tilde{m}} = \frac{Z_o}{2}\left(\frac{1}{\gamma^2}\left(1 + 2\ln\frac{2h}{\pi a}\right) + \left(\frac{h}{\pi R}\right)^2\right), \qquad (4.17)$$

where h is the distance to highly conducting walls which are close to and in the plane of the ring, i.e., $h \ll R$. Thus one wants to put conducting walls as close to the ring as possible and this is done in the experiments. There is a conflict between conducting walls for the negative mass instability and penetration (i.e., non-conducting walls) of the compressor field. Fortunately the frequencies are very different and there is really no conflict. There is, however, a more serious conflict between the negative mass instability (which can be ameliorated by close walls) and the resistive wall instability (which is a strong function of the distance to walls and is least for walls far away). In practice, these two effects must be balanced so that wall design and location is a delicate and important matter.

The resistive wall instability[40] leads to a limit which can be written in the form

$$N_e \leq \frac{2 R Z_o \nu_m \gamma m c^2}{\beta e^2 |Z_T| c} (\Delta S), \qquad (4.18)$$

where Z_T is the transverse beam coupling impedance in ohms/meter and

ΔS is the spread in the collective frequency S (of the first radial mode)

$$S = 2\pi\ (1-\nu_r)\ f_o. \tag{4.19}$$

Both energy spread and amplitude spread contribute to ΔS. The transverse coupling impedance can be expressed in terms of the longitudinal coupling impedance and, roughly,

$$Z_T \simeq \frac{Z\tilde{m}\ 2c}{S\ h^2}\ . \tag{4.20}$$

As has been noted, in practice one must balance the negative mass and resistive wall instabilities in order to accept a large number of particles into a compressor. (Clearly, the situation is most critical at injection.)

Transverse electron-ion oscillations are important even if images are neglected. In that case, and ignoring spreads in beams (a fluid theory), one has[41]

$$\left(\frac{\partial}{\partial t} + \omega_{ce}\frac{\partial}{\partial \phi}\right) x_1 + \omega_{ce}{}^2 Q_e{}^2 x_e = \omega_{ce}{}^2 Q_1{}^2 x_i,$$

$$\frac{\partial^2}{\partial t^2} x_i + \omega_{ce}{}^2 Q_i{}^2 x_i = \omega_{ce}{}^2 Q_i{}^2 x_e, \tag{4.21}$$

where x_e and x_i are the displacements of the electron and ion centers of mass,

$$Q_e{}^2 = \nu^2 + Q_1{}^2, \tag{4.22}$$

and $Q_i\omega_{ce}$ is the ion bouncing frequency:

$$Q_i{}^2 = \frac{e^2 Z N_e R}{M\ c^2\ \pi\ \sigma_r(\sigma_r+\sigma_z)}\ . \tag{4.23}$$

In this formula, the ion has been taken to have charge Ze, the ν is either ν_r or ν_z, $\omega_{ce} = 2\pi f_o$, and the electron bouncing frequency is $Q_1\omega_{ce}$ with

$$Q_1{}^2 = \frac{N_i\ M}{N_e\ m\ \gamma}\ Q_i{}^2\ , \tag{4.24}$$

Modes, $\tilde{m}$, and with frequency, ω, obey the dispersion relation, obtained from Eq. (4.21),

$$\left[Q_e^{\,2} - \left(\tilde{m} - \frac{\omega}{\omega_{ce}}\right)^2 \quad Q_i^{\,2} - \left(\frac{\omega}{\omega_{ce}}\right)^2\right] = Q_i^{\,2} Q_l^{\,2}. \qquad (4.25)$$

The stability diagram which results has been detailed by Koshkarev and Zenkevich[41] and others. In practice, one has to avoid or minimize the effect of instabilities.

The theory needs to be extended to include Landau damping[42] and one then finds a condition for stabilization (applied to the worst; ie, the first radial mode)

$$\left[\Delta_e^{\,2} - \left(\frac{2U}{\nu_r}\right)^2\right] \Delta_i \geq \frac{Q_l^{\,2} Q_i}{\nu_r}, \qquad (4.26)$$

where $\Delta_i \omega_{ce}$ and $\Delta_e \omega_{ce}$ are the spreads in the ion and electron frequencies (FWHM), and U is the coherent frequency shift induced by the collective fields. In practice one can arrange to satisfy Eq. (4.26), but it should be noted that a spread is required in both the ion and electron distributions.

C. Ring acceleration

It is clear that if an electron ring is loaded with only a very small number of ions and then accelerated in an external field the effective charge to mass ratio of the ions is enhanced by the factor $M/m\gamma$. In order to realize this factor we must not load the ring too greatly and, more importantly, the ring must have an electric field which is adequately large to hold the ions while it is being accelerated.

The "holding field," E_H, is close to (but less) than the maximum field, E_{max}, given by Eq. (4.1). The holding field is less than E_{max} because a ring is not uniformly charged and because in accelerating a ring there will be polarization of the ring (the ions, being dragged, will separate from the electrons). In fact, it is to avoid polarization (which can seriously degrade a ring's holding power) that external image focusing is employed. In practice, the limit on the rate of accelerating electron rings is given by the requirement that the holding power not be exceeded; one must not accelerate a ring too quickly.

One can readily employ kinematics to determine the relation between rings at the end of compression (a position designated by the index "c") and rings in a state of motion. To accelerate rings two methods have been considered; namely electric acceleration and magnetic acceleration. Electric methods are more complicated and have not yet been employed in practice. A great deal of theoretical work has gone into this subject, and the Dubna group has devoted many years to building an apparatus to achieve cavity acceleration, but have not yet reported successful use of this approach.

We shall, therefore, consider only magnetic acceleration. The concept is very simple: One arranges a solenoidal field, B_z, which decreases with increasing Z. The electron ring will accelerate (due to the B_r-force) while increasing its radius. The static field does no work on the electrons and their transverse energy (this is all that they have after compression and before acceleration) is converted, in part, into longitudinal (Z) motion.

Let the electron velocity (in units of c) have component $\beta_{\parallel}$ along the Z- direction and β_T in the azimuthal direction. Clearly

$$\beta^2 = \beta_T{}^2 + \beta_{\parallel}{}^2, \tag{4.27}$$

and we can define $\gamma_{\parallel}$ by

$$\gamma_{\parallel}{}^2 = \frac{1}{1-\beta_{\parallel}{}^2} . \tag{4.28}$$

Conservation of energy gives us one desired relation. For the compressed state

$$E = N_e\, mc^2\, \gamma_c + N_i\, Mc^2,$$

and for the accelerated state

$$E = N_e\, mc^2\, \gamma + N_i Mc^2\, \gamma_{\parallel} .$$

Equating these we have

$$\gamma - \gamma_c = \left(\frac{N_i M}{N_e m}\right) (1-\gamma_{\parallel}). \tag{4.29}$$

The other relation which we need comes from the fact that in the expansion, flux will be conserved so that the ring radius, R, is given by BR^2 remaining constant. The cyclotron formula for radius then gives

$$\frac{\gamma^2\beta_T^2}{B} = \frac{\gamma_c{}^2\beta_c^2}{B_c} . \tag{4.30}$$

These two equations can then be solved for $\gamma_{\parallel}$ and γ

$$\gamma_{\parallel} \simeq \frac{1 + g}{\left(\dfrac{B}{B_c}\right)^{1/2} + g}, \tag{4.31}$$

where the mass loading is

$$g = \frac{N_i M}{N_e m \gamma_c} . \tag{4.32}$$

The ion kinetic energy W_i is, hence,

$$W_i \equiv (\gamma_{\parallel} - 1) Mc^2 = \frac{[1-(B/B_c)^{1/2}]}{\left(\frac{B}{B_c}\right)^{1/2} + g} Mc^2 . \tag{4.33}$$

The expansion rate must be limited so that ions remain in the ring. The ion energy gain is

$$\frac{dE_i}{dZ} = Mc^2 \frac{d\gamma}{dZ} ,$$

and this must be not greater than

$$\frac{dE_i}{dZ} = Z_1 E_H ,$$

where

$$E_H = \eta \frac{e N_e}{\pi R (\sigma_r + \sigma_z)} , \tag{4.34}$$

and η is a factor which relates E_H to E_{max} (perhaps $\eta \simeq 1/2$). In the course of expansion the radius, R, and the minor dimensions change, but $BR(\sigma_r+\sigma_z)$ is an invariant. Hence one can determine the magnetic field as a function of position by employing Eq. (4.31) and the last three equations. For small ion loading ($g \sim 0$) we find

$$B_Z = B_c \left[1 + \frac{3}{2} \frac{Z}{\lambda_1}\right]^{-2/3} , \tag{4.35}$$

where the characteristic fall-off length, λ_1, is given by

$$\lambda_1 = \frac{\pi}{2} \frac{R (\sigma_r + \sigma_z) Mc^2}{\eta e^2 N_e} . \tag{4.36}$$

We must also be concerned with the dynamics at the initiation of acceleration. This is a delicate point in the ERA cycle and considerable theoretical effort and experimental effort has been devoted to obtaining a proper transition from a compressor to an expansion column. Clearly, if the transition is not made carefully the spacial variation of B_r will destroy an electron ring. Suffice it to say that the transition has been achieved and we will not go into greater detail here.

An electron ring in the expansion column must not experience a negative mass instability and this puts a serious restriction on the intensity of rings. To estimate this effect we need to evaluate the coupling impedance which has been done by Faltens and Laslett[43] who obtain

$$\frac{Z_{\tilde{m}}}{\tilde{m}} \simeq 300 \ \frac{R-R_c}{R_c} \ (\text{ohms}), \tag{4.37}$$

where R_c is the radius of an inner "squirrel cage" conductor.

Taking $R-R_c \simeq 4\sigma$, and $(\Delta E/E) \simeq 2.4(\sigma/R)$, one can evaluate N_e from Eq. (4.16). From this, with $\eta = 0.8$, one obtains $E_H \simeq 80$ MV/meter for a field of 20 kG (which is a reasonable value). Clearly, E_H could be lower than this, but it is unlikely that E_H can be made greater than (say) 100 MV/meter. Thus we have obtained an upper limit on ERA performance.

D. Experimental situation

Experimental programs, at one time, existed at Dubna, Garching, Berkeley, Maryland, Nagoya, Moscow, Tomsk, Saclay, Karlsruhe, and Bari-Lecce. At the present time the author believes that only the Dubna program is active. Why? Only because the ERA is complicated, for the basic principles seem correct, experimental observations are in all cases in accord with our understanding, and actual acceleration has been achieved by the groups at Dubna and at Garching. As we have seen, there are limits on ERA performance, but these limits are sufficiently high that an ion (in contrast with protons) accelerator would still be very attractive. One can only conclude that the complexity of an ERA is what has discouraged the workers (and funding agencies).

We shall, here, not have space to mention most of the above programs (the interested reader should see the text by Schumacher), but limit ourselves to describing some of the highlights of the Dubna, Garching, and Berkeley programs. Nor can we, here, go into the details of the experimental programs. Suffice it to say that attention to vacuum, field quality, diagnostics, etc., made the work tedious and the resulting devices expensive.

All three groups have produced good rings, two with induction accelerators and one (Garching) with a conventional IREB-producing device. The Dubna parameters are R = 4cm, $N_e \gtrsim 10^{13}$, and minor radii between 2-3 mm. They achieved this by reducing the coupling impedance, in the compressor, to $(Z_{\tilde{m}}/\tilde{m}) \lesssim 6$ ohms.

The Dubna group has accelerated helium, carbon, nitrogen, and heavier ions, at 2-4 MeV/m, to a few MeV per nucleon. The Garching group, in 1974, obtained a holding power $E_H \lesssim 5$ MV/m with which they accelerated hydrogen and helium to about 200 keV/amu.

5. COLLECTIVE FOCUSING ACCELERATOR

The collective focusing accelerator is well-described by Irani and Rostoker[14] (see Fig. 6). We shall follow their notation in this chapter.

They envision employing magnetic confinement, as in the HIPAC,[44] to contain electrons in a torus (see Fig. 6(a)). The electric fields of the electron cloud then contain and focus ions. Acceleration is conventional, i.e., by externally produced toroidal (azimuthal) electric fields. Because the magnetic moment of an electron is much larger than that of ions it is possible to accelerate ions while not accelerating electrons (and hence not "short-out" the betatron).

The magnetic moment of an electron, in a field of magnitude B, is

$$\mu_e = \frac{1}{2}\,\frac{mv_T^{\,2}}{B}\,, \qquad (5.1)$$

where v_T is the electron perpendicular velocity. The electron gyrates, rapidly, about a field line with frequency Ω_e given by

$$\Omega_e = \frac{eB}{mc}\,, \qquad (5.2)$$

while it drifts around the small direction of the torus with frequency

$$\omega_D = \frac{eE_r}{rB}\,. \qquad (5.3)$$

The equation of motion of an electron, if there is an azimuthal electric field, E_Z, present is approximately

$$m\frac{dv_z}{dt} = -\mu_e\,\frac{\partial B}{\partial z} - e\,E_z. \qquad (5.4)$$

Thus, if E_z is less than the magnetic moment force (due to periodic variation in B_z) then electrons will not be accelerated by E_z. In practice, the Maxwell laboratory group[45] has had $E_z \simeq 0.5$ V/cm and negligible toroidal electron current with only a 1% variation in B over a length of 17 cm.

How many electrons can be put into the device? This is, of course, the $64 question. The HIPAC achieved an electron density $n = 4.0\times10^9\,\text{cm}^{-3}$, and the experiments at Maxwell laboratories achieved $n = 10^{10}\,\text{cm}^{-3}$. The HIPAC people saw and studied the diocotron (slipping stream) instability, the magnetron instability and the ion resonance instability. On this basis the group of Rostoker et al. concludes that with proper choice of the electron cloud

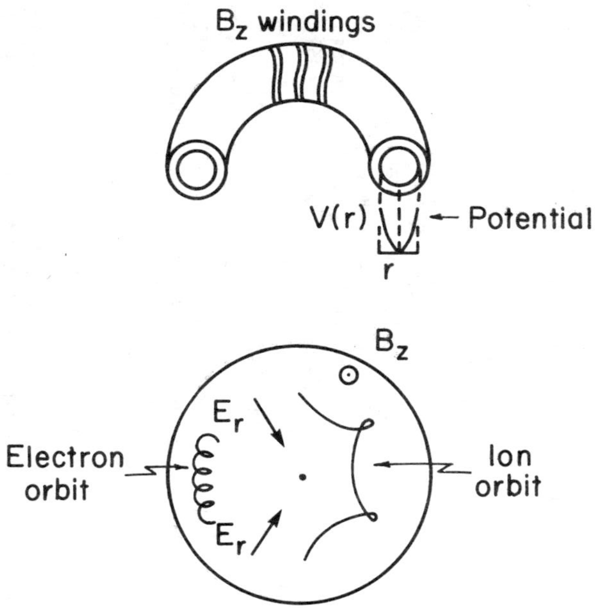

(a). Particle orbits in a toroidal collective accelerator.

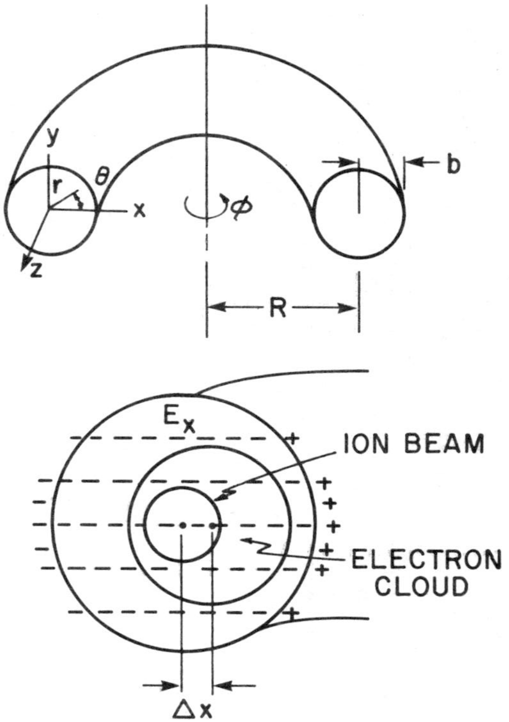

(b). Equilibrium positions of electrons and ions.

Fig. 6. Particle dynamics in a collective focusing accelerator.[14]

parameters one can avoid the diocrotron instability. The magnetron instability is unimportant if

$$\left(\frac{\omega_p}{\Omega e}\right)^2 < 0.05, \qquad (5.5)$$

where ω_p is the electron plasma frequency, and the ion resonance instability (a resonance between the ion bounce frequency and ω_D) can be avoided by choice of parameters and less than 10% loading of the ring with ions. On this basis the estimate that they can achieve $n \simeq 13\ cm^{-3}$.

Typical parameters would be, then, an electron toroidal ring of minor radius $a = 1.0$ cm, major radius $R = 3.3$ m, and a density $n = 1.1\times10^{13} cm^{-3}$. The magnetic field must be large so that Eq. (5.5) is satisfied, and, hence, $B = 50$ kG. The total number of electrons is, then, 7×10^{16}. This number of electrons will produce a peak field

$$\hat{E} = 2\pi nea,$$

which is 10 MV/cm.

Such a large field will, of course, hold ions and Irani and Rostoker observe that it will readily hold 100 GeV uranium if the uranium is 60 times ionized. With an azimuthal field of 10 V/cm they can reach the 100 GeV in a time of ≃ 10 msec and this field is sufficiently small that the electrons won't be accelerated by it. Highly ionized uranium is expected. One of the applications of the HIPAC was, in fact, as an ion source, but of course there will be a spread in ionization states and, hence, a spread in the energy of the accelerated ions.

The ions will have their center-of-mass shifted outward from the electron center-of-mass by an amount, Δx, which is obtained by equating the electric force to the needed centripetal force (see Fig. 6(b)).

$$(\Delta x)2\pi n(Qe) = \frac{MV_z^2}{R} ,$$

where an ion of mass M has velocity V_z, and is ionized Q times. For the example given, $\Delta x = 0.6$ cm which is acceptable.

The electron ring will also shift from the center of the conducting torus and, hence, experience an image force which holds the electrons (which hold the ions). The field needed to do this is about 1 MV/cm and is, automatically, supplied by images. Perhaps surprisingly, the electrons shift inward and force balance can be used to find the magnitude of this shift which, in this example, is about 0.3 cm.

In summary, the collective focusing accelerator will be most interesting if large numbers ($\simeq10^{17}$) of electrons can be contained without instabilites for 10 msec. Only experiments can really tell us whether or not that is possible and such experiments are underway.

6. THE FUTURE

As Niels Bohr said, "Predictions, especially of the future, are difficult." In the spring of 1981 the Department of Energy attempted just that, namely they convened a committee, under the chairmanship of Francis Cole and including the author, to review the situation at present and to give them advice concerning the future of collective accelerators. That committee made predictions concerning the practicality of various approaches, the time scales of experimental studies, the level-of-effort needed in each approach, and, finally, the possible applications of each approach. This author has greatly profited from these deliberations of the DOE committee, but the opinions expressed here are my own. I am expressing opinions on the future because without that no discussion of this subject is complete.

Perhaps not surprisingly, I think that all four collective accelerator approaches, namely a localized space charge in an IREB, a wave accelerator, an ERA, and a collective focusing accelerator, can be made to work. One cannot help but be impressed with the detail of our understanding of the ERA, but one would have to be quite optimistic to think that we shall soon have a useful ERA. Only the Dubna group is now working on the ERA, so progress rests with them alone. A number of groups have worked unsuccessfully on the ERA in the past; this is a significant discouragement to other future researchers.

On the other hand, a number of enthusiastic research groups are now pursuing IREB collective accelerators. The "natural acceleration" has been known for a decade, so the subject is to control, extend, or stage the natural process. The development of space-charge accelerators seems likely to occur soon (within about three years). If so, there are quite a number of uses for these accelerators, and we should see increased effort in the field.

The wave accelerators also seem close to success. The basic difficulty is in enhancing the desired wave and no others. This has been done in at least one case. The ARA Group has also controlled the speed of the desired mode and obtained an accelerating field of 10 MV/m. The acceleration of ions can be expected soon (in three years?). Wave accelerators, because of their diversity and hence the potential variety of applications, are very attractive.

The collective-focusing accelerator is presently being investigated by only one group. If it works (and it is rather close to that stage), I would expect other groups to become interested in it. Thus it could become routinely available in (say) five years.

The present level of U.S. effort on collective accelerators is about 3 M$/yr. This level should be raised somewhat in my opinion, since all of the active groups have marginal resources. On the other hand, I do not think that the field of collective acceleration should take very much more money, for it is not yet ready to go beyond relatively small experiments.

What are the applications for collective accelerators? First, wherever conventional accelerators are now employed, collective acceleration may provide superior approaches. Now, all accelerators which are known widely are large, and most of them produce particles of very high energy (SLAC, Fermilab, AGS, SPS, etc.). But most accelerators are small (and not well-known) and produce isotopes, treat people with medical needs, implant ions, produce radiation damage, produce materials, and produce ions for nuclear physics. It is the latter group which seems most likely to me for collective accelerator use, for they don't demand very high energy, but rather reliability and affordability. There are other uses of accelerators, which have not yet been realized, namely, for the electromagnetic breeding of fuel for light water reactors, as "drivers" for implosion fusion, and as weapons. All of these applications require lots of current and, therefore, seem less likely as applications for collective accelerators. (I think of the current difficulty of producing an IREB with many pulses per second.) Nevertheless, there are a great many potential uses of collective accelerators which we can envision.

Let me be more specific. The space-charge accelerator can be expected to be used as an ion source, as a heavy ion accelerator for nuclear physics, as a pulsed neutron source (for spallation studies), and as a source for medical therapy. The wave accelerators can be expected to be used for material studies and for medical therapy. The collective focusing accelerator has potential use for nuclear physics and as an injector in high energy physics, and the ERA could be used in nuclear physics, in atomic physics (as a source of highly ionized species) and in medical therapy.

In the cliche of the day, "the bottom line" is that I am optimistic about collective accelerators achieving some of their promise in the next few years and that after that I expect that they will be produced, on a routine basis, for a number of important applications. Whether they are, really, "the wave of the future" or not, remains to be seen.

REFERENCES

1. H. Alfven and P. Wernholm, Arkiv, Fysik 5, 175 (1952).
2. S.R.B.R. Harvie, AERE Memorandum G/M 87 (1951).
3. W. Raudorf, Wireless Engineer 28, 215 (1951).
4. V.I. Veksler, CERN Symp. on High Energy Accel., (CERN, Geneva, 1956), Vol. 1, p. 80.
5. G.I. Budker, ibid, p. 68.
6. Ya. B. Fainberg, ibid, p. 84.
7. S.E. Graybill and J.R. Uglum, J. Appl. Phys. 41, 236 (1970).
8. A.A. Plyutto, Sov. Phys. - JETP 12, 1106 (1961).
9. D. Keefe, Particle Accelerators 11, 187 (1981).
10. M. Reiser, IEEE Trans. Nucl. Sci. NS-28, 3355 (1981).
11. Collective Methods of Acceleration, N. Rostoker and M. Reiser, Eds. (Harwood Academic Publishers, New York, 1979).
12. E.L. Olson and U. Schumacher, Collective Ion Acceleration (Springer, New York, 1979), Springer Tracts in Modern Physics, Vol. 84.
13. V.I. Veksler et al., Proc. 6th Int. Conf. on High Energy Accelerators, Cambridge, Mass., USA (1967) (Cambridge Electron Accelerator Report CEALL1-2000, 1967), p. 289.
14. A. Irani and N. Rostoker, Particle Accelerators 8, 107 (1978).
15. U. Schumacher, C. Andelfinger, and M. Ulrich, Phys. Letters 51A, 367 (1975).
16. J.S. Luce, H.L. Sahlin, and T.R. Crites, IEEE Trans. Nucl. Sci. NS-20, 336 (1973).
17. C.L. Olson, IEEE Trans. Nucl. Sci. NS-26, 4231 (1979); C.L. Olson, Proc. IX Int. Conf. High Energy Accelerators, SLAC, Stanford, Cal (1974), p. 272.
18. N. Rostoker, private communication.
19. W.E. Drummond and M.L. Sloan, Phys. Rev. Lett. 31, 1234 (1973).
20. P. Sprangle, A.T. Drobot, and W.M. Manheimer, Phys. Rev. Lett. 36, 1180 (1976).
21. M. Friedman, IEEE Trans. Nucl. Sci. NS-26, 4186 (1979).
22. E.D. Ponets, IEEE Trans. Nucl. Sci. NS-23, 897 (1976) J. Arianer et al, IEEE Trans. Nucl. Sci. NS-26, 3713 (1979).
23. D. Gabor, Nature 160, 89 (1947).
24. T.R. Lockner and M. Friedman, IEEE Trans. Nucl. Sci. NS-26, 4237 (1979).
25. R.M. Mobley, G. Gammel, and A.W. Maschke, IEEE Trans. Nucl. Sci. NS-26, 3112 (1979).
26. L.M. Kazinskii, A.V. Kisletsov, and A.N. Lebedev, Atomnaya Energiya 30, 27 (1970).
27. R.C. Davidson, Theory of Nonneutral Plasmas, (W.A. Benjamin, Reading, Massachusetts, 1974).
28. H.R. Jory and A.W. Trivelpiece, J. Appl. Phys. 40, 3924, (1969).
29. J.W. Poukey and N. Rostoker, Plasma Physics 13, 897, (1971).
30. E. Cornet, H.A. Davis, W.W. Rienstra, M.L. Sloan, T.P. Starke, and J.R. Uglum, Phys. Rev. Lett. 46, 181 (1981).
31. R. Davidson and N. Krall, Phys. Fluids 13, 1543 (1970).
32. R.B. Miller and D.C. Straw, J. Appl. Phys. 48, 1061 (1977).

33. J.A. Nation, D. Aster, and J. Ivers, Bull. Am. Phys. Soc. 25, 945 (1980).
34. R. Adler, J.A. Nation, and V. Serlin, Phys. Fluids 24, 347, (1981).
35. N.C. Christofilos, Phys. Rev. Lett. 22 830 (1969), IEEE Trans. Nucl. Sci. NS-16, 1039 (1969).
36. L.J. Laslett and A.M. Sessler, IEEE Trans. Nucl. Sci. NS-16, 1034 (1969).
37. G. Brantti et al., IEEE Trans. Trans. Nucl. Sci. NS-20, 286 (1973).
38. L.J. Laslett, Report ERAN-200, Lawrence Berkeley Laboratory (1972) (unpublished), I.N. Ivanov, et al., Report P9-4132, JINR, Dubna (1968) (unpublished). (See also Ref. 12).
39. C.E. Nielsen, A.M. Sessler, and K.R. Symon, Int. Conf. on High Energy Accelerators, CERN, Geneva, (1959) p. 239; A.A. Kolomenski, and A.N. Lebedev, ibid p. 115.
40. L.J. Laslett, K. Neil, and A.M. Sessler, Rev. Sci. Instrum. 36, 436 (1965).
41. D. Koshkarev and P. Zenkevich, Particle Accelerators 3, 1 (1972).
42. L.J. Laslett, A.M. Sessler, and D. MD. Mohl, Proc III All-Union National Part. Accelerator Conf., Moscow, (1972); Nucl. Instr. and Methods 121, 517 (1974).
43. A. Faltens and L.J. Laslett, Particle Accelerators 4, 151 (1973).
44. G.S. Janes, R.H. Levy, H.A. Bethe, and B.T. Feld, Phys. Rev. 145, 1925 (1966).
45. W. Clark, P. Korn, A. Mondelli, and N. Rostoker, Phys. Rev. Lett. 37, 592 (1976).

PROBLEMS

1. Give a careful derivation of the stopping distance of an IREB, i.e., present the details of the calculation of Poukey and Rostoker who built on the work of Jory and Trivelpiece.

2. Putnam proposed that a "self-stabilized pinch" could explain the natural acceleration when an IREB is sent into a chamber containing gas. Evaluate the accelerating field due to a pinch. How rapid must the pinch be in order to explain the observed acceleration? Is this likely, given the known rates of ionization?

3. Give the details of a relativistic electron stream propagation in an external, longitudinal magnetic field. In this way work out the characteristics of relativistic Brillouin flow for an IREB.

4. Derive for non-relativistic electron motion through an initially stationary background of ions the resulting longitudinal instability. Thus obtain the critical current and the maximum growth rate of the Buneman analysis of the two-stream instability. Generalize your work for relativistic electrons.

5. Veksler has suggested the "inverse Cherenkov effect" and Lawson has suggested the "inverse Bohr-Bethe stopping-power effect" as acceleration mechanisms. Give formulas for these two mechanisms. Under what circumstances can they give significant accelerating fields (say, 100 MV/m)?

6. Laslett evaluated the contribution to an electron's oscillation frequency due to all of the electrons in a ring. That is he found the effect of "electrons on the other side," which is to say he did not make the usual approximation of a straight beam. Find this "curvature term." When is it important?

7. Pellegrini and I worked out the radiative contribution to the longitudinal coupling impedance of an electron ring. Evaluate this contribution. Is it serious? What should one do to ameliorate the effect?

8. Christofilos, Laslett and I developed the concept of a "static compressor," i.e., an electron ring generator which had (almost) no moving parts and no temporal variation of currents. Work out the design of such a device.

9. Work out, as Koshkarev and Zenkevich did, regions of instability for the transverse electron-ion instability (with no Landau damping). Present plots of your results in the Q_1 vs Q_i plane for modes labelled with $\tilde{m}$ and (say) $\nu = 0$. Evaluate the maximum growth rate for the worst mode. Is it fast or slow, serious or not?

10. Rosenbluth and Furth have noted that in the magnetic expansion column of an electron-ring accelerator one can have both an inner solenoid and an outer solenoid. The inner solenoid, with spacially varying fields has distinct advantages (such as allowing the electron ring to obtain good translational speeds while not expanding its radius). Work out what one can expect for field variation and give some typical design numbers [such as $B_i(Z)$, $B_o(Z)$, $R(Z)$, $\gamma_{11}(Z)$, etc.]. What are the limits to this kind of acceleration?

AIP Conference Proceedings

		L.C. Number	ISBN
No.1	Feedback and Dynamic Control of Plasmas	70-141596	0-88318-100-2
No.2	Particles and Fields - 1971 (Rochester)	71-184662	0-88318-101-0
No.3	Thermal Expansion - 1971 (Corning)	72-76970	0-88318-102-9
No.4	Superconductivity in d-and f-Band Metals (Rochester, 1971)	74-18879	0-88318-103-7
No.5	Magnetism and Magnetic Materials - 1971 (2 parts) (Chicago)	59-2468	0-88318-104-5
No.6	Particle Physics (Irvine, 1971)	72-81239	0-88318-105-3
No.7	Exploring the History of Nuclear Physics	72-81883	0-88318-106-1
No.8	Experimental Meson Spectroscopy - 1972	72-88226	0-88318-107-X
No.9	Cyclotrons - 1972 (Vancouver)	72-92798	0-88318-108-8
No.10	Magnetism and Magnetic Materials - 1972	72-623469	0-88318-109-6
No.11	Transport Phenomena - 1973 (Brown University Conference)	73-80682	0-88318-110-X
No.12	Experiments on High Energy Particle Collisions - 1973 (Vanderbilt Conference)	73-81705	0-88318-111-8
No.13	π-π Scattering - 1973 (Tallahassee Conference)	73-81704	0-88318-112-6
No.14	Particles and Fields - 1973 (APS/DPF Berkeley)	73-91923	0-88318-113-4
No.15	High Energy Collisions - 1973 (Stony Brook)	73-92324	0-88318-114-2
No.16	Causality and Physical Theories (Wayne State University, 1973)	73-93420	0-88318-115-0
No.17	Thermal Expansion - 1973 (lake of the Ozarks)	73-94415	0-88318-116-9
No.18	Magnetism and Magnetic Materials - 1973 (2 parts) (Boston)	59-2468	0-88318-117-7
No.19	Physics and the Energy Problem - 1974 (APS Chicago)	73-94416	0-88318-118-5
No.20	Tetrahedrally Bonded Amorphous Semiconductors (Yorktown Heights, 1974)	74-80145	0-88318-119-3
No.21	Experimental Meson Spectroscopy - 1974 (Boston)	74-82628	0-88318-120-7
No.22	Neutrinos - 1974 (Philadelphia)	74-82413	0-88318-121-5
No.23	Particles and Fields - 1974 (APS/DPF Williamsburg)	74-27575	0-88318-122-3
No.24	Magnetism and Magnetic Materials - 1974 (20th Annual Conference, San Francisco)	75-2647	0-88318-123-1
No.25	Efficient Use of Energy (The APS Studies on the Technical Aspects of the More Efficient Use of Energy)	75-18227	0-88318-124-X

No.26	High-Energy Physics and Nuclear Structure - 1975 (Santa Fe and Los Alamos)	75-26411	0-88318-125-8
No.27	Topics in Statistical Mechanics and Biophysics: A Memorial to Julius L. Jackson (Wayne State University, 1975)	75-36309	0-88318-126-6
No.28	Physics and Our World: A Symposium in Honor of Victor F. Weisskopf (M.I.T., 1974)	76-7207	0-88318-127-4
No.29	Magnetism and Magnetic Materials - 1975 (21st Annual Conference, Philadelphia)	76-10931	0-88318-128-2
No.30	Particle Searches and Discoveries - 1976 (Vanderbilt Conference)	76-19949	0-88318-129-0
No.31	Structure and Excitations of Amorphous Solids (Williamsburg, VA., 1976)	76-22279	0-88318-130-4
No.32	Materials Technology - 1976 (APS New York Meeting)	76-27967	0-88318-131-2
No.33	Meson-Nuclear Physics - 1976 (Carnegie-Mellon Conference)	76-26811	0-88318-132-0
No.34	Magnetism and Magnetic Materials - 1976 (Joint MMM-Intermag Conference, Pittsburgh)	76-47106	0-88318-133-9
No.35	High Energy Physics with Polarized Beams and Targets (Argonne, 1976)	76-50181	0-88318-134-7
No.36	Momentum Wave Functions - 1976 (Indiana University)	77-82145	0-88318-135-5
No.37	Weak Interaction Physics - 1977 (Indiana University)	77-83344	0-88318-136-3
No.38	Workshop on New Directions in Mossbauer Spectroscopy (Argonne, 1977)	77-90635	0-88318-137-1
No.39	Physics Careers, Employment and Education (Penn State, 1977)	77-94053	0-88318-138-X
No.40	Electrical Transport and Optical Properties of Inhomogeneous Media (Ohio State University, 1977)	78-54319	0-88318-139-8
No.41	Nucleon-Nucleon Interactions - 1977 (Vancouver)	78-54249	0-88318-140-1
No.42	Higher Energy Polarized Proton Beams (Ann Arbor, 1977)	78-55682	0-88318-141-X
No.43	Particles and Fields - 1977 (APS/DPF, Argonne)	78-55683	0-88318-142-8
No.44	Future Trends in Superconductive Electronics (Charlottesville, 1978)	77-9240	0-88318-143-6
No.45	New Results in High Energy Physics - 1978 (Vanderbilt Conference)	78-67196	0-88318-144-4
No.46	Topics in Nonlinear Dynamics (La Jolla Institute)	78-057870	0-88318-145-2
No.47	Clustering Aspects of Nuclear Structure and Nuclear Reactions (Winnepeg, 1978)	78-64942	0-88318-146-0
No.48	Current Trends in the Theory of Fields (Tallahassee, 1978)	78-72948	0-88318-147-9
No.49	Cosmic Rays and Particle Physics - 1978 (Bartol Conference)	79-50489	0-88318-148-7

No.50	Laser-Solid Interactions and Laser Processing - 1978 (Boston)	79-51564	0-88318-149-5
No.51	High Energy Physics with Polarized Beams and Polarized Targets (Argonne, 1978)	79-64565	0-88318-150-9
No.52	Long-Distance Neutrino Detection - 1978 (C.L. Cowan Memorial Symposium)	79-52078	0-88318-151-7
No.53	Modulated Structures - 1979 (Kailua Kona, Hawaii)	79-53846	0-88318-152-5
No.54	Meson-Nuclear Physics - 1979 (Houston)	79-53978	0-88318-153-3
No.55	Quantum Chromodynamics (La Jolla, 1978)	79-54969	0-88318-154-1
No.56	Particle Acceleration Mechanisms in Astrophysics (La Jolla, 1979)	79-55844	0-88318-155-X
No. 57	Nonlinear Dynamics and the Beam-Beam Interaction (Brookhaven, 1979)	79-57341	0-88318-156-8
No. 58	Inhomogeneous Superconductors - 1979 (Berkeley Springs, W.V.)	79-57620	0-88318-157-6
No. 59	Particles and Fields - 1979 (APS/DPF Montreal)	80-66631	0-88318-158-4
No. 60	History of the ZGS (Argonne, 1979)	80-67694	0-88318-159-2
No. 61	Aspects of the Kinetics and Dynamics of Surface Reactions (La Jolla Institute, 1979)	80-68004	0-88318-160-6
No. 62	High Energy e^+e^- Interactions (Vanderbilt , 1980)	80-53377	0-88318-161-4
No. 63	Supernovae Spectra (La Jolla, 1980)	80-70019	0-88318-162-2
No. 64	Laboratory EXAFS Facilities - 1980 (Univ. of Washington)	80-70579	0-88318-163-0
No. 65	Optics in Four Dimensions - 1980 (ICO, Ensenada)	80-70771	0-88318-164-9
No. 66	Physics in the Automotive Industry - 1980 (APS/AAPT Topical Conference)	80-70987	0-88318-165-7
No. 67	Experimental Meson Spectroscopy - 1980 (Sixth International Conference , Brookhaven)	80-71123	0-88318-166-5
No. 68	High Energy Physics - 1980 (XX International Conference, Madison)	81-65032	0-88318-167-3
No. 69	Polarization Phenomena in Nuclear Physics - 1980 (Fifth International Symposium, Santa Fe)	81-65107	0-88318-168-1
No. 70	Chemistry and Physics of Coal Utilization - 1980 (APS, Morgantown)	81-65106	0-88318-169-X
No. 71	Group Theory and its Applications in Physics - 1980 (Latin American School of Physics, Mexico City)	81-66132	0-88318-170-3
No. 72	Weak Interactions as a Probe of Unification (Virginia Polytechnic Institute - 1980)	81-67184	0-88318-171-1
No. 73	Tetrahedrally Bonded Amorphous Semiconductors (Carefree, Arizona, 1981)	81-67419	0-88318-172-X
No.74	Perturbative Quantum Chromodynamics (Tallahassee, 1981)	81-70372	0-88318-173-8

No. 75	Low Energy X-ray Diagnostics-1981 (Monterey)	81-69841	0-88318-174-6
No. 76	Nonlinear Properties of Internal Waves (La Jolla Institute, 1981)	81-71062	0-88318-175-4
No. 77	Gamma Ray Transients and Related Astrophysical Phenomena (La Jolla Institute, 1981)	81-71543	0-88318-176-2
No. 78	Shock Waves in Condensed Matter - 1981 (Menlo Park)	82-70014	0-88318-177-0
No. 79	Pion Production and Absorption in Nuclei - 1981 (Indiana University Cyclotron Facility)	82-70678	0-88318-178-9
No. 80	Polarized Proton Ion Sources (Ann Arbor, 1981)	82-71025	0-88318-179-7
No. 81	Particles and Fields - 1981: Testing the Standard Model (APS/DPF, Santa Cruz)	82-71156	0-88318-180-0
No. 82	Interpretation of Climate and Photochemical Models, Ozone and Temperature Measurements (La Jolla Institute, 1981)	82-071345	0-88318-181-9
No. 83	The Galactic Center (Cal. Inst. of Tech., 1982)	82-071635	0-88318-182-7
No. 84	Physics in the Steel Industry (APS.AISI, Lehigh University, 1981)	82-072033	0-88318-183-5
No. 85	Proton-Antiproton Collider Physics - 1981 (Madison, Wisconsin)	82-072141	0-88318-184-3
No. 86	Momentum Wave Functions - 1982 (Adelaide, Australia)	82-072375	0-88318-185-1
No. 87	Physics of High Energy Particle Accelerators (Fermilab Summer School, 1981)	82-072421	0-88318-186-X
No. 88	Mathematical Methods in Hydrodynamics and Integrability in Dynamical Systems (La Jolla Institute, 1981)	82-072462	0-88318-187-8